Cell, Tissue, and Organ Cultures in Neurobiology

ACADEMIC PRESS RAPID MANUSCRIPT REPRODUCTION

Cell, Tissue, and Organ Cultures in Neurobiology

Edited by

S. Fedoroff

L. Hertz

University of Saskatchewan
Saskatoon, Canada

ACADEMIC PRESS

New York San Francisco London 1977
A Subsidiary of Harcourt Brace Jovanovich, Publishers

ACADEMIC PRESS, INC.
111 Fifth Avenue, New York, New York 10003

United Kingdom Edition published by
ACADEMIC PRESS, INC. (LONDON) LTD.
24/28 Oval Road, London NW1

Library of Congress Cataloging in Publication Data

Main entry under title:

Cell, tissue, and, organ cultures in neurobiology.

Proceedings of the international workshop on cell, tissue, and organ cultures in neurobiology held at the University of Saskatchewan, Saskatoon, Canada, March 20–22, 1977.
1. Nerve tissue—Cultures and culture media—Congresses. 2. Cell culture—Congresses. 3. Tissue culture—Congresses. 4. Organs, Culture of—Congresses. I. Fedoroff, S. II. Hertz, L., Date [DNLM: 1. Tissue culture—Congresses. 2. Neurology—Congresses. QS530 C393 1977]
QP363.C44 599'.01'88 77-21066
ISBN 0-12-250450-X

PRINTED IN THE UNITED STATES OF AMERICA

Contents

CELL INTERACTIONS

NUTRITION

USE OF CELL TISSUE AND ORGAN CULTURES IN NEUROBIOLOGY

List of Contributors

Numbers in parentheses indicate the pages on which the authors' contributions begin.

A. S. Aggerwal (539), Department of Anatomy, Faculty of Medicine, University of Ottawa, Ottawa, Ontario K1N 9A9, Canada

Fred Baskin (393), Department of Neurology, Southwestern Medical School, University of Texas, Health Science Center at Dallas, 5323 Harry Hines Blvd., Dallas, Texas 75235

Maria-Helena Benitez (651), Grossherzog-Friedrich-Str. 52, Saarbrucken, Federal Republic of Germany

K. Bennett (485), Department of Anatomy, Northwestern University, Chicago, Chicago, Illinois 60611

B. Betschart (287), Department of Microbiology, University of Connecticut Health Center, Farmington, Connecticut 06032

Elisabeth Bock (407), The Protein Laboratory, 34 Sigurdsgade, Copenhagen N, DK 2200, Denmark

Murray B. Bornstein (141, 651, 667), Department of Neurology, Albert Einstein College of Medicine, Yeshiva University, 1300 Morris Park Avenue, Bronx, New York 10461

G. A. M. Breen (485), Department of Biology, University of California at San Diego, La Jolla, California 92093

A. Carrato-Ibanez (651), Instituto Cajal, Valazquez 144, Madrid 6, Spain

Jaroslava Ciesielski-Treska (593), Centre de Neurochimie du CNRS, 11 rue Humann, Strasbourg F67085, France

J. Cook (287), Roswell Park Memorial Institute, Buffalo, New York

Stanley M. Crain (147, 651, 667), Department of Neuroscience, Albert Einstein College of Medicine, Yeshiva University, 1410 Pelham Parkway South, Bronx, New York 10561

J. de Vellis (485), Department of Anatomy and Mental Retardation Center, Laboratory of Nuclear Medicine and Radiation Laboratory, University of California, Los Angeles, California 90024

J. Diamond (651), Department of Neurosciences, McMaster University, Hamilton, Ontario, Canada

S. Fedoroff (215, 265, 651, 667) Department of Anatomy, College of Medicine, University of Saskatchewan, Saskatoon, Saskatchewan S7N 0W0, Canada

J. Fulcrand (11), Laboratoire de Neurophysiologie, Faculté de Sciences, Montpellier, France

Beatrice B. Garber (515, 651), Department of Biology, University of Chicago, 1103 East 57th Street, Chicago, Illinois 60637

E. Hawrot (651), Department of Neurobiology, Harvard Medical School, 25 Shattuck Street, Boston, Massachusetts

J. J. Hauw (651, 667), Laboratoire de Neuropathologie, La Salpetriere, 47 Blvd. Hopital Paris, Cedex 13 F 75634, France

W. J. Hendelman (539, 667), Department of Anatomy, Faculty of Medicine, University of Ottawa, Ottawa, Ontario K1N 9A9, Canada

L. Hertz (39, 667), Department of Anatomy, College of Medicine, University of Saskatchewan, Saskatoon, Saskatchewan S7N 0W0, Canada

Walther J. Hild (99, 651, 667), Department of Anatomy, University of Texas Medical Branch, Galveston, Texas 77550

S. B. Hrushovetz (667), Kildonian Institute for Geriatric Research, 411 Union Avenue, Winnipeg, Manitoba R2L OC9, Canada

R. K. Hunt (369, 651), Department of Biophysics, Johns Hopkins University, Baltimore, Maryland 21205

D. D. Johnson (667), Department of Pharmacology, College of Medicine, University of Saskatchewan, Saskatoon, Saskatchewan S7N 0W0, Canada

Taiji Kato (223), Brain Research Institute, University of Chicago, 950 East 59th Street, Chicago, Illinois 60637

P. Leveille (485), Department of Anatomy and Mental Retardation Center, Laboratory of Nuclear Medicine and Radiation Laboratory, University of California, Los Angeles, California 90024

Ramon Lim (223, 651), Brain Research Institute and Departments of Surgery and Biochemistry, University of Chicago, 950 East 59th Street, Chicago, Illinois 60637

S. Lin (667), Department of Biophysics, Johns Hopkins University, Charles and 34th Street, Baltimore, Maryland 21212

Katherine M. Lyser (121), Department of Biological Sciences, Hunter College of the City University of New York, P.O. Box 1030, New York, New York

K. McCarthy (485), Laboratory of Nuclear Medicine and Radiation Laboratory, University of California, Los Angeles, California 90024

N. M. McDuffie (651), Department of Physiology, College of Medicine, University of Saskatchewan, Saskatoon, Saskatchewan S7N 0W0, Canada

J. F. McGinnis (485), Department of Anatomy and Mental Retardation Center, Laboratory of Nuclear Medicine and Radiation Laboratory, University of California, Los Angeles, California 90024

Paul Mandel (593, 651, 667), Centre de Neurochimie du CNRS, 11 rue Humann, Strasbourg F67085, France

P. Mancini (287), Department of Microbiology, University of Connecticut Health Center, Farmington, Connecticut 06032

K. C. Marshall (539), Department of Physiology, Faculty of Medicine, University of Ottawa, Ottawa, Ontario KIN 9A9, Canada

J. Mather (619, 667), Department of Biology, University of California at San Diego, La Jolla, California 92093

Irwin S. Mendelson (589), Department of Pediatrics, Alvin Buckwold Centre, University of Saskatchewan, Saskatoon, Saskatchewan S7N 0W0, Canada

Anne Messer (651, 667), Department of Neuroscience, Children's Hospital, 300 Longwood Avenue, Boston, Massachusetts 02115

R. Morris (287), Department of Microbiology, University of Connecticut Health Center, Farmington, Connecticut 06032

I. Munkacsi (667), Department of Anatomy, College of Medicine, University of Saskatchewan, Saskatoon, Saskatchewan S7N 0W0, Canada

Margaret R. Murray (1), Department of Health, Education and Welfare, Public Health Service, National Institutes of Health, Bethesda, Maryland 20014

E. J. H. Nathaniel (667), Department of Anatomy, School of Medicine, 750 Bannatyne Street, University of Manitoba, Winnipeg, Manitoba, Canada

Phillip G. Nelson (347, 651, 667), Behavioral Biology Branch, National Institute of Child Health and Human Development, Bldg. 36, Room 2A21, Bethesda, Maryland 20014

F. T. Oteruelo (651), Department of Anatomy, College of Medicine, University of Saskatchewan, Saskatoon, Saskatchewan, S7N 0W0, Canada

David Pearson (573, 667), Division of Neurosciences, City of Hope, Duarte, California

S. E. Pfeiffer (287, 651, 667), Department of Microbiology, University of Connecticut Health Center, Farmington, Connecticut 06032

J. W. Phillis (73, 667), Department of Physiology, College of Medicine, University of Saskatchewan, Saskatoon, Saskatchewan S7N 0W0, Canada

Kedar N. Prasad (447, 651, 667), Department of Radiology, University of Colorado Medical Center, Denver, Colorado 80220

Nagindra Prashad (393), Department of Neurology, Southwestern Medical School, University of Texas, Health Science Center at Dallas, 5323 Harry Hines Blvd., Dallas, Texas 75235

A. Privat (11, 651), Groupe U. 106 I.N.S.E.R.M., Laboratoire d'Histologie Normale et Pathologique du Systeme Nerveux, 123 Boulevard Port-Royal, Paris F75014, France

Roger N. Rosenberg (393, 667), Department of Neurology, Southwestern Medical School, Health Science Center at Dallas, University of Texas, 5323 Harry Hines Blvd., Dallas, Texas 75235

G. Sato (619), Department of Biology, University of California at San Diego, La Jolla, California 92093

Arne Schousboe (441), Department of Biochemistry A, Panum Institute, Blegdamsvej 3C, University of Copenhagen, Copenhagen DK-2200, Denmark

Bruce K. Schrier (423, 667), Behavioral Biology Branch, National Institutes of Child Health and Human Development, National Institutes of Health, Bethesda, Maryland 20014

B. Scott (651, 667), Services to Developmentally Handicapped Persons, Surrey Place Centre, # 2 Surrey Place, Toronto, Ontario, Canada

F. J. Seil (651, 667), Research Service (151) V.A. Hospital, Portland, Oregon 97207

O. Z. Sellinger (667), MHRI, University of Michigan, 205 Weshtenaw Place, Ann Arbor, Michigan 48109

Monique Sensenbrenner (191, 651, 667), Centre de Neurochimie du CNRS, 11 rue Humann, Strasbourg F67085, France

Vladimir Stefanovic (593), Centre de Neurochimie du CNRS, 11 rue Humann, Strasbourg F67085, France

Shuang S. Troy (223), Brain Research Institute, University of Chicago, 950 East 59th Street, Chicago, Illinois 60637

David E. Turriff (223), Brain Research Institute, University of Chicago, 950 East 59th Street, Chicago, Illinois 60637

Carol K. Vance (393), 1111 Sealy, Galveston, Texas 77550

Silvio Varon (237, 651), Department of Biology, School of Medicine, University of California at San Diego, La Jolla, California 92093

Roderich Walter (589), Department of Physiology, University of Illinois Medical Center, Chicago, Illinois

Charity Waymouth (631), Senior Staff Scientist, The Jackson Laboratory, Bar Harbor, Maine 04609

B. Westermark (651), Wallenberg Laboratory, P.O. Box 562, Uppsala S-75122, Sweden

J. M. Wojtowicz (539), Department of Physiology, University of Ottawa, Ottawa, Ontario KIN 9A9, Canada

Merrill K. Wolf (555), Medical School, University of Massachusetts, Worcester, Massachusetts 01605

J. D. Wood (651, 667), Department of Biochemistry, College of Medicine, University of Saskatchewan, Saskatoon, Saskatchewan S7N 0W0, Canada

Preface

In recent years the application of tissue culture in studies of the nervous system has widely increased. The uses of cell, tissue, and organ cultures in neurobiological research to the present time are reviewed in this fundamental reference book. We thus bring together an interdisciplinary perspective from morphology, biochemistry, pharmacology, endocrinology, embryology, and genetics. Several chapters deal with specific aspects of the nervous system *in vivo*. In order that cultured neural cells and tissues can be assessed as models of their known *in vivo* counterparts, the relevance of various tissue culture preparations for neurobiological investigations are evaluated.

The book will be of interest to people in a wide range of disciplines. It is directed toward tissue culturists concerned with the nervous system, as well as all neurobiologists, cell biologists, and embryologists interested in learning how neural cells and tissues behave in cultures and what has been learned about the nervous system using tissue culture methods, including the applicability of tissue cultures to the study of cell differentiation.

This volume is one outcome of an international workshop held at the University of Saskatchewan in March 1977. Many of the papers presented have been expanded, some considerably, to form chapters in the book; some points made during the discussions have been elaborated on and are included as chapters; the discussion sessions have been summarized and extensively edited to reflect the trends of thought and the interactions between the scientists of various disciplines present at the workshop. The ultimate aim of the volume is to provide a sound basis for the use of nervous tissue culture in the solution of neurobiological problems.

The publication of this volume had the generous support of the College of Medicine and Graduate Studies and the Department of Anatomy, University of Saskatchewan; the Pasadena Foundation for Medical Research; the National Institute on Aging, NIH, U.S. Public Health Service; the Muttart Foundation; the Multiple Sclerosis Society (Canada); and Gibco, Canada.

We are particularly grateful to D. G. Murphy and W. A. Opel for continuing interest in the workshop and to A. Boulton, I. Munkacsi, F. Oteruelo, P. Sulakhe, B. Wenger, K. Seemann, and the staff of the Department of Anatomy, University of Saskatchewan, for help in organizing the workshop.

Elaine Fedoroff, Elna Hertz, B. H. J. Juurlink, Anna Leslie, and Jean MacGregor gave invaluable assistance in the preparation of material for this volume;

Jean MacGregor created the cover design; the staff of the Modern Press in Saskatoon worked diligently in the preparation of the camera ready copy; the staff of Academic Press was patient and helpful in the final stages of publication.

INTRODUCTION

MARGARET R. MURRAY

National Institute of Neurological and Communicative Diseases and Stroke National Institutes of Health Bethesda, Maryland

Nerve tissue culture was born just 70 years ago, when R. G. Harrison, cognizant perhaps of attempts then being made by others to maintain somatic cells *in vitro*, devised a simple and elegant procedure to investigate the genesis of the nerve-fiber. Being an experimental embryologist in the year 1907, he was well acquainted with the principle of "explantation"; but he carried it a step further by transferring a section of neural tube from the frog embryo to an indifferent, non-cellular substrate which consisted of clotted frog-lymph, enclosed in a transparent glass depression-slide. The naked outgrowing nerve fibers with their pseudopodal tips preceded the later-emerging supportive cells by a substantial time — and space — margin, thus strongly indicating that the fibers were not generated as a syncytium or nerve-net by cells along their course. The success of this experiment was spectacular enough to impress on others that the method was worth pursuing, and they took out after it from all directions in this country and abroad.

For these pioneers, practically every observation yielded pay-dirt, and though nearly everyone at some time had a try at growing nerve-tissue or made incidental observations on it, those reports were almost buried in the general accumulation of tissular riches during the ensuing score of years. They were easy to bury because as Harrison's successors developed culture techniques to accommodate other, proliferative, histological types, nervous tissues, especially CNS, for a variety of reasons were found to be refractory. The definitive neurons did not proliferate nor enjoy a long life in culture; they remained viable no more than a week or two, as a rule. Harrison had luck as well as genius going for him in his choice of problem and material: A question that could be answered in short-term (a few days) culture of embryonic amphibian cells which carried with them their own native nutrients in the form of yolk.

Nevertheless, even in its primitive state the culture method could be utilized effectively by others for nerve study. In 1913 Ingebrigtsen, who as a neurosurgeon was interested in the possibility of regeneration, addressed himself with some success to one of the contemporary problems: he cultured cerebellum from the chick and a number of mammals in plasma clots and established that severed CNS axis cylinders of higher vertebrates were able to regenerate — growing out as they did from his explants in typical beaded form, though soon suffering the granular degeneration that (as described by Nauta in 1950) is now well-known to pathologists. The significance of these observations went generally unrecognized and unacclaimed. Then as now, it took unusual showmanship or a unique train of circumstances to breach the habitual thought patterns of medical savants unacquainted with the living cell. Neurobiologists were not far behind them; in 1944 Nonidez felt obliged to publish a review on the status of the Neuronfrage, in a final effort to lay at rest the tenaciously held nerve-net theory. Let us admit that a worker with a large investment in technique or hypothesis does not readily give it up on the publication of an adversary. And in justice it should be said that in the light of more modern technical practices culture conditions then were far from ideal, and results could therefore be suspect. Nevertheless, the principle cannot be enunciated too often: that anything a cell is seen to do in culture must be counted among its potentialities.

In an introduction it would not be possible or desirable to cover even summarily the development of nerve tissue culture in method and substance over the three-score years and ten that have preceded the present sessions designed to illuminate the status quo. Moreover, substantial reviews are now available and these have approached the subject from different angles as the field has become diversified (Murray, 1965, 1971; Sato, 1973; Bunge, 1975; Nelson, 1975; Crain, 1976). The last decade particularly has seen great expansion and multiplication of technical procedures concomitant with the setting of new goal-horizons. Ancillary methods such as light and electron microscopy with their recent advances, also continuing biochemical, neurochemical and pharmacological specializations, have been combined and recombined with enhanced culture procedures in the pursuit of more precisely defined objectives; and since in addition, each experimental run put out by a collaborative group tends to be published as a separate paper with a different senior author, the resultant literature expansion has become unmanageable. As regards the body of nerve tissue culture, the right hand no longer knows what the left hand is doing.

The symposial gathering held in Saskatoon in March 1977 was designed to resolve this dilemma by bringing together practitioners of varied persuasions and goals in order that they might learn from each other and set up a system of cross-fertilization. The group was small but participants were selected on an ecumenical basis and their communications grouped in topical rationale. Provision was made each day at the end of the practical sessions for discussions in depth that might deal with problems of general significance suggested by the reports. No holds were barred, and these lively interchanges provide invaluable information as well as perspective. Now that tissue culture has become established as an adjunct to nearly every discipline, the most urgent general question that faces the research worker today may be the validity of his material as model: its degree of applicability to situations as they occur spontaneously *in situ*. To what extent can principles deduced from observations upon nervous tissues isolated in culture (of whatever type) be presumed to hold in the intact organism? The separation of parts from the whole and of individual factors from a complex is basic to the experimental method; this is accomplished par excellence in tissue culture. Nevertheless, other considerations must also be taken into account in exploring the "Model" problem and these may be of greater or less importance depending on specific circumstances, — as these discussions make abundantly clear.

Following World War II the barriers to neuron longevity in culture were resolved by modifications of medium, substrate and culture-chambers, so as to allow the production of "organotypic" microcosms (Murray, 1965) of differentiating cell communities which with meticulous handling could survive months of isolation in essentially normal form and function. Their LM verisimilitude was amply reinforced by fine-structural and bio-electrical findings. The first region of the nervous system to be brought under control was the sensory ganglion, followed by the cerebellum, the cerebrum, the hypothalamus and spinal cord, and finally the sympathetic ganglia. Time-lapse moving-pictures, preeminently in the hands of Charles Pomerat, greatly accelerated and amplified the general medical understanding of living neurocytology and cellular movements (Hydén, 1967). For qualitative observations of development, specific cytopathology and nerve-cell physiology, the organotypic culture method, competently handled, is unmatched. But the amount of labor and skill involved tends to disqualify it for types of investigation that require more than micro-quantitative procedures, e.g., most problems of analytical neurochemistry.

It is not surprising, therefore, that the possibility of working with established strains of neoplastic neuroblasts and astroblasts attracted the enthusiasm of biochemists and their students, or that a truly enormous literature has burgeoned about these in the last decade, sparked by a high order of ingenuity on the part of their exploiters (Sato, 1973). Here was a source of material that could be propagated in bulk with relatively simple culture manipulations. However, the variability of these lines — even within clones, and a sober appraisal of their neoplastic character led neurobiologists to revert to dissociation methods (long practised by Moscona and his school) for the study of developing normal nervous tissues. By this means, starting with whole brains or regions, one could grow dispersed cells en masse for a substantial period, allow them to reaggregate (Moscona, 1965; Garber and Moscona, 1972) or examine them in monolayers, for whatever purpose. Though the ratio of viable neurons to supporting cells may be small and tends to diminish with age *in vitro*, primitive contacts are formed between them which can be demonstrated by EM and by evoked action potentials. It now appears that with critical timing and treatment (e.g., with anti-mitotics) the relative amounts of various cell types to be cultured can be modified substantially, so as to produce "highly purified" populations of neural or non-neural cells. It should not be overlooked in the study of the *Neuroblastoma* that this tumor is not derived from the CNS, but from sympathetic ganglia or adrenal medulla, as any pathologist will confirm. The spontaneous neuroblastoma is a highly infiltrative tumor which often spreads massively to the spinal cord and quickly metastasizes further. Goldstein (1964), in working with the human type, sometimes finds differentiation in the direction of sympathetic ganglion cells, never of cord or brain neurons.

As these developments were taking place, the promoters of organotypic culture returned to the charge by confronting explants with dissociated cells from other selected regions and thus pinpointing the individual cell contacts on both morphological and physiological and macromolecular grounds. (Bunge *et al.*, 1974). Thus these methods, originally poles apart, have been combined to mutual advantage.

It is from this congeries of approach, method and emphasis that topics were assigned and representative participants were selected for the Saskatoon meeting. Along the way, histologists and embryologists were pressed into service to set forth some basic morphological criteria, electrophysiologists and chemists likewise in their fields, for mutual orientation among this group of sub-specialists.

The participants' communications in all these areas ably speak for themselves. This writer need only direct attention to the architectonics of the compilation. We are dealing here with living nervous tissues removed from their anatomic and histological surroundings *in situ* to an indifferent substrate and unaccustomed medium *in vitro* where questions of varying complexity are asked of them. It is of paramount importance to identify the cell-types which present themselves in a primary culture or its descendents — to recognize neurons, glial cells, Schwann cells and contaminants such as endothelial cells visually as far as possible, though conceding that reliance on other means may be necessary as well as confirmatory. To this end a session was devoted to the morphology, physiology and biochemistry of differentiated nervous tissues, particularly as the major types diverge in development. In the present state of our knowledge, the most generally or overall successful organotypic cultures are explanted in early life from fetal or newborn animals, at a time when morphogenetic movements have ceased and the region under study therefore has approximately its full quota of precursor cells which are sufficiently potent, and at the same time flexible or malleable enough to accommodate to the new environment and continue development. Timing of course varies from species to species and from region to region of the nervous system; for example, cerebellum of the rat is explanted most favorably from the newborn, whilst equivalent material is provided by an 11-day embryo in the chick; retina from a 12-day chick embryo is at essentially the same developmental stage as that from a mouse 9 days post-partum. Relatively little success has been achieved with adult tissues, especially from CNS, though this is a prime desideratum.

As regards glial cell-lines, these are most often of neoplastic origin; even in the normal, chemical and bioelectric methods of identification may be essential.

Tissues also are dissociated and cultured either as reaggregating communities or as single cell types for various purposes in the analysis of normal development or of genetically determined dystrophies or of chemical and pharmacological interdependence among tissues, to name a few research goals. Somatic hybrids can now be made between glia and neuroblasts (usually neoplastic). Methods and pitfalls in the preparation, isolation and identification of components in all such cultures are very usefully discussed in the chapter by Varon; and a handy compendium of the general aspects of cell nutrition and welfare in isolation is contributed in the chapter by Waymouth. Specific considerations for nervous tissues must however be sought in the special articles dealing with the various regions of the NS and methods of culture.

As technical control of nervous tissue maintenance *in vitro* has developed and expanded, the number and variety of feasible research problems has multiplied *pari passu.* An early post-war preoccupation of research workers was with myelin, — which was then first produced *de novo* in culture. (Peterson and Murray, 1955, Hild, 1957). Its fine-structure and developmental mechanics were studied *in vitro* (given a great stimulus by the "jelly-roll" theory of Geren in 1954), and later, the manner of its destruction in the demyelinating diseases (see chapter by Bornstein). All this has made an absorbing topic in itself, with continuing ramifications into the chemistry and immunology of the myelin-forming cell in relation to the neurite which it envelopes, — a field that could only be touched upon in this program, but which can be pursued more fully through the references cited. Furthermore, in neurological mutants of the mouse there is often a developmental myelin deficiency or a myelin lysis, — a condition lending itself to visual monitoring which has opened new vistas to those who would distinguish between local and systemic genetic factors and pursue the pathways through which gene effects are mediated (see chapter by Wolf).

A subject of rising interest and increasing importance to the users of nerve tissue culture concerns the origins and specific functions of the neurotransmitters or neuromodulators, as distinct from the classical neurosecretory substances such as those released by the pineal and certain portions of the hypothalamus, and a new group of "releasing hormones" which operate upon the somatic hormones such as thyroxin and corticosteroids. The capacity of neurons for synaptogenesis *in vitro* and the selectivity of their terminals is now open for analysis (Rees, 1977) also, along with more general aspects of cell recognition. In all these areas the possibility is being explored to bring molecular biology to bear on the mechanisms of a given compound's release and/or its operation upon a target cell membrane, especially through immunochemistry and EM tracers applied in cultures, — thus providing new and exciting extensions of older neurobiological principles.

Finally, an overall concern of neurobiologists must be the regulation of cell and tissue differentiation, which to a degree subsumes all the above topics; and here there is much to be learned from tissue culture and much effort is going into the framing of suitable experiments at various cellular levels, — from the neoplastic monolayers through dissociated and reaggregated cell communities to classic organ cultures. Accounts of the discussion of these problems by this group should be of value to others in the field.

Tissue culture of any sort is a time-consuming, laborious and costly process which should not be undertaken without reasonable expectation that, among possible alternatives, it offers the best hope of solution for a specific problem. The study of isolated yet intact and functioning portions of the nervous system offers a kind of information that can scarcely be gathered in any other way, yet the type of culture method and the nature of the inquiry should be rigorously selected and correlated. Except for the classic "organ cultures", which because of their thickness are microscopically opaque and do not allow sufficient diffusion of gases, nutrients and wastes to support tissue survival for more than a few days or a week, present culture techniques (organotypic, reaggregated, monolayer) have been developed so as to allow repeated microscopic observations of many neuron types and their associated (or dissociated) supporting tissues in stable maintenance for periods of weeks or even months when desired. Especially with organotypic and some surface cultures, — still and time-lapse photographic LM and Nomarski records can be made sequentially at high magnifications for long periods, bioelectric recordings with microelectrodes and modified oscilloscope hookups can be continued for hours, and a variety of pharmacological agents can be administered directly to living cells under continuous monitoring for days if need be. Fluorescence microscopy, TEM, SEM and freeze-fracture procedures are particularly applicable to all types of culture preparations, in which fixation is immediate and artifacts at a minimum; ongoing processes can thus be stopped with precision at a given moment and embalmed in permanent preparations.

On the other hand, normal adult CNS tissues do not lend themselves well to cultivation by any methods yet devised. It therefore may be necessary to carry developing cells in culture until they reach a more or less mature state — which they do in organotypic culture at a slightly slower rate than *in vivo*. On this score, nervous tissues from animals with short life-span and rapid development, such as the small rodents, are preferable to those from man.

The isolation of a region (e.g., cerebellum) or a ganglion (e.g., superior cervical) from afferent and efferent connections obviously may be either an advantage or a disadvantage, depending on the nature of the inquiry. However, in organotypic culture both afferent and efferent terminals can be provided by the introduction, with proper timing, of further explants from whatever source is desired in bioelectric and membrane receptor experiments.

Toxicity studies designed to evaluate effects of chemotherapeutic or carcinogenic agents, industrial pollutants, narcotics and other exogenous additives, have often proved less than satisfactory because of certain critical deficiencies of culture systems vis-a-vis the whole organism. Problems involving the availability of the agent to the tissue *in vivo*, metabolism of the agent by the host, filtration of the agent by the blood-brain barrier or utilization of other routes, systemic toxic effects (as upon vascular or nephric tissue) all tend to reduce correlation between *in vitro* and *in vivo* results. However, it can be of theoretical or suggestive value to observe the *direct*, primary effects of any agent, at physiological concentration, upon the cell or tissue in question. This is particularly applicable to endogenous agents such as those carried in experimental allergic encephalitis (EAE), MS and Guillain-Barré sera and various antitoxins.

Those who work with the many clones and cell lines, "normal" and neoplastic, are convinced, notwithstanding the genotypic and phenotypic variability of their material, that any chemical or physiological process they can analyse and describe as performed by these cells bears on basic genetical and neurological theory; and that from the manifold findings being made available, a mosaic of genetic, developmental and functional laws for the nervous system can be constructed. Certainly much that is of great interest has already been learned, — and this from somatic hybrids also. We must await further developments before evaluating the ultimate applicability of these and other culture results to the overall role of the nervous system *in situ*. The culture system is an imperfect model, but because of the discrepancies revealed by study, it may even be an improvement on the original for analysis.

References

Bunge, P. R. 1975. Changing uses of nerve tissue culture 1950-1975. In: *The Nervous System, Vol. 1: The Basic Neurosciences.* D. B. Tower, ed. Raven Press, N.Y. p. 31.

Bunge, R. P., R. Rees, P. Wood, H. Burton and C.-P. Ko. 1974. Anatomical and physiological observations on synapses formed on isolated autonomic neurons in tissue culture. Brain Res. 66:401-412.

Crain, S. M. 1976. *Neurophysiologic Studies in Tissue Culture.* Raven Press, N.Y.

Garber, B. B. and A. A. Moscona. 1972. Reconstruction of brain tissue from cell suspensions. I. Aggregation patterns of cells dissociated from different regions of the developing brain. Develop. Biol. 27:217-234.

Geren, B. B. 1954. The formation from the Schwann cell surface of myelin in the peripheral nerves of chick embryos. Exp. Cell Res. 7:558-562.

Goldstein, M. N., J. A. Burdman and L. J. Journey. 1964. Longterm tissue culture of neuroblastomas. II. Morphologic evidence for differentiation and maturation. J. Nat. Cancer Inst. 32:165-199.

Harrison, R. G. 1906-1907. Observations on the living, developing nerve fiber. Proc. Soc. Exp. Biol. Med. 4:140-143.

Hild, W. 1957. Myelogenesis in cultures of mammalian central nervous tissue. Z. Zellforsch. 46:71-95.

Hydén, H. (ed.) 1967. *The Neuron.* Chap. III, Dynamic activities of nervous tissue in vitro. Elsevier, Amsterdam. pp. 119-173.

Ingebrigtsen, R. 1913. Regeneration of axis cylinders in vitro. J. Exp. Med. 18:412-415.

Moscona, A. A. 1965. Recombination of dissociated cells and the development of cell aggregates. In: *Cells and Tissues in Culture.* N. Willmer, ed. Academic Press, N.Y. Vol. I. pp. 489-529.

Murray, M. R. 1965. Nervous tissues *in vitro.* In: *The Biology of Cells and Tissues in Culture.* E. N. Willmer, ed. Academic Press, N.Y. Vol. II, pp. 373-455.

Murray, M. R. 1971. Nervous tissues isolated in culture. In: *Handbook of Neurochemistry.* A. Lajtha, ed. Plenum Press, N.Y., Vol. 5, Part A, pp. 373-438.

Nauta, W. H. 1950. Über die sogenannte terminale Degenerazion im Zentralnervensystem und ihre Darstellung durch Silberimprägnation. Arch. Neurol. Psychiat. 66:353-376.

Nelson, P. G. 1975. Nerve and muscle cells in culture. Physiol. Rev. 55:1-61.

Nonidez, J. F. 1944. The present status of the neurone theory. Cambridge Phil. Soc. Biol. Rev. 19:30-40.

Peterson, E. R., and M. R. Murray. 1955. Myelin sheath formation in cultures of avian spinal ganglia. Amer. J. Anat. 96:319-355.

Rees, R. 1977. Structure of cell coats during synapse formation on isolated cultured sympathetic neurons. Brain Res. (in press).

Sato, G. (ed). 1973. *Tissue Culture of the Nervous System.* Plenum Press, N.Y.

Characteristics of Differentiated Cells

NEUROGLIA — FROM THE SUBVENTRICULAR PRECURSOR TO THE MATURE CELL

A. Privat
Laboratoire de Culture de Tissu Nerveux
Paris, France

J. Fulcrand
Laboratoire de Neurophysiologie
Faculte de Sciences
Montpellier, France

I. Introduction

Since the first mention by Dutrochet (1824), and the report of Virchow (1856), neuroglia has been given an identity different from that of the neuron. His postulated (1889) that neurons and neuroglia arise from separate stem cells, while Shaper (1897) suggested that they had a common origin, the apolar and undifferentiated cell. Neuroglia were then considered as a single cell type, and it was only with the introduction of metallic impregnation that Ramon y Cajal (1909) was able to distinguish the astrocyte and, later on, a third element, in addition to the neuron. With the use of the silver carbonate technique, del Rio Hortega (1919) showed that the third element included two separate cell entities, oligodendrocytes and microglia; applying the same technique to embryos and young animals, he concluded that oligodendrocytes and astrocytes originated from neuroectodermal precursors, while microglia came from mesodermal elements. However, the difficulties encountered in achieving satisfactory metallic impregnations of immature cells led to controversy about Hortega's theory. Penfield (1932) attempted to circumvent this difficulty by making use of routinely stained paraffin sections but the identification of the cells remained in doubt. Using the same technique, Kershman (1938, 1939) described the "subependymal layer", a dense sheet of transitional cells located between matrix layer and mantle layer, where he found, besides apolar neuroblasts, the precursors of astrocytes and oligodendrocytes. Like Hortega, he estimated that microglia have a mesenchymal origin, and migrate into the nervous parenchyma from the meninges and the endothelium of blood vessels. The problems of the identification and origin of neuroglia remained in the same state for almost two decades, when the advent of electron microscopy reactivated the old disputes about cell identification, while the use of ^{3}H-thymidine radioautography reopened controversy about their origin. The purpose of the present review is to summarize the present status of these two problems, and to report recent data derived from two different approaches: the freeze-fracturing technique to obtain new criteria for cell identification, and the enucleation of the eye in newborn and young rats to produce Wallerian degeneration in the optic nerve. This dual approach helped in examining the factors which govern glial differentiation.

II. The Subventricular Cells

Subventricular cells, precursors of some neurons and neuroglia, are found around the ventricles in the late fetus, but they are eventually mixed with precursors of neuroblasts. Such is not the case in two situations: the subventricular layer, which persists in the adult animal along the external wall of the lateral ventricle, and the ventricular layer of the optic stalk, are reputed to give rise only to glioblasts.

A. The Subventricular Layer of the Lateral Ventricle

Allen (1912) found that mitotic figures persisted around the lateral ventricles of the adult rat brain. This was later confirmed in other species by Opalski (1934), Kershman (1938), Globus and Kuhlenbeck (1944). Smart (1961) conducted investigations with tritiated-thymidine radioautography and distinguished two cell types in hematoxylin eosin stained paraffin sections. The first type stained lightly and had an irregular nucleus with many indentations and chromatin granules clumped irregularly along the nuclear membrane. The cytoplasm was scarce, and poorly stained. The second type had a smaller, denser elongated nucleus, irregular in shape; coarse chromatin granules gave it a spotted appearance. The cytoplasm appeared as one or two tails at the nuclear poles.

These two cell groups were usually found in small clusters of the same type; they occurred in variable numbers with age, the dark ones being predominant at birth; whereas they occurred in the same numbers in the adult. Tritiated thymidine labeled the dark nucleated cells preferentially, although some light cells were also labeled. Smart tentatively concluded that dark cells gave rise to spongioblasts (subventricular cells), precursors of glial cells, whereas light cells were precursors of neuroblasts (Angevine *et al.*, 1970). Later on, Altman (1966) labeled the subventricular layer of a 13 day old rat with tritiated thymidine, and found that most of the cells migrated into the corpus callosum and the caudate nucleus, where they matured as glial cells. However, the identification of these cells in paraffin sections was only tentative. The advent of electron microscopy, with the pioneer work of Luse (1956) and Farquhar and Hartmann (1957), allowed a clearcut identification of mature glial cells, and prompted the examination of their precursors with the same technique. In the young rat, the

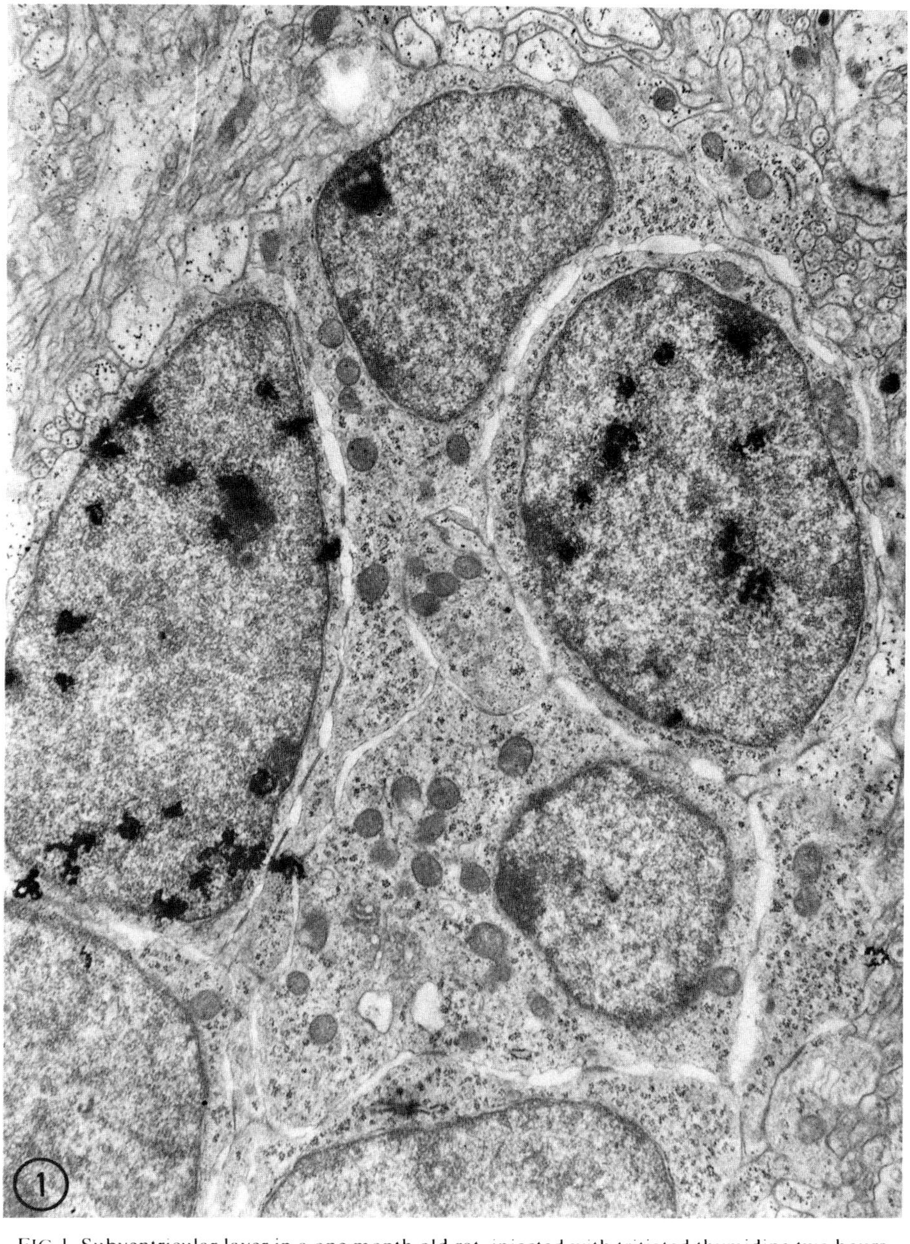

FIG 1. Subventricular layer in a one month old rat, injected with tritiated thymidine two hours before killing. Two cells are labeled: the one on the left has a relatively light, uniform nucleus; whereas the one of the right shows a denser nucleus, with several masses of chromatin. (× 13000).

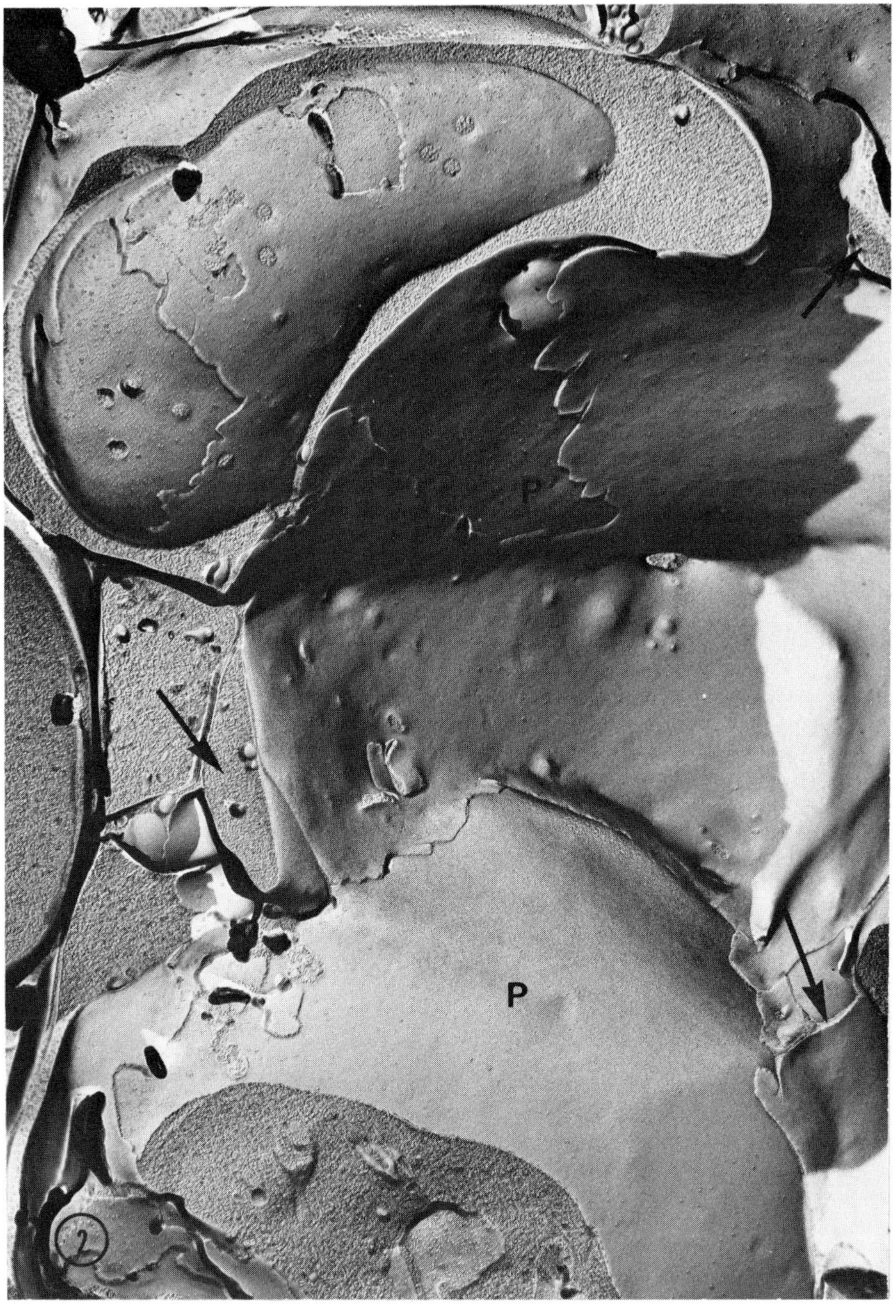
P
P
2

proliferative subventricular layer is concentrated along the external wall of the lateral ventricle, with an extension over the head of the caudate nucleus (Smart, 1961). The cells of the subventricular layer are separated from the ventricular lumen by a single row of ependymal cells. These are identical with those described by Brightman and Palay (1963) in the wall of the third ventricle.

Subventricular cells (Privat and Leblond, 1972) appear to be tightly packed next to the ependyma, without any intervening basement lamina. Their nuclei are, as a rule, smaller and darker than those of ependymal cells. They are spindle-shaped or ovoid, with their long axis parallel to the ventricle surface (Fig. 1). Their nuclear outline is very frequently indented, and their nucleoplasm may vary from relatively light and uniform to a patchy appearance, due to many ill-defined clumps of chromatin. Among the darker, one may identify small cells, usually found as a pair, which represent a late telophase, and also very large cells, with a checkerboard-like nucleoplasm, which likely correspond to early prophase cells. The cytoplasm of subventricular cells is usually restricted to a thin perinuclear rim, with a somewhat larger accumulation at both poles; it contains many free ribosomes, mostly arranged in small polysomes. Cisternae of rough endoplasmic reticulum (RER) are rare, and the few mitochondria contain dense granules. Microtubules are sometimes present, but no filaments may be seen.

These cells often possess long and thin processes which run along their main axis, and usually arise from the two poles (Fig. 2). These processes exhibit a cytoplasmic structure similar to that of the perikaryon. A small number of microglia are also found in the subventricular layer; their features are identical to those of the microglia of the corpus callosum described by Mori and Leblond (1969a).

The subventricular layer proper is separated from the adjacent nervous parenchyma by the so-called intermediate zone. In adult animals, the subventricular layer is divided into isolated small clumps of cells, around which the border area extends. This border area is composed of scattered cells surrounded by a light staining neuropil. These cells include, besides an occasional neuroblast (in rats up to 20 days old), and a few mature neurons, a series of cells similar in some respects to those of the subventricular layer, but differing by having a lighter nucleus and cytoplasm. Chromatin clumps are less numerous, the cytoplasm contains less free ribosomes and more organelles and rough ER cisternae. A few of the cells of the border area display some features of mature neuroglial cells, and they will be considered

FIG. 2. Freeze fracture replica of the subventricular layer of a 20 day old rat. The cells appear closely molded upon each other, and exhibit thin tapering processes (arrow). Notice the relative paucity of intramembranous particles apparent on the P faces (P). (× 15000).

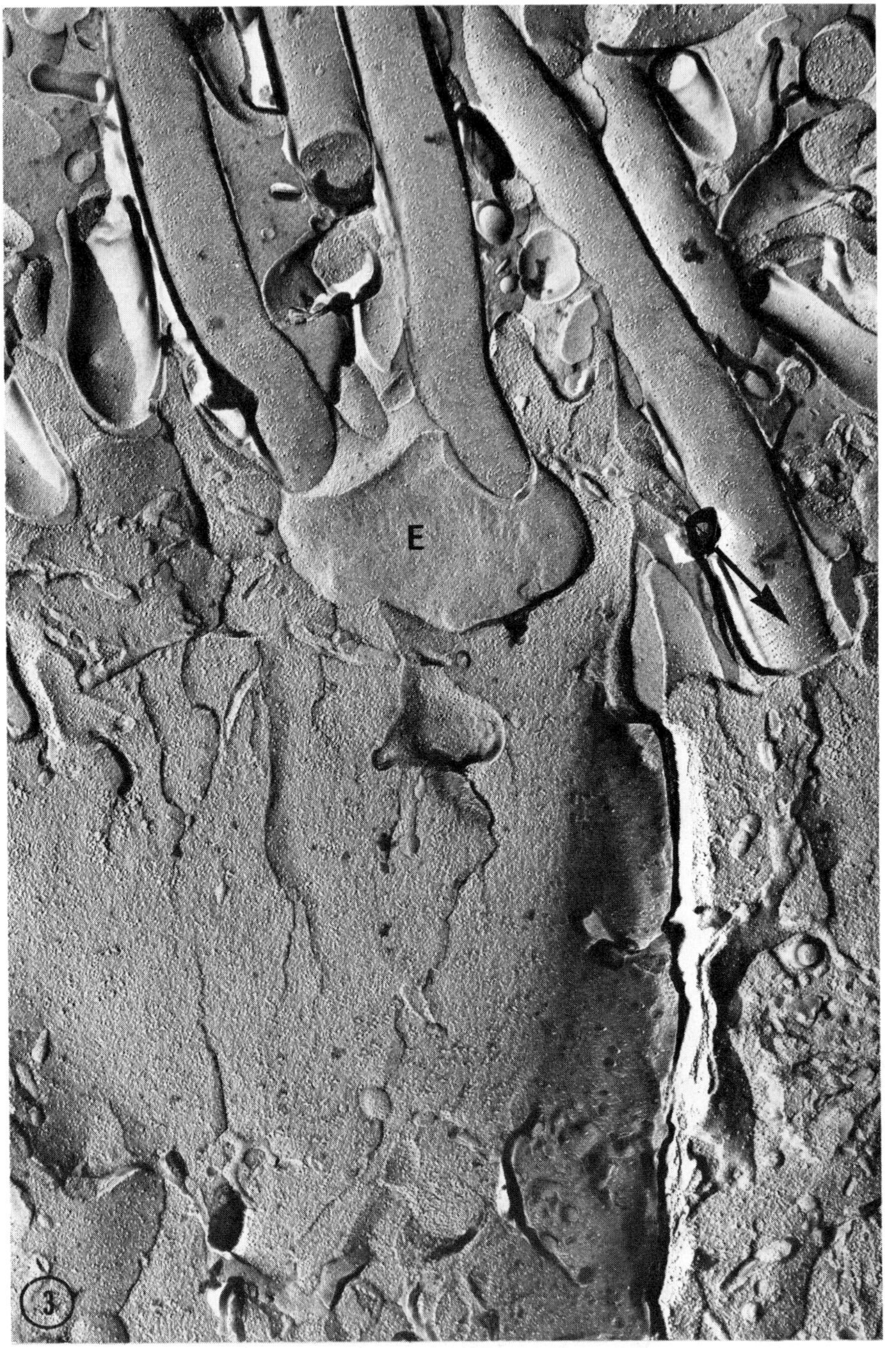
E
3

in the sections relevant to these cell types. Fully mature glial cells are rare, except for an occasional microglia.

Electron microscope radioautographs after tritiated thymidine injection in one month old rats show that two hours after the injection dark and light cells of the subventricular layer are labeled (Fig. 1); whereas ependymal cells are not labeled, nor are most of the cells of the intermediate zone. If one tries to correlate these recent findings with those of Smart (1961), one may admit that the light cells of this author may be a composite group, including the cells of the intermediate zone, which are indeed lighter than those of the subventricular layer, and some of the lighter cells of the subventricular layer itself.

Other information about the subventricular layer came recently from the examination of replicas of freeze-fractured material (Privat, 1977). Subventricular cells were characterized by paucity of intramembranous particles, except for some clustering of particles on the P face of a cell, matching regions of cobbly appearance on the E face of the neighbouring cell. When the irregular appearance of the extracellular space in the subventricular layer is considered, it appears that there are regions of adhesion between cells (Fig. 2).

Another characteristic of subventricular cell membranes was the existence of regions of P faces completely devoid of particles, and corresponding to small, blunt processes. This was compared with the membrane of growth cones, as illustrated by Pfenninger and Bunge (1974) which was similarly devoid of particles.

In addition, it is worth noting that not all the cells of the subventricular layer have the same low density of intramembranous particles. Those located farther from the ependyma usually have a higher density, as more mature cells are expected to do. However, a complete spectrum of densities was found, rather than a clearcut distinction between two classes of cells.

Finally, freeze fracture replicas disclosed the tri-dimensional appearance of cells, and one of the characteristic features of subventricular cells is the existence of very attenuated processes appearing as thin sheaths of cytoplasm sandwiches between perikarya closely molded upon each other (Fig. 2). At variance, ependymal cells were seen to possess membrane characteristics similar to those of astrocytes: the gap junctions already seen on thin sections were readily apparent with freeze fractures, and polygonal assemblies described by Deremietzel (1973) and Landis and Reese (1975) as specific of astrocytes (Fig. 3).

Another set of information was brought by Stensaas and Gilson (1972)

FIG. 3. Freeze fracture replica of the ependyma of a 20 day old rat. Patches of apical surface membrane (E face) show the characteristic polygonal arrays of pits; the ciliary necklace, with its six rows of particles is well demarcated at the base of the cilia (arrow) (× 20000).

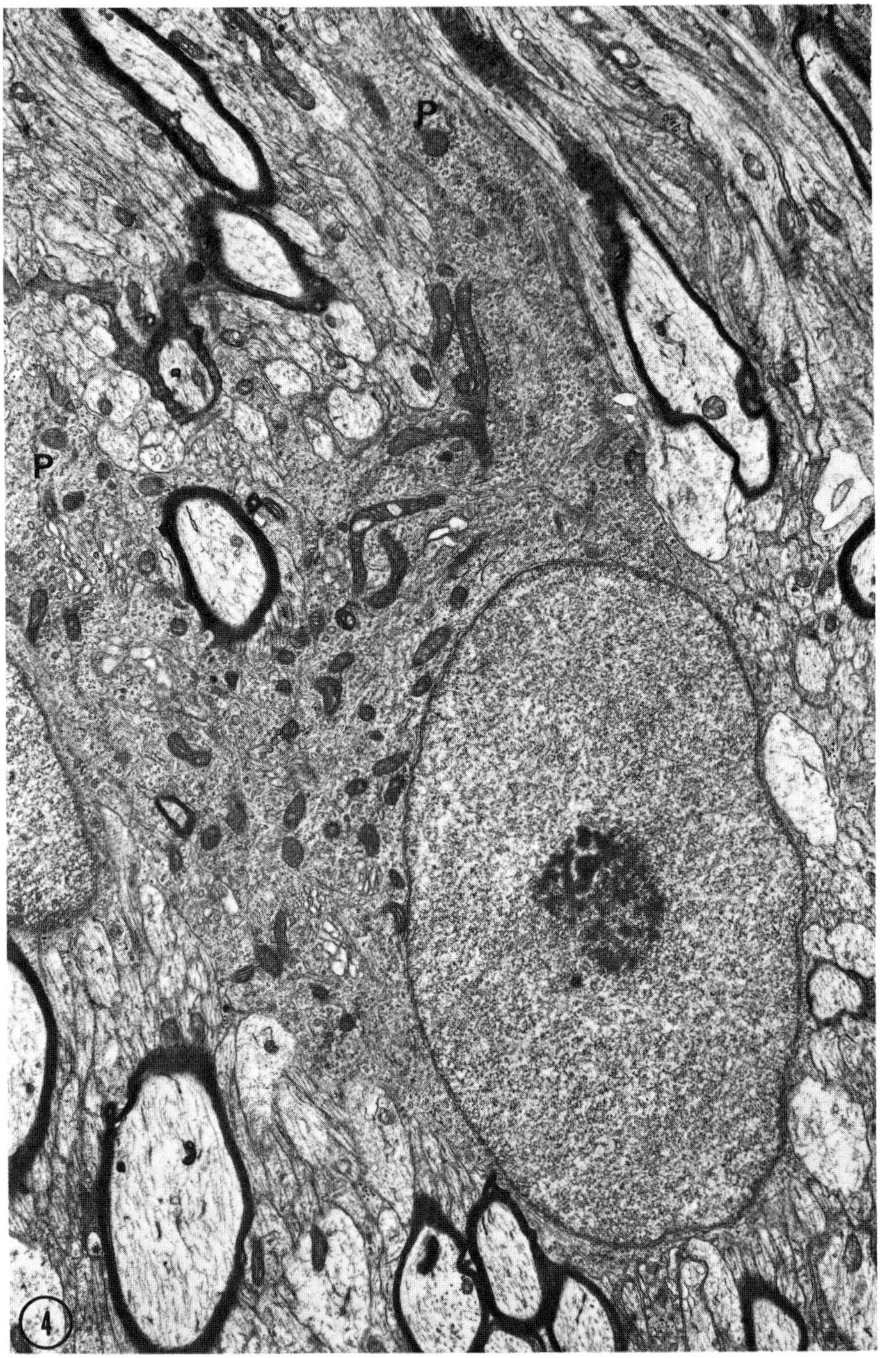
P
P
4

who reconstructed tri-dimensionally several cells of the subventricular layer of the rabbit by electron microscopy serial sectioning; they came to the conclusion that there exists a whole spectrum of cells in the subventricular layer, from the simple, dark apolar cell, to the complex, light, ramified cell; these different shapes and sizes were supposed to represent different stages of the mitotic cycle. It is possible, then, that the subventricular layer of young animals is made up of a single cell type with a high rate of mitosis, which is mixed with the more differentiated cells of the intermediate zone. Work is now in progress, associating radioautography and electron microscopy, to ascertain the validity of this assumption.

B. The Optic Nerve

The optic nerve may be used for the study of gliogenesis in the central nervous system, since the production of its ventricular layer is directed exclusively to gliogenesis. The situation, however, is different from that prevailing in the vicinity of the lateral ventricle, since here neither a true ventricular nor a subventricular layer is present after birth, even though mitoses are still numerous during the first two weeks of postnatal growth. Skoff *et al.* (1976a,b) found no glioblasts and attributed the mitoses to oligodendroblasts and astroblasts.

III. The Oligodendrocyte Line

Following the pilot investigations of del Rio Hortega (1928) the early electron microscope studies described oligodendrocytes as large, light cells (Luse, 1956; Dempsey and Luse, 1958), while others described them as small, dense cells with a dark spherical nucleus (Farquhar and Hartmann, 1957; Palay, 1958). Subsequent workers ignored the work of Luse, but supported the view that these cells were small and dense (De Robertis and Gerschenfeld, 1961; Palay *et al.*, 1962; Malfors, 1963; Mugnaini and Walberg, 1964, Schultz, 1964; Bodian, 1964; Wendell-Smith *et al.*, 1966; Stensaas and Stensaas, 1968; King, 1968).

More recently, other authors (Kruger and Maxwell, 1966; Caley and Maxwell, 1968) estimated that these cells could display a wide range of sizes, and of nuclear and cytoplasmic densities. These apparent discrepancies were cleared up by recent studies of the *maturation* of oligodendrocytes in various regions of the brain. For the sake of clarity, we shall begin with the description of the differentiated oligodendrocytes, and review thereafter the more primitive forms.

FIG. 4. A light oligodendrocyte in the corpus callosum of a 20 day old rat exhibits the typical large, light nucleus, with a prominent nucleolus, abundant cytoplasm, slightly denser than the nucleus, and large processes (P). (× 12500).

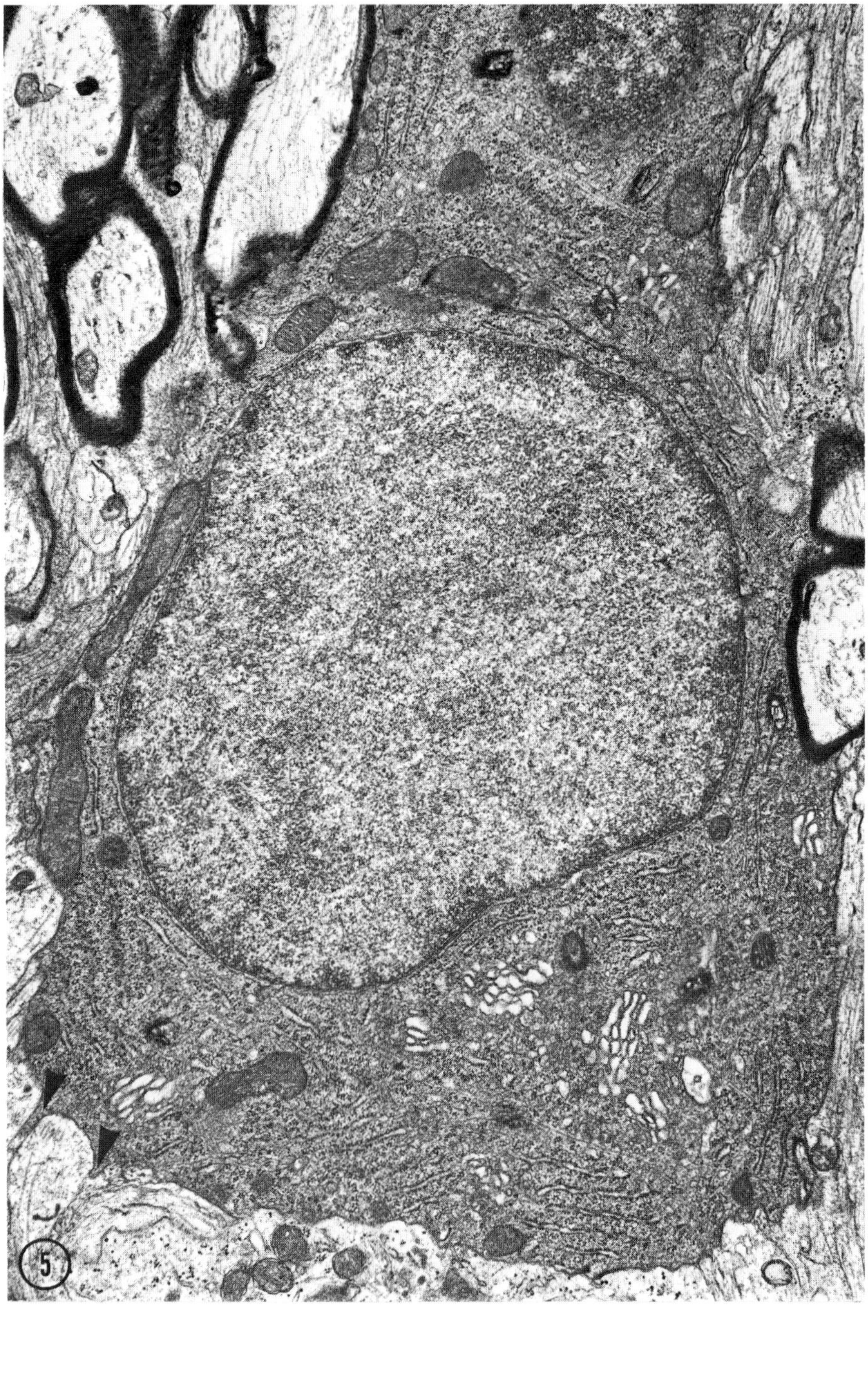
5

A. The Differentiated Oligodendrocytes

1. Corpus callosum. In the corpus callosum of one month old rats, a large spectrum of differentiated oligodendrocytes may be found. This spectrum includes cells at various stages of maturity, but all of those which will be described now are considered as functional, i.e., they are involved in the production or maintenance of the myelinating sheath of axons.

Three basic types were described by Mori and Leblond (1970).

The light oligodendrocyte is characterized by a large, pale nucleus and abundant cytoplasm. The nucleus is as light as the axoplasm of neighbouring axons, has a mean diameter of 7 μm, and a regular outline. There are no clearcut masses of chromatin, and the large nucleolus, most often located centrally (Fig. 4) exhibits fibrillar and granular components. The abundant cytoplasm is usually slightly denser than the nucleus, and contains numerous organelles: the cisternae of rough endoplasmic reticulum are short, with a narrow lumen, and often stacked. The Golgi apparatus is rather prominent, and appears as numerous groups of stacked saccules. Microtubules are abundant in the cytoplasm and may be seen in continuity with those present in the fine processes connecting these cells to the myelin sheaths. These processes are numerous, thin, and tend to be rectilinear; besides the microtubules, they contain only the occasional mitochondria.

The medium oligodendrocyte is denser and/or smaller than, but not fundamentally different from, the light one; there exists in fact a continuous spectrum of sizes and densities. A typical medium oligodendrocyte has a nuclear diameter of 4-7 μm, and the nucleoplasm is denser than the axoplasm of neighbouring axons; it contains a few chromatin masses, stacked along the membrane. The nucleolus is more discrete than in light cells. The cytoplasm is denser with a more prominent Golgi zone and more stacked cisternae. Microtubules are still abundant, and processes are few and thin (Fig. 5).

The dark oligodendrocyte is the smaller and the denser of the series. Again, there exists a large spectrum of shades and sizes, with no clear-cut demarcation from the medium oligodendrocyte. The nucleus, frequently eccentrically located, may be notched; in the scanty cytoplasm, the cisternae of the Golgi stand out as light in a dark ground cytoplasm; the rough ER still appears stacked. Lamellar bodies are consistently found, usually associated with membranous structures of the cells: smooth ER, mitochondria, etc. Processes are no longer seen to arise from these cells.

Systematic counts (Mori and Leblond, 1970) and radiautographic studies (Paterson *et al.*, 1973) have demonstrated that light oligodendrocytes

FIG. 5. A medium oligodendrocyte in the corpus callosum of a 20 day old rat is smaller and denser than the light type; the cisternae of rough E.R. are characteristically stacked, and the cisternae of the Golgi stand out in the dense cytoplasm. (× 16000).

transformed into medium zones, which in turn gave rise to dark oligodendrocytes — the truly mature form. In a preliminary study, Mori and Leblond (1970) estimated that light and medium oligodendrocytes could eventually undergo mitosis, but a more systematic evaluation (Leblond, personal communication) led to the conclusion that they were post-mitotic cells, and must be, as such, regarded as fully differentiated elements. In recent work, it was shown that each type is a stage in the life of the oligodendrocyte; in the rat the light stage was estimated to last about a week, the medium stage about two weeks, and the last stage at least a year (Imamoto and Leblond, 1977).

2. Optic nerve In the optic nerve, at least two types have been recorded by Vaughn (1969), the "active oligodendrocyte", appearing on the fifth postnatal day which seems to be the counterpart of the light oligodendrocytes just described, whereas the "mature" oligodendrocytes correspond to the dark type.

3. Gray matter A uniform situation has been found in the gray matter of the cerebral cortex, cerebellar cortex, and caudate nucleus, where oligodendrocytes are less numerous than in the white matter (Ling and Leblond, 1973) and the maturation seems to be less synchronized than in large myelinated tracts, although the sequence of events is probably the same. Oligodendrocytes are often found in satellite position to neurons, and they were not observed to transit through the light and medium stages. One may suppose that the undifferentiated precursor transforms readily into the mature dark oligodendrocyte.

B. The Precursors of Light Oligodendrocytes

1. Subventricular layer and corpus callosum. It has been indicated by tritiated thymidine labeling that the undifferentiated subventricular cells transformed into young, and then mature oligodendrocytes (Privat, 1970; Paterson *et al.*, 1973). With the electron microscope, cells were seen in the so-called "intermediate zone" (of the subependymal layer) which presented some characters of the oligodendrocyte line (Privat and Leblond, 1972): their cytoplasm was more abundant than that of subventrical cells, their organelles were more numerous, and they exhibited numerous small processes containing microtubules. But, conversely, it must also be noted that both the intermediate zone and the corpus callosum contain undifferentiated cells, isolated or in small clusters, in one month old rats, when the only cell line which has not reached its adult percentage is the oligodendrocyte line. We may then suggest that undifferentiated precursors (glioblasts) migrate from the subventricular layer into the intermediate zone and then into the corpus callosum, where they might undergo some divisions, eventually become specified as oligodendroblasts, undergo their final mitosis, and finally mature through the light-medium-dark sequence.

2. The optic nerve. In a recent series of papers, Skoff *et al.*, (1976a, b) reinvestigated the proliferative capacity of glial precursors and the time of origin of the different cell lines in the rat. They consider that, after birth, oligodendrocytes arise by multiplication of oligodendroblasts; the multiplication of oligodendroblasts appears as a protracted phenomenon, oligodendrocytes being still produced in 30-35 day old rats. Incidentally, the hypothesis of Vaughn (1969), that a small glial precursor, similar to microglia was the common stem cell, is no longer considered valid by Skoff *et al.* (1976a, b).

C. Possible Factors affecting the Differentiation and Maturation of Oligodendrocytes

1. Absence of oligodendrocyte differentiation in the jimpy mouse. The neurological mutant ''jimpy'' was first studied by Sidman and Hayes (1965), then by Hirano *et al.* (1969) who described an almost total absence of myelin. Subsequently, the maturation of oligodendrocytes was investigated in the corpus callosum of jimpy mice by Farkas *et al.* (1972) and Privat *et al.* (1972) who showed that the number of oligodendrocytes was drastically reduced as early as the 14th postnatal day. The few oligodendrocytes present were the light type, whereas the number of free subventricular cells was at least equal or superior to the values of the normal mouse. Very few axons appeared myelinated. In older animals (20 and 29 days postnatal) the number of oligodendrocytes was again dramatically reduced, and all the cells were of the light type. Glioblasts were still present in large numbers, at variance with the control animal of the same age. Astrocytes as well as microglia did not show any evidence of maturation defect, and they appeared slightly reactive, due to a large extent of cell necrosis, possibly of abnormal oligodendrocytes. Thus it appears that the genetic defect leading to the ''jimpy'' condition is restricted to the oligodendrocyte line, and does not affect directly the common precursor of the oligodendrocyte and the astrocyte.

2. Effect of Wallerian degeneration in the optic nerve of the newborn rat on oligodendrocyte differentiation. Wallerian degeneration has been extensively investigated in the optic nerve of the adult rat, and glial reactions were described by Vaughn and Pease (1970), Vaughn *et al.* (1970), and Skoff and Vaughn (1971) among others.

Recently, Privat and Fulcrand (1976), Valat *et al.* (1977) reinvestigated the same pathological process in the young rat, during the period of glial and especially oligodendroglial-maturation, in the hope of shedding some light onto the factors which govern the final fate of undifferentiated glial precursors. It was found that if enucleation of the eye was performed two

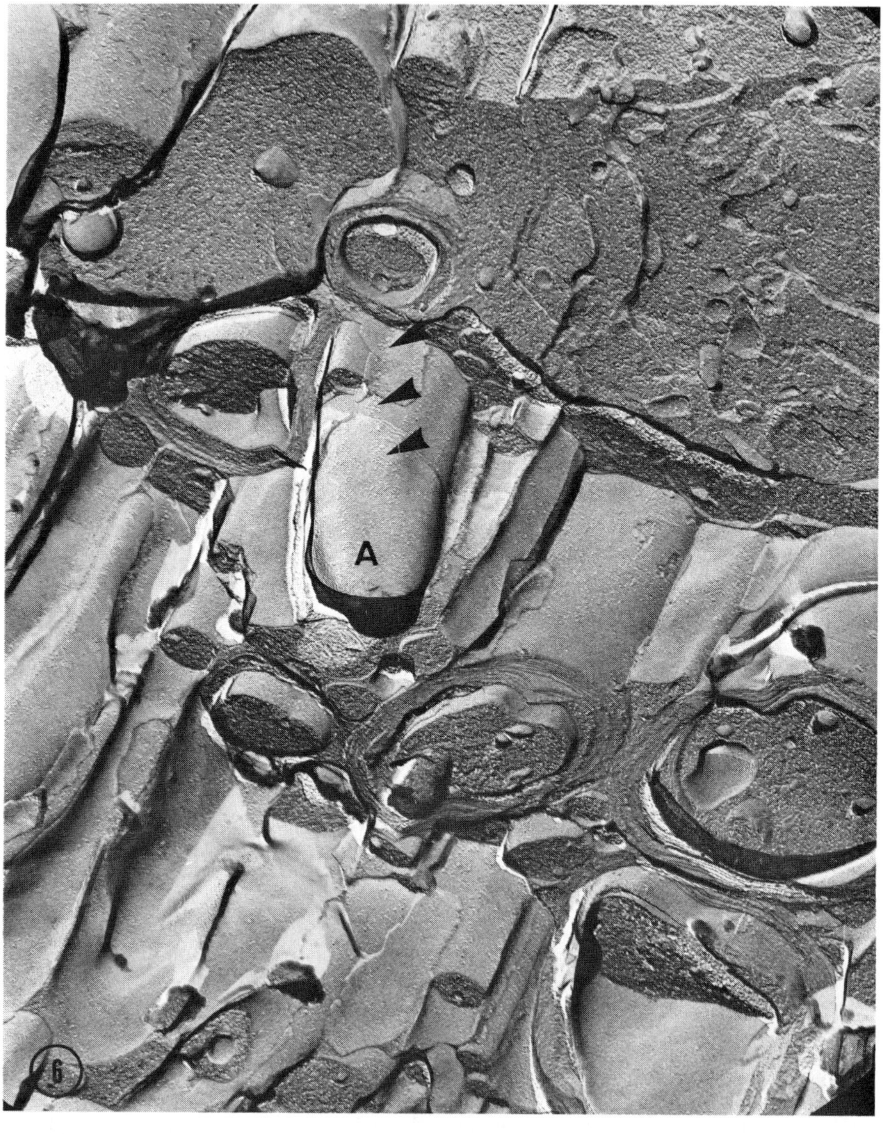

FIG. 6. Freeze fracture replica of the corpus callosum of a 20 day old rat. At upper right is the cross fractured cytoplasm of an oligodendrocyte, which is surrounded by several myelinated axons; on the axon labeled A, notice the correspondence of strands of particles on three subsequent membrane leaflets (arrow). (× 16000).

FIG. 7. Freeze fracture replica of the corpus callosum of a 20 days old rat. The myelinated axon in the center of the field shows a perfect correspondance of tight junction particles from one turn to the other. (Compare with particles on Fig. 6); in the innermost turn the tight junction faces the inner mesaxon. (× 20000).

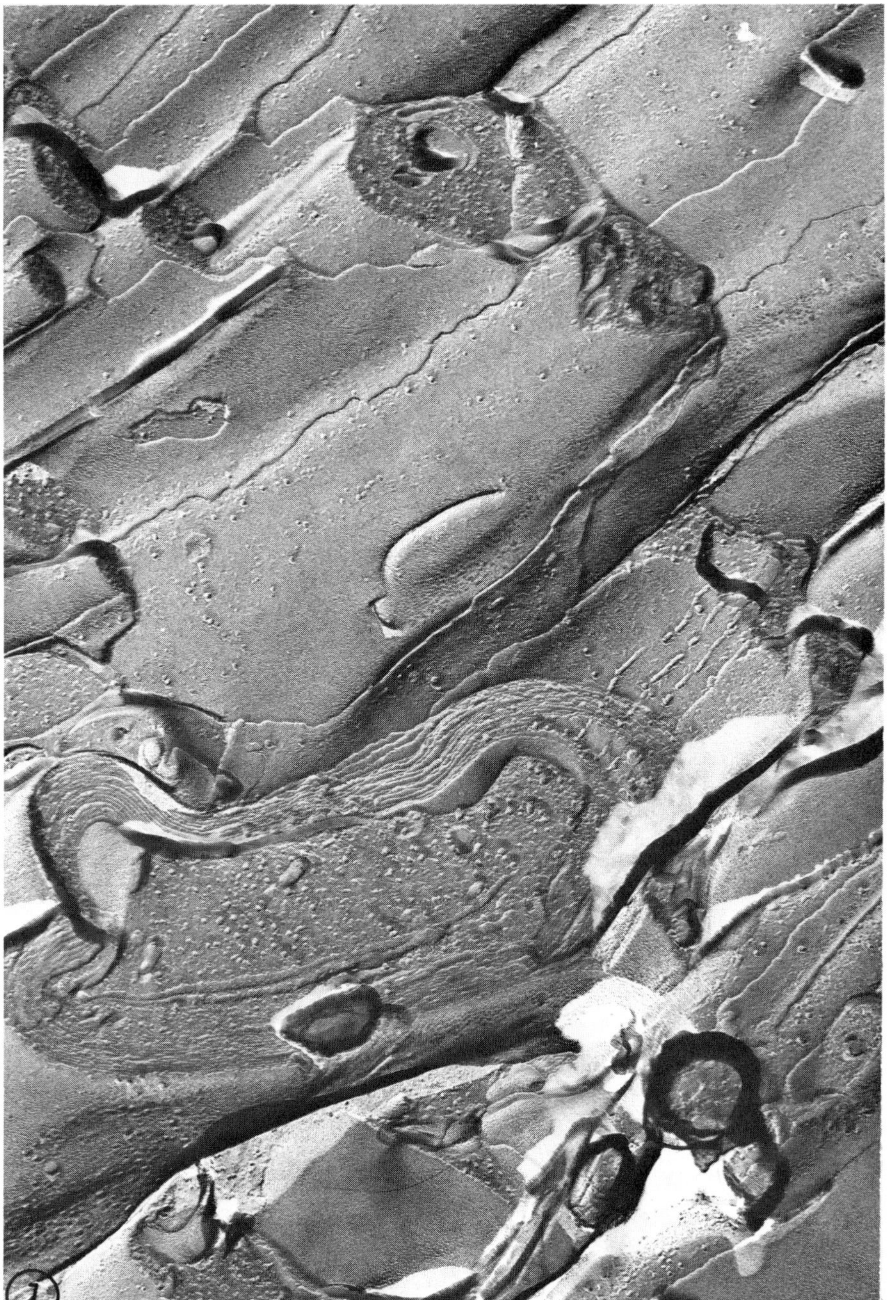
1

days after birth, almost no oligodendrocytes differentiated (3% of the total glial population instead of 70% in the normal animal). When the enucleation was performed five days after birth, a significant number of oligodendrocytes were recognizable in the optic nerve two days later, but they died rapidly, and the same phenomenon occurred when the enucleation was performed 8 days after birth. It was concluded that healthy axons were necessary for the differentiation of oligodendrocytes; in addition, it appeared, especially in the 5 day old animals, that a few cells already specified as oligodendrocytes, but not yet identifiable as mature cells, could progress in their maturation, but that in the absence of axons they died.

3. Formation of tight junctions by oligodendrocytes during myelination. The examination of freeze fracture replicas of the corpus callosum of 20 day old rats allows the disclosure of some of the early events of the myelination process; as soon as the axon is wrapped in the oligodendrocyte cytoplasm, clumping of particles is apparent on the P membrane of the axon, just facing the inner mesaxon, i.e., in the region of the initial contact. As soon as the first turn of glial cytoplasm is completed, tight junctions become apparent between oligodendrocyte membranes. As myelination proceeds, tight junctions are found facing each other at each turn of the myelin (Fig. 6 and 7).

As a temporary conclusion, one can say that oligodendrocytes arise from subventricular cells, through a series of transformations and divisions which may include the division of partly differentiated cells. The proliferation and differentiation of oligodendrocytes is governed by a set of genes independent of those which govern astrocyte proliferation and differentiation. The genetic background may be influenced by the tissue environment, and especially by the absence of healthy axons.

IV. The Astrocyte Line

Since the turn of the century the identification of astrocytes has been based on a metallic impregnation technique, the gold sublimate method of Ramon y Cajal (1913). This method, together with several variants of the Golgi technique, led to the classical distinction between protoplasmic astrocytes, essentially found in the gray matter, and fibrous ones, present in the white matter. Both, however, were provided with vascular endfeet. With ordinary stains applied on paraffin sections, Penfield (1932) identified astrocytes as cells with large, light, irregularly oval nuclei, lighter and larger than those of oligodendrocytes and microglia. The same criteria were used by the early electron microscopists to identify these cells and describe their ultrastructural characteristics.

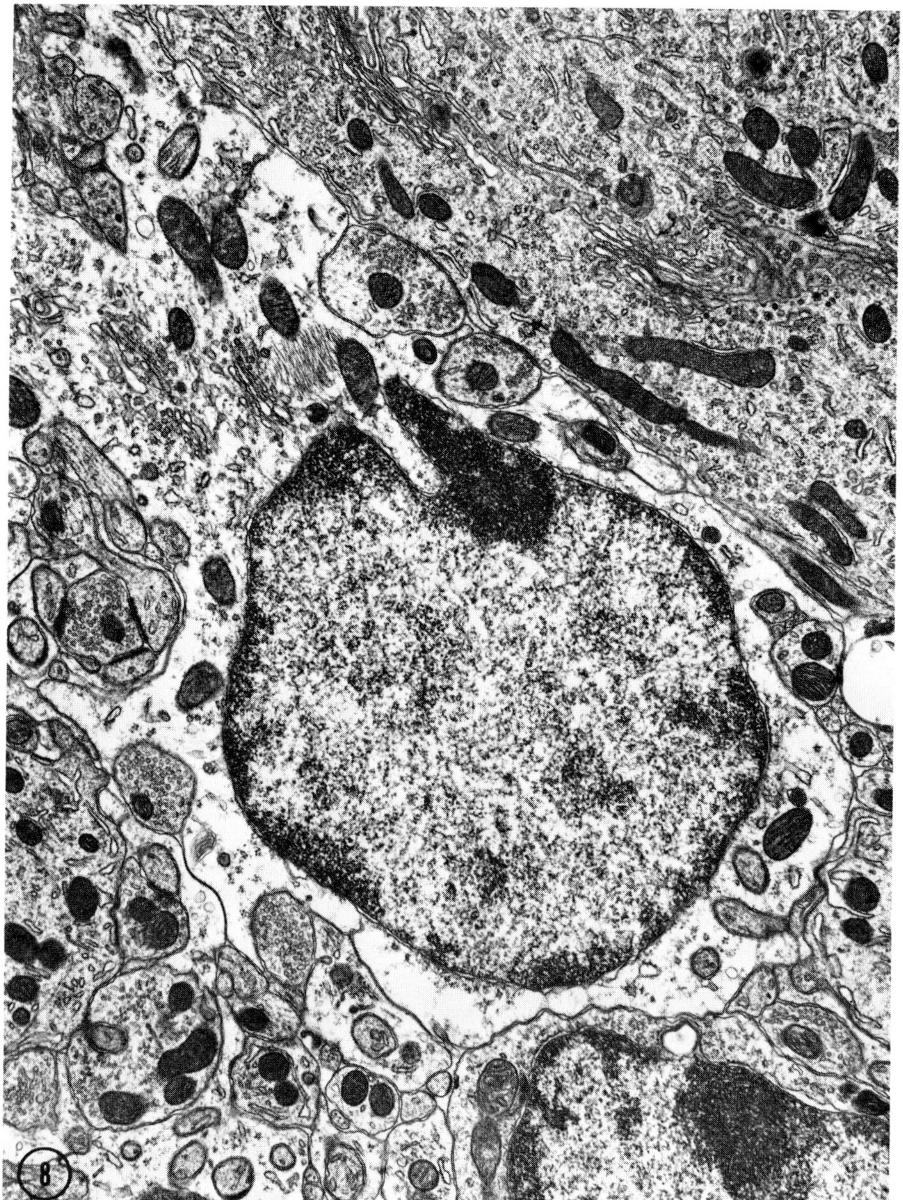

FIG. 8. A typical astrocyte in the cerebellar cortex of a one month old rat exhibits a light nucleus with slight clumping of chromatin along the nuclear membrane, a watery cytoplasm containing bundles of gliofilaments. (× 13000).

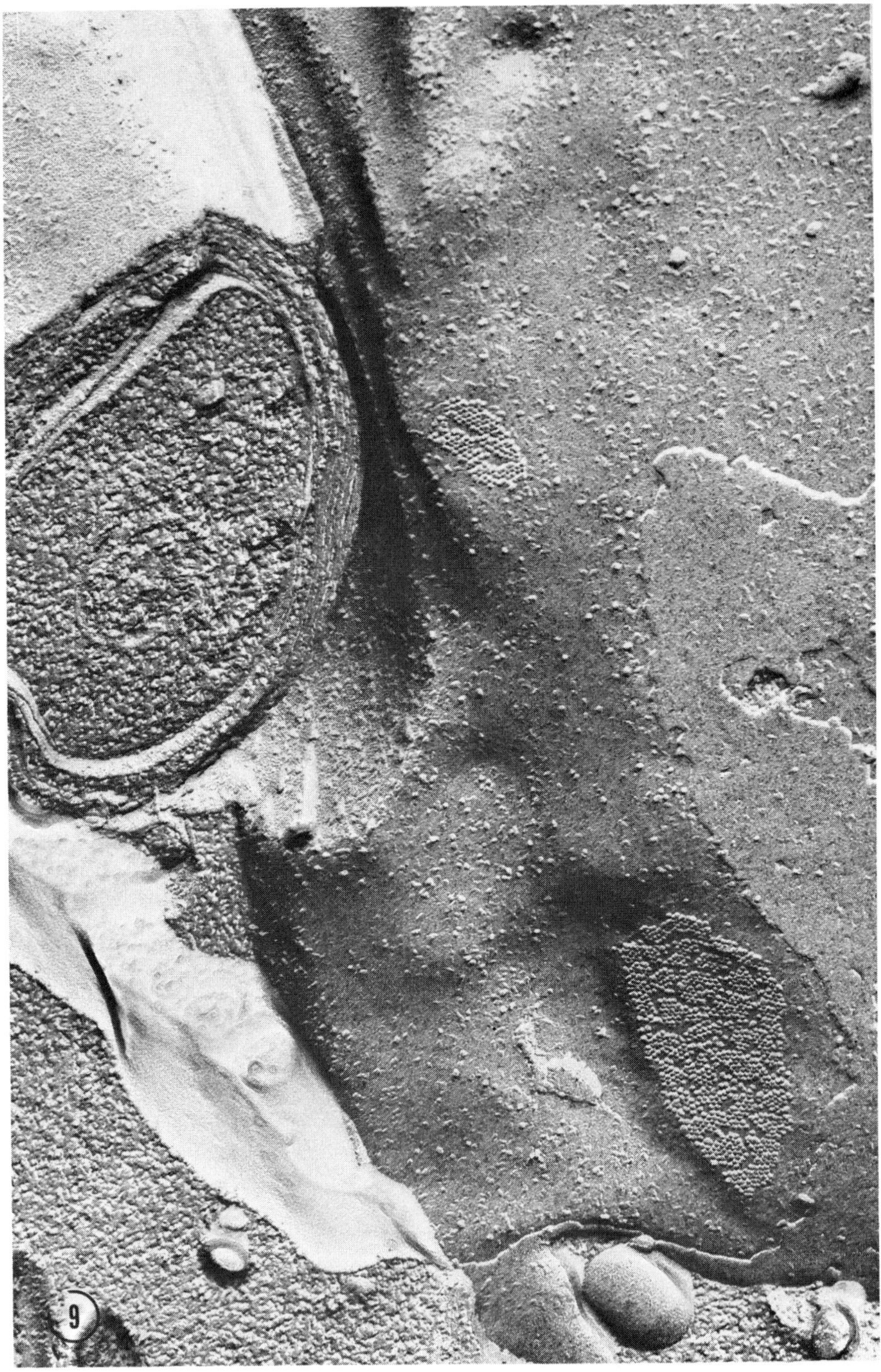
9

A. The Mature Astrocyte

The early description of Palay (1958) mentioned the presence in the astrocyte cytoplasm of bundles of gliofilaments, 8-9 nm in diameter, extending into the processes. This was confirmed by De Robertis and Gerschenfeld (1961), Malmfors (1963), Schultz (1964), Bodian (1964) for fibrous astrocytes, and by Hendron (1964), Wolf (1965), Kruger and Maxwell (1966) for protoplasmic ones. Mori and Leblond (1969b) showed that the filaments were responsible for the gold sublimate staining. Glycogen was also found to be a regular component of the astrocyte cytoplasm (Kruger and Maxwell, 1966; Wendell-Smith *et al.*, 1966; Vaughn and Peters, 1967; Stensaas and Stensaas, 1968; Caley and Maxwell, 1968).

The general consensus can be summarized in describing astrocytes, both protoplasmic and fibrillar, as large cells, with a light nucleus, sometimes notched, having finely dispersed chromatin slightly condensed along the nuclear membrane and a rare nucleolus (Fig. 8). The cytoplasm is usually light, with few organelles. The cisternae of the rough ER are short, their content is usually dense; besides the filaments, an occasional microtubule may be seen, and the other typical inclusion observed after gold sublimate staining is the so-called gliosome, which some authors consider as a lysosomal dense body (Mori and Leblond, 1969) while others relate it to condensed mitochondria. Typical end-feet completely encircle the fused basement membrane of blood vessels and of nervous tissue. The adjacent end-feet, which constitute the glia limitans, are linked together by gap junctions (Fig. 9) and, more rarely, by small attachment plates of the "puncta adherentia" type.

B. The Differentiation and Maturation of the Astrocyte

1. In the subventricular layer and corpus callosum. Radioautographic studies after infusion of tritiated thymidine in the lateral ventricle indicate that in the young rat a small, but significant number of subventricular cells migrate into the corpus callosum where they differentiate into astrocytes (Privat, 1970; Paterson *et al.*, 1973). The electron microscopic examination of the intermediate zone, as well as of the corpus callosum, in one month old rats disclosed some cells similar in many respects to subventricular cells, but already showing some evidence of differentiation into the astrocyte line (Privat and Leblond, 1972); these cells had a light cytoplasm, with few ribosome rosettes, a uniform nucleus, and some isolated cisternae of rough ER. In addition, some of these cells showed discrete bundles of gliofilaments, or a few glycogen granules. In fact, all the intermediate stages may be found between the undifferentiated subventricular cell and the mature astrocyte. Moreover, mitotic cells have been found by Mori and Leblond (1969) to contain bundles of gliofilaments, but these cells were said to be

FIG. 9. Freeze fracture replica of the corpus callosum of a 20 day old rat. Beside a cross fractured myelinated axon, at upper left, is the P face of an astrocyte plasmic membrane, with the characteristic particle assemblies of gap junctions. (A 40000).

immature. At the present time it appears likely that the cells undergoing division are astroblasts rather than astrocytes (Imamoto and Leblond, 1977b). But the possibility that astrocytes can divide has not been eliminated.

2. In the optic nerve. The earliest evidence of differentiation in the astrocyte line is the extension of processes in a plane transversal to the axis of the nerve, which is almost concomitant with the advent of the typical short and dense cisternae of rough ER (Vaughn, 1969). Puncta adherentia are frequent between the radiating processes of the astroblasts, while gap junctions are characteristic of mature forms. Gliofilaments are rarely found in less mature forms, but microtubules are frequent in the processes (Peters and Vaughn, 1967). In a recent radioautographic study, Skoff *et al.* (1976b) showed that the time of origin of astrocytes in the rat optic nerve extends from the 15th fetal day to the 15th postnatal day; after the final mitosis of each astroblast, about two weeks would be necessary for the astrocyte to reach full maturity. While studying Wallerian degeneration in newborn rat optic nerves, Privat and Fulcrand (1976) noticed that young astrocytes were able to actively phagocytose large axonal debris, while fully adult ones could not, and that their morphological maturation was greatly accelerated under this pathological condition. However, they retained for some time their ability to divide, despite their mature appearance (Fulcrand *et al.*, 1977).

3. In the grey matter. In the *cerebellar cortex*, there exists a particular type of astrocyte, the so-called Golgi epithelial cell, or Bergmann glia, whose perikaryon is located in the Purkinje layer, and which extends long, rectilinear processes up to the pia, where the end-feet form the glia limitans (Ramon y Cajal, 1911). According to recent data, (Del Cerro and Swarz, 1976) some of these cells already extend their processes across the whole thickness of the cerebellar cortex before birth. They have been supposed to function as guidelines for the inward migration of granule neuron precursors (Rakic, 1971). However, Das *et al.* (1974) found that Golgi epithelial cells underwent mitosis till the end of the first postnatal week.

In the *cerebral cortex*, Rakic (1972) has found similar relations between long astrocytic processes, reaching the pia, and migrating neuroblasts, well before birth in the monkey; here again, they are supposed to function as guides for the migrating neuroblasts.

In summary, then, astrocytes appear early as differentiated cells throughout the brain. Their time of origin is partly prenatal, and may extend, depending on the regions, to the end of the first postnatal month. Cells of the astrocyte type, referred to as astroblasts, which are already partly differentiated, as testified by the presense of gliofibrils in their cytoplasm, are still able to take up tritiated thymidine and to undergo mitosis. The latter phenomenon is particularly evident in reactive processes such as those secondary to Wallerian degeneration. In the normal adult animal, the mature astrocytes constitute, however, a stable population.

V. The Microglia

Classical microglial cells, identified by Del Rio Hortega (1919) by staining with silver carbonate, were distinguished by their small size, profusely branched processes, and their migratory as well as phagocytic activity. Del Rio Hortega (1930, 1932) found that they appeared in the fetal and perinatal period in the vicinity of the pia and of the ventricles where they represented the so-called microglial fountains, and in his estimation they were mesodermal elements. At variance, Rydberg (1932), Roussy *et al.* (1930), Horstadius (1950) expressed the view that they were ectodermal derivatives, as are astrocytes and oligodendrocytes. With the advent of the electron microscope, new discrepancies appeared in the ultrastructural descriptions and some authors denied the existence of microglia in normal tissue. Early ultrastructural descriptions (Luse, 1956; Farquhar and Hartmann, 1957; Blinzinger and Hager, 1962; Bodian, 1964; Yasuzumi *et al.*, 1961) depicted microglia as small cells with dense nuclei. Kruger and Maxwell (1966), Eager and Eager (1966), King (1968) considered that the only small cells with dense nuclei present in normal tissue were oligodendrocytes. Mori and Leblond (1969a) adopted for the electron microscope the metallic impregnation technique of Del Rio Hortega (1919). They found that one type of cell was selectively impregnated with this technique and described it as an elongated cell of small size, with an angular nucleus, bearing large patches of dark chromatin on a light nucleoplasm, a scanty cytoplasm containing a few, elongated cisternae of rough ER, an inconstantly well developed Golgi zone, and numerous dense bodies (Fig. 10). This cell was present in two locations: the *pericytal* microglia are enclosed within an expansion of the basement lamina of a blood vessel, while the *interstitial* microglia are scattered in the nervous parenchyma. These latter cells bear irregular processes of various calibres, whose density is greater than those of other glial cells and neurons. This description is very similar to that of Blinzinger and Hager (1962) and Stensaas and Stensaas (1968).

Moreover, a similar description was given by Peters *et al.* (1970) for a cell they called the third neuroglial element, which was found in the optic nerve of fetal and postnatal rats, as well as in fully adult animals. These authors considered this cell as an immature multi-potential element, precursor of all glial cells. Recently, Skoff (1976a and b) demonstrated with radioautography that microglia could not be the multipotential precursor in the optic nerve. However, the question remains as to the origin of microglia, and the relationship with oligodendrocytes and astrocytes. Mori and

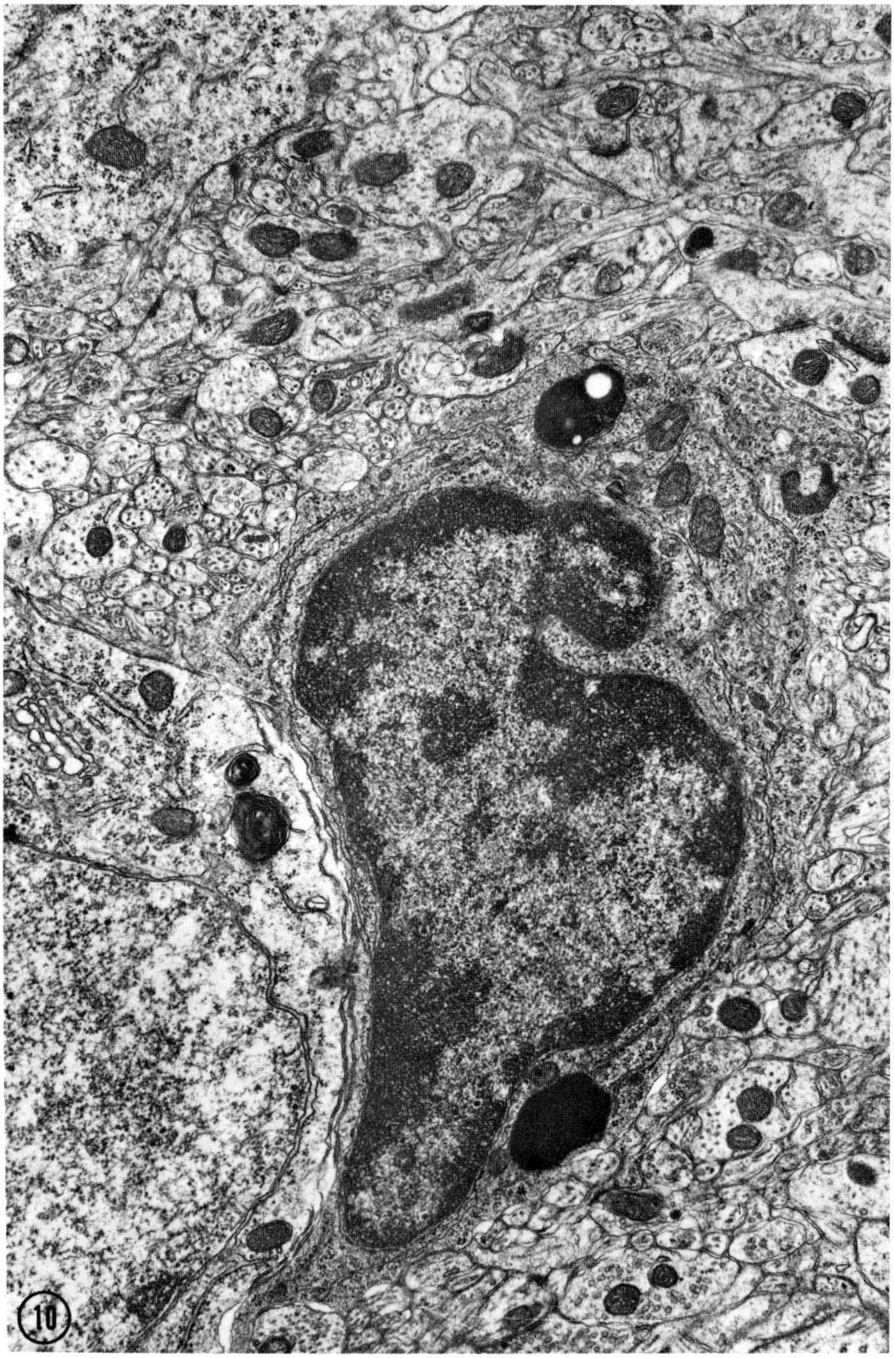
10

Leblond (1969a) found some evidence that pericytes could migrate into the brain parenchyma, and then favored the hypothesis that these cells could be hematogenous extravased elements. Privat and Leblond (1972) found numerous microglial cells in the vicinity of the subventricular layer, or within it, but were unable to find any morphological evidence of a transition between subventricular cells and microglia, as they did with oligodendrocytes and astrocytes. Stensaas and Reichert (1971), Stensaas and Gilson (1972) found two types of silver carbonate stained cells in the newborn rabbit, a round ameboid form and a ramified one (the latter corresponding to what is now recognized as microglia). They were also unable to find transitional forms with subventricular cells, despite an extensive tridimensional reconstruction study. Recently, Privat and Fulcrand (1976) suggested that these two forms could transform into each other, depending on the condition of the surrounding tissue. Ling and Tan (1974) and Imamoto and Leblond (1977a) showed that the ameboid form was a macrophage cell, which within 2 or 3 weeks of birth transformed into typical microglia.

A number of experimental studies were devised in order to decide on the exogenous or endogenous origin of microglia. I have recently (Privat, 1975) summarized the current concepts and available data. To sum up, it must be said that as soon as the endothelium of blood vessels is injured, as happens in a stab wound, hematogenous elements invade the nervous parenchyma (Konigsmark and Sidman, 1963) and at least some of them may take the ultrastructural appearance of microglia (Imamoto and Leblond, 1977a). The same phenomenon occurs in any case where a severe inflammatory process takes place in the brain parenchyma.

In other instances such as retrograde degeneration (Stenwig, 1972) a Wallerian degeneration (Vaughn *et al.*, 1970; Skoff, 1975; Privat and Fulcrand, 1976), the invasion of leucocytes seems to be absent, or minimal, and in this case the proliferative elements which will turn ultimately into phagocytes are indigenous to the brain. Whether they are neuroectodermal or mesodermal derivatives is another issue. In radioautographic experiments, Privat (1970) and Paterson *et al.* (1973) found that numerous microglia were labeled two days after injection of ^{3}H-thymidine in the lateral ventricle. These could have transformed from the initially labeled subventricular cells, but other origins cannot be excluded. Vaughn (1969), Vaughn *et al.* (1970) and Skoff (1975) estimate that in the optic nerve undergoing Wallerian degeneration, the numerous microglia stem from neuroectodermal precursors. It seems reasonable, as advocated by Peters *et*

FIG. 10. In the caudate nucleus of a one month old rat, the satellite microglia is almost completely separated from the neuron by a thin sheath of astroglial cytoplasm; the microglia is easily identified upon the dark chromatin masses on the light nucleoplasm, the long, tortuous cisternae of rough E.R., and the dense bodies (× 20000).

al. (1976) to reserve the term 'microglia' for inactive or minimally loaded cells, whereas the cells reacting to an insult would be called phagocytes, whether they originate from microglia, monocytes, or other exogenous elements.

Conclusions

The morphology and the ultrastructure of glial cells are now well known. The different stages of the maturation of astrocytes and oligodendrocytes are progressively being revealed. The combination of radioautography with electron microscopy will soon clear up some uncertain points. The main problem which faces us now is to discover the mechanisms which govern the differentiation. Purely descriptive studies, combined with experimental approaches, have already given us some insight into these problems. The early interactions between growing axon and immature oligodendrocytes, as well as those between migrating neuroblasts and astroglial processes await further investigation. As to the problem of microglia, considerable work has been accumulated during the last decade.

A general agreement has been reached about the ability of microglia to transform into macrophages. Recent advances indicate that the reverse could also occur, especially during the early postnatal period, in the corpus callosum. The old conception of Del Rio Hortega would thus be confirmed. It remains to be proved that this is true also for other regions of the brain.

Acknowledgments

The freeze-fracturing investigations and some recent radioautographic work have been carried out in the Anatomy Department, McGill University with the support of the Medical Research Council of Canada and of the France-Quebec Exchange Program. The contribution of an INSERM grant is also gratefully acknowledged.

The author is gratefully indebted to Dr. C. P. Leblond for many helpful discussions, and critical reading of the manuscript.

References

Allen, E. 1912. Cessation in mitosis in central nervous system of the albino rat. J. Comp. Neurol. 22:547-568.

Altman, J. 1966. Proliferation and migration of undifferentiated precursor cells in the rat during gliogenesis. Exp. Neurol. 16:263-278.

Angevine, J. B., D. Bodian, A. J. Coulombie, M. V. Edds, V. Hamburger, M. Jacobson, R. M. Lyser, M. C. Prestidge, R. L. Sidman, S. Varon and P. A. Weiss. 1970. Embryonic vertebrate nervous system: revised terminology. Anat. Rec. 166:257-262.

Bodian, D. 1964. An electron microscopic study of the monkey spinal cord. Bull. Johns Hopkins Hospital. 114:13-119.

Brightman, M. W. and S. L. Palay. 1963. The fine structure of ependyma in the brain of the rat. J. Cell Biol. 19:415-439.

Caley, D. W. and D. S. Maxwell. 1968. An electron microscopic study of the neuroglia during postnatal development of the rat cerebrum. J. Comp. Neurol. 133:45-70.

Das, G. D., G. C. Lammert and J. P. McAllister. 1974. Contact guidance and migrating cells in developing cerebellum. Brain Res. 69:13-29.

Del Cerro M. and J. R. Swarz. 1976. Prenatal development of Bergmann glial fibres in rodent cerebellum. J. Neurocytol., 5:669-676.

Del Rio Hortega, P. 1919. El tercer elemento de los centros nervosos. Bol. Soc. Espan. Biol. 9:68-83.

Del Rio Hortega, P. 1928. Tercera apartacion al conosimanto morfologica interpretacion functional de la oligodendroglia. Mem. Real Soc. Espan. Hist. Nat. 14:5-122.

Del Rio Hortega, P. 1930. Concepts historique, morphologique, physiologique et physiopathologique de la microglie. Rev. neurol. 1:956-986.

Del Rio Hortega, P. 1932. Microglia. In: *Cytology and Cellular Pathology of the Nervous System.* W. Penfield, ed. Vol. 2. Harper (Hoeber) New York, pp. 483-534.

Dempsey, E. W. and S. A. Luse. 1958. Fine structure of the neuropile in relation to neuroglial cells. In: *Biology of Neuroglia.* W. F. Windle, ed., C. C. Thomas, Springfield, Ill. pp. 99-108.

Deremietzel, R. 1973. Visualization by freeze-fracturing of regular structures in glial cell membranes. Naturwissenschaften. 60:208.

De Robertis, E. and H. M. Gerschenfeld. 1961. Submicroscopic morphology and function of glial cells. Int. Rev. Neurobiol. 3:1-65.

Dutrochet, H. 1824. Recherches anatomiques et physiologiques sur la structure interne des animaux et des végétaux et sur leur motilité. Paris, Baillière.

Eager, R. P. and P. R. Eager. 1966. Glial responses to degenerating cerebellar cortico-nuclear pathways in the cat. Science. 153:553-554.

Farkas, E., O. Robain, and P. Mandel. 1972. Abnormal glial maturation in the white matter of jimpy mice. An optical study. Acta Neuropath. (Berl.) 21:272-281.

Farquhar, M. G. and J. F. Hartmann. 1957. Neuroglial structure and relationships as revealed by electron microscopy. J. Neuropath. Exp. Neurol. 16:18-39.

Fulcrand, J., J. Valat and A. Privat. 1977. Proliferation of neuroglia in the young rat optic nerve after enucleation, evaluated by ^{3}H thymidine radioautography. Anat. Rec. 187:765.

Globus, J. H. and H. Kuhlenbeck. 1944. Subependymal cell plate (matrix) and its relationship to brain tumors of ependymal type. J. Neuropath. 3:1-35.

Herndon, R. M. 1964. The fine structure of the rat cerebellum. II. The stellate neurons, granule cells and glia. J. Cell Biol. 23:277-293.

Hirano, A., D. S. Sax, H. M. Zimmermann. 1969. The fine structure of the cerebella of Jimpy mice and their normal littermates. J. Neuropath. Exp. Neurol. 28:388-400.

His, W. 1889. Die Neuroblasten und deren Entstehung in embryonalen Mark. Abhandl. d. Math. Phys. Cl. D. Sachs Gesellsch. d. Wissench. 15:311-373.

Horstadius, R. 1950. *The neural crest. Its properties and derivation in the light of experimental research.* Oxf. Univ. Press, London and New York.

Imamoto R. and C. P. Leblond. 1977a. The origin of microglia. J. Comp. Neurol. (in press).

Imamota, R. and C. P. Leblond. 1977b. The life history of oligodendrocytes in the corpus callosum of the rat. J. Comp. Neurol. (in press).

Kershman, J. 1938. The medulloblast and the medulloblastoma. Arch. Neur. Psychiat. 40:937-967.

Kershman, J. 1939. Genesis of microglia in the human brain. Arch. Neurol. Psychiat. 41:24-50.

King, I. S. 1968. A light and electron microscope study of perineuronal glial cells and processes in the rabbit neocortex. Anat. Rec. 161:111-124.

Konigsmark, B. W. and R. L. Sidman. 1963. Origin of brain macrophages in the mouse. J. Neuropath. Exp. Neurol. 22:643-676.

Kruger, L. and D. S. Maxwell. 1966. Electron microscopy of oligodendrocytes in normal rat cerebrum. Am. J. Anat. 118:411-436.

Landis, D. S. and T. S. Reese. 1975. Arrays of particles in freeze fractured astrocytic membranes. J. Cell Biol. 60:316-320.

Ling, E. A. and C. P. Leblond. 1973. Investigation of glial cells in semithin sections. II. Variations with age in the numbers of the various glial cell types in rat cortex and corpus callosum. J. Comp. Neurol. 149:73-82.

Ling, E. A. and C. K. Tan. 1974. Amoeboid microglial cells in the corpus callosum of neonatal rats. Arch. Histol. Jap. 36:265-280.

Luse, S. A. 1956. Electron microscopic observations of the central nervous system. J. Biophys. Biochem. Cytol. 2:531-542.

Malmfors, T. 1963. Electron microscopic description of the glial cells in the nervus opticus in mice. J. Ultrastr. Res. 8:193-212.

Mori, S. and C. P. Leblond. 1969a. Identification of microglia in light and electron microscopy. J. Comp. Neurol. 135:57-80.

Mori, S. and C. P. Leblond. 1969b. Electron microscopic features and proliferation of astrocytes in the corpus callosum of the rat. J. Comp. Neurol. 137:197-226.

Mori, S. and C. P. Leblond. 1970. Electron microscopic identification of three classes of oligodendrocytes and preliminary study of their proliferative activity in the corpus callosum of young rats. J. Comp. Neurol. 139:1-30.

Mugnaini, E. and F. Walberg. 1964. Ultrastructure of Neuroglia. Erg. Anat. Entwicklungs-gesch. 37:194-236.

Opalski, A. 1935. Ueber lokale Unterschiede im Bau der Ventrikelwände beim Menschen. Z. Ges. Neurol. Psychiat. 149:221-254.

Palay, S. L. 1958. An electron microscopical study of neuroglia. In: *Biology of Neuroglia.* W. F. Windle, ed. C. C. Thomas, Springfield, Ill. pp. 24-38.

Palay, S. L., S. M. McGee-Russel, G. Spencer, Jr. and M. A. Grillo. 1962. Fixation of neural tissues for electron microscopy by perfusion with solutions of osmium tetroxyde. J. Cell Biol. 12:385-410.

Penfield, W. 1932. *Cytology and cellular pathology of the nervous system.* Harper (Hoeber), New York.

Paterson, J. A., A. Privat, E. A. Ling, C. P. Leblond. 1973. Investigation of glial cells in semithin sections. III. Transformation of subependymal cells into glial cells, as shown by radioautography after ^{3}H-thymidine injection into the lateral ventricle of the young rat. J. Comp. Neurol. 146:277-302.

Peters, A. and J. E. Vaughn. 1967. Microtubules and filaments in the axons and astrocytes of early postnatal rat optic nerves. J. Cell Biol. 32:113-119.

Peters, A., S. L. Palay and de F. Webster. 1976. *The Fine Structure of the Nervous System*, 2nd ed. Hoeber, New York.

Pfenninger, K. H. and R. Bunge. 1974. Freeze-fracturing of nerve growth cones and young fibers. A study of developing plasma membrane. J. Cell Biol. 63:180-196.

Privat, A. 1970. Sur l'origine des divers types de nevroglie chez le rat. In: *VIth Internt. Congress Neuropath.*, Masson, Paris. pp. 447-448.

Privat, A. 1975. Postnatal gliogenesis in the mammalian brain. Intern. Rev. Cytol. 40:281-323.

Privat, A. 1977. The ependyme and subependymal layer of the young rat. A new contribution with freeze-fracture. Neuroscience, (in press).

Privat, A. and J. Fulcrand. 1976. Glial reactions in the optic nerve of the young rat after unilateral enucleation. Anat. Rec. 184:505.

Privat, A. and C. P. Leblond. 1972. The subependymal layer and neighboring region in the brain of the young rat. J. Comp. Neurol. 146:277-302.

Privat, A., O. Robain and P. Mandel. 1972. Aspects ultrastructuraux du corps calleux chez la souris Jimpy. Acta Neuropath (Berl.) 21:282-295.

Rakic, P. 1971. Neuron-glia relations during granule cell migrations in developing cerebellar cortex. A Golgi and E. M. Study in Macacus Rhesus. J. Comp. Neurol. 141:283-311.

Rakic, P. 1972. Mode of cell migration to the superficial layers of fetal monkey neocortex. J. Comp. Neurol. 145:61-84.

Ramon y Cajal, S. 1909. *Histologie du système* nerveux de l'homme et des certébrés. Maloine, Paris.
Ramon y Cajal, S. 1913. Contribucion al conocimento de la neuroglia del cerebro humano. Trab. Lab. Invest. Biol. Univ. Madr. 11:255-315.
G. Roussy, J. L'Hermitte, Ch. Oberling. 1930. La névroglie et ses réactions pathologiques. Rev. Neurologique. 879-955.
Rydberg, E. 1932. Cerebral injury in newborn children consequent on birth trauma; with an inquiry into the normal and pathological anatomy of neuroglia. Act. Path. Microbiol. Scand. suppl. 10:1-247.
Schaper, A. 1897. Die Frühesten Differenzierungsvorgänge im Centralnervensystem. Arch. Entwickl. Mech. Org. 5:81-132.
Schultz, R. L. 1964. Macroglial identification in electron micrographs. J. Comp. Neur. 122:281-295.
Sidman, R. L. and R. Hayes. 1965. Jimpy: A mouse with inherited sudanophilic leucodistrophy. J. Neuropath. Exp. Neurol. 24:172-191.
Skoff, R. P. 1975. The fine structure of pulse labeled (^{3}H-thymidine) cells in degenerating rat optic nerve. J. Comp. Neurol. 161:595-612.
Skoff, R. P. and J. E. Vaughn. 1971. An autoradiographic study of cellular proliferation in degenerating rat optic nerve. J. Comp. Neurol. 141:133-156.
Skoff, R. P., D. L. Price and A. Stocks. 1976a. Electron microscopic autoradiographic studies of gliogenesis in rat optic nerve. I. Cell Proliferation. J. Comp. Neurol. 169:291-311.
Skoff, R. P., D. L. Price and A. Stocks. 1976b. Electron microscopic autoradiographic studies of gliogenesis in rat optic nerve. II. Time of origin. J. Comp. Neurol. 169:313-334.
Smart, I. 1961. The subependymal layer of the mouse brain and its cell production as shown by radioautography after thymidine H^3 injection. J. Comp. Neurol. 116:325-347.
Stensaas, L. J. and B. C. Gilson. 1972. Ependymal and subependymal cells in the caudate pallial junction in the lateral ventricle of the neonatal rabbit. Z. Zellforsch. Mikr. Anat. 132:297-322.
Stensaas, L. J. and W. H. Reichert. 1971. Round and amoeboid microglial cells in the neonatal rabbit brain. Z. Zellforsch. 119:147-163.
Stensaas, L. J. and S. S. Stensaas. 1968. Astrocytic neuroglial cells, oligendendrocytes and microgliacytes in the spinal cord of the toad. II. Electron Microscopy. Z. Zellforsch. 86:184-213.
Stenwig, A. E. 1972. The origin of brain macrophages in traumatic lesions, wallerian degeneration and retrograde degeneration. J. Neuropath. Exptl. Neurol. 31-696-704.
Valat, J., J. Fulcrand and R. Marty. 1977. Reactions gliales postnatales dans le nerf optique du rat après énucléation. I. Etude quantitative. Acta Anat. (in press).
Vaughn, J. E. 1969. An electron microscopic analysis of gliogenesis in rat optic nerves. Z. Zellforsch. 94:293-324.
Vaughn, J. E. and D. C. Pease. 1970. Electron microscopic studies of wallerian degeneration in rat optic nerves. II. Astrocytes, oligodendrocytes and adventitial cells. J. Comp. Neurol. 140:207-226.
Vaughn, J. E. and A. Peters. 1967. Electron microscopy of the early postnatal development of fibrous astrocytes. Am. J. Anat. 121:131-152.
Vaughn, J. E., P. L. Hinds and R. P. Skoff. 1970. Electron microscopic studies of wallerian degeneration in rat optic nerve. I. The multipotential glia. J.Comp. Neuro. 140:175-206.
Virchow, R. 1856. Zur pathologische Anatomie der Netshaut an des Sehnnerven. Virchow's Arch. 10:170-193.
Wendell-Smith, C. P., M. J. Blunt, and F. Baldwin. 1966. The ultrastructural characterization of macroglial cell types. J. Comp. Neurol. 127:219-239.
Wolf, J. 1965. Elektron mikroskopische Untersuchungen über Struktur und Gestalt von Astrozytenvorsätzen. Z. F. Zellforsch. 66:811-828.
Yasuzmi, G., I. Tsubo, R. Sugihara and Y. Nakai. 1964. Analysis of the development of Japanese B encephalitis (J.B.E.) virus. I. Electron microscopic study of microglia infected with J.B.E. virus. J. Ultrast. Res. 11:213-229.

BIOCHEMISTRY OF GLIAL CELLS

L. HERTZ

Department of Anatomy
University of Saskatchewan
Saskatoon, Saskatchewan, Canada

I. Introduction

Until recently, it was in general tacitly assumed that biochemical and physiological characteristics of brain tissue reflect events occurring in neuronal cells. This concept was probably related to the methodologies available; clinical neurological experience, histochemical techniques (e.g., Wallerian degeneration) and neurophysiological experimentation had already early demonstrated the dramatic, long-range consequences of neuronal activity. Glial cells, in contrast, were mainly known to form scar tissue whenever damage had occurred and were regarded as rather inert cells, glueing neurons together. This concept of a mainly passive role by glial cells in nerve tissue function was challenged by a few authors, most notably Lugaro (1907) who suggested that a major glial function could be to remove toxic substances from neuronal surroundings, and Holmgren (1904) who discussed a nutritive role of glial cells.

The first clear demonstration that non-neuronal cells have a major role to play in an important nerve tissue function seems to be that by Geren (1954) that myelin formation in the periphery is a result of an interaction between a neuronal component (the axon) and a non-neuronal cell (the Schwann cell), in which the latter actually is the cell that produces the myelin. Good evidence is found that oligodendrocytes exert a similar function in the central nervous system (Peters, 1960). Modern morphological and cell dynamic techniques have further expanded the knowledge about glial cell functions and characteristics during development, e.g., the guiding role of 'radial glia cells' (Rakic, 1972) and the severe effect of postnatal undernutrition on glial cell proliferation (Clos *et al.*, 1977). Simultaneously, biochemical and physiological techniques have been introduced which have allowed studies of the biochemistry and physiology of glial cells in the mature central or peripheral nervous system. These include *in vivo* monitoring of membrane potentials and potassium concentrations as well as *in vitro* microtechniques (Lowry 1953; Hydén, 1959), cell separation by gradient centrifugation (e.g., Rose, 1965, 1967; Freysz *et al.*, 1968; Blomstrand and Hamberger, 1969, 1970; Sellinger *et al.*, 1971; Norton and Poduslo, 1971; Nagata *et al.*, 1974; Abe and Norton, 1974) and tissue culture techniques. Each of these methods has its own advantages and disadvantages, a few of which are indicated in Table I, and each of them has been subject to severe criticism (e.g., Utley, 1963; Cremer *et al.*, 1968; Watson, 1974). A mutual cross contamination of bulk prepared glial cells with synaptosomes (Table I) and of synaptosomes with glial cell membranes (Cotman *et al.*, 1971; Henn *et al.*, 1974) may be specially disastrous. Nevertheless a picture of glial biochemistry begins to emerge, though still very incomplete and controversial.

For any discussion of the extent to which the biochemistry of the central nervous system reflects glial characteristics, it is relevant to determine which fraction of the total volume is made up of glial cells. This question has been discussed elsewhere (e.g., Wolff, 1970a,b; Glees, 1972; Hertz and Schousboe, 1975). It will suffice to mention that the glial contribution to the total brain volume is uncertain, but well may be close to 50 per cent in larger species. It varies from region to region and maybe even more from one species to another. This has been clearly demonstrated with respect to the neuronal/glial cell ratio which has been determined with much greater accuracy, and Fig. 1 shows that glial cells outnumber neurons

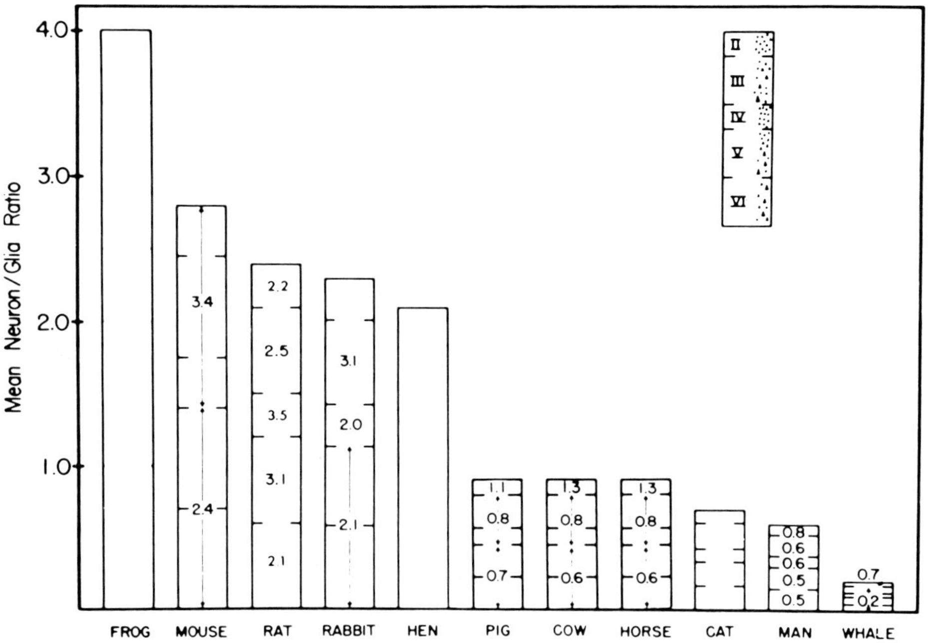

FIG. 1. Neuron/glial cell ratio in different layers (see insert) of the cerebral cortex in different animal species. The total amount of cells was estimated from the content of DNA and the number of neurons was obtained by direct counting. From Bass *et al.*, 1971.

TABLE I
METHODS FOR PREPARATION OF GLIAL CELLS

	Microdissection	Bulk Preparation	Culturing
Preparation time	5 min[a] and upwards	Hours	0 - a few min
Condition during preparation	Physiological	Often highly unphysiological	Physiological
Normality of starting material	Normal	Normal	Questionable
Contamination with neuronal components	5% fibers in glial preparations[b]	About 10-25%[c,d]	Very slight

[a]Hydén and Pigon, 1960; [b]Hamberger, 1963; [c]Hemminki and Holmila, 1971; [d]Sellinger *et al.*, 1971.

by a factor of at least 2 in brains from humans and other large mammals (Van Economo, 1926; Bass *et al.*, 1971; Tower and Young, 1973); astrocytes seem to a large extent to be responsible for this relative increase in cell number (Bass *et al.*, 1971).

If the volume occupied by glial cells in the normal brain is uncertain, the total glial surface area is even less accurately known. The demonstration that astrocytic veils ("velate astrocytes") or lamellae intervene between neuronal cells as well as between oligodendrocytes and neurons (Palay and Chan-Palay, 1974) suggests that the surface/volume ratio for normal astrocytes is very high (Wolff, 1970b). This may be of major importance, e.g., for uptake of transmitter substances and of potassium as well as for other aspects of neuronal-glial interactions, and it may suggest a very active role of glial cells in brain function.

II. Descriptive Biochemistry

A. Macromolecular Compounds (lipids, protein, nucleic acids, glycogen).

Macromolecular compounds have either been observed in histological sections using histochemical techniques or they have been demonstrated and quantitated in isolated glial preparations using conventional biochemical methods or micromethods. The very fact that they are macromolecular may, to some extent, prevent leakage during separation of bulk prepared glial cells. Quantitation in microdissected cells or in histological preparations is methodologically difficult, but some estimates have been made (Table II). Alterations under different conditions have been rather precisely

TABLE II

CONTENTS OF SOLIDS, NUCLEIC ACIDS, PROTEIN AND LIPIDS IN GLIAL CELLS (pg PER CELL AND/OR % OF DRY WT.)

Preparation	Dry Weight pg	DNA pg	RNA pg	Protein pg	Protein %	Lipids pg	Lipids %	Cholesterol %	Phospholipids %
Bulk prepared astrocytes[a,g]	903 ± 77	6.7 ± 0.3[u]		401 ± 21[w]	44.4	294 ± 16	32.6	5.2	23.5[x]
Bulk prepared astrocytes[a,h]			10.3[t]	411[t]					
Bulk prepared astrocytes[a,i]	591	11.2 ± 0.9*	29.1 ± 4.4*	307	51.9	230 ± 42*	38.9		
Bulk prepared astrocytes[a,j]			12.6[t]	767[t]					
Bulk prepared astrocytes[a,k]			11.2 ± 0.2	319 ± 19					
Bulk prepared oligodendroc.[b,l]	25	5.1 ± 0.8*	2.0 ± 0.4*			7 ± 0.8*	29.5		
Bulk prepared oligodendroc.[b,m]					29.2		20.8	6.3	9.8
Bulk prepared oligodendroc.[c,n]		5.3 ± 0.6**	1.8 ± 0.1**	47 ± 3					
Microdissected glial cells[d,o]	500-2000		4-12[v]		28		up to 70		
Subcutaneously grown astroglia[e,p]	239		5.0[t]	117[t]	48.8				
Astrocytes, prim. cultures[f,q]				320-370[v]					
C-6 glioma cell line[a,p]	177[t]		5.7[t]	98[t]	55.2				
C-6 glioma cell line[a,r]							13.5 ± 1.3	2.0	8.8 ± 0.8
C-6 glioma cell line[a,s]		2.0-4.5	7-14	140-250					
NN glial cell line[e,r]							16.1 ± 1.4	1.7	11.8 ± 0.9
Polyoma induced differentiated glioma cell line[e,p]	285 ± 18[t]		5.9 ± 0.5**[t]	167 ± 3**[t]	58.5		8.9		

[a]rat; [b]calf; [c]man; [d]rabbit; [e]hamster; [f]mouse

[g]Freyzs *et al.*, 1968; [h]Nagata *et al.*, 1974; [i]Norton and Poduslo, 1971; [j]Rose and Sinha, 1969; [k]Sellinger *et al.*, 1971; [l]Poduslo and Norton, 1971; [m]Fewster and Mead, 1968; [n]Iqbal Grundke-Iqbal and Wisniewski, 1977; [o]Hydén and Pigon, 1960; [p]Embree *et al.*, 1973; [q]L. Hertz and S. Fedoroff, unpublished; [r]Robert *et al.*, 1976; [s]Vernadakis and Nidess, 1976.

[t]calculated on the basis of a DNA content of 6.5 (or 6.6 (Rose and Sinha, 1969)) pg/cell; this is the amount of DNA per cell found in the rat (confer text) and probably close to that found in the mouse; [u]calculated on the basis of an assumed P content of 10.56%; [v]calculated from 123 pg per glial cell clump containing 10-30 glial cells; [w]calculated on the basis of an assumed N content of 16.0%; [x]calculated on the basis of an average molecular wt. of 775(m)

* S.D. ** S.E.M.

determined, and qualitative studies have been performed with remarkable accuracy. Quantitative determination (e.g., on a protein basis) in cultured cells presents no problems, but the often questionable 'normality' of the material calls for caution in the interpretation.

The macromolecular compounds include lipids, protein, nucleic acids and carbohydrates which together account for, by far, the major part of the solid content. Precise determination of the total amount of solids (as percent of the wet weight or as mass per unit volume) is uncertain because of the difficulties in obtaining a reliable wet weight determination in preparations of isolated cells (on account of adherent fluid) or an accurate mass/volume ratio in histological preparations (e.g., on account of fixation artifacts). Hydén and Pigon (1960) found about 20% solids in microdissected glial cells from the brain stem, i.e., about the same fraction as in the mammalian brain cortex (Katzman and Pappius, 1973; Hertz and Schousboe, 1975). A value of about 22 pg/μm^3 has similarly been calculated from measurements of dry mass in individual glial cells from the cat superior cervical ganglion (Pevzner, 1965); whereas Lodin *et al.* (1967) by interference microscopy determined a dry mass of 0.13 pg/μm^3 (13%) when cell bodies of Bergmann's glia were measured in unfixed preparations, compared to a value of 0.20 pg/μm^3 (20%) after fixation.

Absolute values for dry weights per glial cell fall into several distinct groups (Table II). Bulk preparation of glial cells by gradient centrifugation of dissociated brain tissue leads to a mainly astrocytic (e.g., Norton and Poduslo, 1971; Hamberger *et al.*, 1975) population of cells with a dry weight which consistently has been found to be between 500 and 1000 pg per cell. A rather similar value (500-2000 pg) can be calculated for microdissected perineuronal glial cells and it can be concluded that both the microdissected glial cells and the bulk-prepared astrocytes are relatively large cells. This conclusion is supported by the estimate by Wolff (1970a) that the *in vivo* volume of an astrocyte is 7,000-18,000 μm^3, and by reports that bulk prepared glial cells are 3-10 times larger than bulk prepared neurons isolated under similar conditions (Freysz *et al.*, 1968; Norton and Poduslo, 1971). The content of protein in astrocytes in primary cultures suggests a comparable volume (Table II). Hamster astroglial cells grown subcutaneously or cells from glial cell lines are, however, significantly smaller (177-285 pg) and a solid content of 200 pg/cell is in agreement with a volume of about 1000 μm^3 in C-6 glial cells (Kukes *et al.*, 1976b). Much smaller than any of these cells are the oligodendrocytes (prepared in bulk from white matter) which seem to have a dry wt. of 25-100 pg (Table II).

The major part of the solids is accounted for by the contents of protein and lipids. Microdissected samples of both perineuronal neuropil from Deiters' nucleus [which has been reported to contain less than 5% (vol/vol) of neuronal fibers (Hamberger, 1963], and of neuropil from the spinal cord have a lipid content of more than 50% (Hydén and Pigon, 1960; Berger *et al.*, 1977a), and in bulk-prepared astrocytes the lipids constitute about ⅓ of the total amount of solids (Table II). Even in the latter preparation the lipid content is 50-100% higher than in corresponding samples of bulk-prepared neurons (Freysz *et al.*, 1968; Hamberger and Svennerholm, 1971). The distribution of phospholipids and the fatty acid composition of bulk-prepared astrocytes and neurons is largely identical (Freysz *et al.*, 1968; Hamberger and Svennerholm, 1971), and the lipid composition of bulk-prepared oligodendrocytes resembles that of young myelin (Davison *et al.*, 1966). Bulk-prepared oligodendrocytes have been reported to contain no gangliosides (Fewster and Mead, 1968), but a low ganglioside content has also been reported in bulk prepared astrocytes and neurons, possibly due to a degradation during the isolation procedure (Hamberger and Svennerholm, 1971a). In C-6 and NN glial cells, the lipid contents (in per cent) seem to be much lower than in the bulk prepared cells; this difference may be especially pronounced with respect to cholesterol and possibly also phospholipids (Table II). Otherwise, the lipid composition of NN or C-6 glial cell lines largely resembles that of bulk prepared glial cells (Robert *et al.*, 1976) and of astrocytes in primary culture (Eichberg *et al.*, 1976). A major difference, which may be related to the culturing situation, is the paucity of polyunsaturated fatty acids (Robert *et al.*, 1976) and the simplicity of the ganglioside pattern (Robert *et al.*, 1975)*.

The content of protein in the adult mammalian brain cortex is about 100 mg/g wet wt. or 50% of the dry wt., but both the solid contents and the protein content are lower in the immature brain (Clouet and Gaitonde, 1956). The estimate of a protein content of about 30% in the lipid-rich microdissected glial cells is well below this value, and also in bulk prepared astrocytes or in subcutaneously grown astroglial cells the proteins may constitute slightly less than 50% (Table II). It is in agreement with the concept of a rather low protein content in astrocytes that Freysz *et al.* (1968) found a considerably higher protein content in their neuronal enriched preparation than in their glial-enriched fraction. Protein contents of 70-80 mg/g wet wt. have been reported for C-6 glioma cells (Schrier and Thompson, 1974) and for astrocytes in primary cultures (Schousboe *et al.*, 1975), but these are minimum values since adherent fluid may cause a considerable increase in the wet weight of cellular preparations. On the other hand, Balcar *et al.* (1977) observed a remarkably high protein content (0.15-0.19g/ml tissue volume) in NN glial cells.

Several differences have been observed between glial and neuronal cells with respect to protein composition (Packman *et al.*, 1971; R.N.

---

*For a further discussion of the lipid composition in glial cell lines see the chapter by S. E. Pfeiffer in this volume.

Rosenberg, this volume). Some, often acidic, proteins are exclusively or almost exclusively localized to glial cells, most notably GFA which is confined to astrocytes (Eng *et al.*, 1971), and S100, which is present in much larger amounts in glial cells (both astrocytes and oligodendrocytes (Hartman *et al.*, 1976)) than in neurons, though it seems to be present also in the latter cell type (Hydén and McEwen, 1966; Sviridov *et al.*, 1972). GFA seems to be a major constituent of neuroglial fibers (Eng *et al.*, 1971; Dahl and Bignami, 1973), and in the mature brain and spinal cord, the immunochemically observed distribution of GFA closely resembles that described by Weigert in 1895 using a staining method for astrocytic fibers (Bignami and Dahl, 1974). Ontogenetically, the onset of glial differentiation is accompanied by the appearance of this protein and at a certain stage of development, a radial system of fibers containing GFA is prominent; these fibers are still present (at least in the hippocampus) in the adult rat, but they become widely spaced and thus less prominent than in the immature animal (Bignami and Dahl, 1974). GFA has been observed in immature human astrocytes in primary cultures (Antanitus *et al.*, 1975). Primary, more differentiated astrocyte cultures from the mouse or rat cerebral cortex are highly enriched in this protein (Bock *et al.*, 1975; E. Bock, this volume) and bulk prepared glial cells show a 3-4 fold enrichment (Bock and Hamberger, 1976). The S100 protein is, in contrast, not found in a higher concentration in bulk prepared glial cells than in the cerebral cortex or the whole brain (Olsson *et al.*, 1973; Bock and Hamberger, 1976). Also this protein appears at the onset of glial cell formation and, at least in the cerebral cortex, there is a remarkable parallelism between the increase in the amount of the mainly glial S100 and the neuronal 14-3-2 protein(Cicero *et al.*, 1972); it is found in some, but not all, glial cell lines (Benda *et al.*, 1968; Pfeiffer *et al.*, 1970; Lightbody *et al.*, 1970; Edström *et al.*, 1973; S. E. Pfeiffer this volume).

Nucleic acids are present in glial cells in much lower concentrations than lipids and protein. The content of DNA per cell nucleus (glial or neuronal) in the rat brain has been found to be about 6.4 pg/cell (Santen and Agranoff, 1963). This agrees well with values in nonneuronal cells, and the glial cell numbers on which Fig. 1 is based, were obtained by measurements of total DNA, i.e., total cell numbers (disregarding the small number of polyploid cells as well as the relatively low content of DNA in mitochondria) followed by subtraction of the number of neurons, directly counted in histological preparations (Bass *et al.*, 1971). When contents of DNA per cell have been measured in isolated glial cell preparations, they have in general been found to be reasonably close to the expected value for the species in question (Table II). The low content of DNA (2.0-4.5 pg/cell) reported by Vernadakis and Nidess (1976) is surprising, whereas the somewhat high value (11.2 pg/cell) found by Norton and Poduslo (1971) was explained by a moderate contamination of the bulk prepared cells with broken nuclei. In those studies, cited in Table II, where no values for DNA content per cell were given, an expected value of 6.5 pg (Santen and Agranoff, 1963; Rose and Sinha, 1969) was used to convert DNA/RNA or DNA/protein ratios to RNA or protein content per cell. A similar nuclear contamination in these cases would tend to cause a moderate lowering of the values for protein and RNA contents.

The RNA content in bulk prepared oligodendrocytes seems to be very low, whereas it is considerably higher in cultured and especially in bulk prepared astrocytes, reflecting their larger size (Table II). In microdissected astrocytes the content of RNA is far (10 times) lower than in a corresponding volume of neurons (Hydén and Pigon, 1960). Differences between the patterns of different RNA fractions in neurons and glial cells have been discussed by Jarlstedt and Hamberger (1971), and the nucleotide composition of both glial cells and neurons has been intensively investigated by Hydén and his co-workers. (e.g., Hydén 1967a,b).

Glycogen in the central nervous system has mainly been studied histochemically. It is predominantly, but not exclusively, localized in glial cells as described in detail in a recent review by Ibrahim (1975). Methodological problems seem to exist, since white matter contains practically no histochemically demonstrable glycogen (Ibrahim, 1975); whereas microchemical methods have shown that it contains even larger amounts of glycogen than grey matter (Folbergrova *et al.*, 1970). The total glycogen content in the brain cortex (i.e., glial cells plus neurons) corresponds to about 20 μmol or less than 4 mg glucose/g dry weight (Folbergrova *et al.*, 1970). Even if all the glycogen were localized in the glial cells, their glycogen content would thus not exceed 1-2% indicating that glycogen constitutes roughly the same fraction of the dry weight as DNA or RNA and far less than the content of lipids and protein.

B. Low-molecular Compounds (amino acids, high energy phosphates, inorganic ions).

One of the most remarkable biochemical characteristics of the brain, and especially of the mature cerebral cortex, is the high concentration of several free amino acids (Table III). Already 15-20 years ago it was suggested that a considerable fraction of the glutamine content in brain should be present in non-neuronal elements (Tower, 1960; Margolis *et al.*, 1968; Benjamin and Quastel, 1972) and that GABA may be more or less evenly distributed between neurons and glial cells (Utley, 1963; Margolis *et al.*, 1968). Verification of such a distribution, or determination of neuronal/glial ratios for other amino acids, is difficult since amino acids leak out from bulk prepared cells during the lengthy preparation and since the contents in cultured cells may depend heavily upon the medium

TABLE III

AMINO ACID CONTENTS (μMOL/100 MG PROTEIN OR μMOL/G WET WT.) IN GLIAL CELLS

Preparation	Glutamate	Glutamine	GABA	Aspartate	Glycine	Alanine	Taurine
Bulk prepared astrocytes[a,e]	2.30 ± 0.29*	0.48 ± 0.06*	0.41 ± 0.05*	0.63 ± 0.08*		0.28 ± 0.01*	
Bulk prepared astrocytes[a,f]	2.06 ± 0.54*		0.45 ± 0.11*	0.58 ± 0.12*	0.33 ± 0.07*	0.22 ± 0.05*	
Bulk prepared astrocytes[b,g]	1.84	0.56	0.72	1.00	0.48	0.30	0.60
Microdissection, spinal cord[b,h]	4.29		0.66	5.34	5.18		
Astrocytes, prim. cultures[c,i,n]	5.56 ± 0.40*	9.66 ± 0.47*	0.06 ± 0.01*	3.82 ± 1.97*	0.75 ± 0.12*	1.37 ± 0.15*	
Astrocytes, prim. cultures[c,i,o]	9.92 ± 0.93*	18.94 ± 1.76*	0.06 ± 0.01*	7.44 ± 1.53*	0.71 ± 0.08*	1.67 ± 0.19*	
Astrocytes, prim. cultures[a,c,j,o]	11.70		<0.50				<0.50
Subcutaneously grown astroglia[d,k]	0.35	0.28	0.03	0.13	0.26	0.32	0.35
C-6 glioma cell line[a,l]	9.00						
C-6 glioma cell line[a,k]	3.88	0.77		0.60	5.50	2.70	1.51
SV 40 transformed glioma line[d,k]	3.02	0.95		2.16	5.47	4.07	0.12
Brain cortex, newborn[a,m]	5.78 ± 0.72*	3.16 ± 0.31*	0.40 ± 0.05*	2.33 ± 0.32*	1.31 ± 0.19*	1.34 ± 0.24*	21.60 ± 1.70*
Brain cortex, adult[a,m]	22.54		3.40	4.31			7.71
Whole brain, adult[a,m]	13.60 ± 0.42*	4.40 ± 0.22*	2.32 ± 0.11*	3.66 ± 0.18*	1.66 ± 0.11*	1.07 ± 0.05*	4.84 ± 0.30*

[a]rat; [b]rabbit; [c]mouse; [d]hamster
[e]Rose, 1970; [f]Nagata *et al.*, 1974; [g]Sellström *et al.*, 1975; [h]Berger *et al.*, 1977a; [i]L. Hertz; P. Wu and M. Farmer, unpublished results; [j]Schousboe *et al.*, 1975, 1976b, 1977; [k]Mokrasch, 1971; [l]Passonneau *et al.*, 1976; [m]Cutler and Dudzinski, 1974
[n]plus serum in medium; [o]minus serum in medium
* S.E.M.

concentrations. The latter source of error may be especially important in the case of the essential amino acids. For this reason, only non-essential amino acid contents (μmol/100 mg protein or μmol/g wet wt.) are shown in Table III. In bulk prepared astrocytes, the concentrations of these amino acids can be seen to amount to between ⅓ and ⅛ of their concentrations in the whole brain. In cultured and microdissected glial cells the relation is much more complex: subcutaneously grown hamster astroglia has exceedingly low concentrations of all amino acids (Mokrasch, 1971), whereas mouse astrocytes in primary cultures, glial cell lines, and microdissected neuropil from the spinal cord have high contents of glutamate, glutamine and aspartate (Table III). This seems to suggest, or confirm, that these three amino acids are present in glial cells in concentrations similar to, or maybe even higher, than those in neurons. The distribution of glutamate and aspartate in discrete layers of the monkey retina has similarly led to the conclusion that these two amino acids may be present in especially high concentrations in Muller fibers, ganglion cells, or in both cell types (Berger *et al.*, 1977b).

The high contents of glycine and alanine in the glial cell lines and of glycine in microdissected neuropil samples from the spinal cord (Table III) might indicate that glial cells also contain large amounts of these amino acids. However, the spinal cord is much enriched in glycine (for references, see Berger *et al.*, 1977a). This observation together with the relatively low concentrations of glycine and alanine in bulk prepared astrocytes and in astrocytes in primary cultures, seem to be against the concept of a high concentration of these two amino acids in normal glial cells.

GABA and taurine are notable in being present in lower concentrations in cultured cells (with the exception of the C-6 glioma cell line) than in bulk prepared cells. This may suggest that the content of these amino acids in glial cells is due to uptake from the surroundings rather than to formation in the glial cells. The presence of these two amino acids in C-6 glial cells or in astrocytes in primary cultures does, however, indicate a glial capacity for their formation. Whether this is a sign of metabolic differentiation or dedifferentiation can, at present, not be resolved but the exceedingly high content of taurine in the cortex from neonatal animals may point in the latter direction.

The content of ATP in cultured astrocytes is 2-3 μmol/g wet wt. (Schousboe *et al.*, 1976a) whereas it is much lower in both microdissected and bulk prepared glial cells (for references see Hertz and Schousboe, 1975).

Ten to fifteen years ago, it was repeatedly suggested that glial cells might differ from most other cells in having a high sodium content (for references see Hertz and Schousboe, 1975, Hertz, 1977a). This may be the case in Schwann cells from squid nerve fibers (Villegas *et al.*, 1965; Villegas,

1968), but glial cell preparations from both mammalian and leech central nervous systems have, unless they are very undifferentiated (Trachtenberg *et al.*, 1972), a high potassium content and a low sodium content (Table IV); also microdissected glial cells and cells from the NN cell line contain large amounts of potassium (Hamberger and Röckert, 1964; Gill *et al.*, 1974). The high membrane potentials recorded from glial cells *in vivo* (for references, see Somjen, 1975, Hertz and Schousboe, 1975) support the concept that the intracellular potassium concentration is high.

TABLE IV
ION CONCENTRATIONS IN GLIAL CELLS

Preparation	Species	Concentration µmol/g wet wt.			Authors
		K^+	Na^+	Cl^-	
C-6 glioma cell line	Rat[a]	116	14.4	11.2	Kukes *et al.*, 1976b
Polyoma induced glioma cell line	Hamster[b]	120	12.9		Lees and Shein, 1970
Astroglia grown subcutaneously	Hamster[c]	66	11.2		Lees and Shein, 1970
Astrocytes in primary cultures	Chick[d]	94			Latzkovitz *et al.*, 1974

[a]Calculated from an assumed water content of 80%; [b]calculated from a K^+ content of 0.60 µmol/mg dry wt. and an assumed solid content of 20%; [c]calculated from a K^+ content of 0.33 µmol/mg dry wt. and an assumed solid content of 20%; [d]calculated from an assumed protein content of 100 mg/g wet wt.

III. Dynamic Biochemistry

A. Energy Metabolism.

Before going into a more detailed discussion of different aspects of the dynamic biochemistry of glial cells, it may be advantageous to look at their general metabolic activity as evidenced by rates of oxygen uptake (Table V) and other metabolic parameters. For the sake of comparison, it should be kept in mind that the mean oxygen consumption in the whole normal human brain is 3.6 ml/min per 100 g or close to 100 µmol/hr per g wet wt. (Kety, 1957). In brain cortex slices, the respiratory rate is about half of this value (Elliott and Henderson, 1948; McIlwain, 1953). Regional variations in brain slice respiration have long been known to exist (for references, see Hertz, 1978) and using the elegant 2-deoxy-D-glucose method, Sokoloff and his co-workers have recently obtained evidence for a similar metabolic variation *in vivo* (Reivich *et al.*, 1976). Species differences do occur and higher respiratory rates have been observed in smaller than in larger mammals both *in vivo* and in brain slice experiments (for references see Hertz and Schousboe, 1975).

TABLE V
RATES OF OXYGEN UPTAKE BY GLIAL CELLS[a]

Substrate	Species	Method of Isolation	O_2 Uptake (μmol/hr per g)
Endogenous	Rabbit	Microdiss.	17
Glucose in saline media	Rat, rabbit	Microdiss. Grad. centrif. Culturing	40-187
Glucose in Eagle's M.E.M.	Mouse	Culturing	400
Pyruvate	Rat	Grad centrif.	69
α-Ketoglutarate	Rabbit	Microdiss.	50-172
Glutamate	Rat, rabbit	Microdiss. Grad. centrif.	33-76
Succinate	Rabbit	Microdiss.	90-304
Cytochrome c	Rabbit	Microdiss.	502

[a]This table is condensed from Hertz, 1978. The glucose concentration in the Eagle's M.E.M. (7 mM) is comparable to that in the saline media (6 mM).

Table V shows that the rate of oxygen uptake by glial cells depends heavily upon the substrate used. The oxygen uptake by glial cells incubated in saline media (averaging 100 μmol/hr per g wet wt.) is close to the rate of oxygen uptake in rat brain cortex slices and, together with a corresponding oxygen consumption rate of about 500 μmol/hr per g wet wt. in neuronal perikarya, it was used by Hertz and Schousboe (1975) to deduct that roughly one-third of brain cortex respiration occurs in glial cells. The recent demonstration that astrocytic respiration is a few times more intense (400 μmol/hr per g wet wt.) during incubation in Eagle's MEM medium* (Table V) may suggest that this relative value for glial respiration is too low. Very recently, the 2-deoxy-D-glucose method of Sokoloff has been used at the cellular level, and in agreement with the high respiratory rate in glial cells, it was found that the uptake of glucose in grey matter at several different localizations of the rat central nervous system is only slightly higher in neuronal perikarya than in neuropil (Sharp, 1976). High concentrations of

*In the MEM the main substrate is probably still glucose. The medium does contain 2 mM glutamine which to some extent is degraded to glutamate (to an approximate concentration of 0.2 mM) but from Table V can be seen that glial respiration is rather low with glutamate as the substrate. The Eagle's MEM does not contain pyruvate, succinate or α-ketoglutarate.

potassium have unequivocally been found to cause an increase in the oxygen uptake by glial cells (Hertz, 1966; Bradford and Rose, 1967; Aleksidze and Blomstrand, 1969; Haljamae and Hamberger, 1971; Hertz *et al.*, 1973) together with a corresponding decrease in the content of ATP (Schousboe *et al.*,1970).

The rates of oxygen uptake with cytochrome c, succinate or d-ketoglutarate as the substrate (Table V) have been taken as estimates of, respectively, the cytochrome oxidase, the succinate dehydrogenase and the α-ketoglutarate oxidase activity. The activities of these enzymes are higher in glial cells (from adult animals) than in neuronal perikarya (Hamberger, 1963; Svoboda and Lodin, 1972), but during activity a decrease of the activity in glial cells and an increase of the activity in neurons have been observed (Hamberger, 1963; Hydén, 1967a,b).* Histochemically it has been shown that the activity of the succinate dehydrogenase especially but also of the glucose-6 phosphate dehydrogenase, the malic dehydrogenase, the lactic dehydrogenase and the glutamic dehydrogenase increases in cultured astrocytes when the medium concentration of sodium is raised, and decreases when it is lowered (Friede, 1964). This effect may be behind the fast respiratory decline of microdissected glial cells in a sodium free medium (Hertz, 1966). The succinate dehydrogenase activity in glial cells has (in rats) been found to increase during the first postnatal month and in oligodendrocytes the activity of the oxidative enzymes rises at the time of myelination (Yonezawa *et al.*, 1962).

The end products of oxidative degradation are H_2O and CO_2. In the presence of carbonic anhydrase, CO_2 forms HCO_3^- which may be of major importance for anion transport in the central nervous system. The carbonic anhydrase is almost exclusively localized in glial cells (Giacobini, 1962). Its possible functional importance in the brain has been discussed by Nair and Bau (1971) and by Kimelberg and Bourke (1973). If no oxygen is available, lactate will be formed by anaerobic glycolysis. Evidence that this process mainly occurs in glial cells has been given by Tower and Young (1973).

The uptake of glucose, or rather of the glucose analogue 2-deoxy-D-glucose, into astrocytes in primary cultures has been found to occur with a Vmax of 12.6 nmol/min per mg protein and a Km of below 2 mM (Cummings, 1977); the aerobic breakdown of glucose to lactate under roughly corresponding conditions is relatively slight, i.e., about 6 nmol lactate (corresponding to 3 nmol glucose) per min per mg protein (Hertz, 1978) suggesting that the remaining 9-10 nmol glucose/min per mg protein is oxidized to CO_2 and H_2O. An increase of the potassium concentration to

*The possibility that such a change partly may represent redistribution between neuronal processes and perikarya has been suggested by Derry and Wolfe (1967).

12-15 mM leads to a very substantial increase in Vmax and a decrease in Km for glucose uptake into glial cells (Cummings, 1977), as well as to an increased incorporation of ^{14}C glucose into tricarboxylic acid cycle intermediates (Salem *et al.*, 1975).

A major part of the energy released during oxidative metabolism is utilized via the formation of ATP. This compound is in turn degraded by ATP-ases of which the three main forms are the Ca^{2+}-ATPase, the Mg^{2+}-ATPase and the Na^{+},K^{+}-ATPase. The latter enzyme requires in addition to Na^{+} and K^{+} also Mg^{2+} and is involved in active transport of cations across cell membranes (Skou, 1957, 1965; Schwartz *et al.*, 1975). In several regions of the adult mammalian brain, including the cerebral cortex, the Na^{+}, K^{+}-ATPase activity seems to account for slightly above one half of the total Mg^{2+}-activated ATPase activity (Fahn and Côté 1968; Knox *et al.*, 1972; Stefanovic *et al.*, 1974; cf. however, also Hess *et al.*, 1972), and together the Na^{+},K^{+}-ATPase and the Mg^{2+}-ATPase activity in the cerebral cortex amount to 15-30 μmol phosphate/hr per mg protein. The activity is much lower in cerebral white matter (Fahn and Côté 1968) and in immature brain (Côté, 1964; Samson and Quinn, 1967). Bicarbonate has been found to increase the ATPase activity in brain, but has no significant effect on the Na^{+},K^{+}-ATPase of brain microsomes (Kimelberg and Bourke, 1973).

Histochemical demonstration in intact gray matter has led to the conclusion of a higher Na^{+},K^{+}-ATPase activity in neurons than in glial cells (Stahl and Broderson, 1976)*, but biochemical measurements in isolated cell preparations have repeatedly shown that the activity of the Na^{+},K^{+}-ATPase is higher in glial than in neuronal preparations and that a considerable postnatal increase in activity occurs in the glial cells (Cummings and Hydén, 1962; Henn *et al.*, 1972; Medzihradsky *et al.*, 1972; Nagata *et al.*, 1974 cf. also Hertz, 1976). A general trend seems to be that highest activity is found in bulk prepared glial cells, intermediate activity in glial cells in primary cultures and lowest activity in glial cell lines (Table VI). Table VI shows also that the sum of the Mg^{2+}-ATPase and the Na^{+},K^{+}-ATPase activities in both bulk prepared glial cells and astrocytes in primary cultures comes reasonably close to the activity in the brain cortex (15-30 μmol phosphate/hr per mg protein), but it seems noteworthy that the relative contribution by the Na^{+},K^{+}-ATPase is low in the cultured cells. It is unknown whether this can be related to lack of enzyme induction (by a high potassium concentration) during the culturing period. The sum of the Mg^{2+}-ATPase activity and the Na^{+},K^{+}-ATPase activity is much (5-10

*The activity occurs in radial streaks, which were interpreted as neuronal; it has not been discussed whether these streaks might bear any relation to the radial glial cell system present at early ontogenetic stages (cf. "Introduction").

times) lower in astrocytes grown subcutaneously (Embree *et al.*, 1971; Henn *et al.*, 1972) and in glial or glioma cell lines (Table VI); no correlation seems to exist between the morphological degree of differentiation and the activity of the ATPase (lowest two lines of Table VI).

TABLE VI

SPECIFIC ACTIVITIES (μmol Pi RELEASED /HR PER MG PROTEIN) OF Na^+,K^+-ATPASE AND Mg^{2+}-ATPASE IN GLIAL CELLS

Preparation	Mg^{2+}-ATPase Specific Activity	Na^+,K^+-ATPase Specific Activity[a]	K^+ Conc. (mM)
Bulk prepared astrocytes[b,g]	8.5	3.5	
Bulk prepared astrocytes[b,h]	5	5.7 ± 0.77	10
Bulk prepared astrocytes[b,i]	9	7	20
Bulk prepared astrocytes[c,j]		8[r]	10
Bulk prepared astrocytes[c,j]		5[r]	20
Bulk prepared astrocytes[d,k]	3.50 ± 0.3*	3.7 ± 0.36*	20
Astrocytes, prim. cultures[b,l]		1.4 ± 0.36*	25
Astrocytes, prim. cultures[b,l]		2.4 ± 044*[s]	25
Astrocytes, prim. cultures[b,m]	14.4 ± 1.10*	2.3 ± 0.06*	20
Astrocytes, prim. cultures[b,n]	21.7 ± 0.90*		0
Astrocytes, prim. cultures[b,o]		1.2 ± 0.23*[s,t]	20
Astrocytes, prim. cultures[e,o]		0.7 ± 0.08*[s,t]	20
Subcutaneously grown astroglia[f,p]	2.4	0.4	10
C-6 glioma cell line[b,p]	2.0 ± 0.10*	0.3 ± 0.07*	10
C-6 glioma cell line[b,h]		1.1 ± 0.08*	10
C-6 glioma cell line[b,m]	2.4 ± 0.10*	0.7 ± 0.05*	20
C-6 glioma cell line[b,n]	0.3 ± 0.02*		0
C-6 glioma cell line[b,q]		2.4 ± 0.3	30
NN glial cell line[f,m]	2.8 ± 0.09*	0.9 ± 0.03*	20
NN glial cell line[f,n]	0.4 ± 0.02*		0
Polyoma induced differentiated glioma cell line[f,p]	1.4 ± 0.12*	0.3 ± 0.04*	10
Polyoma induced anaplastic glioma cell line[f,p]	2.6 ± 0.20*	0.9 ± 0.10*	10

[a]The Na^+,K^+-ATPase activity was measured as the difference between the total ATPase activity and that in the absence of Na^+ and K^+ or in the presence of ouabain.
[b]rat; [c]rabbit; [d]beef; [e]mouse; [f]hamster
[g]Nagata *et al.*, 1974; [h]Kimelberg, 1974; [i]Medzihradsky *et al.*, 1972; [j]Henn *et al.*, 1972; [k]Hamberger *et al.*, 1970; [l]Moonen and Franck, 1977; [m]Stefanovic *et al.*, 1975; [n]Stefanovic *et al.*, 1976; [o]Schousboe *et al.*, 1976a; [p]Hess *et al.*, 1972; [q]Cotman *et al.*, 1971;
[r]results obtained using membrane fractions; [s]results obtained after culturing in the presence of dibutyryl cyclic AMP; [t]measured at 30°.
*S.E.M.

Peak activity of the Na^+,K^+-ATPase has often been found at a K^+ concentration of 10-20 mM with higher K^+ concentrations inhibiting the enzyme activity (e.g., Henn *et al.*, 1972). The potassium induced increase of respiration and swelling in brain cortex slices (for references, see Hertz,

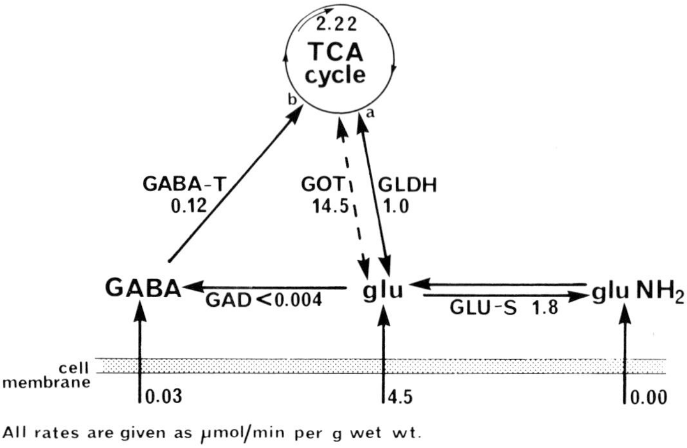

FIG. 2. Rates of energy metabolism and of uptakes and metabolism of GABA, glutamate (glu) and glutamine (glu NH_2) in primary cultures of astrocytes obtained from the dissociated cerebral hemispheres of newborn DBA mice. The turnover rate of the tricarboxylic acid (TCA) cycle has been calculated (1 mol glucose + 6 mol 0_2 → 2 mol acetyl CoA) from the rate of oxygen uptake (Table V), the GABA-T activity and the GABA uptake rate are those observed by Schousboe *et al.* (1977a), the GOT, GDH, GAD and glutamine synthetase activities are unpublished results of A. Schousboe and L. Hertz (cf. A. Schousboe, this volume), the glutamine uptake rate unpublished results of L. Hertz and A. Schousboe, and the glutamate uptake rate that observed by Hertz *et al.* (1977).

a: — α-ketoglutarate; b: — succinate.

1976) requires, however, a higher concentration of potassium, and the possibility of the existence of more than one potassium effect has been discussed (Hertz, 1973, 1976). Recently a potassium induced stimulation of the Na^+,K^+-ATPase with a maximum at 20-50 mM K^+ has, however, been described in cultured astrocytes (Moonen and Franck, 1977).

B. Metabolism and Transport of Amino Acids.

Amino acid metabolism is closely connected to intermediary metabolism since glutamic acid is formed by amination of α-ketoglutaric acid and aspartic acid by transamination of oxaloacetate. Glutamine can be formed from glutamic acid reacting with another molecule of ammonium, and GABA is produced from glutamic acid by decarboxylation (Fig. 2). GABA may in turn, via succinic semialdehyde, be converted to succinate. This "GABA shunt" has in brain slices been estimated to carry a flux corresponding to about 8% of the flux through the tricarboxylic acid cycle (Balazs *et al.*, 1970).

Based upon precursor/product relations in the intact brain *in vivo* or in brain slices after exposure to labelled substrates, a "compartmentation" of glutamate metabolism has been suggested (for references, see Balazs and Cremer, 1972; Berl *et al.*, 1975). Each of the two (or more) glutamic acid compartments seems to be connected with its own tricarboxylic acid cycle, and by computer analysis Van den Berg and Garfinkel (1971) deduced for the mouse brain fluxes of 1.25 and 0.3 μmol/min. per gram wet weight in, respectively, a large and a small compartment,* combined with a bidirectional flux between the two compartments of 0.14 μmol/min per gram wet weight. In the rat, the flux between the two compartments seems to be about half of this value (Dzubow and Garfinkel, 1970), possibly reflecting the species difference in rate of oxygen consumption. Conceivably, the compartmentation could be due to metabolic differences between, for example, mitochondria and cytoplasm, cell bodies, axons and synaptosomes, or neurons and glial cells. Several attempts, including investigations of preparations obtained by gradient centrifugation (Rose, 1970) or structures (e.g., dorsal root ganglia) in which GABA and glutamate are taken up exclusively into satellite cells (Roberts and Keen, 1974; Minchin and Beart, 1975) have been made to distinguish between these possibilities, but it seems at present not possible to correlate with certainty any metabolic compartment in the central nervous system with any specific anatomical structure. However, glutamate metabolism is compartmentalized in bulk prepared glial cells (possibly reflecting the heterogeneity of this fraction) and in dorsal root ganglia (Rose, 1970; Roberts and Keen, 1974; Minchin and Beart, 1975), but maybe not in mouse cerebral astrocytes in primary cultures (L. Hertz and A. Schousboe, preliminary experiments).

It has been suggested that the large compartment should represent neuronal metabolism and the small compartment glial metabolism, and that a release of GABA from the neuronal compartment followed by a subsequent uptake of GABA into the glial compartment should account for the flux in one direction, and a transport of glutamine from glial cells to neurons for that in the opposite direction (e.g., Van den Berg and Garfinkel, 1971). Recently, Benjamin and Quastel (1975) have suggested a somewhat different model according to which glutamate released from the neurons is taken up by glial cells, converted to glutamine and then returned to the neurons, where it is hydrolyzed to glutamate and ammonia. There is no

---

*These two values do not give specific information of *either* compartment sizes *or* turnover rates for either of the two compartments, but they do indicate that the compartment sizes and the turnover rates are such that they *together* lead to fluxes of 1.25 and 0.3 μmol/min per g *total* wet wt. in, respectively, the large and the small compartment. The sum of these two values (1.55 μmol/min per g) is reasonably close to the turnover rate (2.22 μmol/min per g *glial* wet wt.) which can be calculated for mouse glial cells (Fig. 2) on the basis of their rate of oxygen uptake. The similarity between these two values reflects relatively similar rates of oxygen uptake in mouse glial cells (400 μmol/hr per g wet wt. (Table IV) and in the whole mouse brain *in vivo*).

reason that there could not exist a flux of both glutamate and GABA (a putative excitatory and a putative inhibitory transmitter (Krnjevic and Phillis, 1963) from neurons to glial cells. Information about uptake kinetics for glutamate, glutamine and GABA into neurons and glial cells as well as determination of activities of the enzymes involved in the interconversion of these amino acids in glial cells and in neurons will be essential for confirmation or disproval of the neuronal/glial cell interactions outlined above. Only the glial part will be discussed here.

Uptake of amino acids by an efficient, high affinity (i.e., Km < 0.1 mM) uptake is one of the most spectacular characteristics of glial cells. For GABA and glutamic acid, such an uptake was first demonstrated in bulk-prepared glial cells, but no V_{max} values were given (Hamberger, 1971; Henn *et al.*, 1974). A high affinity uptake has since then also been observed in glial cell lines (Hutchison *et al.*, 1974; Schrier and Thomson, 1974; Faivre-Baumann *et al.*, 1974; Balcar *et al.*, 1977), satellite cells in peripheral ganglia (Kelly *et al.*, 1975; Roberts, 1976b), and glial cells in primary cultures (Schousboe *et al.*, 1977; Hertz *et al.*, 1977). The V_{max} values are much higher for uptake into primary astrocytic cultures than for uptake into glial cell lines (A. Schousboe, this volume), and Fig. 2 shows V_{max} values for the high affinity uptake of glutamate, GABA and glutamine into primary cultures of astrocytes. The uptake rate for glutamic acid is extremely high, whereas that for GABA is much lower and no high affinity uptake seems to exist for glutamine.

The enzymes involved in the interconversion between α-ketoglutarate and glutamate, and between glutamate and its metabolites are also indicated in Fig. 2. They are the glutamate dehydrogenase or GLDH (catalyzing the reaction α-ketoglutarate and $NH_4^+ \rightleftarrows$ glutamate), the glutamate oxaloacetate transferase or GOT (catalyzing the transamination between glutamate/α-ketoglutarate and aspartate/oxaloacetate), the glutamine synthetase or GLU-S (catalyzing glutamine synthesis from glutamate and NH_4^+ and requiring ATP), the glutaminase (catalyzing glutamine hydrolysis into glutamate), the glutamic acid decarboxylase or GAD (catalyzing formation of GABA from glutamate) and the GABA transaminase or GABA-T (catalyzing conversion of GABA via succinic semialdehyde to succinate). As in the case of the uptake rates for glutamate and GABA, the activities of these enzymes are higher in primary astrocyte cultures from mouse brain hemispheres than in other glial preparations (A. Schousboe, this volume). It can be seen from Fig. 2 that the activity of the GLDH and the GOT are very high, favouring interconversion between glutamate and α-ketoglutarate; the activity of the glutamine synthetase is also considerable [which is in agreement with a rapid conversion of ^{14}C glutamate into glutamine in both primary astrocytic cultures (A. Schousboe and L. Hertz, unpublished experiments) and glial cells in the dorsal root ganglion (Roberts and Keen, 1974)]: the GAD activity is almost negligible whereas some GABA-T activity is found. When the activities of these enzymes are considered together with the uptake rates for the amino acids, it becomes evident that glial cell properties are consistent with a large flux of glutamic acid from neurons to glial cells and a moderate flux of GABA in the same direction. Thus, these findings support, but do not prove, the concept of a transfer of glutamate and GABA — but not of glutamine — from neurons to glial cells as proposed by Van den Berg and Garfinkel (1971), and Benjamin and Quastel (1975). The uptake and conversion of glutamate are high enough to account for any possible flux connecting neuronal and glial metabolism*. The GABA-T activity is high enough to account for 8% of the metabolic activity in a brain slice, i.e., 8% of 100 μmol O_2 g/hr per g wet wt. (Schousboe, 1977), but not high enough to account for 8% of the actual metabolic degradation *in vivo* or in cultured neurons or glial cells, and the intensity of the uptake is at least 4 times too low to be compatible with the concept (Beart *et al.*, 1974) that the total degradation of GABA passing through the GABA shunt should occur in glial cells.

The uptake of glutamate and GABA into glial cells may obviously be important for termination of transmitter activity (Henn and Hamberger, 1971; Henn *et al.*, 1974; Schrier and Thompson, 1974; Hutchinson *et al.*, 1974; Iversen and Kelly, 1975; McLennan, 1976; Balcar *et al.*, 1977; Schousboe *et al.*, 1977) and this may also apply for uptake of taurine (Schrier and Thompson, 1974; Schousboe *et al.*, 1976b; Borg *et al.*, 1976) and aspartate (Balcar *et al.*, 1977). In this context, it seems of interest that an increased extracellular potassium concentration (which accompanies neuronal excitation *in vivo*) (See "Transport of inorganic ions") enhances the uptake of glutamate (an excitatory transmitter) whereas that of GABA (an inhibitory transmitter) is inhibited (Schousboe *et al.*, 1977; Hertz *et al.*, 1977). Several amino acids without known transmitter function are, however, also accumulated quite efficiently into glial cells, e.g., tryptophan (Bauman *et al.*, 1974) arginine, leucine and serine (Logan, 1976).

Also the release of amino acids has been studied in isolated glial preparations. Addition of non-radioactive GABA during the washout of ^{14}C-GABA from astrocytes previously loaded with labelled GABA has no effect on the efflux, indicating that influx and efflux are not directly connected, but that the uptake occurs as a net accumulation rather than as a homoexchange (L. Hertz, A. Schousboe and P. Wu, unpublished). This is in contrast to the GABA transport in glial cells in peripheral ganglia (Roberts,

*The ultimate consequence of such a neuronal-glial metabolic interaction would be that metabolically interacting neuronal and glial processes constitute one metabolic compartment (e.g., the large compartment) and neuronal perikarya the other (e.g., the small compartment).

1976a). Glutamate release from primary cultures is much slower than the uptake, which again suggests a net uptake (Hertz *et al.*, 1977) and is in accordance with the concept that a large fraction of the glutamate accumulated in glial cells is converted to other compounds, e.g., glutamine. High concentrations of potassium do not increase the release of glutamate or GABA from cultured astrocytes (Hertz, 1976 and unpublished experiments) or of GABA from C-6 glioma cells (B. K. Schrier, this volume) but have been reported to enhance the GABA release from bulk prepared glial cells (Sellström and Hamberger, 1976) and from glial cells in peripheral ganglia (Bowery and Brown, 1972; Minchin and Iversen, 1974; Roberts, 1974; Minchin, 1975).

C. Transport of Inorganic Ions.

The enrichment of Na^+, K^+-ATPase activity in isolated glial cells suggests capacity for transport of sodium and potassium, and the high activity of the carbonic anhydrase, which is localized almost exclusively in glial cells, may be of importance for chloride transport (Kimelberg and Bourke, 1973). Table VII shows that the uptake of ^{42}K occurs at a very much higher rate in primary cultures of mouse astrocytes (150 μmol/min per g wet wt) than in corresponding cultures of chick astrocytes (Latzkovits *et al.*, 1974) or C-6 glioma cells (Kimelberg, 1974; Kukes *et al.*, 1976a), and that the sodium efflux in C-6 cells (Kukes *et al.*, 1976a) is of the same order of magnitude as the potassium fluxes in these cells. The chloride transport into the NN cells was found by Tower and his co-workers (Gill *et al.*, 1974) to be stimulated by an increase of potassium concentration in the medium (with a K_m (for K^+) of 38.5 mM). Similarly an ongoing study in our laboratory of potassium uptake by astrocytes in primary cultures has shown that the K^+ uptake of about 150 μmol/min per g wet wt. at 6 mM extra-cellular K^+ (Table VII) increases to 300 μmol/min per g at 15 mM K^+ and to at least 500 μmol/min per g at 45-50 mM K^+. Further experiments are required to estimate what fractions of these uptakes represent active, net accumulation. From work on brain slices* it is however known that an analogous K^+-induced increase in potassium fluxes depends upon oxidative metabolism (Hertz, 1968). To what extent such a potassium uptake is accompanied by an uptake of chloride and water (Bourke, 1969a, b; Bourke *et al.*, 1970; Franck, 1970; Lund-Andersen and Hertz, 1970; Gill *et al.*, 1974) or by a Na^+,K^+-ATPase-activated sodium transport in the opposite direction (Cummins and Hyden, 1962; Henn *et al.*, 1974; Kukes *et al.*, 1976a) remains unknown. It is, however, well established that high external

*The potassium fluxes are much lower in this preparation (about 10 μmol/min per g wet wt. (for references, see Hertz, 1977a).

concentrations of potassium in the presence of chloride do induce a pronounced swelling (water uptake) in mammalian glial cells without a corresponding effect on neurons (e.g., Bourke and Tower, 1966; Bourke, 1969b; Bourke *et al.*, 1970; Franck, 1970; Lund-Andersen and Hertz, 1970; Lodin *et al.*, 1971; Bourke and Nelson, 1972; Möllgård *et al.*, 1974).

TABLE VII
FLUXES OF POTASSIUM AND SODIUM IN GLIAL CELLS

Preparation	Species	Potassium Influx (μmol/min per g)	Sodium Efflux (μmol/min per g)	External K^+ Conc. (mM)	Authors
Microdissected	Rabbit[a]	1.5		5	Henn *et al.*, 1972
C-6 glioma cell line	Rat[b]	2		5	Kimelberg, 1974
C-6 glioma cell line	Rat[b,c]	3.8±0.5*	7.5±2.0*	4.6	Kukes *et al.*, 1976a
Astrocytes in primary cultures	Chick[b]	5.8		4.7	Latzkovits *et al.*, 1974
Astrocytes in primary cultures	Rat[b]	150-200		6	Hertz, 1977b

[a]Net accumulation; [b]Fluxes determined by aid of radioisotopes; [c]Calculated from surface/volume ratio given by Kukes *et al.*, 1976b.
*S.D.

The transport of potassium, sodium and chloride must be of major importance for the normal function of the central nervous system, since recent *in vivo* experiments using potassium-sensitive electrodes very convincingly have demonstrated an increase of the extracellular potassium concentration in cerebral or spinal grey matter during excitation and even during physiological function*; good concordance is found that the extracellular potassium concentration under normal excitation and even during seizure activity does not exceed 10-12 mM; during spreading depression it may, however, reach the enormous level of 80 mM, and simultaneously the concentrations of both sodium and chloride are drastically decreased (for references, see Somjen, 1975; Katzman and Grossman, 1975; Hertz, 1977a). There is little doubt that extracellular potassium ions actively and efficiently are transported back into cells and that this process is of substantial importance in reducing excess extracellular

*Circumstantial evidence for such an increase had previously been obtained in the leech and necturus nervous system (Kuffler and Nicholls, 1966).

potassium resulting from neuronal activity*. Eventually the potassium ions *must* be returned to neurons and from *in vivo* experiments there is very little evidence that glial cells are involved in potassium homeostasis (cf. however, Grossman and Seregin, 1977). The immense uptake capacity for potassium into astrocytes and the enhancement by excess potassium of this process (cf. above) are, however, highly suggestive that rapid accumulation of potassium ions into glial cells from an extracellular fluid with an increased $K+$ concentration has an important physiological role to play. The pronounced effects of high concentrations of potassium on ATPase activity, ATP content and oxygen uptake in glial cells (cf. "Energy metabolism") support this point of view, and evidence for a transport of potassium between cultured glial cells and neurons has been obtained by Latzkovits *et al.* (1974).

D. *Macromolecular Compounds*

Autoradiographic studies of incorporation of labeled thymidine into DNA have unequivocally shown DNA formation, i.e., cell division in glial cells in brains of postnatal animals (e.g., Altman, 1969; Paterson *et al.*, 1973; Mares and Lodin, 1974; Privat, 1975). Even in young adult mice the DNA synthesis is still so pronounced that continuous exposure to labelled thymidine during the whole second month of life leads to labeling of 25-35% of all oligodendrocytes and 15-30% of all astrocytes (Kraus-Ruppert *et al.*, 1973). These values suggest that an intense DNA synthesis (cell division) continues for a longer time in oligodendrocytes than in astrocytes, and by the reverse technique, i.e., continuous exposure to the radioisotope during the whole prenatal development followed by cessation of the labeling postnatally, it was similarly found that in the 18 week old rat brain cortex only 20% of the oligodendrocytes, compared to more than 50% of the astrocytes and 100% of the neurons, retained the labeling (Haas *et al.*, 1970), i.e., had undergone no (or only few) divisions. In still older animals, the rate of DNA synthesis declines, but even in at least 1 year old mice and rats some synthesis of DNA occurs in the glial cells of the brain (Dalton *et al.*, 1968; Korr *et al.*, 1973; Mares and Lodin, 1974). The proliferation rates are dependent upon external factors, since an increased glial cell proliferation in rats exposed to an intellectually enriched environment has been observed by Altman and Das (1964) and since disease may affect DNA synthesis in glial cells (Kimberlin and Anger, 1969).

---

*The most convincing evidence for such an active uptake of potassium ions from the extracellular space is probably the observation that the decrease in extracellular $K+$ concentration occurs considerably faster than can be explained as a result of diffusion (G. Cordingley and G. Somjen, personal communication).

Biosynthesis of RNA has been studied by administration of labeled adenine, cytidine, uridine, orotic acid, UTP or GTP to microdissected or bulk prepared glial cells or to isolated glial nuclei. A difficulty for the interpretation of such experiments is the unknown magnitude of the precursor pools (e.g., Jarlstedt and Hamberger, 1971; Yanagihara, 1974). A short time (1-4 hrs) after the exposure to the radioactive precursor, a higher specific activity has been found in glial cells than in neurons by some, but not all, authors (Daneholt and Brattgård, 1966; Volpe and Guiditta, 1967; Jarlstedt and Hamberger, 1971). This may be related to differential membrane permeabilities or differences in cytoplasmic precursor pool size, since corresponding experiments using isolated nuclei have shown a faster neuronal RNA synthesis within 20 min of incubation (Banks-Schlegel and Johnson, 1975); after longer exposure to the labeled precursor most authors seem, however, to agree that the specific activity in the RNA from adult animals [but not from neonatal mice (Banks-Schlegel and Johnson, 1975)] is highest in neurons (Volpe and Guiditta, 1967; Flangas and Bowman, 1970, Jarlstedt and Hamberger, 1971; Yanagihara, 1974; Weck and Johnson, 1976: cf. also Lövtrup-Rein, 1970). The largest difference between the two cell types has generally been observed for the heavier types of RNA (Volpe and Guiditta, 1967; Jarlstedt and Hamberger, 1971).

During certain forms of activity the RNA content decreases in glial cells (Hydén, 1967a,b; Pevzner, 1965, 1971), whereas it increases in neurons; for this reason a transfer of RNA from glial cells to neurons has been suggested (Hydén 1967a,b; Pevzner, 1965). In hibernating animals (the ground squirrel) alterations in the opposite direction have been found to occur (Pevzner and Semeshina, 1976). Alterations in glial RNA during a learning experiment have also been observed (Hydén and Egyhazi, 1963).

Most authors seem to agree that the major part of the protein synthesis occurs in neurons (e.g., Blomstrand and Hamberger, 1969, 1970; Blomstrand, 1971; Tiplady and Rose, 1971; Lisy and Lodin, 1973; Blomstrand *et al.*, 1975; Yanagihara, 1976), but Sellinger and his co-workers have repeatedly observed a higher specific activity in the glial proteins (Johnson and Sellinger, 1971; Sellinger *et al.*, 1971, 1973). This difference may partly be explained by the use of different precursor amino acids (Blomstrand and Hamberger, 1970) and maybe also by species or age differences, since the rate of protein synthesis is glial cells (from rats or mice) decreases considerably during the first 4-6 weeks of life (Johnson and Sellinger, 1971; Lisy and Lodin, 1973). The incorporation of leucine into isolated oligodendrocytes or astrocytes is decreased by excess potassium (Takahashi *et al.*, 1970; Blomstrand, 1971). As in the case of RNA, different rates of incorporation of radioactive precursors into proteins may reflect differences in the sizes of the precursor pools as well as true differences in rates of protein synthesis.

Phospholipids account for a major part of the lipid content in glial cells, and the turnover of these compounds has been measured by incorporation of ^{32}P. The incorporation is somewhat slower into bulk prepared glial cells than into bulk prepared neurons (Freysz *et al.*, 1969) but almost identical in C-2 glial cells and neuroblastoma cells (Eichberg *et al.*, 1975). As in the case of protein and RNA, it is difficult to determine whether the observed differences in rates of synthesis are due to a less rapid labeling of precursor pools or to a slower turnover of the lipids.

Conclusions

The biochemical studies of mammalian glial cells, which have been discussed in this chapter seem to have discerned distinctly between two types of cells, i.e., astrocytes and oligodendrocytes, whereas virtually no information seems to exist about biochemical characteristics of microglia.

Most of the information about oligodendroglia relates to the process of myelination and little, if any, evidence is available about other oligodendrocytic functions. Thus astroglia stands out as the glial cell type which is best known from a biochemical point of view. The diverse characteristics of this cell type are spectacular both qualitatively and quantitatively. The uptake rates for glutamate and potassium ions are immense; a preoccupation by astrocytes with functional processes that modify the composition of their surroundings is also suggested by their large surface/volume ratio, which biochemically may be reflected by the high lipid content. If the intensity of the uptake of potassium ions and of excitatory and inhibitory transmitters and the subsequent "fate" of these compounds (metabolic degradation of transmitters? transglial transport and release of potassium or transmitters?) can be regulated, this would provide a mechanism that functionally resembles the slow synaptic transmission, involving active transport mechanisms, which has been described in this volume by J. W. Phillis. In that case, astrocytes and neurons could be equal partners in the function of the central nervous system as suggested by, for example, Galambos (1961), Hertz (1965) and Schrier and Thompson (1974). The role of each cell type for "higher mental functions", as defined by responsiveness to external factors according to previous experience, would depend upon the extent to which cells of each type, and the functional connections they make with other cells, can be modified by incoming stimuli. In this context it seems thought provoking that Golgi type II neurons and astrocytes ontogenetically are closely related (S. Fedoroff, this volume) and that proliferation and migration of astrocytes *mainly* occurs at a young, postnatal stage (A. Privat, this volume) but, as evidenced by DNA synthesis, does continue at a reduced intensity throughout most, or perhaps all, adult life.

Acknowledgements

The financial support to the author by the Medical Research Council of Canada (Grants DG-120, MT-5957) is gratefully acknowledged.

References

Abe, T. and W. T. Norton. 1974. The characterization of sphingolipids from neurons and astroglia of immature rat brain. J. Neurochem. 23:1025-1036.

Aleksidze, N. G. and C. Blomstrand. 1969. Influence of potassium ions on the respiration of the neuron and the neuroglia of the lateral vestibular nucleus of the rabbit. Dokl. Akad. Nauk SSSR 186:1429-1430 or Proc. Acad. Sci. USSR, Biochem. series 186:140-141.

Altman, J. 1969. DNA metabolism and cell proliferation. In: *Handbook of Neurochemistry*, vol. 2. A Lajtha, ed. Plenum Press, New York. pp. 137-182.

Altman, J. and G. D. Das. 1964. Autoradiographic examination of the effects of enriched environment on the rate of glial multiplication in the adult rat brain. Nature (Lond.) 204:1161-1163.

Antanitus, D. S., B. H. Choi and L. W. Lapham. 1975. Immunofluorescence staining of astrocytes *in vitro* using antiserum to glial fibrillary acidic protein. Brain Res. 89:363-367.

Balazs, R, and J. E. Cremer. 1972. *Metabolic Compartmentation in the Brain.* MacMillan Press, London. pp. 1-383.

Balazs, R., Y. Machiyama, B. J. Hammond, T. Julian and D. Richter. 1970. The operation of the γ-aminobutyrate bypath of the tricarboxylic acid cycle in brain tissue *in vitro*. Biochem. J. 116:445-467.

Balcar, V. J., J. Borg and P. Mandel. 1977. High affinity uptake of L-glutamate and L-aspartate by glial cells. J. Neurochem. 28:87-93.

Banks-Schlegel, S. P. and T. C. Johnson. 1975. RNA metabolism in isolated mouse brain nuclei during early postnatal development. J. Neurochem. 24:947-952.

Bass, N. H., H. H. Hess, A. Pope and C. Thalheimer. 1971. Quantitative cytoarchitectonic distribution of neurons, glia, and DNA in rat cerebral cortex. J. comp. Neurol. 143:481-490.

Bauman, A., S. Bourgoin, P. Benda, J. Glowinski and M. Hamon. 1974. Characteristics of tryptophan accumulation by glial cells. Brain Res. 66:253-263.

Beart, P. M., J. S. Kelly and F. Schon. 1974. γ-aminobutyric acid in the rat peripheral nervous system, pineal and posterior pituitary. Biochem. Soc. Trans. 2:266-268.

Benda, P., J. Lightbody, G. Sato, L. Levine and W. Sweet. 1968. Differentiated rat glial cell strain in tissue culture. Science 161:370-371.

Benjamin, A. M. and J. H. Quastel. 1972. Locations of amino acids in brain slices from the rat. Biochem. J. 128:631-646.

Benjamin, A. M. and J. H. Quastel. 1975. Metabolism of amino acids and ammonia in rat brain cortex slices *in vitro:* A possible role of ammonia in brain function. J. Neurochem. 25:197-206.

Berger, S. J., G. Carter and O. H. Lowry. 1977a. The distribution of glycine, GABA, glutamate and aspartate in rabbit spinal cord, cerebellum and hippocampus. J. Neurochem. 28:149-158.

Berger, S. J., M. L. McDaniel, J. G. Carter and O. H. Lowry. 1977b. Distribution of four potential transmitter amino acids in monkey retina J. Neurochem. 28:159-163.

Berl, S., D. D. Clarke and D. Schneider. 1975. Metabolic Compartmentation and Neurotransmission. Relation to Brain Structure and Function. Plenum Press, New York. pp. 1-721.

Bignami, A. and D. Dahl. 1974. Astrocyte-specific protein and neuroglial differentiation. An immunofluorescence study with antibodies to the glial fibrillary acidic protein. J. comp. Neurol. 153:27-38.

Blomstrand, C. 1971. Studies on Protein Metabolism in Neuronal and Glial Cell-enriched Fractions from Brain Tissue. Thesis. University of Göteborg, Sweden.

Blomstrand, C. and A. Hamberger. 1969. Protein turnover in cell-enriched fractions from rabbit brain. J. Neurochem. 16:1401-1407.

Blomstrand, C. and A. Hamberger. 1970. Amino acid incorporation *in vitro* into protein of neuronal and glial cell-enriched fractions. J. Neurochem. 17:1187-1195.

Blomstrand, C., A. Hamberger, Å. Sellström and O. Steinwall. 1975. Protein bound radioactivity in neuronal and glial fractions following intra-carotid ^{3}H-leucine perfusion. Neurobiology. 5:178-187.

Bock, E. and A. Hamberger. 1976. Immunoelectrophoretic determination of brain-specific antigens in bulk-prepared neuronal and glial cells. Brain Res. 112:329-335.

Bock, E., O. S. Jörgensen, L. Dittman and L. F. Eng. 1975. Determination of brain-specific antigens in short term cultivated rat astroglia cells and in rat synaptosomes. J. Neurochem. 25:867-870.

Borg, J., V. J. Balcar and P. Mandel. 1976. High affinity uptake of taurine in neuronal and glial cells. Brain Res. 118:514-516.

Bourke, R. S. 1969a. Evidence for mediated transport of chloride in cat cerebral cortex *in vitro*. Exp. Brain Res. 8:219-231.

Bourke, R. S. 1969b. Studies of the development and subsequent reduction of swelling of mammalian cerebral cortex under isosmotic conditions *in vitro*.Exp. Brain Res. 8:232-248.

Bourke, R. S. and K. M. Nelson. 1972. Further studies on the K^+-dependent swelling of primate cerebral cortex *in vivo:* The enzymatic basis of the K^+-dependent transport of chloride. J. Neurochem. 19:663-685.

Bourke, R. S., K. M. Nelson, R. A. Naumann and O. M. Young. 1970. Studies of the production and subsequent reduction of swelling in primate cerebral cortex under isosmotic conditions *in vivo*. Exp. Brain Res. 10:427-466.

Bourke, R. S. and D. B. Tower. 1966. Fluid compartmentation and electrolytes of cat cerebral cortex *in vitro*. I. Swelling and solute distribution in mature cerebral cortex. J. Neurochem. 13:1071-1097.

Bowery, N. G. and D. A. Brown. 1972. γ-aminobutyric acid uptake by sympathetic ganglia. Nature (Lond.) New Biol. 238:89-91.

Cicero, T. J., J. A. Ferrendelli, V. Suntzeff and B. W. Moore. 1972. Regional changes in CNS levels of the S100 and 14-3-2 proteins during development and aging of the mouse. J. Neurochem. 19:2119-2125.

Clos, J., C. Favre, M. Seleme-Matrat and J. Legrand. 1977. Effects of undernutrition on cell formation in the rat brain and especially on cellular composition of the cerebellum. Brain Res. 123:13-26.

Clouet, D. H. and M. K. Gaitonde. 1956. The changes with age in the protein composition of the rat brain. J. Neurochem. 1:126-133.

Côté, L. J. 1964. Mg^{++}dependent Na^+-K^+-stimulated ATPase in the developing rat brain. Life Sci. 3:899-901.

Cotman, C., H. Herschmann and D. Taylor. 1971. Subcellular fraction of cultured glial cells. J. Neurobiol. 2:169-180.

Cremer, J. E., P. V. Johnston, B. I. Roots and A. J. Trevor. 1968. Heterogeneity of brain fractions containing neuronal and glial cells. J. Neurochem. 15:1361-1370.

Cummins, C. J. 1977. The Effect of Extracellular Potassium on the Intermediary Metabolism of Astrocytes *in vitro*. Thesis, University of Michigan.

Cummins, J. and H. Hydén. 1962. Adenosine triphosphate levels and adenosine triphosphatases in neurons, glia and neuronal membranes of the vestibular nucleus. Biochim. biophys. Acta 60:271-283.

Cutler, R. W. P. and D. S. Dudzinski, 1974. Regional changes in amino acid content in developing rat brain. J. Neurochem. 23:1005-1009.

Dahl, D. and A. Bignami. 1973. Glial fibrillary acidic protein from normal human brain. Purification and properties. Brain Res. 57:343-360.

Dalton, M. M., O. R. Hommes and C. P. Leblond. 1968. Correlation of glial proliferation with age in the mouse brain. J. Comp. Neurol. 134:397-400.

Daneholt, B. and S. O. Brattgård. 1966. A comparison between RNA metabolism of nerve cells and glia in the hypoglossal nucleus of the rabbit. J. Neurochem. 13:913-921.

Davison, A. N., M. L. Cuzner, N. L. Banik and J. Oxberry. 1966. Myelinogenesis in the rat brain. Nature (Lond.) 212:1373-1374.

Derry, D. M. and L. S. Wolfe. 1967. Gangliosides in isolated neurons and glial cells. Science. 158:1450-1452.

Dzubow, L. M. and D. Garfinkel. 1970. A simulation study of brain compartments. II. Atom by atom simulation of the metabolism of specifically labeled glucose and acetate. Brain Res. 23:407-417.

Edström, A., K. G. Haglid, M. Kanje, L. Ronnbäck and E. Walum. 1973. Morphological alterations and increase of S100 protein in cultured human glioma cells deprived of serum. Exp. Cell Res. 83:426-429.

Eichberg, J., H. M. Shein and G. Hauser. 1975. Phospholipid metabolism in cultured neuroblastoma and glioma cells incubated with carbamylcholine and norepinephrine. J. Neurochem. 24:67-70.

Eichberg, J., H. M. Shein and G. Hauser. 1976. Lipid composition and metabolism of cultured hamster brain astrocytes. J. Neurochem. 27:679-685.

Elliot, K. A. C. and N. Henderson. 1948. Metabolism of brain tissue slices and suspensions from various mammals. J. Neurophysiol. 11:471-484.

Embree, L. J., H. H. Hess and H. M. Shein. 1971. Sodium-potassium-ATPase activity of normal and virally transformed hamster astroglia grown subcutaneously. Brain Res. 27:422-425.

Embree, L. J., H. H. Hess and H. M. Shein. 1973. Microchemical studies of lipids, proteins and nucleic acids in polyoma virus-transformed hamster astroglia. J. Neuropath. exp. Neurol. 32:542-551.

Eng, L. F., J. J. Vanderhaeghen, A. Bignami and B. Gerstl. 1971. An acidic protein isolated from fibrous astrocytes. Brain Res. 28:351-354.

Fahn, S. and L. J. Côté. 1968. Regional distribution of sodium-potassium activated adenosine triphosphatase in the brain of the rhesus monkey. J. Neurochem. 15:433-436.

Faivre-Bauman, A., J. Rossier and P. Benda. 1974. Glutamate accumulation by a clone of glial cells. Brain Res. 76:371-375.

Fewster, M. E. and J. F. Mead. 1968. Lipid composition of glial cells isolated from bovine white matter. J. Neurochem. 15:1041-1052.

Flangas, A. L. and R. E. Bowman. 1970. Differential metabolism of RNA in neuronal-enriched and glial-enriched fractions of rat cerebrum. J. Neurochem. 17:1237-1245.

Folbergrova, J. O. H. Lowry and J. V. Passonneau. 1970. Changes in metabolites of the energy reserves in individual layers of mouse cerebral cortex and subjacent white matter during ischaemia and anaesthesia. J. Neurochem. 17:1155-1162.

Franck, G. 1970. Sur la composition ionique des tranches de cerveau de rat. Thesis, University of Liège, Belgium. 1970.

Freysz, L., R. Bieth, C. Judes, M. Sensenbrenner, M. Jacob and P. Mandel. 1968. Quantitative distribution of phospholipids in neurons and glial cells isolated from rat cerebral cortex. J. Neurochem. 15:307-313.

Freysz, L., R. Bieth and P. Mandel. 1969. Kinetics of the biosynthesis of phospholipids in neurons and glial cells isolated from rat brain cortex. J. Neurochem. 16:1417-1424.

Friede, R. L. 1964. The enzymatic response of astrocytes to various ions *in vitro*. J. Cell. Biol. 20:5-15.

Galambos, R. 1961. A glia-neuronal theory for brain function. Proc. Natl. Acad. Sci. U.S.A. 47:129-136.

Geren, B. B. 1954. The formation from the Schwann cell surface of myelin in the peripheral nerves of chick embryos. Exp. Cell Res. 7:558-562.

Giacobini, E. 1962. A cytochemical study of the localization of carbonic anhydrase in the nervous system. J. Neurochem. 9:169-177.

Gill, T. H., O. M. M. Young and D. B. Tower. 1974. The uptake of ^{36}Cl into astrocytes in tissue culture by a potassium-dependent, saturable process. J. Neurochem. 23:1011-1018.

Glees, P. 1972. The neuroglial compartments at light microscopic and electron microscopic levels. In: *Metabolic Compartmentation in the Brain.* R. Balazs and J. E. Cremer, eds. MacMillan Press, London. pp. 209-231.

Grossman, R. G. and A. Seregin. 1977. Glial-neural interaction demonstrated by the injection of Na^+ and Li^+ into cortical glia. Science 195:196-198.

Haas, R. J., J. Werner and T. M. Fliedner. 1970. Cytokinetics of neonatal brain cell development in rats as studied by the 'complex ^{3}H-thymidine labelling' method. J. Anat. 107:421-437.

Haljamäe, H. and A. Hamberger. 1971. Potassium accumulation by bulk prepared neuronal and glial cells. J. Neurochem. 18:1903-1912.

Hamberger, A. 1963. Difference between isolated neuronal and vascular glia with respect to respiratory intensity. Acta physiol. Scand. 58: suppl. 203, 1-52.

Hamberger, A. 1971. Amino acid uptake in neuronal and glial cell fractions from rabbit cerebral cortex. Brain Res. 31:169-178.

Hamberger, A. and H. Röckert. 1964. Intracellular potassium in isolated nerve cells and glial cells. J. Neurochem. 11:757-760.

Hamberger, A. and L. Svennerholm. 1971. Composition of gangliosides and phospholipids of neuronal and glial cell enriched fractions. J. Neurochem. 18:1821-1829.

Hamberger, A., C. Blomstrand and A. L. Lehninger. 1970. Comparative studies on mitochondria isolated from neuron-enriched and glia-enriched fractions of rabbit and beef brain. J. Cell Biol. 45:221-234.

Hamberger, A., H. A. Hansson and Å. Sellström. 1975. Scanning and transmission electron microscopy on bulk prepared neuronal and glial cells. Exp. Cell Res. 92:1-10.

Hartman, B. K., B. W. Moore, W. T. Shearers, K. Fujimoto, S. Kalmbach and H. C. Agrawal. 1976. Biochemical differentiation of brain elements using nervous system specific proteins. Trans. Am. Soc. Neurochem. 7:212.

Hemminki, K. and E. Holmila. 1971. Characterization of neuronal and glial fractions separated in sucrose and ficoll media. Acta physiol. Scand. 82:135-142.

Henn, F. A. and A. Hamberger. 1971. Glial cell function: Uptake of transmitter substances. Proc. Natl. Acad. Sci. U.S.A. 68:2686-2690.

Henn, F. A., H. Haljamäe and A. Hamberger. 1972. Glial cell function: Active control of extracellular K^{+} concentration. Brain Res. 43:437-443.

Henn, F. A., M. N. Goldstein and A. Hamberger. 1974. Uptake of the neurotransmitter candidate glutamate by glia. Nature (Lond.). 249:663-664.

Hertz, L. 1965. Possible role of neuroglia: a potassium-mediated neuronal-neuroglial-neuronal impulse transmission system. Nature (Lond.). 206:1091-1094.

Hertz, L. 1966. Neuroglial localization of potassium and sodium effects on respiration in brain. J. Neurochem. 13:1373-1387.

Hertz, L. 1968. Potassium effects on ion transport in brain slices. J. Neurochem. 15:1-16.

Hertz, L. 1973. *Ion Effects on Metabolism in the Adult Mammalian Brain in vitro.* FADL's Forlag, Copenhagen. pp. 1-207.

Hertz, L. 1976. Potassium effects on transport of amino acids, inorganic ions and water: Ontogenetic and quantitative aspects. In: *Transport Phenomena in the Nervous System, Physiological and Pathological Aspects.* G. Levi, L. Battistin and A. Lajtha, eds. Vol. 69 of Advances in experimental medicine and biology. Plenum Press, New York. pp. 371-383.

Hertz, L. 1977a. Drug-induced alterations of ion distribution at the cellular level of the central nervous system. Pharmacol. Rev. in press.

Hertz, L. 1977b. Potassium transport and energy metabolism in normal astrocytes in cultures. Abstracts, Sixth Internat. Meeting of the Internat. Soc. for Neurochem., Copenhagen. In press.

Hertz, L. 1978. Energy metabolism. In: *Dynamic Properties of Glial Cells.* G. Franck, L. Hertz, E. Schoffeniels and D. B. Tower, eds. Pergamon Press, in press.

Hertz, L. and A. Schousboe. 1975. Ion and energy metabolism of the brain at the cellular level. Int. Rev. Neurobiol. 18:141-211.

Hertz, L., L. Dittmann and P. Mandel. 1973. K^{+} induced stimulation of oxygen uptake in cultured cerebral glial cells. Brain Res. 60:517-520.

Hertz, L., A. Schousboe and S. Fedoroff. 1977. Glutamate uptake metabolism and release in cultured normal astrocytes. Trans. Am Soc. Neurochem. 8:206.

Hess, H. H., L. J. Embree and H. M. Shein. 1972. Enzymatic control of sodium- and potassium-active transport in normal and neoplastic rodent astroglia. Progr. exp. Tumor Res. 17:308-317.

Holmgren, E. 1904. Über die Trophospongien der Nervenzellen. Anat. Anzeiger 24:225-244.

Hutchison, H. T., K. Werrbach, C. Vance and B. Haber. 1974. Uptake of neurotransmitters by clonal lines of astrocytes and neuroblastoma in culture. I. Transport of γ-aminobutyric acid. Brain Res. 66:265-274.

Hydén, H. 1959. Quantitative assay of compounds in isolated, fresh nerve cells and glial cells from control and stimulated animals. Nature (Lond). 184:433-435.

Hydén, H. 1967a. Dynamic aspects on the neuron-glia relationship. A study with micro-chemical methods. In: *The Neuron.* H. Hydén, ed. Elsevier, Amsterdam. pp. 179-219.

Hydén, H. 1967b. RNA in brain cells. In: *The Neurosciences.* G. C. Quarton, T. Melnechuck and F. O. Schmitt, eds. The Rockefeller University Press, New York. pp. 248-266.

Hydén, H. and E. Egyhazi. 1963. Glial RNA changes during a learning experiment in rats. Proc. Nat. Acad. Sci. U.S.A. 49:618-624.

Hydén, H. and B. McEwen. 1966. A glial protein specific for the nervous system. Proc. Natn. Acad. Sci. U.S.A. 55:354-358.

Hydén, H. and A. Pigon. 1960. A cytophysiological study of the functional relationship between oligodendroglial cells and nerve cells of Deiters' nucleus. J. Neurochem. 6:57-72.

Ibrahim, M. Z. M. 1975. Glycogen and its related enzymes of metabolism in the central nervous system. Adv. Anat. Cell Biol. 52:5-89.

Iqbal, K., J. Grundke-Iqbal and H. M. Wisniewski. 1977. Oligodendroglia from human autopsied brain: Bulk isolation and some chemical properties. J. Neurochem. 28:707-716.

Iversen, L. L. and J. S. Kelly. 1975. Uptake and metabolism of γ-aminobutyric acid by neurones and glial cells. Biochem. Pharmacol. 24:933-938.

Jarlstedt, J. and A. Hamberger. 1971. Patterns and labelling characteristics in neuronal and glial RNA. J. Neurochem. 18:921-930.

Johnson, D. E. and O. Z. Sellinger. 1971. Protein synthesis in neurons and glial cells of the developing rat brain: an *in vivo* study. J. Neurochem. 18:1445-1460.

Katzman, R. and R. Grossman. 1975. Neuronal activity and potassium movement. In: *Brain Work.* Alfred Benzon Symposium, vol. 8. D. H. Ingvar and N. A. Lassen, eds. Munksgaard, Copenhagen. pp. 149-166.

Katzman, R. and H. M. Pappius. 1973. *Brain Electrolytes and Fluid Metabolism.* Williams and Wilkins Comp., Baltimore. pp. 1-149.

Kelly, J. S., T. M. Jessell and F. Schon. 1975. Uptake of GABA and glutamate in glial cells in the nervous system. In: *Proceedings of the sixth International Congress of Pharmacology, Vol. 2. Neurotransmission.* L. Ahtee, ed. Pergamon Press, New York. pp. 91-102.

Kety, S. S. 1957. The general metabolism of the brain *in vivo.* In: *Metabolism of the Nervous System.* D. Richter, ed. Pergamon Press, London. pp. 221-237.

Kimberlin, R. H. and H. S. Anger. 1969. DNA synthesis in the glial cells of scrapie-affected mouse brain. J. Neurochem. 16:543-548.

Kimelberg, H. K. 1974. Active potassium transport and (Na^+ + K^+) ATPase activity in cultured glioma and neuroblastoma cells. J. Neurochem. 22:971-976.

Kimelberg, H. K. and R. S. Bourke, 1973. Properties and localization of bicarbonate-stimulated ATPase activity in rat brain. J. Neurochem. 20:347-359.

Knox, W. H., R. G. Perrin and A. K. Sen. 1972. Effect of chronic administration of ethanol on (Na^+ + K^+) activated ATPase activity in six areas of the cat brain. J. Neurochem. 19:2881-2884.

Korr, H., B. Schultze and W. Maurer. 1973. Autoradiographic investigations of glial proliferation in the brain of adult mice. J. Comp. Neurol. 150:169-176.

Krauss-Ruppert, R., J. Laissue, H. Bürki and N. Odartchenko. 1973. Proliferation and turnover of glial cells in the forebrain of young adult mice as studied by repeated injections of ^3H-thymidine over a prolonged period of time. J. Comp. Neurol. 148:211-216.

Krnjevic, K. and J. W. Phillis. 1963. Iontophoretic studies of neurones in the mammalian cerebral cortex. J. Physiol. (Lond.) 165:274-304.

Kuffler, S. W. and J. G. Nicholls. 1966. The physiology of neuroglial cells. Ergeb. Physiol. 57:1-90.

Kukes, G., J. DeVellis and R. Elul. 1976a. A linked active transport system for Na^+ and K^+ in a glial cell line. Brain Res. 104:93-105.

Kukes, G., R. Elul and J. DeVellis. 1976b. The ionic basis of the membrane potential in a rat glial cell line. Brain Res. 104:71-92.

Latzkovits, L., M. Sensenbrenner and P. Mandel. 1974. Tracer kinetic model analysis of potassium uptake by dissociated nerve cell cultures: Glial-neuronal interrelationship. J. Neurochem. 23:193-200.

Lees, M. B. and H. M. Shein. 1970. Sodium and potassium content of normal and neoplastic rodent astrocytes in cell culture. Brain. Res. 23:280-283.

Lightbody, J., S. E. Pfeiffer, P. L. Kornblith and H. Herschman. 1970. Biochemically differentiated clonal human glial cells in tissue culture. J. Neurobiol. 1:411-417.

Lisy, V., H. Kovaru, J. Faltin and Z. Lodin. 1971. Activity of succinate dehydrogenase and acetylcholinesterase in synaptic endings isolated from neuropil and cerebral cortex. Physiol. Bohemoslov. 20:229-234.

Lisy, V. and Z. Lodin. 1973. Incorporation of radioactive leucine into neuronal and glial proteins during postnatal development. Neurobiology 3:320-326.

Lodin, Z., J. Hartman, M. Kazachashvili, K. A. Chaubal and J. Müller. 1967. The dry mass of the cerebellar cortex of the rat and rabbit. Results and methodological problems of determining the dry mass with the use of the interference microscope. Brain Res. 6:489-509.

Lodin, Z., J. Hartman, M. P. Kage, P. Korinkova and J. Booher. 1971. Potassium-induced hydration in cultured neural tissue. Neurobiology 1:69-85.

Logan, W. J. 1976. Amino acid transport by two glial cell lines and by proliferating glia. Exp. Neurol. 53:431-443.

Lövtrup-Rein, H. 1970. Synthesis of nuclear RNA in nerve and glial cells. J. Neurochem. 17: 853-863.

Lowry, O. H. 1953. The quantitative histochemistry of the brain: Histological sampling. J. Histochem. Cytochem. 1:420-428.

Lugaro, E. 1907. Sulle funzioni delle nevroglia. Rivist. Patol. nerv. ment. 12:225-233.

Lund-Andersen, H. and L. Hertz. 1970. Effects of potassium and glutamate on swelling and on sodium and potassium content in brain cortex slices from adult rats. Exp. Brain Res. 11:199-212.

Mares, V. and Z. Lodin. 1974. An autoradiographic study of DNA synthesis in adolescent and adult mouse forebrain. Brain Res. 76:557-561.

Margolis, R. K., A. Heller and R. Y. Moore. 1968. Effects of changes in cellular composition following neuronal degeneration on amino acids in brain. Brain Res. 11:19-31.

McIlwain, H. 1953. Substances which support respiration and metabolic response to electrical impulses in human cerebral tissues. J. Neurol. Neurosurg. Physchiat. 16:257-266.

McLennan, H. 1976. The autoradiographic localization of L-[^{3}H] glutamate in rat brain tissue. Brain Res. 115:139-144.

Medzihradsky, F., O. Z. Sellinger, P. S. Nandhasri and J. C. Santiago. 1972. ATPase activity in glial cells and in neuronal perikarya of rat cerebral cortex during early postnatal development. J. Neurochem. 19:543-545.

Minchin, M. C. W. 1975. Factors influencing the efflux of ^{3}H-gamma-aminobutyric acid from satellite glial cells in rat sensory ganglia. J. Neurochem. 24:571-577.

Minchin, M. C. W. and P. M. Beart. 1975. Compartmentation of amino acid metabolism in the rat dorsal root ganglion; a metabolic and autoradiographic study. Brain Res. 83:437-449.

Minchin, M. C. W. and L. L. Iversen. 1974. Release of ^{3}H-gamma amino-butyric acid from glial cells in rat dorsal root ganglia. J. Neurochem. 24:533-540.

Mokrasch, L. C. 1971. Free amino acid content of normal and neoplastic rodent astroglia. Brain Res. 25:672-676.

Möller, M., H. Lund-Andersen, K. Möllgård and L. Hertz. 1974. Concordance between morphological and biochemical estimates of fluid spaces in rat brain cortex slices. Exp. Brain Res. 22:299-314.

Moonen, G. and G. Franck. 1977. Potassium effect on Na^+-K^+-ATPase activity of cultured new-born rat astroblasts during differentiation. Neuroscience Letters 4:263-267.

Nagata, Y., K. Mikoshiba and Y. Tsukada. 1974. Neuronal cell body enriched and glial cell

enriched fractions from young and adult rat brains: Preparation and morphological and biochemical properties. J. Neurochem. 22:493-503.
Nair, V. and D. Bau. 1971. Studies on the functional significance of carbonic anhydrase in central nervous system. Brain Res. 31:185-193.
Norton, W. T. and S. E. Poduslo. 1971. Neuronal perikarya and astroglia of rat brain: Chemical composition during myelination. J. Lipid. Res. 12:84-90.
Olsson, J. E., C. Blomstrand and K. G. Haglid. 1973. Cellular distribution of beta-trace protein in CNS and brain tumors. J. Neurol. Neurosurg. Psychiat. 37:302-311.
Packman, P. M., C. Blomstrand and A. Hamberger. 1971. Disc electrophoretic separation of proteins in neuronal, glial and subcellular fractions from cerebral cortex. J. Neurochem. 18:1-9.
Palay, S. L. and V. Chan-Palay. 1974. *Cerebellar Cortex: Cytology and Organization.* Springer, New York. pp. 1-348.
Passonneau, J. V., W. D. Lust and S. K. Crites. 1976. GABA shunt metabolites and enzymes in glioma and neuroblastoma cell lines. Trans. Am. Soc. Neurochem. 7:192.
Paterson, J. A., A. Privat, E. A. Ling and C. P. Leblond. 1973. Investigation of glial cells in semi-thin sections III. Transformation of subependymal cells into glial cells as shown by radioautography after ^{3}H-thymidine injection into the lateral ventricle of the young rat. J. Comp. Neur. 149:83-102.
Peters, A. 1960. The formation and structure of myelin sheaths in the central nervous system. J. Biophys. Biochem. Cytol. 8:431-446.
Pevzner, L. Z. 1965. Topochemical aspects of nucleic acid and protein metabolism within the neuron-neuroglia unit of the superior cervical ganglion. J. Neurochem. 12:993-1002.
Pevzner, L. Z. 1971. Topochemical aspects of nucleic acid and protein metabolism within the neuron-neuroglia unit of the spinal cord anterior horn. J. Neurochem. 18:895-907.
Pevzner, L. Z. and T. M. Semeshina. 1976. Topochemical and cytochemical studies on RNA content in the central nervous system of the active and hibernating ground squirrel, *citellus erythrogenys.* Brain. Res. 108:205-211.
Pfeiffer, S. E., H. R. Herschman, J. Lightbody and G. Sato. 1970. Synthesis by a clonal line of rat glial cells of a protein unique to the nervous system. J. cell. Physiol. 75:329-340.
Poduslo, S. E. and W. T. Norton. 1971. Isolation and some chemical properties of oligodendroglia from calf brain. J. Neurochem. 19:727-736.
Privat, A. 1975. Postnatal gliogenesis in the mammalian brain. Internat. Rev. Cytol. 40:281-323.
Reivich, M., L. Sokoloff, C. Kennedy and M. des Rosiers. 1976. An autoradiographic method for the measurement of local glucose metabolism in the brain. In: *Brain Work.* The Coupling of Function Metabolism and Blood Flow in the Brain. Alfred Benzon symposium VIII. A. D. H. Ingvar and N. A. Lassen eds. Munksgaard, Copenhagen. pp. 377-384.
Rakic, P. 1972. Mode of cell migration to the superficial layers of fetal monkey neocortex. J. Comp. Neurol. 145:61-84.
Robert, J., L. Freysz, M. Sensenbrenner, P. Mandel and G. Rebel. 1975. Gangliosides of glial cells: A comparative study of normal astroblasts in tissue culture and glial cells isolated on sucrose-ficoll gradients. FEBS letters 50:144-146.
Robert, J., P. Mandel and G. Rebel. 1976. Neutral lipids and phospholipids from cultured astroblasts. J. Neurochem. 26:771-777.
Roberts, P. J. 1974. Amino acid release from isolated dorsal root ganglia. Brain Res. 74:327-332.
Roberts, P. J. 1976a. Gamma-aminobutyric acid homoexchange in sensory ganglia. Brain Res. 113:205-209.
Roberts, P. J. 1976b. Amino acid transport in spinal and sympathetic ganglia. In: *Transport Phenomena in the Nervous System, Physiological and Pathological aspects.* G. Levi, L. Battistin, and A. Lajtha, eds. Vol. 69 of Advances in Experimental Medicine and Biology. Plenum Press, New York. pp. 165-178.
Roberts, P. J. and P. Keen. 1974. [^{14}C] Glutamate uptake and compartmentation in glia of rat dorsal sensory ganglion. J. Neurochem. 23:201-209.
Rose, S. P. R. 1965. Preparation of enriched fractions from cerebral cortex containing isolated, metabolically active neuronal cells. Nature (Lond.). 206:621-622.

Rose, S. P. R. 1967. Preparation of enriched fractions from cerebral cortex containing isolated, metabolically active neuronal and glial cells. Biochem. J. 102:33-43.

Rose, S. P. R. 1970. The compartmentation of glutamate and its metabolites in fractions of neuron cell bodies and neuropil; studied by intraventricular injection of [U-^{14}C] glutamate. J. Neurochem 17:809-816.

Rose, S. P. R. and A. K. Sinha. 1969. Some properties of isolated neuronal cell fractions. J. Neurochem. 16:1319-1328.

Sålem, R. D., R. Hammerschlag, H. Bracho and R. K. Orkand. 1975. Influence of potassium ions on accumulation and metabolism of [^{14}C]glucose by glial cells. Brain Res. 86:499-503.

Samson, F. E. and D. J. Quinn. 1967. Na^+-K^+-activated ATPase in rat brain development. J. Neurochem. 14:421-427.

Santen, R. J. and B. W. Agranoff. 1963. Studies on the estimation of deoxyribonucleic acid and ribonucleic acid in rat brain. Biochim. biophys. Acta. 72:251-262.

Schousboe, A. 1977. Metabolism and Function of GABA in the mammalian brain. Possible role of glia cells. Thesis, University of Copenhagen, Denmark.

Schousboe, A., H. Fosmark and B. Formby. 1976a. Effect of serum withdrawal on Na^+-K^+ ATPase activity in astrocytes cultured from dissociated brain hemispheres. J. Neurochem. 26:1053-1055.

Schousboe, A., J. Booher and L. Hertz. 1970. Content of ATP in cultivated neurons and astrocytes exposed to balanced and potassium-rich media. J. Neurochem. 17:1501-1504.

Schousboe, A., H. Fosmark and L. Hertz. 1975. High content of glutamate and of ATP in astrocytes cultured from rat brain hemispheres: Effect of serum withdrawal and of cyclic AMP. J. Neurochem. 25:909-911.

Schousboe, A., H. Fosmark and B. Formby. 1976a. Effect of serum withdrawal on Na^+-K^+ ATPase activity in astrocytes cultured from dissociated brain hemispheres. J. Neurochem. 26:1053-1055.

Schousboe, A., H. Fosmark and G. Svenneby. 1976b. Taurine uptake in astrocytes cultured from dissociated mouse brain hemispheres. Brain Res. 116:158-164.

Schousboe, A., L. Hertz and G. Svenneby. 1977. Uptake and metabolism of GABA in astrocytes cultured from dissociated mouse brain hemispheres. Neurochem. Res. 2:217-229.

Schrier, B. K. and E. J. Thompson. 1974. On the role of glial cells in the mammalian nervous system. J. Biol. Chem. 249:1769-1780.

Sellinger, O. Z., J. M. Azcurra, D. E. Johnson, W. G. Ohlsson and Z. Lodin. 1971. Independence of protein synthesis and drug uptake in nerve cell bodies and glial cells isolated by a new technique. Nature (Lond.) New Biol. 230:253-256.

Sellinger, O. Z., D. E. Johnson, J. C. Santiago and V. Idoyaga-Vargas. 1973. A study of biochemical differentiation of neurons and glia in rat cerebral cortex. Prog. Brain Res. 40:331-347.

Schwartz, A., G. E. Lindenmayer and J. C. Allen. 1975. The sodium-potassium adenosine triphosphatase: Pharmacological, physiological and biochemical aspects. Pharmacol. Rev. 27:3-134.

Sellström, Å., and A. Hamberger. 1977. Potassium stimulated γ-aminobutyric acid release from neurons and glia. Brain Res. 119:189-198.

Sellström, Å., L. B. Sjöberg and A. Hamberger. 1975. Neuronal and glial systems for γ-aminobutyric acid metabolism. J. Neurochem. 25:393-398.

Sharp, F. R. 1976. Relative cerebral glucose uptake of neuronal perikarya and neuropil determined with 2-deoxyglucose in resting and swimming rat. Brain Res. 110:127-139.

Skou, J. C. 1957. The influence of some cations on an adenosine triphosphatase from peripheral nerves. Biochim. Biophys. Acta. 23:394-401.

Skou, J. C. 1965. Enzymatic basis for active transport of Na^+ and K^+ across cell membrane. Physiol. Rev. 45:596-617.

Somjen, G. G. 1975. Electrophysiology of neuroglia. Ann. Rev. Physiol. 37:163-190.

Stahl, W. L. and S. H. Broderson. 1976. Localization of Na^+, K^+-ATPase in brain. Fed. Proc. 35:1260-1265.

Stefanovic, V., A. Ebel, J. C. Hermetet and P. Mandel. 1974. Na^{+}-K^{+}-activated ATPase and K^{+}-activated p-nitrophenylphosphatase activities: a comparison of their regional localizations in the C_{57} mouse brain. J Neurochem. 23:895-897.

Stefanovic, V., J. Ciesielski-Treska, M. Ledig and P. Mandel. 1975. (Na^{+} + K^{+})-activated ATPase and K^{+}-activated p-nitrophenyl-phosphatase activities of the nervous system cells in tissue culture. Experientia 31:807-808.

Stefanovic, V., J. Ciesielski-Treska, A. Ebel and P. Mandel. 1976. Neuroblast-glial interaction: The effect of co-cultivation upon ecto-ATPase activity of neuroblastoma and glioma cells. Exptl. Cell Res. 98:191-203.

Sviridov, S. M., L. I. Korochkin, V. N. Ivanov, E. I. Maletskaya and T. K. Kakhtina. 1972. Immunohistochemical studies of S-100 protein during postnatal ontogenesis of the brain of two strains of rats. J. Neurochem. 19:713-718.

Svoboda, P. and Z. Lodin. 1972. Postnatal development of some mitochondrial enzyme activities of cortical neurons and glial cells. Physiol. Bohemoslov 21:457-465.

Takahashi, Y., C. S. Hsü and S. Honma. 1970. Potassium and glutamate effects on protein synthesis in isolated neuroglial cells. Brain Res. 23:284-287.

Tiplady, B. and S. P. R. Rose. 1971. Amino acid incorporation into protein in neuronal cell body and neuropil fractions *in vitro*, J. Neurochem. 18:549-558.

Tower, D. B. 1960. The neurochemistry of asparagine and glutamine. In: *Neurochemistry of Nucleotides and Amino Acids*. R. O. Brady and D. B. Tower, eds. pp. 173-204.

Tower, D. B. and O. M. Young. 1973. The activities of butyrylcholinesterase and carbonic anhydrase, the rate of anaerobic glycolysis, and the question of a constant density of glial cells in cerebral cortices of various mammalian species from mouse to whale. J. Neurochem. 20:269-278.

Trachtenberg, M. C., P. L. Kornblith and J. Häuptli. 1972. Biophysical properties of cultured human glial cells. Brain Res. 38:279-298.

Utley, J. D. 1963. Gamma aminobutyric acid and 5-hydroxytryptamine concentrations in neurons and glial cells in the medial geniculate body of the cat. Biochem. Pharmac. 12:1228-1230.

Van den Berg, C. J. and D. Garfinkel. 1971. A simulation study of brain compartments. Metabolism of glutamate and related substances in mouse brain. Biochem. J. 123:211-218.

Van Economo, C. 1926. Ein Koeffizient für die Organisationshöhe der Grosshirnrinde. Klin. Wschr. 5:593-595.

Vernadakis, A. and R. Nidess. 1976. Biochemical characteristics of C-6 glial cells. Neurochem. Res. 1:385-402.

Villegas, J. 1968. Transport of electrolytes in the Schwann cell and location of sodium by electron microscopy. J. Gen. Physiol. 51:61-71.

Villegas, J., L. Villegas and R. Villegas. 1965. Sodium, potassium and chloride concentrations in the Schwann cell and axon of the squid nerve fiber. J. Gen. Physiol. 49:1-7.

Volpe, P. and A. Giuditta. 1967. Biosynthesis of RNA in neuron- and glia-enriched fractions. Brain Res. 6:228-240.

Watson, W. E. 1974. Physiology of neuroglia. Physiol. Rev. 54:245-271.

Weck, P. K. and T. C. Johnson. 1976. Nuclear-cytosol interactions that modulate RNA synthesis and transcript size of mouse brain nuclei. J. Neurochem. 27:1367-1374.

Wolff, J. R. 1970a. Quantitative aspects of astroglia, VIth Intern. Congr. Neuropathology, Paris. pp. 327-333, 352.

Wolff, J. R. 1970b. The astrocyte as link between capillary and nerve cell. Triangle. 9:153-164.

Yanagihara, T. 1974. RNA metabolism in rabbit brain: Study with neuron-glia and subcellular fractions J. Neurochem. 23:833-837.

Yonezawa, T., B. Bornstein, R. Peterson and M. R. Murray. 1962. A Histochemical study of oxidative enzymes in myelinating cultures of central and peripheral nervous tissue. J. Neuropat. Exp. Neurol. 21: 479-487.

Yanagihara, T. 1976. Cerebral anoxia: Effect on neuron-glia fractions and on polysomal protein synthesis. J. Neurochem. 27:539-543.

PHYSIOLOGICAL AND PHARMACOLOGICAL STUDIES ON CENTRAL SYNAPTIC TRANSMISSION

J. W. PHILLIS

Department of Physiology
University of Saskatchewan
Saskatoon, Saskatchewan
Canada

I. Introduction

The concept that the nervous system is composed of discrete units or nerve cells was first proposed by His and Forel and then independently by Ramón y Cajal. It was Cajal's elegant histological studies which, above all others, established that each nerve cell or neuron is a structural entity bounded by a plasma membrane and that the functional connections between nerve cells are effected by close contacts and not by continuity in a syncytial network, as had been proposed in the rival reticular hypothesis of Gerlach and Golgi. The name 'synapse' was given to these functional connections between nerve cells by Sherrington (1906). More recently, electron microscope studies have confirmed the existence of a cytoplasmic discontinuity between nerve cells and have revealed additional facets of the extraordinary variety of morphological features that differentiate the synaptic contacts between nerve cells.

At the majority of synapses in vertebrates, the presynaptic element produces its effect on the postsynaptic structure by the secretion of a specific chemical substance which in turn alters ion fluxes across the postsynaptic membrane and thus evokes a change in the membrane potential. Such junctions are known as chemical synapses. At many synapses in invertebrates, and at some in vertebrates, the presynaptic impulse generates a flow of current across the postsynaptic membrane in the absence of the release of a chemical mediator. Such junctions are called electrical synapses and they possess distinctive morphological features which enable them to be identified as such. Some junctions may possess the ability to utilize both electrical and chemical transmission.

Some of the general features of synaptic ultrastructure are illustrated in Fig. 1. At chemically transmitting synapses the pre- and postsynaptic structures are separated by a synaptic cleft of 20-30 nm, which is continuous with the intercellular space. Both pre- and post-synaptic membranes may

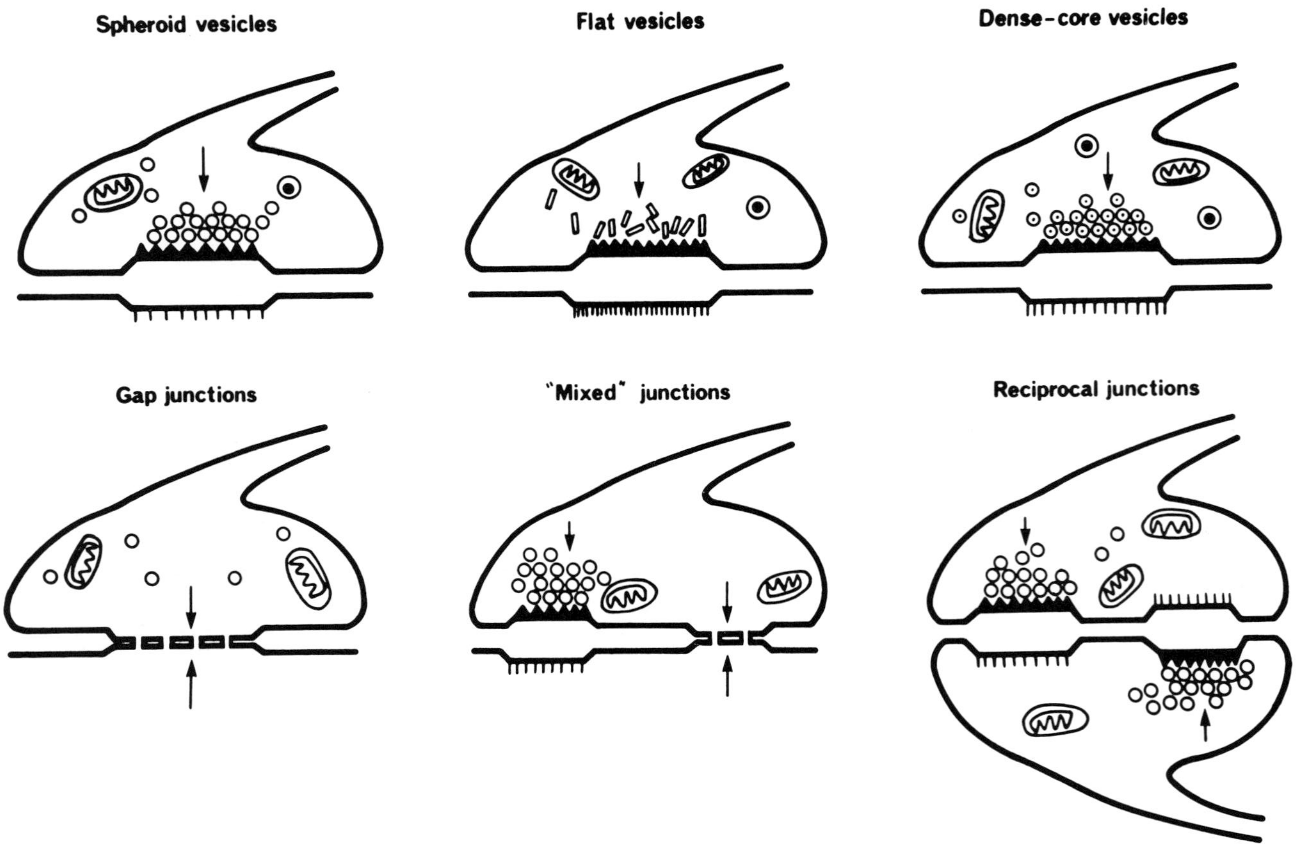

FIG. 1. Schematic representation of variations of synaptic morphology. (Redrawn after Bodian, 1972).

contain specialized regions of electron-dense material, and areas of enhanced electron-density may also be present in the synaptic cleft. A characteristic feature of chemically transmitting synapses is the presence in the presynaptic terminal of numerous small vesicles (20-80 nm in diameter) called synaptic vesicles, which are thought to contain the transmitter chemical. The presynaptic terminal commonly contains a large number of mitochondria.

The majority of chemically transmitting synapses can be assigned to either of the two major classes (Types I and II) of asymmetric synaptic contacts (Gray, 1969; Shepherd, 1974; Palay and Chan-Palay, 1976). Type I synapses are predominantly axodendritic in their distribution and are thought to be excitatory in function. The synaptic vesicles in the presynaptic terminal are rounded in aldehyde-fixed material and are usually aggregated in the vicinity of the electron dense material bordering the synaptic cleft. Type II synapses are commonly axosomatic and contain flattened vesicles in aldehyde-fixed tissue. Type II synapses are thought to be inhibitory in their action. At some synapses in the olfactory bulb and retina, known as reciprocal or symmetric junctions, there are no clear-cut pre- or post-synaptic elements, rather at one point along the interface transmission appears to be in one direction, whereas elsewhere it may be in the reverse direction.

"Tight" junctions, where the membranes of adjacent neurons come into very close apposition, have been observed in the central nervous system. In electron micrographs the outer layers of the membranes of the two cells appear to be fused and the combined membranes present a five-layered appearance. At "gap" junctions the external surfaces of the plasma membranes remain separated by a 2-3 nm gap and present a seven-layered appearance. In several cases, the presence of these "gap" junctions has been correlated with the finding of a low resistance electrical pathway between two neurons (Reese and Brightman, 1970; Bennet, 1973), and on this basis they have been categorized as electrical synapses. Electrical junctions are a common form of inter-neuronal connection in lower vertebrates (Bennet, 1972), but have been described less frequently in higher vertebrates (Llinas, 1975).

Chemically transmitting synapses can be loosely subdivided into two categories. At one type, of which the mammalian skeletal neuromuscular junction is a prime example, the synaptic transmitter acts on the postsynaptic membrane to rapidly increase its permeability to certain ions. At other types of junction, the transmitter induces alterations in the excitability of the postsynaptic element by inducing membrane potential alterations in the absence of any decrease in membrane resistance; indeed there are reports of increase in membrane resistance. (These effects are frequently slower in onset than those described above). In at least some of these instances the transmitter may exert its effect by altering the activity of

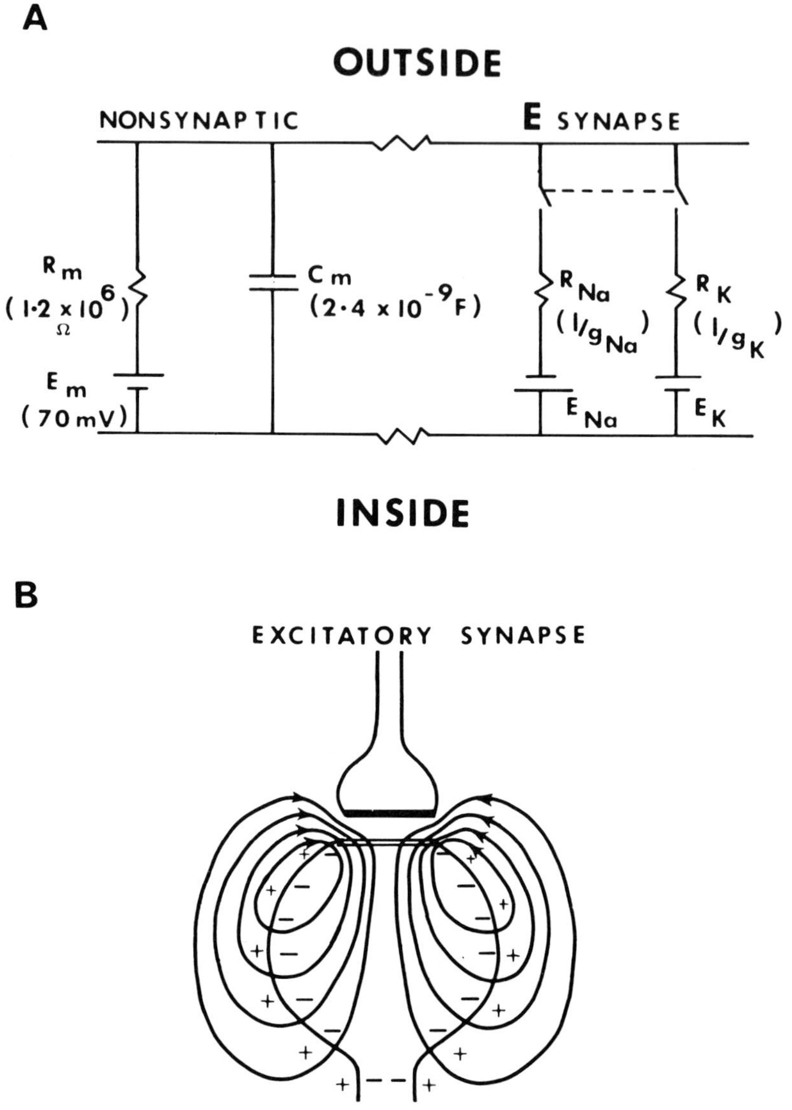

FIG. 2. *A*. Formal electrical diagram of the membrane of a motoneuron. On the right are the synaptically activated conductances of potassium (g_K) and sodium (g_{Na}), in series with their respective equilibrium potentials E_{Na} and E_K (shown as emf's). On the left, the membrane resistance (R_m) of the nonsynaptic membrane is in series with the membrane potential (emf, E_m) and in parallel with the membrane capacitance (C_m). *B*. Current flow induced by activation of an excitatory synapse.

metabolic systems in either the presynaptic terminals or postsynaptic neuron, and, conceivably, both structures may be affected simultaneously.

II. Physiological Studies in Synaptic Transmission

A. Fast Excitatory Synaptic Potentials

In the mammalian central nervous system (CNS) the mechanisms of synaptic transmission have been studied most extensively on spinal motoneurons. The excitatory postsynaptic potential (EPSP) generated by activation of Ia afferent terminals is a result of an inward flow of current, largely carried by sodium ions, across the subsynaptic membrane (Fig. 2A). The outward current (Fig. 2B), which must be of equal magnitude, is at first capacitative as the membrane capacity is discharged, and then some ionic currents (carried by potassium and perhaps chloride ions) pass across the membrane.

In some experiments it has been possible to demonstrate a reversal of the EPSP when the charge inside the motoneuron was reversed so as to be positive with respect to the exterior, but in others there has been no clearcut reversal of the EPSP or reversal only of its late phase (Coombs *et al.*, 1955a; Smith *et al.*, 1967; Kuno and Llinas, 1970; Werman and Carlen, 1976; Edwards *et al.*, 1976). Hyperpolarization of motoneurons causes little change in the peak amplitude of the EPSP, an observation which has been attributed to the shunting effect of anomalous rectification (Nelson and Frank, 1967), or to a failure of the synaptic current to increase (Edwards *et al.*, 1976).

In many motoneurons the expected decrease in membrane resistance during the generation of an EPSP has not been detectable (Smith *et al.*, 1967), an observation which can be accounted for if it is assumed that many of the active synapses are on dendrites and distant from the recording microelectrode. Such a distribution of excitatory endings could also account for the failure of many EPSP's to reverse when the charge within the cell soma is reversed (Smith *et al.*, 1967; Rall *et al.*, 1967).

Edwards *et al.* (1976), studying somatically generated EPSPs, calculated that the reversal potentials should not be less positive than the accepted equilibrium for either Na or Ca ions. These investigators re-examined the suggestion of Rall *et al.* (1967) that transmission between Ia afferents and motoneurons might be electronically mediated, but were unable to come to a definite conclusion regarding the differentiation between electrical and chemical transmission. Werman and Carlen (1976) have attempted to reconcile some of the difficulties inherent in the interpretation of the experimental data by suggesting that the Ia-motoneuron synapse operates by both electrical and chemical transmission.

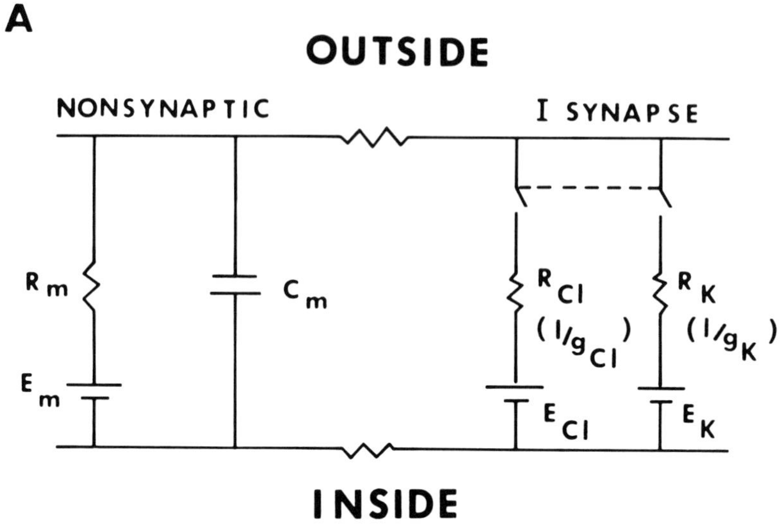

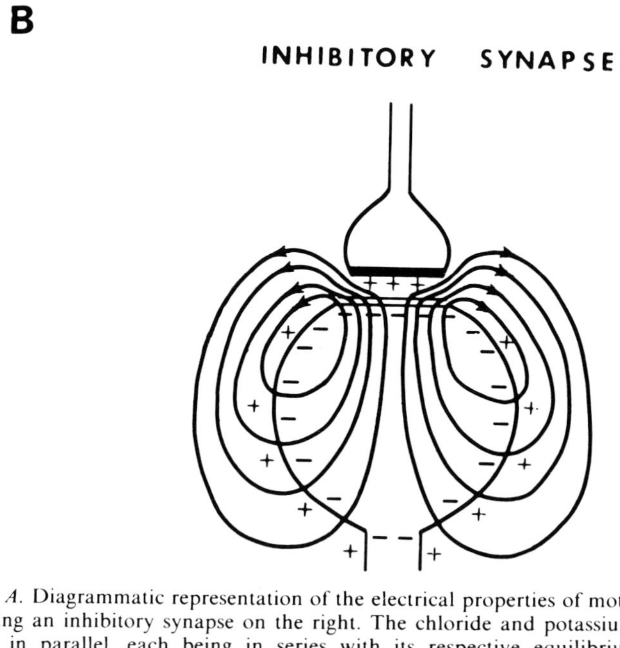

FIG. 3. *A*. Diagrammatic representation of the electrical properties of motoneuron membrane including an inhibitory synapse on the right. The chloride and potassium conductances are shown in parallel, each being in series with its respective equilibrium potential (emf). Nonsynaptic membrane is on the left. *B*. Current flows induced by activation of an inhibitory synapse.

B. Fast Inhibitory Synaptic Potentials

The inhibitory synaptic action on antagonist motoneurons of Ia afferents to the spinal cord is generated by a brief hyperpolarization of the postsynaptic membrane, the inhibitory postsynaptic potential (IPSP). The IPSP is the result of an outwardly directed current flux across the subsynaptic membrane of the activated inhibitory synapse, with an inwardly directed component across the remainder of the membrane (Fig. 3). The ionic mechanism underlying the generation of the IPSP in spinal motoneurons was first studied by Coombs *et al.* (1955b), who concluded that both potassium and chloride ions were carrying the current for the generation of the inhibitory potential (Fig. 3A), since the equilibrium potential for the IPSP was approximately midway between the calculated equilibria for the K and Cl ions (chloride was assumed to be distributed passively across the membrane).

More recently, Lux has proposed the existence of an outwardly directed chloride pump in spinal motoneurons which maintains a chloride gradient with an equilibrium potential which would be more negative than the resting potential (Lux *et al.*, 1970; Lux, 1971). Blockage of the chloride pump with externally applied ammonium ions reduced the amplitude of IPSP's and altered their reversal potential, suggesting that alterations in chloride conductance are of major significance in the generation of the IPSP. Another example of a chloride-dependent IPSP has been revealed in cat trochlear motoneurons, where the IPSP has an equilibrium potential that is even more negative than the potassium-dependent action potential after-hyperpolarization (Llinas and Baker, 1972). These IPSP's are reversed by intracellular chloride injection. Ammonium ions reduce the IPSP amplitude by altering its equilibrium potential rather than by affecting the magnitude of the postsynaptic conductance change (Llinas *et al.*, 1974).

The permeability changes induced by the inhibitory transmitter have been studied by intracellular injection of various anions. Various small anions, including chloride, will reverse the IPSP, and it has been estimated that the diameter of the conductance channel in motoneurons must be at least 1.14 times the size of a hydrated K ion (Ito *et al.*, 1962). The channels in cerebral cortical neurons appear to be substantially larger, as intracellular injection of a range of univalent organic and inorganic anions reversed the IPSP's in these neurons (Kelly *et al.*, 1969).

C. Presynaptic Inhibition

As originally postulated by Eccles (1963), presynaptic inhibition was effected by a reduction in the amount of transmitter released from a pre-synaptic nerve terminal during its invasion by an action potential, when the terminal had itself been depolarized by a transmitted released from another nerve terminal. Although depolarization makes the affected terminal more excitable, it has the effect of reducing the amplitude of invading action potentials and thus the amount of transmitter released by the action potential (Schmidt, 1971).

Presynaptic inhibition has been a controversial phenomenon, however, and other investigators have interpreted their findings in a different vein, arguing that the inhibition actually occurs postsynaptically at the remote sites on the dendrites (remote dendritic inhibition; Kellerth and Szumski, 1966; Diamond, 1968; Carlen *et al.*, 1976). Regardless of its size of occurrence, inhibition of this type is antagonized by picrotoxin and bicuculline, which also block the actions of γ-aminobutyric acid (GABA). GABA depolarizes presynaptic nerve terminals and hyperpolarizes neurons, and is therefore a likely candidate for the role of transmitter of the "presynaptic" inhibitory transmitter.

Somewhat different inhibitory presynaptic actions may be associated with other putative transmitter agents. Acetylcholine (ACh) reduces the release of ACh by nerve terminals in skeletal muscle and sympathetic ganglia (Ciani and Edwards, 1963; Hubbard *et al.*, 1965; Nishi, 1970; Ginsborg, 1971). At both sites the presynaptic ACh receptors are nicotinic. In the CNS, ACh depresses its own release as well as that of dopamine and norepinephrine by an action on muscarinic receptors (Szerb and Somogyi, 1973; Reader *et al.*, 1976).

Catecholamines act presynaptically in increasing ACh release from motor nerve terminals (Krnjević and Miledi, 1958; Jenkinson *et al.*, 1968), but depress ACh release from preganglionic nerve terminals in sympathetic ganglia (Weir and McLennan, 1963; Dun and Nishi, 1974). Adenosine, another compound with potent depressant effects on central neurons, depresses ACh release from sympathetic nerve terminals (Ginsborg and Hirst, 1972; Hedqvist and Fredholm, 1976). Adenosine and the adenine nucleotides inhibit the contractile responses of guinea pig ileum to electrical stimulation, and reduce ACh release by this preparation (Mori *et al.*, 1973; Kazić and Milosavljević, 1976; Vizi and Knoll, 1976). The contractile responses of the preparation to exogenously applied ACh, histamine and potassium are unaffected by concentrations of the adenine derivatives which inhibit the nerve mediated contractions (Mori *et al.*, 1973; Kazić and Milosavljević, 1976) indicating that the inhibition occurs at a presynaptic site, with a reduction in ACh release. Increases in external calcium can prevent this action of the adenine nucleotides (Kazić and Milosavljević, 1976). The inhibitory effects of adenosine on postsynaptic potentials in olfactory cortical slices can also be reversed by elevated calcium levels (Kuroda *et al.*, 1976), and may also be due to a reduction in transmitter release.

D. Slow Synaptic Events

The most fully documented examples of synapses at which transmission occurs in the absence of a decrease in membrane resistance are found in the sympathetic ganglia of mammals and amphibia. The slow EPSP generated at muscarinic cholinergic receptors on the ganglion cell membrane occurs either with no change in resistance (rabbit, frog; Kobayashi and Libet, 1968; Nishi *et al.*, 1969), or in conjunction with an increased resistance (frog; Weight and Votava, 1970). The slow IPSP in mammalian sympathetic ganglion cells evoked by synaptically released catecholamines is also generated without a change in membrane resistance (Kobayashi and Libet, 1968, 1970). The slow IPSP in amphibian ganglia may be associated with an increase in membrane resistance (Weight and Padjen, 1973a), but there is some uncertainty as to whether the inhibitory transmitter is ACh (Weight and Padjen, 1973b) or a catecholamine (Libet and Kobayashi, 1974).

Nishi and Koketsu (1968) have proposed that the slow IPSP in frog sympathetic ganglion cells is generated by an electrogenic sodium pump, as ouabain, metabolic inhibitors and low extracellular potassium depress the slow IPSP. Weight and Padjen (1973a) dispute this hypothesis and propose as an alternative mechanism that the hyperpolarization is generated by a decrease in resting sodium conductance of the ganglion cell membrane. The slow EPSP in amphibian ganglia may be due to a decrease in membrane potassium conductance (Weight and Votava, 1970).

Examples of slow synaptic transmission within the CNS have been observed in the cerebellar cortex and hippocampus. Fibres from the nucleus locus coeruleus hyperpolarize both Purkinje cells and hippocampal pyramidal cells by a mechanism which is associated with an increase in membrane resistance (Hoffer *et al.*, 1973; Segal and Bloom, 1974; Oliver and Segal, 1974). The ionic mechanism underlying these hyperpolarizations, or those evoked by norepinephrine of hippocampal neurons, has still to be determined.

III. Studies with Putative Neurotransmitters on Cerebral Cortical Neurons

A. Monoamines

The inhibitory actions of iontophoretically applied catecholamines, serotonin and histamine on central neurons were initially described by Curtis and Davis (1962), who recorded from lateral geniculate neurons, and by Krnjević and Phillis (1963), studying cerebral cortical neurons.

All three amines hyperpolarize spinal motoneurons, reducing the amplitude of evoked EPSP's and IPSP's and causing failure of the invasion of orthodromically and antidromically propagating action potentials (Figs.

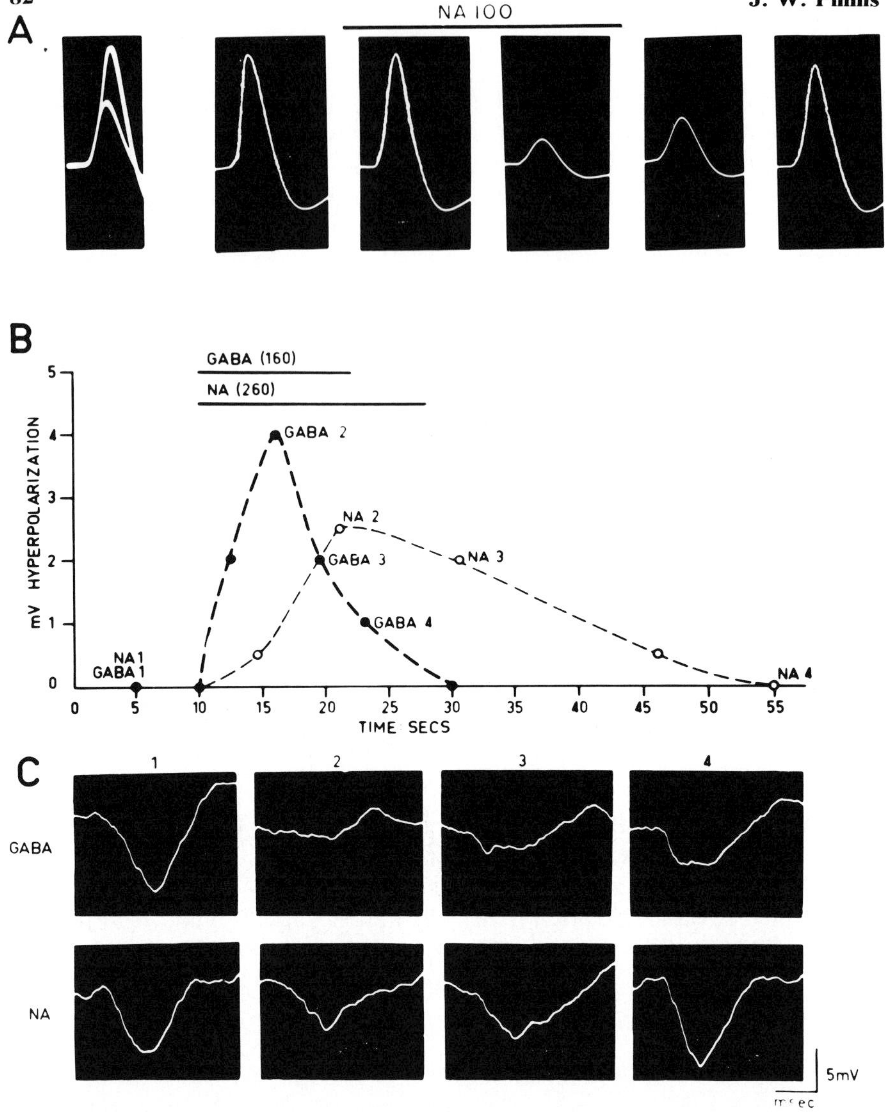

FIG. 4. Intracellular records showing inhibitory effects of noradrenaline (NA) and γ-amino-butyric acid (GABA). A. Antidromic spike of a biceps-semitendinosus (BST) motoneurone evoked by stimulation of ventral root L7. The first record shows partial failure of the soma-dendritic (SD) spike during repetitive stimulation at 20/sec. An application of NA (100 nA) blocked the SD spike and reduced the size of the initial segment (IS) spike within a few sec. Recovery occurred within 60 sec after the application was terminated. B. Graph showing changes in membrane potential of another BST motoneurone (resting potential 50mV) during the application of γ-amino-butyric acid (GABA) and NA. The latter was applied 1 min after the end of GABA application. C. NA and GABA depression of IPSPs evoked by stimulation of the sural nerve in the same motoneurone represented in B. The relative times of recording of each IPSP shown in C are indicated by the corresponding numbers in graph B. (Phillis *et al.*, 1968).

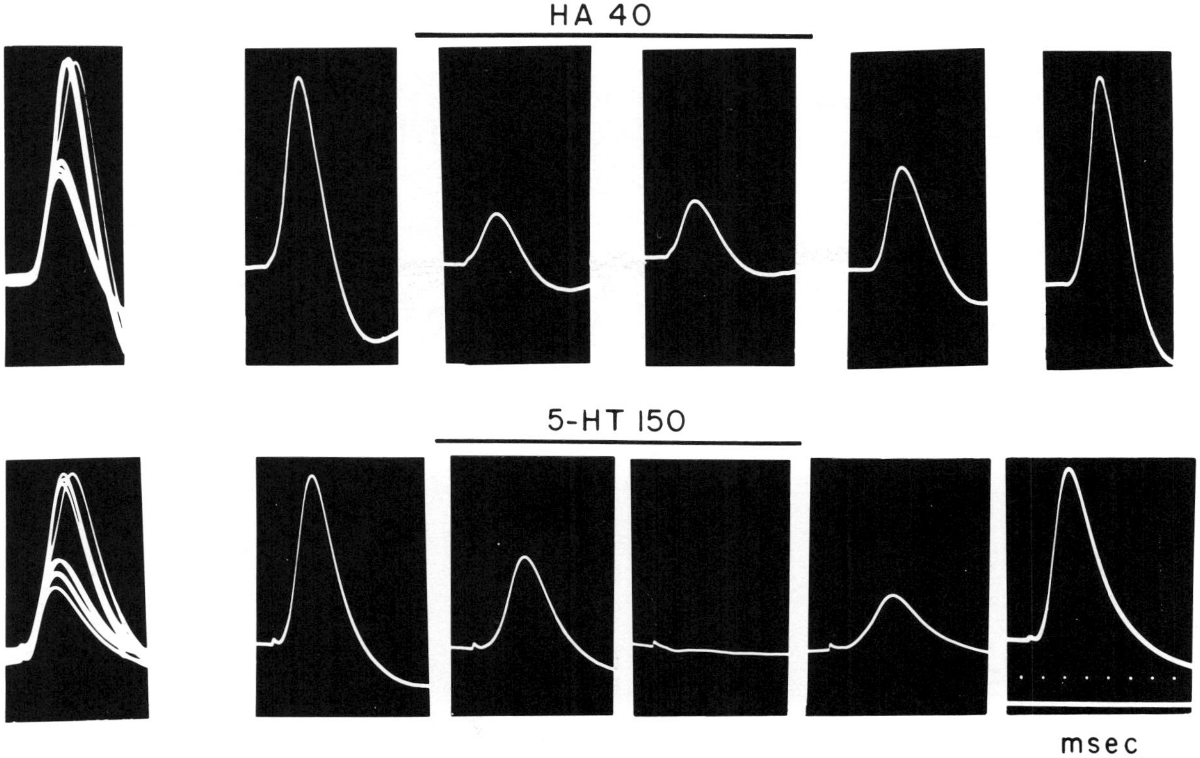

FIG. 5. Intracellular records from two motoneurons showing failure of antidromic spike invasion during applications of histamine (HA, 40 nA) and serotonin (5-HT, 150 nA). Both neurons were hyperpolarized by 3-5 mV during the amine applications. The initial record in each instance shows partial failure of the soma-dendritic spike during repetitive stimulation at 20/sec. (Phillis *et al.*, 1968).

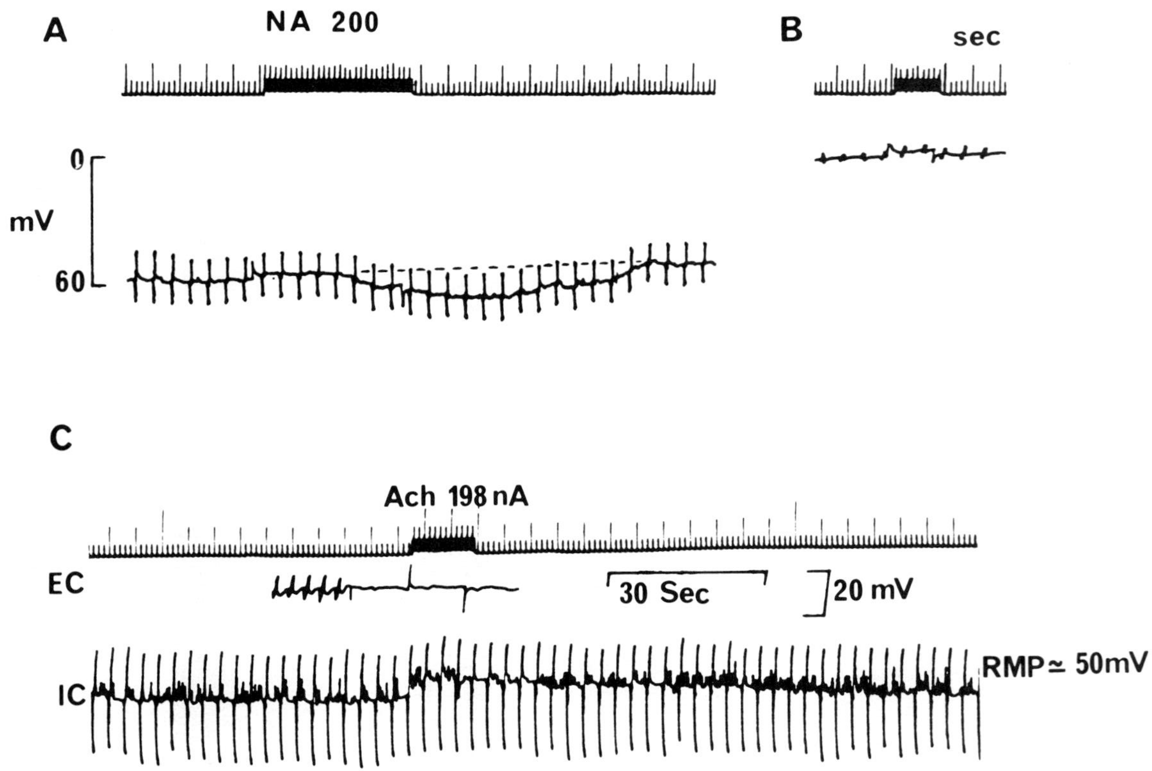
A
NA 200
0
mV
60
B
sec
C
Ach 198 nA
EC
30 Sec
20 mV
IC
RMP≃ 50mV

4 and 5; Phillis *et al.*, 1968). A concurrent increase in motoneuron membrane resistance has also been recorded (Engberg *et al.*, 1974).

Intracellular records have demonstrated a hyperpolarizing action of norepinephrine on cat cerebral cortical neurons (Fig. 6A), but concurrent alterations in membrane resistance have not been a consistent finding. Thus, norepinephrine (200 nA) hyperpolarized the neuron illustrated in Fig. 6A by several millivolts without any associated change in the membrane resistance (demonstrated by the lack of alteration in the amplitude of the voltage changes evoked by brief hyperpolarizing or depolarizing current pulses). The decreases in EPSP amplitudes which can be observed during the application of monoamines onto cerebral cortical and spinal neurons (Edstrom and Phillis, unpublished observations; Phillis *et al.*, 1968) suggest that these substances may be able to depress the release of excitatory transmitters from presynaptic terminals. Such an action could account for at least part of the post-synaptic hyperpolarization and would also explain the increase in membrane resistance observed by some investigators during the application of monoamines.

At least two separate observations indicate that the monoamines have a direct postsynaptic depressant action. Iontophoretic applications of magnesium in amounts which should have been sufficient to block on-going presynaptic release of excitatory transmitter (Nicoll, 1971) failed to modify norepinephrine- and serotonin-evoked depressions of cerebral cortical neurons (Yarbrough *et al.*, 1974). This finding also makes it unlikely that the amines depress by evoking the release of an endogenous inhibitory transmitter. A postsynaptic action of the monoamines is further implied by the observation that these substances are potent depressants of amino acid- (L-glutamate, L-aspartate, etc.) evoked firing of central neurons. These excitant amino acids have a direct, postsynaptic action on central neurons, as they can depolarize motoneurons in amphibian spinal cord preparations in which synaptic transmission has been blocked by 20 mM magnesium (Barker and Nicoll, 1973).

FIG. 6. Intracellular potentials of two cerebral cortical neurons recorded on a chart recorder. A. Norepinephrine (NA, 200 nA) hyperpolarized the neuron. 0.5 nA pulses of inward and outward current lasting 250 msec were used to test the neuron's input impedance. This did not change during the norepinephrine-induced hyperpolarization. B. Recording made after the microelectrode had been withdrawn from the neuron showing the almost negligible effects of electrode polarization and bridge imbalance. Time calibrations: 1 and 5 sec. C. (IC) Intracellular potential record of a cerebral cortical neuron on a chart recorder. Acetylcholine (ACh, 198 nA) depolarized the neuron in the absence of any marked alteration in membrane resistance. 0.5 nA pulses of 250 msec duration of inward and outward current were used to test input impedence. (EC) Recording made after the electrode had been withdrawn from the neuron showing the small amount of bridge imbalance and the lack of effect of the current application on the recorded potential. (J. P. Edstrom and J. W. Phillis, unpublished observations).

Evidence from skeletal muscle and ganglion experiments implies that the postsynaptic actions of the amines may be a result of activation of an electrogenic sodium pump. Catecholamines can hyperpolarize avian and amphibian skeletal muscles by a process which is ouabain and potassium ion-sensitive (Somlyo and Somlyo, 1969; Bressler *et al.*, 1975; Koketsu and Ohta, 1976). Frog skeletal muscle contains a sodium pump which is stimulated by epinephrine, greatly increasing sodium efflux. This action of epinephrine is blocked by strophanthidin (Hays *et al.*, 1974). Koketsu and his colleagues have presented evidence that the hyperpolarizing actions of norepinephrine and serotonin on amphibian ganglion cells are generated by activation of an electrogenic pump (Koketsu *et al.*, 1973; Watanabe and Koketsu, 1973; Koketsu and Shirasawa, 1974; Koketsu and Nakamura, 1976).

Na^+, K^+-ATPase activity, the biochemical substrate for the sodium pump, in mammalian brain is stimulated by various amines including norepinephrine, dopamine, serotonin, histamine and ACh (Table I) (Schaefer *et al.*, 1972; Yoshimura, 1973; Godfraind *et al.*, 1974; Gilbert *et al.*, 1975; Sulakhe *et al.*, 1977; Desaiah and Ho, 1976; Kometiani *et al.*, 1976). Stimulation by norepinephrine is antagonized by the α-adrenoreceptor blocker, phentolamine (Iwangoff *et al.*, 1974; Gilbert *et al.*, 1975). Na^+, K^+-ATPase in the CNS is known to be associated with synaptic membranes (Hosie, 1965; Kurokawa *et al.*, 1965). In rat cerebral cortex, the enzyme has been localized histochemically in neuronal processes, including axons and dendrites (Stahl and Broderson, 1976).

Iontophoretic studies on rat cerebral cortical neurons support the hypothesis that amines hyperpolarize by activating an electrogenic sodium pump. Thus ouabain and lithium ions antagonize the depressant actions of norepinephrine on cerebral cortical neurons (Phillis, 1974b, 1976; Phillis and Limacher, 1974), and it has recently been reported that ouabain antagonizes the depressant actions of norepinephrine and dopamine on cerebellar Purkinje cells and caudate neurons, respectively (Yarbrough, 1976). Vizi has suggested that catecholamines may depress the release of acetylcholine at both peripheral and central synapses by stimulating Na^+, K^+-ATPase (Vizi, 1972, 1973, 1975). According to this concept, inhibition of Na^+, K^+-ATPase by calcium, which enters the nerve terminal during an action potential, is responsible for the release of acetylcholine. Catecholamines may decrease the calcium influx and thus reduce the extent of inhibition of the enzyme.

An involvement of calcium in the sequence of events which couples the amine receptors to the pump is suggested by the finding that a series of so-called calcium antagonists including polyvalent metallic cations, local anaesthetics, neomycin, ruthenium red and verapamil reduce or abolish the depressant effects of the monoamines on cerebral cortical neurons (Yarbrough *et al.*, 1974; Phillis, 1974a). Two calcium antagonist metallic ions, lanthanum and lead, block norepinephrine-evoked inhibitions of cerebellar Purkinje cells (Nathanson *et al.*, 1976; Taylor *et al.*, 1976). These experiments with "calcium antagonists" have been subject to variable interpretations, since the antagonists themselves may affect other enzymes, including adenylate cyclase. However, the only action which is presently known to be common to all these compounds remains that of calcium antagonism.

TABLE I
EFFECTS OF BIOGENIC AMINES ON THE ATPASE ACTIVITIES OF SYNAPTOSOMES OF RAT CEREBRAL CORTEX[a]

Drug	ATPase Activity as Percentage of Control Na^+,K^+-ATPase	 Ca^{2+},Mg^{2+},-ATPase[b]
Control	100%[c]	100%
Noradrenaline (mM)		
0.01	132.5 ± 11.0% (5)†	58.3 ± 7.5% (3)
0.05	154.0 ± 8.1% (5)†	111.1 ± 21.7% (3)
0.1	170.0 ± 8.8% (5)†	97.2 ± 13.0% (3)
0.15	160.0 ± 10.2% (5)†	127.7 ± 6.8% (3)
Dopamine (mM)		
0.01	128.3 ± 13.6% (2)	97.2 ± 18.8% (2)
0.05	150.1 ± 11.8% (3)†	100.0 ± 19.0% (3)
0.1	159.8 ± 3.6% (3)†	102.8 ± 32.7% (3)
0.15	163.5 ± 9.3% (3)†	163.8 ± 16.1% (3)
5-HT (mM)		
0.05	111.5 ± 7.3% (5)	122.2 ± 13.0% (5)
0.1	125.5 ± 9.2% (4)†	91.7 ± 24.0% (4)
0.2	134.0 ± 8.6% (5)†	80.6 ± 16.5% (5)
0.5	132.2 ± 9.0% (5)†	113.8 ± 24.2% (4)
Histamine (mM)		
0.05	110.8 ± 2.6% (2)	---
0.2	110.3 ± 16.0% (3)	64.3 ± 7.2% (3)
0.5	126.6 ± 9.7% (5)*	102.9 ± 9.2% (4)
2.0	129.5 ± 6.9% (5)*	105.8 ± 57.2% (3)

[a] Each value is the mean ± S.E.M. of the number of measurements given in parenthesis.
[b] The incubation medium contained 0.02 mM $CaCl_2$ and 1.0 mM ouabain.
[c] **The specific activities (μmoles Pi per mg protein per hour) for the control Na^+,K-ATPase and Ca^{2+}, Mg^{2+}-ATPase are 21.6 and 3.6 respectively.**
† Significantly different ($p < 0.02$) from control.
* Significantly different ($0.02 < p < 0.05$) from control.

From: S. L. Lee and J. W. Phillis, unpublished observations.

A characteristic finding of the experiments with ouabain and the calcium antagonists has been a simultaneous block of the actions of histamine, serotonin and norepinephrine, indicating a common final mode of action of the three amines. The recent discovery of specific antagonists which can discriminate between the depressant actions of histamine (metiamide; Phillis *et al.*, 1975) or serotonin (metergoline; Sastry and Phillis, 1977) and those of the other amines on cerebral cortical neurons implies the existence of specific receptors for each type of amine. Each receptor is presumably able to activate a similar or common mechanism which evokes the membrane hyperpolarization. The currently available adrenergic antagonists appear to be less specific in their ability to discriminate between the different types of amine (Frederickson *et al.*, 1972; Dray *et al.*, 1976).

Evidence has now been obtained in this laboratory for a norepinephrine-releasing inhibitory pathway from the locus coeruleus to the rat cerebral cortex, where the inhibited cortical neurons included identified cortico-spinal neurons (Phillis and Kostopoulos, 1977). A serotonin-releasing inhibitory pathway projects from the raphé nuclei to the rat cerebral cortex (Sastry and Phillis, 1977), and a histamine-releasing inhibitory pathway to the cerebral cortex has been identified at the level of the ventro-lateral hypothalamus (Sastry and Phillis, 1976). The nucleus of origin of the histaminergic pathway has not been identified.

These pathways have much in common with the norepinephrine, serotonin and histamine-releasing pathways which project to the hippocampus and/or cerebellar cortex (Hoffer *et al.*, 1973; Segal and Bloom, 1974; Segal, 1975; Haas and Wolf, 1977).

B. Acetylcholine

ACh depresses many of the more superficially located cerebral cortical neurons (Phillis and York, 1968; Jordan and Phillis, 1972) and excites deeper cells, including corticospinal neurons (Krnjević and Phillis, 1963). The depolarizing excitant action of ACh on these neurons has been attributed to a fall in membrane potassium permeability, and an increase in neuronal membrane resistance has been reported as occurring in conjunction with the depolarization (Krnjević *et al.*, 1971). However, experiments in other laboratories have failed to demonstrate a consistent increase in membrane resistance during ACh excitation. Thus Zieglgänsberger and Reiter (1974) observed increases in membrane resistances in only a small proportion of the spinal motoneurons which they depolarized with ACh, nor were increases in membrane resistance a noticeable feature of ACh-induced depolarizations of caudate neurons (Bernardi *et al.*, 1976).

Our own findings (Edstrom and Phillis, unpublished observations) indicate that increases in membrane resistance do not occur consistently in the cerebral cortical neurons that are depolarized by ACh (Fig. 6B). ACh has a depressant effect on the release of transmitters by nerve terminals, and the disfacilitation accompanying such an action could conceivably account for those increases in membrane resistance that have been observed.

C. Purines and Pyrimidines

An extensive range of purine and pyrimidine nucleotides has been tested on identified corticospinal and deep spontaneously active cerebral cortical neurons in the rat sensorimotor cortex (Table II). Adenine nucleotides and adenosine were observed to have a pronounced depressant action, guanine nucleotides a weak depressant action, and little or no depressant effects were observed with cytidine, inosine, thymidine, uridine and xanthine nucleotides (Phillis *et al.*, 1974). Further studies have confirmed the potent depressant actions of 2′-, 3′-, and 5′-adenosine monophosphates and 2′, 3′-cyclic monophosphate, and the relatively weak depressant action of 3′, 5′-cyclic AMP (Phillis *et al.*, 1975). A list of the adenine derivatives which have been tested and their relative depressant potencies is presented in Table III.

Excitant actions have been observed with the various purine and pyrimidine di- and especially tri-nucleotides, as well as with 3′, 5′-cyclic AMP. These effects have been more pronounced when the compounds were tested on non-spontaneously active neurons and are comparable to the excitant effects of 5′-ATP or 3′, 5′-cyclic AMP that have been reported in the cerebellar cortex (Siggins *et al.*, 1971) and dorsal column nuclei (Galindo *et al.*, 1967). Excitant actions of the nucleotides have been tentatively attributed to their calcium chelating abilities (Galindo *et al.*, 1967; Siggins *et al.*, 1971), but this explanation remains to be either proven or disproven.

Methylxanthines such as caffeine and theophylline antagonize the depressant actions of adenosine or 5′-AMP on cerebral cortical neurons (Phillis and Kostopoulos, 1975). On account of its sparing solubility in aqueous solutions, theophylline was administered intravenously in doses of 10-100 mg/kg. Caffeine has been effective as an antagonist when administered either iontophoretically or intravenously. The antagonism exerted by either compound can be overcome by increasing the adenosine or 5′-AMP application current through the appropriate microelectrode barrel, implying that the methylxanthines act as competitive antagonists. Theophylline also antagonized the depressant actions of 3′, 5′-cyclic AMP, indicating that this substance exerts its effects on an identical receptor to adenosine or 5′-AMP, possibly after its conversion to adenosine.

TABLE II
DEPRESSANT POTENCY OF VARIOUS PURINES AND PYRIMIDINES ON CORTICOSPINAL CELLS

Pronounced	Weak	Very Weak or Nil
Adenosine	Adenosine 3′,5′-cyclic monophosphate	Adenine
Adenosine 2′-monophosphate	Guanosine 5′-monophosphate	Cytidine
Adenosine 3′-monophosphate	Guanosine 5′-diphosphate	Cytidine 3′(2′)-monophosphate
Adenosine 2′,3′-cyclic monophosphate	Guanosine 5′-triphosphate	Cytidine 5′-monophosphate
Adenosine 5′-monophosphate	Guanosine 3′, 5′-cyclic monophosphate	Cytidine 5′-triphosphate
Adenosine 5′-diphosphate		Guanosine
Adenosine 5′-triphosphate		Hypoxanthine
Adenosine 5′-tetraphosphate		Inosine
		Inosine 5′-monophosphate
		Inosine 5′-triphosphate
		Purine riboside
		Thymidine
		Thymidine 5′-monophosphate
		Thymidine 5′-triphosphate
		Uridine
		Uridine 3′ (2′)-monophosphate
		Uridine 5′-monophosphate
		Uridine 5′-triphosphate
		Xanthine
		Xanthosine
		Xanthosine 5′-monophosphate
		Xanthosine 5′-triphosphate

The significance of adenosine deaminase in terminating the action of adenosine has been investigated with two potent inhibitors of this enzyme, deoxycoformycin and erythro-9-(2-hydroxy-3-nonyl) adenine (Phillis and Edstrom, 1976). Both substances caused a pronounced potentiation of the duration of the depressant effects of adenosine and 5′-AMP. As uptake into adjacent cellular elements represents an alternative route of removal of extracellular adenosine, various uptake blockers were tested, including papaverine, hexobendine and 2-hydroxy-5-nitrobenzylthioguanosine (Phillis and Kostopoulos, 1975; Phillis and Edstrom, 1976). All three compounds potentiated the actions of adenosine and 5′-AMP.

Intracellular recordings have shown that 5′-AMP depresses by hyperpolarizing cerebral cortical neurons in the absence of any significant alterations in membrane resistance, but with a pronounced reduction in the number and amplitude of excitatory postsynaptic potentials. The mechanism by which 5′-AMP produces these effects is still uncertain, but the findings would be compatible with some form of inhibition of transmitter release from presynaptic terminals.

TABLE III
STRUCTURE-ACTIVITY RELATIONSHIPS FOR ADENOSINE ANALOGS

Substance	Depressant Activity on Cerebral Cortical Neurons
Adenine	-
Adenosine	- - - -
Inosine	- -
Purine Derivatives of Adenosine	
2-Chloroadenosine	- - - - -
2-Fluoroadenosine	- - - - -
2-Bromoadenosine	- -
2-Aminoadenosine	- - -
2-Hydroxyadenosine	- - -
8-Bromoadenosine	0
Ribose Derivatives of Adenosine	
2′-Deoxyadenosine	- - -
3′-Deoxyadenosine	- -
5′-Deoxyadenosine	- - -
Adenine xylofuranoside	-
Adenosine N′-oxide	-
Adenosine 5′-sulphate	0
Adenosine 5′-mononicotinate	- - -
7-Deazaadenosine (tubercidin)	- -
Mononucleotides	
Adenosine 5′-monophosphate	- - - -
Adenosine 5′-diphosphate	- - - -
Adenosine 5′-triphosphate	- - - -
Adenosine 5′-tetraphosphate	- - - -
Adenosine 2′-monophosphate	- - - -
Adenosine 3′-monophosphate	- - - -
Adenosine 3′, 5′-cyclic monophosphate	- -
Adenosine 2′, 3′-cyclic monophosphate	- - - -
Adenosine 5′-imidodiphosphate	- - - -
Homoadenosine 6′-phosphonic acid	0
α,β-Methylene 5′-ATP	-
β,γ-Methylene 5′-ATP	- - -
α,β-Methylene 5′-ADP	-
8-Parachlorophenylthio 3′, 5′-cAMP	-
8-Benzylthio 3′, 5′-cAMP	0
2′-Deoxyadenosine 5′-monophosphate	- -
Dinucleotides	
Nicotinamide adenine dinucleotide phosphate	- - -

Adenosine and 5′-AMP have been tested on neurons in several other regions of the brain, including the cerebellar cortex, olfactory bulb, thalamus, hippocampus, caudate nucleus and superior colliculus (Kostopoulos *et al.*, 1975; Kostopoulos and Phillis, 1977). Depression of neuronal firing was observed in each region tested.

Adenosine and the adenine nucleotides are released by the cerebral cortex (Berne *et al.*, 1974; Sulakhe and Phillis, 1975). Release is enhanced when the brain becomes ischemic (Berne *et al.*, 1974), and adenosine may account for the reported effects of anoxia on cerebral cortical neurons, which include an initial hyperpolarization with a decrease and then disappearance of postsynaptic potentials (Grossman and Williams, 1971).

Conclusions

The actions of the established neurotransmitters, norepinephrine, serotonin, histamine and acetylcholine, on central neurons, and in particular on cerebral cortical neurons, are discussed. Evidence is presented for the existence of norepinephrine, serotonin and histamine-mediated ascending inhibitory pathways to the cerebral cortex. These amines may depress neuronal firing by both pre- and postsynaptic actions, and both actions could be the result of stimulation of an electrogenic sodium pump.

The recently discovered depressant action of adenosine and the non-cyclic nucleotides on central neurons is described. The actions of these compounds are antagonized by the well known CNS stimulants, caffeine and theophylline. The possibility that these methylxanthines exert their stimulant effects by antagonizing the actions of endogenously released adenosine and the adenine nucleotides has therefore to be considered. In this context, it is interesting to note that the tranquilizers, valium and librium, may also act through adenosine, since these compounds can block the adenosine uptake system in brain slices (Mah and Daly, 1976).

Acknowledgments

This research was supported by the Medical Research Council and the University of Saskatchewan.

References

Barker, J. L. and R. A. Nicoll. 1973. The pharmacology and ionic dependency of amino acid responses in the frog spinal cord. J. Physiol. (Lond.) 228:259-277.

Bennett, M. V. L. 1972. A comparison of electrically and chemically mediated transmission. In: *Structure and Function of Synapses*. G. D. Pappas and D. P. Purpura, eds. Raven Press, New York. pp. 221-256.

Bennett, M. V. L. 1973. Function of electrotonic junctions in embryonic and adult tissues. Fed. Proc. 32:65-75.

Bernardi, G., V. Floris, M. G. Marciani, C. Morocutti and P. Stanzione. 1976. The action of acetylcholine and L-glutamic acid on rat caudate neurons. Brain Res. 114:134-138.

Berne, R. M., R. Rubio and R. Curnish. 1974. Release of adenosine from ischaemic brain. Circulat. Res. 35:262-272.

Bodian, D. 1972. Neuron junctions: A revolutionary decade. Anat. Record 174:73-82.

Bressler, B. H., J. W. Phillis and W. Kozachuk. 1975. Noradrenaline stimulation of a membrane pump in frog skeletal muscle. Europ. J. Pharmac. 33:201-204.

Carlen, P. L., Y. Yaari and R. Werman. 1976. Postsynaptic conductance changes during presynaptic inhibition. Neuroscience Abstracts 2:1000.

Ciani, S. and C. Edwards. 1963. The effect of acetylcholine on neuromuscular transmission in the frog. J. Pharmac. exp. Ther. 142:21-23.

Coombs, J. S., J. C. Eccles and P. Fatt. 1955a. Excitatory synaptic action in motoneurones. J. Physiol. (Lond.) 130:374-395.

Coombs, J. S., J. C. Eccles and P. Fatt. 1955b. The specific ionic conductances and the ionic movements across the motoneuronal membrane that produce the inhibitory post-synaptic potential. J. Physiol. (Lond.) 130:326-373.

Curtis, D. R. and R. Davis. 1962. Pharmacological studies upon neurones of the lateral geniculate nucleus of the cat. Br. J. Pharmac. Chemother. 18:217-246.

Desaiah, D. and I. K. Ho. 1976. Effect of biogenic amines and GABA on ATPase activities in mouse tissues. Europ. J. Pharmac. 40:255-261.

Diamond, J. 1968. The activation and distribution of GABA and L-glutamate receptors on goldfish Mauthner neurones: an analysis of dendritic remote inhibition. J. Physiol. (Lond.) 194:669-723.

Dray, A., T. J. Gonye and N. R. Oakley. 1976. Effects of α-flupenthixol on dopamine and 5-hydroxytryptamine responses of substantia nigra neurones. Neuropharmac. 15:793-796.

Dun, N. and S. Nishi. 1974. Effects of dopamine on the superior cervical ganglion of the rabbit. J. Physiol. 239:155-164.

Eccles, J. C. 1963. Postsynaptic and presynaptic inhibitory actions in the spinal cord. Progr. Brain Res. 1:1-18.

Edwards, F. R., S. J. Redman and B. Walmsley. 1976. The effect of polarizing currents on unitary Ia excitatory post-synaptic potentials evoked in spinal motoneurones. J. Physiol. (Lond.) 259:705-723.

Engberg, I., J. A. Flatman and K. Kadzielawa. 1974. The hyperpolarization of motoneurones by electrophoretically applied amines and other agents. Acta. Physiol. Scand. 91:2-4A.

Frederickson, R. C. A., L. M. Jordan and J. W. Phillis. 1972. A reappraisal of the actions of noradrenaline and 5-hydroxytryptamine on cerebral cortical neurons. Comp. gen. Pharmac. 3:443-456.

Galindo, A., K. Krnjević and S. Schwartz. 1967. Micro-iontophoretic studies on neurones in the cuneate nucleus. J. Physiol. (Lond.) 192:359-377.

Gilbert, J. C., M. G. Wyllie and D. V. Davison. 1975. Nerve terminal ATPase as possible trigger for neurotransmitter release. Nature (Lond.) 255:237-238.

Ginsborg, B. L. 1971. On the presynaptic acetylcholine receptors in sympathetic ganglia of the frog. J. Physiol. (Lond.) 216:237-246.

Ginsborg, B. L. and G. D. S. Hirst. 1972. The effect of adenosine on the release of the transmitter from the phrenic nerve of the rat. J. Physiol. (Lond.) 224:629-645.

Godfraind, T., M.-C. Koch and N. Verbeke. 1974. The action of EGTA on the catecholamines stimulation of rat brain Na-K-ATPase. Biochem. Pharmacol. 23:3505-3511.

Gray, E. G. 1969. Electron microscopy of excitatory and inhibitory synapses: a brief review. Progr. Brain Res. 31:141-155.

Grossman, R. G. and V. F. Williams. 1971. Electrical activity and ultra-structure of cortical neurons and synapses in ischemia. In: *Brain Hypoxia*. J. B. Brierley and B. S. Meldrum, eds. W. Heinemann, London. pp. 61-75.

Haas, H. L. and P. Wolf. 1977. Central actions of histamine: microelectrophoretic studies. Brain Res. 122:269-279.

Hays, E. T., T. M. Dwyer, P. Horowicz and J. G. Swift. 1974. Epinephrine action on sodium fluxes in frog striatal muscle. Amer. J. Physiol. 227:1340-1347.

Hedqvist, P. and B. B. Fredholm. 1976. Effects of adenosine on adrenergic neurotransmission; prejunctional inhibition and postjunctional enhancement. Arch. Pharmacol. 293:217-223.

Hoffer, B. J., G. R. Siggins, A. P. Oliver and F. E. Bloom. 1973. Activation of the pathway from locus coeruleus to rat cerebellar Purkinje neurons: pharmacological evidence of noradrenergic central inhibition. J. Pharmacol. exp. Therap. 184:553-569.

Hosie, R. J. A. 1965. The localization of ATPase in morphologically characterized subcellular fractions of guinea pig brain. Biochem. J. 96:404-412.

Hubbard, J. I., R. F. Schmidt and T. Yokota. 1965. The effect of acetylcholine upon mammalian motor nerve terminals. J. Physiol. (Lond.) 181:810-829.

Ito, M., P. G. Kostyuk and T. Oshima. 1962. Further study on anion permeability in cat spinal motoneurones. J. Physiol. (Lond.) 164: 150-156.

Iwangoff, P., A. Enz and A. Chappuis. 1974. Effect of adrenergic blockers on the activation of brain ATPase by noradrenaline. Experientia 30:688.

Jenkinson, D. H., B. A. Stamenović and B. D. L. Whitaker. 1968. The effect of noradrenaline on the end-plate potential in twitch fibres of the frog. J. Physiol. (Lond.) 195:743-754.

Jordan, L. M. and J. W. Phillis. 1972. Acetylcholine inhibition in the intact and chronically isolated cerebral cortex. Br. J. Pharmac. 45:584-595.

Kazić, T. and D. Milosavljević. 1976. Influence of adenosine, cAMP and db-cAMP on responses of the isolated terminal guinea-pig ileum to electrical stimulation. Arch. int. Pharmacodyn. 223:187-195.

Kellerth, J. -O. and A. J. Szumski. 1966. Effects of picrotoxin on stretch-activated post-synaptic inhibitions in spinal motoneurones. Acta. Physiol. Scand. 66:146-156.

Kelly, J. S., K. Krnjević, M. E. Morris and G. K. W. Yim. 1969. Anionic permeability of cortical neurones. Exp. Brain Res. 7:11-31.

Kobayashi, H. and B. Libet. 1968. Generation of slow post-synaptic potentials without increases in ionic conductance. Proc. Nat. Acad. Sci. U.S.A. 60:1304-1311.

Kobayashi, H. and B. Libet. 1970. Actions of noradrenaline and acetylcholine on sympathetic ganglion cells. J. Physiol. (Lond.) 208:353-372.

Koketsu, K. and M. Nakamura. 1976. The electrogenesis of adrenaline-hyperpolarization of sympathetic ganglion cells in bullfrogs. Jap. J. Physiol. 26:63-77.

Koketsu, K. and Y. Ohta. 1976. Acceleration of the electrogenic Na^+ pump by adrenaline in frog skeletal muscle fibres. Life Sci. 19: 1009-1014.

Koketsu, K. and Y. Shirasawa. 1974. 5-HT and the electrogenic sodium pump. Experientia 30:1034-1035.

Koketsu, K., T. Shoji and S. Nishi. 1973. Slow inhibitory postsynaptic potentials of bullfrog sympathetic ganglia in sodium-free media. Life Sci. 13:453-458.

Kometiani, Z. P., T. Ya. Dzhariashvili and L. G. Tsakadze. 1976. Effect of acetylcholine on Na,K-ATPase of synaptosomes. Biochemistry U.S.S.R. 40:885-891.

Kostopoulos, G. K., J. J. Limacher and J. W. Phillis. 1975. Action of various adenine derivatives on cerebellar Purkinje cells. Brain Res. 88:162-165.

Kostopoulos, G. K. and J. W. Phillis. 1977. Purinergic depression of neurons in different areas of the rat brain. Exp. Neurol. 55:719-729.

Krnjević, K. and R. Miledi. 1958. Some effects produced by adrenaline upon neuromuscular propagation in rats. J. Physiol. (Lond.) 141:291-304.

Krnjević, K. and J. W. Phillis. 1963. Actions of certain amines on cerebral cortical neurones. Br. J. Pharmac. Chemother. 20:471-490.

Krnjević, K., R. Pumain and L. Renaud. 1971. The mechanism of excitation by acetylcholine in the cerebral cortex. J. Physiol. (Lond.) 215:247-268.

Kuno, M. and R. Llinas. 1970. Alterations of synaptic actions in chromatolysed motoneurones of the cat. J. Physiol. 210:823-838.

Kuroda, Y., M. Saito and K. Kobayashi. 1976. High concentrations of calcium prevent the inhibition of postsynaptic potentials and the accumulation of cyclic AMP induced by adenosine in brain slices. Proc. Jap. Acad. 52:86-89.

Kurokawa, M., T. Sakamoto and M. Kato. 1965. Distribution of Na-plus-K-stimulated ATPase activity in isolated nerve ending particles. Biochem. J. 97:833-844.

Libet, B. and H. Kobayashi. 1974. Adrenergic mediation of slow inhibitory postsynaptic potential in sympathetic ganglia of the frog. J. Neurophysiol. 37:805-814.

Llinas, R. 1975. Electrical synaptic transmission in the mammalian central nervous system. In: *Golgi Centennial Symp. Proc.* M. Santini, ed., Raven Press, New York. pp. 379-386.

Llinas, R. and R. Baker, 1972. A chloride-dependent inhibitory postsynaptic potential in cat trochlear motoneurons. J. Neurophysiol. *35*, 484-492.

Llinas, R., R. Baker and W. Precht. 1974. Blockage of inhibition by ammonium acetate action on chloride pump in cat trochlear motoneurons. J. Neurophysiol. 37: 522-532.

Lux, H. D. 1971. Ammonium and chloride extrusion: Hyperpolarizing synaptic inhibition in spinal motoneurons. Science 173:555-557.

Lux, H. D., C. Loracher and E. Neher. 1970. The action of ammonium on post-synaptic inhibition of cat spinal motoneurons. Exp. Brain Res. 11:431-447.

Mah, H. D. and J. W. Daly. 1976. Adenosine-dependent formation of cyclic AMP in brain slices. Pharmac. Res. Commun. 8:65-79.

Mori, M., S. Yamada, S. Takamura and E. Hayashi. 1973. Effect of purine nucleotides on acetylcholine output from cholinergic nerves in guinea-pig ileum. Jap. J. Pharmac. 23: Suppl. 124, Abstr. No. 217.

Nathanson, J. A., R. Freedman and B. J. Hoffer. 1976. Lanthanum inhibits brain adenylate cyclase and blocks noradrenergic depression of Purkinje cell discharge independent of calcium. Nature (Lond.) 261:330-332.

Nelson, P. G. and K. Frank. 1967. Anomalous rectification in cat spinal motoneurones and effect of polarizing currents on excitatory postsynaptic potential. J. Neurophysiol. 30:1097-1113.

Nicoll, R. A. 1971. Pharmacological evidence for GABA as the transmitter in granule cell inhibition in the olfactory bulb. Brain Res. 35:137-149.

Nishi, S. 1970. Cholinergic and adrenergic receptors at sympathetic pre-ganglionic nerve terminals. Fedn. Proc. 29:1957-1965.

Nishi, S. and K. Koketsu. 1968. Analysis of slow inhibitory post-synaptic potential of bullfrog sympathetic ganglion. J. Neurophysiol. 31:717-728.

Nishi, S., H. Soeda and K. Koketsu. 1969. Unusual nature of ganglionic slow EPSP studied by a voltage clamp method. Life Sci. 8:33-42.

Oliver, A. P. and M. Segal. 1974. Transmembrane changes in hippocampal neurons: hyperpolarizing actions of norepinephrine, cyclic AMP and locus coeruleus. Proc. Soc. Neurosci. 361.

Palay, S. L. and V. Chan-Palay. 1976. A guide to the synaptic analysis of the neuropil. Cold Spring Harbor Symposia on Quantitative Biology 40:1-16.

Phillis, J. W. 1974a. The role of calcium in the central effects of biogenic amines. Life Sci. 14:1189-1201.

Phillis, J. W. 1974b. Neomycin and ruthenium red antagonism of monoaminergic depression of cerebral cortical neurones. Life Sci. 15:213-222.

Phillis, J. W. 1976. An involvement of calcium and Na, K-ATPase in the inhibitory actions of various compounds on central neurons. In: *Taurine.* R. Huxtable and A. Barbeau, eds. Raven Press, New York. pp. 209-223.

Phillis, J. W. and J. P. Edstrom. 1976. Effects of adenosine analogs on rat cerebral cortical neurons. Life Sci. 19:1041-1094.

Phillis, J. W. and G. K. Kostopoulos. 1975. Adenosine as a putative transmitter in the cerebral cortex. Studies with potentiators and antagonists. Life Sci. 17: 1085-1094.

Phillis, J. W. and G. K. Kostopoulos. 1977. Activation of a noradrenergic pathway from the brain stem to rat cerebral cortex. Gen. Pharmac. (in press).

Phillis, J. W. and J. J. Limacher. 1974. Effects of some metallic cations on cerebral cortical neurones and their interactions with biogenic amines. Can. J. Physiol. Pharmacol. 52:566-574.

Phillis, J. W. and D. H. York. 1968. Pharmacological studies on a cholinergic inhibition in the cerebral cortex. Brain Res. 10:297-306.

Phillis, J. W., A. K. Tebecis and D. H. York. 1968. Depression of spinal motoneurones by noradrenaline, 5-hydroxytryptamine and histamine. Europ. J. Pharmac. 7:471-475.

Phillis, J. W., G. K. Kostopoulos and J. J. Limacher. 1974. Depression of corticospinal cells by various purines and pyrimidines. Can. J. Physiol. Pharmacol. 52:1226-1229.

Phillis, J. W., G. K. Kostopoulos and J. J. Limacher. 1975. A potent depressant action of adenine derivatives on cerebral cortical neurones. Eur. J. Pharmac. 30:125-129.

Rall, W., R. E. Burke, T. G. Smith, P. G. Nelson and K. Frank. 1967. Dendritic location of synapses and possible mechanisms for the monosynaptic EPSP in motoneurons. J. Neurophysiol. 30:1169-1193.

Reader, T. A., J. De Champlain and H. Jasper. 1976. Catecholamines released from cerebral cortex in the cat; decrease during sensory stimulation. Brain Res. 111:95-108.

Reese, T. S. and M. W. Brightman. 1970. Olfactory surface and central olfactory connections in some vertebrates. In: *CIBA Foundation Symposium on Taste and Smell in Vertebrates*. G. E. W. Wolstenholme and J. Knight, eds. Churchill, London. pp. 115-149.

Sastry, B. S. R. and J. W. Phillis. 1976. Evidence for an ascending inhibitory histaminergic pathway to the cerebral cortex. Can. J. Physiol. Pharmacol. 54: 782-786.

Sastry, B. S. R. and J. W. Phillis. 1977. Inhibition of cerebral cortical neurones by a 5-hydroxytryptaminergic pathway from median raphé nucleus. Can. J. Physiol. Pharmacol. (in press).

Schaefer, A., G. Unyi and A. K. Pfeifer. 1972. The effects of a soluble factor and of catecholamines on the activity of adenosine triphosphatase in subcellular fractions of rat brain. Biochem. Pharmac. 21:2289-2294.

Schmidt, R. F. 1971. Presynaptic inhibition in the vertebrate central nervous system. Ergebn. Physiol. 63:20-107.

Segal, M. 1975. Physiological and pharmacological evidence for a serotonergic projection to the hippocampus. Brain Res. 94:115-131.

Segal, M. and F. E. Bloom. 1974. The action of norepinephrine in the rat hippocampus. II. Activation of the input pathway. Brain Res. 72:99-114.

Shepherd, G. M. 1974. *The Synaptic Organization of the Brain. An Introduction.* Oxford University Press, London.

Sherrington, C. S. 1906. *The Integrative Action of the Nervous System.* Yale University Press. New Haven.

Siggins, G. R., B. J. Hoffer and F. E. Bloom. 1971. Studies on norepinephrine-containing afferents to Purkinje cells of rat cerebellum. III. Evidence for mediation of norepinephrine effects by cyclic 3′, 5′-adenosine monophosphate. Brain Res. 25:535-553.

Smith, T. G., R. B. Wuerker and K. Frank. 1967. Membrane impedance changes during transmission in cat spinal motoneurones. J. Neurophysiol. 30:1072-1096.

Somlyo, A. P. and A. V. Somlyo. 1969. Pharmacology of excitation-contraction coupling in vascular smooth muscle and in avian slow muscle. Fedn. Proc. 28:1634-1642.

Stahl, W. L. and S. H. Broderson. 1976. Localization of Na^+, K^+-ATPase in brain. Fedn. Proc. 35:1260-1265.

Sulakhe, P. V., S. -H. Jan. and S. J. Sulakhe. 1977. Studies on the stimulation of (Na^+-K^+) ATPase of neural tissues by catecholamines. Gen. Pharmac. 8:37-41.

Sulakhe, P. V. and J. W. Phillis. 1975. The release of ^3H-adenosine and its derivatives from cat sensorimotor cortex. Life Sci. 17:551-556.

Szerb, J. C. and G. T. Somogyi. 1973. Depression of acetylcholine release from cerebral cortical slices by cholinesterase inhibition and by oxotremorine. Nature (Lond.) New Biol. 241:121-122.

Taylor, D., J. Nathanson, B. Hoffer, L. Olson and A. Seiger. 1976. Lead blockade of noradrenergic inhibition in cerebellar Purkinje neurons. Neuroscience Abstracts II, (1), 507.

Vizi, E. S. 1972. Stimulation by inhibition of (Na^+-K^+-Mg^{2+})-activated ATPase, of acetylcholine release in cortical slices from rat brain. J. Physiol. (Lond.) 226:95-117.

Vizi, E. S. 1973. Does stimulation of Na^+-K^+-Mg^{2+}-activated ATPase inhibit acetylcholine release from nerve terminals. Br. J. Pharmacol. 48: 346-347.

Vizi, S. E. 1975. The role of Na^+-K^+-activated ATPase in transmitter release: acetylcholine release from basal ganglia and its inhibition by dopamine and noradrenaline. In: *Subcortical Mechanisms and Sensorimotor Activities.* T. L. Frigyesi, ed. Hans Huber Publishers, Bern. pp. 63-87.

Vizi, E. S. and J. Knoll. 1976. The inhibitory effect of adenosine and related nucleotides on the release of acetylcholine. Neuroscience 1:391-398.

Watanabe, S. and K. Koketsu. 1973. 5-HT hyperpolarization of bullfrog sympathetic ganglion cell membrane. Experientia 29: 1370-1372.

Weight, F. F. and A. Padjen. 1973a. Slow synaptic inhibition in sympathetic ganglion cells: evidence for synaptic inactivation of sodium conductance in sympathetic ganglion cells. Brain Res. 55:219-224.

Weight, F. F. and A. Padjen. 1973b. Acetylcholine and slow synaptic inhibition in frog sympathetic ganglion cells. Brain Res. 55:225-228.

Weight, F. F. and J. Votava. 1970. Slow synaptic excitation in sympathetic ganglion cells: Evidence for synaptic inactivation of potassium conductance. Science 170:755-758.

Weir, M. C. L. and H. McLennan. 1963. The action of catecholamines in sympathetic ganglia. Can. J. Biochem. Physiol. 41:2627-2636.

Werman, R. and P. L. Carlen. 1976. Unusual behavior of the Ia EPSP in cat spinal motoneurons. Brain Res. 112:395-401.

Yarbrough, G. G. 1976. Ouabain antagonism of noradrenaline inhibitions of cerebellar Purkinje cells and dopamine inhibitions of caudate neurones. Neuropharmac. 15:335-338.

Yarbrough, G. G., N. Lake and J. W. Phillis, 1974. Calcium antagonism and its effect on the inhibitory actions of biogenic amines on cerebral cortical neurones. Brain Res. 67:77-88.

Yoshimura, K. 1973. Activation of Na-K activated ATPase in rat brain by catecholamine. J. Biochem. (Tokyo) 74:389-391.

Zieglgänsberger, W. and Ch. Reiter. 1974. A cholinergic mechanism in the spinal cord of cats. Neuropharmacol. 13:519-527.

CHARACTERISTICS OF NEURONS IN TISSUE CULTURE

WALTHER J. HILD

The University of Texas
Medical Branch
Galveston, Texas

It appears appropriate to remember, at least with a short thought, that 70 years have passed since Ross G. Harrison (1907) launched the endeavors with which we are still concerned, when he explanted fragments of the neural tube of 3 mm frog embryos shortly after closure of the neural folds. The significance of his experiments and their impact on the problems and questions in neurocytology of those days are common knowledge. It is satisfying to observe that his technique of explanting living tissue, modified and refined by many investigators over the decades, still today provides us with a powerful tool to investigate properties and capabilities of nervous tissue yielding results that would be unobtainable with other techniques.

It is not the purpose of this paper to give a review of the growth and modifications of the culture techniques and to enumerate the important contributions by experts in the field. Rather, I would like to refer to Murray's (1965) review and would like to give here only an account of the cytological, histological and general morphological features of cultures, taken as primary explants from various portions of the neuraxis, and thereby provide a background against which the experimental work of several other contributors to this volume can be understood.

It should be kept in mind that the preparations from which the illustrations are taken were maintained in a culture situation called "flying coverslip technique"; i.e., a coverslip, to which the tissue fragment adheres by means of a plasma clot or reconstituted collagen gel, is free to move in a slowly rotating roller tube containing a measured amount of nutrient medium (for details see Hild and Tasaki, 1962, 1964). This method favors rapid thinning of the explants by outward migration of certain cell types thereby resulting in a rather flat tissue layer in which single elements are conveniently visualized in the microscope under various optical conditions. Preparations of this kind may differ markedly from cultures in which a more three-dimensional arrangement is favored with its inherent advantages in other aspects, such as organotypic maintenance or establishment of functional interconnections between different cells. Cultures of the latter kind will be described in other chapters of this volume.

Fig. 1 depicts the margin of a primary explant of rat retina seven days after explantation. Numerous nerve fibers have grown into the cell-free medium. Their advancing tips, in the form of so-called growth cones with their finger-like extensions, can be seen at a somewhat higher magnification in Fig. 2. Such naked nerve fibers without glial ensheathment have a relatively short life span. Therefore, during the second week *in vitro* one can observe their rapid degeneration and dissolution. On the other hand, those nerve fibers which remain within the confines of the original explant or which will become associated with migrating glial cells will remain intact and can be seen to form networks within the culture, possibly becoming myelinated and establishing synaptic connections.

As flattening of the cultures proceeds, with time *in vitro* neuronal perikarya present themselves with increased clarity. Fig. 3 shows a very thin spread of neuronal cell bodies from one of the motor nuclei of the rat brain stem. Their dendrites and their as yet unmyelinated axons can only be ill defined in the neuropil. An extremely flattened area containing a multipolar neuron is seen in Fig. 4. This cell is probably derived from the lateral vestibular nucleus. Preparations of this kind are very suitable for experiments involving the introduction of microelectrodes, microsurgical instruments, or application of chemical compounds because cellular reactions can be directly observed, recording sites can be exactly determined, or injury phenomena can be followed with either still or moving picture photography.

Frequently, but not always and certainly not with great accuracy, neuronal cell bodies in cultures retain their morphological characteristics sufficient for the identification of their derivation. For example, the neuron depicted in Fig. 5 is clearly recognizable as a Purkinje cell from the kitten cerebellar cortex. The main dendritic stem with some of the major branches is seen at one extremity of the perikaryon while the axon originates from the other side. A similar Purkinje cell in a silver impregnated whole mount of a 40 day old culture is seen in Fig. 6. Axonic and dendritic poles of the cell body are at opposite ends. The dendritic tree is rather short and stubby; it is possibly underdeveloped because of the lack of interaction with parallel fibers which may never have been present in this culture or which may have degenerated. This example may well serve to explain many of the differences which can be observed in specific cell types if compared in the *in situ* and the *in vitro* condition.

More primitive Purkinje cells are represented in Fig. 7. The dendritic ramifications are still coming off the cell body at separate sites and a common stem has not yet formed. These two cells are members of a row of similar cells lined up in a manner resembling a layer of Purkinje cells *in situ*.

Another easily identifiable cell type in cerebellar explants is seen in

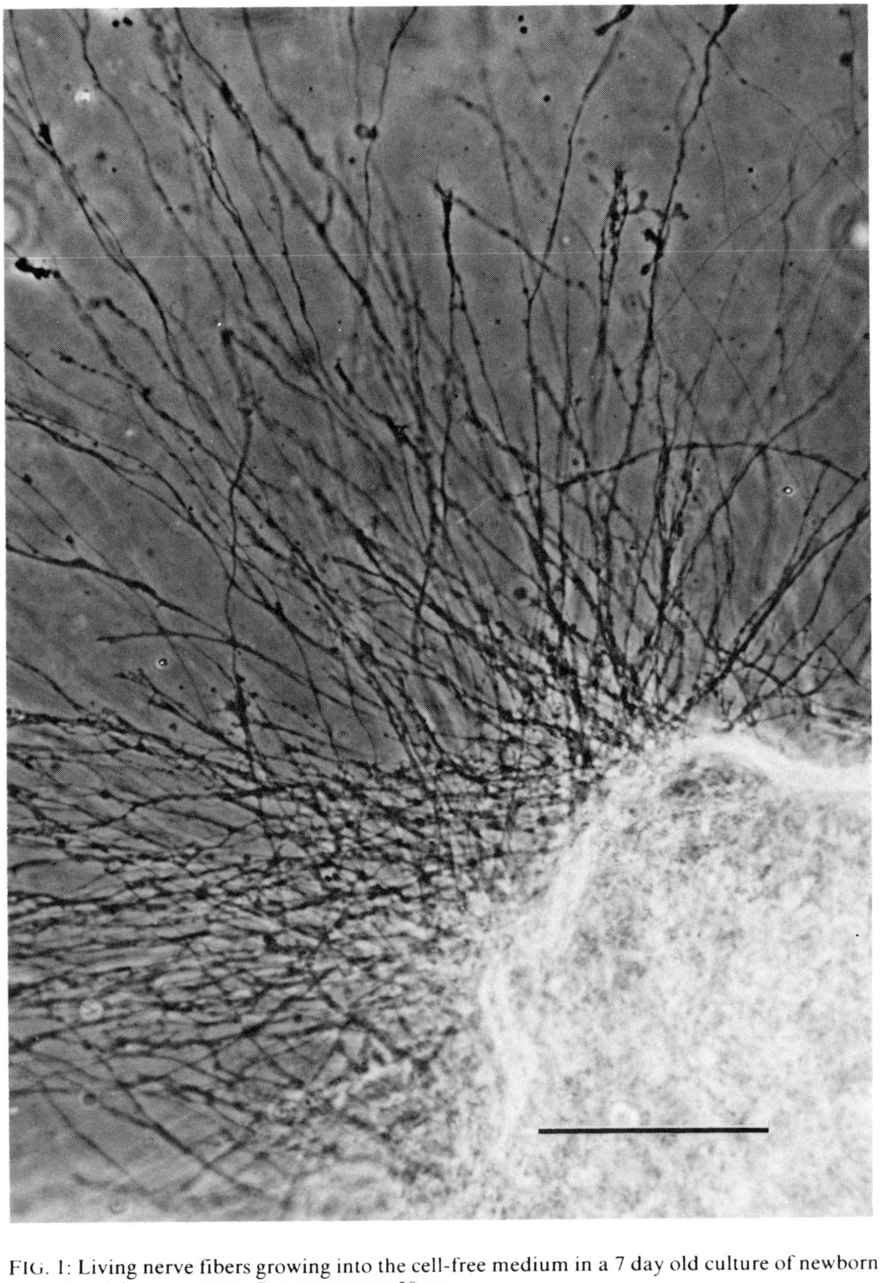

FIG. 1: Living nerve fibers growing into the cell-free medium in a 7 day old culture of newborn rat retina. Phase contrast. Bar represents 50 μm.

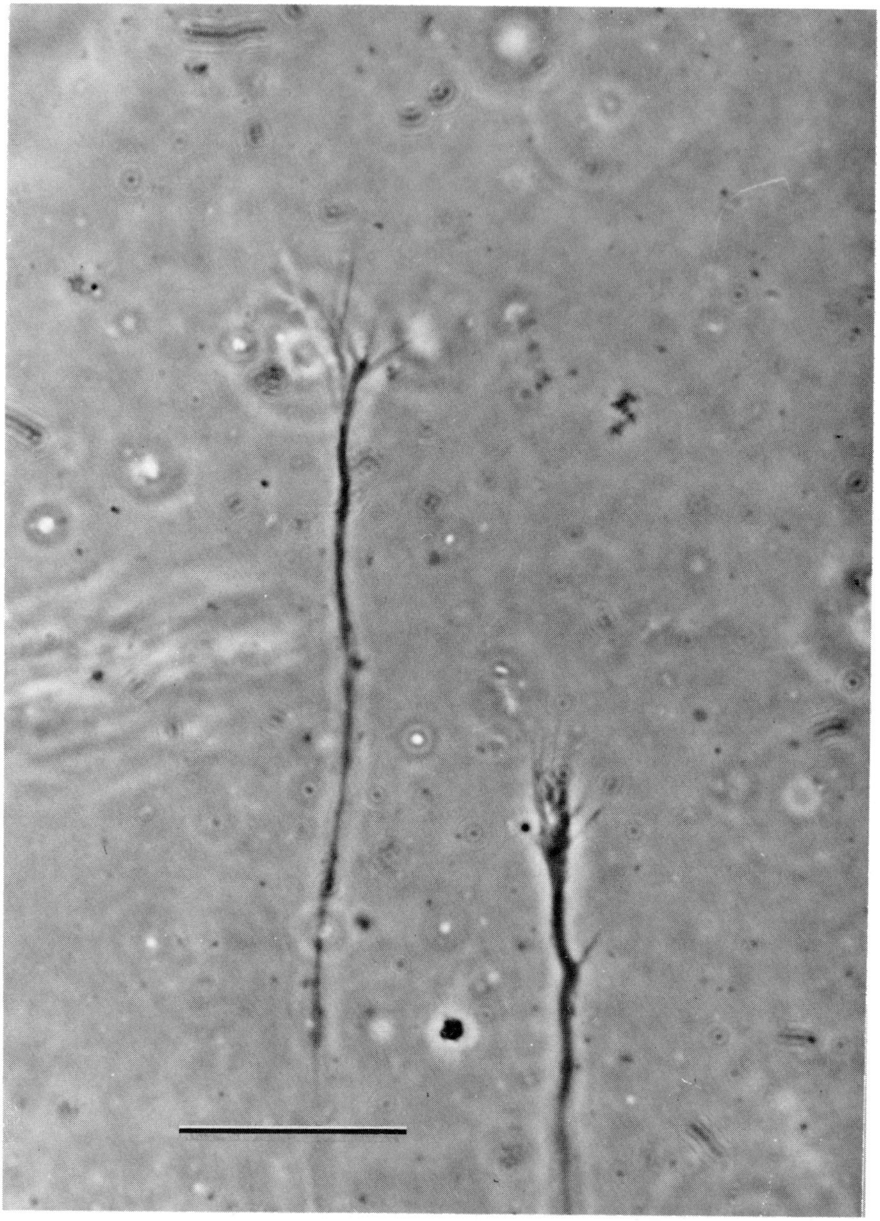

FIG. 2: Higher magnification photograph of two growth cones in a 4 day old culture of rat retina. Note the extremely fine finger-like extensions at the advancing tip of the nerve fibers. With time-lapse photography these can be observed in continuous motion, retraction and new emergence. Phase contrast. Bar represents 20 μm.

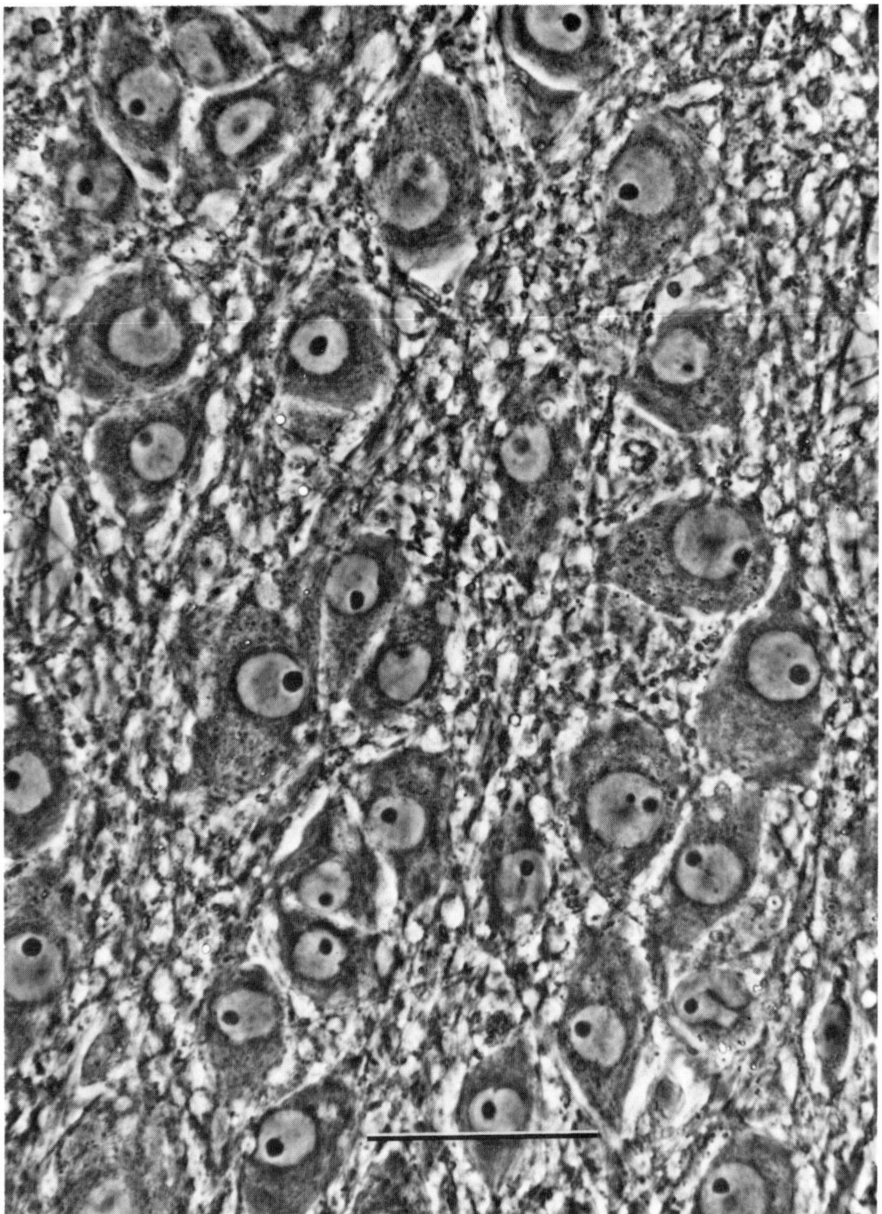

FIG. 3: Living neurons in a 35 day old culture of rat brain stem. Cell size and shape may indicate the derivation of these cells from one of the motor nuclei of the brain stem but it is impossible to determine from which one in particular. Cell processes cannot be clearly distinguished because of considerable light reflexion and formation of light haloes between the perikarya. Phase contrast. (From Hild, 1966). Bar represents 50 μm.

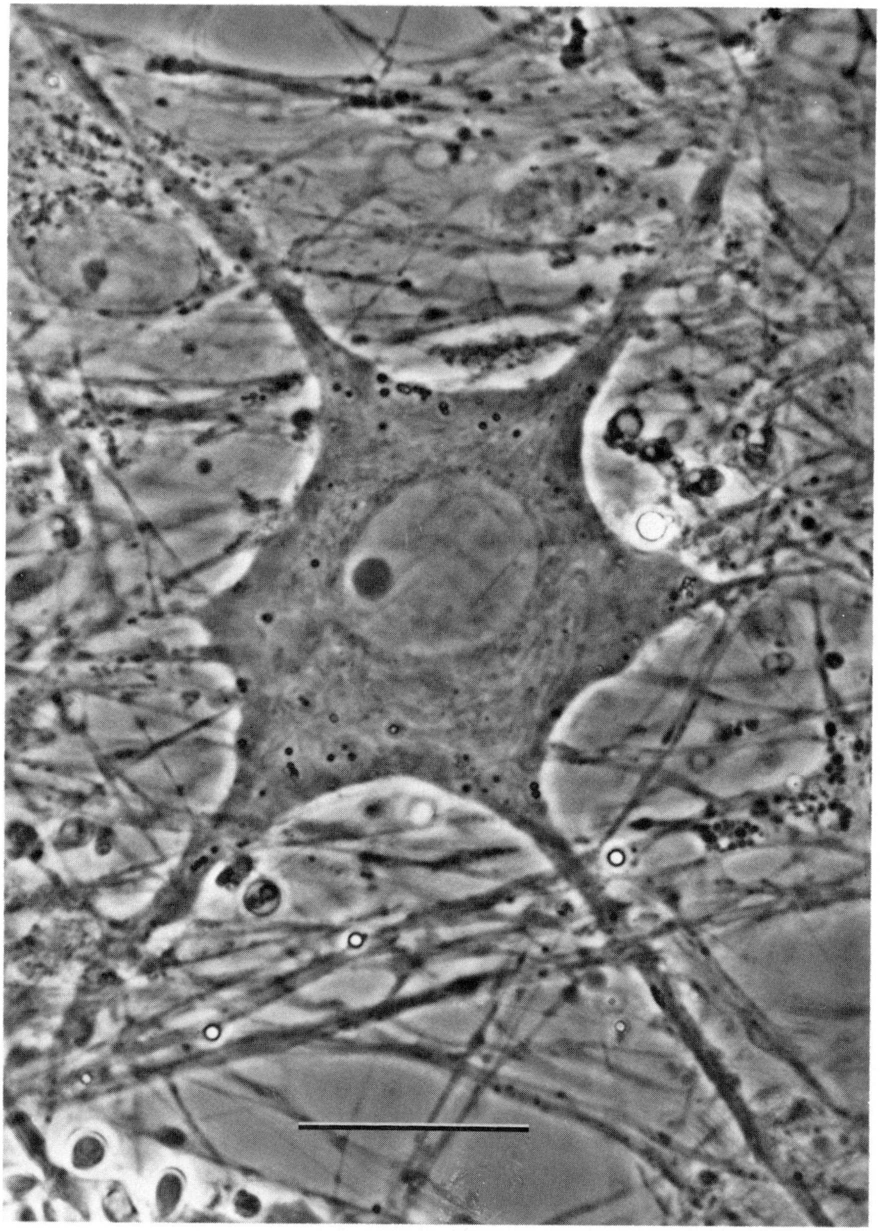

FIG. 4: Well isolated living neuron in a 17 day old culture of rat brain stem. Size and shape are suggestive of the derivation of this cell from the lateral vestibular nucleus. The thickness of this area is not more than one cell layer. Glial fibers form a loose network around the neuron. Phase contrast. (From Hild, 1966). Bar represents 20 μm.

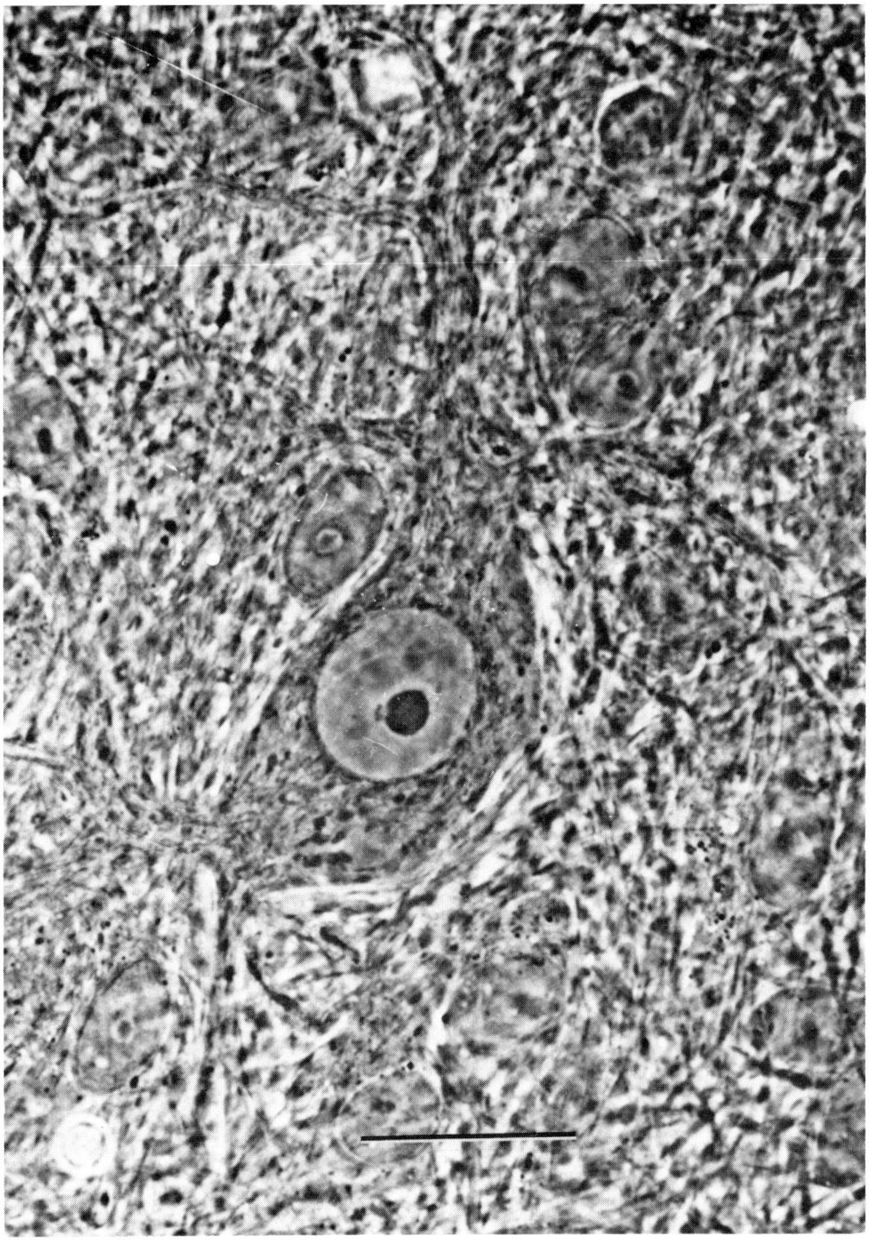

FIG. 5: Living Purkinje cell in a 21 day old culture of kitten cerebellum. Phase contrast. (From Hild, 1959). Bar represents 20 μm.

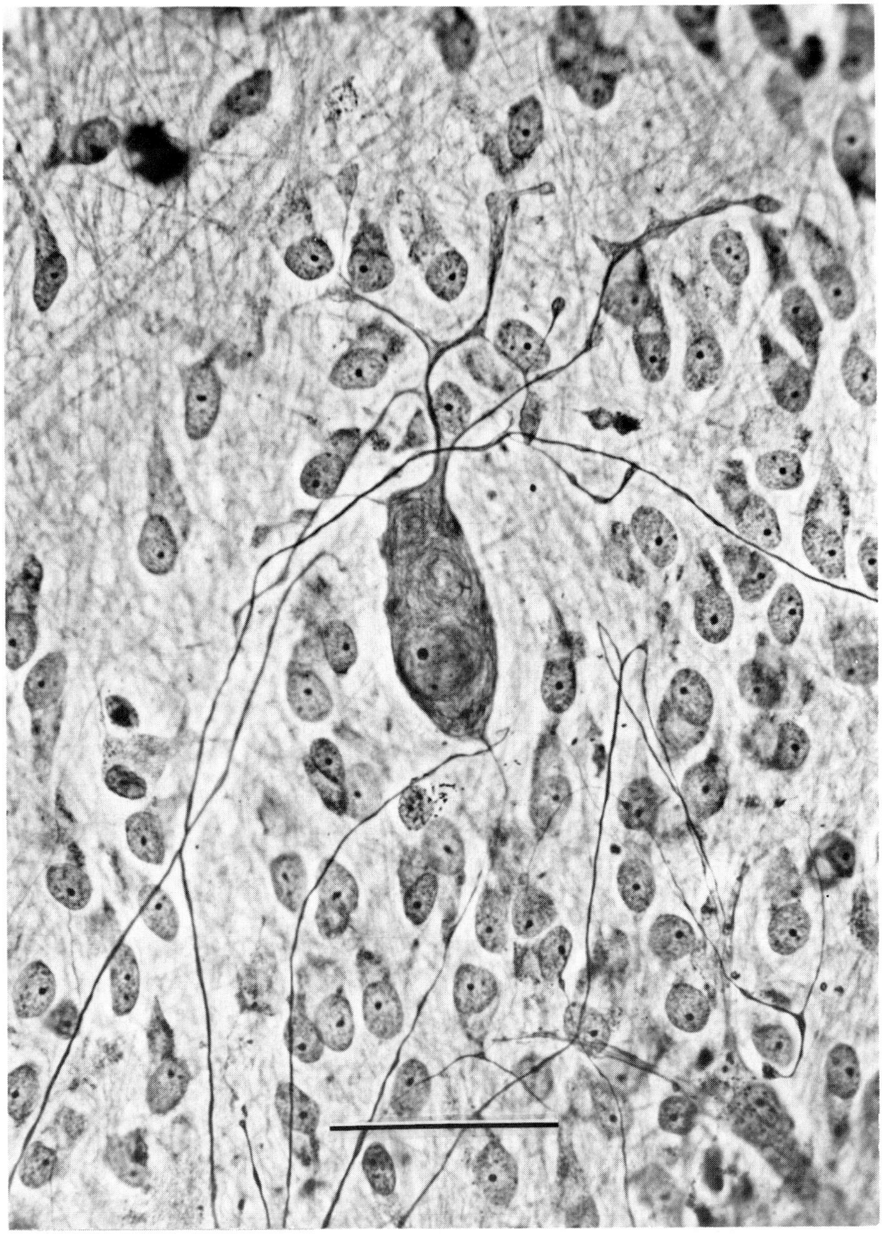

FIG. 6: Purkinje cell in a 40-day culture of kitten cerebellum. The dendritic stem of this cell extends upward. The arborization is very much reduced in complexity. Instead of numerous finely drawn-out tips there are only a few blunt club-shaped endings in which some fibrillar pattern is recognizable. The axon springs from the lower circumference of the cell body. Bodian silver impregnation. (From Hild, 1966). Bar represents 50 μm.

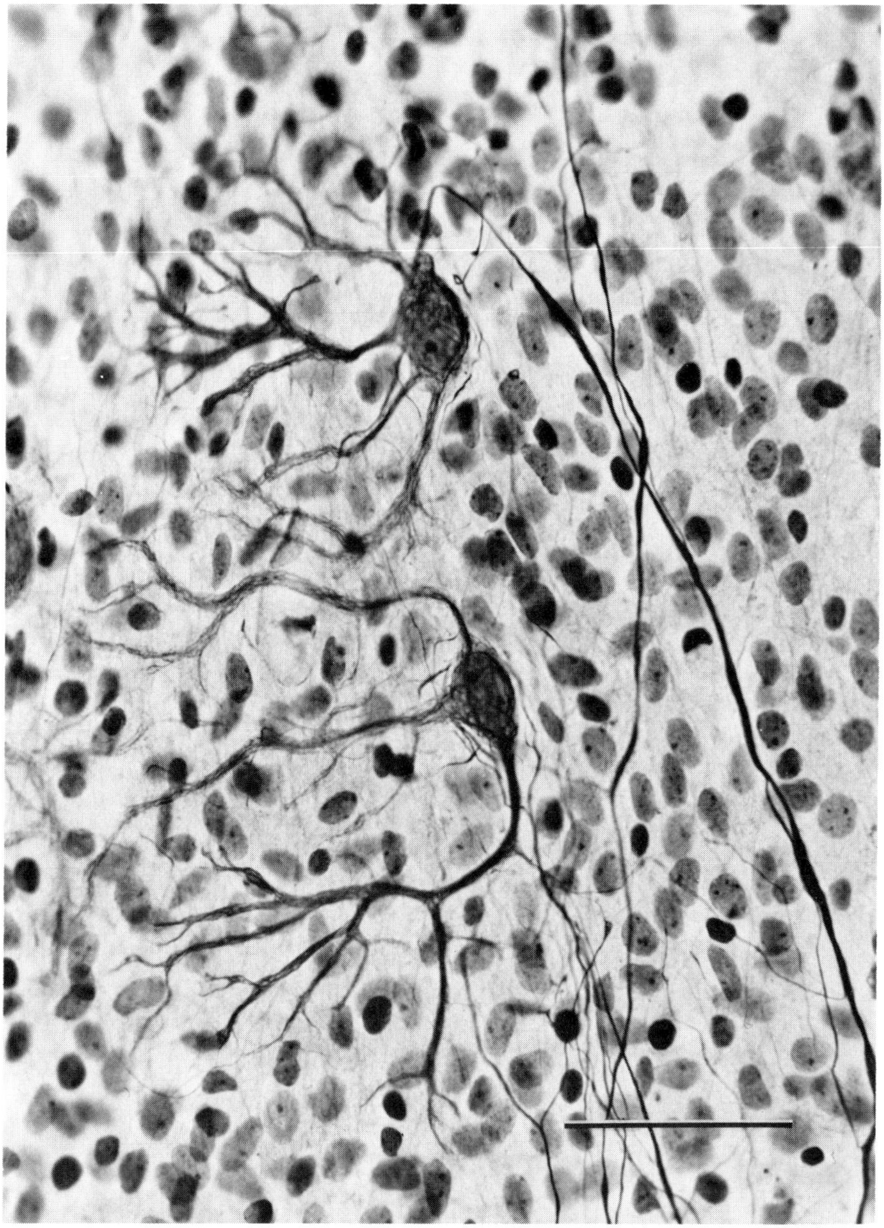

FIG. 7: Immature Purkinje cells in a 13 day old culture of rat cerebellum. Holmes silver impregnation. (From Hild, 1966). Bar represents 50 μm.

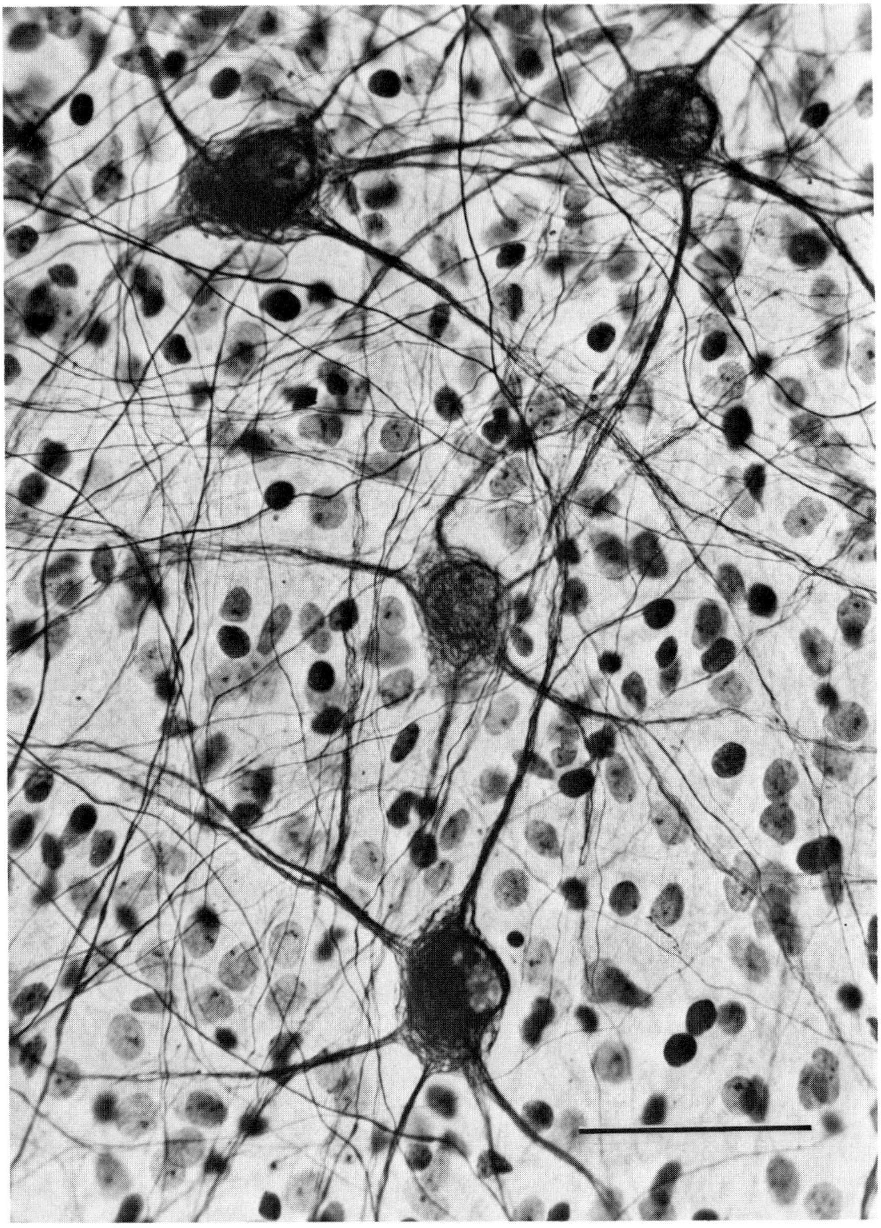

FIG. 8: Multipolar neurons from the deep cerebellar nuclei in a 12 day old culture of rat cerebellum. Holmes silver impregnation. (From Hild, 1966). Bar represents 50 μm.

Fig. 8. These neurons from the deep cerebellar nuclei have retained their multipolar shape. In many explants, which contain both Purkinje neurons and neurons from the deep nuclei, the axons of Purkinje cells can be seen to enter the area occupied by multipolar cells suggesting the maintenance of original conditions and perhaps functional interrelationships.

Unipolar sensory neurons, as from dorsal root ganglia or from the mesencephalic nucleus of the trigeminal nerve, are among those which retain their original shape most faithfully. Fig. 9 shows two neurons from the mesencephalic nucleus of the trigeminal nerve in a silver impregnated culture. Most of the time, these cells are unipolar with their extensions breaking up into two or more branches some distance away from the perikaryon. In some instances however, the cells may be bipolar or even tripolar.

Large multipolar neurons in cultures taken from the lateral areas of the floor of the fourth ventricle are suggestive of their derivation from the lateral vestibular nucleus. One of these neurons is seen in Fig. 10. It has seven large dendrites from one of which the axon takes its origin. This same feature can be seen frequently in multipolar neurons that are commonly found in brain stem cultures but which cannot be classified with certainty as belonging to a particular nucleus. One of these neurons is seen in Fig. 11. The dendrites of these cells are very long, sometimes several hundred μm but still easily distinguishable from the axon which usually can be followed for several mm.

In Fig. 12, another distinct neuronal cell type is shown. This is predominantly found in explants from the cerebral cortex. These cells are multipolar and small. In some instances, an axon can easily be recognized, in others it is impossible to distinguish an axon among the multitude of cell extensions.

In our material, it has been a general observation that small perikarya give rise to shorter and thinner axons than large perikarya. Sometimes, however, small perikarya send out unusually large axons. Fig. 13 shows such a situation. The insert shows the cell body of origin of the massive fiber convolution consisting of the axon and several of its collaterals. At its origin from the cell body this axon is very thin, but it then broadens out into a complicated convolution with a total mass many times that of the cell body.

Primary explant cultures maintained with modern techniques are capable of developing myelin sheathes around axons. Peterson and Murray (1955) have observed this phenomenon first in cultures of peripheral ganglia, and Hild (1957) was first to observe myelin formation in cultures of mammalian central nervous tissue. Fig. 14a and b shows myelinated living axons as seen under phase contrast and polarized light illumination.

It is clearly evident that such cultures can profitably be used for quite a variety of experiments and that the culture situation is unique in that it offers the advantage of continuous observation simultaneous with experimental procedures.

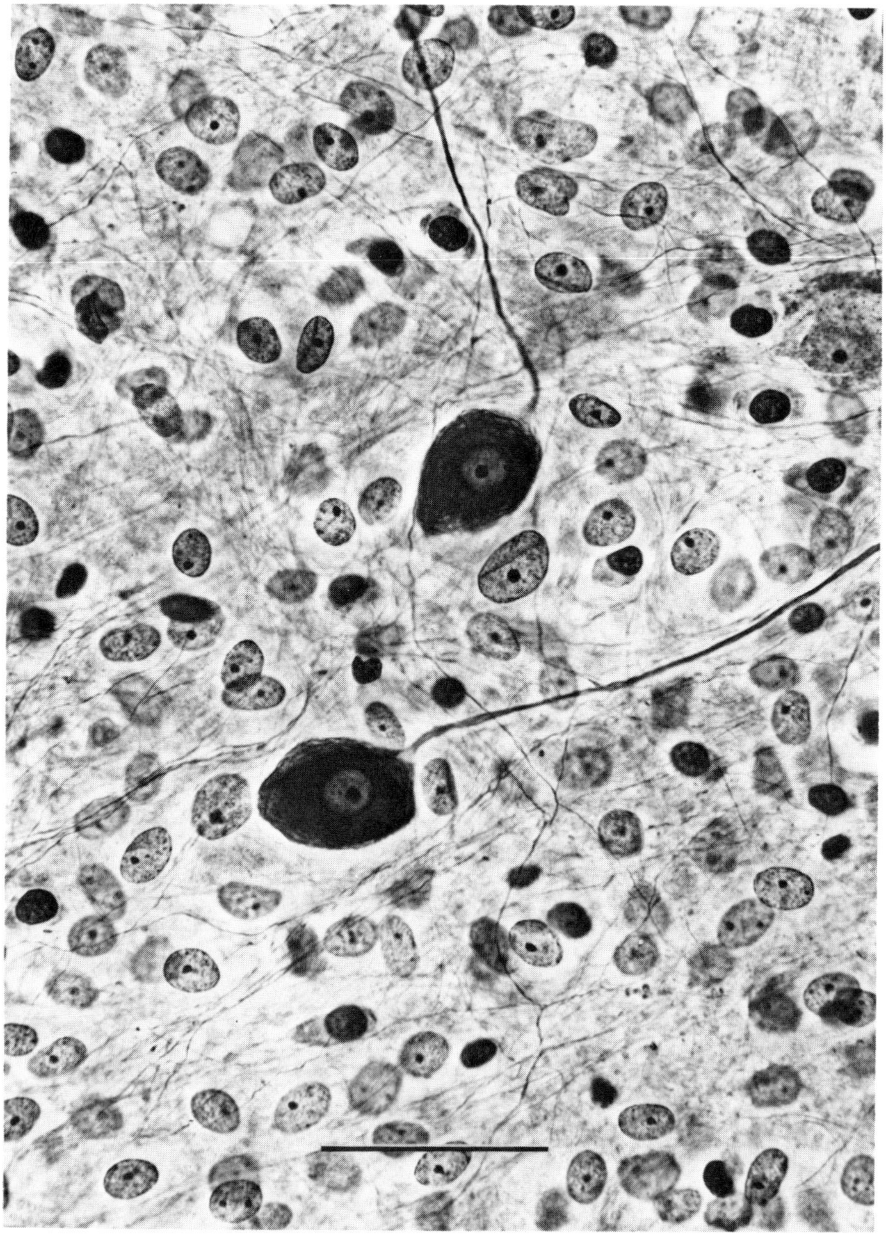

FIG. 9: Neurons from the mesencephalic V_{th} nucleus in a 20 day old culture of rat midbrain. These cells are mostly unipolar but may sometimes be bipolar or tripolar. Bodian silver impregnation. (From Hild, 1966). Bar represents 50 μm.

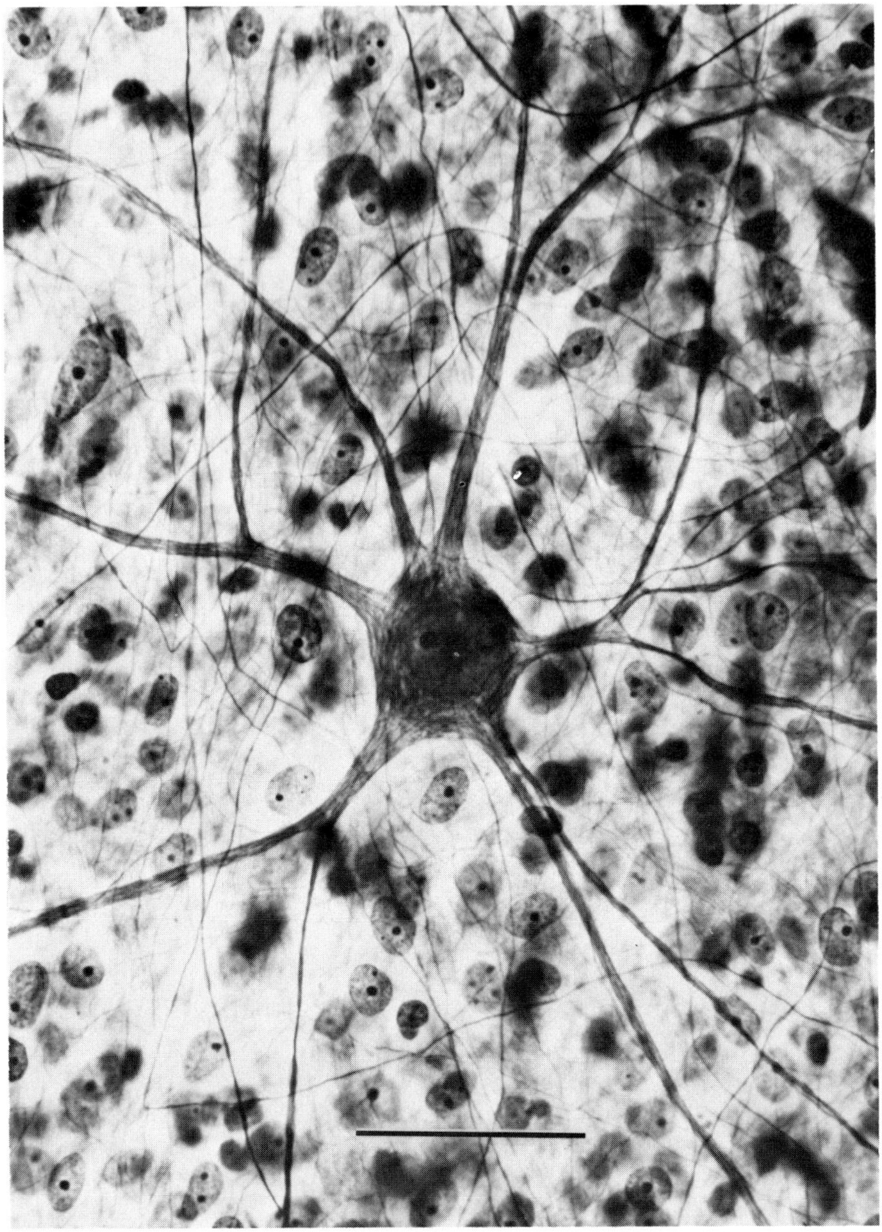

FIG. 10: Large multipolar neuron in a 17 day old culture from the lateral area of the floor of the 4th ventricle of a rat. Size and shape suggest its derivation from the lateral vestibular nucleus. The axon originates from one of the seven large dendrites, and some of the thinner nerve fibers traversing the field are collaterals of the axon of this cell. Bodian silver impregnation. Bar represents 50 μm.

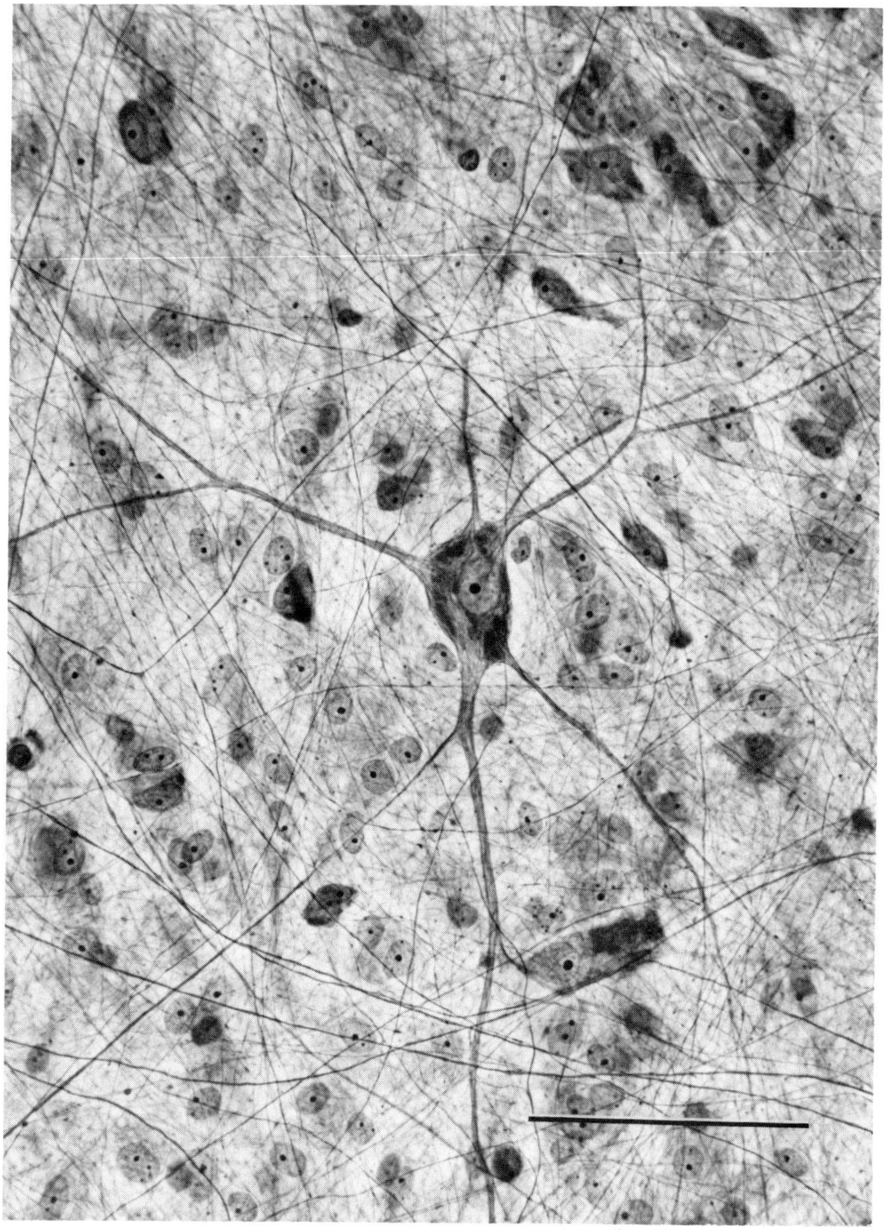

FIG. 11: Medium-sized neuron in a 29 day old culture of rat brain stem. The axon springs from one of the dendrites. Cells of this kind may be derived from the reticular formation. Bodian silver impregnation. (From Hild, 1966). Bar represents 100 μm.

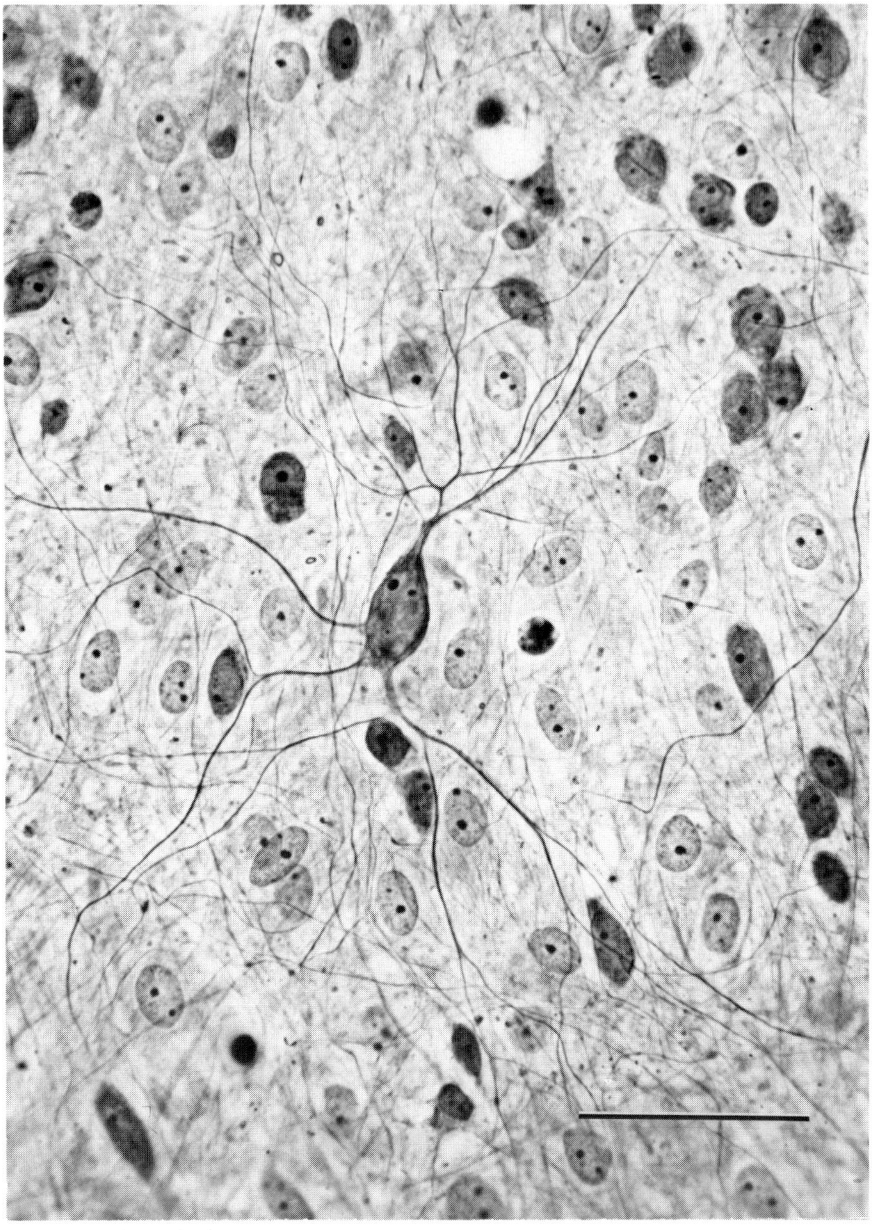

FIG. 12: Small neuron in a 30 day old culture of rat cerebral cortex. Dendrites originate from both pointed ends of the cell body, while the axon springs from the left long side. Bodian silver impregnation. (From Hild, 1966). Bar represents 50 μm.

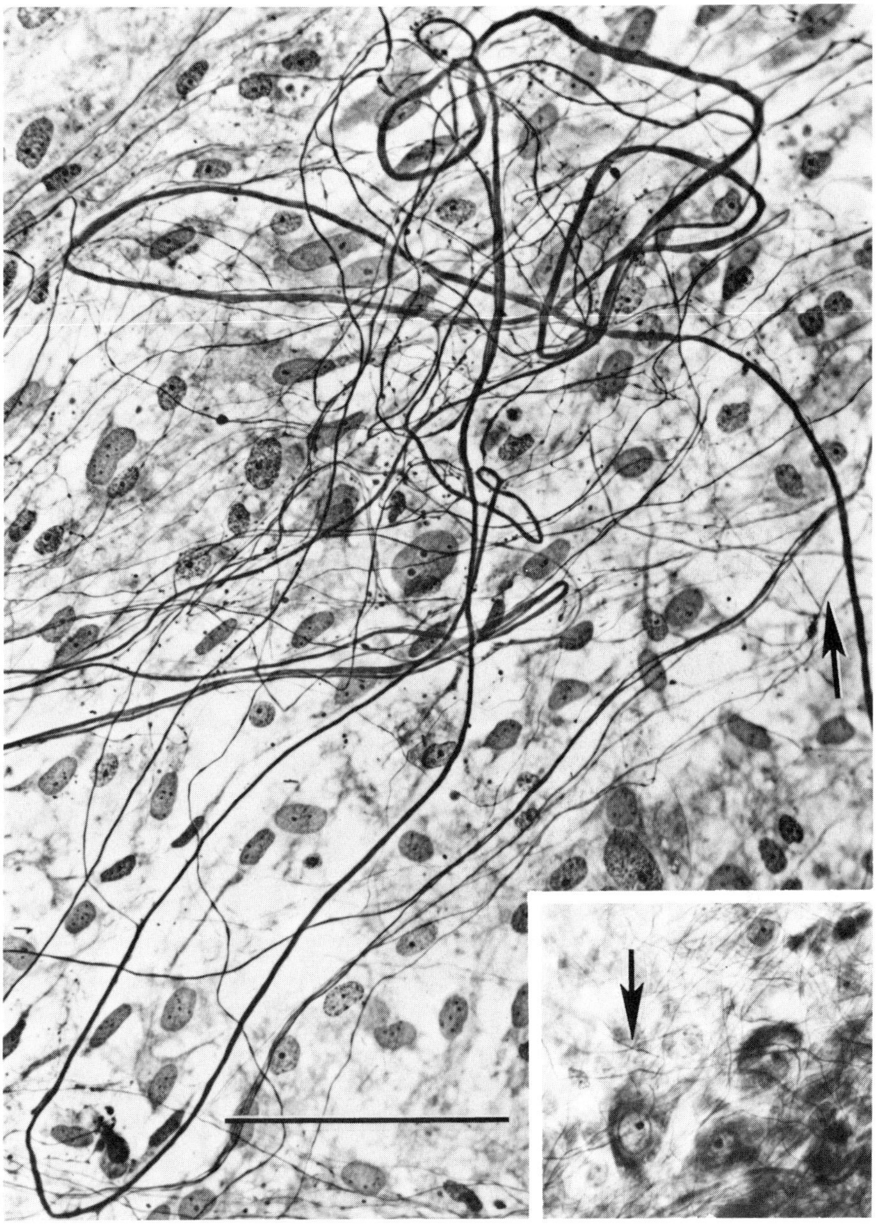

FIG. 13: Fiber convolution in a 41 day old culture of rat cerebellum. The cell of origin of this fiber is pointed out (arrow) in the insert photograph. Both photographs are taken at the same magnification which emphasises the tremendous size difference between axon and perikaryon of this neuron. The point where the axon enters the photographic field on the right side (arrow) is about 1½ mm distant from the cell body. Bodian silver impregnation. (From Hild, 1966). Bar represents 100 μm.

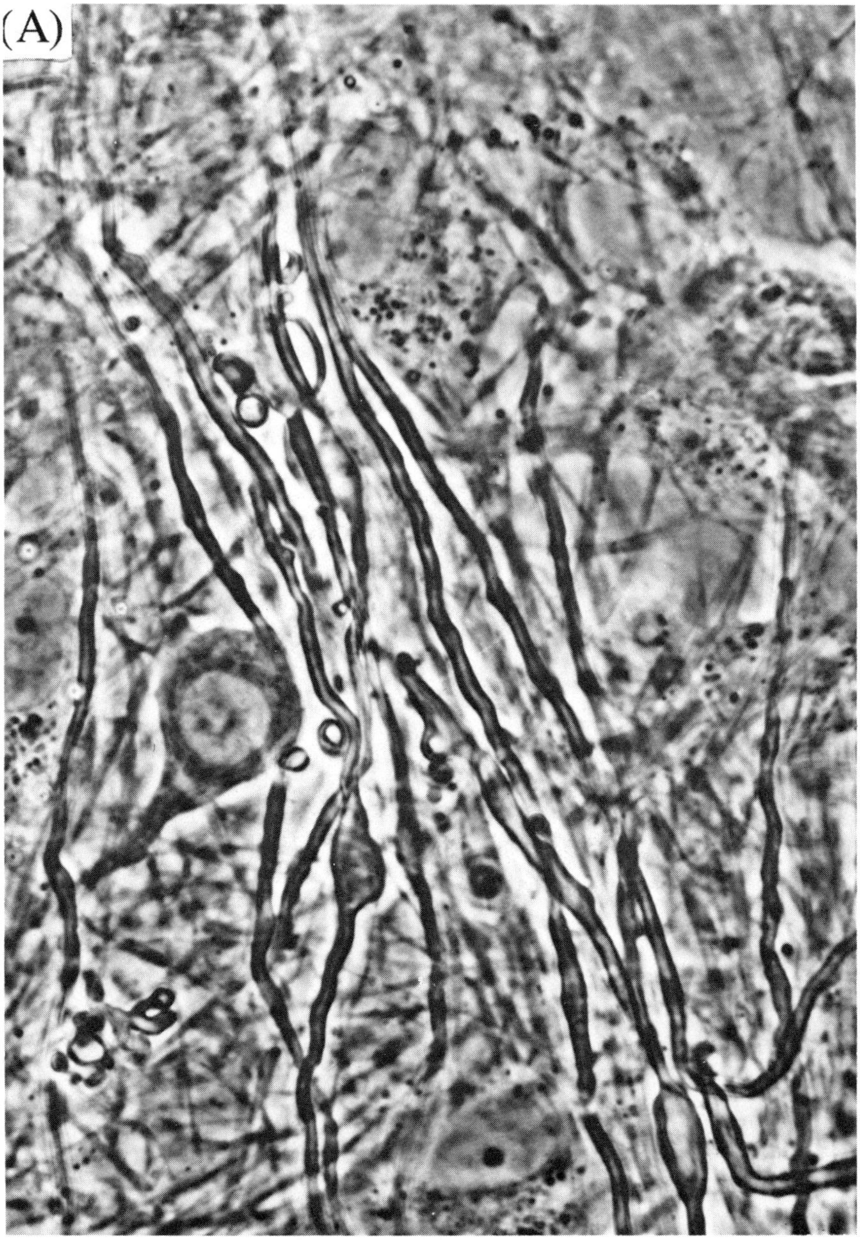

FIG. 14: a and b. Living myelinated axons in a 17 day old culture of rat midbrain. Phase contrast (a) and polarized light illumination (b). Bar represents 30 μm.

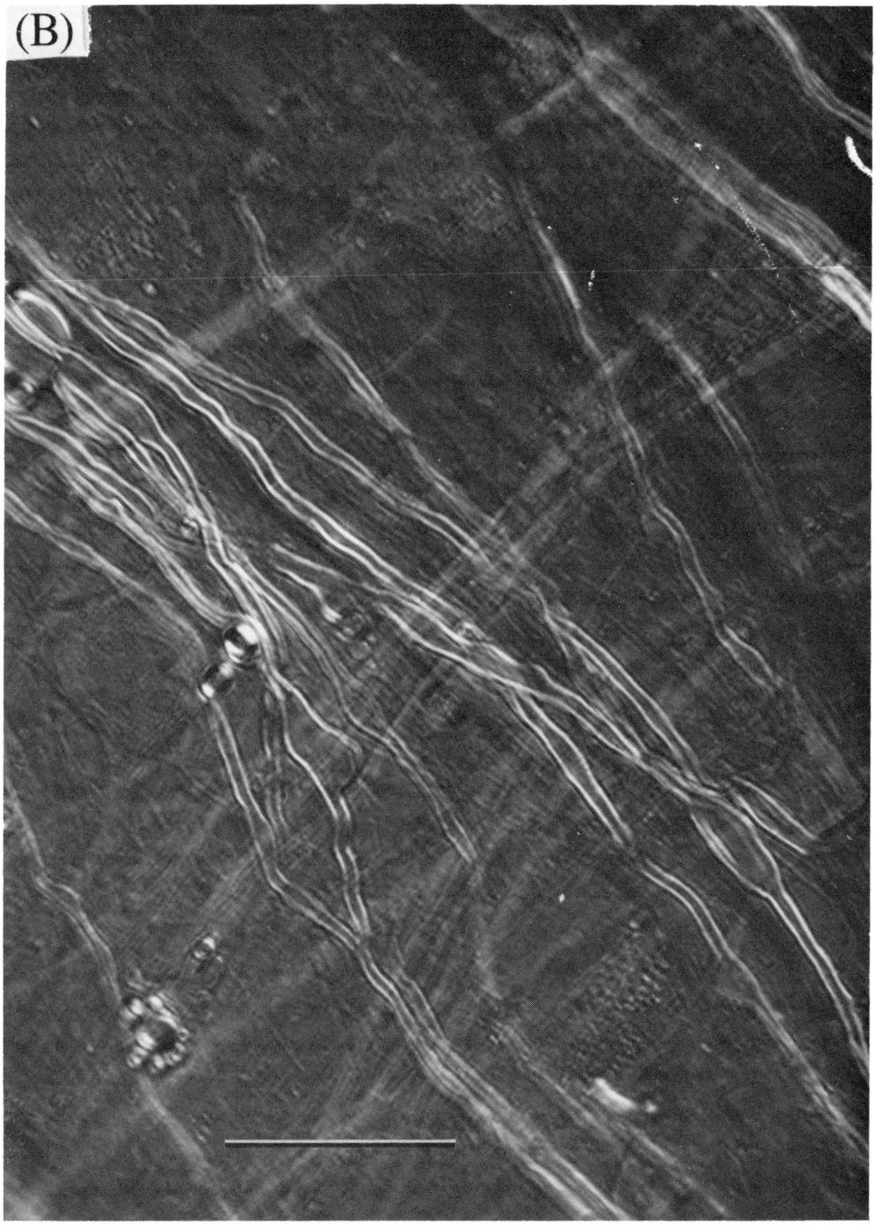

FIG. 14b

References

Harrison, R. G. 1907. Observations on the living developing nerve fiber. Anat. Rec. 1:116-118.

Hild, W. 1957. Myelogenesis in cultures of mammalian central nervous tissue. Zeitschr. f. Zellforsch. 46:71-95.

Hild, W. 1959. Das Neuron. In: *Handbuch der mikroskopischen Anatomie des Menschen.* Vol. IV, Part 4. Springer Verlag, New York-Berlin. pp. 184.

Hild, W. 1966. Cell types and neuronal connections in cultures of mammalian central nervous tissue. Zeitschr. f. Zellforsch. 69:155-188.

Hild, W. and I. Tasaki. 1962. Morphological and physiological properties of neurons and glial cells in tissue culture. J. Neurophysiol. 25:277-304.

Hild, W. and I. Tasaki. 1964. The use of tissue cultures in neurophysiological research. In: *Methods in Medical Research.* Vol. 10, Section IV, Chapter III. Year Book Medical Publishers, Chicago. pp. 327-334.

Murray, M. R. 1965. Nervous tissues *in vitro.* In: *Cells and Tissues in Culture*, Vol. 2. E. N. Willmer (ed.) Academic Press, London and New York. pp. 373-455.

Peterson, E. R. and M. R. Murray, 1955. Myelin sheath formation in cultures of avian spinal ganglia. Amer. J. Anat. 96:319-355.

Differentiation of Cells in Primary Cultures

DIFFERENTIATION OF CELLS IN ORGAN CULTURES

KATHERINE M. LYSER

Hunter College
City University of New York
New York, New York

I. Introduction

A number of methods have been devised for *in vitro* study of the nervous system. Primary cultures consisting of fragments of central nervous system and/or peripheral ganglia are generally included in the category 'organ culture' (Fedoroff, 1967). In such cultures, basic nervous system tissue architecture, and sometimes specific organotypic arrangements, are maintained, and neuron and glial cell differentiation are promoted.

Harrison's classical experiments with primary explants in lymph clots were essentially of this sort (Harrison, 1907, 1910). Clot techniques and then collagen coated coverslips in Maximow assemblies, as developed in Margaret Murray's laboratory and widely used with various modifications and adaptations, have been the foundation of the study of organized, differentiating nervous tissue *in vitro* (Murray, 1965, 1971). Other arrangements include Rose chambers; growing explants at the medium-gas interface using various 'rafts' to support the tissues at the surface of a liquid or semi-solid medium; and culture on collagen, or, more recently, on plastic and various other surfaces in small Petri-type or similar dishes.

Though much work has been done using explants of partially differentiated neural tissues, a wide range of stages can be used, depending on the aims of the project and the phases of development upon which attention is to be focused. To begin this discussion, organization and cellular differentiation of spinal cord will be considered as an illustration of the potentials for differentiation in organ culture.

II. Spinal Cord as an Example of Differentiation in Organ Culture

Spinal cord has been explanted from embryos at early neural tube stages through fetal stages. The early neural tube consists mainly of ventricular cells which are still undergoing mitotic division and are arranged in a pseudostratified columnar epithelium (Boulder Committee, 1970; Watterson, 1965). Though occasional cells undergo their last DNA synthesis (S phase) even earlier (McConnell and Sechrist, 1976), the

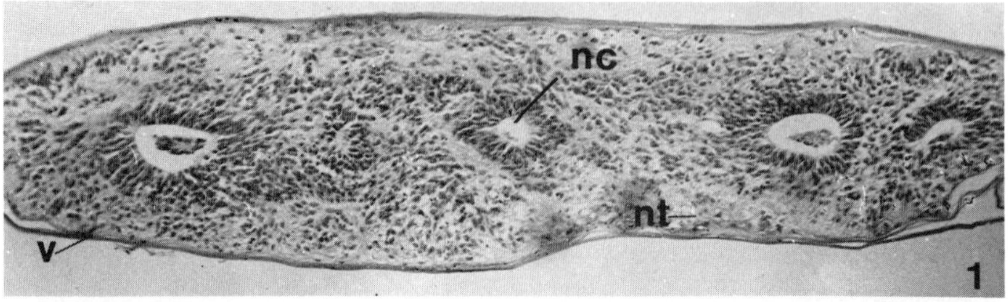
nc
nt
v
1

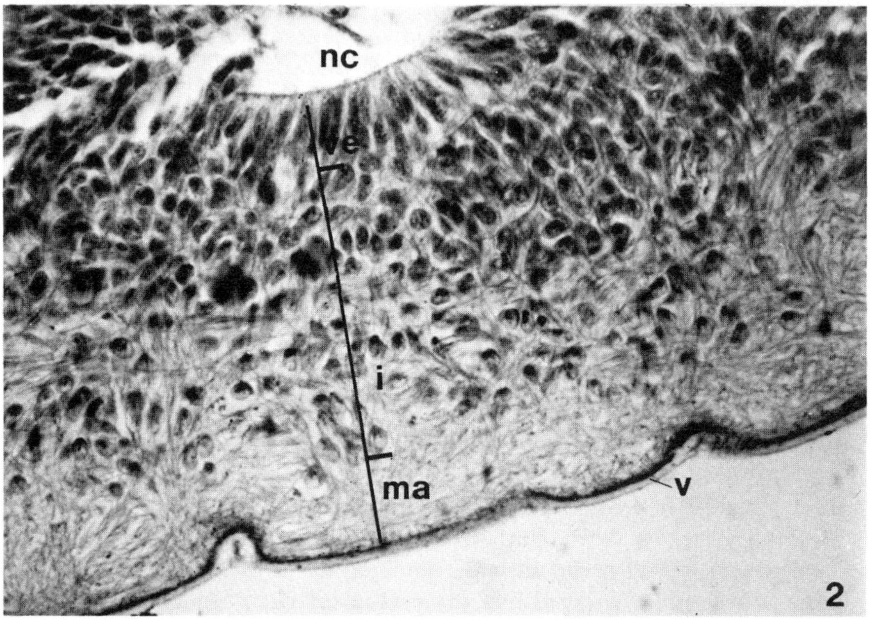
nc
ve
i
ma
v
2

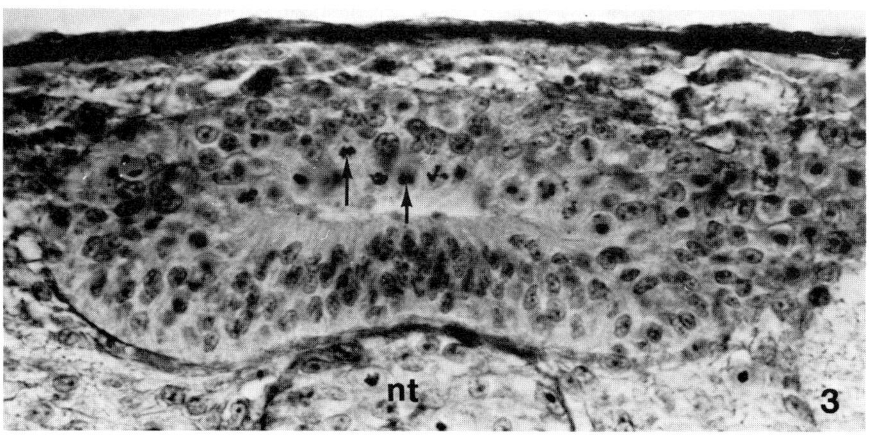
nt
3

differentiation of nerve cells generally begins during the second day of incubation in the chick and embryonic day 11 in the rat. The differentiating nerve cells are scattered near the outer edge of the neural tube, recognizable as monopolar or primitive bipolar cells with axons in silver-stained sections (17-somite chick embryo,* Tello, 1923) or cells with round nuclei which are unlabeled following continuous exposure to ^{3}H-thymidine (20-somite chick embryos, Martin and Langman, 1965). Electron microscopic observations of the same stages are consistent with light microscopic observations (Lyser, 1964, 1968; Tennyson, 1970). The height of proliferative activity occurs at 3 days in the basal plate and 6 days in the alar plate in the chick (Hamburger, 1948). The 'birthdays', i.e., period of DNA synthesis preceding the last mitotic division of the precursor cells, are between 2 and 8 days for the majority of chick spinal cord neurons (Langman and Haden, 1970). By day 6 in the chick (Barron, 1946) or day 13 in the rat (Windle and Baxter, 1936) multipolar motor neurons are present, spinal ganglion neurites have entered the cord to begin the formation of the dorsal funiculus, and interneurons have processes. Synapses are seen as early as 3 days in the chick (Oppenheim *et al.*, 1975); and 13.5 days in the rat (Vaughn and Grieshaber, 1973). The origin of glial cells from the ventricular cell population and the beginning of their overt differentiation generally follows that of neurons in a particular area of the nervous system (see Boulder Committee, 1970).

The potential for *early phases of neuron differentiation in vitro* may be studied in spinal cord explants from neural tube stages. We have cultured spinal cord of 12- to 27-somite chick embryos (stages 11 to 15) in a fold of vitelline membrane from an unincubated chicken egg (Wolff, 1961). The cultures are supported by pieces of millipore filter on stainless steel grids and maintained in Falcon organ culture dishes (Lyser, 1966, 1971a, 1971b,

*17-somite chick embryos correspond to Hamburger and Hamilton (1951) stage 12, about 45 to 49 hours of incubation; 20-somite embryos, stages 13 to 14, 48 to 53 hours.

FIGS. 1-3. Light micrographs of explants from 2 day chick embryos cultured in a fold of vitelline membrane in Falcon organ culture dishes for one week.

FIG. 1. Three lengths of isolated neural tube (with some notochord). Four 'neurocoels' (nc) lined with ventricular-type cells are present. Areas similar to intermediate zone with cell bodies and neurities extend to the outer edge of the culture in many places. v, vitelline membrane; nt, very small cluster of notochord cells. From 17 and 19 somite embryos, levels of somites 10-17, 11-18 and 19 and posterior to somites. H and E stain. × 125. (From Lyser, 1971a, Fig. 4)

FIG. 2. Three lengths of isolated neural tube. In this area ventricular zone cells (ve) lining the neurocoel (nc), an intermediate zone with cells and neurites (i), and an outer marginal zone (ma) are seen. v, vitelline membrane. From 17 and 18 somite embryos, levels of somites 10-17 and posterior to somites. Holmes silver stain. × 520. (From Lyser, 1971a, Fig. 5)

FIG. 3. Neural tube explanted with adjacent mesoderm, notochord, ectoderm and endoderm. Cultured 6 days, with 2.7×10^{-7} Colcemid in the medium during the last day *in vitro*. Cell division is still taking place in the spinal cord as indicated by arrested mitoses, seen particularly in the dorsal ventricular zone (arrows). nt, notochord. From a 17 somite embryo, level of somites 12-17 and posterior to somites. H and E stain. × 460.

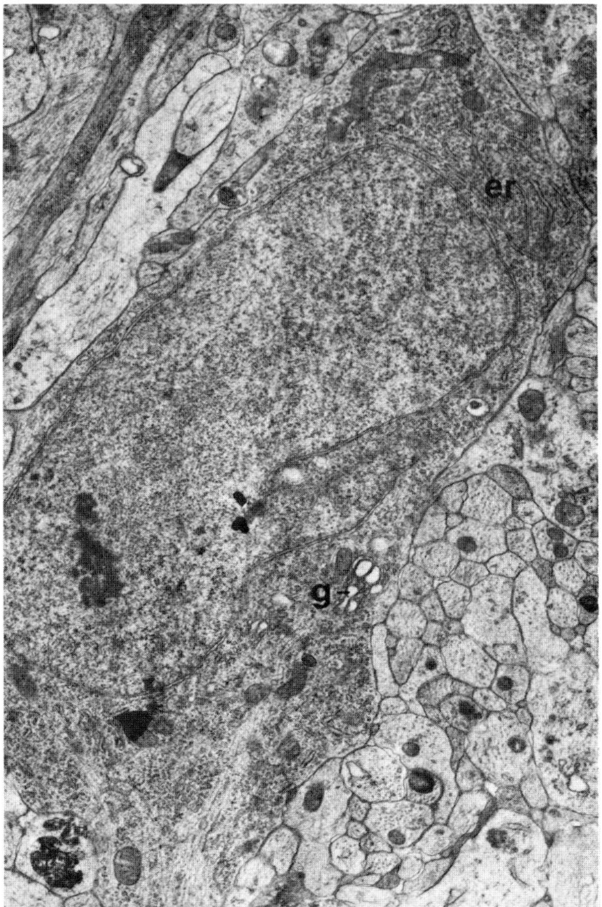
er
g
4

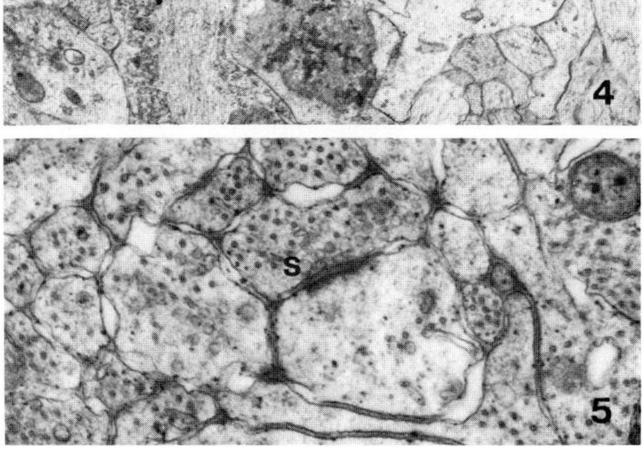
s
5

1972, 1976). Neural tube can be freed of other tissues, or a length of neural tube with adjacent somites and other mesoderm, notochord, endoderm and ectoderm can be explanted. In vitelline membrane cultures, the explants are somewhat flattened in the plane of the support, and may taper to a thin layer at the edges, but do not exhibit the 'explant' and surrounding 'outgrowth' zones typical of many primary explant tissue cultures.

During the initial period in culture, mitosis continues and a number of nerve cells undergo the first steps of their overt differentiation. In cultures of isolated neural tube, histological organization is maintained to an appreciable extent, so that after about a week *in vitro* ventricular, intermediate and marginal zones may be recognized in some areas (Figs. 1 and 2). Many nerve cells with processes can be seen in sections stained by silver methods. Neurites extend in various directions among the cell bodies, as well as constituting the marginal zone. When several lengths of neural tube are cultured together, the pieces fuse into one mass; original dorsal-ventral orientation is not obvious, and cultures tend to be symmetrical with respect to upper and lower surfaces. In other areas of the cultures the pattern may be less typical, for example, central canals branching or disappearing, or indistinct intermediate and marginal zones.

In cultures of neural tube with adjacent tissues, spinal cord and other structures are found in approximately their normal relationships after about a week *in vitro*. The spinal cord is rounded or oval in transverse section, and dorsal-ventral orientation is usually indicated by the position of the notochord. There is some variation in thickness of the spinal cord and arrangement of cells. In the areas showing better differentiation, ventricular, intermediate and marginal zones are present; sections of neurites and cells with processes can be seen. Occasionally 'spinal nerves' — bundles of fibers extending into surrounding tissues — are present.

Although we have not adequately studied mitotic patterns in vitelline membrane cultures, it might be mentioned that mitosis does continue *in vitro*. First, the number of cells increases considerably, particularly in isolated neural tube, though possibly not to the extent that occurs *in vivo*. Also, a few mitotic figures are seen in sections, and blocked mitoses can be accumulated by addition of Colcemid to the medium (Fig. 3).

FIGS. 4,5. Electron micrographs of chick neural tube after one week *in vitro*.

FIG. 4. Differentiating neuron from a vitelline membrane culture of three lengths of isolated neural tube. Note granular endoplasmic reticulum (er) and the base of a process with microtubules (toward the bottom of the figure). g, Golgi complex. From 18 somite embryos, levels of somites 10-18 and posterior to somites. × 10,000. (From Lyser, 1971a, Fig. 12)

FIG. 5. Early synapse (s) from a culture of neural tube explanted with adjacent tissues. Note also sections of processes with microtubules. From a 21 somite embryo, level of somites 17-21 and posterior to somites, cultured directly on millipore filter support in a Falcon organ culture dish. × 26,000. (From Lyser, 1971a, Fig. 10)

v
6

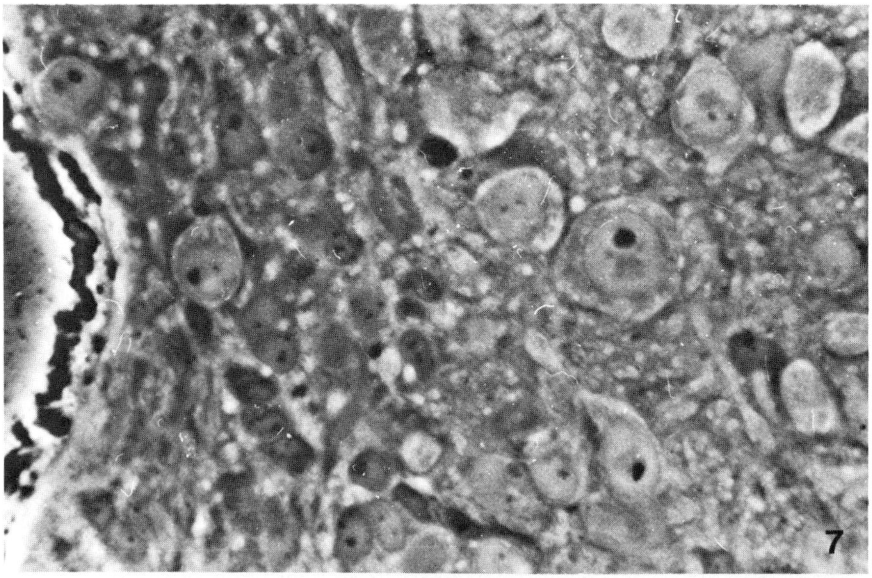
7

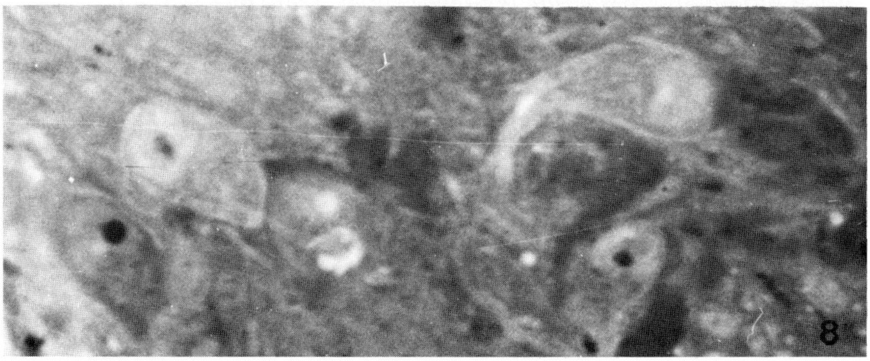
8

Cell bodies of differentiating neurons contain many free ribosomes, some microtubules, moderately developed Golgi complexes, and sometimes several elongated, stacked profiles of granular endoplasmic reticulum, the beginning of establishment of Nissl bodies (Fig. 4). In a few instances, synapses have been seen in electron micrographs (Fig. 5). However, cell body size and development of organelles are considerably behind the more advanced nerve cells of 8-day embryos (see Lyser, 1964). There is thus a definite lag in development in these early explants, perhaps proportionately greater than in explants from older embryos. There may be some small differences in the sequence of cytological changes: differentiating neurons in cultures of neural tube plus adjacent tissues tend to have more microtubules in the cell bodies, and less granular endoplasmic reticulum than is seen in those of isolated neural tube *in vitro* or during early stages of cytological differentiation *in vivo*. However, development from these initial steps to mature neurons and glia when cultures are maintained for longer periods, indicates that the initial changes are essentially part of a normal developmental sequence.

Later stages of development proceed *in vitro* with continued maintenance of early explants or from nervous system explanted at partially differentiated stages. Neurons and glia with cytological features comparable to those of cells *in vivo* develop; nervous system tissue architecture is elaborated. The specific patterns of neurons and fiber tracts may be modified to a greater or lesser extent; some of the differences seem to be correlated with the size and orientation of the explant relative to the substrate, and physical and nutritional constraints of the culture conditions.

The patterns are perhaps most easily compared to the *in vivo* counterpart in transverse slices of spinal cord from intermediate stage embryos placed with the cut transverse surface on the supporting surface as seen in Maximow double-coverslip assemblies (Peterson *et al.*, 1965; R. P. Bunge *et al.*, 1965; Sobkowicz *et al.*, 1968; Guillery *et al.*, 1968). In cultures of spinal cord only (meninges and ganglia removed) from 16- to 18-day rat or 14- to 16-day mouse embryos, the initial explant becomes thinner due to some spreading and to loss of cells, and is surrounded by an outgrowth zone

FIGS. 6-8. Light micrographs of isolated neural tube from 2 day chick embryos after 3-4 weeks *in vitro*.

FIG. 6. Culture with intermingled neurons, glial cells, and neuronal and glial processes. Culture is thicker at the center, but no 'outgrowth' zone is formed. v, vitelline membrane. From 19 and 21 somite embryos, levels of somites 10-19 and 11-20. 20 days *in vitro*. × 230.

FIG. 7. Same culture as figure 6, with larger neurons and smaller, dense glial cells. v, vitelline membrane. × 1,000.

FIG. 8. Neurons with adjacent glia and neuropil from another culture. From 18 and 23 somite embryos, levels of somites 13-20, and 15-18 and posterior to somites. 31 days *in vitro*. × 1,200.

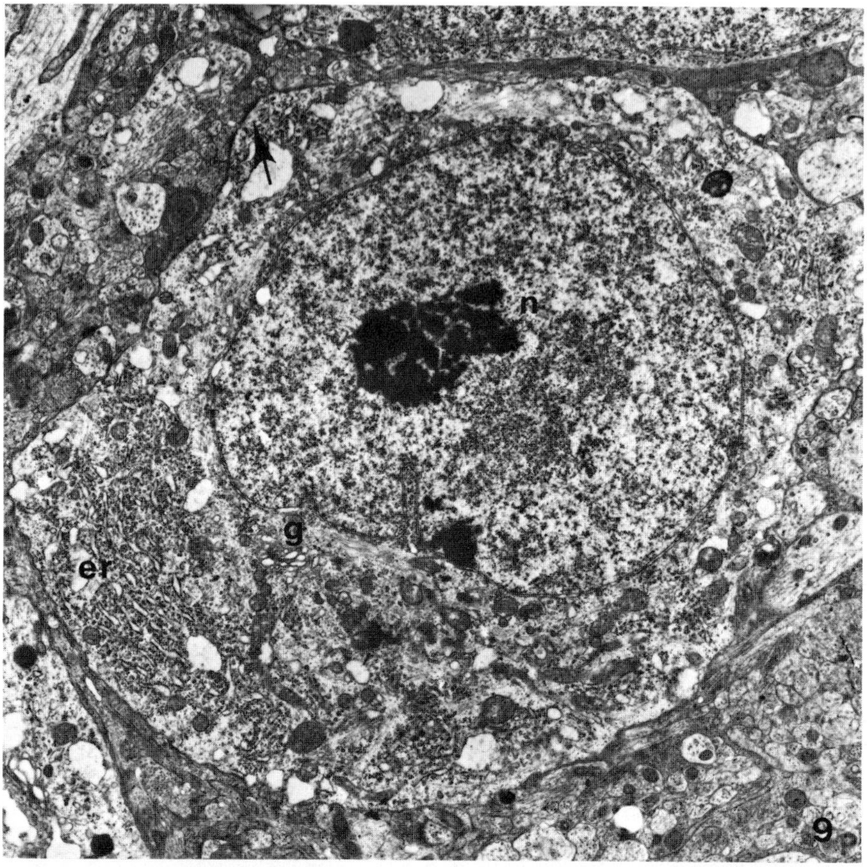
n
g
er
9

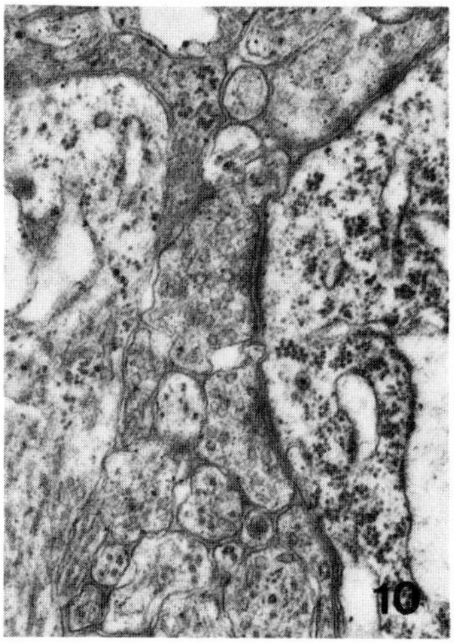
10

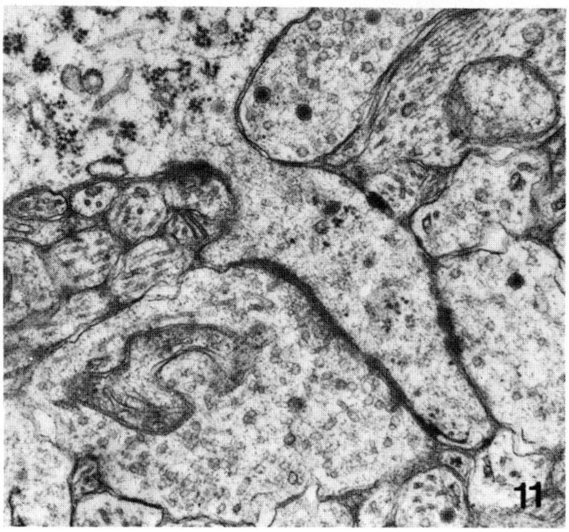
11

of glial cells and neurites. In the primary explant, the main areas of the spinal cord can be seen in some cases (see especially Sobkowicz *et al.*, 1968): ependymal cells surrounding a central canal, dorsal and ventral horn areas with neurons and glial cells. The longitudinal fiber columns disappear; instead, bands of fibers oriented circumferentially parallel to the plane of the coverslip are usually present, along with bundles extending into the outgrowth zone dorsally and ventrally. The latter 'ventral roots' contain axons of large ventral multipolar — motor — neurons, hence are comparable to ventral roots *in vivo.* The dorsal bundles contain neurites of smaller neurons, which usually loop back into the primary explant.

In other cases, especially where spinal cord is cultured with the longitudinal axis parallel to the substrate surface, specific patterns of the *in vivo* cord may not be recognizable. This has been observed in explants from early neural tube maintained for longer periods *in vitro.* Vitelline membrane cultures of isolated early neural tube remain about 200 μm thick. In many areas they consist of intermingled medium-sized and small neurons, glial cells, and neuropil (Figs. 6, 7, 8). There may be a border of glial cells and neuronal and glial processes, but no recognizable patterns of neurons or fiber tracts have been seen. When necrotic zones occur, they tend to be between the center and the lower (toward the medium) side of the culture. Sometimes a more patterned arrangement of cells is present, though still appearing morphologically not to be a reproduction of the *in vivo* organization. For example, larger neurons may be clustered near the bottom of the culture, with thick processes extending toward the lower surface, and eccentric nuclei toward the upper ends of the cells. Ependymal cells and neurocoel are infrequently observed; ependymal zone and neurocoel tend to be flattened parallel to the surfaces of the culture. Sometimes there is a layer of neurites between the ependyma and neurons.

Multipolar neurons and abundant neurites are also seen after 21 days *in vitro* in explants from 2½ to 3 day old chick embryo spinal cord cultured on collagen coated coverslips in roller tubes. In contrast to explants from older embryos, some neuron cell bodies are seen in the outgrowth zone of these explants (Fisher *et al.*, 1975).

Neurons typically become quite mature cytologically, though cell bodies are usually smaller than those of corresponding cells *in vivo.* Chick

FIGS. 9-11. Electron micrographs of explants of 2 day chick embryo spinal cord, after 3-4 weeks in vitelline membrane cultures.

FIG. 9. Neuron and adjacent neuropil. Note large nucleus with prominent nucleolus (n), granular endoplasmic reticulum — forming Nissl bodies (er), Golgi complex (g). Axo-somatic synapses (arrow) are seen at higher magnification in Fig. 10. From the same culture as Fig. 8. × 7,200.

Fig. 10. Synapses on soma of neuron in Fig. 9 × 23,000.

FIG. 11. Synapses on a spine arising from the soma of a neuron (edge of soma at upper left). From the same culture as Fig. 6. × 25,000.

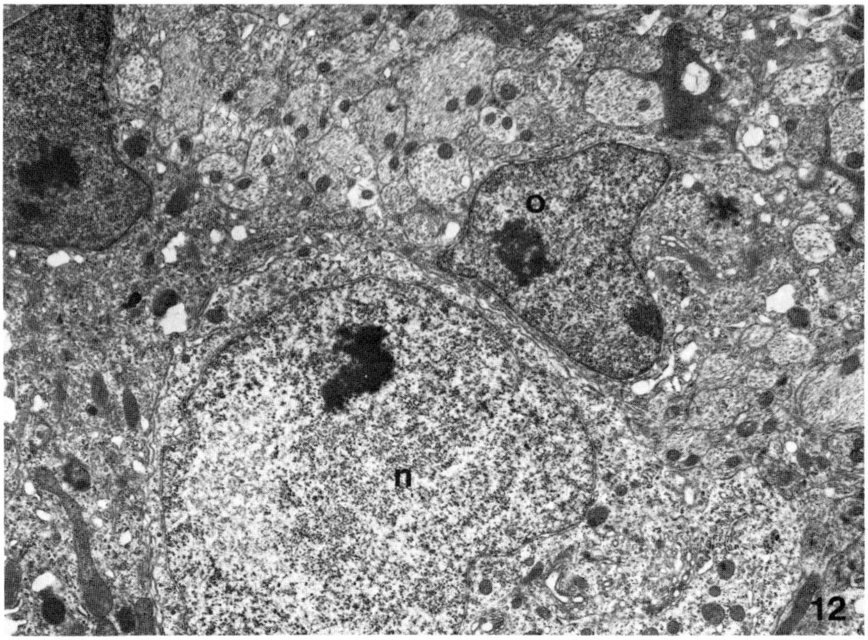
o
n
12

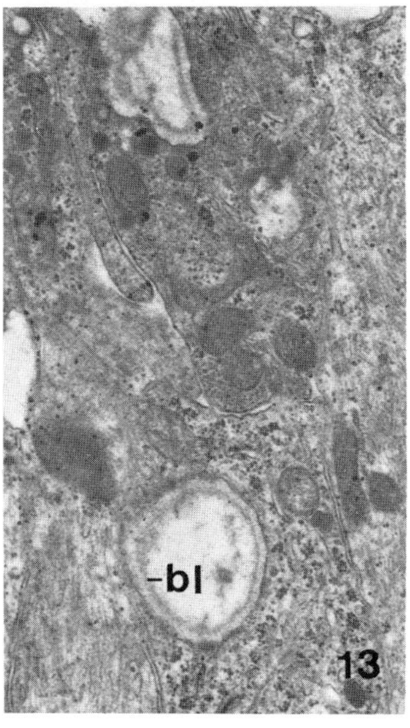
-bl
13

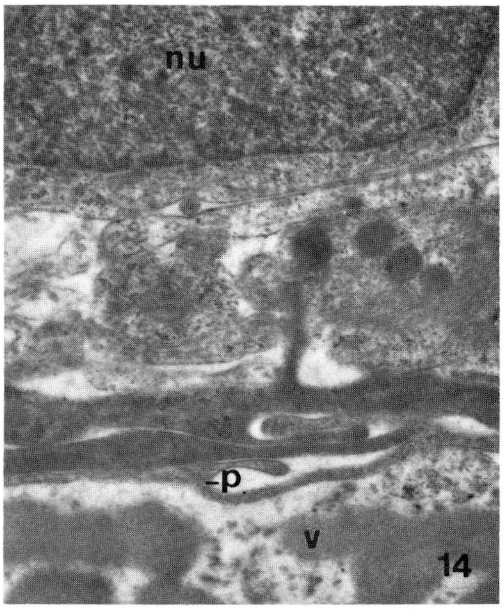
nu
-p
v
14

and mouse neurons and their processes are argentophilic. In vitelline membrane cultures of explants from early chick neural tube the larger neuron somas after about 3 weeks *in vitro* measure about 10 × 20 μm (compared to 24 × 37 μm for cervical motor neurons of hatched chicks) and tend to have eccentric nuclei (Figs. 7, 8). Nuclei are typically large, 'vesicular', with prominent nucleoli. Cytoplasmic organelles include fairly well developed Golgi complexes, granular endoplasmic reticulum which may have the configuration of Nissl bodies, mitochondria, some microtubules and filaments. Axo-dendritic and some axo-somatic synapses with the usual fine structure, including membrane densities and synaptic vesicles, are present (Figs. 9-12).

Neurons in cultures using other methods and a range of embryonic and fetal stages are comparable as to differentiation of organelles, and may become a little more mature with sufficient time *in vitro*. Large multipolar cells with extensive cytoplasm and centrally located nuclei have been seen in some cultures of partially differentiated fetal rat and mouse cord (R. P. Bunge *et al.*, 1965; Sobkowicz *et al.*, 1968). Dendrites and long, branching axons extending via ventral roots through the outgrowth zone have been followed in thin collagen-coated coverslip cultures stained by silver methods (Guillery *et al.*, 1968).

Glial cells also differentiate considerably *in vitro*. Cells with the characteristics of *in vivo* oligodendrocytes, astrocytes and ependymal cells have been described in various spinal cord cultures. Oligodendrocytes have rather dense cytoplasm and are sometimes seen in a satellite relation to neurons (Fig. 12). Their processes occasionally can be seen to form myelin sheaths (Guillery *et al.*, 1968). At least some axons are typically myelinated in well differentiated spinal cord cultures, though the extent of myelination varies in explants from different species as well as with culture conditions. Astrocytes have less dense cytoplasm, particles which are presumably glycogen, and typical filaments. Ependymal cells with cilia, sometimes observed to beat in living cultures (Guillery *et al.*, 1968) and abundant microvilli line neurocoels (Fig. 16). Glial cells also form a limiting layer at edges of explants, sometimes comparable to glia limitans *in vivo*. In vitelline membrane cultures of isolated spinal cord, glial processes, similar in fine

Figs. 12-14. Electron micrographs of explants of 2 day chick embryo spinal cord after 4 weeks in vitelline membrane cultures.

FIG. 12. Neuron (n) with adjacent oligodendrocytes (o) and neuropil. Same culture as Fig. 8. × 7,300.

FIG. 13. An area within the same culture where glial processes surround intercellular space lined with basal lamina (bl). × 15,500.

FIG. 14. The edge of the same culture, where neural tissue is bounded by thin glial processes (p) but no basal lamina is seen between them and the vitelline membrane (v). nu, nucleus of glial cell. × 16,500.

structure to those of the glia limitans of normal baby chicks, form the borders of the culture adjacent to the vitelline membrane. In this case basal lamina is not observed, though occasionally a layer of basal lamina material is seen adjacent to glial processes lining an extracellular space within the culture (Figs. 13, 14). A similar border of glial processes is found at the surface of coverslip cultures facing the medium (Guillery *et al.*, 1968). In contrast, a basal lamina, with the appearance of that seen *in vivo*, is consistently found just outside the glia limitans in cultures of spinal cord explanted with surroundings tissues (Lyser, 1972).

III. Explants from Other Parts of the Nervous System

Other parts of the central nervous system, particularly those with well defined specific patterns of organization of various neuronal types, illustrate further that specific organization, as well as differentiation of various neuronal and glial cell types, can proceed *in vitro*. Neonatal mouse cerebellum contains immature Purkinje neurons and dividing external granular cells. When pieces of cerebellum are explanted at this stage, the external granular cells complete their mitotic phase and migrate *in vitro*, so that molecular, granular and Purkinje layers may evolve (Wolf, 1970; see chapters by Hendelman and by Wolf, this volume).

Spinal and sympathetic ganglia, explanted at partially differentiated stages, illustrate the development of peripheral nervous system in organ culture. Spinal ganglia from 5 to 15 day old chick embryos or 16½ to 19 day old rat fetuses may be cultured in Maximow assemblies in plasma clots or on collagen (Peterson and Murray, 1955; Peterson *et al.*, 1965; M. B. Bunge *et al.*, 1967). After an initial chromatolytic reaction in the soma and axon regeneration, neurons become cytologically quite well differentiated. Nuclei with prominent nucleoli are centrally located in larger nerve cells; characteristic Nissl bodies of granular endoplasmic reticulum, Golgi complexes, subsurface cisternae, neurofilaments and microtubules are present. Histotypic organization and accessory cell differentiation are seen in satellite sheaths which surround neuron somas and in fascicles of axons with the typical arrangement of peripheral nerves. The latter contain mixtures of small unmyelinated axons, several associated with one Schwann cell, and larger axons, each myelinated by a single Schwann cell at any one level.

Explants of sympathetic ganglia (chick, 14 to 16 day old embryos, to 8 days post hatching; rat, neonatal to 20 days post partum; guinea pig, neonatal) behave in a similar way (Chamley *et al.*, 1972a,b; Benitez *et al.*, 1974). Large principal neurons, and sometimes other neuron types, are distinguishable in cultures, and are comparable to neurons seen *in vivo*. The principal neurons show moderate formaldehyde-induced fluorescence, and well differentiated cytoplasmic organelles including granular endoplasmic reticulum, and receive axo-somatic and axo-dendritic synapses. Smaller, brightly fluorescent cells, the interneurons (SG or SIF cells), may be present, particularly in mammalian ganglia. Accessory cells are associated with the neurons.

IV. Variations among Cultures and Discrepancies between Development in Vivo and in Vitro

It is perhaps useful to review some of the variations in cell differentiation and patterns of organization among cultures and in comparison with *in vivo* development, in consideration of cultures as 'models' of normal nervous systems, and in looking for clues to developmental controls by exploiting differences.

Development often proceeds at a slower pace *in vitro* than in the embryo, in tissues explanted at partially differentiated as well as early stages. While the delay is seen in most features, including differentiation of neuronal and glial organelles, myelinization, etc., there may be some differences among various aspects of differentiation. Synapses, for example, seem to develop with less lag than some other cytological features. Synapses have been described in spinal cord explants from 14 day rat fetuses after 3 days *in vitro*, which is close to the additional time required *in vivo* (M. B. Bunge *et al.*, 1967), and occasionally in chick cord explanted at 2 days after 7 days in culture.

The observations that neuron cell bodies tend to be smaller than *in vivo* and are not completely mature after considerable periods *in vitro* may in part reflect continuing lags in rate of development, as well as restriction of development by lack of appropriate connections. The presence or absence of other tissues affects the way neural explants develop, but not necessarily in a manner such that differences can be easily attributed to target organ or other synaptic interactions. In spite of variations in organization within each group, characteristic differences are seen in spinal cord of 2 day chick embryo isolated from other tissues, and that explanted with adjacent somites and other mesoderm, notochord, ectoderm and endoderm. In contrast to isolated neural tube, cultures of neural tube with adjacent tissues tend to have prominent ependymal cells lining a conspicuous neural canal and fewer neurons (Figs. 15, 16 and Lyser, 1971b). The latter cultures, after 3 to 4½ weeks *in vitro*, contain masses of cartilage, striated muscle and/or whorls of epidermis, in addition to spinal cord. A regular pattern of spinal nerves has not been recognized, but bundles of neurites extending to other tissues, particularly muscle masses, are sometimes found. Even in these areas, larger neurons are much less numerous than in cultures of isolated neural tube, though a few are found that are at least as well differentiated as

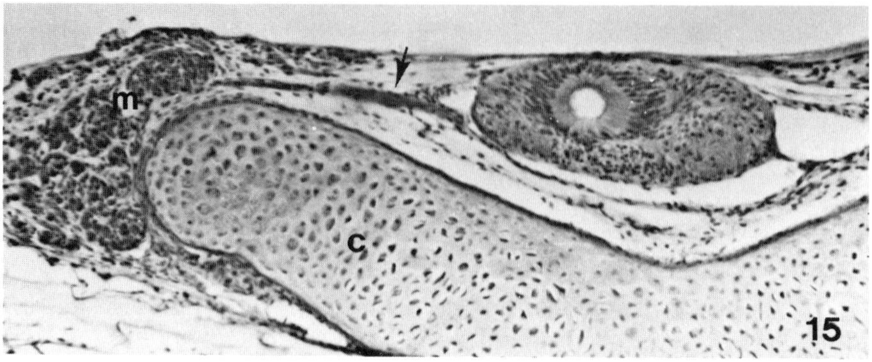

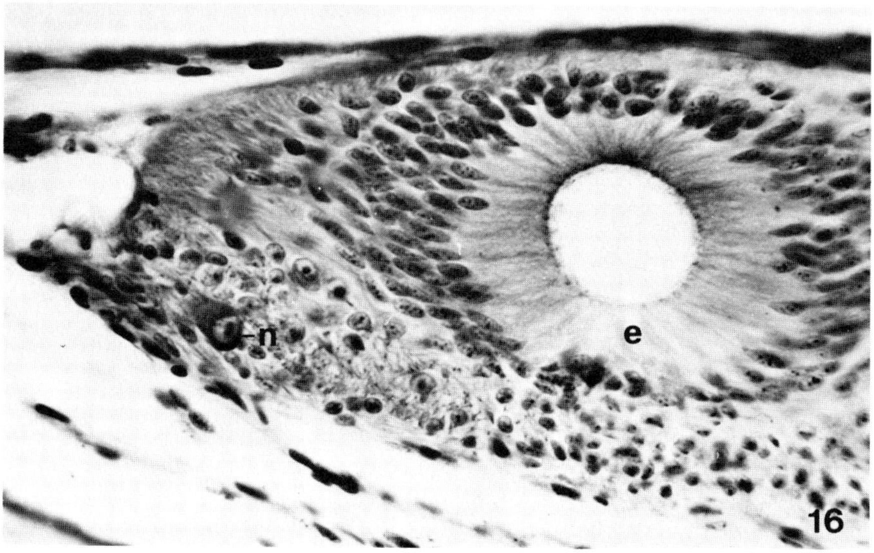

FIGS. 15, 16. Light micrographs of neural tube with adjacent tissues from 2 day chick embryo after 3 weeks *in vitro*..

FIG. 15. Area of the culture where spinal cord, cartilage (c) and muscle (m) can be seen. A bundle of neurites (arrow) extends from the spinal cord to muscle tissue (connects to spinal cord and muscle in adjacent sections, not shown). From a 27 somite embryo, level of somites 15-21, 21 days *in vitro*. Holmes silver stain. × 140.

FIG. 16. Another section from the same culture as figure 15. Prominent ependymal zone (e) lines neurocoel; a large neuron is seen in the ventral lateral area of the spinal cord (n). × 530.

those in isolated spinal cord cultures (Fig. 16). Occasionally, there are regions of the culture where other tissues are scanty, and spinal cord appears as a larger mass similar to cultures of isolated neural tube. Thus, if there is a trophic effect of target organs, it is overshadowed by other factors which seem to be less favorable for formation or maintenance of larger numbers of neurons. One has the impression that other tissues might physically constrain the spinal cord. This is particularly noticeable where there is extensive cartilage, but begins before much cartilage forms, as seen in 7 day cultures.

Another example of apparent changes affecting specific cell types is that of longitudinally-oriented 14½ to 15½ day fetal rat cord (R. P. Bunge and Wood, 1973). In sagittal halves or whole lengths opened along the dorsal midline, usually placed with the ependymal surface facing the substrate, dorsal-ventral differences disappear, and large motor neurons seem to be lost gradually *in vitro*.

Changes from the *in vivo* situation, and differences among cultures, in total numbers of neural cells, as well as proportions of various neuronal and glial types, could result from a number of causes, which are not easy to sort out. Non-specific degeneration, due to deficiencies in nutrient, etc., reaching central areas of tissue, has been widely observed. Disproportions in numbers of certain types of cells could be due to differential survival under certain environmental conditions, as has been observed with dissociated cell cultures with various combinations of media and gas phases (Mains and Patterson, 1973). However, little is known about possible effects, general or specific, on mitotic patterns, initiation of cell differentiation, or relations between mitosis, differentiation, cell death, and cell maturation.

As suggested above, even though more or less specific interactions among neurons and between neurons and end organs are undoubtedly very important in later stages of differentiation and maintenance of neurons, if not during early stages *in vivo* (see Smith and Kreutzberg 1976), such effects are not at all clear *in vitro*. For example, large, well differentiated multipolar cells, which must be motor neurons, are seen after extended periods in some spinal cord cultures, but disappear in others, in both instances in the absence of muscle in the culture.

V. Importance of Histotypic Organization

Another question upon which organ culture, together with cell cultures, might shed some light is the importance of histotypic organization for development of neural cells. Though much more needs to be done, the available results suggest that histotypic organization is more important during early phases of neuron differentiation than for later phases.

As discussed above, spinal and autonomic ganglia explanted at partially differentiated stages develop fairly mature neurons and satellite and Schwann cells. Ganglia from these same stages of development can also be dissociated, and neurons regenerate neurites and continue their development in cell cultures.

However, at earlier developmental stages, dispersion of spinal ganglion cells seems to be unfavorable for neuronal differentiation but encourages differentiation of melanocytes. Peterson and Murray (1955) found that in explants from 4 day chick embryos, there was considerable differentiation of pigment cells. These explants spread more than those from 5 day or older embryos. Cowell and Weston (1970), on the basis of further studies, suggest that the melanocytes were probably not derived from presumptive pigment cells which happened to be in the vicinity of the ganglia at the time of explantation, as suggested earlier, but from 'pluripotent' cells which would become ganglion neurons or accessory cells under normal conditions. Also, 4 day spinal ganglia cultured in a fold of vitelline membrane with 8 to 9 day mesonephros develop recognizable neurons after a week *in vitro;* the ganglia remain rounded masses and pigment cells have not been observed. Though the limited nature of these observations makes them inconclusive, they are consistent with the idea that dispersion of cells favors initial stages of melanocyte differentiation, and closer cell-to-cell contacts favors initial neuronal differentiation.

Studies of neural crest give some further indications of the same sort, though the picture is complicated by the possibility that inductive interactions may be involved. The explants in these studies consist of neural tube, or dorsal half of the neural tube, excised just before the neural crest cells have begun their migration. *In vitro* the neural crest cells migrate out from the neural tube, which can then be removed from the culture if desired, and the development of neural crest cells followed. In petri-type tissue culture dishes (on collagen), neurons have been seen only after the culture becomes overgrown; otherwise, pigment cells are the only identifiable cell type observed (Cohen and Konigsberg, 1975). In organ cultures on millipore filters, early sympathetic neurons (identified by formaldehyde-induced fluorescence) are found only when the culture contains somites contiguous to neural crest cells, and ventral cord, which may be on the opposite side of the filter (Norr, 1973). The latter observations suggest an inductive type of interaction, especially since similar observations have been made in chorio-allantoic grafts (Cohen, 1972).

Such results, and the difficulties met by various workers in attempting to use culture methods suitable for partially differentiated neural cells with earlier material, indicate that requirements for initial stages of cell differentiation differ from those for axon regeneration, cytological maturation, maintenance, i.e., later stages of differentiation. However, in some types of cultures, dissociated cells from early central nervous system differentiate very nicely. Kim and Wenger (1973), placing 1×10^4 to 4×10^4 cells from spinal cords of 2½ to 3 day chick embryos in a drop of medium on collagen-coated coverslips in Maximow assemblies, observed bipolar and primitive multipolar neurons after 3 to 5 days *in vitro*, and rather mature multipolar neurons with central nuclei beginning at about 3 weeks. Perhaps a more precisely controlled microenvironment is required by early neural cells, and is provided by 'conditioning' of the immediate areas in organ cultures or of a relatively small volume of medium in Maximow double coverslip cultures, or some more complex 'factor' is maintained in sufficiently high concentrations. This may very well be an area where *in vitro* studies could contribute much to our understanding of conditions necessary for various phases of neural differentiation, and ultimately to mechanisms of progressive sequences of control of cellular activities in development.

Conclusions

Nervous tissue explanted before neuronal and glial differentiation are apparent, or at partially differentiated stages, continues to develop in organ culture. Neurons may become quite mature cytologically, and form abundant synapses. The various types of glial cells also differentiate in a way comparable to their development *in vivo*. Relations among cell types result in a general pattern of tissue architecture characteristic of the central or peripheral nervous system. Specific patterns of organization, particularly of neurons, resemble those *in vivo* in some instances; in other cases they are more or less distorted by conditions in culture.

These points are illustrated by cultures of spinal cord using various methods and beginning with explants of early embryonic through fetal stages. The same sorts of observations have been made of explants from a number of areas of the central nervous system and from peripheral ganglia.

The degree to which specific patterns of organization resemble those in the normal animal varies considerably, even among cultures of one part of the nervous system, and appears to be the result of numerous factors affecting the development *in vitro*. Initial phases of differentiation may be more demanding in requirements than later stages. Closer comparisons of cultures to one another and to *in vivo* development may provide a way to explore factors controlling various aspects and steps of neural development.

Acknowledgments

Work of the author has been supported in part by an NIH Biomedical Research Support Program grant and a PSC-BHE Research Award (11407) from the Research Foundation of the City University of New York.

I wish to acknowledge also the collaboration of Elizabeth Barnes, Donald Burmeister, Joanne Carroll, Helene Jaffa, Margaret Pieczonka, Margaret Sulivan, Geraldin Van Dorn and Gail Zeevalk.

References

Barron, D. H. 1946. Observations on the early differentiation of the motor neuroblasts in the spinal cord of the chick. J. Comp. Neurol 85:149-169.

Benitez, H. H., E. B. Masurovsky and M. R. Murray. 1974. Interneurons of the sympathetic ganglia in organotypic culture. A suggestion as to their function, based on three types of study. J. Neurocytol. 3:363-384.

Boulder Committee. 1970. Embryonic vertebrate central nervous system: Revised terminology. Anat. Rec. 166:257-262.

Bunge, M. B., R. P. Bunge, E. R. Peterson and M. R. Murray. 1967. A light and electron microscope study of long-term organized cultures of rat dorsal root ganglia. J. Cell Biol. 32:439-466.

Bunge, R. P. and P. Wood. 1973. Studies on the transplantation of spinal cord tissue in the rat. I. The development of a culture system for hemisections of embryonic spinal cord. Brain Res. 57:261-276.

Bunge, R. P., M. B. Bunge and E. R. Peterson. 1965. An electron microscope study of cultured rat spinal cord. J. Cell Biol. 24:164-191.

Chamley, J. H., G. E. Mark and G. Burnstock. 1972a. Sympathetic ganglia in culture. II. Accessory cells. Z. Zellforsch. 135:315-327.

Chamley, J. H., G. E. Mark, G. R. Campbell and G. Burnstock. 1972b. Sympathetic ganglia in culture. I. Neurons. Z. Zellforsch. 135:287-314.

Cohen, A. M. 1972. Factors directing the expression of sympathetic nerve traits in cells of neural crest origin. J. Exp. Zool. 179:167-182.

Cohen, A. M. and I. R. Konigsberg. 1975. A clonal approach to the problem of neural crest determination. Develop. Biol. 46:262-280.

Cowell, L. A. and J. A. Weston. 1970. An analysis of melanogenesis in cultured chick embryo spinal ganglia. Develop. Biol. 22:670-697.

Fedoroff, S. 1967. Proposed usage of animal tissue culture terms. J. Nat. Cancer Inst. 38:607-611.

Fisher, K. R. S., S. Fedoroff and E. L. Wenger. 1975. Effect of osmotic pressure on neurogenesis in cultures of chick embryo spinal cords. In Vitro. 11:329-337.

Fujita, S. 1963. The matrix cell and cytogenesis in the developing central nervous system. J. Comp. Neurol. 120:37-42.

Guillery, R. W., H. M. Sobkowicz and G. L. Scott. 1968. Light and electron microscopical observations of the ventral horn and ventral root in long term cultures of the spinal cord of the fetal mouse. J. Comp. Neurol. 134:433-476.

Hamburger, V. 1948. The mitotic patterns in the spinal cord of the chick embryo and their relation to histogenetic processes. J. Comp. Neur. 88:221-283.

Hamburger, V. and H. L. Hamilton. 1951. A series of normal stages in the development of the chick embryo. J. Morph. 88:49-92.

Harrison, R. G. 1907. Observations on the living developing nerve fiber. Proc. Soc. Biol. Med. 4:140-143.

Harrison, R. G. 1910. The outgrowth of the nerve fiber as a mode of protoplasmic movement. J. Exp. Zool. 9:787-846.

Kim, S. U. and E. L. Wenger. 1973. Morphological and enzyme histochemical studies of dissociated chick neural tube cultured *in vitro*. J. Neurobiol. 4:513-523.

Langman, J. and C. C. Haden. 1970. Formation and migration of neuroblasts in the spinal cord of the chick embryo. J. Comp. Neurol. 138:419-432.

Lyser, K. M. 1964. Early differentiation of motor neuroblasts in the chick embryo as studied by electron microscopy. I. General aspects. Develop. Biol. 10:433-466.

Lyser, K. M. 1966. Difféenciation du tube neural de l'embryon de poulet en culture organotypique. Arch. Anat. Microscop. Morph. Exp. 55:37-54.

Lyser, K. M. 1968. Early differentiation of motor neuroblasts in the chick embryo as studied by electron microscopy. II. Microtubules and neurofilaments. Develop. Biol. 17:117-142.

Lyser, K. M. 1971a. Early differentiation of the chick embryo spinal cord in organ culture: Light and electron microscopy. Anat. Rec. 169:45-64.

Lyser, K. M. 1971b. Further differentiation of early neural tube explants in organ culture as studied by electron microscopy. Wilhelm Roux' Arch. 168:269-281.

Lyser, K. M. 1972. The differentiation of glial cells and the glia limitans in organ cultures of chick spinal cord. In Vitro. 8:77-84.

Lyser, K. M. 1976. Organotypic culture of early embryonic nervous system. In: *Tissue Culture Association Manual.* V. J. Evans, V. P. Perry and M. M. Vincent, eds. Vol. I:121-126.

Mains, R. E. and P. H. Patterson. 1973. Primary cultures of dissociated sympathetic neurons. I. Establishment of long-term growth in culture and studies of differentiated properties. J. Cell Biol. 59:329-345.

Martin, A. and J. Langman. 1965. The development of the spinal cord examined by autoradiography. J. Embryol. Exp. Morph. 14:25-35.

McConnell, J. A. and J. W. Sechrist. 1976. A comparative radiographic study of early neuron origin in the mouse and chick. Neurosci. Abst. 2:201.

Murray, M. R. 1965. Nervous tissues *in vitro.* In: *Cells and Tissues in Culture: Methods, Biology and Physiology.* E. N. Willmer, ed. Academic Press, London and N. Y. Vol. 2, pp. 373-455.

Murrray, M. R. 1971. Nervous tissues isolated in culture. In: *Handbook of Neurochemistry.* A. Lajtha, ed. Plenum Press, N. Y. Vol. V (Pt. A) pp. 373-438.

Norr, S. C. 1973. *In vitro* analysis of sympathetic neuron differentiation from chick neural crest cells. Develop. Biol. 34:16-38.

Oppenheim, R. W., I. W. Chu-wang and R. F. Foelix. 1975. Some aspects of synaptogenesis in the spinal cord of the chick embryo: A quantitative electron microscopic study. J. Comp. Neur. 161:383-418.

Peterson, E. R., S. M. Crain and M. R. Murray. 1965. Differentiation and prolonged maintenance of bioelectrically active spinal cord cultures (rat, chick and human). Z. Zellforsch. 66:130-154.

Peterson, E. R. and M. R. Murray. 1955. Myelin sheath formation in cultures of avian spinal ganglia. Am. J. Anat. 96:319-355.

Smith, B. H. and G. W. Kreutzberg. 1976. Neuron-target cell interactions. Neuroscience Res. Prog. Bull. 14:211-453.

Sobkowicz, H. M., R. W. Guillery and M. B. Bornstein. 1968. Neuronal organization in long-term cultures of the spinal cord of the fetal mouse. J. Comp. Neur. 132:365-396.

Tello, J. F. 1923. Les differenciations neuronales dans l'embryon du poulet, pendant les premiers jours de l'incubation. Trabajos Lab. Invest. Biol. (Madrid) 21:1-93.

Tennyson, V. M. 1970. The fine structure of the developing nervous system. In: *Developmental Neurobiology.* W. A. Himwich, ed. Charles C. Thomas,Springfield, Illinois. pp. 47-116.

Vaughn, J. E. and J. A. Grieshaber. 1973. A morphological investigation of an early reflex pathway in developing rat spinal cord. J. Comp. Neur. 148:177-210.

Watterson, R. L. 1965. Structure and mitotic behavior of the early neural tube. In: *Organogenesis.* R. L. DeHaan and H. Ursprung, eds. Holt, Rinehart and Winston, New York. pp. 129-159.

Windle, W. F. and R. E. Baxter. 1936. Development of reflex mechanisms in the spinal cord of albino rat embryos. Correlations between structure and function, and comparisons with the cat and the chick. J. Comp. Neur. 63:189-209.

Wolf, M. K. 1970. Anatomy of cultured mouse cerebellum. II. Organotypic migration of granule cells demonstrated by silver impregnation of normal and mutant cultures. J. Comp. Neur. 140:281-298.

Wolff, E. 1961. Utilisation de la membrane vitelline de l'oeuf de poule en culture organotypique. I. Technique et possibilités. Develop. Biol. 3:767-786.

DIFFERENTIATION OF CELLS IN PRIMARY CULTURES: MYELINATION

MURRAY B. BORNSTEIN

The Saul R. Korey Department of Neurology and the Department of Neuroscience The Albert Einstein College of Medicine of Yeshiva University Bronx, New York

Peterson and Murray (1955) were the first to report myelinogenesis in cultured tissues. This pioneering study utilized chick embryo dorsal root ganglia maintained in long-term culture in the Maximow slide assembly. The Maximow slide technique offers a number of advantages, and some disadvantages, to the investigator. Among the former is the observation that the method favors differentiation and maturation of the cultured tissues. Secondly, it permits high power light microscopic examinations of cellular events at all times during the tissue's entire *in vitro* existence. Consequently, this first report emphasized a sequence of maturational events of the interacting cellular components, the neuron and the Schwann cell, which were apparently necessary for the occurrence of myelination *in vitro:* the migration of neuronal nuclei to central locations within their perikarya, the development of Nissl substance, the increasing diameter of the axon, and the alignment of Schwann cells along and their close apposition to the developing axon. By direct observations and photographic recording all these maturational developments were demonstrated to be associated with myelinogenesis of the chick peripheral nerve tissue *in vitro*. Experimental manipulations of the chemical or physical environments of the cultures and their effects on the program of inter-related differentiation were demonstrated to be sufficient to interfere with myelinogenesis (Peterson and Murray, 1960). In those days, it was important to establish the phenomena as observed in the living unstained tissues as representing the formation of myelin sheaths and, at the same time, the faithfulness of the *in vitro* model system to the processes as they were known to occur *in vivo*. This was proven by a number of techniques available; namely, examination of the living sheaths for their characteristic birefrigence under polarized light and the staining characteristics of the fixed cultures. Most recent electron microscopic observations of the cultured PNS fragments (Bunge *et al.*, 1967) have fully confirmed the original conclusion that the observed phenomena did, in fact, represent myelinogenesis *in vitro*.

Hild (1957) established the second landmark in the chronicle of *in vitro* myelinogenesis by demonstrating its development in CNS tissue as represented by fragments of newborn kitten cerebellum. This technique, as practiced in Pomerat's laboratory, was somewhat different from the Maximow slide assembly. A "flying" coverslip to which the fragments were attached, was maintained in a roller tube. This method did not permit sequential observations of individual cultures. When, after a few weeks, the coverslips were removed from the tube and viewed with a light microscope, the fragments were noted to have myelinated. Possibly as a consequence of the limited accessibility of these cultures for observation, and the apparent lack of structural features which might characterize the involved cells as oligodendroglia, Hild suggested that, in these cultures, the myelin sheath might be the product of the axon proper. In view of the established oligodendroglial - axonal relationship in myelinogenesis in *vivo,* this was unfortunate since it brought into question the faithfulness of the *in vitro* model system. However, this impression soon yielded to the results obtained in our extending Hild's use of the newborn cerebellum to similar tissues from kittens and rats maintained in the Maximow slide assembly (Bornstein and Murray, 1958). The further examination of the myelinated cerebellar cultures by means of electron microscopy (Ross *et al.*, 1962) established beyond doubt that myelinogenesis in cultured CNS tissue involves the participation of a neuroglial (oligodendroglial) cell contributing its external membrane to the formation of the multi-lamellated myelin sheath.

At the time of these studies, two theories were being proposed concerning the intimate details of myelinogenesis in the CNS. Most investigators subscribed to the hypothesis that the oligodendroglial cell in some (still) unexplained fashion wrapped its processes around the axon in a spiral fashion similar to that proposed by Geren (1956) for the Schwann cell and the peripheral axon, essentially an extracytoplasmic process. However, De Robertis *et al.* (1958), for example, presented the possibility that vesicles within the glia would approach the axon, and fuse with each other to form the lamella of myelin, an intracytoplasmic process. This idea had already been proposed by Luse (1956) in her electron microscopic studies of rodent brain and spinal cord. The ultrastructural examinations of cultures chosen by direct observation during the sequence of *in vitro* myelination (Ross *et al.*, 1962) unequivocally supported the former theory that myelin is composed of modified neuroglial cell membrane.

Another suggestion arising from the early light and electron microscopic observations of myelinating cerebellar cultures was that more than one cell process and, by extension, more than one oligodendroglial cell might contribute to the multilamellated structure along any particular segment of the sheath. Maturana (1960), Peters (1960) and Bunge *et al.*, (1961), however, favored a unitary theory, that a single oligodendroglial cell is responsible for the myelin at any particular internode. This question is still open and the possibility of multicellular involvement may be of some consequence in the light of our present knowledge of CNS remyelination *in vitro* and *in vivo* (Bunge *et al.*, 1961; Bornstein, 1963).

The possession of an *in vitro* model invites biochemical analyses of both substrate and enzyme components as they may relate to the process of myelination. The relatively small amount of tissue available, however, acted as a deterrent to these studies for some time. However, Lehrer and his co-workers successfully applied the microchemical techniques developed by Lowry to the cultured material (Lehrer *et al.*, 1970 a, b). Detailed analyses of the enzymes associated with carbohydrate metabolism (hexokinase, lactic dehydrogenase, and malic dehydrogenase) lent further support to the faithful character of the model system, in that, for the most part, the enzyme activities during the periods of maturation and differentiation resembled those found to occur *in vivo* at similar times. Differences, however, were found in the activities of β-glucuronidase and glucose-6-phosphate dehydrogenase. The greatest rate of change was noted to occur between days 8 and 15, i.e., during the onset of myelinogenesis.

A number of interesting insights concerning the biology of myelination have been generated by a series of experiments designed to explore the immunopathology of the demyelinating disorders, multiple sclerosis (MS) and experimental allergic encephalomyelitis (EAE), by employing the organotypic myelinated mammalian CNS cultures (Bornstein and Appel, 1961).

One of the first observations was that CNS tissue, which had been demyelinated by exposure to anti-myelin antiserum in the presence of complement, possessed the capacity to remyelinate when removed from the influence of the abnormal serum (Bornstein, 1963; Raine and Bornstein, 1970). These observations, which were based on the direct observation and photographic recording of particular axons of the stages of myelination, demyelination and remyelination, confirmed the report of Bunge *et al.* (1961) that remyelination in the adult cat spinal cord followed a traumatically induced demyelination (CSF barbotage). Many confirmatory reports have appeared since then, demonstrating remyelination in the CNS of patients with MS (Feigin and Popoff, 1966), and animals with EAE (Lampert, 1965).

The immunological inhibition of myelination is another useful laboratory procedure for the *in vitro* analysis of the biology of myelin. It had first been demonstrated that the continued presence of the anti-CNS serum would prevent remyelination. As an extension of those observations, an experiment was carried out to examine the effects of antiserum on primary myelin formation (Bornstein and Raine, 1970). Low concentrations of antiserum plus complement were introduced into the nutrient medium of fragments of 13-14 day old mouse embryo spinal cord, rodent CNS, at the time of explantation and continuously maintained thereafter. Whereas the cultures maintained in their normal nutrient environment were observed to myelinate by about the tenth day *in vitro,* those exposed to the influence of the antiserum showed no evidence of myelin formation for periods of time ranging up to 60 days. Light and electron microscopic examinations of the inhibited cultures demonstrated a total absence of differentiated oligodendroglia and no evidence of myelin sheaths. Neuronal perikarya, axons, synapses and astrocytes developed normally, apparently unaffected by the presence of the EAE serum and complement. When the inhibited cultures were washed free of the antiserum and returned to their normal nutrient solution, oligodendroglia promptly differentiated and myelin sheaths were formed around the axons. Thus, in this combination of immunological and tissue culture techniques, we are in possession of a system by which myelinogenesis can be turned off and on, or inhibited and disinhibited, at will. The inhibition - disinhibition model has since been employed to include antisera against cerebroside which have also been found to inhibit myelinogenesis (Fry *et al.,* 1974). In other studies, biochemical analyses of the inhibited state have been performed. Thus, the incorporation of sulfate into cerebroside sulfate does not proceed in the inhibited state, but promptly occurs when the tissue is disinhibited (Fry *et al.,* 1972). Another series of analyses revealed the development of cyclic 2′3′ nucleotide 3′ phosphohydrolase (CNP) to be abolished in inhibited cultures whereas a number of enzymes related to carbohydrate metabolism were unaffected (Fry *et al.,* 1973). Again, the cultures promptly produced CNP when removed from the presence of the antiserum. The capacity to permit a burst of myelinogenesis in culture tissues at will is finding other interesting applications. Other biochemical analyses are currently being applied to the inhibited and disinhibited cultures in attempts to characterize details of cerebroside synthesis during myelination.

Further applications of the tissue culture-immunological techniques contain some promise of insights into the intimate structure of myelin. The studies began with the observation that increased birefrigence of myelin sheaths preceded the whole CNS antiserum-induced pattern of demyelination in investigations on experimental allergic encephalomyelitis (EAE) (Bornstein and Appel, 1961). The phenomenon was transient, however, being rapidly followed by the breakdown of the myelin. Recently, we (Bornstein and Raine, 1976) noted that the antiserum without complement could induce the increased refringence in cultured CNS but that demyelination did not ensue. All myelin sheaths in the exposed cultures could be brought to and maintained at this stage. Electron microscopic examinations revealed that the width of the myelin lamellae had been doubled, explaining the apparent increase in birefringence noted in the light microscope. In addition, the intraperiod line had disappeared, its usual double linear image having been replaced by a quadruplet. In its early stages, the swelling of the myelin sheaths was reversible.

This reproducible alteration in the ubiquitous pattern of the myelin sheath by immunological means presents a model by which one may be able to analyse its physico-chemical relationships. In view of the significant involvement of myelin basic protein (MBP) in the induction of EAE and its occurrence in high concentration in myelin, it is possible that the observed changes in myelin may in some way result from a modification in the quantity or the quality of MBP. These possibilities are now under investigation.

Another observation in these last experiments may lead to further analyses in culture of the oligodendroglial-neuronal relationship during myelinogenesis. At this period of development, exposure to the antiserum without complement induces a bizarre transformation of oligodendroglia. They appear activated to produce a large number of cell processes which surround the oligodendrocyte. Where the aberrant processes chance to meet, they form abnormal, swollen lamellae. If the heat inactivated antiserum without complement is introduced into the nutrient medium at the time of explantation and continued during the period of myelination, similar but more pronounced, aberrant oligodendroglial membranes are produced and wander aimlessly around the oligodendrocytes (Diaz *et al.*, 1977). Very few of these processes appear capable of finding and encircling axons to form myelin. Therefore, one might suggest that the axon sends at least two messages to the oligodendroglia: first, "make myelin" and, second, "come and find me". The first appears to operate. The cells differentiate and produce vast numbers of cell processes. The second message is interfered with in some manner, for, they are incapable for the most part of contacting and surrounding axons.

In these experiments, the tissue culture model system continues as a valuable tool for the analysis and understanding of myelinogenesis.

Acknowledgments

Investigations in the author's laboratory were supported by grants NS 06735 and NS 11920 and a grant from the Alfred P. Sloan Foundation.

References

Bornstein, M. B. 1963. A tissue culture approach to demyelinative disorders. N.C.I. Monogr. No. 11:197–214.

Bornstein, M. B. and S. H. Appel. 1961. The application of tissue culture to the study of experimental allergic encephalomyelitis. I. Patterns of demyelination. J. Neuropath. Exp. Neurol. 20:141–157.

Bornstein, M. B. and M. R. Murray. 1958. Serial observations on patterns of growth, myelin formation, maintenance and degeneration in cultures of newborn rat and kitten cerebellum. J. Biophys. Biochem. Cytol. 4:499-504.

Bornstein, M. B. and C. S. Raine. 1970. Experimental allergic encephalomyelitis: Antiserum inhibition of myelination *in vitro*. Lab. Invest. 23:536-542.

Bornstein, M. B. and C. S. Raine. 1976. The initial structural lesion in serum-induced demyelination *in vitro*. Lab. Invest. 35:391-401.

Bunge, M. B., R. P. Bunge and H. Ris. 1961. Ultrastructural study of remyelination in an experimental lesion in adult cat spinal cord. J. Biophys. Biochem. Cytol. 10:61-94.

Bunge, M. B., R. P. Bunge, E. R. Peterson and M. R. Murray. 1967. A light and electron microscope study of long-term organized cultures of rat dorsal root ganglia. J. Cell Biol. 32:439-466.

Diaz, M., M. B. Bornstein and C. S. Raine. 1977. Dissociation of myelinogenesis in tissue culture by anti-CNS antiserum. J. Neuropath. Exp. Neurol. Abstract. 36:594.

Feigin, I. and N. Popoff. 1966. Regeneration of myelin in multiple sclerosis. The role of mesenchymal cells in such regeneration and in myelin formation in the peripheral nervous system. Neurology 16:364.

Fry, J. M., G. M. Lehrer and M. B. Bornstein. 1972. Sulfatide synthesis in CNS tissue culture and its inhibition by experimental allergic encephalomyelitis (EAE) serum. Science 175:192-193.

Fry, J. M., G. M. Lehrer and M. B. Bornstein. 1973. Experimental inhibition of myelination in spinal cord tissue cultures: Enzyme assays. J. Neurobiol. 4:453-459.

Fry, J. M., S. Weissbarth, G. M. Lehrer and M. B. Bornstein. 1974. Cerebroside antibody inhibits sulfatide synthesis and myelination and demyelinates in cord tissue cultures. Science 183:1540-542.

Geren, B. B. Structural studies of the formation of the myelin sheaths in peripheral nerves. In: *Cellular Mechanisms in Differentiation and Growth.* D. Rudnick (ed.) Princeton Univ. Press, Princeton, 1956. pp. 213-220.

Hild, W. 1957. Myelogenesis in cultures of mammalian central nervous tissue. Z. Zellforsch. 46:71.

Lampert, P. W. 1965. Demyelination and remyelination in experimental allergic encephalomyelitis. Further electron microscopic observations. J. Neuropath Exp. Neurol. 24:371.

Lehrer, G. M., M. B. Bornstein, C. Weiss, M. Furman and C. Lichtman. 1970a. Enzymes of carbohydrate metabolism in the rat cerebellum developing *in situ* and *in vitro*. Exp. Neurol. 27:410-425.

Lehrer, G. M., M. B. Bornstein, C. Weiss and D. J. Silides. 1970b. Enzymatic maturation of mouse cerebral neocortex *in vitro* and *in situ*. Exp. Neurol. 26:595-606.

Luse, S. A. 1956. Formation of myelin in the central nervous system of mice and rats as studied with the electron microscope. J. Biophys. Biochem. Cytol. 2:777.

Maturana, H. C. 1960. The fine anatomy of the optic nerve of anurans - an EM study. J. Biophys. Biochem. Cytol. 2:107.

Peters, A. 1960. The formation and structure of myelin sheaths in the central nervous system. J. Biophys. Biochem. Cytol. 8:431.

Peterson, E. R. and M. R. Murray. 1955. Myelin sheath formation in cultures of avian spinal ganglia. Am. J. Anat. 96:319-356.

Peterson, E. R. and M. R. Murray. 1960. Modification of development in isolated dorsal root ganglia by nutritional and physical factors. Develop. Biol. 2:461-476.

Raine, C. S. and M. B. Bornstein. 1970. Experimental allergic encephalomyelitis. A light and electron microscope study of remyelination and "sclerosis" *in vitro*. J. Neuropath. Exp. Neurol. 29:552-574.

de Robertis, E. H., E. H. Gerschenfeld and F. Wald. 1958. Cellular mechanisms of myelination in the central nervous system. J. Biophys. Biochem. Cytol. 4:651-658.

Ross, L. L., M. B. Bornstein and G. M. Lehrer. 1962. Electron microscopic observations on myelin formation in tissue cultures of developing rat cerebellum. J. Cell Biol. 14:19-30.

SYNAPSE FORMATION IN NEURAL TISSUE CULTURES

STANLEY M. CRAIN

Departments of Neuroscience and Physiology
Albert Einstein College of Medicine
New York, New York

I. Introduction

Techniques are now available for preparation of neuronal arrays in cultures of varying complexity, ranging from intact explants containing a few thousand closely packed neurons and glial cells to monolayers of completely separated neurons. Synaptogenesis can occur in all of these types of cultures derived from embryonic CNS tissues, but intact explants have so far been more reliable as model systems for studying mature synaptic networks with reproducible organotypic bioelectric activities and characteristic pharmacologic sensitivities (Crain, 1976). Emphasis has therefore been placed on analyses of mechanisms underlying development of the functional integrity of CNS explants as a foundation for systematic correlative cytologic and electrophysiologic studies of various types of simpler dissociated neuronal arrays. The latter preparations facilitate direct microscopic observation, microelectrode manipulations, and other biophysical analyses (Fischbach and Dichter, 1974; Godfrey *et al.*, 1975; Nelson, 1975, 1976; Peacock *et al.*, 1973; Ransom and Nelson, 1975; see also Section IIC2. and Nelson, this volume). However, studies of some of the uniquely interesting cerebral functions related to memory and learning, important trophic interactions of CNS cells, and mechanisms underlying rhythmic behavior patterns may require more complex cellular assemblies in culture.

Pharmacologic analyses of the complex bioelectric activities of intact CNS explants provide, moreover, a powerful tool for linking selective neuropharmacologic effects in the whole organism and direct molecular actions on individual cells. This review therefore includes recent evidence demonstrating the remarkable degree to which characteristic pharmacologic sensitivities can develop and be maintained in identified regions within mammalian CNS explants in relation to their capacity to generate specific synaptically-mediated organotypic bioelectric network discharges (see Section IIB and III).

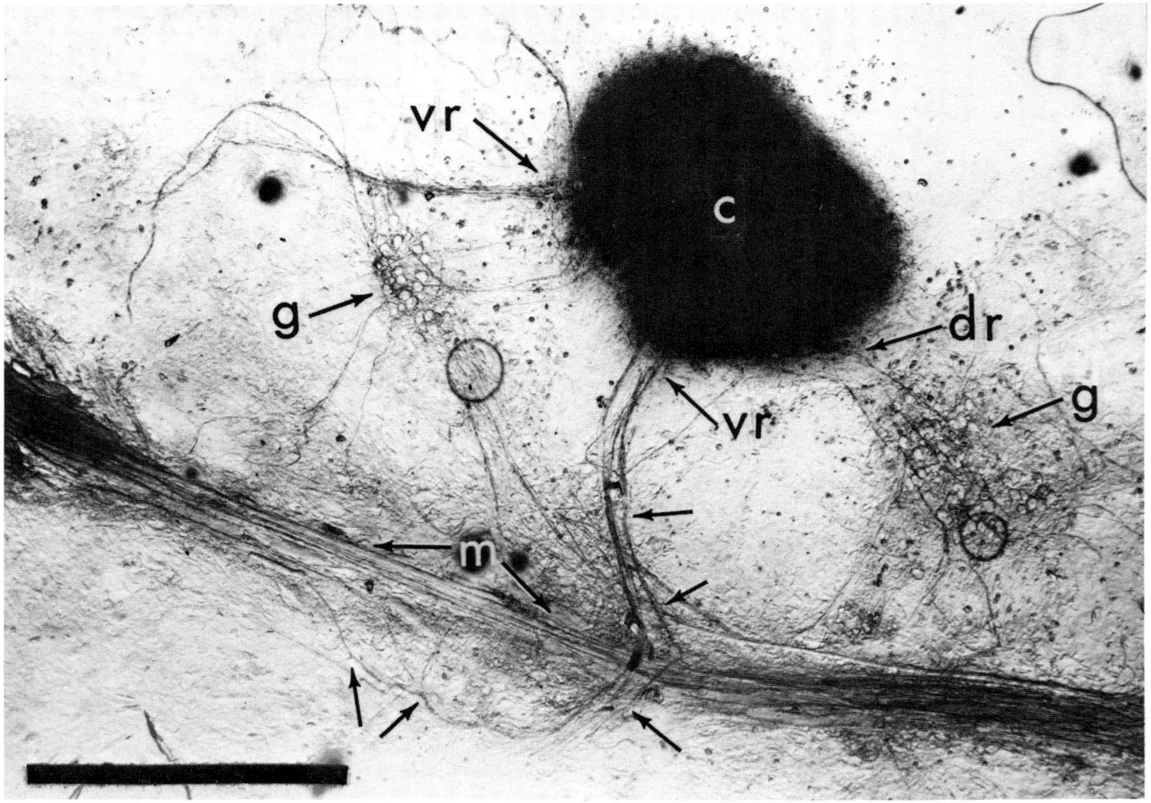
vr
c
g
dr
vr
g
m

II. Synaptogenesis in Embryonic Spinal Cord Explants

Electrophysiologic studies of explants of embryonic spinal cord (rat, chick, and human) provided the first clear-cut demonstration that "nerve cells in cultures may maintain, for months *in vitro*, not only the capacity to propagate impulses along their neurites but also a remarkable degree of functional organization resembling the activity of synaptic networks of the central nervous system" (Crain and Peterson, 1963, 1964).

A. Internuncial Neuronal Networks

Cross sections (0.5-1 mm thick) of spinal cord from 13- to 14-day rat and mouse fetuses and from smaller groups of human (6-week) and chick (9-day) embryos have been explanted for various developmental neurobiology studies [see Peterson *et al.* (1965) for details of the spinal cord culture technique and morphologic development; also reviews in Crain, 1976 and Bornstein, this volume]. The cord segment is frequently explanted together with its attached dorsal root ganglia (DRGs) and associated meningeal tissues (Figs. 1, 6). A large number of cord neurons survive and mature *in vitro*. Many of the axons become myelinated in both the explant and the outgrowth zones (Fig. 1). Although the neuronal somas generally remain thickly invested in a framework of neurites and glia, the explants may become thin enough (100-300 μm) to permit at least some of the perikarya to be visible in the living state at high magnification (Peterson *et al.*, 1965). A dense neuropil (*ca.* 100 μm thick) containing fine arborizing neurites (and glial cells) often covers the main body of the explant (Bunge *et al.*, 1965) and extends irregularly outward onto the collagen gel as the clearly visible "neuritic growth zone" (see also Guillery *et al.*, 1968).

Cultures selected after serial microscopic examinations during days or weeks of incubation in Maximow slides are transferred to a micrurgical chamber of an inverted microscope for bioelectric study. Openings in the chamber walls permit manipulation of microelectrodes under direct visual control at high magnification for focal recording and stimulation within an explant (Crain, 1973). Extracellular microelectrode recordings are made

FIG. 1. Long-term culture of fetal mouse spinal cord explant, with attached dorsal root ganglia, after innervation of mouse skeletal muscle (2 months *in vitro;* muscle added to 4-day cord culture). Low-power view of living culture. The spinal cord cross section (c) shows distinct dorsal (dr) and ventral root (vr) fibers emerging from the explant. Note well-organized, myelinated motor nerve fascicle (arrows) penetrating the central region of the long muscle strip. Muscle fibers (m) in central region have regenerated *in vitro* after contact with cord neuritic outgrowth (residual "parent" muscle fibers are present toward the ends of the strip). Note the clusters of dorsal root ganglion perikarya (g) and their myelinated peripheral branches arborizing diffusely throughout the area. (Dorsal root ganglion on left is located in aberrant position due to slight asymmetry of original hand-sliced cord cross section.) Scale: 1 mm. (From Crain and Peterson, 1974a.)

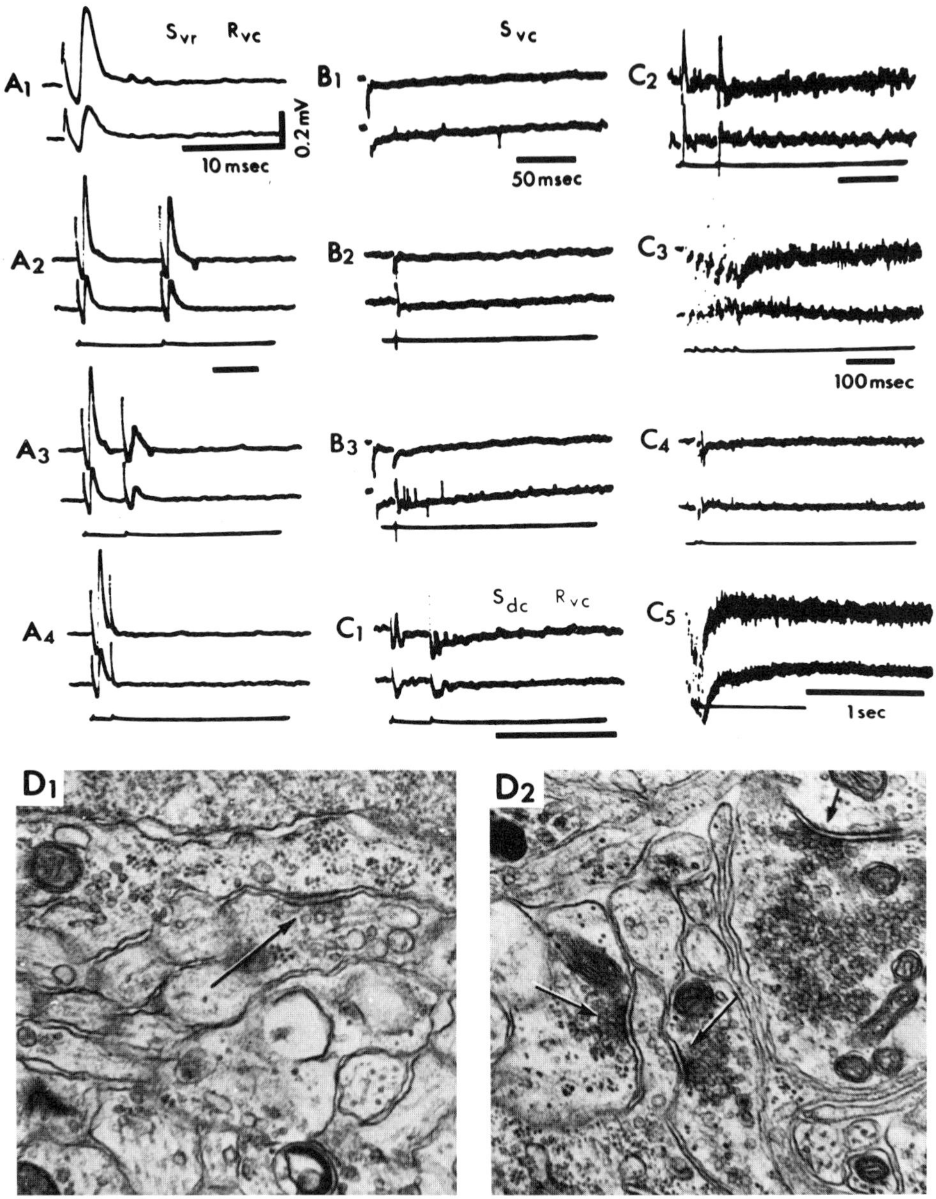
A1
A2
A3
A4
Svr
Rvc
0.2 mV
10 msec
B1
B2
B3
Svc
50 msec
C1
Sdc
Rvc
C2
C3
C4
C5
100 msec
1 sec
D1
D2

with chloridized silver electrodes via saline-filled pipettes with tip diameters of 1-5 μm, and electric stimuli (0.1-0.5 ms duration and up to 50 μA in strength) are applied through pairs of similar pipettes with 10-μm tips. A large chloridized silver wire in the periphery of the culture bath serves as a ground electrode. Bioelectric signals are recorded with high-input impedance preamplifiers and a 4-channel oscilloscope (passband generally from 0.2 Hz to 10 kHz). RF stimulus-isolation circuits are used to minimize shock artifacts. Intracellular recordings are made via potassium citrate-filled pipettes with tips of about 0.1 μm. Miniaturized magnetically coupled micromanipulators now permit the incorporation of an entire array of microelectrodes and micromanipulators within a *sealed* chamber, using magnetic controls *external* to the chamber for positioning each electrode in the tissue. This sealed chamber array facilitates far more rigorous control of the physicochemical environment of the culture, and sterilized assemblies are being developed for longitudinal bioelectric recordings over days or weeks, instead of the usual limit of less than 1 day (Crain, 1973, 1976).

During the first 2 days after explantation of 13- to 14-day fetal rodent spinal cord (Fig. 2A) at essentially "presynaptic" stages (see below), generally only simple spike potentials are detected in response to single or tetanic electric stimuli applied to these immature cord explants. The spikes

FIG. 2: Transition from simple to complex bioelectric responses in spinal cord tissue cultured for 2-4 days after explantation from a 14-day fetal rat. A and B: Two days *in vitro*. A_1: Simultaneous extracellular recordings of simple spike potentials evoked in two sites of ventral cord (300 μm apart) by a stimulus applied to ventral root (400 μm from edge of explant). Application of paired stimuli demonstrates relative ($A_{2,3}$) and absolute (A_4) refractoriness to test stimulus. (Responses to dorsal cord, or root, stimulus were much smaller in amplitude, in both dorsal and ventral cord.) $B_{1,2}$: Similar simple spike potentials evoked in another region of ventral cord by stimulus applied nearby, within the cord explant. B_3: Paired stimuli, at a 15-msec test interval, elicit a multiple spike burst of small amplitude lasting more than 35 msec (barely detectable in most explants at 2 days *in vitro*, even with tetanic stimuli and strychnine). C: Four days *in vitro*. C_1: Longer-lasting afterdischarge (small amplitude) in ventral cord evoked by paired stimuli to dorsal cord. C_2: Strychnine (10 μg/ml) now enhances barrage response even at test intervals of 30 msec. C_{3-5}: Brief tetanic stimulation (20/sec) produces a further increase in duration and amplitude of after-discharges. (*Note:* In these extracellular recordings and all others in subsequent figures, upward deflection indicates negativity at active recording electrode. Time and amplitude calibrations, and specification of recording and stimulating sites, apply to all succeeding records, *until otherwise noted*. Onset of stimuli is indicated, where necessary, by a sharp pulse (or arrow) or a break in the baseline of the third sweep.) All recordings were made in Simms' BSS unless otherwise noted. D_1: Electron micrograph of cultured spinal cord tissue approximately 4 days after explantation as a "synapse-free" slice from a 14.5-day fetal rat. Note the typical, newly formed synapse (at arrow) with moderate synaptic membrane density and intervening cleft substance. A few synaptic vesicles are clustered on the presynaptic side of this axosomatic junction (near arrowhead). D_2: Similarly prepared electron micrograph of fetal rat cord explant after 10 weeks *in vitro*. Synaptic membrane density and cleft substance are more pronounced at the axodendritic (black arrow) and the two axosomatic (white arrows) synapses. Many more synaptic vesicles have accumulated at the presynaptic side of all three junctions. (A-C from Crain and Peterson, 1967. D from Bunge *et al.*, 1965, 1967a.)

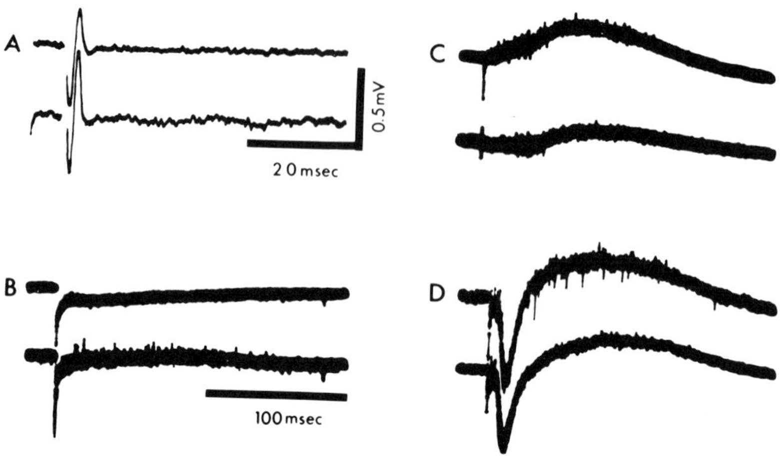

FIG. 3. Onset of complex synaptic network functions in fetal mouse spinal cord explants (13–14 days *in utero;* 1–3 days *in vitro*). A: Simple early-latency spike potentials evoked in two sites of cord explant (at 1 day *in vitro*) by a single stimulus applied to the third site. No signs of complex discharges were detected following these spikes, either in BSS or after addition of pharmacologic agents (see below). B: Long-lasting repetitive spike barrages and small-amplitude slow-wave responses elicited in two sites of cord explant at 2 days *in vitro,* but only after introduction of bicuculline (10^{-5}M) and caffeine (10^{-3}M). (See also Figs. 1, 13) Note the change in time calibration; afterdischarges lasted more than 1-2 sec. C: Similar long-lasting repetitive spike barrages, but these are accompanied by much larger-amplitude negative slow-wave responses, evoked in two sites of another cord explant at 2 days *in vitro* (tissue was obtained from a slightly more mature fetus than that in B). These complex discharges were observed only after introduction of strychnine (10^{-5}M) and caffeine (10^{-3}M). D: Still more elaborate diphasic slow-wave potentials and spike barrages evoked at two sites in a cord explant at 3 days *in vitro* after addition of strychnine (10^{-5}M). Smaller-amplitude slow-wave potentials and shorter spike barrages (as in C) could also be elicited in this explant in BSS (prior to strychnine). (From Crain, 1976.)

(Fig. $2A_1$) are similar to those observed in neurons of cultured DRGs, where impulses can propagate along the conductile portions of the neurons, but no bioelectric evidence of transmission from one neuron to another has been detected. Application of paired stimuli reveals only refractory properties characteristic of electrically excitable, conductile membranes (Fig. $2A_{2\text{-}4}$). Both unit and summated spike potentials can be recorded depending on the size of the recording electrode tips and their location relative to the active neurons, as well as other features of the volume-conductor recording conditions (Crain, 1964a, 1956).

On the other hand, by 3 days *in vitro* facilitation effects can be demonstrated in the cord explants with paired stimuli at long test intervals, and long-lasting spike barrages and "slow-wave" potentials may be evoked with single or brief tetanic stimuli at critical thresholds (Fig. 2B,C). Complex discharges have also been detected in some 13- to 14-day fetal rat and mouse spinal cord explants after only 2 days *in vitro* (Figs. 3,4) (Crain and Peterson, 1967; Crain, 1976; see also Pappas *et al.*, 1975). The appearance of these complex bioelectric activities indicates that polysynaptic networks are beginning to function in the cultured cord tissue (Crain and Peterson, 1963, 1964, 1967). Although intracellularly recorded postsynaptic potentials (PSPs) have not yet been obtained from cultured CNS explants during the first week *in vitro* (see however, Section IVC), electrophysiologic analysis of extracellular recordings provides excellent, although less direct, evidence of activities mediated through synaptic pathways. The responses in 3-day and older cultures can often be enhanced with strychnine, picrotoxin, or bicuculline (*ca.* 10^{-6}M), whereas no significant effects are produced by the application of these drugs during the first 2 days *in vitro*. Furthermore, all of the complex bioelectric discharges can be rapidly blocked by raising the Mg^{2+} concentration of the medium from the usual 1 mM level to 5-10 mM whereas short-latency spike potentials can still be evoked. This selective interference with complex discharges by increased Mg^{2+} levels provides further support for synaptic mediation of these activities (Crain and Pollack, 1973; Crain, 1974a, b). In contrast, introduction of low concentrations of tetrodotoxin (*ca.* 10^{-8} g/ml), a drug that selectively blocks active Na^{+} channels in conductile membranes, rapidly abolishes all action potentials in these spinal cord (and other neural) explants, thereby also blocking all of the synaptic network discharges which require conduction through numerous internuncial axon pathways between the synapses (Crain, 1976).

The complex spike barrage and slow wave responses in cord explants are quite labile, often requiring rest periods of 1 to 10 seconds between stimuli, in contrast to the more rapid recovery of spike potentials which could be reliably evoked at repetition rates of 10-100/sec. The lability of these complex discharges and their relatively long and variable latencies provide strong additional evidence for their mediation by sequential activation through polysynaptic networks (Crain and Peterson, 1963, 1964). The lability and "fatigability" properties are even more pronounced during the first few days after onset of complex network responses, and post-discharge recovery periods may be longer than 1 minute.

The marked sensitivity of 13- to 14-day rodent spinal cord explants to strychnine, picrotoxin, and bicuculline by 3 days *in vitro* suggests, moreover, that inhibitory circuits may already be functioning quite early during CNS synaptogenesis (see Section IVC).

During the next few days of culture beyond the critical 2- to 3-day period, the afterdischarges increase in amplitude and complexity; and they can be elicited at lower thresholds with fewer stimuli (Fig. 5A-C). These

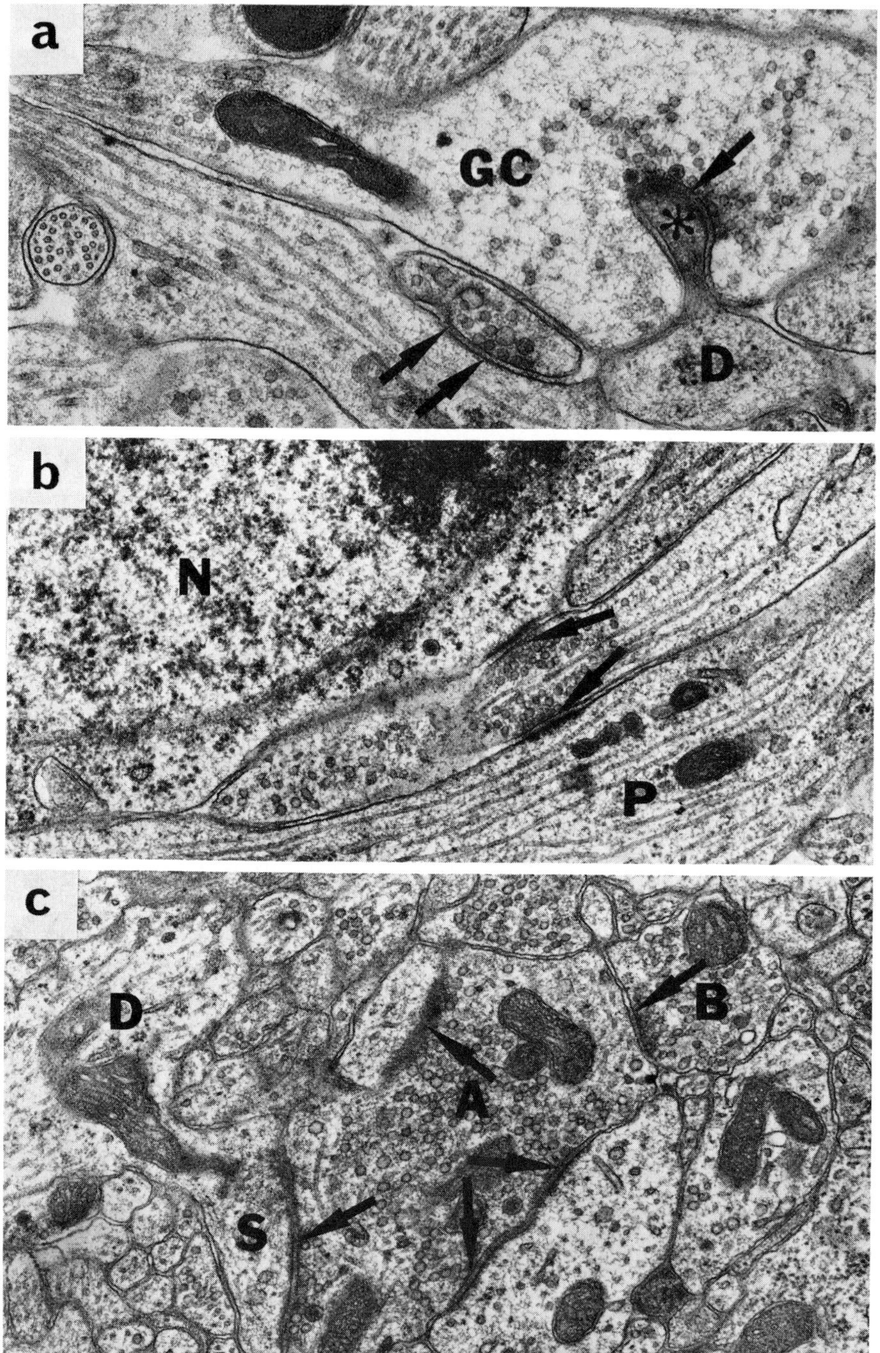
a
GC
*
D
b
N
P
c
D
B
A
S

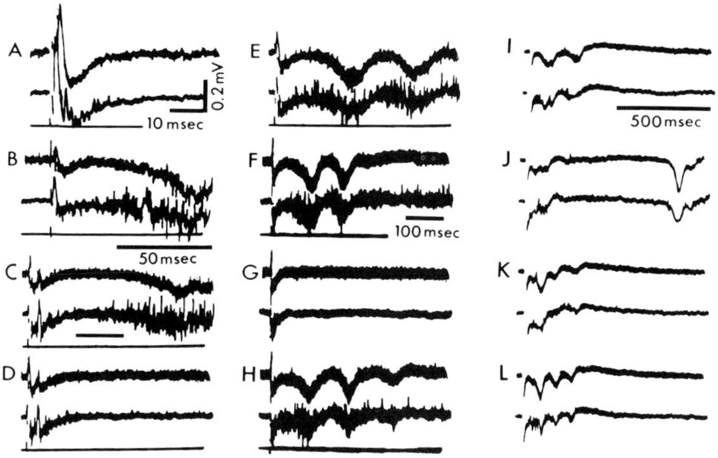

FIG. 5. Oscillatory afterdischarges evoked by single, brief, electric stimuli in a long-term culture of rat spinal cord (3 months *in vitro*). A-C: Simultaneous recordings of typical repetitive, spike-barrage responses evoked in two regions of the explant (400 μm apart) by a stimulus applied 500 μm from both recording sites. D: Short-duration response when stimulus was applied 1 sec after spike barrage in C. E,F, and H: Long-lasting oscillatory sequences (up to 400 msec) evoked shortly after records A-D. Note the regularity of the pattern of these 7-15/sec diphasic potentials and the synchronization of the responses between the two recording sites. G: Control record (subthreshold stimulus). I-L: Similar oscillatory afterdischarges at a slower sweep rate. Note the second sequence in J, which appears after a silent period of 500 msec. (From Crain and Peterson, 1964; Crain, 1966.)

FIG. 4. Electron micrographs of 14-day fetal mouse spinal cord explants during early stages of synaptogenesis and after maturation in culture. *a* and *b:* two days *in vitro* (see bioelectric network discharges in Fig. 3B,C). *a:* Portion of a growth cone (GC) which envelops and forms a synaptic junction with a filopodium (☆) arising from a presumed dendritic (D) process containing ribosomes. Another immature synaptic contact appears nearby (double arrow). Note clusters of presynaptic vesicles at both junctions. *b:* A neuritic shaft forms synaptic contact (arrows) with a cell body (nucleus N) and another process (P). *c:* Dorsal horn region of cord with attached DRGs, 6 weeks *in vitro.* A large presynaptic terminal (A) forms junctions with three neuritic elements (arrows), including a spine (S) arising from a dendritic shaft (D). This presynaptic terminal also receives synaptic input from another neurite (B, at arrow). ×26,000. (*a* and *c* from Crain, 1976; provided by E. B. Masurovsky and G. D. Pappas. *b* From Pappas *et al.*, 1975.)

long-lasting "slow-wave" and oscillatory potentials are probably extracellularly recorded field potentials representing summated excitatory and inhibitory postsynaptic potentials (EPSPs and IPSPs respectively) generated in complex sequences during activation of the synaptic networks in these cord explants. Still more elaborate repetitive discharges often occur in older rodent spinal cord explants (Fig. 1) — e.g., long-lasting rhythmic (7-15/sec) oscillatory afterdischarge sequences (Fig. 5E-L) — and these patterned repetitive sequences have also been evoked in long-term explants of *human* embryo spinal cord (Crain and Peterson, 1963; Crain, 1966). The afterdischarges appear in all-or-none fashion at a critical stimulus intensity, and they may also be evoked following facilitation of paired subthreshold stimuli at long test intervals. The oscillatory afterdischarge patterns are often relatively stereotyped, but variations in response to single stimuli of constant strength may occur. The repetitive discharges may continue uninterrupted for many seconds after the stimulus. They often occur synchronously over widespread regions of the explant, even after a weak stimulus localized to a few neurites close to a 10-μm stimulating electrode. Nevertheless, the bioelectric responses evoked in 4-day cultures of 14-day fetal rodent spinal cord tissue already include the basic organotypic properties characteristic of mature explants of cord and other CNS tissues (Crain, 1966, 1970b, 1974a).

The 2- to 3-day lag in onset of functional interneuronal transmission within these immature spinal cord tissues in culture cannot be attributed simply to depression of activity following surgical trauma at explantation, since complex, long-lasting afterdischarges can already be evoked within 1 day after explantation of 18-day fetal rat cord. It should be noted, moreover, that the 18-day cord tissue appears to undergo more degeneration during the first few days *in vitro* than does the 14-day cord, since a larger fraction of the tissue from the older fetus is composed of longitudinal fiber tracts which are transected during preparation of these 0.5- to 1-mm cord cross sections. The 2- to 3-day lag in onset of complex bioelectric activity in the 14-day fetal cord explants can therefore be considered to be due primarily to the immature state of the neuronal networks developing during the first few days in culture, rather than merely to depression of activity following explantation trauma.

The onset of complex bioelectric activity in 14-day fetal rodent spinal cord explants after 2-3 days *in vitro* correlates closely with the earliest time at which clear-cut synaptic profiles, mainly axodendritic, are regularly seen in electron micrographs of these cultured tissues (Figs. $2D_1$; 4a, b) (in rat: Bunge *et al.*, 1967a; Crain *et al.*, 1964b; in mouse: Pappas *et al.*, 1975). Furthermore, formation of functional synaptic networks by 2 days after explantation of 13- to 14-day fetal mouse cord and onset of DRG-evoked cord discharges by 3 days *in vitro* (Crain, 1976) correlates well with evidence of spinal reflex activity by 15-16 days in the fetal mouse *in utero* (Vaughn *et al.*, 1975). The marked increase in number of synapses as the cord explants mature in culture [as many as nine synaptic junctions have been found on a single neuron in a thin electron microscopic section (Bunge *et al.*, 1965)] and the concomitant developments in synaptic membrane density, cleft substances, and presynaptic vesicle accumulation (Figs. $2D_2$; 4C) (Bunge *et al.*, 1965; Pappas *et al.*, 1975; see also Guillery *et al.*, 1968) provide important morphologic parameters which probably underlie the sequential developments in bioelectric activities seen during the first week *in vitro* and thereafter. Although collateral fiber sprouts and neuritic arborizations are also developing in abundance during this period, these are not likely to constitute a critical parameter since similar elaborate neuritic processes develop in explants of DRGs (Levi and Meyer, 1938; Lumsden, 1951; Peterson and Murray, 1955) where no such complex bioelectric activities have been detected (Crain, 1956) in the absence of synaptic junctions (Bunge *et al.*, 1967b; see however, Crain, 1971).

B. Sensory Inputs from Dorsal Root Ganglia

Initial electrophysiologic studies of 13- to 14-day fetal rat cord cross-sections with attached pairs of DRGs demonstrated that synaptically mediated cord responses could be evoked by DRG stimuli by 6 days after explantation of the "presynaptic" tissues (Crain and Peterson, 1967). More recent extension of this work to 13- to 14-day fetal mouse cord-DRG cultures indicates, however, that DRG stimuli may be effective in eliciting complex cord discharges as early as 2-3 days after explantation when tested in the presence of appropriate pharmacologic agents, e.g. strychnine and bicuculline (e.g., Fig. 3) (Crain, 1976). Onset of sensory-evoked cord network responses in 13- to 14-day fetal cord DRG explants at 2-3 days *in vitro* correlates quite closely with evidence of spinal reflex activity as early as 15-16 days in the mouse fetus *in situ* (Vaughn *et al.*, 1975), as in the rat fetus (Windle and Baxter, 1935-1936; Narayanan *et al.*, 1971; see also May and Biscoe, 1975).

Although DRG-evoked cord network responses could be obtained in some explants even after months in culture, we were concerned that a substantial fraction (up to 90%) of the DRG neurons in these embryonic rodent cord-ganglion explants atrophied or degenerated during the first few days in our control culture medium (Fig. 6A, B). In earlier studies with 15- to 18-day fetal rat cord-DRG explants, on the other hand, a much larger fraction of the ganglion cells generally survived and matured in control cultures (Peterson *et al.*, 1965; Bunge *et al.*, 1967b; Sobkowicz *et al.*, 1973), possibly due to decreased dependence of more mature DRG neurons on exogenous nerve growth factor (NGF) (Levi-Montalcini and Booker, 1960) or other trophic factors. Introduction of NGF at explantation, in

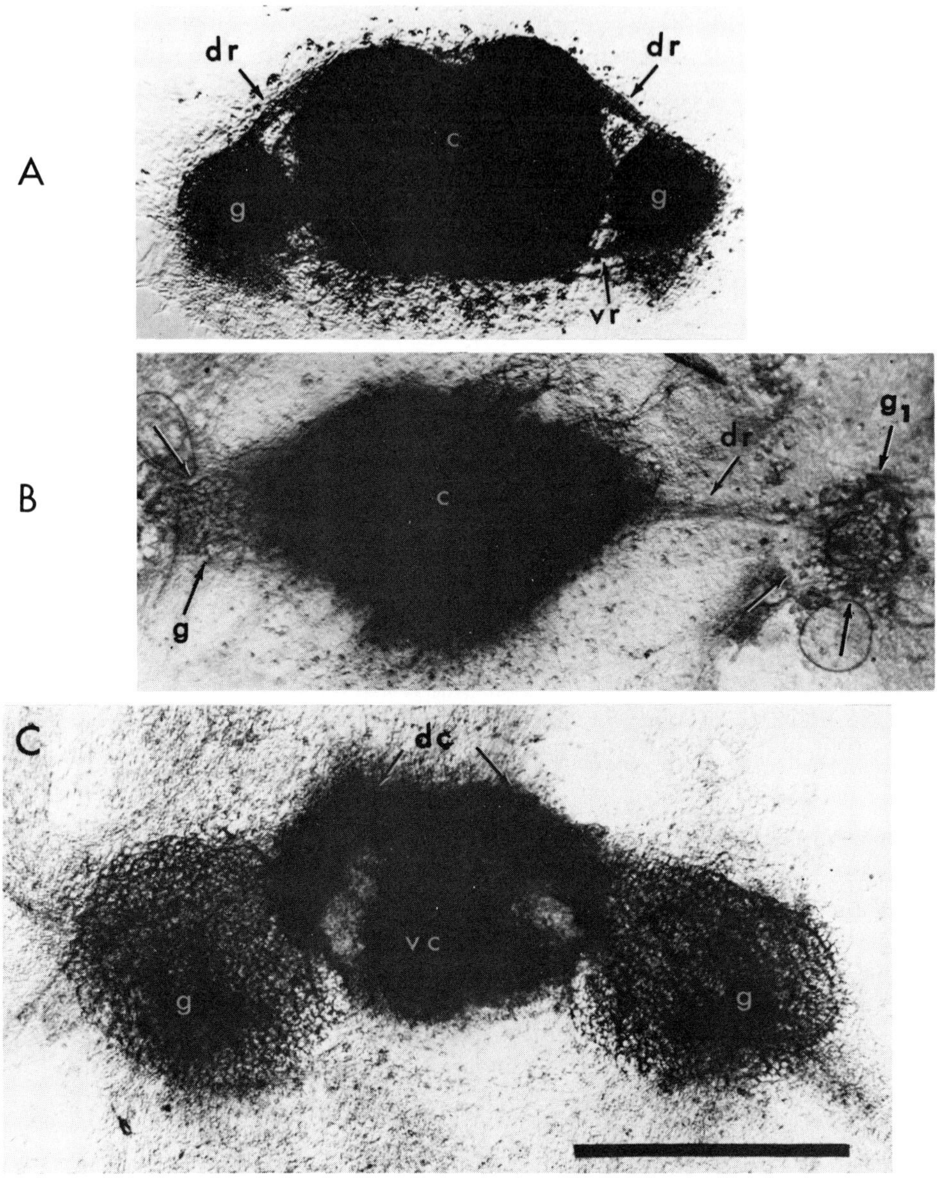
A
dr
dr
c
g
g
vr
B
c
dr
g1
g
C
dc
vc
g
g

concentrations ranging from 10^{-8} to 10^{-5} g/ml [i.e., 1-1,000 biologic units (BU) ml] led to a higher survival rate of DRG neurons (Crain and Peterson, 1974b) in addition to the characteristic dose-dependent enhancement of neuritic outgrowth (Levi-Montalcini, 1958; Levi-Montalcini and Angeletti, 1968a, b). Vastly increased neuronal survival occurred in higher concentrations of NGF (100-1,000 units/ml) concomitant with more abundant neuritic outgrowths from the DRG cells, as in cultures of NGF-"hypertrophied" mouse sympathetic ganglia (Crain *et al.*, 1964a). These *in vitro* effects resemble the remarkable selective hypertrophy of sensory DRGs and sympathetic ganglia in 7- to 10-day chick embryos *in ovo* within 3-4 days after daily injection of microgram quantities of NGF (Levi-Montalcini, 1958; Levi-Montalcini and Cohen, 1960). At the higher NGF levels, massive DRGs containing many hundreds of neuron perikarya developed during the first week *in vitro* and often remained densely packed in close proximity to the spinal cord tissue (Fig. 6C). This is in sharp contrast to the sparse, irregular array of several dozen DRG neuron perikarya (rarely more than 100) in control cultures (Figs. 1, 6B), which generally "migrate" approximately 0.5-1 mm away from the cord (Peterson *et al.*, 1965; Sobkowicz *et al.*, 1973) retaining characteristic dorsal root connections (Peterson and Crain, 1972; Crain and Peterson, 1974b). The NGF-stimulated DRG neurons continued to mature during the following weeks in culture even though no additional NGF was introduced beyond that included in the medium at explantation. Some of the DRGs actually became comparable in volume to their associated spinal cord segment (cf. Figs. 6C and 6A). Examination of the living cultures at high magnification indicates that the enlarged DRGs were composed of many layers of healthy neuron perikarya; and normal neuronal and supporting cell morphology has been observed in preliminary histologic analyses at the light- and electron-microscopic level (Peterson *et al.*, 1974).

FIG. 6. Photomicrographs of 14-day fetal mouse spinal cord explants (cross sections) with attached DRGs; these are living, unstained cultures. Scale: 1mm. A: Shortly after explantation (1 day *in vitro*). Note size of DRGs (g) relative to cord tissue (c); also dorsal (dr) and ventral (vr) roots. B: One month in normal culture medium. Many of the ganglion cells degenerated during the first few days *in vitro*, leaving a small thinly spread array of DRG neurons (g), which have matured and retained characteristic (myelinated) dorsal root (dr) connections to the cord (see also Fig. 1). Note that the DRGs are of similar size, although only one (g_1) shows the characteristic "migration" away from the cord. Most of the other control cultures showed even lower survival of DRG neurons (see text). C: Another cord-DRG explant after 1 month in the same culture medium, but NGF was added at explantation (1,000 BU/ml). Note remarkable enlargement of DRGs (g) relative to their initial size at explantation (A) and in contrast to the control culture (B). Many hundreds of ganglion cells form densely packed clusters close to the cord. (The major DRG volume increase was reached by the second week *in vitro*.) Relatively dense appearance of dorsal cord (dc) is due to large numbers of myelinated axons which represent central branches of DRG neurons (see text). Dense region in ventromedial cord (vc) is due primarily to a "necrotic core" which generally develops in both treated and control explants. (From Crain and Peterson, 1974b.)

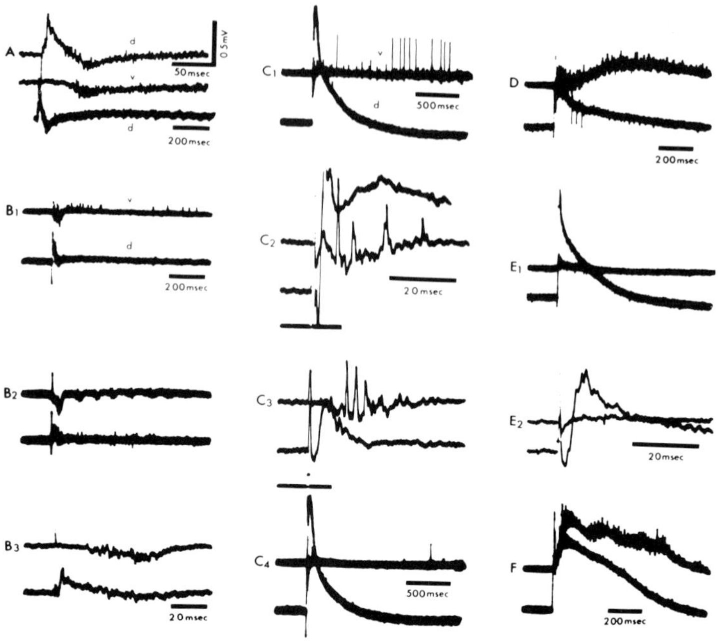

FIG. 7. Enhanced responses evoked in dorsal regions of fetal mouse spinal cord explants by stimuli applied to NGF-hypertrophied DRG. A: Control culture, 4 weeks *in vitro* (Fig. 6B). Early-latency negative slow-wave potential (resembling PAD — see text) is evoked in dorsal cord (d) by single DRG stimulus (via focal 10 μm tip electrode), and is followed by a positive slow-wave concomitant with a high-frequency spike barrage (dorsal cord response is shown at slower sweep rate on lowest record). Ventral cord response (v) begins after longer latency and involves primarily a positive slow-wave and spike barrage. B_1: Smaller PAD elicited in dorsal cord (d) of another control explant (2 weeks *in vitro*); ventral cord response (v) again consists of a primary positive slow-wave and a repetitive spike barrage. $B_{2,3}$: After introduction of strychnine (10^{-6}M), the ventral cord discharge becomes larger and more complex, but PAD is relatively unchanged although it is now followed by a long spike barrage. Fast sweep (B_3) shows that PAD begins shortly after DRG stimulus, whereas ventral cord response occurs at longer latency (as A). C_1: Similar explant, 2 weeks *in vitro*, but NGF was added at explantation (1,000 BU/ml). A PAD-like potential evoked in the dorsal cord (d) by a single stimulus to NGF-hypertrophied DRG (Fig. 6C) is much larger in amplitude and longer in duration (cf. A and B_1), whereas the ventral cord response (v) is similar to the control pattern. C_2: An early-latency, sharp rising phase and the complexity of the PAD response are seen at a faster sweep rate. C_3: Tenfold reduction of DRG stimulus intensity evoked smaller but still prominent PAD, and the ventral cord discharge now begins after a longer latency, during the falling phase of PAD (as in A and B_3). C_4: Larger DRG stimulus (as in C_1) again elicits characteristic large PAD just before drug application. D: Introduction of bicuculline (10^{-5}M) leads to a marked decrease in amplitude of PAD concomitant with the onset of a convulsive negative slow-wave and repetitive-spike discharge in the ventral cord (v). E_1: After transfer to 10^{-3} GABA, the large PAD response is restored (*cf.* C_4) in contrast to almost complete block of the ventral cord discharge. E_2: Tenfold reduction of DRG stimulus intensity still evokes relatively large PAD (*cf.* C_3, in BSS). F: Return to bicuculline (10^{-5}M) leads to partial depression of PAD and appearance of the secondary longer-lasting negative slow-wave in the dorsal cord, concomitant with the onset of a huge negative slow-wave and oscillatory discharge in the ventral cord. (From Crain and Peterson, 1974b.)

Correlative electrophysiologic experiments have been carried out on these fetal mouse cord-DRG explants to determine whether the NGF-stimulated growth and development of DRG neurons leads to enhancement of organized synaptic relationships with spinal cord neurons or if the additional neuritic bundles and aborizations are merely unregulated growth processes leading to "deadends" or to aberrant functional connections — as occurs in some types of NGF-stimulated sympathetic ganglion cells *in situ* (e.g., Olson, 1967).

In control cultures application of a focal stimulus close to the DRG or dorsal root (via a 10-μm electrode) often evokes a small negative slow-wave potential in the dorsal regions of the spinal cord, with a rapid rising phase and a slower falling phase (Fig. 7A, B). The amplitude and duration of this potential are quite variable, ranging up to approximately 0.5 mV and 100 msec, depending on the stimulus intensity, geometry of the cellular array, and proximity of the stimulating and recording electrodes to the excitable neuronal elements. This negative slow-wave response appears after a latency of 2-3 msec when the stimulus is applied to the DRG perikarya or the dorsal root (Fig. 7A, B_3). Simultaneous recordings in the ventral regions of the spinal cord segment generally show longer-latency (>5-10 msec) responses, often including a positive slow wave potential (*ca.* 100 msec) and a spike barrage of variable complexity (Fig. 7A, B).

Application of similar stimuli to NGF-hypertrophied DRGs evoked comparable negative slow-wave responses in the dorsal cord with similar latency and a sharp rising phase, but the amplitude and duration were now remarkably larger (Fig. 7C) (Crain and Peterson, 1974b, 1975a; see also Crain, 1976). The cord potentials often reached 2 mV in response to a single large DRG stimulus and a prominent negative slow wave could be evoked even after a 10-fold reduction in stimulus strength to these low-threshold neurons (Fig. $7C_3,E_2$). Furthermore, the duration of the DRG-triggered dorsal cord responses was often more than 500 msec (Fig. 7C) — far longer-lasting than generally observed in control cultures. Single stimuli applied to peripheral branches of NGF-stimulated DRG cells, as much as 2-3 mm beyond the explant zone (e.g., Figs. 1, 6C), were also effective and elicited responses comparable to those in Fig. $7C_3,E_2$, but with longer latencies (*ca.* 5-10 msec).

The temporal patterns of the dorsal cord potentials evoked by stimulation of NGF-hypertrophied DRGs *in vitro* contain components that are remarkably similar to those characteristic of the primary afferent depolarization (PAD) response in mammalian spinal cord *in situ* (Eccles, 1964). The PAD-like component recorded with extracellular microelectrodes in the cord explants appears to be a field potential produced by summated excitatory postsynaptic potentials (EPSPs) generated following DRG activation of dorsal cord circuits (see below). The weak PAD-like response generally observed with focal DRG stimulation in control cultures may be due partly to the lower density of excitable DRG neurons developing in the absence of added NGF, as well as to less extensive (or less effective) synaptic connections of each DRG cell with cord neurons. In contrast to the dramatic enhancement of the 'PAD' potentials in the dorsal cord of NGF-treated explants, no substantial alterations in the response patterns of *ventral* cord regions have been observed (cf. Figs. 7A and 7C).

Stimuli to nearby ventral cord regions (100-200 μ away) evoked, on the other hand, only early-latency spikes in dorsal cord, whereas large PAD-like potentials could be readily elicited with stimuli to distant DRGs (> 1 mm away).

Marked temporal facilitation of 'PAD' responses in explants occurs following brief application of 100/sec DRG volleys at low stimulus strength — resembling characteristic facilitation of PADs evoked by repetitive afferent stimuli *in situ* (Eccles, 1964). Furthermore, whereas strychnine showed relatively little effect on 'PAD' potentials in dorsal cord [even at concentrations (*ca.* 10^{-5}M) which greatly enhanced complex long-latency spike-barrage and slow-wave discharges in both dorsal and ventral cord (cf. Figs. $7B_2$ & $7B_1$)], bicuculline and picrotoxin (10^{-5}M) produced marked attenuation of the 'PADs' concomitantly with the onset of the convulsive discharges, especially in the ventral cord regions (Fig. 7D, F). On the other hand, after introduction of 10^{-3}M γ-aminobutyric acid (GABA) into the culture bath the 'PAD' responses in dorsal cord were generally maintained or even *augmented* (Fig. 7E), in contrast to the rapid and sustained depression of almost all detectable synaptically mediated discharges in ventral cord regions as well as long-latency discharges in dorsal cord (Fig. 7E; upper sweep). GABA enhancement of 'PADs' occurred in control as well as in NGF-treated cord explants and was especially marked during brief 100/sec DRG volleys. Generation of the 'PADs' by Ca^{2+} -dependent synaptic transmitter release is suggested by the rapid and complete block of 'PAD' potentials after increasing the Mg^{2+} concentration from 1 to 10mM, whereas spikes could still be directly evoked (Crain, 1974b, c, 1976).

Although the 'PADs' recorded in our explants probably include field potentials due to summated EPSPs in the dorsal cord interneurons triggered by DRG collaterals, the marked selective attenuation of these responses in bicuculline and picrotoxin suggests that a major component is generated by GABA-mediated EPSPs, possible at DRG terminals as *in situ* (Davidoff, 1972a,b; Barker and Nicoll, 1973; Benoist *et al.*, 1974). Depolarization of DRG terminals (by these GABA interneurons triggered by DRG collaterals) presumably decreases both the amplitude of the presynaptic action potential and quantity of transmitter released, and this presynaptic inhibition provides a potent regulatory mechanism that depresses the central excitatory actions of many primary afferent fibers in the mammalian

CNS (Eccles, 1964; Wall, 1964; cf. Curtis *et al.*, 1971). Analyses of dorsal and ventral cord responses to selective single and paired stimuli applied to various neural elements in these cord-DRG explants provide additional evidence that the observed 'PADs' are associated with inhibitory mechanisms. The longer-latency negative slow-wave and oscillatory potentials elicited by DRG stimuli in bicuculline probably represent summated postsynaptic potentials synchronously and sequentially activated after interference with GABA-mediated inhibitory postsynaptic potentials (IPSPs) of cord neurons involved in these complex network responses (Crain, 1974b, 1975c,d; Crain and Bornstein, 1974; see also Zipser *et al.*, 1973). Introduction of 10^{-3}M GABA, on the other hand, may block convulsive network discharges by enhancing GABA-mediated IPSPs of cord neurons as well as GABA-mediated EPSPs (or other depolarization) of DRG terminals.

This study demonstrates that at least some of the abundant additional neurites which develop after exposure of fetal DRG cells to NGF can proceed to make characteristic long-term synaptic relationships with *specific* types of spinal cord neurons. The data also indicate that responses resembling PADs can be generated in organized spinal cord-DRG explants with remarkable mimicry of these specialized synaptic network functions *in situ.* Moreover, if further analyses of the greatly enhanced 'PADs' generated in spinal cord explants in response to the unusually large input from NGF-hypertrophied DRGs do indeed demonstrate their relationship to presynaptic inhibitory functions, this may be evidence of an intrinsic CNS regulatory system, involving development of compensatory (homeostatic) inhibitory circuits proportional to the magnitude of the excitatory synaptic input (Crain, 1974a).

Neurites from isolated DRGs can also grow across gaps of 0.5-1 mm on a collagen-film substrate (in high NGF) and invade separate spinal cord explants. During the first week of co-culture, abundant DRG neurites grow into slabs of *dorsal* cord tissue, in contrast to relatively sparse invasion of similarly apposed *ventral* cord slabs. Large 'PADs' are evoked by DRG stimuli in co-cultures with dorsal, but not ventral, cord (Crain and Peterson, 1975b). Although longer-latency positive slow-waves or spike barrages are occasionally evoked in the latter explants, these responses are rapidly blocked in 10^{-3}M GABA, whereas the DRG-evoked 'PADs' in dorsal cord are unaffected or enhanced at this GABA concentration. Furthermore, in cultures where the cut end of longitudinal slabs of *whole* spinal cord is apposed to DRG clusters, DRG neurites often form prominent fascicles directly towards dorsal regions of the cord. Stimuli to these "*de novo*" dorsal roots evoke large 'PADs', whereas none were detected with stimuli to the adjacent neuritic growth zone of ventral cord regions.

Under the equivalent geometric conditions provided by these explant arrays, the apparently selective growth of DRG neurites toward specific regions of CNS tissue rich in sensory target neurons provides further support for the theory of "chemoaffinity in the orderly growth of nerve fiber patterns and connections" (Sperry, 1963), which may now be accessible to more direct analysis under rigidly controlled physicochemical conditions in CNS tissue cultures. Correlative histologic studies after selective labeling of the DRG cells innervating cord cross-sectional explants may clarify whether the central DRG neurites arborize *diffusely* throughout the cord tissue (as they often do when growing on homogeneous collagen gel substrates) — making initial transitory contacts with all types of neurons including the final target cells — or whether they do indeed tend to be restricted to the dorsal horn regions as suggested by the microelectrode mapping experiments described above. Furthermore, if the DRG fibers are in fact concentrated primarily in the dorsal cord, is this due to "repulsion" by ventral cord tissue (see above) or to "tropic" (or trophic) effects exerted by target cells in the dorsal cord?

C. Motorneuron Outputs

1. Innervation of skeletal muscle. Tissue culture studies of neuromuscular relationships have been the subject of intensive investigations in many laboratories during the past few decades (see reviews by Murray, 1965a, b, 1972; Shimada and Fischman, 1973; Nelson, 1973, 1975; Crain and Peterson, 1974a; Fischbach, 1974; Fischbach *et al.*, 1974a, b, 1976; Harris, 1974a, b; Crain, 1976).

Direct electrophysiologic recordings from cord-myotome cultures were first begun in 1964 using mouse tissues (Crain, 1964b; see also Crain, 1970a, and morphologic correlates by Bornstein *et al.*, 1968). Cross sections (0.5 mm thick) of the dorsal half of 12- to 13-day mouse embryos were explanted, each fragment containing spinal cord, DRG, paravertebral masses of myoblasts, and various connective tissues, with relatively little disruption of the normal tissue array. Microscopically visible contractions of large numbers of muscle fibers were generally evoked in these myotome cultures by selective stimulation of spinal cord neurons. Muscle action potentials could be detected with extracellular microelectrodes, and they occurred with latencies ranging from 1 to 15 msec. Cord-evoked muscle responses were generally triggered at a critical stimulus threshold and occurred only with careful positioning of the stimulating microelectrode in the cord tissue. Muscle responses with earliest latencies were evoked with minimal stimulus intensities at critical sites in the ventral cord (or ventral root). Together with observations in *in vitro*-coupled and cord-muscle explants (see below), the data suggested that innervation of the muscle fibers in these cultures was provided by motor neurons located in the ventral horn regions of the cord cross sections.

The effects of *d*-tubocurarine and eserine on the bioelectric activities of cord-myotome explants indicated that neuromuscular transmission was mediated by cholinergic synaptic mechanisms as *in situ*. Introduction of *d*-tubocurarine (0.3-10 μg/ml) produced a characteristic rapid block of cord-evoked muscle responses, while cord afterdischarges were unaffected (as in Fig. 8). Presynaptic neural spikes could still be elicited and the muscle fibers could still be activated by *direct* electric stimuli. The anticholinesterase agent eserine, on the other hand, produced characteristic excitatory effects of myotome explants at low concentrations (0.1-0.5 μg/ml). Single cord stimuli, which evoked single muscle responses in control media, now began to elicit repetitive muscle twitches even though the cord discharges showed no enhancement (Crain, 1970a). Strychnine also produced "convulsive" muscle contractions in the myotome cultures, but in this case the effects were mediated by action of the *cord* tissue. Long series of repetitive muscle spikes (and contractions) were now associated with long-lasting cord afterdischarges evoked by single stimuli or occurring spontaneously (as in Fig. 8A_3).

These studies provided the first direct demonstration that immature neurons of mammalian spinal cord retain the capacity *in vitro* to form *functional* neuromuscular junctions with skeletal muscle fibers. The organotypic bioelectric properties of cord-myotome explants are in good agreement with histologic evidence of characteristic cholinesterase-loaded neuromuscular synapses on cross-striated muscle fibers (Bornstein *et al.*, 1968). It should be emphasized that neuromuscular junctions have probably not yet formed in the immature 12- to 13-day mouse embryo tissue explanted for these experiments. Clear-cut junctions in the rat fetus have not been detected by electron microscopy until 17-18 days *in utero* (Teräväinen, 1968; Kelly and Zacks, 1969), in agreement with physiologic studies (Windle *et al.*, 1935; Diamond and Miledi, 1962). Critical developmental processes associated with formation of neuromuscular junctions must therefore have occurred after explantation of 12- to 13-day cord-myotomes in culture.

Techniques were then developed to permit innervation *in vitro* of isolated skeletal muscle explants positioned within 1 mm of the ventral edge of previously explanted fetal rodent spinal cord (Fig. 1) (Peterson and Crain, 1970, 1972). Electrophysiologic studies demonstrated that characteristic neuromuscular transmission developed not only after innervation *in vitro* of fetal rodent muscle (Crain, 1968, 1970a) but also after innervation of muscle fibers regenerating in *adult* rodent and even *human* muscle explants (Crain *et al.*, 1970). After coupling periods of 2-7 weeks *in vitro*, selective stimulation of ventral horn regions of the spinal cord explants (or ventral

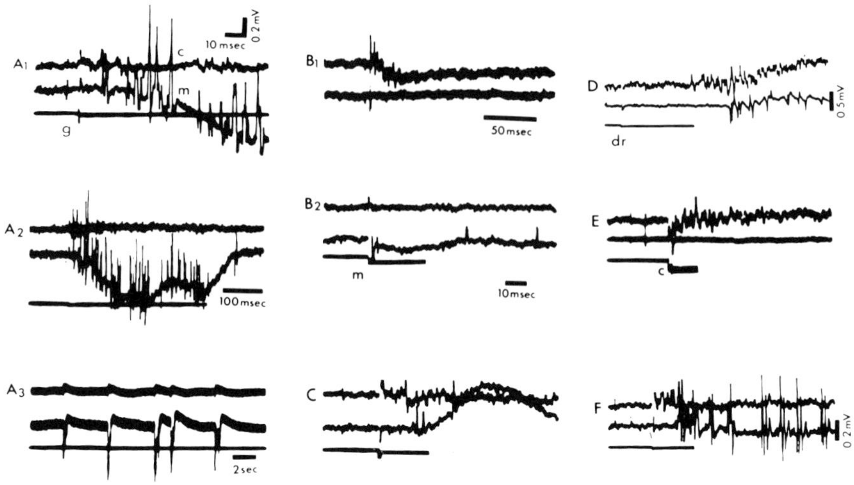

FIG. 8. Neuromuscular transmission between separate explants of fetal mouse spinal cord and *adult* mouse skeletal muscle in 6- and 8-week cultures; the muscle tissue was "presented" to the isolated cord explant at 4 days *in vitro*. $A_{1,2}$: Simultaneous recordings from cord and muscle explants (8-week culture) show a brief burst of spike potentials in cord (first sweep, c), evoked by a single stimulus to attached DRG (third sweep, g), and a much longer-lasting barrage of spikes superimposed on a large slow wave in muscle (second sweep, m). The muscle discharge in A_2 actually included approximately 10 additional spikes of large amplitude (up to 1 mV) — similar to the first four extending above the first sweep. They occurred in groups of two or three, every 20-40 msec, as in A_1, but were barely visible at the slower sweep rate in A_2 (only the first few were retouched, to avoid undue complexity). Synchronous contractions of large numbers of muscle fibers occurred concomitantly with the slow wave (see text). A_3: Cord-muscle discharge sequence (and contractions) occurs repetitively at 2- to 5-sec intervals following the initial responses evoked by a single ganglion stimulus. B_1: Neurally evoked muscle discharge (and contraction) is blocked under *d*-tubocurarine (10 μg/ml) while the spike burst is still elicited in cord explant by ganglion (or cord) stimuli. B_2: Muscle spike potential (and contraction) can still be evoked by *direct* electric stimulus to muscle tissue (third sweep, m). C: After return to normal medium, characteristic muscle discharges (and contractions) can again be elicited with cord (or ganglion) stimuli. D: Cord and muscle discharges evoked by dorsal root stimulus in another culture (6 weeks *in vitro*) showing much longer latencies — in this case approximately 30 msec for the onset of cord barrage and an additional delay of 20 msec before the appearance of the muscle response. *d*-Tubocurarine produces a selective block of cord-evoked muscle discharge (E), and characteristic responses can again be elicited after a return to normal medium (cf. F and A_1). (From Crain, Alfei and Peterson, 1970.)

root loci) evoked widespread coordinated contractions in all of these types of muscle fibers (e.g., Fig. $8A_1$, C). Simultaneous microelectrode recordings of ventral cord and muscle responses to focal cord stimuli show that muscle action potentials (and contractions) generally occurred with latencies of several milliseconds after onset of the cord discharges. The cord potentials appear to be due to activation of motoneurons, which then propagate impulses toward the muscle fibers (Crain, 1970a). Similar temporal relations were often seen during spontaneous rhythmic discharges of the coupled cord and muscle tissues. Long series of repetitive discharges at 2- to 5-sec intervals may occur synchronously between these cord and muscle explants in response to single cord (or DRG) stimuli, and they may also appear spontaneously. As in the cord-myotome cultures, *d*-tubocurarine (1-10 μg/ml) selectively and reversibly blocked neuromuscular transmission (Fig. 8B, E) and eserine accelerated recovery of normal function. Repetitive cord and muscle discharges were greatly augmented after introduction of strychnine. Complex rhythmic oscillatory (*ca.* 10/sec) afterdischarges generated in strychninized cord explants led to similarly patterned muscle discharges (and contractions) which may also occur in normal medium (Fig. 8A).

The organotypic bioelectric properties of these cultured neuromuscular tissues correlate well with histologic evidence of silver-impregnated axons arborizing among cross-striated muscle fibers and terminating at characteristic neuromuscular junctions with marked cholinesterase localization (Peterson and Crain, 1970, 1972) and typical motor endplate ultrastructure (Fig. 9) (Pappas *et al.*, 1971a, b). Some of these cord-innervated muscle cultures have remained in healthy condition for more than 1 year, suggesting that neurotrophic effects may play an important role in maintaining the structural and functional integrity of the muscle fibers (Crain and Peterson, 1974a). This view is supported by our demonstration of serious muscle atrophy within 2-4 weeks after surgical denervation of these *in vitro*-matured muscle fibers. Furthermore, muscle denervation atrophy after cord extirpation is markedly postponed if sympathetic (but not dorsal root) ganglia were previously explanted close to the muscle fibers. Although these neurotrophic effects were observed in the apparent absence of direct synaptic connections between sympathetic ganglion and muscle cells, some of the muscle fibers in these cultures appeared to have become innervated by the sympathetic neurites which arborized over the muscle tissue (Crain and Peterson, 1974a). Similarly unexpected formation of connections in culture between dissociated sympathetic ganglion cells and skeletal myotubes were recently demonstrated by Nurse and O'Lague (1975) (see also Crain, 1976).

2. Innervation of sympathetic ganglion cells. Coordinated electrophysiologic and cytologic analyses have been carried out during the formation and development of synapses between neurons in fetal rat spinal cord explants and dissociated sympathetic ganglion cells (Bunge *et al.*, 1974; Bunge, 1976a, b; Rees *et al.*, 1976). Two types of synapses were observed by electron microscopy of these cultures. One type consisted of adrenergic dense-core vesicles (Rees and Bunge, 1974). The other type contained only

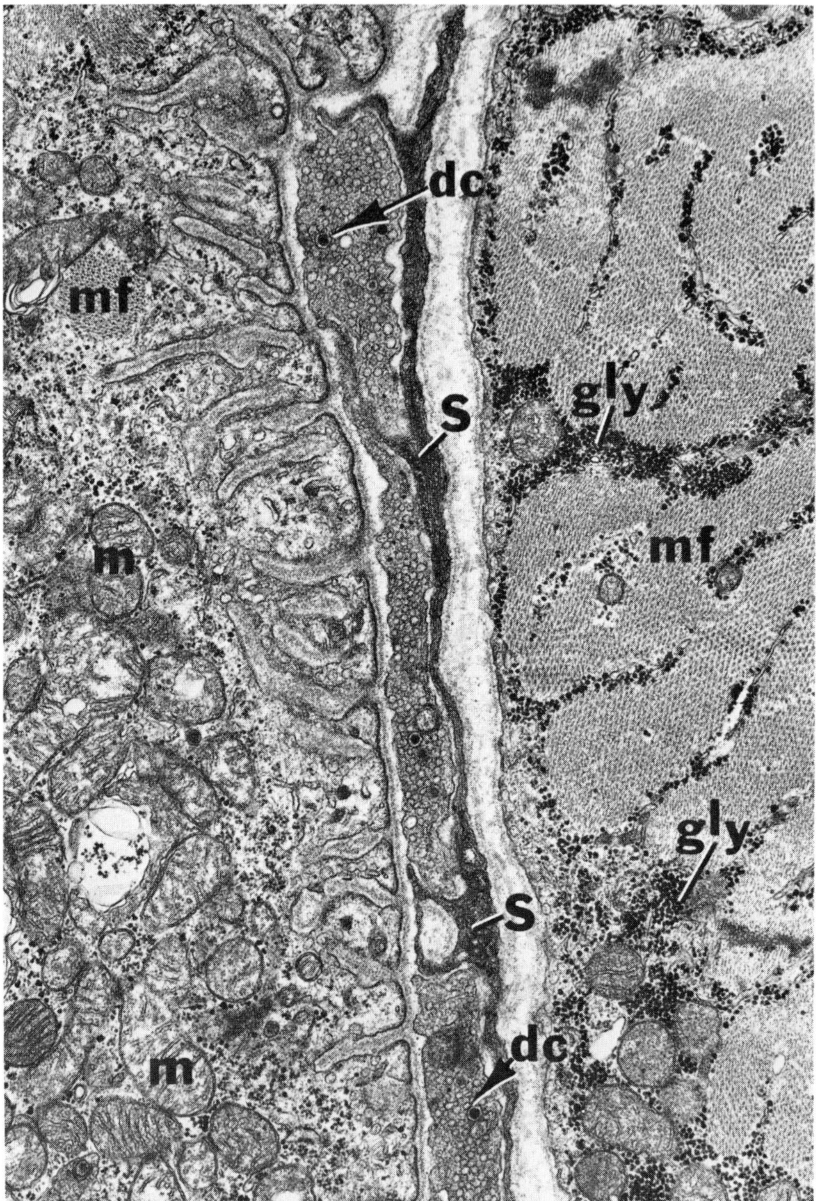
dc
mf
gly
S
mf
m
gly
S
dc
m

clear vesicles, and these cholinergic-like synapses degenerated selectively 2-3 days after extirpation of the spinal cord explant. Intracellular recordings from the dissociated sympathetic neurons showed characteristic membrane resting and action potentials (see also O'Lague *et al.*, 1974). In ganglion cells grown near spinal cord explants, small local depolarizations frequently occurred spontaneously, sometimes associated with the generation of action potentials. No such spontaneous activity was detected in superior cervical ganglion cells cultured in the absence of spinal cord. The synaptic nature of the local potentials was suggested by increases in their amplitude during hyperpolarization of the postsynaptic membranes and their reduction in amplitude and frequency when the Mg^{2+} concentration was increased to 15 mM. The spontaneous depolarizations of the sympathetic neurons generally occurred at rates below 5/sec, but at times bursts of 25/sec were observed. Bunge *et al.* (1974) suggested that these depolarizations may have been due to spontaneous discharges generated in the spinal cord explants leading to chemical transmitter activity at the cord-ganglion synapses, as *in situ.* These interpretations were recently confirmed by electrically stimulating ventral horn regions of the spinal cord explant with extracellular microelectrodes (Ko *et al.*, 1975, 1976a) and recording intracellular dissociated superior cervical ganglion neurons. These synaptic potentials could be mimicked by iontophoretic application of ACh. Both the ACh-evoked potential and the EPSP elicited by spinal cord stimuli were shown to be blocked by nicotinic blocking agents (*d*-tubocurarine, mecamylamine or hexamethonium at 10^{-5}-10^{-4}M) and the EPSPs could be selectively depressed by increasing the Mg^{2+} concentration or reducing the Ca^{2+} concentration. On the other hand, adrenergic blocking agents, e.g., phenoxybenzamine, were ineffective. These data demonstrate that the spinal cord explants form a cholinergic nicotinic synapse *de novo* in culture on the sympathetic neurons that is functionally similar to the synapse made in the intact animal (see also Ko *et al.*, 1976b).

The experiments with dissociated sympathetic ganglion cells presented to explants of spinal cord demonstrate the potentialities of combining the advantages of the two types of culture techniques, i.e., the enhanced visibility and accessibility of dissociated neurons with the regionally

FIG. 9. Electron micrograph of a motor endplate that has developed on a regenerated adult mouse skeletal muscle fiber after innervation by a motoneuron in a fetal mouse spinal cord-ganglion explant, 3 months in culture. (Low-power view is shown in Fig. 1.) The axon terminals are filled with presynaptic vesicles, some of which are of the dense-core variety (dc). The postsynaptic infoldings of the sarcolemma, characteristic of mature neuromuscular junctions, extend along the entire region of synaptic contact. The axon terminals are ensheathed by Schwann cell processes (S), except at the synaptic contact surface. The myofilament lattices (mf) are seen here in cross section; note also the mitochondria (m), glycogen particles (gly), and ribosomes. Scale 1 μm. (From Pappas, *et al.*, 1971a.)

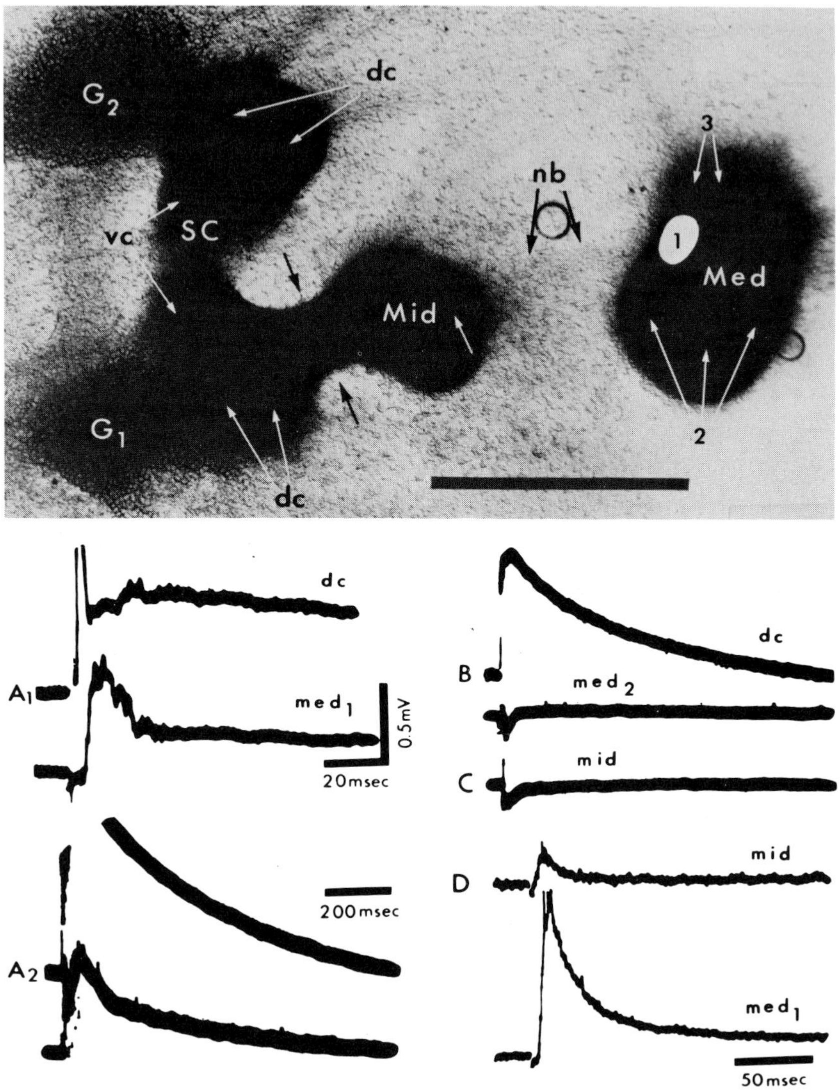
dc
G2
3
nb
vc
SC
1
Med
Mid
G1
2
dc
dc
A1
med1
0.5mV
20msec
dc
B
med2
mid
C
200msec
A2
mid
D
med1
50msec

organized neuronal arrays in CNS explants. Introduction of dissociated DRG cells, muscle cells, etc. near spinal cord explants should also provide valuable preparations for correlative cytologic and electrophysiologic analyses of specific types of synapses during development in culture.

Explants of fetal rat sympathetic ganglia co-cultured with spinal cord explants have also provided significant data on the formation of specific synaptic connections *in vitro* (Olson and Bunge, 1973). Electron microscopic analyses of these cultured ganglia demonstrated development of abundant synapses as *in situ*, whereas the number of synapses decreased drastically following removal of the cord explant from the culture. In contrast, no significant increase in synapses was detected in sympathetic ganglion explants when co-cultured with cerebral cortex explants (see also Bunge, 1976a, b).

III. Sensory Innervation of Specific Target Regions in Medulla Explants

After demonstrating that DRG neurons could establish characteristic functional synaptic networks in dorsal regions of spinal cord explants, we proceeded to prepare model systems suitable for analyzing further development of central DRG neurites up the neuraxis. Our earlier electrophysiologic studies showed that complex synaptic interactions could develop between neurons in explants of fetal rodent spinal cord and brainstem tissues after growth of neurites across gaps of 0.5-1 mm, but no signs of selective synaptic connections between specific types of neurons were detected (Crain and Peterson, 1966; Crain *et al.*, 1968).

FIG. 10. Negative slow-wave potentials resembling PADs evoked in spinal cord and brainstem explants by DRG stimuli (14-day fetal mouse tissues; 14 days in culture). Top: Photomicrograph shows spinal cord cross section (SC) with attached DRG ($G_{1,2}$). The dorsal edge of one-half of the cord is "fused" to the midbrain (Mid) explant (black arrows); a thinly spread array of neurites and glial cells have formed a bridge (nb) to the medulla (Med) explant (barely visible at this low magnification). Midline section of dorsal cord (see text) resulted in lateral displacement of the dorsal horns and DRGs at either end of the ventral cord (vc). Scale 1 mm. Bottom: Simultaneous recordings of PADs in dorsal cord (dc, lower left arrow in photomicrograph; see also Fig. 7C) and medulla (Med, site 1) in response to a single DRG stimulus (G_1). Note abrupt onset and long duration of these potentials; also longer latency of medulla response. A_2: Brief 100/sec DRG volley (at smaller, near-threshold stimulus strength) elicits much larger and longer-lasting PADs in both cord and medulla. B: Large DRG_1 stimulus evoked only small, positive slow-wave responses in other regions of medulla explant (Med_2); the dorsal cord PAD is still large. Microelectrode mapping showed no signs of medulla PADs in response to DRG stimuli, except in the zone indicated in white around site 1, and none in the entire midbrain (Mid) explant (e.g., at white arrow; record C). D: After adding 10^{-3}M GABA, PAD in medulla (med_1) shows marked increase in amplitude and duration (as in Fig. 11B), and it can now also be evoked in adjacent region 3; the polarity of the midbrain response (Mid) has now become negative, although still relatively small. (From Crain and Peterson, 1975a.)

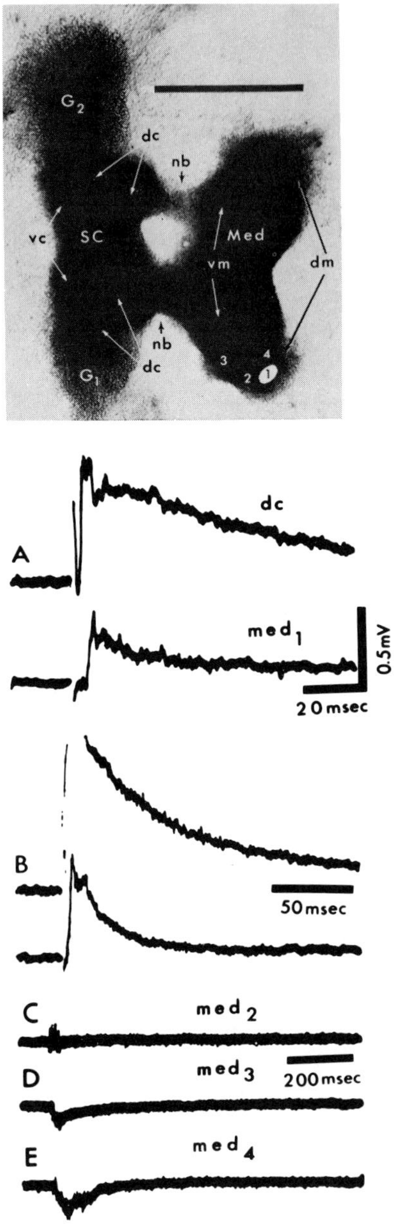

FIG. 11. PAD responses evoked in dorsal region of spinal cord (SC) and medulla (Med) explants (complete cross sections) by DRG stimuli (14-day fetal mouse tissues; 14 days in culture). Top: Medulla cross section is at the level of the cuneate and gracilis nuclei; dorsal closure has not yet occurred at this fetal stage so that the dorsal medulla tissues (dm) are laterally displaced. Note the bridges (neurites and glia) which have formed between the dorsal edge of the cord (dc) and the ventral edge of the medulla (vm) explants in two regions (nb); DRGs ($G_{1,2}$) are located more laterally in this explant (*cf.* Fig. 10) further away from the ventral cord (vc). Scale: 1 mm. Bottom: A: Simultaneous recordings of PADs in dorsal cord (dc, lower left arrow in photomicrograph) and dorsal medulla (Med, site 1) in response to a single DRG_1 stimulus. B: After adding 10^{-3}M GABA, PADs at this site in medulla (Med_1) and in dorsal cord are augmented (as in Fig. 10D). C: A large DRG_1 stimulus evokes only a spike burst at site 2 in the medulla (Med_2) and small positive slow-wave responses at $Med_{3,4}$. D,E: Systematic mapping of entire medulla explant showed no signs of PADs in response to DRG stimuli except in a small zone indicated in white around Med_1 (*ca.* 01 × 0.2 mm), even during GABA exposure. (Mapping was not attempted in this culture with stimuli to DRG_2) (From Crain and Peterson, 1975a).

In more recent experiments aimed at facilitating *direct* DRG innervation of medulla target neurons, a midline section of the spinal cord cross-section was made from the central canal through the dorsal cord and meninges prior to explantation of fetal cord-DRG explants (Fig. 6). In the younger fetuses only the meninges required cutting since dorsal closure of the cord was not yet completed. This ensured outgrowth of CNS neurites and glial cells, including DRG fibers that have passed through the cord, comparable to dorsal column axons. Fetal mouse brainstem explants were carefully positioned near the dorsal edge of the cord cross sections so as to be in the path of the outgrowing "dorsal column" (Figs. 10, 11, 12) (Crain and Peterson, 1975a). Tissues from various regions of the medulla and midbrain were presented to the cord-DRG explant, including complete cross sections through the medulla at the level of the cuneate and gracilis nuclei (Figs. 11, 12).

'PADs' similar to those evoked in dorsal cord explants were detected in small regions of medulla explants connected to cord with NGF-hypertrophied DRGs. The medulla 'PADs' evoked by single DRG stimuli ranged up to 1 mV and arose after longer (*ca.* 3-10 msec) latencies (Figs. 10, 11). The large amplitude of these 'PADs' indicates that relatively large numbers of DRG terminals probably make synaptic connections with target neurons in the medulla explants. In cord-DRG cultures without added NGF, where only a few dozen DRG neurons may survive (Fig. 6B), dorsal cord 'PADs' are often much smaller than these medulla 'PADs' in spite of the abundant dorsal cord neurons available for establishing sensory synaptic networks with ingrowing DRG neurites (e.g., Fig. 7A, B) (Crain and Peterson, 1974b, 1975b). Introduction of 10^{-3}M GABA generally augmented the brainstem and dorsal cord 'PADs' (Figs. 10D, 11B), whereas various cord-evoked brainstem network discharges were seriously depressed, as were ventral cord responses. Moreover, in cases where a midbrain explant was positioned between the cord and medulla, prominent DRG-evoked 'PADs' were detected only in the latter explant, even when it was located more than 1 mm distal to the interposed midbrain tissue (cf. Fig. 10A and 10C). Weak 'PADs', however, could be detected in some midbrain regions, especially when medulla target neurons were relatively distant (Fig. 10D) or absent. Using this pharmacologic marker technique, we have been able to map more than 20 DRG-cord-brainstem cultures, and in most cases 'PADs' were sharply localized to one or two small zones (*ca.* 100-300 μm) in each medulla explant. In 10 cultures where cross sections of the entire medulla at the level of the cuneate and gracilis nuclei were presented to the DRG-cord explant with controlled orientation, prominent 'PADs' were evoked only in the *dorsal* medulla regions (Fig. 11), precisely where dorsal column sensory fibers normally terminate and lead to PADs *in situ* (Anderson *et al.*, 1964; Eccles, 1964; Wall, 1964; Davidson and Southwick, 1971). Similar results

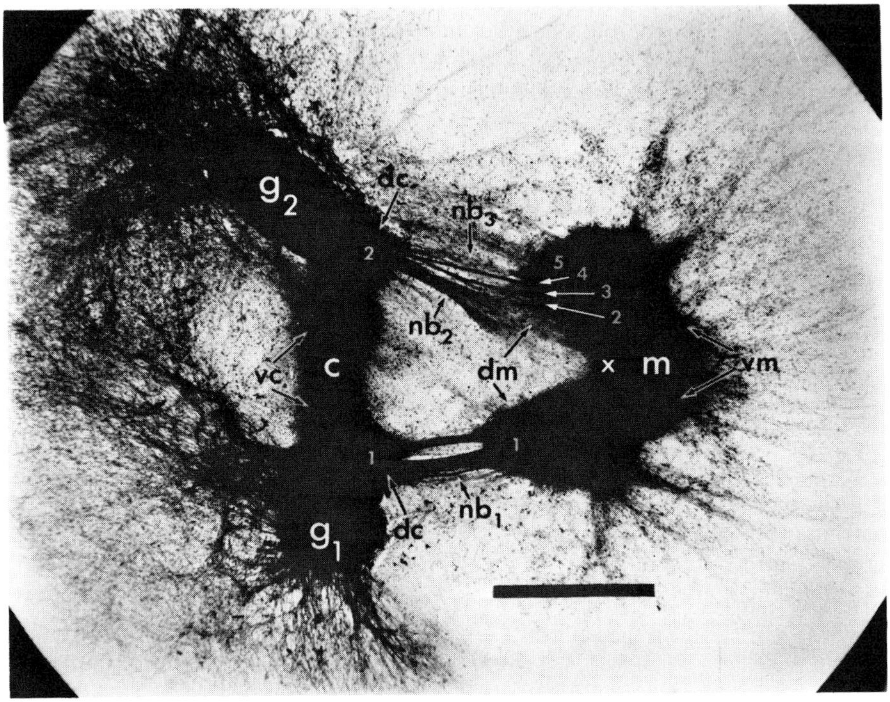

FIG. 12. Photomicrograph of a spinal cord-DRG explant co-cultured with a medulla explant (as in Fig. 11 but separated by a larger gap), showing details of the neuritic bridges that formed between the two explants (14-day fetal mouse tissue). Medulla explant (**m**) consists of a cross section at the level of the cuneate and gracilis nuclei and it is positioned so that the dorsal edge faces the dorsal edge of the cord explant (**c**) (*cf.* Fig. 11 where the ventral edge of the medulla faces the dorsal cord). The cord-DRG explant was added to a 1 week-old medulla culture, and the paired array was then maintained for two additional weeks *in vitro*. Note the prominent neuritic bridges ($nb_{1,2,3}$) which formed between the dorsolateral regions of the cord and medulla explants (including "dorsal column" fibers; see text). Scale: 1 mm. Large PAD responses were evoked in area 1 in the lower part of the medulla explant by stimuli to the lower DRG (g_1; as in Fig. 11A), and similar PADs were elicited in area 2-3 by stimuli to the upper DRG (g_2). All of these DRG-evoked PAD responses were maintained or enhanced during exposure to 10^{-3}M GABA (as in Figs. 10,11). Furthermore, no PADs could be detected in other regions of the medulla explant, e.g. medial (between 1 and 2); ventral (vm); and lateral, near 5, even with maximal stimuli to g_2 or g_2.

were obtained in cultures where the medulla cross section was rotated 90° or 180° (Fig. 12) with respect to the axis of the cord explant.

These *in vitro* experiments demonstrate that DRG neurites, after passing through spinal cord tissue, can grow across a homogeneous collagen-film substrate and, in mimicry of dorsal column fibers *in situ*, establish characteristic functional synaptic networks with programmed target neurons in brainstem explants, even in the presence of a variety of alternative CNS neurons with abundant synaptogenic receptor sites (see Olson and Bunge, 1973; Section IIC2). Furthermore, although the initial neuritic and glial spinal cord outgrowth in relation to these nearby medulla explants was comparable to that extending toward nontarget CNS tissues, preliminary analyses in suitably arrayed cultures suggest that prominent fascicles of "dorsal column fibers" may become organized toward the target neuron zones in the medulla (Fig. 12) (Crain, 1976; Crain and Peterson, 1977). In more recent studies with isolated DRGs positioned near medulla explants (see Section IIB), we have also demonstrated that DRG neurites can, under favorable NGF stimulation, grow directly into medulla cross sections. Focal stimulation of DRG neurites located 1-2 mm from the medulla explant could evoke characteristic 'PADs' in dorsolateral medulla target zones, whereas similar stimuli applied within nearby ventral of medial regions of the medulla explant (100-200 μm away) were often ineffective. The remarkable degree of characteristic pharmacologic sensitivity and regional specificity of these sensory-evoked spinal cord and brainstem networks provides the basis for a powerful new model system to analyze mechanisms underlying formation and development of specific synaptic connections in the mammalian CNS (Sperry, 1963, 1965) under isolated conditions in culture (see also Weiss, 1955, 1966; Jacobson, 1970).

Significant electrophysiologic and morphologic studies have also been carried out during the development of specific synaptic circuits in fetal rodent cerebellar explants (see reviews by Hendelman and Wolf, this volume).

IV. Synaptogenesis in Fetal Cerebral Explants

A. Early Functions

Explants of late fetal or newborn mouse cerebral neocortex (frontal region) show a sequence of developmental changes in bioelectric properties similar to those in 13- to 14-day fetal rodent spinal cord cultures during the first week after explantation (Crain, 1964a, 1976; Crain and Bornstein, 1964, 1974; morphologic correlates in Bornstein, 1964; Kim, 1972; Seil *et al.*, 1974).

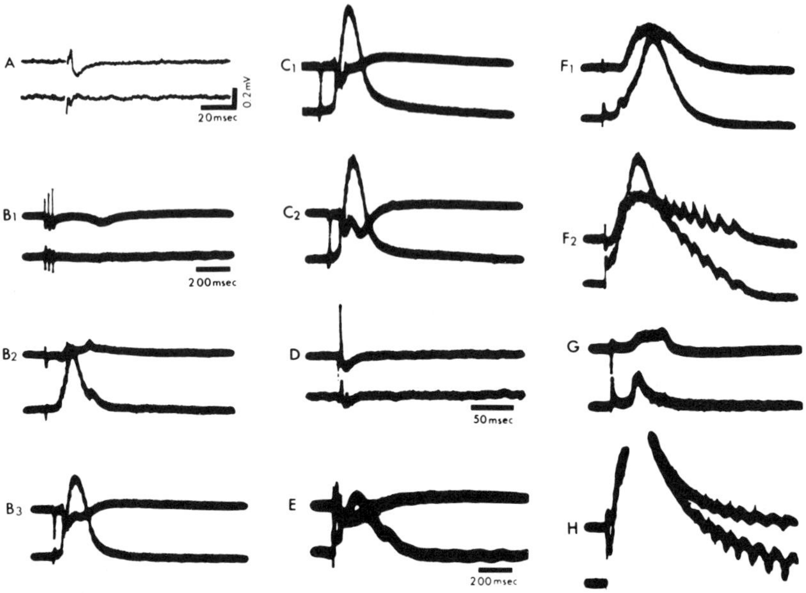

FIG. 13. Complex bioelectric discharges evoked in explants of 18-day fetal mouse hippocampus after 3-5 days in culture. A: Only simple, early-latency spike potentials could be elicited in this 3-day explant in response to single or repetitive stimuli in normal medium (BSS). (All recordings were made with electrodes at two sites approximately 0.5 mm apart in response to a stimulus applied to a third region of the explant about 0.5 mm away.) B_1: Shortly after introduction of strychnine (*ca.* 10^{-5}M), positive slow-wave potentials with small amplitude and long latency could now be evoked at one site (upper sweep) with repetitive stimuli (three or four at 50/sec). B_2: Several minutes later negative slow-wave potentials with larger amplitudes and shorter latencies could be elicited at both sites, even with single stimuli. These responses were quite labile, and rest periods of approximately a minute were generally required even with large stimuli. B_3: Similar but larger-amplitude responses were evoked occasionally after a return to BSS (lower sweep). C_1: Within 1 min after replacing regular BSS with a chloride-free BSS (propionate substituted for all chloride ions), the amplitude of the negative slow-wave response at one recording site (lower sweep) increased by nearly 50%. C_2: Shortly afterward, the response at other recording site (upper sweep) became more complex, involving an additional positive phase followed by an enhanced, longer-lasting negative phase. D: In another hippocampal explant from the same batch, only simple spike potentials could be evoked in BSS even after 5 days *in vitro*. E: Introduction of strychnine (*ca.* 10^{-5}M) led to the appearance of prominent slow-wave responses as in the 3-day explant (*cf.* $B_{2,3}$). F: Removal of chloride ions by introducing propionate-BSS dramatically enhanced the evoked responses, i.e., doubling of the amplitude of the negative slow-wave potential (lower sweep), inversion of the positive slow-wave potential at the other recording site (upper sweep), and onset of long-lasting repetitive oscillatory (*ca.* 10/sec) potentials at both sites (F_2). G: Responses became much smaller again after returning to the regular BSS. H: Huge primary negative slow-wave potentials and secondary oscillatory discharges can be evoked again after removal of the chloride ions, as in F. (From Crain and Bornstein, 1974.)

Only simple spike potentials could be evoked during the first 2 days after explantation of newborn mouse cerebral neocortex, similar to those evoked in immature fetal spinal cord explants (Figs. 2A, 3A; see also Fig. 13A). Within 3 days after explantation from newborn mice, much more complex response patterns may be evoked in cerebral cultures under the same stimulus and recording conditions. A long-duration (*ca.* 400 msec) negativity may arise, for example, with a latency of approximately 100 msec after the early superficial spike and a long-duration positivity, with a still greater latency after the early deep spike (Crain, 1964a; Crain and Bornstein, 1964). The slow waves and superimposed rhythmic, repetitive discharges may appear spontaneously in bursts (*ca.* 1 sec) occurring sporadically several times per minute. The amplitude and duration of these complex discharges were greatly enhanced by strychnine, resembling the onset of strychnine sensitivity in electrocorticographic recordings of 4-day-old rat cerebal neocortex *in situ* (Crain, 1952, 1970b, 1975a).

Explants of *fetal* mouse cerebral tissues have been used in more recent studies to facilitate direct microelectrode recordings and pharmacologic manipulations during the critical developmental period just before birth (Crain and Bornstein, 1974; Crain *et al.*, 1975). Only simple early-latency spike potentials could be elicited by large single, or repetitive, electric stimuli in 18-day fetal hippocampus and neocortex explants after 1-2 days *in vitro* (Fig. 13A). On the other hand, by 3 days *in vitro*, although most 18-day fetal explants still showed only simple spike responses to electric stimuli when bathed in normal medium (Fig. 13A), much more complex discharges could be evoked after introducing relatively low concentrations of strychnine (Fig. 13B), picrotoxin, or bicuculline (*ca.* 10^{-5}M), or by transfer to a chloride-free medium (Fig. 13C). Negative slow-wave potentials reached unusually large amplitudes, especially in a chloride-free medium. Generation of the drug-induced slow-wave discharges by Ca^{2+} -dependent synaptic transmitter release is indicated by the rapid and complete block of these complex potentials after increasing the Mg^{2+} concentration from 1 to 5-10 mM, whereas spikes could still be directly evoked (Crain, 1974a,b, 1976). The complex network discharges often occurred synchronously over widespread regions of the explant, even at these early stages of synaptogenesis.

B. Morphologic Correlates of Early Network Discharges

No typical synapses were detected in fresh fragments of 18-day fetal mouse cerebral neocortex or hippocampus, or in explants sampled after 1-2 days *in vitro*, in agreement with the absence of complex slow-wave discharges at these stages. Membrane thickenings suggestive of synaptic junctions were indeed seen often, but *not* in association with clusters of vesicles. In other cases vesicle arrays were seen in the absence of characteristic membrane thickenings. On the other hand, in the same 18-day fetal cerebral neocortex and hippocampus explants which showed synaptic network discharges at 3 days *in vitro*, ultrastructural analyses showed that at least some characteristic, although immature, synaptic complexes could be detected in each thin section (Crain *et al.*, 1975). Both axosomatic and axodendritic synapses were observed. Similar types of synapses were also noted in explants of 18-day fetal rat hippocampus after 3 days in culture (Wenzel *et al.*, 1973). The synaptic junctions showed morphologic features comparable to those observed during early stages of synaptogenesis in other CNS tissues *in situ* and *in vitro* (Bunge *et al.*, 1967a; Johnson and Armstrong-James, 1970; Model *et al.*, 1971; LaVail and Wolf, 1973). Only small numbers of vesicles were clustered near the electron-dense thickening of the presynaptic membrane, and mature synapses were quite sparse in these 18-day fetal cerebral explants after 3 days in culture.

The data therefore indicate that in spite of the sparse distribution of synapses in these immature cerebral explants, prominent, organotypic slow-wave discharges can be recorded with extracellular microelectrodes. They further suggest that some types of complex synaptic networks may already be functioning during late fetal or newborn stages in mouse cerebral neocortex and hippocampus — assuming a gestation period of 21 days (Crain and Bornstein, 1974; Crain, 1975a; cf. Armstrong-James and Williams, 1963, 1964).

C. *Early Onset of Inhibition*

The primitive cerebral synaptic networks appear, moreover, to be under tonic inhibition at these early stages of synaptogenesis *in vitro*, since the pharmacologic procedures which revealed or greatly augmented the complex discharges in immature explants have been shown to produce selective interference with CNS inhibitory mechanisms *in situ*, e.g., picrotoxin, bicuculline, strychnine, or chloride-free medium (Fig. 13) (Crain and Bornstein, 1974). Simple cell assemblies connected by excitatory synaptic connections probably develop first during CNS synaptogenesis (Purpura, 1972; Crain, 1974a), but tonic inhibitory mechanisms may appear shortly thereafter which maintain a restraining influence on these primitive networks until they reach critical stages of development in forward reference to their ultimate functions (Crain, 1974a, 1976; Pollack and Crain, 1972).

Correlative extra- and intracellular recordings in older hippocampal explants (Zipser *et al.*, 1973) provide strong additional evidence that the large negative slow waves evoked in these explants after 3-5 days *in vitro*, represent, or at least include, summated EPSPs (Crain, 1966, 1975b, c). Furthermore, the large "primary" positive slow-wave components may represent summated IPSPs. The early tonic inhibition that appears to occur in these CNS explants could be produced not only by sustained inhibitory interneuronal impulse activity at synaptic junctions, but also by spontaneous leakage of inhibitory transmitter from nerve terminals to neighboring extrajunctional receptor sites (Crain, 1976).

Electrophysiologic studies of fetal mouse olfactory bulb explants suggest a similar development of GABA-mediated inhibitory synaptic circuits, as well as complex excitatory synaptic networks, during the first week *in vitro* (Corrigall *et al.*, 1976).

V. Formation of Functional Synaptic Connections in Co-Cultures of Embryonic Mouse Retinal and Tectal Explants

Explants of retina and superior colliculi (whole tecta) from 14-day mouse embryos have now been successfully cultured on collagen-coated coverslips in Maximow slide chambers for up to 6 weeks (Smalheiser *et al.*, 1977). Retinal explants developed characteristic "rosettes", laminar arrays of neurons and glia, and neuritic outgrowth (e.g., LaVail and Hild, 1971). Retinal recordings (with 3-5μ electrodes) showed complex spontaneous and electrically-evoked repetitive-spike and slow-wave patterns which were enhanced by strychnine and bicuculline (10^{-5}M). Tectal explants also developed synaptic-network activities, as in cultures of cerebral explants (e.g., Crain and Bornstein, 1964, 1974). The tectal responses consisted of early-latency spikes, followed by more complex spike and slow-wave potentials, and they were strongly enhanced by *d*-tubocurarine, strychnine and bicuculline (10^{-6}-10^{-5}M). In retinotectal co-cultures, single focus stimuli in the retinal explants (with 5-10μ electrodes) evoked characteristic tectal responses at stimulus strengths much lower than needed to backfire tectal neurites. Stimuli applied to retinal neurites in the outgrowth far from the tectum also evoked similar tectal responses. These data (and other controls) indicate that functional retinotectal connections developed after growth of retinal neurites into the tectal explant. Microelectrode mapping is being used to determine if retinal neurites may project to specific regionally localized sites in explants of the whole tectum (as occurs with dorsal root ganglion neurites in relation to sensory target zones in co-cultures with spinal cord and medulla explants) (Crain and Peterson, 1975a,b; see Section IIB, III). Preliminary electrophysiologic analyses suggest that groups of axons arising from ganglion cells in a small region of a retinal explant tend to arborize within a restricted zone of the tectal explant, and axons from another retinal locus tend to be associated with a different tectal zone. Topographic mapping of retinal and tectal explants with controlled orientation has not yet been attempted. These electrophysiologic studies of retinotectal cultures should provide a valuable new model system to supplement and extend to the cellular level *in situ* studies of mechanisms underlying formation of specific patterned synaptic networks. In view of the complexities involved in current experiments *in situ*, and the ambiguities and controversies engendered by these analyses (e.g., Hunt and Jacobson, 1974; Meyer and Sperry, 1974), more direct studies with simpler cellular arrays, under more flexibly controlled conditions in culture, may provide valuable clues to basic neurospecificity mechanisms (see also Keating, 1976; Meyer and Sperry, 1976).

Recent studies by Yoon (1976) indicate, moreover, that even small fragments of adult goldfish tectum retain their "original topographic polarity [as determined by electrophysiologic mapping of the visual projection] . . . regardless of the orientation of [surgical] reimplantation after either a rotation or even an inversion." It will be of interest to determine if tectal explants can also develop and/or maintain "topographic polarity" under isolated conditions *in vitro* (see also Jacobson and Levine, 1975a,b; Jacobson, 1976). Evidence of "patterned migration *in vitro* of embryonic DRG perikarya attached to a spinal cord explant — forming organized dorsal roots — and outgrowth of orderly fascicles of ventral root fibers onto a homogeneous collagen substrate in the absence of target tissues" led Crain (1976) to suggest "that neurons in organotypic cultures can grow complex 3-dimensional neuritic arborizations with patterns that are determined by genetic programs which develop in relation to the position of the neuron perikaryon in 'body-space' (see also Diamond *et al.*, 1976). The concept of body-space in a culture may be useful in cases of organotypic CNS explants which include sufficient components to provide a coded representation of the original body axes in its geometrical array of neurons and glial cells." Perhaps "*sets* of neurons in suitably prepared CNS explants *may* retain in culture complex properties which determine, not only the formation of connections between specific types of neurons (i.e., phenotypic specificity), but also the development of an organized spatial framework in which the neurons make synaptic connections at particular positions within the cell population, leading to functional patterns related to the 3-dimensional body axes, i.e., locus specificity (Hunt and Jacobson, 1974; see also Sperry, 1965). If this high degree of order can occur in culture it would permit experimental analyses of some of the basic principles which regulate development of the intricate spatial organization and precise regional localizations in the central nervous system. . . . Analyses of these complex types of tissue cultures may provide clues to some of the fundamental aspects of spatial relationships between arrays of CNS neurons that are invariant to drastic environmental transformation" (Crain, 1976).

Conclusions

Synapse formation has been demonstrated to occur in cultures prepared from almost all types of embryonic mammalian central and peripheral neurons. Correlative electrophysiologic and electron microscopic studies of cultured CNS tissues have emphasized that development of many organotypic synaptic network structures and functions appear to be so tightly coupled to genetic factors that the explants continue to organize in rather stereotyped fashion in spite of wide variations in environmental conditions — even after chronic exposure to pharmacologic agents which block all nerve impulses (Crain *et al.*, 1968; Model *et al.*, 1971), as well as after random dispersion of immature dissociated CNS neurons in a relatively homogeneous, unpatterned culture environment (e.g., Crain and Bornstein, 1972; Bornstein and Model, 1972; Nelson and Peacock, 1973). Similarly, functional neuromuscular junctions can form in cultures containing high concentrations of *d*-tubocurarine and other cholinergic blocking agents (e.g., Crain and Peterson, 1971, 1974a; Cohen, 1972; Obata, 1977) as well as during chronic exposure to levels of tetrodotoxin which block all nerve action potentials (Obata, 1977).

The formation and long-term maintenance of synaptic junctions, involving membrane specializations in the postsynaptic cells, are clearly neurotrophic effects; their occurrence *in vitro* in the absence of neural impulse activity suggests mediation by neurotrophic chemical agents. Maturation and long-term maintenance of organotypic, triggerable synaptic networks in a typical CNS explant therefore involves a complex *set* of neurotrophic interactions between large numbers (e.g., hundreds) of component neurons and glial cells (Crain and Peterson, 1974a; see also Guth and Windle, 1970).

Tissue cultures provide a valuable model system for analyses of the molecular mechanisms underlying these neurotrophic effects in relatively simple neuromuscular synapses as well as in more complex polysynaptic CNS networks. Biochemical studies, however, must be coordinated with electrophysiologic analyses which provide clear-cut means of determining if a neural assembly is operating normally in a particular group of cultures — morphologic or chemical tests demonstrate only that essential, but not necessarily *sufficient,* components are present.

In a recent study on synapse formation between clonal neuroblastoma x glioma hybrid cells (NG108-15) and dissociated striated muscle cells in culture, Nelson *et al.*, (1976) noted: "Establishment of synapses between nerve and muscle poses a paradox, for on the one hand, muscle movements are highly coordinated, which suggests that neuromuscular synapses and other synapses in the neural circuits are assembled with high precision, whereas the demonstrated ability of autonomic neurons of the vagus

(Landmesser, 1971), sympathetic ganglion neurons (Nurse and O'Lague, 1975), and clonal (NG108-15) hybrid cells to synapse with striated muscle cells suggest that functional synapses can form that may not be dependent upon highly specific cell recognition molecules. . . . The molecular nature of the cell interactions that lead to synapse formation is not known. . . . The results suggest that synapse formation and the efficiency of transmission are regulated *in vitro,* apparently by independent processes. It may be possible to find conditions for the conversion of the early form of synapse to the late, mature form." Puro and Nirenberg (1976) demonstrated, moreover, that "clonal NG108-15 hybrid cells form synapses with cells from different muscles and from different organisms such as chick, mouse and rat. . . . These results with clonal cells confirm and extend the findings of Crain *et al.* (1970) which show that explanted mouse spinal cord neurons form synapses with mouse, rat, or human muscle cells." On the basis of these and related data, Puro and Nirenberg (1976) postulate "that much of the specificity of the normal neuromuscular synapse is acquired after the synapses form by a process of selection that reduces the number of synapses and that is dependent upon effective transmission across the synapse, rather than by a process of matching complementary molecules on neurons and muscle cells that code for different synaptic connections" (see also Changeux and Danchin, 1976). These hypotheses need to be tested by more systematic experiments with critical arrays of neurons and target cells in culture, especially in regard to their relevance for CNS synaptic network formation. It should be emphasized that the apparent lack of specificity of synapse formation in many of the types of cultures noted above may really be due to deficiencies in the *in vitro* environment which preclude phenotypic expression of a genotypic neuronal cell-recognition code (see also Varon, this volume). Judicious presentation of arrays of target and appropriate non-target cells to neurons growing under more *organotypic* culture conditions should help to clarify whether specific cell recognition molecules may indeed play a significant role in preferential formation of at least some types of specific synaptic connections. Our recent studies of patterned neuritic growth of isolated dorsal root ganglion cells in relation to synaptic network formation with regionally localized target neurons in explants of spinal cord and medulla (Crain and Peterson, 1975a,b; 1976; Crain, 1976) provide a valuable model system for further analyses of this fundamental problem in neurobiology.

Correlation of electrophysiologic studies on cord-DRG cultures with *in situ* analyses suggest, moreover, that formation of DRG synapses on dorsal cord neurons leads to sequential development of GABA-mediated interneuron feedback loops to DRG terminals (see Section IIB). It will be of interest to determine if these types of networks can also develop in cultures of dissociated cord-DRG neurons (e.g., Peacock *et al.*, 1973), or whether formation of patterned neuritic feedback circuits requires the existence of orderly neuronal-glial arrays which maintain positional information regarding body-space coordinates (Crain, 1976). In addition to development of GABA-mediated interneuron feedback systems, recent studies of the effects of opiates and their antagonists on cord-DRG explants suggest that endorphinergic interneuron loops to DRG terminals may also mediate some of the inhibitory controls on sensory-evoked network responses in dorsal horn regions of these organotypic cultures (Crain, 1977; Crain *et al.* 1977). This interesting new evidence is based on selective enhancement by naloxone of DRG-evoked negative slow-wave responses in dorsal cord loci of our explants (following introduction of low concentrations, 10^{-8}-10^{-6}M, comparable to levels which reverse analgesic doses of morphine *in situ*). In contrast, low concentrations of morphine (10^{-7}-10^{-6}M) or etorphine (10^{-8}M) produce selective depression of these sensory-evoked dorsal-horn network responses. The opiate-depressant effects in cord-DRG cultures can be prevented by prior introduction of naloxone, as *in situ.* It is of interest that high levels of opiate receptors develop in the same neuroblastoma x glioma hybrid cells which form synapses with muscle cells (Klee and Nirenberg, 1974). Although marked attenuation of cyclic AMP levels occurs rapidly after exposure of these cells to morphine (Klee *et al.* 1975), preliminary electrophysiologic analyses have not detected any specific depression of electric excitability at concentrations of 10^{-6}-10^{-5}M (P. G. Nelson, personal communication; this volume). These observervations emphasize the importance of developing tissue culture systems which permit faithful phenotypic expression of basic nerve cell assemblies encoded within the CNS genetic program. In addition to providing useful preparations for *in vitro* analyses of complex CNS functions, organotypic explants also constitute a valuable bridge to establish realistic links between studies in the intact organism and simple arrays of dissociated nerve cells.

Acknowledgments

This work was supported by research grants NS-06545 and NS-12405 from the National Institute of Neurological and Communicative Disorders and Stroke, grant BMS75-03728 from the National Science Foundation, and a grant from the Alfred P. Sloan Foundation.

I wish to express my appreciation to Edith R. Peterson and Murray B. Bornstein, Depts. of Neuroscience and Neurology, at Einstein, for providing the neural cultures used in these studies (see also Bornstein, this volume).

Thanks are also due to Bea Crain for skillful technical assistance in carrying out many of the electrophysiologic experiments.

References

Anderson, P., J. C. Eccles, R. F. Schmidt and T. Yokoto. 1964. Slow potential waves produced in cuneate nucleus by cutaneous volleys and by cortical stimulation. J. Neurophysiol. 27:78-91.

Armstrong-James, M.A. and T. D. Williams. 1963. Post-natal development of the direct cortical response in the rat. J. Physiol. (Lond.) 168:19-20.

Armstrong-James, M. A. and T. D. Williams. 1964. Differences in the direct cortical-response to unifocal stimuli during the post-natal development of the rat cerebral cortex. J. Physiol. (Lond.). 170:15P-16P.

Barker, J. L. and R. A. Nicoll. 1973. The pharmacology and ionic dependency of amino acid responses in the frog spinal cord. J. Physiol. (Lond.). 228:259-278.

Benoist, J. M., J. M. Besson and J. R. Boissier. 1974. Modification of presynaptic inhibition of various origins by local application of convulsant drugs on cat's spinal cord. Brain Res. 71:172-177.

Bornstein, M. B. 1964. Morphological development of neonatal mouse cerebral cortex in tissue culture. In: *Neurological and Electroencephalographic Correlative Studies in Infancy.* P. Kellaway and I. Petersen, eds. Grune and Stratton, New York. pp. 1-11.

Bornstein, M. B. and P. G. Model. 1972. Development of synapses and myelin in cultures of dissociated embryonic mouse spinal cord, medulla and cerebrum. Brain Res. 37:287-293.

Bornstein, M. B., H. Iwanami, G. M. Lehrer and L. Breitbart. 1968. Observations on the appearance of neuromuscular relationships in cultured mouse tissues. Z. Zellforsch. 92:197-206.

Bunge, M. B., R. P. Bunge and E. R. Peterson. 1967a. The onset of synapse formation in spinal cord culture as studied by electron microscopy. Brain Res. 6:728-749.

Bunge, M. B., R. P. Bunge, E. R. Peterson and M. R. Murray. 1967b. A light and electron microscope study of long-term organized cultures of rat dorsal root ganglia. J. Cell Biol. 32:439-466.

Bunge, R. P. 1976a. The expression of neuronal specificity in tissue culture. In: *Neuronal Recognition.* S. H. Barondes, ed. Plenum Press, New York. pp. 109-128.

Bunge, R. P. 1976b. Changing uses of nerve tissue culture 1950-1975. In: *The Nervous System. I. Basic Neurosciences.* R. O. Brady, ed. Raven Press, New York. pp. 31-42.

Bunge, R. P., M. B. Bunge and E. R. Peterson. 1965. An electron microscope study of cultured rat spinal cord. J. Cell Biol. 24:163-191.

Bunge, R. P., R. Rees, P. Wood, H. Burton and C.-P. Ko. 1974. Anatomical and physiological observations on synapses formed on isolated autonomic neurons in tissue culture. Brain Res. 66:401-412.

Changeux, J. P. and A. Danchin. 1976. Selective stabilisation of developing synapses as a mechanism for the specification of neuronal networks. Nature (Lond.). 264:705-712.

Cohen, M. W. 1972. The development of neuromuscular connexions in the presence of *d*-tubocurarine. Brain Res. 41:457-463.

Corrigall, W. A., S. M. Crain and M. B. Bornstein. 1976. Electrophysiological studies of fetal mouse olfactory bulb explants during development of synaptic functions in culture. J. Neurobiol. 7:521-536.

Crain, S. M. 1952. Development of electrical activity in the cerebral cortex of the albino rat. Proc. Soc. Exper. Biol. Med. 81:49-51.

Crain, S. M. 1956. Resting and action potentials of cultured chick embryo spinal ganglion cells. J. Comp. Neurol. 104:285-330.

Crain, S. M. 1964a. Development of bioelectric activity during growth of neonatal mouse cerebral cortex in tissue culture. In: *Neurological and Electroencephalographic Correlative Studies in Infancy.* P. Kellaway and I. Petersen, eds. Grune and Stratton, New York. pp. 12-26.

Crain, S. M. 1964b. Electrophysiological studies of cord-innervated skeletal muscle in long-term tissue cultures of mouse embryo myotomes. Anat. Rec. 148:273.

Crain, S. M. 1966. Development of "organotypic" bioelectric activities in central nervous tissues during maturation in culture. Internat. Rev. Neurobiol. 9:1-43.

Crain, S. M. 1968. Development of functional neuromuscular connections between separate explants of fetal mammalian tissues after maturation in culture. Anat. Rec. 160:466.

Crain, S. M. 1970a. Bioelectric interactions between cultured fetal rodent spinal cord and skeletal muscle after innervation *in vitro*. J. Exper. Zool. 173:353-370.

Crain, S. M. 1970b. Tissue culture studies of developing brain function. In: *Developmental Neurobiology*. W. A. Himwich, ed. C. C. Thomas, Springfield, Ill. pp. 165-196.

Crain, S. M. 1971. Intracellular recordings suggesting synaptic functions in chick embryo spinal sensory ganglion cells isolated *in vitro*. Brain Res. 26:188-191.

Crain, S. M. 1973. Microelectrode recording in brain tissue cultures. In: *Methods in Physiological Psychology, 1. Bioelectric Recording Techniques:* Cellular Processes and Brain Potentials. R. F. Thompson and M. M. Patterson, eds. Academic Press, New York. pp. 39-75.

Crain, S. M. 1974a. Tissue culture models of developing brain functions. In: *Studies on the Development of Behavior and the Nervous System, 2. Aspects of Neurogenesis*. G. Gottlieb, ed. Academic Press, New York. pp. 69-114.

Crain, S. M. 1974b. Selective depression of organotypic bioelectric activities of CNS tissue cultures by pharmacologic and metabolic agents. In: *Drugs and the Developing Brain*. A Vernadakis and N. Weiner, eds. Plenum Publ., New York. pp. 29-57.

Crain, S. M. 1974c. Paroxysmal discharges after release from tonic inhibition in mouse CNS tissue cultures during acute exposure to chloride-free medium. Fed. Proc. 33:285.

Crain, S. M. 1975a. Onset of inhibitory functions during synaptogenesis in fetal mouse brain cultures. In:*Golgi Centennial Symposium*. M. Santini, ed. Raven Press, New York. pp. 625-634.

Crain, S. M. 1975b. Tissue culture models of brain functions. In: *Methods in Brain Research*. P. B. Bradley, ed. John Wiley & Sons, New York. pp. 379-411.

Crain, S. M. 1975c. Development of complex synaptic functions in 'simple' neuronal arrays in culture. In: *'Simple' Nervous Systems*. P. N. R. Usherwood and D. R. Newth, eds. Edward Arnold Ltd., England. pp. 67-117.

Crain, S. M. 1975d. Physiology of CNS tissues in culture. In: *Metabolic Compartmentation and Neurotransmission. Relation to Brain Structure and Function*. S. Berl, D. D. Clarke and D. J. Schneider, eds. Plenum Press, New York. pp. 273-303.

Crain, S. M. 1976. *Neurophysiologic Studies in Tissue Culture*. Raven Press, New York.

Crain, S. M. 1977. Selective depressant effects of opiates and GABA on sensory-evoked synaptic networks in dorsal-horn regions of spinal cord cultures. In: *Iontophoresis and Transmitter Mechanisms in the Mammalian Central Nervous System*. J. S. Kelly, ed. Elsevier-North Holland, Amsterdam, in press.

Crain, S. M. and M. B. Bornstein. 1964. Bioelectric activity of neonatal mouse cerebral cortex during growth and differentiation in tissue culture. Exper. Neurol. 10:425-450.

Crain, S. M. and M. B. Bornstein. 1972. Organotypic bioelectric activity in cultured reaggregates of dissociated rodent brain cells. Science 176:182-184.

Crain, S. M. and M. B. Bornstein. 1974. Early onset in inhibitory functions during synaptogenesis in fetal mouse brain cultures. Brain Res. 68:351-357.

Crain, S. M. and E. R. Peterson. 1963. Bioelectric activity in long-term cultures of spinal cord tissues. Science 141:427-429.

Crain, S. M. and E. R. Peterson. 1964. Complex bioelectric activity in organized tissue cultures of spinal cord (human, rat and chick). J. Cell. Comp. Physiol. 64:1-15.

Crain, S. M. and E. R. Peterson. 1966. Formation of functional interneuronal connections between explants of rat brainstem and spinal cord during development in culture. Fed. Proc. 25:701.

Crain, S. M. and E. R. Peterson. 1967. Onset and development of functional interneuronal connections in explants of rat spinal cord-ganglia during maturation in culture. Brain Res. 6:750-762.

Crain, S. M. and E. R. Peterson. 1971. Development of paired explants of fetal spinal cord and adult skeletal muscle during chronic exposure to curare and hemicholinium. In Vitro 6:373.

Crain, S. M. and E. R. Peterson. 1974a. Development of neural connections in culture. Ann. N.Y. Acad. Sci. 228:6-34.

Crain, S. M. and E. R. Peterson. 1974b. Enhanced afferent synaptic functions in fetal mouse spinal cord-sensory ganglion explants following NGF-induced ganglion hypertrophy. Brain Res. 79:145-152.

Crain, S. M. and E. R. Peterson. 1975a. Development of specific sensory-evoked synaptic networks in fetal mouse cord-brainstem cultures. Science 188:275-278.

Crain, S. M. and E. R. Peterson. 1975b. Selective growth of neurites from isolated fetal mouse dorsal root ganglia toward specific CNS target tissues. Soc. Neuroscience, 5th Ann. Mtg., New York, p. 783.

Crain, S. M. and E. R. Peterson. 1976. Formation of specific synaptic networks between dorsal root ganglia and longterm-deafferented explants of fetal mouse spinal cord and medulla. Soc. Neuroscience, 6th Ann. Mtg., Toronto, Canada. p. 1018.

Crain, S. M. and E. R. Peterson. 1977. Development of specific synaptic networks in cultures of CNS tissues. In: *Reviews of Neuroscience.* S. Ehrenpreis and I. Kopin, eds. Raven Press, New York. in press.

Crain, S. M. and E. D. Pollack. 1973. Restorative effects of cyclic AMP on complex bioelectric activities of cultured fetal rodent CNS tissues after acute Ca^{++}-deprivation. J. Neurobiol. 4:321-342.

Crain, S. M., H. Benitez and A. E. Vatter. 1964a. Some cytologic effects of salivary nerve-growth factor on tissue cultures of peripheral ganglia. Ann. N.Y. Acad. Sci. 118:206-231.

Crain, S. M., R. P. Bunge, M. B. Bunge and E. R. Peterson. 1964b. Bioelectric and electron microscopic evidence for development of synapses in spinal cord cultures. J. Cell Biol. 23:114-115A.

Crain, S. M., E. R. Peterson and M. B. Bornstein. 1968. Formation of functional connections between explants of various mammalian central nervous tissues during development *in vitro.* In: *Ciba Found. Symposium, Growth of the Nervous System.* G. E. W. Wolstenholme and M. O'Connor, eds. Churchill, London. pp. 13-31.

Crain, S. M., L. Alfei and E. R. Peterson. 1970. Neuromuscular transmission in cultures of adult human and rodent skeletal muscle after innervation *in vitro* by fetal rodent spinal cord. J. Neurobiol. 1:471-489.

Crain, S. M., C. S. Raine and M. B. Bornstein. 1975. Early formation of synaptic networks in cultures of fetal mouse cerebral neocortex and hippocampus. J. Neurobiol. 6:329-336.

Crain, S. M., E. R. Peterson, B. Crain and E. J. Simon. 1977. Selective opiate depression of sensory-evoked synaptic networks in dorsal-horn regions of spinal cord cultures. *Brain Res.,* (in press).

Curtis, D. R., A. W. Duggan, D. Felix and G. A. R. Johnston. 1971. Bicuculline, an antagonist of GABA and synaptic inhibition in the spinal cord of the cat. Brain Res. 32:69-96.

Davidoff, R. A. 1972a. Gamma-aminobutyric acid antagonism and presynaptic inhibition in the frog spinal cord. Science. 175:331-333.

Davidoff, R. A. 1972b. The effects of bicuculline on the isolated spinal cord of the frog. Exper. Neurol. 35:179-193.

Davidson, N. and C. A. P. Southwick. 1971. Amino acids and presynaptic inhibition in the rat cuneate nucleus. J. Physiol (Lond.). 219:689-708.

Diamond, J. and R. Miledi. 1962. A study of foetal and new-born rat muscle fibres. J. Physiol. (Lond.). 162:393-408.

Diamond, J., E. Cooper, C. Turner and L. Macintyre. 1976. Trophic regulation of nerve sprouting. Neuron-target interactions and spatial relations control sensory nerve fields in salamander skin. Science 193:371-377.

Eccles, J. C. 1964. *The Physiology of Synapses.* Springer-Verlag, Berlin.

Fischbach, G. D. 1974. Some aspects of neuromuscular junction formation. In: *Cell Communication.* R. P. Cox, ed. John Wiley, New York. pp. 43-66.

Fischbach, G. D. and M. A. Dichter. 1974. Electrophysiologic and morphologic properties of neurons in dissociated chick spinal cord cell cultures. *Devel. Biol.* 37:100-116.

Fischbach, G. D., S. A. Cohen and M. P. Henkart. 1974a. Some observations on trophic interaction between neurons and muscle fibers in cell culture. *Ann. N. Y. Acad. Sci.* 228:35-46.

Fischbach, G. D., M. R. Henkart, S.A. Cohen, A. C. Breuer, J. Whysner and F. M. Neal.

1974b. Studies on the development of neuromuscular junctions in cell culture. In: *Soc. Gen. Physio. 26th Symp:* Raven Press, New York. pp. 259-283.

Fischbach, G. D., D. K. Berg, S. A. Cohen and E. Frank. 1976. Enrichment of nerve-muscle synapses in spinal cord-muscle cultures and identification of relative peaks of ACh sensitivity at sites of transmitter release. In: *Cold Spring Harbor Symp. on Quantitative Biol. XL. The Synapse,* pp. 347-357.

Godfrey, E. W., P. G. Nelson, B. K. Schrier, A. C. Breuer and B. R. Ransom. 1975. Neurons from fetal rat brain in a new cell culture system: a multidisciplinary analysis. Brain Res. 90:1-21.

Guillery, R., H. M. Sobkowicz and G. L. Scott. 1968. Light and electron microscopical observation of the ventral horn and ventral root in long-term cultures of the spinal cord of the fetal mouse. J. Comp. Neurol. 134:433-476.

Guth, L. and W. F. Windle. 1970. The enigma of central nervous regeneration. Exper. Neurol. (Suppl. 5) 28:1-43.

Harris, A. J. 1974a. Inductive functions of the nervous system. Ann. Rev. Physiol. 36:251-305.

Harris, A. J. 1974b. Role of acetylcholine receptors in synapse formation. In: *Soc. Gen. Physiol. 26th Symp.* Raven Press, New York. pp. 315-337.

Hunt, R. K. and M. Jacobson. 1974. Neuronal specificity revisited. In: *Current Topics in Developmental Biology.* A. Moscona and A. Monroy, eds. Academic Press, New York. pp. 203-259.

Jacobson, M. 1970. *Developmental Neurobiology.* Holt, Rinehart and Winston, New York.

Jacobson, M. 1976. Neuronal recognition in the retinotectal system. In: *Neuronal Recognition.* S. H. Barondes, ed. Plenum Press, New York. pp. 3-23.

Jacobson, M. and R. L. Levine. 1975a. Plasticity in the adult frog brain: filling the visual scotoma after excision of translocation of parts of the optic tectum. Brain Res. 88:339-345.

Jacobson, M. and R. L. Levine. 1975b. Stability of implanted duplicate tectal positional markers serving as targets for optic axons in adult frogs. Brain Res. 92:468-471.

Johnson, R. and M. Armstrong-James. 1970. Morphology of superficial postnatal cerebral cortex with special reference to synapses. Z. Zellforsch. 110:540-558.

Keating, M. J. 1976. The formation of visual neuronal connections: An appraisal of the present status of the theory of "neuronal specificity". In: *Studies on the Development of Behavior and the Nervous System. 3. Neural and Behavorial Specificity.* G. Gottlieb, ed. Academic Press, New York. pp. 59-110.

Kelly, A. M. and S. I. Zacks. 1969. The fine structure of motor endplate morphogenesis. J. Cell Biol. 42:154-169.

Kim, S. U. 1972. Light and electron microscope study of mouse cerebral neocortex in tissue culture. Exper. Neurol. 35:305-321.

Klee, W. E. and M. Nirenberg. 1974. A neuroblastoma x glioma hybrid cell line with morphine receptors. Proc. Natl. Acad. Sci. 71:3474-3477.

Klee, W. E., S. K. Sharma and M. Nirenberg. 1975. Opiate receptors as regulators of adenylate cyclase. Life. Sci. 16:1869-1874.

Ko, C.-P., H. Burton and R. P. Bunge. 1975. Cholinergic synapses between spinal cord and sympathetic neurons in tissue culture. *Soc. Neuroscience,* 5th Ann. Mtg., New York. p. 816.

Ko, C. -P., H. Burton and R. P. Bunge. 1976a. Synaptic transmission between rat spinal cord explants and dissociated superior cervical ganglion neurons in tissue culture. Brain Res. 117:437-460.

Ko, C. -P., H. Burton, M. I. Johnson and R. P. Bunge. 1976b. Synaptic transmission between rat superior cervical ganglion neurons in dissociated cell cultures. Brain Res. 117:461-485.

Landmesser, L. 1971. Contractile and electrical responses of vagus-innervated frog sartorius muscles. J. Physiol. (Lond.). 213:707-725.

LaVail, M. M. and W. Hild. 1971. Histotypic organization of the rat retina *in vitro.* Z. Zellforsch. 114:557-579.

LaVail, J. H. and M. K. Wolf. 1973. Postnatal development of the mouse dentate gyrus in organotypic cultures of the hippocampal formation. Amer. J. Anat. 137:47-66.

Levi, G. and H. Meyer. 1938. Presentation de cultures d'un nombre restreint d'éléments nerveux avec quelques considérations sur les rapports d'interdépendence entre les neurones. C. R. Assoc. Anat. 33:312-328.

Levi-Montalcini, R. 1958. Chemical stimulation of nerve growth. In: *Symposium on the Chemical Basis of Development.* W. D. McElroy and B. Glass, eds. Johns Hopkins Press, Baltimore. pp. 646-664.

Levi-Montalcini, R. and P. U. Angeletti. 1968a. Biological aspects of the nerve growth factor. In: *Ciba Found. Symposium, Growth of the Nervous System.* G. E. W. Wolstenholme and M. O'Connor, eds. Churchill, London. pp. 126-147.

Levi-Montalcini, R. and P. U. Angeletti. 1968b. Nerve Growth Factor. Physiol. Rev. 48:534-569.

Levi-Montalcini, R. and B. Booker. 1960. Excessive growth of the sympathetic ganglia evoked by a protein isolated from mouse salivary glands. Proc. Natl. Acad. Sci. 46:373-384.

Levi-Montalcini, R. and S. Cohen. 1960. Effects of the extract of the mouse submaxillary salivary glands on the sympathetic system of mammals. Ann. N.Y. Acad. Sci. 85:324-341.

Lumsden, C. E. 1951. Aspects of neurite outgrowth in tissue culture. Anat. Rec. 110:145-179.

May, M. K. and T. J. Biscoe. 1975. An investigation of the foetal rat spinal cord. II. An ultrastructural study on the development of synapses with the aid of observations on some electrophysiological properties. Cell and Tiss. Res. 158:251-268.

Meyer, R. L. and R. W. Sperry. 1974. Explanatory models for neuroplasticity in retinotectal connections. In: *Plasticity and Recovery of Function in the Central Nervous System.* D. G. Stein, J. J. Rosen and N. Butlers, eds. Academic Press, New York. pp. 45-63.

Meyer, R. L. and R. W. Sperry. 1976. Retinotectal specificity: chemoaffinity theory. In: *Studies on the Development of Behavior and the Nervous System, 3. Neural and Behavioral Specificity.* G. Gottlieb, ed. Academic Press, New York. pp. 111-149.

Model, P. G., M. B. Bornstein, S. M. Crain and G. D. Pappas. 1971. An electron microscopic study of the development of synapses in cultured fetal mouse cerebrum continuously exposed to xylocaine. J. Cell Biol. 49:362-371.

Murray, M. R. 1965a. Muscle tissues *in vitro.* In: *Cells and Tissues in Culture,* 2. E. N. Wilmer, ed. Academic Press, New York. pp. 311-372.

Murray, M. R. 1965b. Nervous tissues *in vitro.* In: *Cells and Tissues in Culture,* 2. E. N. Wilmer, ed. Academic Press, New York. pp. 373-455.

Murray, M. R. 1972. Skeletal muscle in culture. In: *Structure and Function, 1,* (2nd ed.). G. H. Bourne, ed. Academic Press, New York. pp. 237-299.

Narayanan, C. H., M. W. Fox and V. Hamburger. 1971. Prenatal development of spontaneous and evoked activity in the rat (Rattus Norvegicus Albinus). Behaviour 40:100-134.

Nelson, P. G. 1973. Electrophysiological studies of normal and neoplastic cells in tissue culture. In: *Tissue Culture of the Nervous System.* G. Sato, ed. Plenum Press, New York. pp. 135-160.

Nelson, P. G. 1975. Nerve and muscle cells in culture. Physiological Rev. 55:1-61.

Nelson, P. G. 1976. Central nervous system synapses in cell culture. In: *Cold Spring Harbor Symp. on Quantitative Biol. XL. The Synapse.* pp. 359-371.

Nelson, P. G. and J. H. Peacock. 1973. Electrical activity in dissociated cell cultures from fetal mouse cerebellum. Brain Res. 61:163-174.

Nelson, P. G., C. Christian and M. Nirenberg. 1976. Synapse formation between clonal neuroblastoma x glioma hybrid cells and striated muscle cells. Proc. Natl. Acad. Sci. 73:123-127.

Nurse, C. A. and P. H. O'Lague. 1975. Formation of cholinergic synapses between dissociated sympathetic neurons and skeletal myotubes of the rat in cell culture. Proc. Natl. Acad. Sci. 72:1955-1959.

Obata, K. 1977. Development of neuromuscular transmission in culture with a variety of neurons and in the presence of cholinergic substances and tetrodotoxin. Brain Res. 119:141-153.

O'Lague, P. H., K. Obata, P. Claude, E. J. Furshpan and D. D. Potter. 1974. Evidence for cholinergic synapses between dissociated rat sympathetic neurons in cell culture. Proc. Natl. Acad. Sci. 71:3602-3603.

Olson, L. 1967. Outgrowth of sympathetic adrenergic neurons in mice treated with a nerve-growth factor (NGF). Z. Zellforsch. 81:155-173.

Olson, M. J. and R. P. Bunge. 1973. Anatomical observations on the specificity of synapse formation in tissue culture. Brain Res. 59:19-33.

Pappas, G. D., G. Q. Fox, E. B. Masurovsky, E. R. Peterson and S. M. Crain. 1975. Differentiation in neuronal growth cone relationships in the mammalian CNS. In: *Advances in Neurology, 12. Physiology and Pathology of Dendrites.* G. W. Kreutzberg, ed. Raven Press, New York. pp. 163-180.

Pappas, G. D., E. R. Peterson, E. B. Masurovsky and S. M. Crain. 1971a. The fine structure of developing neuromuscular synapses *in vitro.* Ann. N.Y. Acad. Sci. 183:33-45.

Pappas, G. D., E. R. Peterson, E. B. Masurovsky and S. M. Crain. 1971b. The fine structure of developing neuromuscular synapses *in vitro.* Anat. Rec. 169:395-396.

Peacock, J. H., P. G. Nelson and M. W. Goldstone. 1973. Electrophysiologic study of cultured neurons dissociated from spinal cords and dorsal root ganglia of fetal mice. Devel. Biol. 30:137-152.

Peterson, E. R. and S. M. Crain. 1970. Innervation in cultures of fetal rodent skeletal muscle by organotypic explants of spinal cord from different animals. Z. Zellforsch. 106:1-21.

Peterson, E. R. and S. M. Crain. 1972. Regeneration and innervation in cultures of adult mammalian skeletal muscle coupled with fetal rodent spinal cord. Exper. Neurol. 36:136-159.

Peterson, E. R. and M. R. Murray. 1955. Myelin sheath formation in cultures of avian spinal ganglia. Amer. J. Anat. 96:319-356.

Peterson, E. R., S. M. Crain and M. R. Murray. 1965. Differentiation and prolonged maintenance of bioelectrically active spinal cord culture (rat, chick and human). Z. Zellforsch. 66:130-154.

Peterson, E. R., E. B. Masurovsky and S. M. Crain. 1974. Enhanced survival and selective 'hypertrophy' of dorsal root ganglia after exposure of fetal rodent spinal cord-ganglion explants to nerve growth factor. J. Cell Biol. 63:265a.

Pollack, E. D. and S. M. Crain. 1972. Development of motility in fish embryos in relation to release from early CNS inhibition. J. Neurobiol. 3:381-385.

Puro, D. G. and M. Nirenberg. 1976. On the specificity of synapse formation. Proc. Natl. Acad. Sci. 73:3544-3548.

Purpura, D. P. 1972. Ontogenetic models in studies of cortical seizure activities. In: *Experimental Models of Epilepsy.* D. P. Purpura, J. K. Penry, D. Tower, D. M. Woodbury and R. Walter, eds. Raven Press, New York. pp. 531-556.

Ransom, B. R. and P. G. Nelson. 1975. Neuropharmacological responses from nerve cells in tissue culture. In: *Handbook of Psychopharmacology, 2.* L. L. Iversen, S. D. Iversen and S. H. Snyder, eds. Plenum Press, New York. pp. 101-127.

Rees, R. and R. P. Bunge. 1974. Morphological and cytochemical studies of synapses formed in culture between isolated rat superior cervical ganglion neurons. J. Comp. Neurol. 157:1-12

Rees, R. P., M. B. Bunge and R. P. Bunge. 1976. Morphological changes in neuritic growth cone and target neuron during synaptic junction development in culture. J. Cell Biol. 68:240-263.

Seil, F. J., J. M. Kelly and A. I. Leiman. 1974. Anatomical organization of cerebral neocortex in tissue culture. Exper. Neurol. 45:435-450.

Shimada, Y. and D. A. Fischman. 1973. Morphological and physiological evidence for the development of functional neuromuscular junctions *in vitro.* Develop. Biol. 31:200-225.

Smalheiser, N. R., S. M. Crain and M. B. Bornstein. 1977. Development of functional synaptic networks in cultures of embryonic mouse retinal and tectal explants. Anat. Rec., 187:780.

Sobkowicz, H. M., H. A. Hartmann, R. Monzain and P. Desnoyers. 1973. Growth, differentiation and ribonucleic acid content of the fetal rat spinal ganglion cells in culture. J. Comp. Neurol. 148:249-284.

Sperry, R. W. 1963. Chemoaffinity in the orderly growth of nerve fiber patterns and connections. Proc. Nat. Acad. Sci. 50:703-710.

Sperry, R. W. 1965. Embryogenesis of behavioral nerve nets. In: *Organogenesis.* R. L. DeHaan and H. Ursprung, eds. Holt, Rinehart and Winston, New York. pp. 161-186.

Teräväinen, H. 1968. Development of the myoneural junction in the rat. Z. Zellforsch. 87:249-265.

Vaughn, J. E., C. K. Henrikson, C. R. Chernow, J. A. Grieshaber and C. C. Wimer. 1975. Genetically-associated variations in the development of reflex movements and synaptic junctions within an early reflex pathway of mouse spinal cord. J. Comp. Neurol. 161:541-554.

Wall, P. D. 1964. Presynaptic control of impulses at the first central synapse in the cutaneous pathway. Prog. Brain Res. 12:92-118.

Weiss, P. 1955. Nervous system. In: *Analysis of Development.* H. Willier, P. Weiss and V. Hamburger, eds. Saunders, Phila, Pa. pp. 346-401.

Weiss, P. 1966. Specificity in the neurosciences. In: *Neuroscience Research Symposium Summeries, 1.* F. O. Schmitt and T. Melnechuk, eds. M. I. T. Press, Cambridge, Mass. pp. 179-212.

Wenzel, J., M. Wenzel, G. Gross and W. Kirsche. 1973. Elektroenmikroskopische untersuchungen zur entwicklung der Synapsen in Explantatkulturen des Hippocampus bei Rattenembryonen. Z. mikrosk. anat. Forsch. 87:195-217.

Windle, W. F. and R. E. Baxter. 1935-36. Development of reflex mechanism in the spinal cord of albino rat embryos. Correlation between structure and function, and comparisons with the cat and chick. J. Comp. Neurol. 63:189-210.

Windle, W. F., W. L. Minear, M. F. Austin and D. W. Orr. 1935. The origin and early development of somatic behaviour in the albino rat. Physiol. Zool. 8:156-185.

Yoon, M. G. 1976. Topographic polarity of the optic tectum studied by reimplantation of the tectal tissue in adult goldfish. In: *Cold Spring Harbor Symp. on Quantitative Biol. XL. The Synapse.* pp. 503-519.

Zipser, B., S. M. Crain and M. B. Bornstein. 1973. Directly evoked "paroxysmal" depolarizations of mouse hippocampal neurons in synaptically organized explants in longterm culture. Brain Res. 60:489-495.

DISSOCIATED BRAIN CELLS IN PRIMARY CULTURES

MONIQUE SENSENBRENNER

Centre de Neurochimie du CNRS
Strasbourg, France

I. Introduction

Primary cultures of cells dissociated from neural tissue are now extensively used in neurobiological studies. They are extremely useful in the investigations of specific neurobiological aspects of neuronal and glial cells: growth, differentiation, effect of growth factors, hormones and drugs, neuronal-glial interactions and metabolic activities.

Started by Nakai (1956) with spinal ganglia, the dissociation procedures used by the various authors initially produced cultures consisting of heterogeneous populations of dispersed neurons and glial cells. Most experiments have been performed with dissociated spinal ganglia (Levi-Montalcini and Angeletti, 1963; Scott *et al.*, 1969; Sensenbrenner *et al.*, 1969; Ciesielski-Treska *et al.*, 1970; Lodin *et al.*, 1970, 1973a, b; Varon and Raiborn, 1971; Blood, 1973; Contestabile *et al.*, 1973; Luduena, 1973; Peacock *et al.*, 1973; Varon *et al.*, 1973; Shahar *et al.*, 1975). Morphological, biochemical and physiological characteristics of these cells have been documented. Dissociated cells from brain (Sensenbrenner *et al.*, 1971; Booher and Sensenbrenner, 1972; Grosse and Lindner, 1972; Shapiro and Schrier, 1973; Breen and De Vellis, 1974; Greenham *et al.*, 1974; Godfrey *et al.*, 1975; Wiesmann *et al.*, 1975; Bonkowski and Dryden, 1976), from spinal cord (Cavanaugh, 1955; Kim, 1972; Bird and James, 1973, 1975; Contestabile *et al.*, 1973; Kim and Wenger, 1973; Peacock *et al.*, 1973; Fischbach and Dichter, 1974) and from cerebellum (Lasher and Zagon, 1972; Contestabile *et al.*, 1973; Nelson and Peacock, 1973; Burry and Lasher, 1975) can be maintained in culture and thus specific properties of the central nervous system (CNS) can be investigated. Relatively few reports of dissociated sympathetic ganglia have been published (Levi-Montalcini and Angeletti, 1963; Hermetet *et al.*, 1970; Varon and Raiborn, 1972; Mains and Patterson, 1973a, b, c; Chalazonitis *et al.*, 1974).

Neurobiological studies would be facilitated by cell culture systems consisting of either pure neuronal or pure glial cell cultures. Primary cultures of only astroblasts have recently been described (Booher and Sensenbrenner, 1972; Lim *et al.*, 1973; Massarelli *et al.*, 1974; Moonen *et al.*,

1975). It is generally accepted that neurons in culture require intimate contact with glial cells for their survival and growth. However, neurons in the virtual absence of non-neuronal cells can be grown as well. Several methods were developed for growing neurons in the absence of other cell types from spinal and sympathetic ganglia (Okun, 1972; Varon and Raiborn, 1972; Mains and Patterson, 1973a; Lazarus *et al.*, 1976; McCarthy and Partlow, 1976). These isolated neurons could be maintained in long-term culture because they were grown in the presence of nerve growth factor. Some attempts have also been made with neurons from the CNS (Varon and Raiborn, 1969; Jaros *et al.*, 1975; Sensenbrenner *et al.*, 1975). However, these neurons, up to now, cannot survive any longer than 10 days.

Due to the enormously expanded research efforts in the field of dissociated cell cultures from nervous tissue this discussion will concentrate on dissociated brain cell cultures.

The interested reader is referred for general information on peripheral and central nervous tissues in cell culture to the following reviews: Herschman, 1973; Richelson, 1975; Varon, 1975; Mandel *et al.*, 1976).

II. Differentiation of Dissociated Brain Cells From Chick Embryos

The procedures employed to obtain dispersed brain cells involve enzymatic (pronase, trypsin) or mechanical (nylon sieve, pipet, syringe) dissociation or a combination of the two. The dissociated cells can be successfully cultivated either on a collagen substrate in various culture chambers (Maximow slides, Rose chambers, roller tubes, plastic flasks or Petri dishes) or directly on plastic surfaces. Various chemically defined culture media, different kinds and amounts of serum and special supplements are used as nutrient media.

A. Undifferentiated Neural Cells from Young Embryos

Dissociated cells from the neural plate of chick embryos at a very early stage (36 hours) remain round and undifferentiated under the usual culture conditions (Touzet *et al.*, 1975). However, when these cells are cultivated on glial cell layers they are able to differentiate into neurons (N. R. Touzet and M. Sensenbrenner, unpublished data). On the other hand, the undifferentiated brain cells obtained from older embryo (48 and 72 hours) are capable of differentiation and mature into neuronal and glial cells (Fig. 1a).

FIG. 1. Dissociated brain cells from chick embryos after 2 weeks in culture: a mixed neuronal-glial population. Phase contrast ×250. (a) from 3 day old chick embryo, (b) from 5 day old chick embryo. (c) from 7 day old chick embryo.

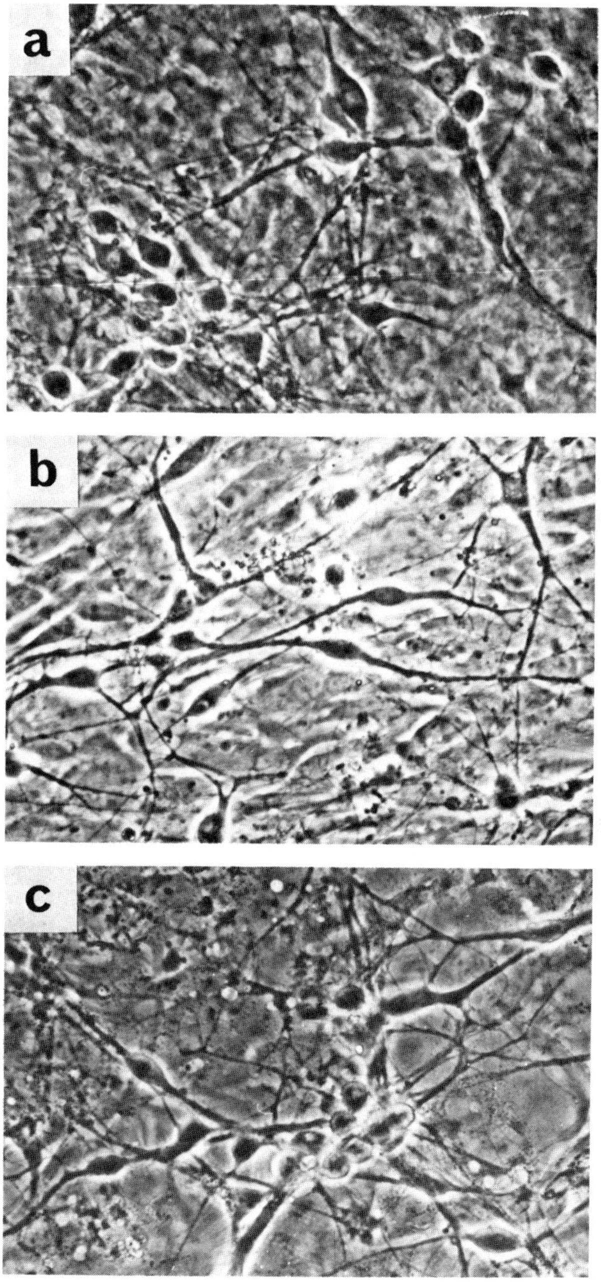
a
b
c

These results seem to indicate that the 36th hour is a critical stage during brain embryogenesis. Once this stage has been passed, the cells are capable of differentiation in culture.

A recent immunofluorescent study has shown that brain specific proteins appeared *in vivo* between 36 and 48 hours in the chick embryo brain (Touzet *et al.*, 1977). Therefore, the inability of the 36 hour old neural cells to differentiate in dissociated culture and without contact with other cell types may be related to the absence of the neurospecific antigens.

B. Differentiation of Dissociated Brain Cells from Older Embryos

1. Morphological differentiation. Dissociated cells from nervous tissue of 5 to 12 day old chick embryos undergo maturation in culture and develop into a mixed population of neuronal and glial cells (Booher and Sensenbrenner, 1972; Grosse and Lindner, 1970, 1972; Sensenbrenner *et al.*, 1971). The flat polygonal glial precursor cells multiply actively and form, after two weeks, layers of glial cells. Bipolar and multipolar neurons with many processes of varying length and thickness are dispersed upon the glial cell layer (Fig. 1b and 1c).

When the dissociated cells are derived from 14 to 19 day old chick embryos, the neuronal cells degenerate during the first week of cultivation and only the non-neuronal cells (glial cell precursors) survive. After 2 weeks in culture, a nearly pure glial cell population has developed (Booher and Sensenbrenner, 1972; Massarelli *et al.*, 1974) (Fig. 2).

Ultrastructural studies have shown the presence of synapses and dense core synaptic vesicles in the mixed neuronal-glial cell population in cultures originating from dissociated brains of 7 day old chicks (Fig. 3). Neurotubules and microfilaments accumulate in the neurons and the glial cells respectively (M. Miehe, M. Sensenbrenner, A. Ponte and P. Mandel, unpublished data). However, myelin formation was never observed in such dissociated cell cultures. Nissl bodies and neurofibrils, revealed by histochemical staining, develop with increasing differentiation of the neurons (Sensenbrenner *et al.*, 1972, 1973) (Fig. 4a,b).

We were able to establish that brain extracts or mesenchymal tissue extracts obtained from chick embryos of different ages stimulate the morphological differentiation of the neuronal and the glial cells when compared to control cultures or cultures treated with total embryonic extract (Sensenbrenner *et al.*, 1972; Athias *et al.*, 1974; Cam *et al.*, 1975).

The extract prepared from the surrounding mesenchyme of 8 day old chick embryo brains enhances the differentiation of the still undifferentiated neuronal cells from 5 day old chick embryos. Later on, when the neuroblasts have become pyriform and have already started their differentiation process (7 day old chick embryo) they no longer respond to the

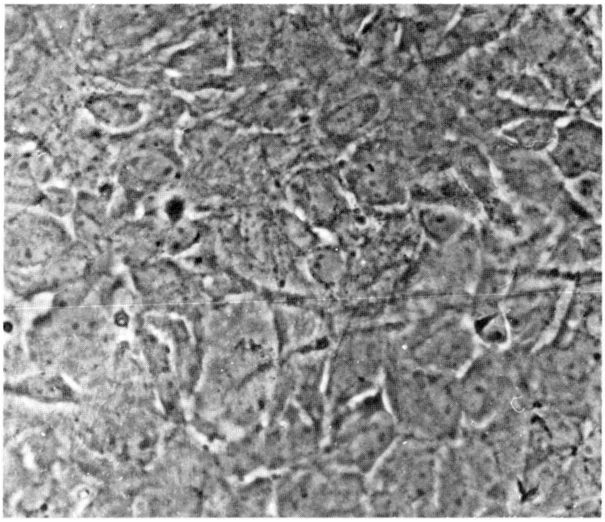

FIG. 2. A glial cell (astroblast) culture from 15 day old chick embryo brain after 2 weeks in culture. Phase contrast ×250.

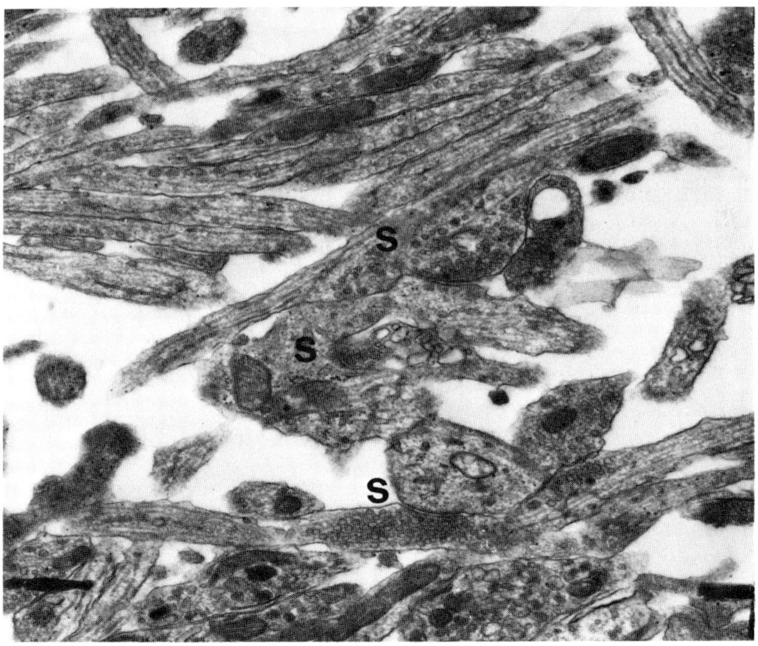

FIG. 3. Electron micrograph of a 10 day old culture of dissociated 7 day old chick embryo brain cells (S indicates synapses). ×22,100.

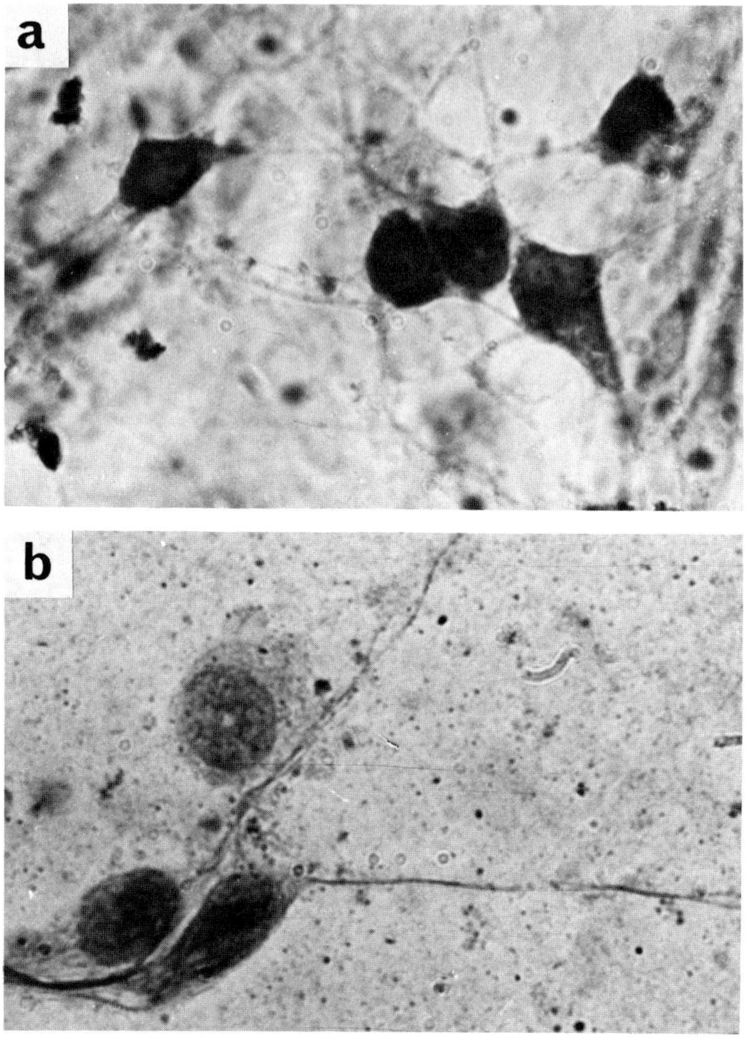

FIG. 4. Dissociated brain cells from 7 day old chick embryos after 1 week in culture. ×1,000. (a) Neurons with Nissl bodies (thionine staining). (b) Neurons with neurofibrils (Holmes silver impregnation).

extract from mesenchymal tissue but are only stimulated to differentiate by brain extract from 8 and 12 day old chick embryos. In the presence of these extracts large multipolar as well as differentiated pyramidal neurons develop. An increase in Nissl bodies and neurofibrils was observed. The whole fiber network was more developed forming many ramifications and thicker fascicles (Fig. 5a, b, c and d).

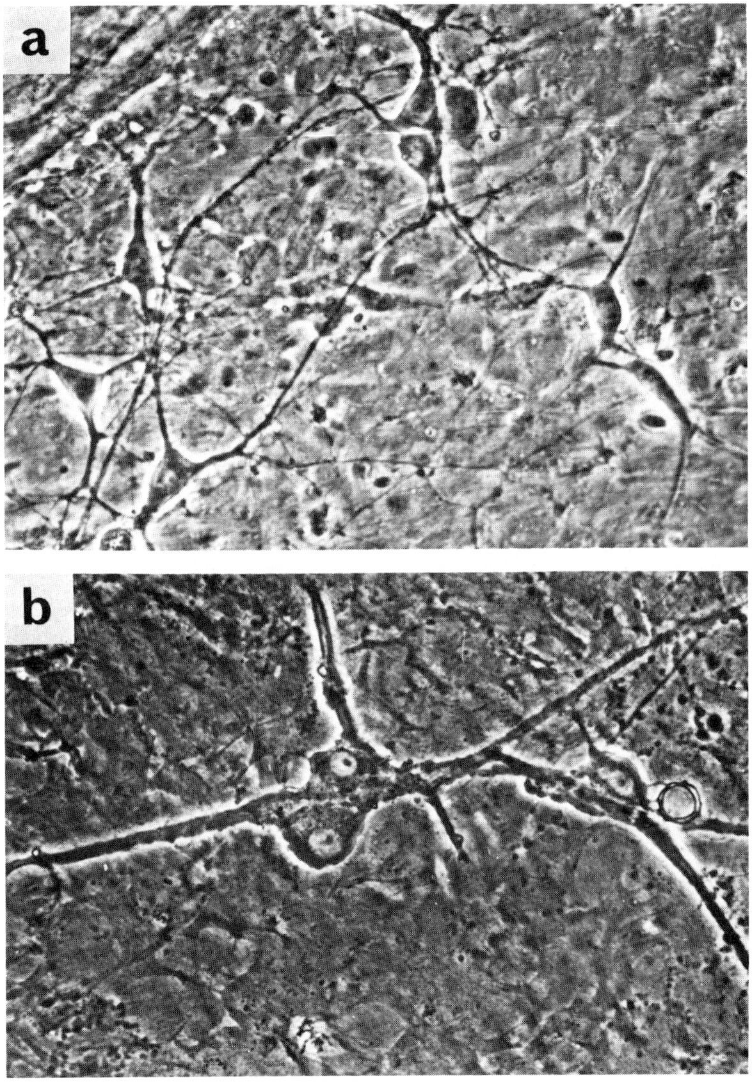

FIG. 5(a)(b). Dissociated brain cells from 7 day old chick embryos cultivated during 2 weeks in presence of brain extract. Phase contrast ×250. (a) Multipolar neurons. (b) Large neuronal cells.

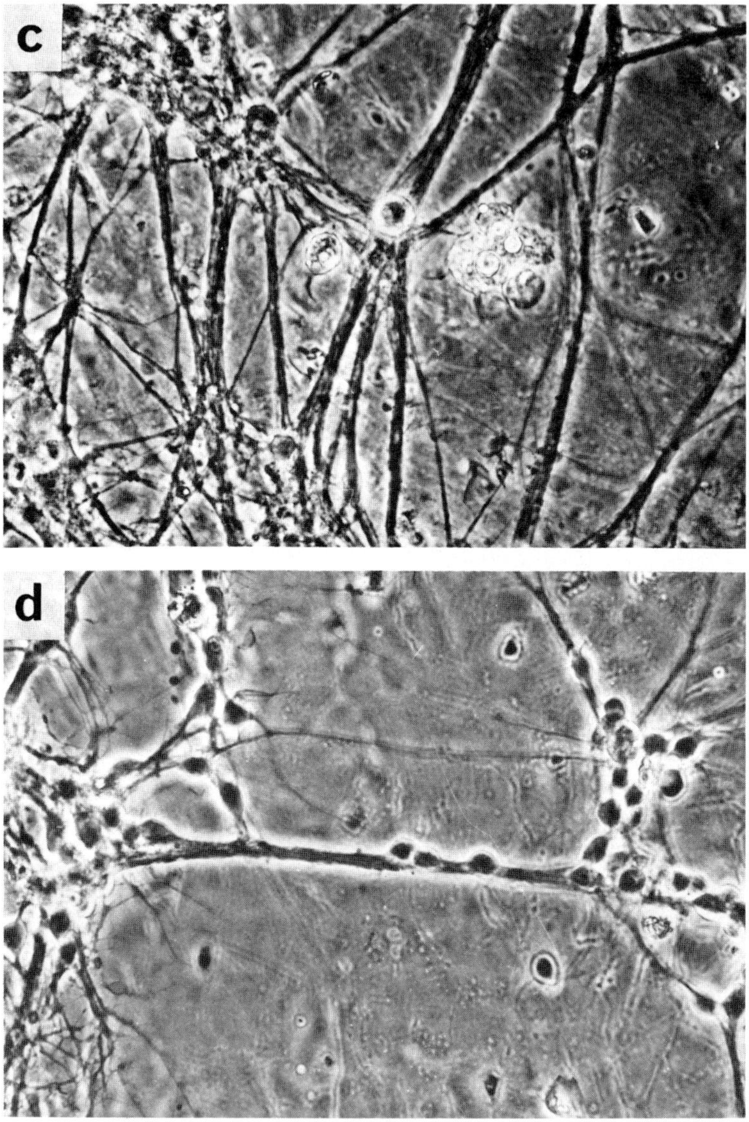

FIG. 5. (continued). (c) Fiber bundles. (d) Small cells along the fibers.

The flat polygonal glial precursor cells are markedly influenced by brain extracts prepared from 17 to 19 day old embryos and change to a more astrocytic appearance (Fig. 6). A similar morphological alteration was obtained by addition of dibutyryl cyclic AMP (dBcAMP) (Moonen and Sensenbrenner, 1976).

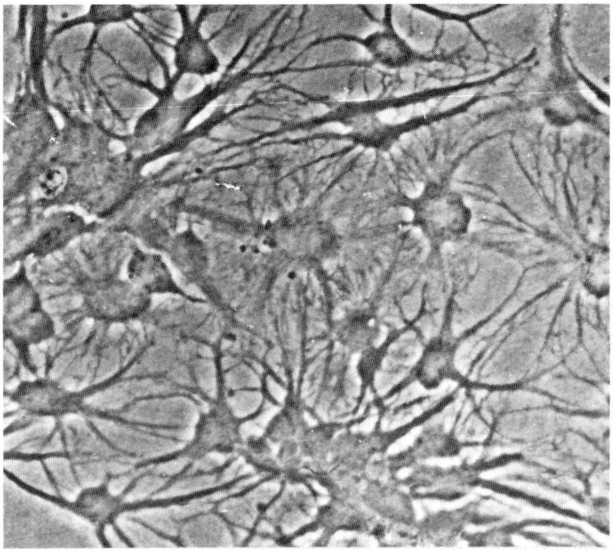

FIG. 6. A glial culture (astrocytes) from 15 day old chick embryos cultivated for 3 weeks in presence of brain extract. (Compare to Fig. 2). Phase contrast ×250.

These extracts have no stimulatory effects on synaptogenesis and do not initiate the formation of myelin.

The active components of the brain extracts which stimulate the differentiation of neuronal and glial cells have been partially identified. They are proteins of low and high molecular weight, which enhance the maturation of the neurons and the glial cells respectively (Y. Cam, M. Ledig, M. Sensenbrenner and P. Mandel, unpublished data).

The differentiation of the neuronal cells can also be enhanced in the presence of glial conditioned medium (Sensenbrenner *et al.*, 1977a). Furthermore, neuroblasts seeded on a glial cell layer adhere very quickly to this layer, remain scattered over it and are more differentiated than those plated on other culture substrates (Sensenbrenner and Mandel, 1974) (Fig. 7a, b, c).

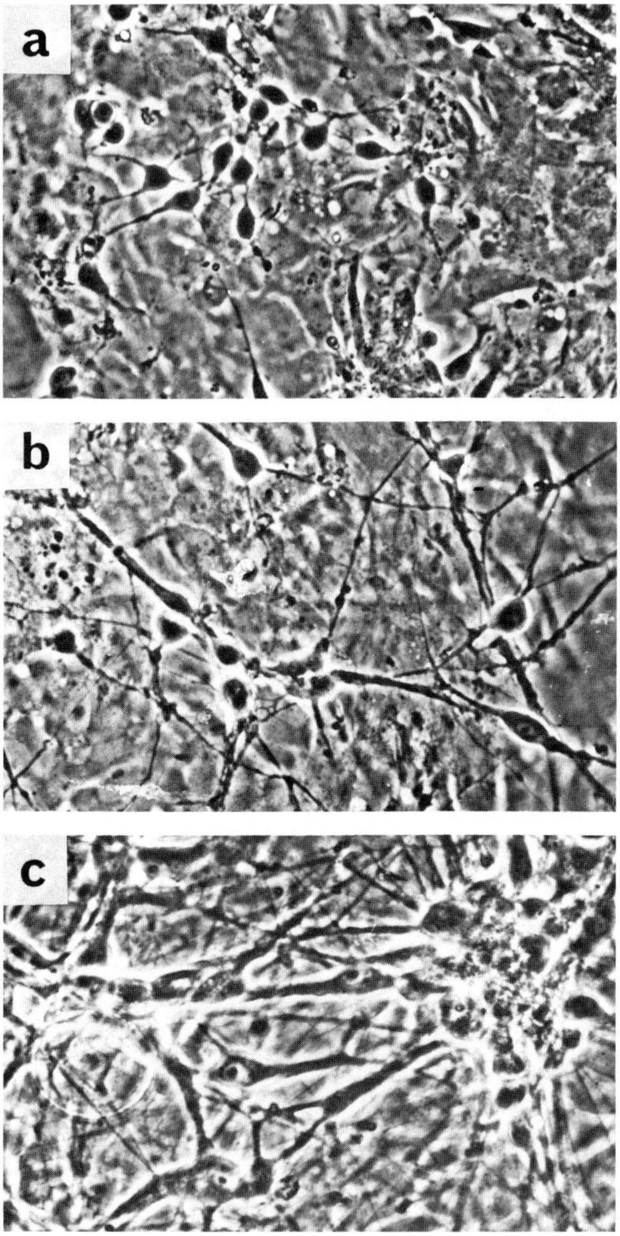

FIG. 7. Dissociated nerve cells from 7 day old chick embryo brains cultured on glial cell layers. Phase contrast ×250. (a) 24 h in culture. Dispersed neuroblasts on the glial layer. (b) 4 days in culture. (c) 14 days in culture (from Sensenbrenner and Mandel, 1974).

From these results the question arises whether neurons in culture require intimate contact with glial cells for their survival and growth. Some attempts have been made to grow neurons in the virtual absence of glial cells (Varon and Raiborn, 1969; Jaros *et al.*, 1975; Sensenbrenner *et al.*, 1975). More recently a pure neuronal culture was obtained from dissociated 7 day old chick embryo brain, by culturing the cell suspension on a polylysine substrate (M. Sensenbrenner, unpublished data). The cells grew processes which formed a network. Bipolar and multipolar neurons developed. However, these neurons, could not survive any longer than 10 days. These results seem to indicate that contacts with glial cells are necessary for neuronal survival *in vitro*. Further studies will be needed to test different growth factors which may prove favorable for the maintenance of isolated neurons in long-term culture.

2. Biochemical differentiation. Acetylcholinesterase (AChE) and choline acetyltransferase (ChAc), two enzymes closely associated with cholinergic transmission, have been investigated in dissociated brain cell cultures.

Histoenzymologic studies demonstrated the presence of AChE in the cell body and the fibers of the neuronal cells. In the presence of brain extracts this activity was increased. Weak or no activity was observed in the non-neuronal cells (Sensenbrenner *et al.*, 1972, 1973) (Fig. 8a, b).

Biochemical determinations have shown that cultured neurons possess AChE and ChAc (Werner *et al.*, 1971; Peterson *et al.*, 1973; Ebel *et al.*, 1974). AChE increased during the first days in culture, when neuronal precursor cells actively proliferated followed by neuronal maturation, and then declined during massive glial cell proliferation. In contrast, ChAc increased during the whole cultivation period and paralleled the development of synapses. The pattern of increase of the ChAc/AChE ratio was similar to that seen *in vivo* in developing brain (Fig. 9a, b, c). Recently, it has been shown that brain extracts enhance AChE and ChAc activities (Y. Cam, A. Ebel, M. Sensenbrenner, unpublished data). The flat glial cells possess neither of these activities, but have butyrylcholinesterase activity and a choline independent acetylase activity that may be carnitine acetyltransferase (Peterson *et al.*, 1973).

Acetylcholine as well as catecholamines are present in the cultivated neuronal cells (M. Sensenbrenner, A. Ebel, M. Miehe, unpublished data).

Guanyl cyclase activity has been detected in all mammalian organs studied and has been found to be high in the brain and overwhelmingly concentrated in neurons. Cultures containing neurons in substantial numbers always have higher guanyl cyclase activities than those consisting mainly of glial cells (Goridis *et al.*, 1974). These results provide further evidence that brain guanyl cyclase essentially is localized in neuronal cells. An increase of this activity was seen during the outgrowth of nerve fibres,

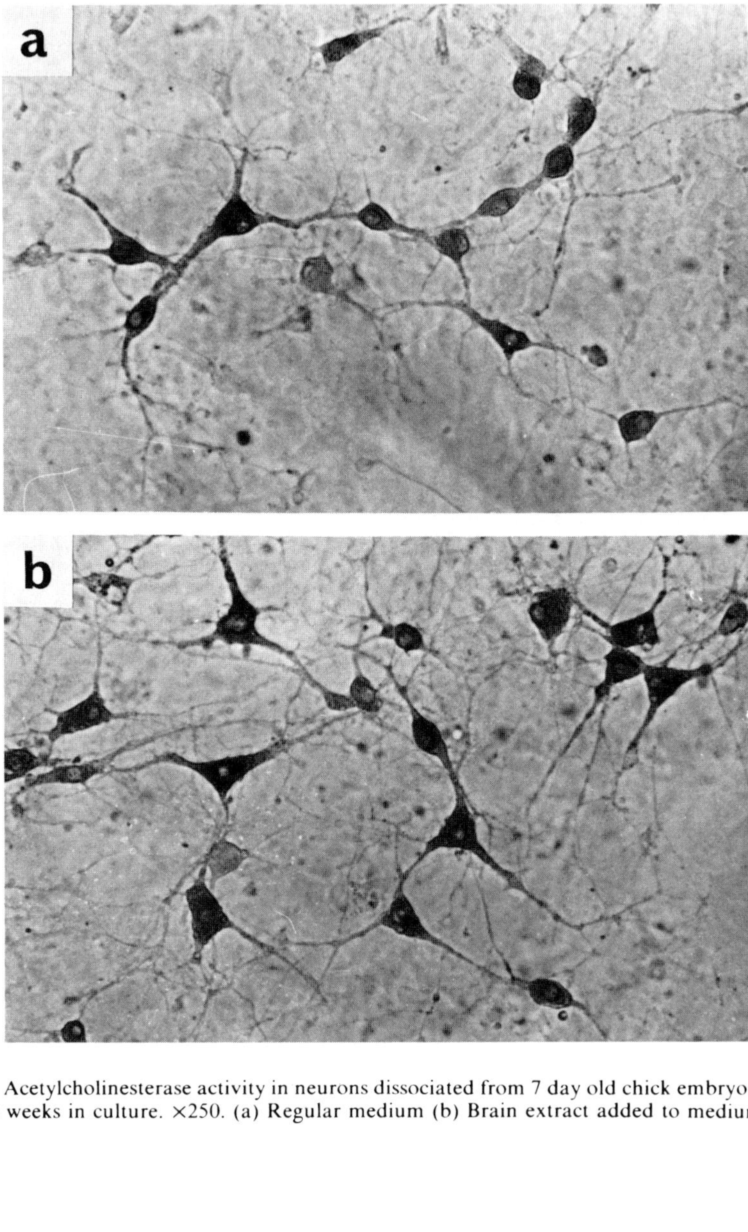

FIG. 8. Acetylcholinesterase activity in neurons dissociated from 7 day old chick embryo brains after 2 weeks in culture. ×250. (a) Regular medium (b) Brain extract added to medium.

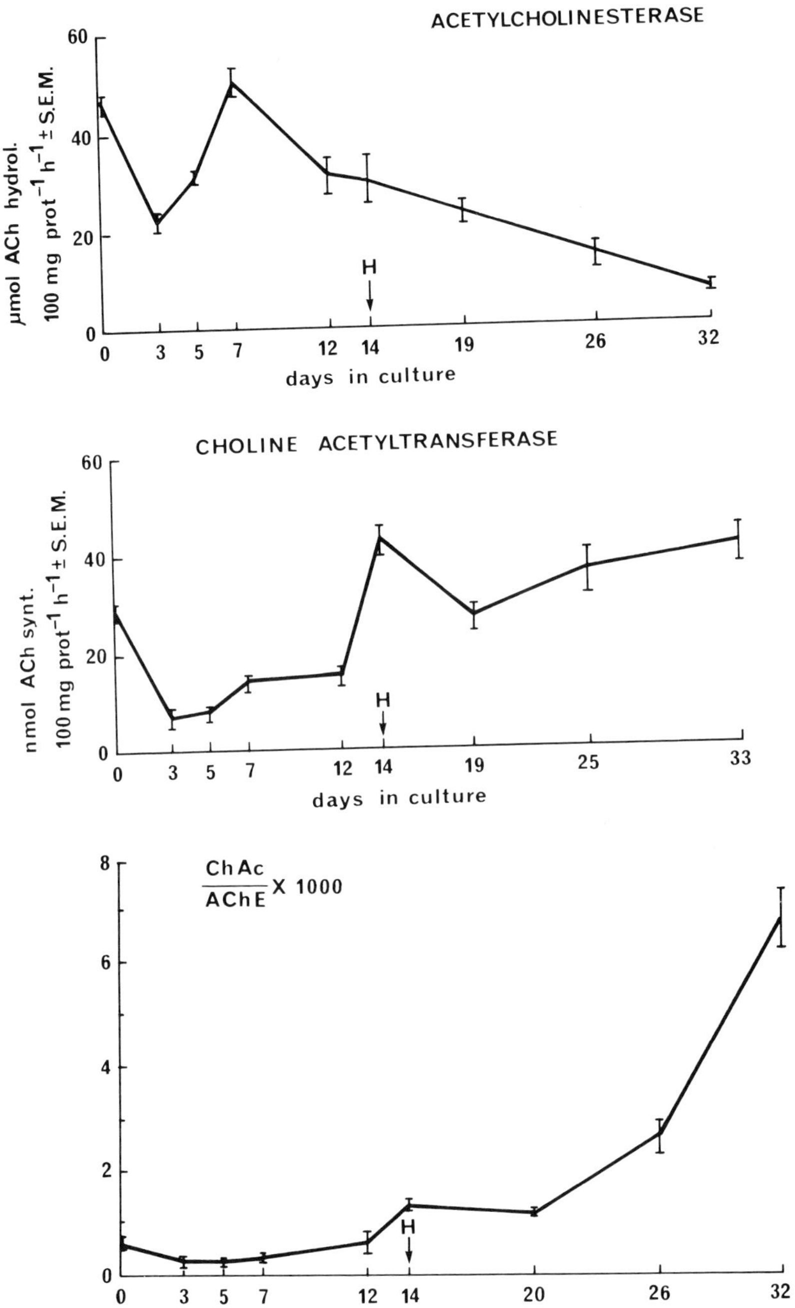

FIG. 9. Activity levels of AChE and ChAc in cultured brain cells from 7 day old chick embryos. Hatching point is designated by H. (a) AChE (b)CHAc (c) ChAc/AChE activity ratios(from Ebel *et al.*, 1974).

and the first steps of differentiation of cultured neurons (Fig. 10). Guanyl cyclase activity is stimulated in the presence of brain extracts (Y. Cam, M. Ledig, M. Sensenbrenner and P. Mandel, unpublished data).

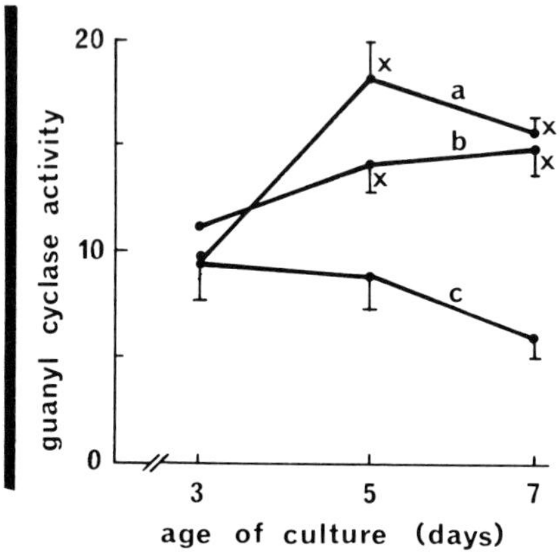

FIG. 10. Guanyl cyclase activity in cultured chick embryo brain cells derived from embryos of different ages. (a) 8 day old. (b) 12 day old. (c) 14 day old embryos. Guanyl cyclase activity is expressed as pmol cyclic GMP formed/min/mg protein. Mean ± S.E.M. for 8-10 cultures is shown. ˣ$P<0.01$ when compared with the corresponding value for c (from Goridis *et al.*, 1974).

Some information about the energy metabolism in cultured brain cells during differentiation is available. Dittman *et al.* (1973b) have shown that the rate of oxygen uptake increases as a function of the age of the culture in parallel with the increase in the rate of oxygen uptake observed in the embryo. Furthermore, brain extract increases the O_2 consumption (Dittman *et al.*, 1973a) (Fig. 11).

The activities of total ATPase, Mg^{2+} - and Na^+, K^+-ATPases have been studied as a function of maturation in cultures of dissociated brain cells. These activities increase with the cultivation time and parallel the evolution of these enzymes *in vivo* (Latzkovits *et al.*, 1975).

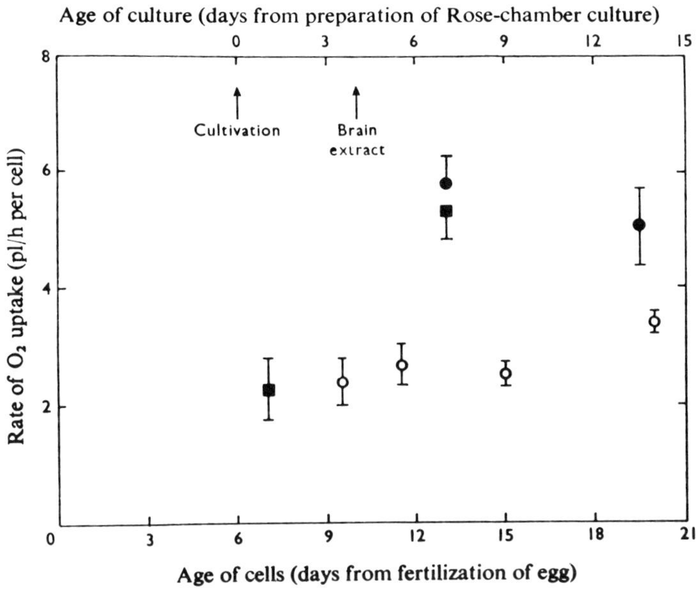

FIG. 11. Respiratory rates of fresh chick embryo brain cells and of brain cells cultured in the absence or the presence of brain extract.
■, fresh cells. O, cells cultured in the absence of brain extract.
●, cells cultured in the presence of brain extract (from Dittmann *et al.*, 1973a).

III. Differentiation of Mammalian Brain Cells

A. Mixed Neuronal-glial Cell Population

Recent studies have demonstrated that brain cells from mice and rats can be cultured following mechanical or enzymatic dissociation and that they undergo morphological and biochemical differentiation (Booher and Sensenbrenner, 1972; Rosenberg, 1972; Shapiro and Schrier, 1973; Greenham *et al.*, 1974; Grosse and Lindner, 1974, 1975; Godfrey *et al.*, 1975). In the rat, it has been shown that 16-18 day old embryos provide the best brain tissue for obtaining a mixed neuronal-glial cell population. The dissociated cells have been cultivated on different substrates (collagen or plasma film) or directly on a plastic or glass surface. Dissociated mammalian brain cells have also been grown on a polylysine-coated surface, which is said to favor the attachment of the cells (Yavin and Yavin, 1974).

Morphological observations have shown that the dissociated cells tend to reaggregate and very few cells remain isolated (Booher and Sensenbrenner, 1972; Schrier, 1973; Greenham *et al.*, 1974; Yavin and Yavin, 1974; Grosse and Lindner, 1974, 1975; Bonkowski and Dryden, 1976). The cells grow fibers during the first 48 hours and a fiber network develops within 5 days. Bipolar and multipolar, small and large neurons are observed after 2 weeks *in vitro*. Oligodendroglia can be identified in such cultures. Flat polygonal glial precursor cells proliferate and rapidly form a carpet upon which the neuronal cells are observed (Fig. 12). This layer includes some areas of ependymal cells with actively beating cilia.

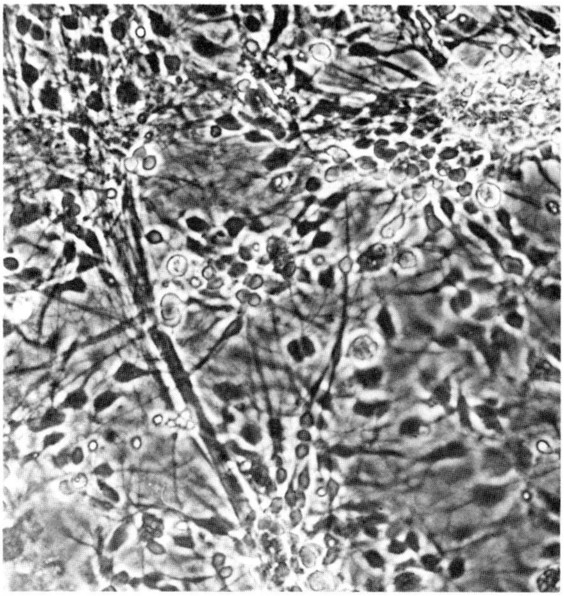

FIG. 12. Dissociated brain cells from 16 day old rat embryo cultivated during 14 days on a polylysine-coated surface. Phase contrast ×200.

Godfrey *et al.* (1975) have developed a new culture system for dissociated mammalian brain cells. This modification involves the use of fluorodeoxyuridine, which prevents an overgrowth of the glial cells and thus indirectly allows more complete development of the neurons. These cultures have been shown to be very suitable for morphological, biochemical and electrophysiological studies.

Electron microscopy has revealed that synaptogenesis occurs during the development of the mammalian brain cell culture and myelin becomes visible during the 3rd week (Bornstein and Model, 1972).

Some authors succeeded in culturing dissociated cells from a specific area of the brain, namely the hypothalamus (Wilkinson *et al.*, 1974; Benda *et al.*, 1975). These cultures represent a useful system for the investigation of neurosecretory cells. These cells undergo morphological differentiation *in vitro* as shown by phase contrast and electron microscopy.

Several laboratories have investigated biochemical characteristics of cultured dissociated mammalian brain cells. Protein and DNA content, as well as activities of AChE, ChAc, glutamate decarboxylase and catechol *O*-methyltransferase have been determined (Rosenberg, 1972; Wilson *et al.*, 1972; Shapiro and Schrier, 1973; Schrier, 1973). These enzymes have been selected because they are important in neurotransmitter metabolism and are considered to be markers of neuronal differentiation. All the activities, and particularly glutamate decarboxylase, have been found to be higher in the cultured brain cells than in non-neuronal cell lines (Wilson *et al.*, 1972). The activities of ChAc are high in cultures derived from young (18 day) fetuses rich in neuronal cells; whereas in cultures from older (21 day) fetuses containing very few neurons, the ChAc activity is low (Shapiro and Schrier, 1973). ChAc and glutamate decarboxylase activities increase with the development of the culture, but AChE remains at a constant level (Rosenberg, 1972). Schrier (1973) describes a marked increase of ChAc activity during the cell maturation and this development is stimulated by subculturing.

The study of Breen and De Vellis (1974) on the regulation of glycerol phosphate dehydrogenase (GPDH) by hydrocortisone brings further evidence that dissociated brain cells in culture differentiate biochemically. GPDH is known to be regulated by glucocorticoids in nervous tissues, while other tissues are not responsive. The level of GPDH remains low and unchanged during the evolution of the culture, but the induction of GPDH increases with time and reaches a maximum around 21 days. This parallels the developmental increase of GPDH *in vivo*.

Recently, Wiesmann *et al.* (1975) have shown that cerebroside, a myelin precursor, increases with the development of the culture and also parallels in some respects the *in vivo* maturation.

B. Pure Glial Cell Cultures

Dissociated brain cells from newborn rats and mice cultivated on a plastic surface develop into a pure glial cell population. The neuronal cells degenerate during the first two days and some oligodendroblasts survive for about one week. On the other hand, the flat polygonal shaped cells multiply actively and form a monolayer within 2 weeks (Booher and Sensenbrenner, 1972; Shapiro, 1973; Lim *et al.*, 1973; Moonen *et al.*, 1975) (Fig. 13a). In an attempt to characterize these cells the cultures have been examined by

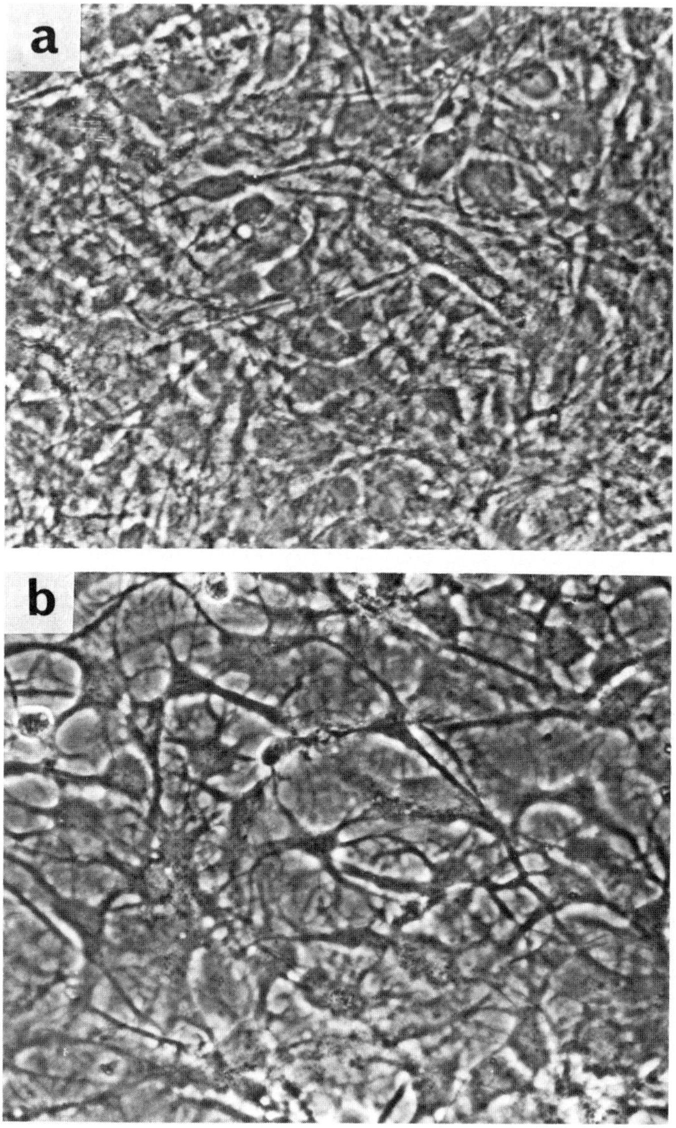

FIG. 13. Glial cultures from newborn rat brain after 3 weeks. Phase contrast ×250. (a) Regular medium: glial precursor cells (b) Medium with brain extract: astrocytes.

electron microscopy (Moonen *et al.*, 1976). A few randomly dispersed microfilaments are present in these cells after 7-10 days in culture, but they become very abundant after 25 days and are then organized in bundles similar to those observed in fibrillary astrocytes. This morphological differentiation occurs in parallel with the biochemical development of the glial fibrillary acidic (GFA) protein, a specific marker for fibrillary astrocytes (E. Bock, M. Möller, C. Nissen and M. Sensenbrenner, unpublished data). Other specific glial markers such as protein S100 and carbonic anhydrase have also been found in these cells (Lim *et al.*, 1977; Sensenbrenner *et al.*, 1977b).

These polygonal astrocyte precursor cells assume a star-like aspect and become more like mature astrocytes under the effect of dBcAMP (Shapiro, 1973; Lim *et al.*, 1973; Moonen *et al.*, 1975). The same morphological change was observed after the addition of brain extract (Lim *et al.*, 1973, 1976; Lim and Mitsunobu, 1974; Sensenbrenner *et al.*, 1977b) (Fig. 13b) or pentobarbital (Roth-Schechter *et al.*, 1976) into the medium. Recently, this morphological alteration of the glial precursor cells has been related to changes of biochemical glial markers, indicating a simultaneous biochemical differentiation (Lim *et al.*, 1977; Sensenbrenner *et al.*, 1977b).

A factor, isolated from those brain extracts which stimulate the differentiation of the cultured glial cells has been partially purified and characterized. It is a protein of high molecular weight (Lim and Mitsunobu, 1974, 1975).

Conclusions

Dissociated brain cells from the chick as well as from mammalian animals can be maintained in long-term cultures. These cultures appear to contain a mixed population of neuronal and glial cells. The proportion of the two elements varies with the age of the brain tissue used. Thus, cultures rich in neuronal cells are obtained from young chick embryos and young mammalian fetuses while brains from older animals can be used to produce pure glial cultures. Furthermore, by changing the culture conditions pure neuronal cultures can be developed.

Morphological as well as biochemical differentiation takes place in dissociated brain cell cultures, analogous in some respects to the *in vivo* situation. Therefore, primary cultures of brain cells can be considered an excellent model system in the pursuit of neurobiological studies.

References

Athias, P., M. Sensenbrenner and P. Mandel. 1974. The behaviour of dissociated chick embryo brain cells in long-term cultures in presence and absence of brain extracts. Differentiation 2:99-106.

Benda, P., F. De Vitry, R. Picart and A. Tixier-Vidal. 1975. Dissociated cell cultures from fetal mouse hypothalamus. Patterns of organization and ultrastructural features. Exp. Brain Res. 23:29-47.

Bird M. M. and D. W. James. 1973. The development of synapses *in vitro* between previously dissociated chick spinal cord neurons. Z. Zellforsch. 140:203-216.

Bird, M. and D. W. James. 1975. Myelin formation in cultures of previously dissociated mouse spinal cord. Cell Tissue Res. 162:93-105.

Blood, L. A. 1973. A preliminary cinemicrographic study on the effect of NGF on the growth of chick embryonic sensory neurites *in vitro*. Neurobiology 3:53-65.

Bonkowski, L. and W. F. Dryden. 1976. The effects of putative neurotransmitters on the resting membrane potential of dissociated brain neurons in culture. Brain Res. 107:69-84.

Booher, J. and M. Sensenbrenner. 1972. Growth and cultivation of dissociated neurons and glial cells from embryonic chick, rat and human brain in flask cultures. Neurobiology 2:97-105.

Bornstein, B. M. and P. G. Model. 1972. Development of synapses and myelin in cultures of dissociated embryonic mouse spinal cord, medulla and cerebrum. Brain Res. 37:287-293.

Breen, G. A. M. and J. De Vellis. 1974. Regulation of glycerol phosphate dehydrogenase by hydrocortisone in dissociated rat cerebral cell cultures. Developmental Biol. 41:255-266.

Burry, R. W. and R. S. Lasher. 1975. Uptake of GABA in dispersed cell cultures of postnatal rat cerebellum: an electron microscope autoradiographic study. Brain Res. 88:502-507.

Cam, Y., M. Sensenbrenner and P. Mandel. 1975. A comparative study of the effects of brain extracts and mesodermal membrane extracts on nerve cell differentiation. Experientia 31:1430-1431.

Cam, Y., M. Sensenbrenner, M. Ledig and P. Mandel. 1977. Characterization of a brain extract that stimulates nerve cell differentiation in culture. Neuroscience in press.

Cavanaugh, M. W. 1955. Neuron development from trypsin-dissociated cells of differentiated spinal cord of the chick embryo. Exp. Cell Res. 9:42-48.

Chalazonitis, A., L. A. Greene and M. Nirenberg. 1974. Electrophysiological characteristics of chick embryo sympathetic neurons in dissociated cell culture. Brain Res. 68:235-252.

Ciesielski-Treska, J., J. C. Hermetet and P. Mandel. 1970. Histochemical study of isolated neurons in culture from chick embryo spinal ganglia. Histochemie 23:36-43.

Constestabile, A., G. Minelli and F. Ciani. 1973. Enzymic activities in dissociated neurons differentiated in cultures *in vitro*. J. Neurochem. 20:691-698.

Dittmann, L., L. Hertz, H. Fosmark, M. Sensenbrenner and P. Mandel. 1973a. Development of respiration in fresh and cultivated brain cortex cells. Trans. Biochem. Soc. 1:136-138.

Dittmann, L., L. Hertz, A. Schousboe, H. Fosmark, M. Sensenbrenner and P. Mandel. 1973b. Energy metabolism of nerve cells during differentiation. Exp. Cell Res. 80:425-431.

Ebel, A., R. Massarelli, M. Sensenbrenner and P. Mandel. 1974. Choline acetyl-transferase and acetylcholinesterase activities in chicken brain hemispheres *in vivo* and in cell culture. Brain Res. 76:461-472.

Fischbach, D. and A. Dichter. 1974. Electrophysiological and morphological properties of neurons in dissociated chick spinal cord cell cultures. Developmental Biol. 37:100-116.

Godfrey, E. W., P. G. Nelson, B. K. Schrier, A. C. Breuer and B. R. Ransom. 1975. Neurons from fetal rat brain in a new cell culture system: a multidisciplinary analysis. Brain Res. 90:1-21.

Goridis, C., R. Massarelli, M. Sensenbrenner and P. Mandel. 1974. Guanyl cyclase in chick embryo brain cell cultures: evidence of neuronal localization. J. Neurochem. 23:135-138.

Greenham, L. W., T. J. Hill, C. A. Moss and D. B. Peacock. 1974. Preparation of disaggregated mouse brain cultures, and observations on early neurite and axon formation therein. Exp. Cell Res. 84:287-299.

Grosse, G. and G. Lindner. 1970. Untersuchungen zur Differenzierung isolierter Nerven und Gliazellen des zentralnervösen Gewebes von Hühnerembryonen in der Zellkultur. J. Hirnforsch. 12:207-215.

Grosse, G. and G. Lindner. 1972. Untersuchungen zur Differenzierung disaggregierter Zellen des Telencephalon vom Hühnerembryo in Langzeit Zellkulturen. Z. mikrosk. anat. Forsch. Leipzig 85:438-448.

Grosse, G. and G. Lindner. 1974. Zytodifferenzierung des Hippocampus fetaler Ratten in der Zellkultur. Z. mikrosk. anat. Forsch. 88:705-712.

Grosse, G. and G. Lindner. 1975. Morphologische Untersuchungen der Entwicklung des Hippocampus in der Zell- und Explantat-Kultur. J. Hirnforsch. 16:451-456.

Hermetet, J. C., J. Treska and P. Mandel. 1970. Histochemical study of isolated neurons in culture from chick embryo sympathetic ganglia. Histochemie 22:177-186.

Herschman, H. R. 1973. Tissue and cell culture as a tool in neurochemistry. In: *Proteins of the Nervous System.* D. J. Schneider, R. H. Angeletti, R. A. Bradshaw, A. Grasso and B. W. Moore, eds. Raven Press, New York, pp. 95-115.

Jaros, G. G., G. Moonen, M. Sensenbrenner and P. Mandel. 1975. Cultures de cellules dissociées d'hémisphères cérébraux à composante essentiellement neuronale. C. R. Acad. Sci. Paris série D 280:327-330.

Kim, S. U. 1972. Formation of synapses and myelin sheaths in cultures of dissociated chick embryonic spinal cord. Exp. Cell Res. 73:528-530.

Kim, S. U. and E. L. Wenger. 1973. Morphological and enzyme histochemical studies of dissociated chick neural tube cultured *in vitro.* J. Neurobiol. 4:513-523.

Lasher, R. S. and I. S. Zagon. 1972. The effect of potassium on neuronal differentiation in cultures of dissociated newborn rat cerebellum. Brain Res. 41:482-488.

Latzkovits, L., A. Juhasz, A. Rimanoczy and I. Szentistvanyi. 1975. Adenosine triphosphatase in cultures of dissociated brain cells as compared with that in the developing chick brain. 5th Intern. Meeting Intern. Soc. Neurochem. Barcelona Abstr. p. 233.

Lazarus, K. J., R. A. Bradshaw, N. R. West and R. P. Bunge. 1976. Adaptive survival of rat sympathetic neurons cultured without supporting cells or exogenous nerve growth factor. Brain Res. 113:159-164.

Levi-Montalcini, R. and P. U. Angeletti. 1963. Essential role of the nerve growth factor in the survival and maintenance of dissociated sensory and sympathetic embryonic nerve cells *in vitro.* Developmental Biol. 7:653-659.

Lim, R. and K. Mitsunobu. 1974. Brain cells in culture: morphological transformation by a protein. Science 185:63-66.

Lim, R. and K. Mitsunobu. 1975. Partial purification of a morphological transforming factor from pig brain. Biochim. Biophys. Acta 400:200-207.

Lim, R., K. Mitsunobu and W. K. P. Li. 1973. Maturation-stimulating effect of brain extract and dibutyryl cyclic AMP on dissociated embryonic brain cells in culture. Exp. Cell Res. 79:243-246.

Lim, R., D. E. Turriff and S. S. Troy. 1976. Response of glioblasts to a morphological transforming factor: cinematographic and chemical correlations. Brain Res. 113:165-170.

Lim, R., D. E. Turriff, S. S. Troy, B. W. Moore and L. F. Eng. 1977. Glia maturation factor: effect on chemical differentiation of glioblasts in culture. Science 195:195-196.

Lodin, Z., J. Booher and F. H. Kasten. 1970. Long-term cultivation of dissociated neurons from embryonic chick dorsal root ganglia in the Rose chamber. Exp. Cell Res. 59:291-298.

Lodin, Z., J. Faltin, J. Booher, J. Hartman and M. Sensenbrenner. 1973a. Fiber formation and myelinization of cultivated dissociated neurons from chicken dorsal root ganglia. Neurobiology 3:66-87.

Lodin, Z., J. Faltin, J. Booher, J. Hartman and M. Sensenbrenner. 1973b. Formation of intercellular contacts in cultures of dissociated neurons from embryonic chicken dorsal root ganglia. Neurobiology 3:376-390.

Luduena, M. A. 1973. Nerve cell differentiation *in vitro.* Developmental Biol. 33:268-284.

McCarthy K. D. and L. M. Partlow. 1976. Preparation of pure neuronal and non-neuronal cultures from embryonic chick sympathetic ganglia: a new method based on both differential cell adhesiveness and the formation of homotypic neuronal aggregates. Brain Res. 114:391-414.

Mains, R. E. and P. H. Patterson. 1973a. Primary cultures of dissociated sympathetic neurons. I. Establishment of long-term growth in culture and studies of differentiated properties. J. Cell Biol. 59:329-345.

Mains, R. E. and P. H. Patterson. 1973b. Primary cultures of dissociated sympathetic neurons. II. Initial studies on catecholamine metabolism. J. Cell Biol. 59:346-360.

Mains, R. E. and P. H. Patterson. 1973c. Primary cultures of dissociated sympathetic neurons. III. Changes in metabolism with age in culture. J. Cell Biol. 59:361-366.

Mandel, P., J. Ciesielski-Treska and M. Sensenbrenner. 1976. Neurons *in vitro*. In: *Molecular and Functional Neurobiology*. W. H. Gispen, ed. Elsevier, Amsterdam, pp. 111-157.

Massarelli, R., M. Sensenbrenner, A. Ebel and P. Mandel. 1974. Kinetics of choline uptake in mixed neuronal-glial and exclusively glial cultures. Neurobiology 4:414-418.

Moonen, G. and M. Sensenbrenner. 1976. Effects of dibutyryl cyclic AMP on cultured brain cells from chick embryos of different ages. Experientia 32:40-41.

Moonen, G., Y. Cam, M. Sensenbrenner and P. Mandel. 1975. Variability of the effects of serum-free medium, dibutyryl cyclic AMP or theophylline on the morphology of cultured new-born rat astroblasts. Cell Tiss. Res. 163:365-372.

Moonen, G., E. Heinen and G. Goessens. 1976. Comparative ultrastructural study of the effects of serum-free medium and dibutyryl cyclic AMP on newborn rat astroblasts. Cell Tiss. Res. 167:221-227.

Nakai, J. 1956. Dissociated dorsal root ganglia in tissue culture. Am. J. Anat. 99:81-103.

Nelson, P. G. and J. H. Peacock. 1973. Electrical activity in dissociated cell cultures from fetal mouse cerebellum. Brain Res. 61:163-174.

Okun, L. H. 1972. Isolated dorsal root ganglion neurons in culture: cytological maturation and extension of electrically active processes. J. Neurobiol. 3:111-151.

Peacock, J. H., P. G. Nelson and M. W. Goldstone. 1973. Electrophysiologic study of cultured neurons dissociated from spinal cords and dorsal root ganglia of fetal mice. Developmental Biol. 30:137-152.

Peterson, G. R., G. W. Webster and L. Shuster. 1973. Characteristics of choline acetyltransferase and cholinesterase in two types of cultured cells from embryonic chick brain. Developmental Biol. 34:119-134.

Richelson, E. 1975. Tissue culture of the nervous system: applications in neurochemistry and psychopharmacology. In: *Handbook of Psychopharmacology*. L. L. Iversen, S. D. Iversen and S. H. Snyder, eds. Plenum Press, New York. pp. 101-135.

Rosenberg, R. N. 1972. Neuronal and glial enzyme studies in cell culture. *In Vitro* 8:194-206.

Roth-Schechter, B., M. Sensenbrenner and P. Mandel. 1976. Effet du pentobarbital sur la morphologie et la respiration des cellules gliales en culture primaire. C. R. Acad. Sci. Paris série D 283:1333-1335.

Schrier, B. K. 1973. Surface culture of fetal mammalian brain cells: effect of subculture on morphology and choline acetyltransferase activity. J. Neurobiol. 4:117-124.

Scott, B. S., V. E. Engelbert and K. C. Fisher. 1969. Morphological and electrophysiological characteristics of dissociated chick embryonic spinal ganglion cells in culture. Exp. Neurology 23:230-248.

Sensenbrenner, M. and P. Mandel. 1974. Behaviour of neuroblasts in the presence of glial cells, fibroblasts and meningeal cells in culture. Exp. Cell Res. 87:159-167.

Sensenbrenner, M., Z Lodin, J. Treska, M. Jacob, M. P. Kage and P. Mandel. 1969. The cultivation of isolated neurons from spinal ganglia of chick embryo. Z. Zellforsch. 98:538-549.

Sensenbrenner, M., J. Booher and P. Mandel. 1971. Cultivation and growth of dissociated neurons from chick embryo cerebral cortex in the presence of different substrates. Z. Zellforsch. 117:559-569.

Sensenbrenner, M., N. Springer, J. Booher and P. Mandel. 1972. Histochemical studies during the differentiation of dissociated nerve cells cultivated in the presence of brain extracts. Neurobiology 2:49-60.

Sensenbrenner, M., J. Booher and P. Mandel. 1973. Histochemical study of dissociated nerve cells from embryonic chick cerebral hemispheres in flask cultures. Experientia 29:699-701.

Sensenbrenner, M., G. G. Jaros, G. Moonen and P. Mandel. 1975. Effects of synthetic tripeptide on the differentiation of dissociated cerebral hemisphere nerve cells in culture. Neurobiology 5:207-213.

Sensenbrenner, M., G. G. Jaros, G. Moonen, E. J. Meyer and P. Mandel. 1977a. Effects of conditioned media on nerve cell differentiation *in vitro*. S. Afr. J. Sci. in press.

Sensenbrenner, M., G. Moonen, J. P. Delaunoy, E. Bock, P. Poindron and P. Mandel. 1977b. Morphological and biochemical differentiation of rat astroblasts in primary culture. Trans. Am. Soc. Neurochem. 8:43.

Shahar, A., Y. Grunfeld, M. Y. Spiegelstein and R. Mouzain. 1975. Myelinization in long-term cultures of dissociated mammalian neurons. Brain Res. 88:44-51.

Shapiro, D. L. 1973. Morphological and biochemical alterations in foetal rat brain cells cultured in the presence of monobutyryl cyclic AMP. Nature (Lond.) 241:203-204.

Shapiro, D. L. and B. K. Schrier. 1973. Cell cultures of fetal rat brain. Growth and marker enzyme development. Exp. Cell Res. 77:239-247.

Touzet, N., R. Jeanmaire-Zylberberg and M. Chaminade. 1977. Etude par immunofluorescence de l'apparition d'antigènes neurospécifiques d'adulte chez l'embryon de poulet. J. Embryol. exp. Morph. in press.

Touzet, N., M. Sensenbrenner, Th. Lender and P. Mandel. 1975. Cultivation and differentiation of dissociated cells of chick embryo encephalon. Differentiation 4:183-187.

Varon, S. 1975. Neurons and glia in neural cultures. Exp. Neurol. 48:93-134.

Varon, S. and C. W. Raiborn, Jr. 1969. Dissociation, fractionation, and culture of embryonic brain cells. Brain Res. 12:180-199.

Varon, S. and C. W. Raiborn. 1971. Excitability and conduction in neurons of dissociated ganglionic cell cultures. Brain Res. 30:83-98.

Varon, S. and C. Raiborn. 1972. Dissociation, fractionation and culture of chick embryo sympathetic ganglionic cells. J. Neurocytol. 1:211-221.

Varon, S., C. Raiborn and E. Tyszka. 1973. *In vitro* studies of dissociated cells from newborn mouse dorsal root ganglia. Brain Res. 54:51-63.

Werner, I., G. R. Peterson and L. Shuster. 1971. Choline acetyltransferase and acetylcholinesterase in cultured brain cells from chick embryos. J. Neurochem. 18:141-151.

Wiesmann, U. N., K. Hofmann, Th. Burkhart and N. Herschkowitz. 1975. Dissociated cultures of newborn mouse brain. I. Metabolism of sulfated lipids and mucopolysaccharides. Neurobiology 5:305-315.

Wilkinson, M., C. J. Gibson, B. H. Bressler and D. R. Inman. 1974. Hypothalamic neurons in dissociated cell culture. Brain Res. 82:129-138.

Wilson, S. H., B. K. Schrier, J. L. Farber, E. J. Thompson, R. N. Rosenberg, A. J. Blume and M. W. Nirenberg. 1972. Markers for gene expression in cultured cells from the nervous system. J. Biol. Chem. 247:3159-3169.

Yavin, E. and Z. Yavin. 1974. Attachment and culture of dissociated cells from rat embryo cerebral hemispheres on polylysine-coated surface. J. Cell Biol. 62:540-546.

TRACING GLIAL CELL LINEAGES BY COLONY FORMATION IN PRIMARY CULTURES*

S. FEDOROFF

Department of Anatomy
University of Saskatchewan
Saskatoon, Saskatchewan
Canada

Booher and Sensenbrenner (1972) demonstrated that glial precursor cells can be obtained from brain of embryos provided the embryos are selected at developmental stages at which most neurons are already in the post-mitotic state of differentiation but at which glial cell precursors are still proliferating. In cultures such glial precursor cells form monolayers and when dibutyryl cyclic AMP (dBcAMP) is added (see chapter by Sensenbrenner, this volume) many transform into cells which morphologically and biochemically resemble mature astrocytes (see chapter by Hertz, this volume) and contain large amounts of the astrocyte specific GFA protein (see chapter by Bock, this volume). The procedure used by Booher and Sensenbrenner (1972) leads to confluent glial cell cultures within a couple of weeks but the cultures are too dense to study glial precursor cells as they differentiate into mature glial cells. We modified this technique by using a small inoculum of dissociated cells so that *colonies*, rather than a monolayer, formed. The hope was that precursor cells from various stages of the glial cell lineages would be segregated into the discrete colonies.

A suspension of dissociated cells was made by gently forcing mouse cerebral hemispheres, free of meninges, through sterile Nitex nets of mesh size 75 or 120 μm, depending on the age of the mice. The cells were suspended in a growth medium consisting of Eagle's MEM containing a fourfold concentration of vitamins, a double concentration of amino acids, except glutamine (kept at 2 mM level), 7.5 mM (135 mg/100 ml) glucose with 20% (v/v) fetal calf serum, and antibiotics (penicillin and streptomycin). The larger aggregates were allowed to settle and the supernatant was collected. Cell viability was determined as the ability to exclude the dye Nigrosine. The supernatant was serially diluted and cells were planted in 60 mm plastic Falcon petri dishes. Each petri dish contained 2.5 ml of cell

*This chapter was presented as part of the discussion of cell and tissue differentiation.

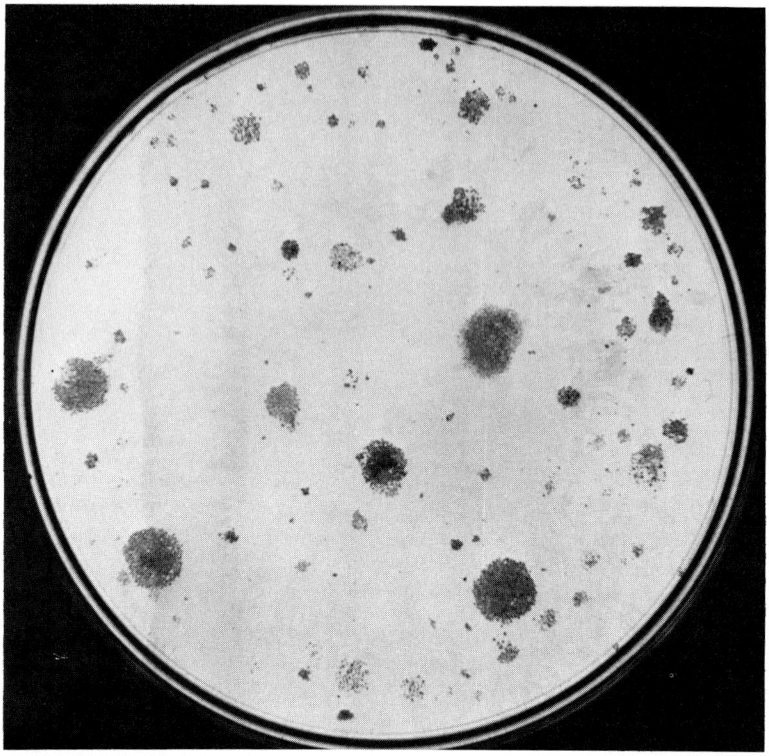

FIG. 1. Culture from dissociated 15-day mouse embryo cerebral hemispheres grown for 14 days in 60 mm Falcon petri dish. Cells form discrete colonies varying in size, density and morphology.

suspension, containing 0.4-3.6 × 10^4 viable cells/ml and was incubated at 37°C in an atmosphere of 5% (v/v) CO_2 in air.

After 3 days of culture, round refractile cells could be seen attached to the plastic dish and many single cells, cell clumps and cell debris were floating in the medium. The medium was removed, cell debris and non-attached cells were washed out and fresh medium was added. The cultures were then incubated for a further 11 days without disturbance or feeding. (This was possible because of the small cell inoculum.) After a total of 14 days of incubation, the medium was removed, the cultures were washed with warm Hanks' balanced salt solution (37°C) and fixed in Bouin's fixative or in absolute methanol. Cultures were then examined with darkfield and phase contrast optics. Subsequently, cultures were stained with 2.4% (w/v) Coomasie Brilliant Blue for 5 minutes and then examined with light field microscopy.

The dissociated cells had, indeed, formed discrete colonies which varied in size, density and overall morphology (Fig. 1). The number of

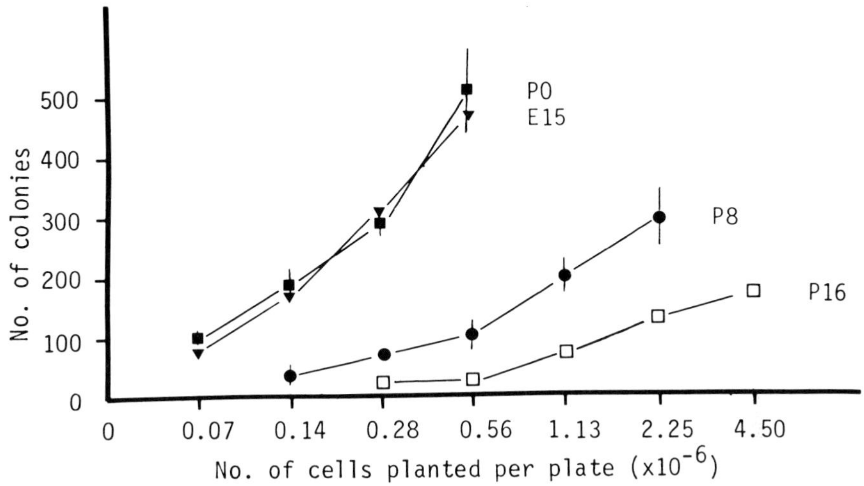

FIG. 2. Correlation between number of colonies, number of viable cells originally planted and age of animal. Dissociated cells from mouse cerebral hemisphere were planted in different concentrations in Falcon petri dishes. The number of colonies formed was determined after 14 days in culture. There is a relationship between the number of cells planted and the number of resulting colonies. Cells from cerebral cortex from 15 day old embryos (E15) and newborn (P0) mice formed the same number of colonies. Ability to form colonies declined postnatally. Cells from 16 day old mice (P16) formed fewer colonies than cells from 8 day old mice (P8).

colonies formed depended on the amount of viable cells initially planted as well as on the age of the mice (Fig. 2). There was no significant difference in the number of colonies when the same number of viable cells were obtained from 15-day embryos (E15) or from newborn mice(P0), but fewer colonies developed from cell suspensions obtained from 8 day old mice (P8) and even fewer from 16 day old mice (P16). This seems to be a reflection of the declining intensity of glial cell proliferation *in vivo* with increasing age.

Under low power darkfield illumination the differences between colonies were pronounced. Based on the compactness of the colonies and the pattern of shininess in darkfield illumination, six distinct types of colonies were identified and designated A to F (Figs. 3-8). The type A colony always appeared compact and uniformly shiny (Fig. 3). Under the phase contract microscope (Fig. 9), these colonies were seen to consist of closely apposed epithelial cells, the smaller ones toward the centre. Nuclei tended to be oriented toward the centre of the colony, rather than being centrally placed in the cells. The peripheral cells had much thinly spread cytoplasm, outlining the colony boundaries. The type B colony (Fig. 4),

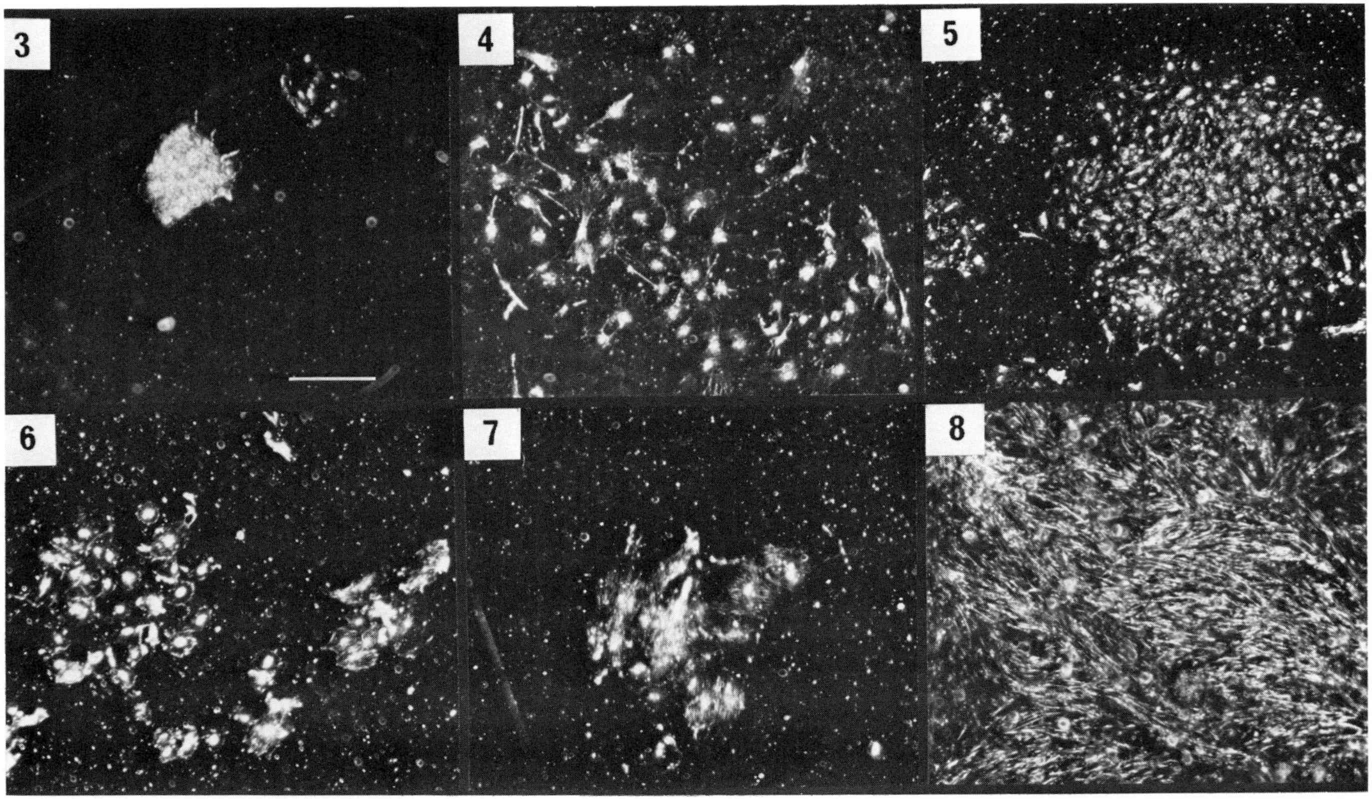

FIGS. 3-8. Darkfield illumination of the six types of colonies. Fig. 3: Type A colony; Fig. 4: Type B colony; Fig. 5: Type C colony; Fig. 6: Type D colony; Fig. 7: Type E colony; Fig. 8: Type F colony. For details see text. Scale = 500 μm.

consisted of loosely-arranged cells. Nuclei shone brightly and the cytoplasm was faintly visible. Under phase contrast, numerous irregular cytoplasmic processes were seen. The nuclei were centrally placed within the cells. The type C colony appeared compact (Fig. 5). Only the cell nuclei were bright. They were not uniformly illuminated but exhibited small, shiny dots. With phase contrast microscopy (Fig. 11), the cells were seen to be apposed by many cytoplasmic processes and although the colonies were compact, spaces could be seen between the cells. Nuclei were centrally placed. The type D colony consisted of loosely-arranged cells with brightly shining nuclei, and cytoplasm shining as a thin continuous line (Fig. 6). In phase contrast microscopy, large round cells, some with cytoplasmic processes issuing from one region, were seen. The type E colony had brightly illuminated cell nuclei and dimly shining fibrils in the cytoplasm gave these cells a "misty" appearance (Fig. 7). Under phase contrast, very large cells, often with two or more nuclei, were seen. The cytoplasm contained bundles of fibers going in one direction and, less commonly, in several directions. The type F colony contained brightly illuminated cells and showed a directional arrangement (Fig. 8). Such colonies were rarely seen. With phase contrast, large colonies consisting of spindle-shaped cells, attenuated at each end by long thin processes were seen.

The study of type A colonies using timelapse cinematography revealed that cells of type A colonies (Fig. 9) gave rise to cells forming type C colonies (Fig. 11) and that cells of type C colonies, on addition of 0.25 mM dBcAMP, formed morphologically typical fibrous astrocytes (Fig. 12). Such astrocytes were rich in acidic glial protein and stained with Cajal's gold chloride sublimate.

All six types of colonies were observed in the cultures irrespective of the age of the animal (E10-P16); only the proportion of the different types of colonies varied, and the relative number of type A colonies decreased with increasing age. Type A colonies formed by cells from young embryos (E13-E16), as a rule, contained a large number of small bipolar cells (shiny in darkfield) which clustered on the surface of the type A colony epithelial cells (Fig. 10). Type A colonies formed by cells from newborn mice had only a few such small cells and colonies formed from postnatal mice were free of these cells. Thus, the association of small bipolar cells with type A colonies is age specific and probably reflects developmental potential of the type A colony cells. In timelapse cinematography the small bipolar cells appeared to be motile and survived in cultures for 2-3 weeks. It is likely that these cells represent a cell population among which are neuronal cell precursors, since some of them developed long, branching processes which in all respects resembled neurites.

When cultures were initiated by cells from different parts of the cerebrum, the largest number of type A colonies were obtained from cells

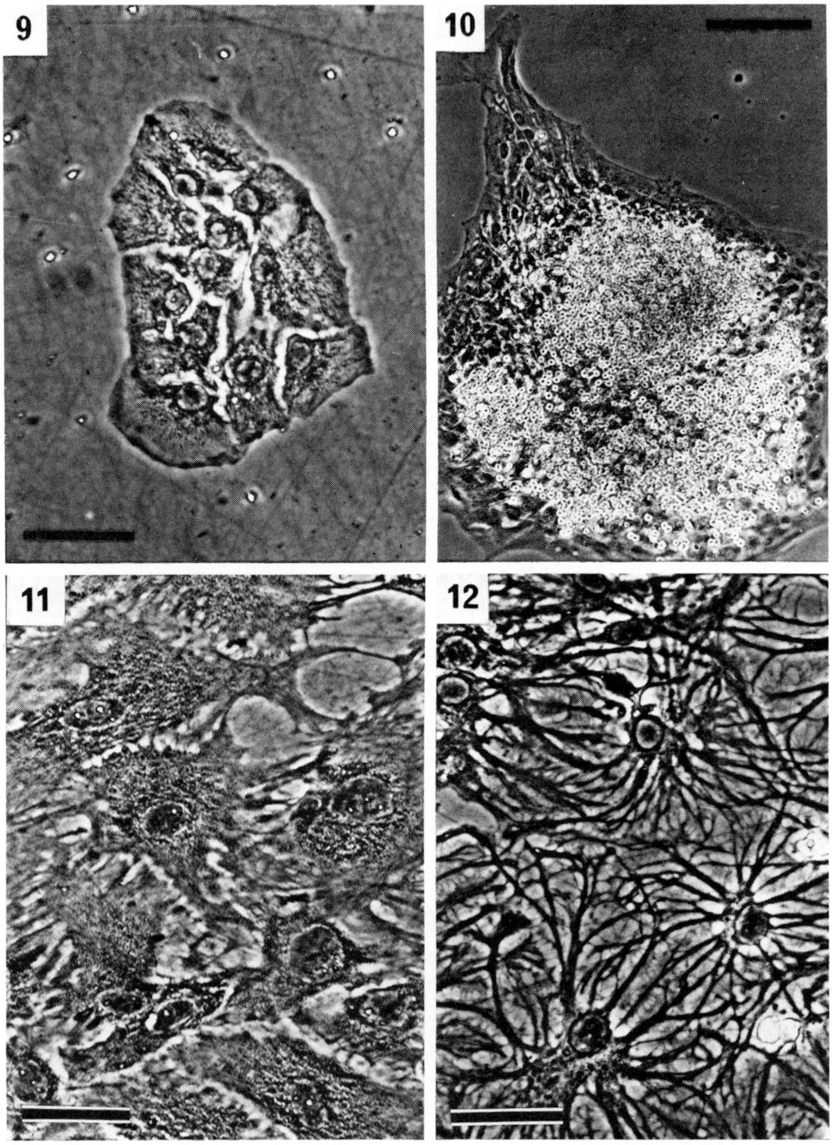
9
10
11
12

from the part of the brain containing the subventricular zone (B. Juurlink and S. Fedoroff, unpublished observation). Because of this observation and because of the epithelial nature of the cells and their resemblance in electromicrographs to subventricular cells (Oteruelo *et al.*, 1977), it seems reasonable to assume that type A colonies are indeed composed of immature subventricular cells which *in vivo* can form astrocytes and oligodendrocytes (see chapter by Privat, this volume) as well as neurons (Sidman, 1970).

The exact relationship between type A colony cells and the small bipolar cells as well as the interrelationship and potential of the other types of colonies observed in the cultures are still under study. It seems, however, that because it is possible to identify *various* undifferentiated cells according to their ability to form specific types of colonies in culture, it should be possible by this technique to sort out the different developmental stages of various cell lineages in the developing nervous system in a manner which is not feasible *in vivo*.

Acknowledgments

This work was supported by MRC (Canada) grats MA and MT 4235. I am grateful to Drs. L. Hertz and B. Juurlink and Mr. C. Hall for their help and criticisms.

References

Booher, J. and M. Sensenbrenner. 1972. Growth and cultivation of dissociated neurons and glial cells from embryonic chick, rat and human brain in flask cultures. Neurobiology 2:97-105.

Sidman, R. L. 1970. Cell proliferation, migration, and interaction in the developing mammalian central nervous system. In: *The Neurosciences: Second Study Program.* F. O. Schmitt, ed. Rockefeller Univ. Press, New York. pp. 100-107.

Oteruelo, F. T., S. Fedoroff and L. Hertz. 1977. Ultrastructure of glial precursor cells in tissue culture. Proc. Canad. Fed. Biol. Soc. 20:47.

FIG. 9. Typical type A colony from a newborn (P0) mouse cerebrum. Smaller cells in center surrounded by larger cells with eccentric nuclei. Phase contrast, scale = 50 μm.

FIG. 10. Typical type A colony from 16 day old (E16) mouse embryo cerebrum. Large number of small bipolar cells adhere to the surface of type A colony epithelial cells. Phase contrast, scale = 100 μm.

FIG. 11. Typical type C colony cells. Cells with multiple processes abut against neighbouring cells. Phase contrast, scale = 50 μm.

FIG. 12. Type C colony cells transformed into fibrous astrocytes on addition of dBcAMP to the culture medium. Phase contrast, scale = 50 μm.

DIFFERENTIATION OF GLIOBLASTS UNDER THE INFLUENCE OF GLIA MATURATION FACTOR*

RAMON LIM, DAVID E. TURRIFF, SHUANG S. TROY AND TAIJI KATO

Brain Research Institute and Departments of Surgery (Neurosurgery) and Biochemistry, University of Chicago Chicago, Illinois

I. Introduction

For the past few years (Lim *et al.*, 1972, 1973; Lim and Mitsunobu, 1974) our laboratory has engaged in the study of dissociated fetal brain cells grown in surface culture. Emphasis is placed on the detection of agents that can promote the development and differentiation of these cells. In so doing we have detected the presence of a protein factor, endogenous to the adult brain, which is capable of reversibly promoting the morphological and chemical maturation of glioblasts. We have previously referred to this factor as the Morphological Transforming Factor (Lim and Mitsunobu, 1975; Lim *et al.* 1976, 1977b) based on the striking morphological effect on the cells. In this chapter we designate it as the Glia Maturation Factor (GMF). The use of this term does not preclude the possibility of other functions that the factor may have, nor does it imply that it is the only factor needed for the differentiation of glial cells. In this paper, evidence is presented in favor of the glial nature of the cells under study and in favor of their differentiation in the presence of the factor. The effects of cyclic nucleotides are included for comparison.

II. Isolation of Glioblasts for GMF Study

Cerebrums and cerebellums, freshly dissected from 17-day fetal rats, are combined and dissociated with trypsin into single cells. The majority of the cells are round and about 10 μm in diameter. The cells are left undisturbed at 37°C for 24 hours, at which time they group themselves into numerous aggregates of about 200 to 300 cells each. (Cells dissociated from other organs of the same embryos do not form aggregates under similar conditions). These aggregates adhere to the flask surface as the cells start to

*This paper was presented as part of the discussion dealing with development of neuroglia cells from precursor cells.

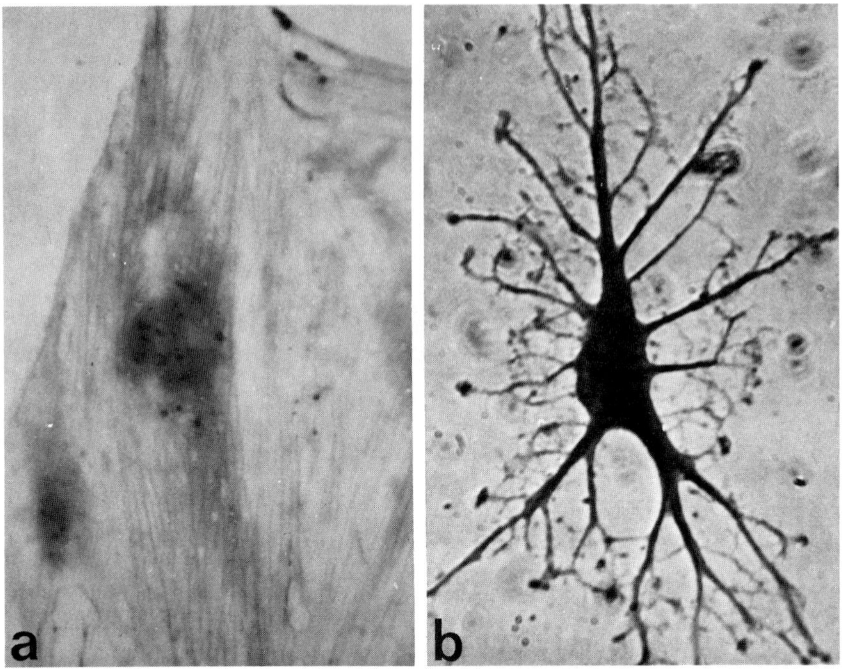

FIG. 1. Effect of GMF on glioblast morphology. (a) Before addition of GMF. Note the spread-out appearance and the tenuous cytoplasm. Only the nuclear and perinuclear regions are clearly visible. Borders between adjacent cells are ill-defined. (b) Twenty four hours after addition of GMF. Note the presence of multipolar branched processes with knob-like terminals which resemble astrocytic feet. The cells were fixed with 2% glutaraldehyde: 1% paraformaldehyde in 0.1 M phosphate buffer, pH 7.4, and stained with hematoxylin and eosin. × 900.

migrate out. The first outgrowing cells are the glioblasts. They assume an epithelial morphology and tend to grow in sheets. They eventually cover the entire flask in the form of a continuous cell carpet. The second outgrowth of cells is composed of bipolar neuroblasts with a phase-dark cell body. These cells stay above the glioblasts and never attach directly to the plastic or glass surface. The neuroblasts contain many microtubules in the processes and occasionally show synaptic connections. Some necrotic cells appear in the center of the aggregates, probably because of limited accessibility of nutrients. The surviving cells are approximately 50% glioblasts and 50% neuroblasts. The neuroblasts are eliminated during the second passage by means of differential attachment (by changing the medium within 24 hours after subculture, the neuroblasts are deprived of a chance to adhere to the substratum), leaving a homogeneous population of epithelial glioblasts. That these cells are glial in nature is evidenced by a high level of the glial

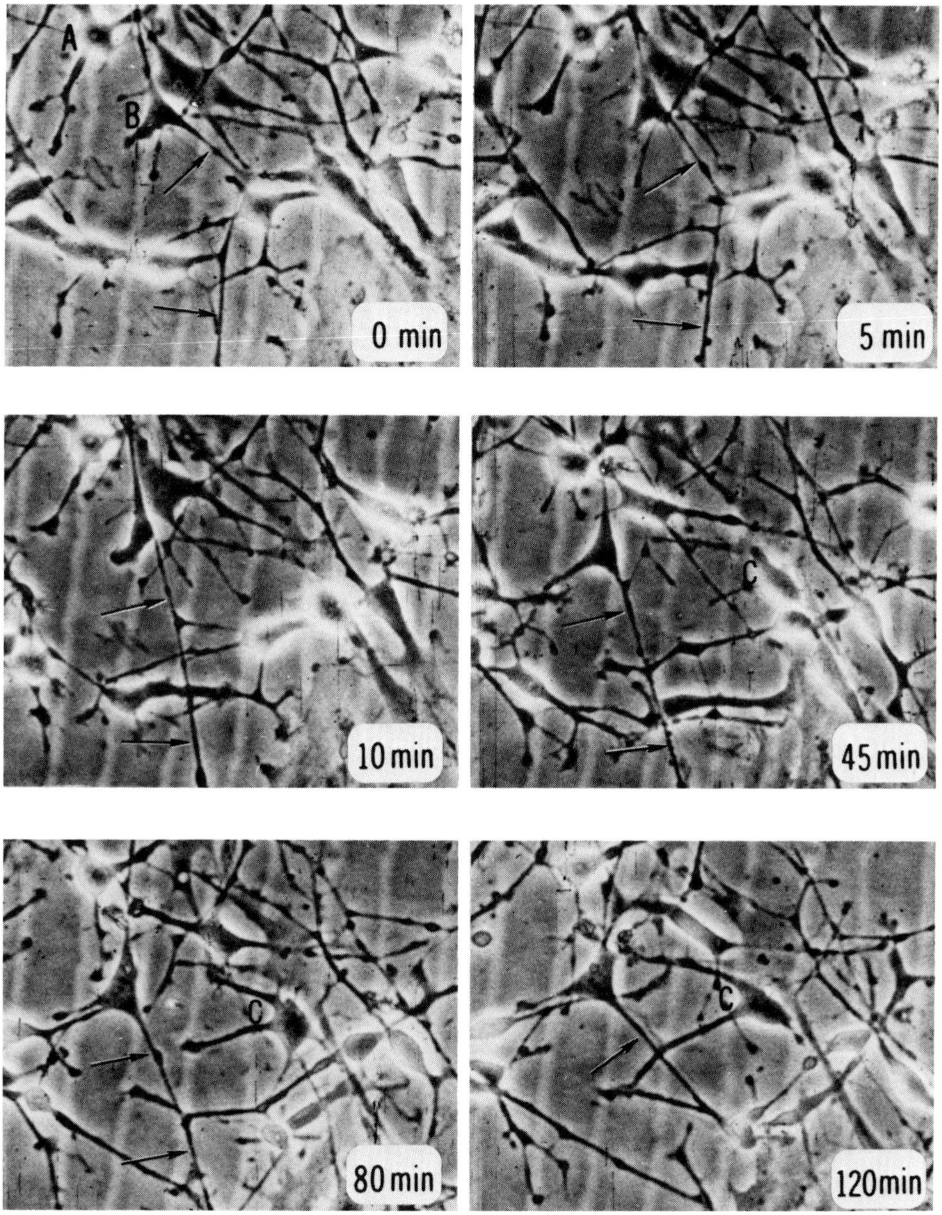

FIG. 2. Time-lapse photography of glioblasts 24 hours after exposure to GMF, showing movement of the cell processes. The times indicated are intervals with respect to the first picture of the series. Note that cell "A" has a process pointing downward at zero minutes, which is retracted at 45 minutes and re-extended at 120 minutes. Cell "B" has a process (arrows) pointing downward which assumes different angulations with time as a result of pulling among different cells. Cell "C" (at 45 minutes) extends a process toward the lower left of the field. (Vertical lines are manufacturer's markings on the plastic flask surface.) Live phase-contrast. × 250.

fibrillary acidic protein (GFAP) (Bock *et al.*, 1975). Under our culture conditions, with or without subsequent passage, the cells can survive for weeks or even months without showing signs of morphological differentiation.

For our studies we routinely use glioblasts in the second passage at the confluent or near-confluent stage.

III. Effect of GMF on Glioblasts

The addition of GMF to a monolayer of glioblasts results in morphological changes after a lag period of about 12 hours (Fig. 1). The sequence of events is best observed with time-lapse photography (Lim *et al.*, 1976). First, the cells tend to round up without detaching from the flask. They then send out processes which are usually multipolar and branched. The processes adhere tightly to the flask surface and reach out to make contacts with processes from other cells. Tugging among cells is frequently observed (Fig. 2). In the steady state there is a constant making and breaking of connections as different processes are being extended and retracted at the same time (Fig. 2). While this goes on, the cell bodies remain relatively localized, rarely moving from one place to another. The overall pattern is that of an undulating cell net (best seen between 24 and 48 hours) which is in great contrast to the original quiescent state of a cell carpet. The dynamic pattern differs from that of fibroblast cultures where the cells are usually engaged in random movement, without extruding processes and making connections with one another.

Withdrawal of the factor from the medium leads to the reversal of the differentiated morphology in one or two days. However, if the factor is renewed, the processes continue to elongate and overlap with one another, conveying the impression of criss-crossing tree branches. (The cell bodies, on the other hand, do not overlap.) This is most strikingly seen between the 4th and 7th day. At this time the tugging activity subsides.

Another property characteristic of glial cells usually appears at 3 to 5 days, when many of the cells pulsate vigorously, some with a periodicity of 5 to 10 minutes per beat (Lim *et al.*, 1976). This pulsatile movement has never been reported for neuroblasts or fibroblasts. Myoblasts do contract but they have a completely different rhythm.

The morphological changes are most dramatically demonstrated in a confluent culture, but glioblasts in sparse cultures also respond to the factor. Glioblasts in the primary and secondary cultures react equally well. Increasing the passage of the cells leads to a progressive decrease of responsiveness (Lim *et al.*, 1977d), probably because of the loss of receptors from the cell surface. Neuroblasts, on the other hand, never react to the factor.

In addition to the morphological effect, GMF also stimulates cell division in glioblasts, as can be observed directly with cinematography and indirectly with thymidine incorporation (Lim *et al.*, 1976). However, the surge in cell multiplication is limited to the first two days. Thereafter mitosis is inhibited in the continued presence of the factor. This mitogenic effect persists with the purification of GMF. Nonetheless, it remains to be proved if a single polypeptide mediates both the mitogenic and the morphologic effects.

The surge in DNA synthesis is preceded by an increase in intracellular cyclic GMP, which occurs a few hours after exposure to the factor, and is the earliest detectable chemical event (D. E. Turriff and R. Lim, unpublished data).

The change in cell shape is correlated with a restructuring of the fine components of the cell, as revealed by electron microscopy (Lim *et al.*, 1977b). The epithelial glioblasts contain numerous microfilaments (5 nm) and only a few gliofilaments (10 nm) and microtubules (20 nm), the last two being randomly oriented. Exposure of the cells to GMF for a week leads to the complete disappearance of the microfilaments, while the gliofilaments greatly increase in number, filling most of the cell processes. The microtubules remain sparse, but they are now arranged parallel to the long axis of the processes, as are the gliofilaments.

The epithelial glioblasts are exclusively lined with desmosome junctions, which are so numerous that they frequently appear in series (Lim *et al.*, 1977b). The differentiated cells do not contain any desmosomes. Instead, they contain the *puncta adhaerentia* type of junction at points of contact between processes (Lim *et al.*, 1977b). *Puncta adhaerentia* are normally found among glial cells and among neurons in the brain.

The morphologically differentiated glioblasts in surface culture do not contain glycogen granules, as do those grown in an aggregate (Lim *et al.*, 1977b). This difference could be due to the relatively anoxic state of cells under the latter condition.

A 10-fold increase in the glia-specific protein S100 occurs four days after stimulation by GMF (Fig. 3), a fact strongly in favor of glial cell differentiation (Lim *et al.*, 1977c). Concomitantly, intracellular cyclic AMP increases three-fold (Lim *et al.*, 1977c), as shown in Fig. 4. For reference, we observed that cyclic AMP in fetal (17-day) and adult rat brains are 30 pmoles and 88 pmoles/mg protein, respectively. Thus, the rise in cyclic AMP as a result of exposure to GMF is consistent with cell maturation. Furthermore, the decrease in cell division and motility at a time when cyclic AMP increases could be explained by the known suppressive effect of the cyclic nucleotide on these activities.

The question arises as to whether the increase in S100 and cyclic AMP is secondary to cell density. This is not the case since total DNA content per

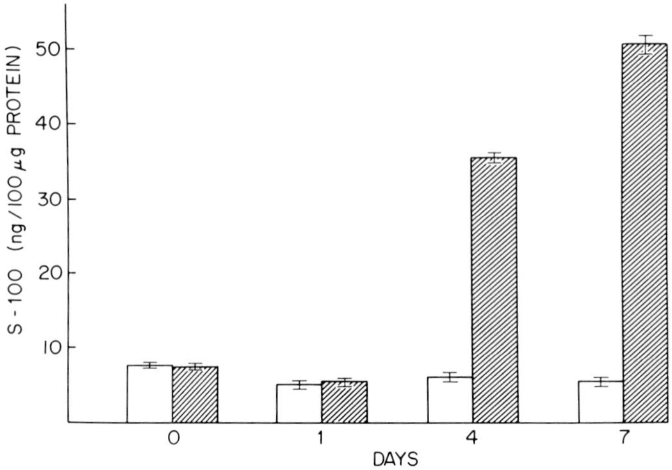

FIG. 3. Effect of GMF on S100 protein levels in glioblasts. Empty bars, control cells; hatched bars, cells exposed to GMF. Values are means ± S.D. from 4 pools of two culture flasks each. Days indicated are the lengths of time after the initial exposure to the factor.

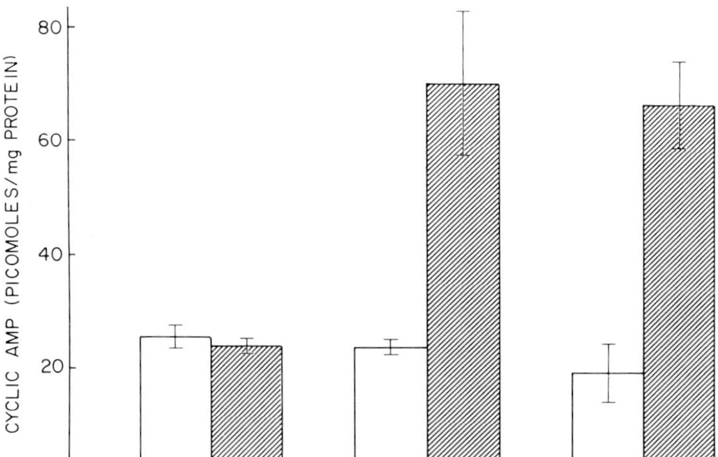

FIG. 4. Effect of GMF on cyclic AMP levels in glioblasts. For explanations see legend to Fig. 3.

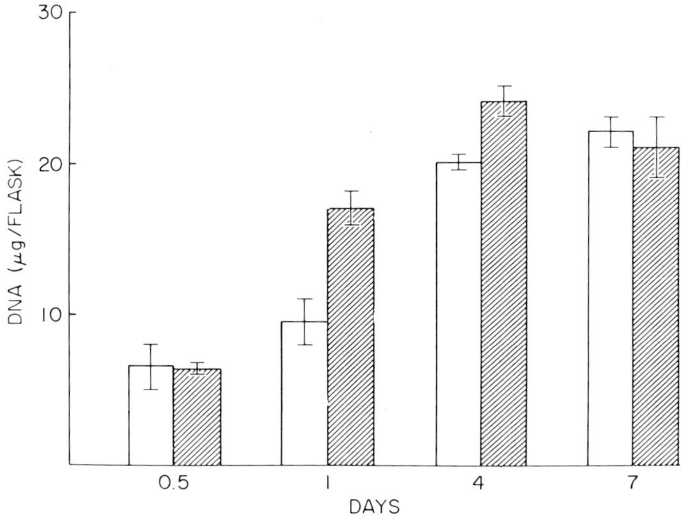

FIG. 5. DNA content of glioblasts grown in the absence and presence of GMF. Surface area of the flask was 75 cm^2. For other explanations see legend to Fig. 3.

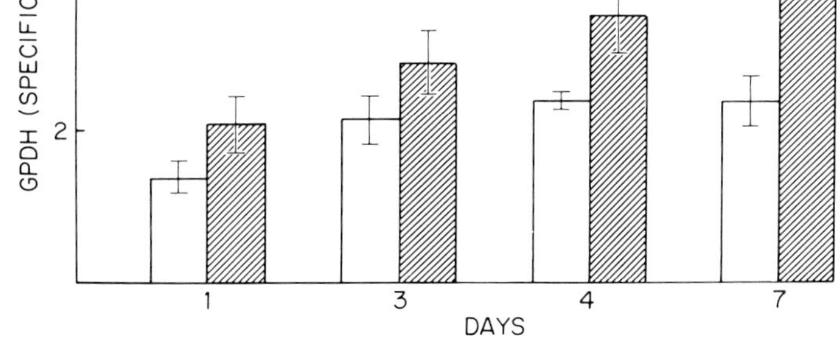

FIG. 6. Effect of GMF on glycerophosphate dehydrogenase (GPDH) in glioblasts. Specific activity is defined as that amount of enzyme catalyzing one nmole of substrate per minute per mg total cell protein. Values are means of 4 flasks ± S.D. For other explanations see legend to Fig. 3.

flask (an indication of cell density) is about the same in experimental and control cultures (Fig. 5) when such increase is detected. Apparently, while cell division in the GMF-stimulated cells is sequentially turned on and off, that in the control cells goes on at a slow but steady rate.

To further rule out the role of cell density effect, we conducted a "cell reversion" experiment. In this experiment the glioblasts were exposed to GMF for two days and then switched to the control medium. Cells subjected to this treatment went through the initial phase of morphological changes and the spurt of cell division but failed to show an increase in S100 and cyclic AMP at 4 and 7 days. This indicates that the continued presence of GMF is needed to trigger chemical differentiation.

GMF does not increase the neuron-specific protein antigen α, also known as "14-3-2", which is present at a uniformly low level in the glioblasts (G. S. Bennett and R. Lim, unpublished data).

GMF induces the cytoplasmic enzyme glycerophosphate dehydrogenase (EC 1.1.1.8) in the glioblasts by as much as three-fold (Fig. 6). Hydrocortisone alone also causes a two-fold increase. However, hydrocortisone and GMF added together do not produce a further increase over the value obtained by GMF alone (Lim *et al.*, 1977a). By contrast, another cytoplasmic enzyme, lactate dehydrogenase, does not respond to either agent, indicating that the inductive effect is selective (Lim *et al.*, 1977a).

Intracellular recording of the morphologically differentiated glioblasts failed to detect any action potential (D. Gruol and R. Lim, unpublished data).

In summary, the glioblasts undergo a biphasic change in response to GMF. The first stage (first three days) consists of: a rise in cyclic GMP; rounding up of cell bodies; outgrowth of short processes which are highly motile; a surge in cell division. The second stage (after three days) consists of: elongation of the processes whose motility is now suppressed; inhibition of cell division; onset of chemical maturation.

IV. Effect of Cyclic Nucleotides on Glioblasts

This section provides a comparison between GMF and the cyclic AMP (cAMP) analogues. Of all the known compounds tested, only a few analogues of cAMP and molecules known to raise the intracellular levels of cyclic nucleotides can elicit the formation of cell processes in the glioblasts under study.

Dibutyryl cyclic AMP (dB-cAMP) causes morphological changes and process formation in as early as five minutes. After reaching a peak at about four hours, the cells gradually revert to the "flat" morphology (Lim *et al.*, 1976). Very few retain the differentiated appearance after 24 hours. The reversion can be retarded by theothylline. (By contrast, theophylline does not potentiate the effect of GMF). That the morphological effect is indeed due to the cAMP moiety is borne out by the abilitv of 8-bromo-cAMP but not sodium butyrate to simulate the effect. Unlike dB-cAMP, the 8-bromo analogue has a prolonged effect, which can be explained by the stability of the molecule inside the cell. Unsubstituted cAMP, on the other hand, has no effect when added to the medium, a fact commonly attributed to its poor penetration across the cell membrane. Cyclic GMP and its analogues have no morphological effect. The results are summarized in Table I.

TABLE I
EFFECT OF CYCLIC NUCLEOTIDES ON GLIOBLASTS[a]

	Time after addition of agents		
	2 hours	4 hours	24 hours
Control (no addition)	—	—	—
GMF (60 μg/ml medium)	—	—	+++
cAMP (1.0 mM)	—	—	—
dB-cAMP (1.0 mM)	+++	++++	+
Theophylline (0.1 mM)	—	—	—
dB-cAMP (1.0 mM) + theophylline (0.1 mM)	+++	++++	+++
8-Bromo-cAMP (1.0 mM)	++	+++	+++
Na butyrate (1.0 mM)	—	—	—
cGMP (1.0 mM)	—	—	—
dB-cGMP (1.0 mM)	—	—	—
8-Bromo-cGMP (1.0 mM)	—	—	—

[a]Degrees of morphological differentiation are expressed as follows (% of cells that form processes with respect to total cell population): — = 0-5%; ± = 5-10%; + = 10-25%; ++ = 25-50%; +++ = 50-75%; ++++ = 75-95%. All experiments were carried out in F10 nutrient with 5% fetal calf serum.

Differences exist between the effects of dB-cAMP and GMF. First, the formation of processes in response to dB-cAMP is mainly due to the retraction of the epithelial glioblasts at selected points, whereas the events in response to GMF consist of active extension of processes as well as retraction of cell bodies. Secondly, the processes formed in the presence of dB-cAMP are filamentous, resembling tree roots, whereas those in the presence of GMF have more volume and are more rigid, resembling tree branches. Thirdly, it takes only a few minutes for the effect of dB-cAMP to show up, whereas a lag period of at least 12 hours exists for GMF. Lastly, dB-cAMP suppresses cell division and motility whereas GMF stimulates these activities.

That cAMP is not a mediator of GMF is evidenced by the absence of its increase at the time when GMF starts to reorganize the cell pattern, even though cAMP level does increase at a much later period (Lim *et al.*, 1977c).

A somewhat similar situation is observed with nerve growth factor, where cAMP is not a mediator (Frazier *et al.*, 1973), yet dB-cAMP stimulates neuritic outgrowth in the dorsal root ganglia (Roisen *et al.*, 1972).

Isoproterenol, and to some extent norepinephrine, stimulate process formation in glioblasts, with a morphological feature and time course identical to that of dB-cAMP. This is probably due to the known effect of the drugs to raise the intracellular cAMP level of glial cells (Gilman and Nirenberg, 1971). That the β-blocker dichloroisoproterenol, but not the α-blocker phentolamine, inhibits this morphological change indicates that the glioblasts possess β-adrenergic receptors. The results are summarized in Table II.

TABLE II
EFFECT OF β-ADRENERGIC AGENTS ON GLIOBLASTS[a]

	Time after addition of agents 4 hours	24 hours
Control (no addition)	—	—
GMF (60 μg/ml medium)	—	+++
l-Norepinephrine (0.1 mM)	+	±
l-Isoproterenol[b] (0.01 mM)	+++	±
Dichloroisoproterenol (0.1 mM)	—	—
Phentolamine[c] (0.1 mM)	—	—
l-Isoproterenol (0.01 mM) + dichloroisoproterenol[d] (0.1 mM)	±	—
l-Isoproterenol (0.01 mM) + phentolamine[d] (0.1 mM)	+++	±

[a]See footnote to Table I for scoring of morphological differentiation.
[b]Trade name Isuprel (Winthrop Laboratories).
[c]Trade name Regitine (Ciba Corporation).
[d]The cells were pre-incubated with the inhibitors for 15 minutes before the addition of isoproterenol.

V. Chemical Nature of GMF

The strongest evidence in favor of a protein nature of the factor comes from its susceptibility to pronase (Lim and Mitsunobu, 1974, 1975). Other pieces of evidence include its thermolability and inactivation by extremes of pH and denaturing agents such as 8M urea. However, GMF is resistant to trypsin, a proteolytic enzyme whose action is highly selective. The factor is resistant to reducing agents such as dithiothreitol, and to alkylating agents such as iodoacetamide. It is not affected by DNAase, RNAase and periodate oxidation, nor is the activity eliminated by lipid extractants such as ethanol and butanol. The isoelectric point is below neutrality. The factor in the brain extract is a large molecule (apparent molecular weight 350,000) which can be dissociated into smaller nondialyzable subunits. The large molecule is the stable form, while the active subunit easily loses activity. One contributing factor to the instability of the subunit is the requirement of a dialyzable component. We have purified this factor down to three protein bands in the polyacrylamide gel. Efforts to achieve complete homogeneity are currently under way.

GMF appears to be distinct from other growth factors reported (Gospodarowicz and Moran, 1976). Nerve growth factor (NGF) has a molecular weight of 140,000 for the complex form and 13,000 for the active subunit which is a basic polypeptide. NGF activity is very low in the brain and there is no biological cross-reactivity between NGF and GMF. Epidermal growth factor (EGF) has a molecular weight of 70,000 for the complex form and 6,000 for the active chain. It differs from GMF in being heat and acid stable (Starkey *et al.*, 1975). Fibroblast growth factor (FGF) is distinct from GMF since it is a basic protein. It has a molecular weight of 13,000 and has not been found in any larger form. Furthermore, EGF and FGF are primarily mitogenic rather than maturation factors. The so-called "Cell Surface Protein" or LETS is a large glycoprotein (MW 250,000) found on the cell surface (Yamada *et al.*, 1976). It differs from GMF in being highly trypsin-sensitive and in having an opposite morphological effect (cell flattening in fibroblasts).

VI. Biological Significance of GMF

The reversibility of GMF effect (Lim *et al.*, 1977d) and the fact that the factor exists at a high level in the adult brain but low in the embryonic brain (Lim and Mitsunobu, 1974; Lim *et al.*, 1977d) suggests a role in glial cell differentiation which is primarily maintenance.

It is well known that mammalian cells exhibit high degrees of interdependency. When grown individually *in vitro*, they either fail to survive or frequently lose their differentiated characteristics. In a living organism, there are three possible mechanisms by which cell integrity can be maintained through mutual influence: (1) by humoral factors mediating long-range communicatons between cells in different organs, as exemplified by the well known hormones; (2) by diffusible factors acting locally between cells in the same organ, as exemplified by the ionic interactions across the gap junctions; (3) by non-diffusible or membrane-bound molecules on cell surfaces acting on adjacent cells. GMF is not a humoral factor (Lim *et al.*, 1977d). Its unusually large molecular weight and our failure to demonstrate the activity in the conditioned medium (Lim *et al.*, 1977d) lead us to speculate that GMF is bound to the surface membrane, surviving as a "solid state" intercellular message when cells are in physical contact (assuming that the receptors are also present on the membrane) but not when they are separated (presuming that self-stimulation is hindered by an unfavorable steric relationship between the factor and its receptors on the same cell). The addition of GMF to a monolayer of glioblasts would satisfy the unoccupied receptor sites, partially reconstructing the *in vivo* environment of the cells. This hypothesis is supported by our recent observation that a homogeneous population of glioblasts undergoes spontaneous differentiation if the cells are grown in an aggregate (Lim *et al.*, 1977b).

GMF is not species specific: brain extracts taken from various animals are equally effective on rat glioblasts. GMF does not induce morphological changes in cells taken from other fetal rat organs, nor does it promote process outgrowth in several cell lines. However, time has not permitted us to explore the chemical changes in these cells. Extracts prepared from several adult organs other than the brain seem to have some stimulating activity on the outgrowth of glial processes (Lim *et al.*, 1977d). This peculiarity in distribution suggests that GMF could be but one member of a general class of macro-molecules participating in the universal phenomenon of short-range cellular interaction. These molecules need not be identical but may exhibit a certain degree of cross-reactivity.

It is also plausible that neurons may contain high levels of GMF which can then promote and maintain the differentiation of the glial cells in the brain. This hypothesis can also explain the onset of gliosis following neuronal injury. *A priori*, there is no reason why the neuronal role of GMF and the general cellular role of GMF-like molecules are mutually excluding possibilities.

In any event, although the exact biological significance of GMF remains to be explored, it seems likely that GMF promises to explain some of the phenomena in developmental and cellular neurobiology. The study of this factor is potentially important in the understanding and management of such disorders as mental retardation, aging, brain tumors and gliosis.

Acknowledgments

This work was supported by U.S. Public Health Service grants No. NS-09228, NS-07376, CA-14599 and CA-19266. Drs. Turriff, Troy and Kato are Research Associates in Dr. Lim's laboratory. Dr. Turriff is an NIH Postdoctoral Fellow (NS-05017). Dr. Kato is on leave from the Department of Biochemistry, Nagoya City University, Nagoya, Japan.

Sincere appreciation is due to Dr. Silvio S. Varon, who gave us moral support since the inception of this work.

References

Bock, E., O. S. Jörgensen, L. Dittmann and L. F. Eng. 1975. Determination of brain-specific antigens in short term cultivated rat astroglial cells and in rat synaptosomes. J. Neurochem. 25:867-870.

Frazier, W. A., C. E. Ohlendorf, L. F. Boyd, L. Aloe, E. M. Johnson, J. A. Ferrendelli and R. A. Bradshaw. 1973. Mechanism of action of nerve growth factor and cyclic AMP on neurite outgrowth in embryonic chick sensory ganglia. Proc. Nat. Acad. Sci. U.S.A. 70:2448-2452.

Gilman, A. and M. Nirenberg. 1971. Effect of catecholamines on the adenosine 3′:5′ cyclic monophosphate concentrations of clonal satellite cells of neurons. Proc. Nat. Acad. Sci. U.S.A. 68:2165-2168.

Gospodarowicz, D. and J. S. Moran. 1976. Growth factors in mammalian cell cultures. Ann. Rev. Biochem. 45:531-558.

Lim, R., and K. Mitsunobu. 1974. Brain cells in culture: morphological transformation by a protein. Science. 185:63-66.

Lim, R., and K. Mitsunobu. 1975. Partial purification of a morphological transforming factor from pig brain. Biochim. Biophys. Acta (Amst.). 400:200-207.

Lim, R., W. K. P. Li and K. Mitsunobu. 1972. Morphological transformation of dissociated embryonic brain cells in the presence of brain extracts. Abstr. of 2nd Ann. Meeting Soc. Neurosci. Houston, Texas, p. 181.

Lim, R., K. Mitsunobu and W. K. P. Li. 1973. Maturation-stimulating effect of brain extract and dibutyryl cyclic AMP on dissociated embryonic brain cells in culture. Exp. Cell Res. 79:243-246.

Lim, R., D. E. Turriff and S. S. Troy. 1976. Response of glioblasts to a morphological transforming factor: cinematographic and chemical correlations. Brain Res. 113:165-170.

Lim, R., S. S. Troy and D. E. Turriff. 1977a. Induction of glycerophosphate dehydrogenase by glia maturation factor. Trans. Am. Soc. Neurochem. 8:106.

Lim, R., S. S. Troy and D. E. Turriff. 1977b. Fine structure of cultured glioblasts before and after stimulation by a glia maturation factor. Exp. Cell Res. (in press).

Lim, R., D. E. Turriff, S. S. Troy, B. W. Moore and L. F. Eng. 1977c. Glia maturation factor: effect on chemical differentiation of glioblasts in culture. Science. 195:195-196.

Lim, R., D. E. Turriff, S. S. Troy and K. Mitsunobu. 1977d. Morphological transforming factor: influence on histotypic organization of epithelial brain cells. In: *Cell Culture and its Applicaton.* R. T. Acton, ed. Academic Press, New York, (in press).

Roisen, F. J., R. A. Murphy and W. G. Braden. 1972. Neurite development *in vitro.* I. Effects of adenosine 3′5′-cyclic monophosphate. J. Neurobiol. 4:347-368.

Starkey, R. H., S. Cohen and D. N. Orth. 1975. Epidermal growth factor: identification of a new hormone in human urine. Science 189:800-802.

Yamada, K. M., S. S. Yamada and I. Pastan. 1976. Cell surface protein partially restores morphology, adhesiveness, and contact inhibition of movement to transformed fibroblasts. Proc. Nat. Acad. Sci. U.S.A. 73:1217-1221.

NEURAL CELL ISOLATION AND IDENTIFICATION

SILVIO VARON

Department of Biology
School of Medicine
University of California, San Diego
La Jolla, California

I. Introduction

Cellular heterogeneity, a conspicuous feature of neural tissue, is a serious obstacle to the recognition and study of chemical constituents and biochemical activities characteristic of each neural cell type. At the same time, this heterogeneity underlines the concept that the nervous system must develop, maintain itself, and operate through a tightly controlled interaction of its diverse classes of cells. To study these interactions, one must be able to investigate neural cells in contexts which are different, and possibly simpler, than the normal situation *in situ* and which, in addition, can be manipulated with regard to both humoral and cellular components. The investigation of neural tissue *in vitro* provides several ways to attack the problem. The rapid progress made in recent years with these approaches and the increasing number of researchers adopting them testify to their present and potential usefulness (Varon, 1970, 1975b; Varon and Saier, 1975). In parallel with these rapid advances, the need has also become increasingly pressing for neuronal and glial preparations of different types, separated from one another yet viable enough to be maintained in an *in vitro* culture system.

In this chapter, no effort will be made to review bulk isolation techniques, usually developed for the main purpose of biochemical analyses, but without an obligatory concern with subsequent viability (cf. Poduslo and Norton, 1975). Rather, the focus will be on the separation of neural cells retaining the *capability to survive* as segregated or recombined populations. Several attempts have been made in different laboratories to obtain viable populations of purified neurons and glial cells. However, the systematic effort has not yet been made to develop a general technology for neural cell separation. Such technology should include: 1) efficient dissociation techniques and supportive culture conditions, 2) effective segregation procedures, and 3) definable criteria to distinguish cell classes, before and after their segregation (Varon and Saier, 1975). This paper will first review general principles and methodologies and then illustrate some of them using recent examples of neural cell separation systems.

II. Methodology

A. Tissue Dissociation and Culture Conditions.

1. Dissociation. With few exceptions, the source tissue must be dissociated into component cells before a separation of different cell classes can be attempted. Dissociation of peripheral neural tissue is achieved effectively by enzymatic treatment, while use of enzymes on central neural tissues has given uncertain results thus far. With or without enzymatic pretreatment, cell dispersion is obtained by straining the tissue through metal or nylon sieves or, more often, by multiple aspirations in and out of a small bore pipet (trituration). The manner in which the tissue is dissociated will influence in a number of ways the outcome of subsequent attempts to cell separation. The number of cells yielded by the dissociation sets a ceiling to the size of pure populations to be obtained, a limitation particularly important for neurons in view of their inability to proliferate. Besides immediate destruction, cell viability can be jeopardized by less obvious damage to the cell membrane. Materials released by cell damage may offset the advantages of the dissociation by promoting rapid cell reaggregation. The resolving efficiency of several cell separation procedures may be further reduced by cell surface alterations that are not critical for viability, or by unfavorable ratios of nonneurons to neurons in the dissociate (see part III for examples).

Almost any detail of the dissociation procedure, no matter how seemingly trivial, is likely to affect the product of the dissociation (cf. Varon and Raiborn, 1972a; Varon and Saier, 1975). With respect to enzyme treatments, for example, relevant aspects will be choice, storage conditions and concentration of the enzyme, but also composition of the fluid and duration of exposure, volume-to-tissue ratios, divalent cations present before enzyme treatment, and the washing procedure to remove or inhibit enzyme residues. With respect to cell dispersion, important factors are orifice diameters, the pressure applied to force through the tissue, the number of passes, the composition of the dispersion medium and, again, the volume-to-tissue ratio. It is not surprising, therefore, that dissociates prepared according to the "same" dissociation procedure — at times, even by the same investigator — often exhibit gross variations in cell yield or composition, or more subtle differences in subsequent cell behavior. To minimize such variations, and also to facilitate the transmission of effective information to other experimenters, it is essential that procedures be defined in most minute details, with explicit attention to parameters already recognized as hazardous. It is also important to characterize routinely the product of each dissociation run by a minimum of easy determinations, such as total and differential cell counts, dye exclusion tests and, eventually, cell culture behaviors.

2. Culturing. Culturing of the dissociated cells is aimed at serving three main purposes: a) to verify and quantitate cell viability, b) to provide morphological information for classification and identification of the several cell classes, and c) to offer approaches to their separation.

a). The number of cells that will *attach* to a culture substratum over the first several hours and the number of cells that will *survive* on a longer term are reflections of both the quality of the dissociate that was seeded and the suitability of the several components of the culture system. Thus, a carefully standardized and reproducible set of culture conditions will permit a quantitative evaluation of neural dissociates from repeated preparations, or different preparators, and also set the stage for later assessment of the products of cell separation procedures. Conversely, a dissociate of consistent composition and qualities will allow to carry out a comparative evaluation of different sets of culture conditions. Among culture components to be evaluated are nutritional support (see Waymouth, this volume), special trophic factors (see Sato and Mather, this volume; also Varon, 1975a; Varon and Bunge, 1977), but also the choice and conditions of the culture substratum (Varon and Tayrien, 1977 and unpublished data).

b). Culture substratum and medium composition are also important determinants of cell shape and motility (Letourneau, 1975). Therefore, they must be explicitly controlled if one intends to classify diverse cell subpopulations by their morphologies and intercellular organization patterns. In addition, cell-cell adhesion and cell-substratum adhesion have been variably implicated in the regulation of cell proliferation and differentiated behaviors (cf. Varon and Tayrien, 1977). Thus, one must expect that biochemical markers may also be modulated differently in culture systems using different — or inadequately defined — substrata.

c). Cell separation, in several approaches, is based on differential interactions of different cell classes with a common culture substratum (see next section). Consequently, cell separation also may be affected by the choice of substratum and medium. In addition, neurons and glia pose different problems because of their different replicative abilities and, thus, are not equally subject to control by a given set of culture conditions. Purified neuronal populations will benefit from conditions that inhibit cell proliferation and prevent any remaining nonneuronal cells from gradually taking over. In contrast, purified glial populations could be expanded, and further enriched, if culture conditions are geared to favor their replication.

B. Segregation Procedures

Approaches to cell separation may be categorized according to the cellular *property* on which they are based, or according to the *process* chosen to exploit a given property. Table I illustrates some relationships between the two criteria.

1. *Differential filtration* attempts to take advantage of differences in cell size, a property that distinguishes large principal neurons from Golgi type II neurons and glial cells, but not the latter from one another. During an exploration of this approach with embryonic cerebral cells (Varon and Raiborn, 1969), drawbacks have been noted in terms of both the plasticity of the larger cells (capable of squeezing through three-fold narrower pores) and the ease with which cells may coat or clog the filters (even those providing uniform, cylindrical pores rather than tortuous, irregular channels).

2. *Differential sedimentation* exploits differences in both size and density of the cells (cell buoyancy) sedimenting through solutions of different densities. Bulk fractionation procedures use mainly this approach (Poduslo and Norton, 1975). Live cells, however, react to the changing osmotic properties of the environment through which they migrate by altering their own fluid content, hence their own buoyancy. In addition, they may be adversely affected by either the tonicity or the toxicity of the solutions used. Sucrose gradients usually involve considerable, possibly excessive osmotic consequences. Ficoll gradients have been reported both as tolerated (e.g., Barkley *et al.*, 1973) and as harmful (e.g., Varon and Raiborn, 1969) environments. Gradients generated with bovine serum albumin have revealed no adverse consequences (Varon and Raiborn, 1969; Wilkin *et al.*, 1976; Campbell *et al.*, 1977). Despite the several attempts to exploit differential sedimentation, examples of documented success are still rare. One way for the future may be to increase or even impose artificial differences in sedimentation properties, for example by loading selected cells with erythrocyte rosettes via an immunochemical ligand (Schachner and Hammerling, 1974).

TABLE I

CELL PROPERTIES AND SELECTIVE PROCESSES INVOLVED IN MOST CELL SEPARATION APPROACHES

	Size	Density	Adhesion	Survival Requirements	Drug Sensitivity	Special Surface Constituents
Filtration	×					
Sedimentation	×	×				×
Electronic sorting						×
Affinity binding						×
Attachment			×			
Migration			×			
Destruction				×	×	×

3. *Fluorescence-activated cell sorting* is an ingenious approach developed for the separation of different classes of lymphoid cells (e.g., Kreth and Herzenberg, 1974). Treatment with cell-specific antibody provides the corresponding cells with a fluorescent tag. The cell suspension is allowed to flow through the cell-sorting apparatus, where optic monitors recognize the tagged cells and divert their flow to different collectors. Very recently, the electronic cell sorting approach has been applied to dissociates from 10-day old mouse cerebellum, tagged with antiserum to corpus callosum, and has yielded a probably viable, glial-enriched subpopulation (Campbell *et al.*, 1977).

4. *Affinity binding* also takes advantage of special cell surface constituents, but exploits them to attach the cell to corresponding ligands properly anchored to an insoluble substratum. Much attention has been given in recent years to specific binding properties of cell surfaces, such as hormonal receptors (Cuatrecasas, 1974; Frazier *et al.*, 1974), neurotransmitter receptors (Alexander *et al.*, 1975; Maguire *et al.*, 1976; Greene, 1976), and lectin-binding sites (Nicolson, 1974; Barondes, 1976). Moreover, considerable efforts are expended in several laboratories to obtain antisera specifically directed to cell surface antigens characteristic of certain but not all neural cells (e.g., Schachner *et al.*, 1975; Stallcup and Cohn, 1976; Dorfman *et al.*, 1976; Akeson and Seeger 1977; see also Bock, this volume). Hormones, neurotransmitters (or their analogs), lectins, and immune globulins could be used as cell-selective ligands to sort out the cells possessing the corresponding binding sites on their surface. The approach has been recently illustrated with C6 glioma cells, using glass or sepharose beads derivatized with iodothyronine, corticotropin or isoproterenol (Venter *et al.*, 1976). Additional aspects will be discussed further on (see part III).

5. *Differential attachment* to substrata uses cell surface properties which are much less defined than receptors or even surface antigens, namely those responsible for differences in cell surface adhesiveness. The approach has been successfully applied to dissociates from cerebrum (Varon and Raiborn, 1969), sympathetic ganglia (Varon and Raiborn, 1972b; McCarthy and Partlow, 1976a) and dorsal root ganglia (Okun *et al.*, 1972; Varon *et al.*, 1973). It also has the great merit of involving a minimum of external intervention beyond the dissociation stage and the timely removal of the unattached cells, and of retaining the potential to yield all cell subpopulations over sequential attachment steps. Effectiveness and/or selectivity of the individual steps might be increased by appropriate manipulations of the attachment surface, the seeding medium, the composition and the post-dissociation qualities of the mixed cell population, and the time course of the attachment (see part III for more detailed discussion). Moreover, the attached cells can be directly characterized at least by morphological

features and culture behavior. Finally, differential adhesiveness applies to cell-cell as well as cell-substratum associations and, in special cases, selective reaggregation within the original cell suspension might yield purified subpopulations (cf. McCarthy and Partlow, 1976a).

6. *Differential migration* also exploits differences in cell-substratum adhesiveness, in addition to differences in cellular motility itself (Letourneau and Wessells, 1974). In explant cultures using locomotion-permissive substrata, fibroblastic elements usually migrate earlier and further out than glial cells, glial cells move out alone or in association with outgrowing neurites, and neuronal somata remain within, or at the margins of, the explant itself. It should be possible to enhance, or better control, the differential out-migration of certain cell classes, remove the explant, and use it as the source tissue for another round of cell migration (under circumstances favoring a different set of cells) or for dissociation and recovery of the unmigrated population(s). Similarly, in dissociated cell cultures, cell migration leads to selective re-association patterns (Varon, 1970) — particularly between neurons and glia — and the opportunity to collect off special groups of cultured elements (Fine and Bray, 1971; Peterson *et al.,* 1973; J. Martner and S. Varon, unpublished. Both approaches will be discussed further in Part III.

7. *Differential destruction* relies on different requirements for survival, or different susceptibilities to toxic treatments, by different cell classes. The notable disadvantage is, of course, the loss rather than the segregation of the "vulnerable" subpopulations. Nevertheless, the approach has been effectively used in a number of systems. Witholding nerve growth factor leads to an enrichment of the small highly fluorescent neurons in sympathetic ganglionic cultures (Jacobowitz and Greene, 1974), or to neuron-free nonneuronal subpopulations from dorsal root or sympathetic ganglia (Varon *et al.*, 1973; Mains and Patterson, 1973). Conversely, antimitotic agents (aminopterin, cytosine arabinoside, 5-fluorodeoxyuridine) not only suppress proliferation but may favor destruction of replication-competent elements, such as fibroblasts or glial cells (Fischbach, 1972; Mains and Patterson, 1973). Drugs selective for certain cell types (e.g., 6-hydroxydopamine for sympathetic neurons) will eventually be employed in a similar manner. Equally effective will be the use of surface-specific antibodies to elicit complement-dependent cell lysis. Culture substrata (Varon and Tayrien, 1977, and unpublished data) and media ingredients (e.g., Lasher and Zagon, 1972; Holley and Kiernan, 1975) can also be manipulated for either cytostatic or cytotoxic cell-selective effects. Finally, a judicious choice of source tissue ages and selective culture conditions can lead to the obtention of purified populations of astrocytes (see Sensenbrenner, this volume).

C. Criteria for Cell Classification and Identification

The emphasis of this chapter being on viable cells, it is appropriate to discuss criteria for *viability* before addressing the question of cell type characterization. Attachment and persistance in a standardized surface culture system are necessary features of viability, even though they also depend on the choice of appropriate culture conditions. They are not, however, sufficient and unequivocal criteria until morphological or biochemical evidence is added that the attached and persisting cells do function as living elements. Examination of surface cultures provides the opportunity to evaluate all three such aspects, again depending on adequate culture conditions. One must note, for example, that knowledge about special "trophic" requirements of neural cells is still very limited (Varon, 1977; Varon and Bunge, 1978). Consequently, purified cell preparations should be tested for viability in coculture with their traditional partners, and not only in isolation.

The distinction between cell *classification* and cell *identification* is an important one (cf. Varon and Raiborn, 1969; Varon, 1970). In a mixed cell culture, it is always feasible to distinguish several cell categories, based on morphological and optical features displayed under phase contrast as well as on cytochemical features revealed by fixation and staining. However, it is not always possible to determine whether such "classification" reflects distinctive cell types or variable behaviors by the same kind of cell. Even more importantly, the *expression* of most, if not all, cell traits is likely to be subject to regulation by a variety of extrinsic influences, few of which are currently known. Presumably, display of a differentiated trait requires the concurrence of three components: an already acquired intrinsic program, environmental cues eliciting or modulating its execution, and permissive conditions so that ancillary cell activities do not limit such execution. Thus, absence of a trait manifestation may be due to "inappropriate" circumstances rather than lack of the trait itself. These considerations are likely to apply *in vivo* as well as *in vitro*. Indeed, a major concept emerging from the *in vitro* studies has been that, in the normal tissue, neurons and glial cells may be expressing only a portion of the "neuronal" or "glial" programs acquired during development.

Usually, the ultimate goal of cell characterization *in vitro* is viewed as the *attribution* of an identity defined by traditional criteria for neural cells *in situ*. Two major difficulties derive from the *in situ*, rather than the *in vitro*, situation. One is that individual cell-specific properties remain to be adequately identified even *in vivo* — a major reason for the need of purified neural cell preparations, in the first place. The other problem is that traditional criteria for neural cell classification often depend heavily on regional and contextual references, hence rely on "circumstantial" features more than on intrinsic ones. Consequently, neural cell "markers" must be qualified by the conditions under which they have been observed and, ultimately, by the conditions required for their manifestation. Neural cell separation and culture approaches are again needed to i) define under permissive conditions the entire programmed repertoire of different classes of neurons and glia, and ii) identify the various sets of extrinsic influences that specifically call forth "normotypic" and "aberrant" neuronal or glial behaviors. Until this is achieved, identification of an isolated neural cell population must avoid the pitfall of negative attributions, as well as that of positive attributions based on traits not yet properly established as cell type-specific.

Chemical and biochemical markers, as well as immunological markers, are discussed elsewhere in this volume. Here, one only needs to stress the importance of developmental age differences, both *in vivo* and *in vitro*, in view of the fact that most of the *in vitro* primary neural systems have been obtained, thus far, from prenatal or perinatal tissues. A few additional comments may be appropriate with regard to *morphological features*. In general, recognition problems are not likely to arise between most neurons and nonneurons (Varon and Saier, 1975), but rather among subclasses of neurons or of glial cells, between small neurons (and neuroblasts) and oligodendrocytes or Schwann cells, or between glial and mesenchymal cells (connective sheaths, vascular elements). Cell morphology is subject to extensive modulation by culture conditions. Humoral modulators may involve specific, though as yet poorly understood agents (Sensenbrenner *et al.*, 1972; Lim *et al.*, 1976), a distinctive example being nerve growth factor itself (Varon, 1975a). The importance of the culture substratum for cell shape and motility has already been stressed (Letourneau, 1975), as have its potential implications to biological regulation (Varon and Tayrien, 1977, and unpublished observations). Extent and patterns of proliferation are critical in determining social organization (Varon, 1970; Varon and Raiborn, 1972b), and the "conditioning" properties of the medium with respect to other cells in the culture (e.g., Patterson *et al.*, 1976; Giller *et al.*, 1977).

III. Experimental Examples

A. Sympathetic Ganglionic Cells

Varon and Raiborn (1972a) reported that 11-day chick embryo sympathetic ganglia can yield a dissociate that i) contains about equal numbers of neurons and nonneurons, and ii) can be fractionated by differential attachment into viable neuronal and nonneuronal subpopulations. More recently, McCarthy and Partlow (1976a) described an attractive improvement to the separation procedure. The improvement involves the

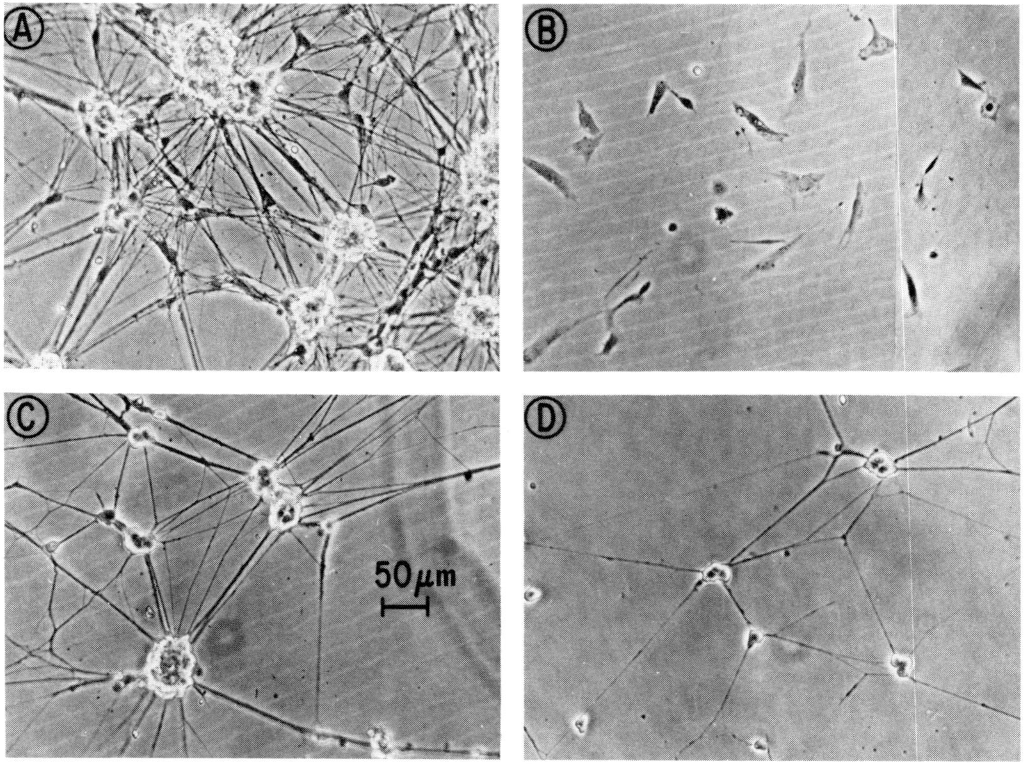

FIG. 1. Sympathetic ganglia dissociates, 24 hours in culture on a collagen-coated surface. A. Cells seeded as a dispersed suspension and attached without mechanical interventions. (Mixed cell culture). B. Vibration pulses were applied for 4 hours after seeding, the unattached materials removed, and the remaining cells rinsed several times. (Nonneuronal cell subpopulation). C. The unattached material, collected after the 4 hour vibration, was depleted of single cells and small clumps by filtration, and transferred to a fresh collagen-coated dish. (Neuronal aggregates). D. The same material (as in C) was redissociated before reseeding, yielding a pure neuronal subpopulation.

administration, during the first few hours after seeding, of periodic bursts of vertical vibrations. The vibrations interrupt any incipient attachment of the neurons and, thus, sharpen the resolution of the system. Vibration also causes the unattached cells to aggregate.

Fig. 1 illustrates the kind of preparations currently obtained in our laboratory by use of the vibration-inclusive procedure (Bottenstein and Varon, unpublished data). The sympathetic dissociate, left to attach undisturbed to a collagen-coated dish, yields a mixed cell culture (Fig. 1 A), in which neuronal aggregation occurs only after attachment and over a 1 to 5 day span (cf. Varon and Raiborn, 1972b). However, if seeding is followed by 4 hours of periodic vibrations, the neurons are made to aggregate in suspension within the seeding medium. At 4 hours, the unattached material can be removed, leaving behind a ganglionic subpopulation which is essentially nonneuronal (Fig. 1 B). Repeated washes, and further incubation without nerve growth factor, can reduce neuronal contamination to less than 1% of the attached cells. Conversely, the unattached material, collected after 4 hours of vibration, comprises almost exclusively neuronal elements. It can be fractionated over a nylon mesh into a filtrate, containing the few unattached single cells and small neuronal clusters, and a retentate, made up exclusively by larger neuronal aggregates. The latter can be cultured directly (Fig. 1 C), or they can be redissociated to yield a dispersed, viable subpopulation of pure neurons (Fig. 1 D).

Effective segregation of neurons by the vibration method requires selective reaggregation properties of the unattached cells (neuron-to-neuron, but not neuron-to-nonneuron), in addition to differential attachment to the substratum. Self-aggregation may not occur with equal thoroughness under any and all circumstances, nor with any type of neural dissociate. For example, chick embryo sympathetic neurons tend to self-aggregate in surface cultures as well, but the extent to which they do so is apparently regulated by the density of the underlying nonneuronal mono-layer (Varon and Raiborn, 1972b). The number of nonneurons also appears to control reaggregation in surface cultures of chick embryo dorsal root ganglionic cells, but here the reaggregates include nonneurons as well as neurons (Varon 1970, 1971). Thus, attempts to extend the usefulness of this approach to dissociates from other tissues may encounter, and have to deal with, the possibility that unacceptable numbers of nonneurons be involved, or trapped, in vibration-promoted reaggregates.

Fig. 2 represents the kinetics of neuronal (open circles) and non-neuronal (full circles) cell attachments from different sympathetic ganglia dissociates, and illustrates another instructive aspect of these studies. The first graph (Fig. 2 A), redrawn from the earlier study by Varon and Raiborn (1972b), shows a very small overlap between the attachment patterns of the two cell classes — thereby offering an excellent opportunity for their

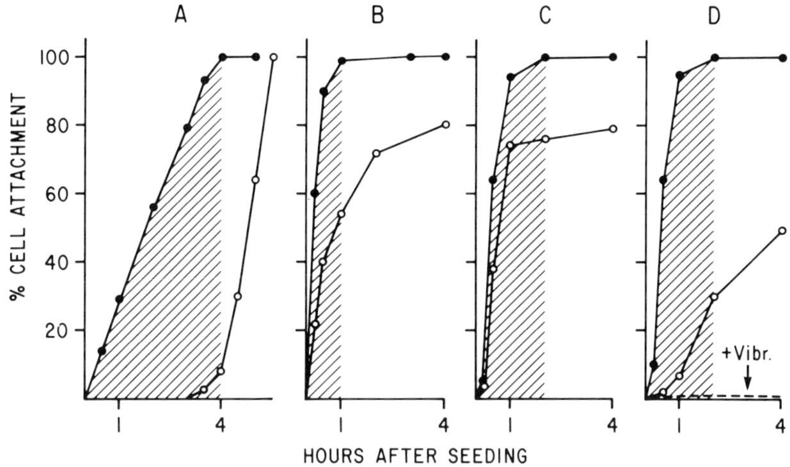

FIG. 2. Attachment curves of sympathetic ganglia cells obtained by different procedures. Ordinate: neurons (o) and nonneurons (●) as % of final attachment levels. A. Ganglia collected in Ca^{++}, Mg^{++}-free saline and dissociated by trypsin, cells seeded on ammonia-treated collagen (Varon and Raiborn, 1972b). B. Collection in presence of divalent cation, dissociation by collagenase. C. Collection in presence of divalent cations, dissociation by trypsin. D. Collection in Ca^{++}, Mg^{++}-free saline, dissociation by trypsin. B, C and D: cells seeded on hypertonic saline-treated collagen (adapted from McCarthy and Partlow, 1976a).

separation even without vibration. The other three diagrams are reconstructed from McCarthy and Partlow's report (1976a), and describe the behavior of sympathetic dissociates obtained in three different manners and seeded on the same collagen surface, albeit one prepared differently from that used in the earlier study. Specifically, one preparation (Fig. 2 B) resulted from ganglia collected in the presence of divalent cations and dissociated by collagenase; a second preparation (Fig. 2 C) derived from similarly collected ganglia but by the use of trypsin; and the third one (Fig. 2 D) involved both collection in a divalent cation-free solution and use of trypsin — much as had been done by the earlier investigators (see Fig. 2 A). Clearly different attachment kinetics are displayed in the four cases by both nonneurons and neurons, presumably in response to changed conditions for tissue collection (cf. C and D), enzyme treatment (cf. B and C), or collagen coating (cf. A and B, C or D). More explicit correlations between attachment kinetics and procedural parameters are currently being sought in this laboratory, as is the definition of reproducible cell yields at the several stages of the separation procedure.

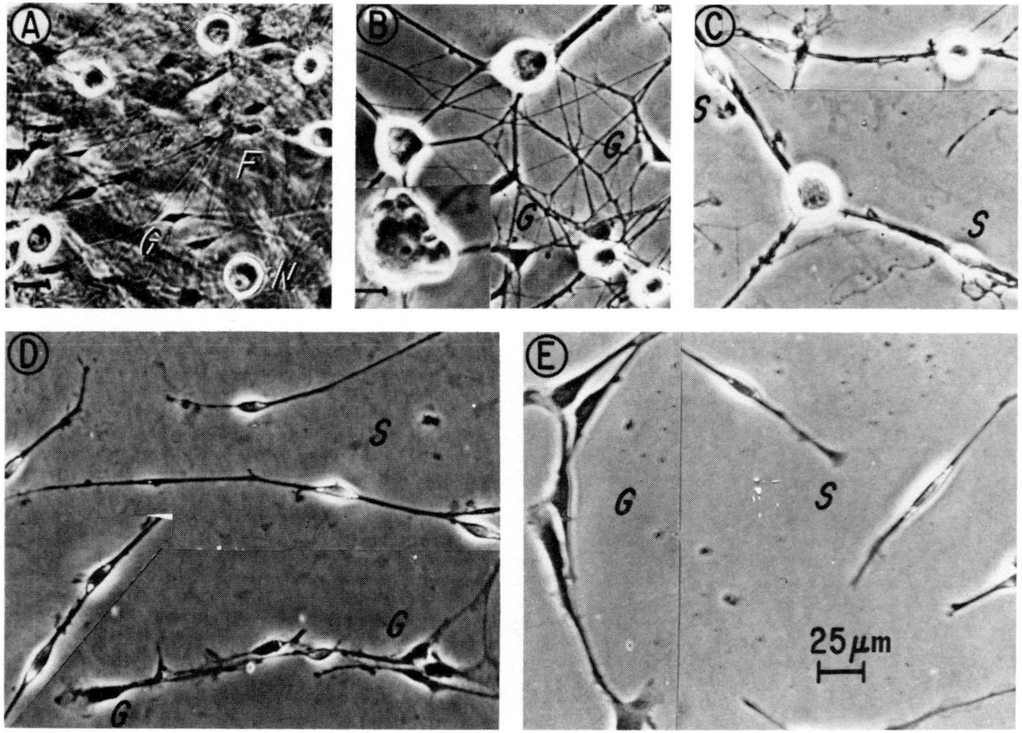

FIG. 3. Culture morphology of mouse peripheral glia. A. Week-old primary DRG culture showing neurons (N), glial cells (G), and background flat cells (F). B, C. One day old DRG cultures, displaying typical Schwann cells (S) in addition to G cells. D. Morphologies of both the G and S Cells are exhibited by cells migrated out of a sciatic nerve explant in three day old cultures. E. Satellite cells released by neuronal fractions deprived of nerve growth factor. Cells have morphology of G and S cells.

B. Dorsal Root Ganglionic (DRG) Cells

Neurons prefer the surface of nonneuronal cells (or, possibly, microexudates laid down by them) to the conventional culture substrata such as tissue culture plastic or collagen (cf. Varon, 1975b; Varon and Saier, 1975). If viable nonneurons are preponderant in the cell dissociate to be seeded, their early attachment may offset any differential attachment condition provided by the system. Dissociates from embryonic chick sympathetic ganglia owe their successful separation in large part to their low nonneuron-to-neuron ratio (near unity) — and some of the differences in attachment behaviors discussed in Fig. 3 may reflect altered ratios in the cell dissociates. Chick embryo DRG dissociates, on the other hand, usually are obtained with ratios substantially greater than unity (1.5 to 3.0), and cell separation by differential attachment has not been achieved on a practical level (Varon and Raiborn, 1971, 1972a; Varon *et al.*, 1974a; however, see Okun *et al.*, 1971). Similar difficulties were encountered with DRG dissociates from neonatal rat, obtained with a nonneuron-to-neuron ratio of 7 (Varon, *et al.*, 1974a). In fortunate contrast, DRG from neonatal mice have yielded dissociates with nearly equal numbers of nonneurons and neurons, and cell separation by differential attachment was effectively achieved (Varon *et al.*, 1973). Even with the mouse DRG, however, higher nonneuron-to-neuron ratios, and correspondingly less satisfactory cell separations, were readily imposed by modest (and all-too-often unnoticed) alterations in the dissociation process. The procedure was extensively used in the investigation of the nerve growth factor-like properties of ganglionic nonneurons (e.g., Burnham *et al.*, 1972; Varon *et al.*, 1974b).

The mouse DRG material provides an opportunity to illustrate morphological modulation and identification problems of cultured peripheral glia (S. Varon and C. Raiborn, unpublished data). Fig. 3A shows a week-old primary culture, typically displaying: round, birifringent neurons (N); dark, compact glial cells (G) strung along neurite bundles; and background flat cells (F), forming an underlying carpet more epithelioid than fibroblastic in character (Varon *et al.*, 1973). The glial elements, and their special relationship to the neurites, are particularly evident in the absence of an underlying carpet, as occurs after only 24 hours in culture (Fig. 3 B, C). Clearly, their appearance can vary considerably, from opaque or glowing triangular elements (Fig. 3B) to highly birifringent, classical Schwann-like (S) cells (Fig. 3C). Neuronal subpopulations from mouse DRG retain satellite cells tightly associated with neuronal somata (e.g., left bottom corner of Fig. 3B). If such subpopulations are seeded without nerve growth factor, most neurons die and the satellite cells attach and stretch out on their own, revealing the morphology of both G and S cells (Fig. 3E). Cells that have migrated out of peripheral nerve explants (Fig. 3D), also

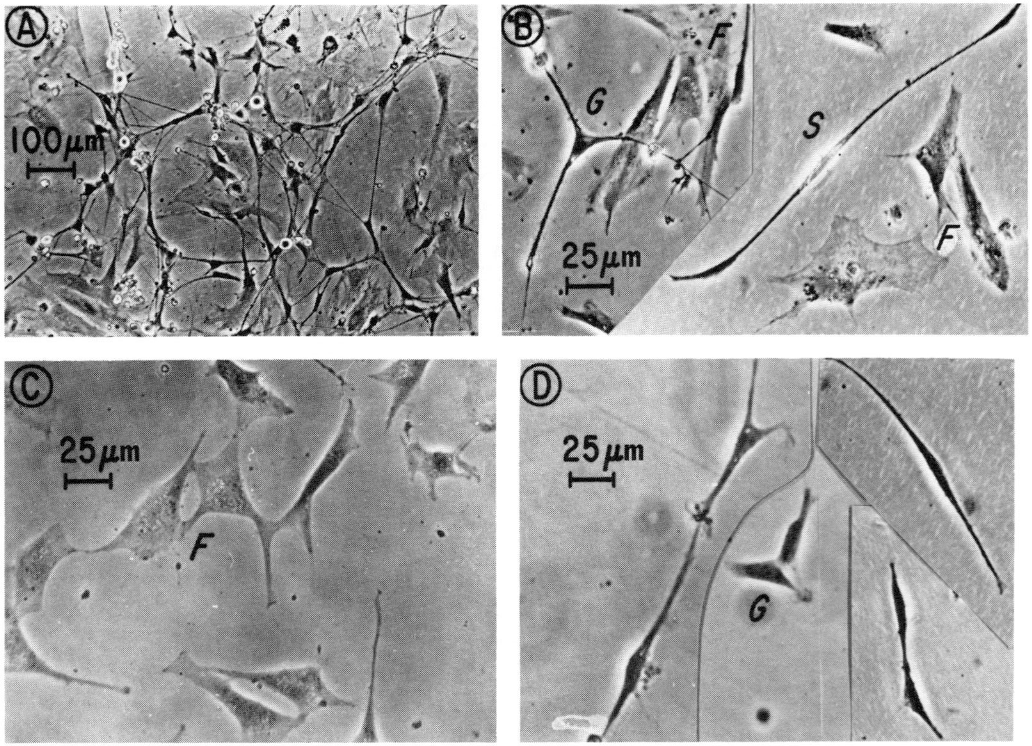

FIG. 4. Separation of neuronal-glial networks from cultures of 8 day old chick embryo DRG. Primary dissociate was seeded on culture plastic, and cultured for 3 days with serum- and NGF-supplemented medium in a bicarbonate-CO_2 system. A, B. Primary culture has well developed neuronal networks, and both G and S-like cells, mainly associated with the networks. C. Cells, left in primary culture after manual removal of networks, then removed by trypsin and reseeded, show the typical morphology of F cells. D. The networks, redissociated by trypsin, yield a nearly pure population of G and S (glia) cells.

display the morphology of G and S cells. Thus, cultured satellite and Schwann cells cannot be distinguished morphologically. In addition, in all three preparations, G cells are also found considerably more flattened out than usual, particularly where there is no neuritic association (e.g., Fig. 3 D, E) and may become poorly distinguishable from undersized F-type cells.

The selective tendency of neurons and glia to associate with one another and form extended, heterocellular networks (cf. Varon, 1970) can be exploited for cell separation purposes. Sympathetic networks have been lifted off from primary cultures after neuronal aggregation had occurred (Fine and Bray, 1971), or from recombined cultures where heavy nonneuronal supplements have prevented neuronal reaggregation (Varon and Raiborn, 1972b). Networks have been similarly dislodged by gentle irrigation from primary dissociated cultures of embryonic chick cerebral cells (Peterson *et al.*, 1973). A similar procedure, currently under investigation with respect to chick embryo DRG cells (J. Martner, and S. Varon, unpublished), is illustrated in Fig. 4. Primary cultures are allowed to establish well-developed networks (Fig. 4A), where the nonneuronal elements are undistinguishable from typical G and S cells. (Fig. 4B). The network is lifted off manually, and redissociated by trypsin into a secondary suspension of "network" cells. The residual cells, trypsinized off the primary culture after network removal and reseeded as a secondary population, yield a typical F population (Fig. 4C), with few recognizable G or S cells and no neurons. Conversely, network cells, on reseeding, reveal essentially the morphology of G or S cells, with less than 2% F-like cells (Fig. 4D). No neurons have been recovered after network redissociation, as yet. The distinction between the two secondary nonneuronal subpopulations is also striking in terms of DNA labeling and cell proliferation — high for the F cells and practically undetectable for the network cells. However, comparable levels of labeled DNA are found in both populations when collected from pre-pulsed primary cultures. This behavior is in keeping with the recently demonstrated dependance of ganglionic glial proliferation on the presence of their neuronal partners (Wood and Bunge, 1975; McCarthy and Partlow, 1976b), and studies are under way of the trophic and mitogenic requirements of the network cell subpopulation.

An entirely different procedure for ganglionic cell separation has been developed by Wood and Bunge (Wood and Bunge, 1975; Wood, 1976), using DRG from perinatal rat. The procedure combines the use of several approaches already discussed (see Part II), chiefly among them cell migration, glial cell proliferation, and capsule cell (fibroblast) destruction. The culture system and its separated products are depicted by the microphotographs of Figure 5, for whose availability we are most grateful to Drs. P. M. Wood and R. P. Bunge. The excised ganglion is decapsulated (to minimize the ratio of capsule to glial cells) and explanted on collagen in a

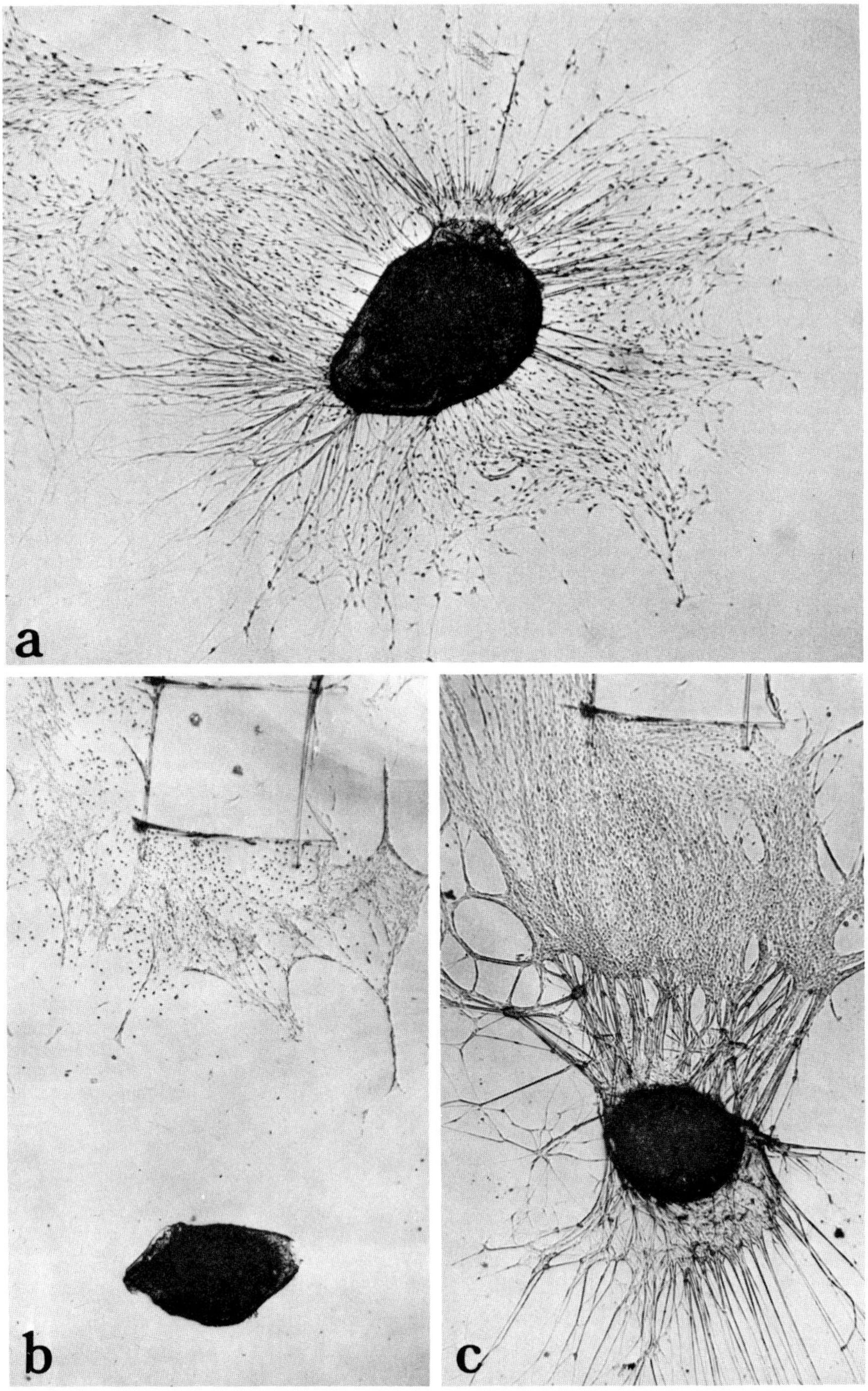
a
b
c

"rich" medium containing nerve growth factor, where neurite outgrowth and emigration of both fibroblasts and glial cells proceed rapidly. The culture is subjected to daily alternations of the "rich" medium and an "inhibitory" one, containing antimitotic drugs (as well as nerve growth factor). This treatment results in suppression and progressive loss of the fibroblasts, while Schwann cells still manage to recover and proliferate under the stimulus of the growing neurites. After 2-3 such cycles, the culture displays an outgrowth of neurites accompanied only by Schwann cells (Figure 5A). The residual explant, containing the neuronal somata, can then be excised off the culture and transferred to fresh collagen, in a new or in the same dish (Figure 5B, bottom). In the outgrowth left behind, neurites will degenerate, resulting in a pure population of Schwann cells (Figure 5B, top). The transplanted ganglion demonstrates its pure neuronal composition by the ability to develop a new set of neurites, unaccompanied by any cell emigration (Figure 5C, bottom). In this plate, part of the neuritic outgrowth was allowed to reach into a pure Schwann cell bed. Interaction of neurites and Schwann cells results in a vast increase in proliferation of the glial cells. If this culture is carried out for several more weeks, Schwann cells will continue to proliferate, ensheath all the bare neurites in the entire culture, and subsequently initiate myelination (R. P. Bunge, personal communication). As yet, no attempt has been reported on the dissociation of the "pure neuronal ganglia", to obtain a pure population of dispersed rat DRG neurons.

C. Central Neural Tissues

1. Cerebral cortex. Several investigators have described surface cultures of cells dissociated from rat, mouse or chick *cerebrum* (cf. Varon, 1975b; Sensenbrenner, this volume; Hertz; this volume). Two reports deal with the actual fractionation of viable cerebral cells, both using chick embryo cerebral hemispheres and mechanical, rather than enzymatic, dissociation procedures. One of these studies (Varon and Raiborn, 1969) was, in fact, the first description of CNS cell separation by selective attachment, although it also described for comparison the results of two

FIG. 5. Separation of Schwann cells and neurons from explant cultures of perinatal rat DRG. A. A decapsulated ganglion, explanted on collagen and subjected to alternations of "rich" medium (NGF + serum) and proliferation-inhibiting medium (NGF + antimitotic drugs), exhibits typical outgrowth of neurites and Schwann cells but few, if any, fibroblasts. B. Excision of the "residual ganglion" (bottom) leaves behind an essentially pure population of Schwann cells (top). C. The "residual ganglion" shows neuritic outgrowth but no further nonneuronal out-migrations, indicating its pure neuronal composition (bottom). If the pure neurites are allowed to grow into a pure Schwann cell population, vast proliferation of the Schwann cells is triggered (top). Courtesy of P. M. Wood and R. P. Bunge.

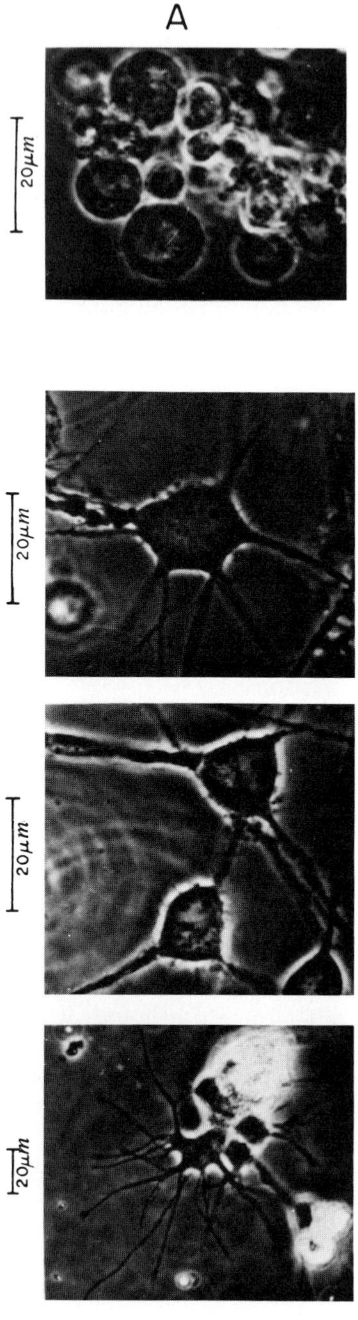

FIG. 6. Main cell categories in a dissociated culture from 11 day chick embryo cerebrum. A. Large, oligo- or multipolar neurons. B. Small, uni- and tripolar neurons. C. Small, extensively spread-out nonneuronal cells. Culture age, 24 hours. (Varon and Raiborn, 1969).

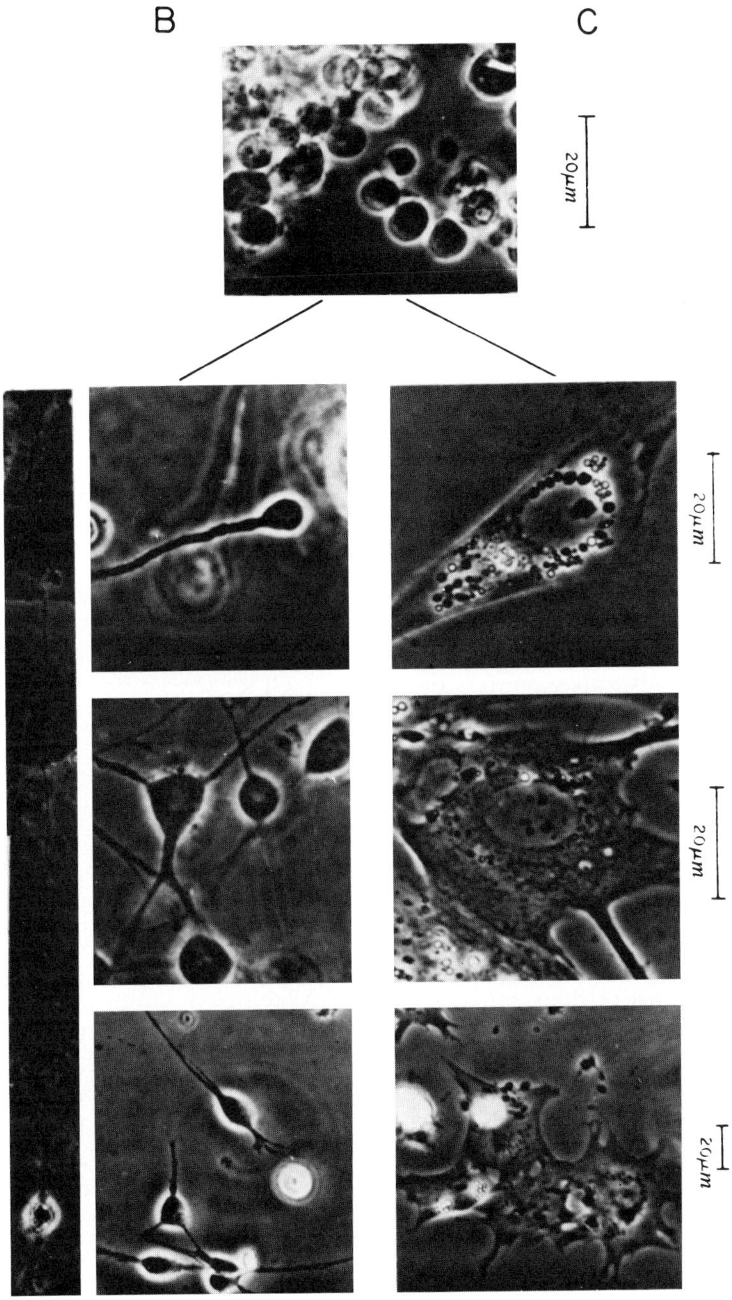
B
C
20μm
20μm
20μm
20μm

other approaches (filtration; sedimentation on albumin gradients). Three cell classes were distinguished (Fig. 6): large neurons (A), small, uni- or bipolar neurons (B), and nonneuronal cells (C) — this last class characterized by high mobility, active proliferation and a flat, epithelioid morphology. Both the A and the C classes were obtained with 90% or better purity, while B cells displayed a high and specific adhesiveness to C cells and could only be segregated in combination with them. No further work was done with these preparations, and a resumption of their investigation by use of present day biochemical tools might well be worthwhile. The second report of cerebral cell separation (Peterson *et al.*, 1973), as has already been mentioned, was based on the mechanical collection of culture reaggregates (mainly neuronal) linked into a network and overlying the confluent carpet of nonneuronal elements. Network material and nonneuronal cells could, then, be subjected to separate biochemical analyses, for example with respect to choline acetyltransferase and acetylcholinesterase activities. Purified cultures of astroblasts and astrocytes have also been obtained from late embryonic cerebral tissue by dissociation and culture conditions that allowed no survival of neurons or oligodendrocytes (see Sensenbrenner, this volume).

2. Cerebellar cortex is an attractive source for neural cell separation studies, given its relatively few, well characterized cell classes. Dissociated cerebellar cultures have also been studied by several investigators (cf. Varon, 1975b; Wolf, this volume), and so have cell separation approaches. Barkley *et al.* (1973) dissociated newborn mouse cerebellum by use of trypsin and DNAase (to reduce reaggregation in suspension), and sedimented the cells on a Ficoll gradient at unit gravity (to exploit largely size differences). The procedure removes cellular debris, intact nuclei and dead (trypan blue-positive) cells. The region of trypan blue-excluding cells was collected fractionally and analysed by autoradiography for nuclei prelabeled by *in vivo* pulses at selected dates. The fractions obtained were still heterogenous, partly because size differences were not yet fully established, partly because the size of external granule cells apparently varies in the course of their generation cycle. Viability in culture was not verified beyond the ability of each fraction to form randomly organized aggregates. Wilkin *et al.* (1976) point out that most procedures applied to dissociation and fractionation of cerebellar cells may have not preserved structural integrity, in particular continuity of plasma membranes. They described a method for the isolation of well preserved "cell bodies" from up to 2 week old rat cerebella. The metabolic integrity of the "cell bodies" was also retained as measured by glucose metabolism, potassium accumulation, protein synthesis and patterns. No report was provided at that time on the viability of such "cells" in culture, or on their separation into distinct classes.

Of considerable interest with regard to both methodology and cell identification is the already mentioned, recent study by Campbell *et al.* (1977) with 10 day old mouse cerebella. The tissue was dissociated by a two-step (0.05% then 0.005%) trypsin-DNAase treatment followed by trituration, with an overall yield of 10^7 cells/cerebellum. Sedimentation on albumin gradients produced two fractions of debris-free, trypan blue-excluding cells (overall yield=30-50% of load). Analysis of glial "markers" (CNPase, S100, GFA protein, NS-1 antigen) suggested that one fraction (B, equal to about one fourth of the recovered cells) was enriched for glial cells, and also was more reactive to a rabbit antiserum against bovine corpus callosum. The other fraction (C) was low in glial cells by both the chemical and the immunoreactive tests. Fraction C, treated with this antiserum and with fluorescein-conjugated goat antirabbit IgG, was used to explore the electronic cell sorter approach to cell separation (see Part II). The processing of 30-40 $\times$ 10^6 cells (in about 5 hours) yielded half of the loaded cells. Of the cells recovered, 6% had been sorted out as antigen-positive, and were subsequently confirmed to be enriched in the glial "markers", as well. Cell fractions obtained by either approach were 95% viable by trypan blue exclusion, but none of them was tested for viability in culture, as yet.

3. Spinal cord has also been dissociated and the cells cultured in several laboratories. No report is available on separation of viable cell subpopulations from such dissociates. Dissociates from chick embryo spinal cord, as well as other cell systems, are currently used in our laboratory to explore the technology for affinity-binding neural cell separation, and particularly the use of cell-specific antibodies — as recommended by the wide options they offer in cell separation approaches (see Table I). The project involves the development of several tools and procedures:

1) surface cultures of peripheral and central neural cells and the procedure to use them for quantitative determination of complement-dependent cytotoxicity of an immune serum.
2) application of the techniques to a screening of existing, or new, antisera and the definition of their target-specificities.
3) immobilization of immunoglobulin on a semisolid matrix or a solid surface by covalent coupling (derivatization), with particular concern for maximal conservation of the immunoglobulin and of its reactivity.
4) modalities for an effective presentation of a mixed cell suspension to the derivatized material, so that it captures and retains the antigen-positive cells while allowing for mechanical removal of the negative ones.
5) procedures to recover the sequestered cells from the derivatized material.
6) use of the initial surface culture systems to assay quantitatively the viability of the separated cells and to help verify their classification.

Conclusion

From the general discussion of principles and methodology, and even more so from the survey of some successful separation systems, it is safe to predict that future investigations will increasingly exploit a *multi-step* approach to neural cell separation, one in which a modest success by one technique can be amplified or multiplied by subsequent application of other techniques. For example:

i) a small number of pure glial cells, obtained by a first technique, could be scaled up to bulk preparations if conditions are established for exclusive and controlled glial proliferation.

ii) if population A is obtained pure by selective destruction of other cell classes, immune sera could be developed against it, the sera could be used to remove population A from the original heterotypic cell system and the other cell classes would also be made available in a purified condition.

iii) immune sera developed against population A in the preceding example could also be applied to fractionate other heterotypic cell populations containing A-like cells, and obtain both its A-like and its A-unlike subpopulations in viable and bulk forms.

Of the individual approaches (Table I), the two most promising are clearly those exploiting cell surface properties, i.e., differential adhesiveness and/or affinity binding via special cell surface constituents. Besides minimizing physical stresses and abnormal changes in the chemical environment, they are most likely to take advantage of fine distinctions between different cell classes and thus have the greatest resolving potential. In addition, they can be put to use via a variety of technical approaches, particularly in the case of the immunochemical ligands. Lastly, the success of either approach is undoubtedly and critically influenced by the state of the cell surface (dissociation conditions), the substratum and the medium used for attachment, the relative number of "attaching" versus "non-attaching" cells. Learning to control, and possibly select, such influences should give a great impetus to the state of the art. Given the increasing attention paid to these aspects, one may confidently look forward to increasing numbers of successful cell separation systems for both central and peripheral neural tissues.

Acknowledgments:

Research described in this paper has been supported by USPHS grants NS-07606 and NS-12893 from the National Institute of Neurological and Communicative Disorders and Stroke.

References

Akeson, R. and R. C. Seeger. 1977. Interspecies neural membrane antigens on cultured human and murine neuroblastoma cells. J. Immunology (in press).

Alexander, R. W., J. N. Davis and R. J. Lefkowitz. 1975. Direct identification and characterization of β-adrenergic receptors in rat brain. Nature (Lond.) 258:437-440.

Barkley, D. S., L. L. Rakic, J. K. Chaffee and D. L. Wong. 1973. Cell separation by velocity sedimentation of postnatal mouse cerebellum. J. Cell. Physiol. 81:271-280.

Barondes, S. 1976 *Neuronal Recognition.* Plenum Press, New York.

Burnham, P., C. Raiborn and S. Varon. 1972. Replacement of Nerve Growth Factor by ganglionic nonneuronal cells for the survival in vitro of dissociated ganglionic neurons. Proc. Nat. Acad. Sci. USA 69:3556-3560.

Campbell, F. LeM., M. Schachner and S. Sharrow. 1977. Isolation of glial cell-enriched and -depleted populations from mouse cerebellum by density gradient centrifugation and electronic cell sorting. Brain Res. (in press).

Cuatrecasas, P. 1974. Membrane receptors. Ann. Rev. Biochem. 43: 169-214.

Dorfman, S. H., H. Holtzer, and D. H. Silberberg. 1976. Effect of 5-bromo-2-deoxyuridine or cytosine-D-arabinofuranoside hydrochloride on myelination in newborn rat cerebellum cultures following removal of myelination-inhibiting antiserum to whole cord or cerebroside. Brain Res. 104:283-294.

Fine, R. E. and D. Bray. 1971. Actin in growing nerve cells. Nature (Lond.) New Biol. 234:115-118.

Fischbach, G. D. 1972. Synapse formation between dissociated nerve and muscle cells in low density cell cultures. Devel. Biol. 28:407-429.

Frazier, W. A., L. F. Boyd, A. Szutowicz, M. W. Pulliam and R. A. Bradshaw. 1974. Specific binding sites for ^{125}I-Nerve Growth Factor in peripheral tissues and brain. Biochem. Biophys. Res. Comm. 57:1096-1103.

Giller, E. L. Jr., J. H. Neale, P. N. Bullock, B. K. Schrier and P. G. Nelson. 1977. Choline acetyltransferase activity of spinal cord cell cultures increased by co-culture with muscle and by muscle conditioned medium. J. Cell Biol. (in press).

Greene, L. A. 1976. Binding of α-bungarotoxin to chick sympathetic ganglia: properties of the receptor and its rate of appearance during development. Brain Res. 111:135-146.

Holley, R. W. and J. A. Kiernan. 1975. Control of the initiation of DNA synthesis in 3T3 cells: low-molecular weight nutrients. Proc. Nat. Acad. Sci. USA 71:2942-2945.

Jacobowitz, D. and L. A. Greene. 1974. Histofluorescence study of chromaffin cells in dissociated cell cultures of chick embryo sympathetic ganglia. J. Neurobiol. 5:65-83.

Kreth, H. W. and L. A. Herzenberg. 1974. Fluorescence-activated sorting of human T and B lymphocytes. I. Direct evidence that lymphocytes with a high density of membrane-bound immunoglobulin are precursors of plasmacytes. Cell Immunol. 12:396-406.

Lasher, R. S. and I. S. Zagon. 1972. The effect of potassium on neuronal differentiation in cultures of dissociated newborn rat cerebellum. Brain Res. 41:482-488.

Letourneau, P. C. 1975. Possible roles for cell-to-substratum adhesion in neuronal morphogenesis. Devel. Biol. 44:77-91.

Letourneau, P. C. and N. K. Wessells. 1974. Migratory cell locomotion versus nerve cell elongation. Differences based on the effects of lanthanum ion. J. Cell. Biol. 61:56-69.

Lim, R., D. E. Turriff and S. S. Troy. 1976. Response of glioblasts to a morphological transforming factor from adult brain: cinematographic and chemical correlations. Brain Res. 113:165-170.

Maguire, M. E., R. A. Wiklund, H. J. Anderson and A. G. Gilman. 1976. Binding of [^{125}-I] iodohydroxybenzylpindolol to putative β-adrenergic receptors of rat glioma cells and other cell clones. J. Biol. Chem. 251:1221-1231.

Mains, R. E. and P. H. Patterson. 1973. Primary cultures of dissociated sympathetic neurons. I. Establishment of long-term growth in culture and studies of differentiated properties. II. Initial studies on catecholamine metabolism. III. Changes in metabolism with age in culture. J. Cell Biol. 59:329-366.

McCarthy, K. D. and L. M. Partlow. 1976a. Preparation of pure neuronal and nonneuronal cultures from embryonic chick sympathetic ganglia. A new method based on both differential cell adhesiveness and the formation of homotypic neuronal aggregates. Brain Res. 114:391-414.

McCarthy, K. D. and L. M. Partlow. 1976b. Neuronal stimulation of ^{3}H-thymidine incorporation by primary cultures of highly purified nonneuronal cells. Brain Res. 114:415-426.

Nicolson, G. L. 1974. The interactions of lectins with animal cell surfaces. Int. Rev. Cytol. 39:89-190.

Okun, L. M., F. K. Ontkean, and C. A. Thomas. 1972. Removal of nonneuronal cells from suspensions of dissociated embryonic dorsal root ganglia. Exp. Cell. Res. 73:226-229.

Patterson, P. H., L. F. Reichardt and L. L. Y. Chun. 1976. Biochemical studies on the development of primary sympathetic neurons in cell culture. Cold Spring Harbor Symp. Quant. Biol. 40:389-397.

Peterson, G. R., G. W. Webster and L. Shuster. 1973. Characteristics of choline acetyltransferase and cholinesterases in two types of cultured cells from embryonic chick brain. Devel. Biol. 34:119-134.

Poduslo, S. E. and W. T. Norton. 1975. Isolation of specific brain cells. In: *Methods of Enzymology*. Vol. 35, part B. S. P. Colowick and N. O. Kaplan, eds. Academic Press, New York. pp. 561-578.

Schachner, M. and U. Hammerling. 1974. The postnatal development of antigens on mouse brain cell surfaces. Brain Res. 73:362-371.

Schachner, M., K. A. Wortham, L. D. Carter and J. K. Chaffee. 1975. NS-4 (nervous system antigen-4), a cell surface antigen of developing and adult mouse brain and mature sperm. Devel. Biol. 44:313-325.

Sensenbrenner, M., N. Springer, J. Booher and P. Mandel. 1972. Histochemical studies during the differentiation of dissociated nerve cells cultivated in the presence of brain extracts. Neurobiology 2:49-60.

Stallcup, W. B. and M. Cohn. 1976. Correlation of surface antigens and cell type in cloned cell lines from the rat central nervous system. Exp. Cel. Res. 98:285-297.

Varon, S. 1970. *In vitro* study of developing neural tissue and cells: past and prospective contribution. In: *The Neurosciences Second Study Program*. F. O. Schmitt, ed.-in-chief. Rockefeller Univerity Press, New York, pp. 83-99.

Varon, S. 1971. The investigations of neural development by experimental *in vitro* techniques. In: *Cellular Aspects of Neural Growth and Differentiation*. D. Pease, ed. UCLA Forum Med. Sci., No. 14, University of California Press, Los Angeles. pp. 223-242.

Varon, S. 1975a. Nerve Growth Factor and its mode of action. Exp. Neurol. 48:75-92.

Varon, S. 1975b. Neurons and glia in neural cultures. Exp. Neurol. 48:93-134.

Varon, S. 1977. Neural growth and regeneration: a cellular perspective. Exp. Neurol., 54:1-6.

Varon, S. and R. P. Bunge. 1978. Trophic mechanisms in the peripheral nervous system. Ann. Rev. Neuroscience 1. (in press).

Varon, S. and C. Raiborn. 1969. Dissociation, fractionation and culture of embryonic brain cells. Brain Res. 12:180-199.

Varon, S. and C. Raiborn. 1971. Excitability and conduction in neurons of dissociated ganglionic cell cultures. Brain Res. 30:83-98.

Varon, S. and C. Raiborn. 1972a. Dissociation of chick embryo spinal ganglia and effects on cell yields by the mouse 7S Nerve Growth Factor protein. Neurobiol. 2:196-222.

Varon, S. and C. Raiborn. 1972b. Dissociation, fractionation and culture of chick embryo sympathetic ganglionic cells. J. Neurocytol. 1:211-221.

Varon, S. and M. Saier. 1975. Culture techniques and glial-neuronal interrelationships *in vitro*. Exp. Neurol. 48:135-162.

Varon, S. and M. Tayrien. 1977. Influence of substratum on ganglionic cells in surface culture. Trans. Am. Soc. Neurochem. 7:90.

Varon, S., C. Raiborn and E. Tyszka. 1973. *In vitro* studies of dissociated cells from newborn mouse dorsal root ganglia. Brain Res. 54:51-63.

Varon, S., C. Raiborn and P. A. Burnham. 1974a. Selective potency of homologous ganglionic nonneuronal cells for the support of dissociated ganglionic neurons in culture. Neurobiol. 4:231-252.

Varon, S., C. Raiborn and S. Norr. 1974b. Association of antibody to Nerve Growth Factor with ganglionic nonneurons (glia) and consequent interference with their neuron-supportive action. Exp. Cell Res. 88:247-256.

Venter, B. R., J. C. Venter and N. O. Kaplan. 1976. Affinity isolation of cultured tumor cells by means of drugs and hormones covalently bound to glass and sepharose beads. Proc. Nat. Acad. Sci. USA 73:2013-2017.

Wilkin, G. P., R. Balász, J. E. Wilson, J. Cohen and G. R. Dutton. 1976. Preparation of cell bodies from the developing cerebellum: structural and metabolic integrity of the isolated "cells". Brain Res. 115:181-200.

Wood, P. M. 1976. Separation of functional Schwann cells and neurons from normal peripheral nerve tissue. Brain Res. 115:361-376.

Wood, P. M. and R. P. Bunge. 1975. Evidence that sensory axons are mitogenic for Schwann cells. Nature (Lond.) 256:662-664.

Cell Lines and Cell Strains

PRIMARY CULTURES, CELL LINES AND CELL STRAINS: TERMINOLOGY AND CHARACTERISTICS

S. FEDOROFF

Department of Anatomy
University of Saskatchewan
Saskatoon, Saskatchewan
Canada

I. Introduction

The various cultures used nowadays in biomedical research differ according to the species and tissue of origin, their biological age, behavior and function, the culture regimen and the various special treatments they may have received during their life *in vitro.* All this and more must be communicated and, as in other fields, a specific terminology has evolved. In this chapter, I propose to discuss the historical background of the terminology related to cell cultures and also the context in which it is used today. Certain terms were short-lived and others changed in meaning as Tissue Culture developed. The reader should not be misled by the early usage of some of the terms.

II. Historical Review of Origin of Terms

It is 70 years since Ross Granville Harrison performed his famous experiment at Johns Hopkins in which the medullary plate of a frog was cultured and the outgrowth of neurites was observed (Harrison 1906-1907, 1910, 1912; Harvey, 1975). This experiment not only settled the argument about the origin of nerve fibers but also set in motion events which led to the development of the field of Tissue Culture. The term "Tissue Culture", according to Philip White (1954), was coined by Thomas Montrose Burrows, a medical student at Johns Hopkins, who was the link between Harrison and Alexis Carrel, another major figure in early tissue culture work (Burrows, 1913).

Tissue culture preparations at first consisted of explants of small fragments of tissue and the outgrowth of cells from them (Parker, 1938; Ebeling, 1942; White, 1954). At that time the terminology was based on the method of culturing rather than on cell behavior or function, e.g., Carrel flask cultures, hanging drop cultures, Maximow slide cultures, etc.

Although the early work dealt with cultures grown for a short period of time, before long attempts were made to transplant fragments from one culture vessel to another serially. The first such continuous culture was reported by Carrel, who started it on January 7, 1912 from heart tissue of a 7-day chick embryo and serially transplanted the cells from flask to flask (Carrel, 1914). According to Parker (1964) this culture survived in continuous propagation for 34 years, although questions have been raised as to whether it was an authentic continuous culture or whether at each feeding with embryonic extract fresh cells were introduced (Hayflick, 1972). This suspicion persists because although many have attempted, no one, except Carrel and his co-workers (Ebeling, 1942), has succeeded in propagating chick cells in continuous cultures for any appreciable length of time. It was Carrel who introduced the term "strain" into the field of tissue culture by calling his continuous culture of fibroblasts from heart tissue a "strain of connective tissue" (Carrel, 1914). As this culture became a "celebrity" it became known as "old strain" (Ebeling, 1942) or "Carrel's old strain". Following Carrel's success with chick heart fibroblasts, many more "strains" were developed in his own and other laboratories. On serial transplantation, the outgrowth from the explants appeared to be fairly uniform; consequently, the term "pure strain of cells" appeared more and more often in the literature (Carrel, 1923; for further information see the 5 pages of references dealing with pure cultures in the *Tissue Culture Bibliography* by Murray and Kopeck, 1953). Willmer argued that the so-called "pure strains of fibroblasts" are not pure and that in the light of observations made in parallel work with bacteria and on the basis of the growth rates and cell modulations in tissue cultures, and because of the very inadequate means of distinguishing between the different types of fibroblasts, it was unreasonable to expect to isolate pure cell populations with methods which existed at that time (Willmer, 1935).

Parker (1932a,b, 1933, 1934) compared "strains" of fibroblasts from different tissues and organs and from corresponding parts of embryos of different ages as to their rates of growth, the relative amount of free acid accumulating in the medium, survival in abnormally acid medium and the extent to which they could digest fibrin. He found distinct differences between the strains. In order to indicate that although all the 9 strains he developed were composed of fibroblasts which appeared to be similar, they were, in fact, different and the differences depended on the tissue and region from which they were isolated. Parker introduced the term "race" to distinguish the variations between strains of fibroblasts. Willmer and Parker both used the term "race" (Willmer, 1960); however, it has never been very popular among tissue culturists.

With the development of L and HeLa cells a new era began in tissue culture. In 1940 Earle at the U.S. National Cancer Institute in Bethesda explanted subcutaneous tissue from the abdominal wall of a male C_3H

mouse (Andervont substrain) and subcultured it continuously for 291 days. A number of subcultures were then treated with 20-methylcholanthrene. At various intervals after the start of treatment, cultures were separated from the treated ones and grown in ordinary medium. In this manner a series of cultures was obtained which had been exposed to 20-methylcholanthrene for varying periods of time (Earle 1943; Earle and Nettleship 1943). Earle designated the treated cultures by letters and called them "strains". One such "strain" which emerged after treatment with the carcinogen for 111 days was called "strain L" (Earle 1943). This "strain" could be propagated as a single cell suspension and on injection into C_3H mice it formed tumors (Earle and Nettleship, 1943; Earle *et al.*, 1950; Sanford *et al.*, 1956). Sanford *et al.* (1948) isolated a single cell from a culture of strain L cells. This cell proliferated to form a cell population which could be continuously subcultured and was called NCTC clone 929 (Strain L). Thus new terms were introduced, the term "clone" and the term "clonal strain", meaning origin from a single cell and implying homogeneity of a culture.

The other famous "cell strain" was developed in Gey's laboratory in Johns Hopkins in 1951 when tissue from a biopsy from a cervical carcinoma was grown in cultures (Jones *et al.*, 1971), and gave rise to a culture which could be continuously propagated. Gey called the resulting cultures "HeLa strain" (Gey *et al.*, 1952; Gey, 1956; Hsu *et al.*, 1976). It was not surprising that the many cell cultures which were developed thereafter were referred to as "strains". A few workers, however, used the term "line" and possibly it was White who used it first. In his book he refers to Carrel's "strain" of connective tissue as a "single-tissue pure-line" (White, 1954). By 1960 the terms "cell line" and "cell strain" were used interchangeably. For example, Berman *et al.* (1957) in describing the morphology of a number of cultures which they developed, referred to them, in the same paper, as "Detroit strains" and as "cell lines". Hsu and Moorhead (1957) wrote "Comparing strain S-3 with its mother line, HeLa, it is rather obvious that the 2 strains possessed similar stemline elements", and Krooth *et al.* (1968) in a review on genetics of cultured mammalian cells stated that they used the words "strain" and "line" interchangeably. As information accumulated about the origin and fate of cell "strains" (cell lines) it became clear that certain cell "strains" (cell lines) survive only for a certain number of subcultures; others undergo alteration in karyotype, i.e., their chromosome numbers increase, abnormal types of chromosomes appear and tumors form on injection of such cells into animals (e.g., Levan and Biesele, 1958; Hsu, 1961; Hayflick and Moorhead, 1961), and that cells from some animal species are more prone to cell alteration in cultures than others (Hsu, 1963; Macpherson, 1970). In order to emphasize the importance of these differences, Hayflick and Moorhead in 1961 proposed that cells in culture pass through three distinct phases. "Phase I" or the early growth phase, constitutes that period when the cells have been freed from the intact tissue and are just establishing themselves on glass (primary culture). Phase I ends with the formation of the first confluent sheet, at which time the culture is ready for its first subcultivation and entry into "Phase II" (Hayflick and Moorhead, 1961), described as a period during which rapid cell multiplication takes place necessitating frequent subculturing. Hayflick and Moorhead proposed that cultures in this phase be called *cell strains.* At the end of this phase the cells begin to degenerate, their mitotic activity decreases and the cells enter Phase III, or the terminal phase, described as having slowing mitotic activity, progressive cell degeneration, the appearance of cell debris and cells with bizarre nuclear forms and sizes (Hayflick and Moorhead, 1961). During Phase II "cell strains" may undergo alterations and thus escape cell death (Phase III) to acquire the potential for infinite life.

Hayflick and Moorhead proposed that after such alterations "cell strains" be called "cell lines", and listed 11 characteristics by which human "cell strains" differ from "cell lines". "Cell strains" were supposed to be diploid, to retain sex chromatin, histotypical differentiation, inadaptability to growth in suspension culture, non-malignant characteristics *in vivo*, finite limit to cultivation, a virus spectrum similar to primary tissues, cell morphology similar to primary tissue, increased acid production compared to "cell lines", retention of Coxsackie A9 receptor substance, and retention of the ease with which "strains" can be developed (Hayflick and Moorhead, 1961). They also proposed that such terms as "established cell lines" and "stable cell lines" be avoided and that the term "cell alteration" be preferred to "cell transformation", the latter being a term which has specific implications in bacteriology.

The term "cell alteration" is used when suddenly an island of cells with somewhat different morphology appears in a culture. The cells are more refractile than regular cells and are rapidly proliferating, and eventually repopulate the culture (see e.g., Parker 1954; Parker *et al.*, 1957; Hayflick, 1961; Todaro and Green 1963; Sanford, 1967). The cells usually have abnormal chromosome numbers as well as some morphologically abnormal chromosomes. [Later it was found that many so-called alterations resulted from contaminations with cells from other well-established cultures, (Roothfels *et al.*, 1959)]

The Hayflick-Moorhead proposals did not eliminate the existing confusion in terminology. The time at which a "cell line" originates from a "cell strain", as well as the time when a culture can no longer be considered to be diploid, had a wide range of interpretations. The Tissue Culture Association responded to the situation by setting up an *ad hoc* committee with R. S. Chang of Harvard University as chairman. The committee, after 3 years of deliberation, prepared a report on tissue culture terminology. The report was based, in principle, on the Hayflick-Moorhead proposals

(Hayflick was a member of the committee). The report elicited a number of criticisms by members of the Tissue Culture Association. In 1964 the committee was dissolved and a new committee was formed consisting of V. J. Evans of the National Cancer Institute in Bethesda and T. C. Hsu of the M. D. Anderson Hospital in Houston, with myself as chairman. The committee prepared a report on terminology and by means of thorough discussion with groups of scientists on this continent and abroad, the report underwent five major revisions. The fifth draft of proposed usage of animal Tissue Culture terms was accepted by the Tissue Culture Association at its Annual Meeting on June 3, 1966 in San Francisco. It was published in a number of journals and during the subsequent two years it was accepted internationally (Fedoroff, 1966). The following excerpt from the report outlines the proposed usage of the term *cell line* and *cell strain:* "In everyday usage 'cell line' and 'cell strain' have sometimes been considered interchangeable. However, according to Roget's Thesaurus the terms are not synonymous. The word 'line' is used most properly with the meaning of an *uninterrupted sequence.* In this context it is synonymous with such words as procession, series and succession. In relation to paternity, 'line' is synonymous with tree, pedigree, descent and family. In relation to posterity, 'line' is synonymous with straight descent. 'Strain', on the other hand, is generally used with the meaning of *relationships of kindred.* In this context 'strain' is synonymous with race, stock, clan and tribe. Therefore it is proposed that the term 'cell line' in the field of Tissue Culture should imply an uninterrupted sequence of cell growth, whereas the term 'cell strain' should imply a certain relationship of the cells, i.e., that they all have one or more common properties or markers for which these cells were specifically selected." (Fedoroff, 1967). These definitions are somewhat different to those implied by Hayflick and Moorhead and therefore require a more detailed examination of the meaning of the terms primary culture, cell line and cell strain and their major characteristics.

In the sections to follow terms will be used according to the recommendations of the Tissue Culture Association (Fedoroff, 1967).

III. Primary Cultures

The committee defined "primary culture" in this way: "This term implies a culture started from cells, tissues or organs taken directly from organisms [and grown for more than 24 hr]. It does not include cultures started from explants of tumors developed by injecting cultured cells into animals. Such cultures would be considered more properly as continuations of the injected cell line or strain. A primary culture may be regarded as such until it is subcultured for the first time. It then becomes *a cell line.*" (Fedoroff, 1966). This definition does not differ from that in the Hayflick-Moorhead proposal.

Cells in primary cultures are direct descendants of cells *in situ* and provide a continuation of the *in situ* cell lineage. Therefore, behaviour and function of the cells will be greatly influenced by their past life history *in situ* and the conditions *in vitro*. Cultures differ markedly when initiated by cells from embryos at various stages of development and animals of various ages. In each case the cell population in the primary culture will be different, leading to differences in the capacity for differentiation, in functions and in survival time in culture. During embryogenesis when the developmental processes proceed with astonishing rapidity at certain stages, only a few hours may make a considerable difference in the type of cells which can be isolated from the same tissue. For example, in the mouse embryo by day 7 the primitive streak has developed. Half a day later the neural plate has formed and blood circulation has begun to develop outside the embryo. By day 8 some post-mitotic neurons have formed, the neural tube has begun to take shape and heart rudiments appear. By 8½ days the heart is beating, blood circulation is established and on day 9 the major subdivisions of the brain are visible grossly. The large cervical spinal neurons are already in the post-mitotic stage of differentiation. All these changes have occurred in 2 days. Cultures from the brain of 10 day mouse embryos will have multi-potential stem cells (ventricular cells) which proliferate and can form both neuronal and glial cells (Sidman, 1970). Cultures from perinatal brain will also have cells which proliferate in cultures (subventricular cells); most of them, however, can no longer form neurons but can still form glial cells (Paterson *et al.*, 1973).

Whether all the cell types comprising the initial explant cell population will survive in culture is another question. Survival depends on the age of the tissue, methods of explantation, culture milieu and culture regimen. For example, when prenatal, perinatal and postnatal mouse brain are dissociated by the same method, the resulting cell suspensions will have different percentages of viable cells (as determined by dye exclusion). Suspensions from 10- or 12-day mouse embryos, in our experience, have approximately 60-70% viable cells; suspensions from perinatal brains have 25% and those from postnatal brain have only about 10-15% viable cells. The procedure for dissociating cells also has an effect on the cells and preferentially selects certain cells of the population; even minor alterations in technique may bring about a somewhat different selection.

A cell suspension with high cell viability, however does not guarantee that all the cells will survive in culture. The culture milieu or the properties of the culture vessel surface exert additional and different selective pressures; as a result certain cells may die in the first few days of culture. For example, in our laboratory suspensions prepared from mouse spleens have 95-98% viable cells. When such cells are planted into Marbrook flasks (Marbrook, 1967), only 25-30% of the cells remain viable after 24 hours

(Fedoroff *et al.*, 1975) and cell death occurs regardless of what culture medium is used. *In vivo*, spleen cells are functionally interrelated, but in culture where the proper proportion of cells within a population is altered due to the death of large numbers of cells, the population of cells will not function in the same coordinated way as did the original population *in vivo*. Indeed, R. White in this laboratory observed that within 6 days in culture, mouse spleen cells (type B lymphocytes) become cytotoxic and kill syngeneic tumor cells even though the spleen cells have never been exposed to such tumor cells (Fedoroff *et al.*, 1975). This unusual phenomenon may be the result of the disruption of the interdependence which the cells had *in vivo*. Cells which performed an inhibitory function *in vivo* may have been lost during the first 24 hours of culturing, allowing the spleen cells to become cytotoxic cells.

The culture medium — its osmotic pressure, amino acid and vitamin composition and in particular, its serum component — may have a pronounced effect on one cell type but not on another, thus exerting a strong selective pressure. For example, cells of dissociated prenatal and perinatal mouse brain form discrete colonies in cultures. Six types of colonies (A-F) have been identified (Fedoroff *et al.*, 1977). When such dissociated cells are planted in appropriate concentrations in medium containing 5% horse serum, many type C colonies but very few type A colonies will be present after 14 days. If, however, the serum concentration is increased, the proportion of type A and type C colonies is reversed.

It has become customary to refer to cultures by the organ of their origin, e.g., kidney cell cultures, brain cell cultures, spleen cell cultures, etc. Such designations are very misleading. Kidney tissue in culture will not contain all the cell types present in kidney tissue, nor will the culture necessarily mimic the function of kidney tissue *in vivo*. Most likely various cell types will be present in different proportions and some types of kidney cells will be absent totally. Individual cells which do survive may indeed have functions similar or identical to those *in vivo* equivalent to the kidney tissue *in situ*. Therefore, one must not be satisfied with knowing the organ of tissue origin, but must attempt to identify the types of cells present in the culture, as well as their relative proportions.

IV. Cell Lines

According to the recommendation of the Tissue Culture Association, the term "cell-line" should imply an uninterrupted sequence of cell growth. The cell-line arises from a primary culture at the time of the first subculture and, as the term implies, cultures arising consist of numerous lineages of cells originally present in the primary culture. In order to describe a cell line more precisely, the following adjectives are used: diploid, established heteroploid, and cloned (Fedoroff, 1966, 1968).

A. Diploid Cell Lines

The recommended usage of the term "diploid cell line" is as follows: ". . . a cell line in which, arbitrarily, at least 75% of the cells have the same karyotype as the normal cells of the species from which the cells were originally obtained. It should be noted that a diploid chromosome number is not necessarily equivalent to the diploid karyotype, because there are situations in which a cell may lose one type of chromosome and acquire another type of chromosome. Thus the karyotype of the cell has changed but the diploid number of chromosomes remains the same. Such cells should be referred to as 'pseudo-diploid'." (Fedoroff, 1966).

Diploid cell lines originate from primary cultures and correspond to the "cell strains" of Hayflick and Moorhead (1961), the homonuclear cell lines of Krooth *et al.* (1968) and the hyperplastid cell lines of Martin *et al.* (1970). Hayflick and Moorhead (1961) listed 11 characteristics by which diploid and established cell lines differ, and Martin and Sprague (1973) list 21 such criteria.

The life of diploid cell lines in cultures is finite (Hayflick and Moorhead, 1961; Hayflick, 1965). It is not related to chronological time but to the number of times the cultures can be serially passaged or, more precisely, the number of times the cell population doubles (population doubling number). Originally, it was suggested that there is a relationship between population doubling number and the life span of the animal species from which the cells were obtained (Hayflick, 1973). The maximum human life span is estimated to be 110 years and the population doubling number is 50-70; chicken life span is 30 years and the population doubling number is 15-35; mouse life span is 3.5 years and population doubling number 14-28 (Schwarzacher and Schnedl, 1965; Hayflick, 1972, 1973; Thompson and Holliday, 1973; Nichols *et al.*, 1977). However, a survey of ten additional species, although done on a small number of animals and using both embryonic and adult animals as sources of tissue, indicated no relationship between maximum life span of an animal and the doubling number of fibroblast-like cells in cultures (Stanley *et al.,* 1975). In cultures of diploid fibroblast-like cells and glial cells, at least those of human origin, cells are not synchronously dividing either after a few or after many population doublings. At all population doublings some cells have long and some have short interphase times (Merz and Ross, 1969, Pontén, 1973; Smith and Hayflick, 1974; Marcieira-Coelho, 1974; Absher *et al.,* 1974, 1975). Therefore, cultures of diploid cell lines cannot be considered to be homogeneous cell populations. Although most experience has been gained with diploid fibroblast-like cells, there is evidence that population doubling number is not the same for all cell types, even from the same species. For example, human fetal diploid fibroblast-like cells from the lung (HEL) begin to exhibit signs of senescence at 45-65 population doublings; human fetal diploid epithelial-like cells from kidneys (HEK) begin to show senescence at 8-10 population doublings (R. C. Miller, W. W. Nichols, J. Potlash and M. M. Aronson, unpublished) and human glial cells begin senescence at 15 to 30 population doublings (Pontén and Macintyre, 1968).

During their life *in vitro* diploid cell lines change as the number of population doublings increases. As cultures age, there are increases in mean cell volume (Bowman and Daniel, 1975; Mitsui and Schneider, 1976), mean nuclear size (Mitsui and Schneider, 1976), interdivision time (Macieira-Coelha, 1974; Absher *et al.,* 1974, 1975), percentage of chromosome aberrations (Sax and Passano, 1961; Todaro and Martin, 1967; Thompson and Holliday, 1973; Kodanka *et al.,* 1973; Schneider and Mitsui, 1976), RNA content (Cristofalo, 1970), the number of lysosomes and autophagic vacuoles in the cells (Lipetz and Cristofalo, 1972), and changes in DNA metabolism (Mattern and Cerotti, 1975), as well as a decline in cloning ability (Merz and Ross, 1969; Orgel, 1973). However, diploid cell lines throughout their life remain non-tumorigenic, i.e., they do not produce tumors when injected into appropriate hosts such as genetically similar hosts (inbred strains), or into Syrian hamsters' cheek pouch (Foley *et al.*, 1962) or antilymphocytic serum (ALS) treated animals (Stanbridge and Perkins, 1969).

Every culture of a diploid cell line will have a small number of chromosomally abnormal cells. The percentage can fluctuate within a certain range. The Tissue Culture Association proposed that, in normal diploid cell lines, cells with abnormal chromosomes should not exceed 25% (Fedoroff, 1966). However the Cell Culture Committee of the International Association of Microbiological Societies recommends more stringent requirements for monitoring diploid cell lines for virus production. Its recommendation is based on specified sampling procedures. Examples of the allowable chromosome abnormalities are: 3.5% chromosome breaks, 0.37% hyperdiploid cells, 2.07% polyploid cells. (For more details, see Cell Culture Committee, 1971.)

Cell lines are serially transplanted; obviously the more rapidly proliferating cells will be preferentially selected and the non-dividing cells will be overgrown by dividing cells and on subsequent transplantations, will be lost from the cultures. Furthermore, when the culture of origin (primary culture) has more than one cell type, each with a different finite life span, then on serial passages cells with shorter finite life spans will be lost from the culture before the others. For example, when cells from pancreas are explanted into primary cultures and a diploid cell line is developed from such cultures, it is likely that epithelial cells from the Langerhans' islands or the excretory ducts will be lost from the culture before the fibroblast-like cells derived from the connective tissue of the pancreas, because the latter have a longer finite life span.

It is obvious that the definition of the "diploid cell line" is an operational one and that it in no ways defines the extent to which the cells are similar or dissimilar from the cells of the tissue origin; it only defines a culture with specific characteristics.

B. Established Cell Lines

A large number of established cell lines are available and are used extensively in all aspects of biomedical research. Culture of established cell lines, however, is not without its problems and considerable skill is required for their proper maintenance. Their definition is beset with a number of difficulties.

Established cell lines originate from primary cultures or diploid cell lines, either spontaneously (Earle and Nettleship, 1943; Sanford *et al.*, 1950; Parker, 1954; Todaro and Frean, 1963; Rothfels *et al.*, 1963; Sanford, 1965, 1974), or by transformation by means of viruses (Kaplan, 1974; Kurstack and Maramorosh, 1974; Rapp, 1974; Butel and Estes, 1975; Studzinski, 1977), chemicals (Heidelberger, 1973, 1975; Casto and DiPaolo, 1973; Van Duuren *et al.*, 1974; Mishra and Di Mayorca, 1974; Diamond and Baird, 1977), or radiation (Borek and Sachs, 1966, 1967; Borek and Hall, 1973, 1974). The term "transformation" is a confusing one because of its many meanings. Transformation induced by viruses has been defined by Rapp and St. Jeorr (1974) as the acquisition of new properties by normal cells which are passed in a stable, heritable form to all progeny cells. Transformation has also been defined as "the stable acquisition of those characteristics which increase the cell's oncogenic potential" (Studzinski, 1977). Diamond and Baird (1977) defined transformation as "the acquisition by cells in culture of heritable morphological and growth properties that are generally associated with tumorigenic cells but may not include tumorigenicity. 'Spontaneous transformation' refers to transformation that occurs in the absence of known carcinogenic stimuli."

The important point is that in order for primary cultures or diploid cell lines to become established cell lines, by a process not yet understood a large number of new cell characteristics are acquired and, in general, cell behavior is changed. This change of many characteristics and in behavior is then referred to by the "umbrella" term, transformation. Transformation of a culture does not occur suddenly but rather over a period of time. Not every cell type nor cells of every species are equally prone to transformation. For example, murine cells transform fairly easily but human cells are very resistant (Macpherson 1970).

The situation does not become less complicated once transformation has occurred. At what precise point does a diploid cell line become an established cell line? How can the newly formed line be defined? Unfortunately, there is no single criterion by which these questions can be answered. Instead, established cell lines can be defined only by a multitude of criteria, and a particular established cell line may not fulfil all of these. The lists of criteria to differentiate established cell lines are many. One prepared by Diamond and Baird (1977) is as follows: altered cell and colony morphology, lack of contact inhibition of locomotion, lack of density-dependent inhibition of cell multiplication, increased saturation density, loss of anchorage dependence, high fibrinolytic activity, high agglutinability by plant lectins, expression or re-expression of cellular antigens, the capability to undergo an unlimited number of cell divisions, the ability to grow in soft agar medium, and the ability to produce progressively growing tumors when inoculated into syngeneic or immunosuppressed hosts. Auersperg and Finnegan (1974) suggest that established lines which originate from tumors differ from those which originate from normal cultures in their greater capacity to form multicellular, organized structures in the absence of heterologous tissues, in their greater capacity to maintain high levels of differentiation, and in the relative ease with which the cells derived from malignant tissues adapt to long-term growth in suspension cultures.

Levan and Biesele (1958), after careful study of the transformation of murine cell lines in cultures, concluded that the established cell lines, both of normal and of cancerous origin have similar chromosome behaviour: the mitotic mechanisms show instability, with variations in chromosome number and structure, and the stemline chromosome set differs from the normal karyotype of the species. Hsu *et al.* (1961) also followed the pattern of murine cell transformations leading to heteroploidy and concluded that the process of transformation can proceed by at least 2 routes: diploidy-tetraploidy-hypotetraploidy, or, diploidy-hypodiploidy-hypotetraploidy.

Cell lines transform gradually in cultures and the many new characteristics emerge at different times; therefore, there is no definite point in the life history of the cell line when one can say that yesterday it was a diploid cell line but today and from now on it is an established cell line.

In struggling with this difficult situation, the Terminology Committee of the Tissue Culture Association came up with the proposal that as for human fibroblast-like cell lines, they must have undergone at least 70 subculturings before they can be pronounced to be established cell lines (Fedoroff 1967). Obviously, this is an arbitrary decision based on experience with human fibroblast-like diploid cell line (WI-38). It is in no way applicable to other species or cell types.

A number of misconceptions about established cell lines persist. The first is that all established cell lines are characterized by a wide range of chromosome numbers in their cells and therefore the terms heteroploid cell line and established cell line are often used interchangeably. It is true that many established cell lines have extensive heteroploidy but this is not true for all established cell lines. For example, many established cell lines which originated from blood or lymphatic tissue are diploid or pseudodiploid, (e.g., Hamerton, 1973; Bloom *et al.,* 1974; Moore, 1975); two epithelial established cell lines which originated from cervical carcinoma, maintained hypodiploid chromosome numbers (Auersperg and Hawryluk, 1962; Auersperg, 1964); epitheloid cell line RPMI-2650 which originated from a human anaplastic squamous cell carcinoma maintained a distinct mode of 46 chromosomes (Moorhead, 1965) and SV40 transformed mouse cells produced a pseudodiploid cell line (Kelley, 1975). Therefore, the comment by Biedler (1975) that extensive heterploidy is neither a prerequisite nor a necessary consequence of *in vitro* cultivation of human cells in established cell lines is correct. It is obvious, then, that the degree of chromosome abnormalities in established cell lines will depend on species and tissue origin as well as on the mechanism by which the cells were transformed into an established cell line.

The second misconception is that all established cell lines are tumorigenic. Studzinski (1977) suggests that all established cell lines capable of indefinite proliferation *in vitro* can give rise to tumors if the right conditions prevail. The "right conditions" would be either injection of the cell into "nude" mice which are deficient in thymus-dependent immunological functions (Freedman and Shin, 1974); intraocular implantation (Evans *et al.,* 1964); transplantation into chick blastoderm (Marsel *et al.,* 1975), into an ALS pretreated syngeneic host (Stanbridge and Perkins, 1969), into hamster cheek pouch (Foley *et al.* 1962), or into an irradiated host (Murphy and Morton, 1915; Taliaferro, 1957). It can be argued that formation of a nodule at the site of injection of cells of an established cell line when the site is a specially selected and prepared "ideal" site, is not a valid test for neoplastic properties as defined in human pathology. Such test procedures do however, demonstrate the degree of ability of cells to survive and proliferate in a host and give a measure of certain cell characteristics which are accepted by tissue culturists as constituting what is usually referred to as the "tumorigenicity" of the cells (Fedoroff, 1966, 1968). The degree of tumorigenicity can change with the progression of time in a culture. For example, the incidence of tumors formed at the site of injection of line L cells into normal C_3H mice (Andervont strain) was 46%, 68%, 42%, 2% and 1%, respectively, in the years 1942, 1943, 1944, 1945 and 1946 (Earle *et al.*, 1950). Subsequently, a single cell was isolated from a culture of line L cells which gave rise to a line designated as clone 929 (Sanford *et al.*, 1948). After 4 to 5 years in culture, these cells were retested for tumorigenicity. They produced sarcomas in 15% of non-irradiated C_3H mice and in 64% of x-ray irradiated C_3H mice (Sanford *et al.*, 1956). More recently Fedoroff *et al.* (1974), on injection of 1×10^6 line L cells (clone 929) into normal non-irradiated C_3H mice, obtained tumors in 2% of the animals, whereas cells from a cell strain isolated from a culture of line L cells (clone 929) designated as L_E cells, produced tumors in 58% of injected mice. In x-ray irradiated mice, both line L cells (clone 929) and L_E cells produced tumors in nearly 100% of the mice (depending on the dosage of irradiation). There are many other instances in which the tumorigenicity of cells markedly decreased or was completely lost on prolonged culturing, e.g., MB mouse lymphoma cells (deBruyn and Gey, 1952); rat hepatoma cells (Hsu and Klatt, 1959); mouse sarcoma 180 cells (Foley and Drolet, 1964); and Ehrlich ascites cells (Moore *et al.*, 1968). A whole series of tissue culture lines established from Ehrlich, TA_3, Ehrlich-Lettré and $6C_3HED$ mouse ascites tumors lost their transplantability after 1 to 5 years of propagation in cultures (Morgan *et al.*, 1970).

Another misconception is that established cell lines consist of uniform cell populations. As mentioned previously, most established cell lines are heteroploid. Idiograms of the frequencies of cells with various chromosome numbers provide profiles of cell lines. Such profiles are not stable but change under various culturing conditions. For example, a different batch of serum or heat inactivation of serum can alter the profile of a cell line (Saxén and Penttinen, 1956, 1965; Saksela *et al.*, 1961; Saksela and Saxén, 1962; Saksela, 1962). In addition many types of cell mutants, e.g., cells resistant to certain drugs, nutritional auxotrophs, temperature sensitive conditioned lethal mutants, or mutants in which certain specialized cell functions are altered, have been isolated from a number of established cell lines (Thompson and Baker, 1973). This indicates that established cell lines consist of genetically heterogeneous proliferative cell populations and that the genetic heterogeneity is due to rearrangements in chromosome numbers and structure as well as accumulation of gene mutations. As such, the cell populations are very adaptive to changing environmental conditions due to the continuous selection of cells with the genotype most compatible with the culture milieu. Such genetic drifts affect the stability of the chemical and biological characteristics of some cell lines; however, certain characteristics remain relatively stable despite genetic drifts. Consequently, careful selection of an established cell line (with as little genetic heterogeneity as possible), rigidly controlled culturing conditions and regular selection of the cells is required to secure stability of a cell line with desired characteristics.

It is often assumed that established cell lines consist of dedifferentiated cells which have deviated considerably from the cells of the tissue of origin. In many instances, this assumption is true. There are, however, established cell lines which have retained some of their specialized functions. In many cases, cultures from tumors transform into established cell lines without the loss of the differentiated state they had *in situ*. One of the most dramatic examples is that of human thyroid carcinoma cells, which secrete thyrocalcitonin, and which continue such secretion when explanted in tissue culture, even after being transformed by SV-40 virus (Grimley *et al.*, 1969). Another example is that in established cell lines from neuroblastoma, cells may express a number of the characteristics of differentiated neurons (Giller *et al.*, 1975). In the chapter by P. G. Nelson, (this volume) acquisition of some properties of neuronal cells by neuroblastoma cell lines, and in the chapter by S. E. Pfeiffer (this volume), the retention of many characteristics of normal glial cells in glial established cell lines are discussed.

Yet another misconception is the idea that established cell lines are easy to culture and therefore do not require much skill. Established cell lines are propagated in cultures in a continuous state of proliferation over a very long period of time. During their *in vitro* life, the cells become exposed to a variety of sera from many animals of the same species or even from other species, as well as media impurities, non-degradeable materials and, most important, to microorganisms and viruses of the immediate environment. In time, all these contaminants may accumulate intracellularly. The use of antibiotics in the media may mask some infections and even facilitate intracellular contamination. Furthermore, laboratories in which more than one cell line is carried are exposed to the risk of a possible cross-contamination by cells of other lines (Rothfels *et al.*, 1959; Gartler, 1967; Nelson-Rees *et al.*, 1974; Nelson-Rees and Flandermeyer, 1976, 1977; Lavappa, 1977), as has been the bitter experience of so many laboratories in the past.

In order to maintain an established cell line over a long period of time with minimal alterations, it must be periodically cloned or must be treated with selective media to eliminate mutants. In addition, cell lines periodically must be surveyed with microbiological, cytogenetic, morphological and biochemical methods (Fogh, 1973; O'Brian *et al.*, 1977). All this can become extremely time-consuming. The over-riding concern of tissue culturists these days must be provision not only of a microorganism-free environment for the cultures, but also avoidance of the possible biohazards to the tissue culturists themselves (Berg *et al.*, 1975; U.S. National Institutes of Health, 1976; Report of the Working Party on the Practice of Genetic Manipulation, 1976; MRC (Canada), 1977).

C. Clonal Lines

The term "clone" denotes a population of cells derived from a single cell by mitosis (Fedoroff, 1966). Clonal lines can be initiated from primary cultures, diploid cell lines or established cell lines. There are many techniques for isolating a single cell from a culture and from that cell to develop cultures which can be serially propagated (Ham, 1972). These are the methods by which cells with specific characteristics can be isolated from a culture and by periodic cloning or by use of selecting media can be maintained with the desired characteristics in culture in stable form. It should be understood that the term "clonal line" does not necessarily imply an homogeneous cell line. Even the progeny from a single cell develop heterogeneity and the longer they are maintained in culture after cloning, the more heterogeneous the line may become (Ruddle, 1961). Therefore, only frequent cloning and selection of cells with specific characteristics can secure a minimum of heterogeneity in the culture.

V. Cell Strains

In the introduction to this chapter, the accepted definition of the term cell strain was discussed. Accordingly, strain has the meaning of *relationship and kindred.* Therefore, the term "cell strain" implies a certain relationship between the cells, i.e., they all have one or more common characteristics or markers for which these cells were originally selected (Fedoroff, 1966).

A cell strain can be derived from either a primary culture or a diploid or established cell line by the selection or cloning of cells having specific characteristics or markers. The characteristics or markers must persist during subsequent culturing (Fedoroff, 1966). For example, a cell line may consist of many different types of cells, characterized morphologically or by their karyotypes, or by biochemical markers or physiological activities. Such a line can be cloned and the clones can be serially propagated. The resulting cultures are often referred to as clonal lines; however, if such clonal lines have certain specific characteristics or markers which persist on subculturing, the lines are properly referred to as cell strains. The term "clonal line" implies only that a given cell line has been cloned and should be used unless and until its characteristics have been proven stable. On the other hand, not all cell strains are clonal lines. There are means other than cloning by which cell variants can be selected from a cell population to obtain a cell strain. If the strain loses its characteristics or markers, it should no longer be designated as a cell strain. In other words, it reverts to a cell line. In describing a cell strain, therefore, its specific features must be defined, e.g., a cell strain with a certain marker chromosome, a cell strain with acquired resistance to a certain drug or virus, etc.

Conclusions

All cell cultures fall into one of three basic types: primary cultures, diploid cell lines and established cell lines. These types differ considerably and each has certain advantages and disadvantages. All such cultures are at present readily accessible to the investigator, but one should be very critical in selecting the right type of culture for an experiment, in the hope of answering specific questions.

A serious weakness in the field of tissue culture is that in only a few instances have the cells been identified as cells which occur *in vivo*. In most cases, we work with heterogeneous cell populations, whether they are primary, diploid or established cell lines; experimenters must take this into consideration.

At first glance, it seems a hopeless task to work with cell cultures but, on reflection, it can be seen that the variability can be an advantage. The investigator must therefore delineate his spectrum of inquiry, and master the methodology to the extent that cells behave predictably and experiments are repeatable.

References

Absher, P. M., R. G. Absher and W. D. Barnes, 1974. Genealogies of clones of diploid fibroblasts. Exp. Cell Res. 88:95-104.

Absher, P. M., R. G. Absher and W. D. Barnes, 1975. Time-lapse cinemicrophotographic studies of cell division patterns of human diploid fibroblasts (WI-38) during their *in vitro* lifespan. In: *Cell Impairment in Aging and Development.* V. J. Cristofalo and E. Holečková, eds. Plenum Press, New York and London. pp. 91-105.

Auersperg, N. and A. P. Hawryluk, 1962. Chromosome observations on three epithelial-cell cultures derived from carcinomas of the human cervix. J.Natl. Cancer Inst. 28:605-627.

Auersperg, N. 1964. Long-term cultivation of hypodiploid human tumor cells. J. Natl. Cancer Inst. 32:135-148.

Auersperg, N. and C. V. Finnegan. 1974. The differentiation and organization of tumors *in vitro.* In: *Neoplasia and Cell Differentiation.* G. V. Sherbet, ed. S. Karger, Basel-Müenchen-Paris, London, New York, Sydney. pp. 279-318.

Berg, P., D. Baltimore, S. Brenner, R. O. Roblin and M. F. Singer. 1975. Summary statement of the Asilomar Conference on recombinant DNA molecules. Science 188:991-994; Nature 255; 442-444; Proc. Natl. Acad. Sci. U.S.A. 72:1981-1984.

Berman, L., C. S. Stulberg, and F. H. Ruddle. 1957. Human cell culture. Morphology of the Detroit strains. Canc. Res. 17:668-676.

Biedler, J. L. 1975. Chromosome abnormalities in human tumor cells in culture. In: *Human Tumor Cells in Vitro.* J. Fogh, ed. Plenum Press, New York, London. pp. 359-394.

Bloom, A. D., J. A. McNeill, and F. T. Nakamura. 1974. Epstein-Barr virus infection of lymphoid cells and the cytogenetics of established human lymphocyte cell lines. In: *Chromosomes and Cancer.* J. German, ed. John Wiley and Sons, New York, London, Sydney, Toronto. pp. 565-599.

Borek, C. and E. J. Hall. 1973. Transformation of mammalian cells in vitro by low doses of X-rays. Nature (Lond.) 243:450-453.

Borek, C. and E. J. Hall. 1974. Effect of split dose of X-ray on neoplastic transformation of single cells. Nature (Lond.) 252:499-501.

Borek, C. and L. Sachs. 1966. *In vitro* transformation by X-irradiation. Nature (Lond.) 210:276-278.

Borek, C., and L. Sachs. 1967. Susceptibility to transformation by X-irradiation and fixation of the transformed state. Proc. Natl. Acad. Sci. U.S.A. 57:1522-1527.

Bowman, P. D. and C. W. Daniel. 1975. Characteristics of proliferative cells from young, old and transformed WI-38 cultures. In: *Cell Impairment in Aging and Development.* V. J. Cristofalo and E. Holečková, eds. Plenum Press, New York and London. pp. 107-122.

Burrows, M. T. 1913. The tissue culture as a physiological method. Trans. Congr. Am. Phys. Surg. 9:77-94.

Butel, J. S. and M. K. Estes. 1975. Properties of cells transformed by DNA tumor viruses. In Vitro. 11:142-150.

Casto, B. C. and J. A. DiPaolo. 1973. Virus, chemicals and cancer. Prog. Med. Virol. 16:1-47.

Carrel, A. 1914. Present condition of a strain of connective tissue twenty-eight months old. J. Exp. Med. 1-2.

Carrel, A. 1923. Les cultures pures de cellules en physiologie. Soc. biol. Compt. rend. 89:972-974.

Cell Culture Committee: Permanent section on microbiological standardization: Minutes of the eighth meeting of the committee 1971. International Association Biological Standardization. Cell Culture Committee, Medical Research Council, Hampstead Laboratories, London, England.

Cristofalo, V. J. 1970. Metabolic aspects of aging in diploid human cells. In: *Aging in Cell and Tissue Culture.* E. Holečková and V. J. Cristofalo, eds. Plenum Press, New York and London. pp. 83-119.

deBruyn W. M. and G. O. Gey. 1952. Further studies in vitro of transplantable mouse lymphosarcoma MB(T86157), Acta. Un. Intern. contra. Cancrum. 7:772-775.

Diamond, L. and W. M. Baird. 1977. Chemical carcinogenesis in vitro. In: *Growth, Nutrition, and Metabolism of Cells in Culture,* Vol. III. G. H. Rothblat and V. J. Cristofalo, eds., Academic Press, New York, San Francisco, London. pp. 421-470.

Earle, W. R. 1943. Changes induced in a strain of fibroblasts from a strain C_3H mouse by the action of 20-methylcholanthrene. J. Nat. Cancer Inst. 3:555-558.

Earle, W. R., and A. Nettleship. 1943. Production of malignancy *in vitro.* Results of injection of cultures into mice. U.S. J. Natl. Cancer Inst. 4:213-227.

Earle, W. R., E. Shelton, and E. L. Schilling. 1950. Production of malignancy *in vitro.* XI Further results from reinjection of *in vitro* cell strains into strain C_3H mice. U.S. J. Nat. Cancer Inst. 10:1105-1113.

Ebeling, A. H. 1942. Dr. Carrel's immortal chicken heart. Present, authentic facts about this oft-falsified scientific "celebrity". Scientific Am. 166:22-24.

Evans, V. J., G. A. Parker, and T. B. Dunn. 1964. Neoplastic transformation in C_3H mouse embryonic tissue *in vitro* determined by intraocular growth. I Cells from chemically defined medium with and without serum supplement. U.S. J. Natl. Cancer Inst. 32:89-121.

Fedoroff, S. 1966. Proposed usage of animal tissue culture terms. *In Vitro* 2: 155-159; 1967. J. Genet. Cytol. 9:412-418; Hereditas 58:86-94; Exp. Cell Res. 46:648-672; Cytogenetics (Basel) 6:161-167; J. Nat. Cancer Inst. 38:607-611; 1976, TCA Manual 1:53-57.

Fedoroff, S. 1968. Definition of terms used in the conference on "Cell Cultures for Virus Vaccine Production". U.S. Nat. Cancer Inst. Monograph 29:587-595.

Fedoroff, S., J. A. Boechler and R. V. White. 1974. Interactions between host (C_3H mice) and line L cells. *In Vitro* 9:382-383.

Fedoroff, S., B. Juurlink, L. Hertz and F. Oteruelo. 1977. Formation of cell colonies in cultures from mouse cerebral cortex. *In Vitro* 13:154.

Fedoroff, S., R. V. White and J. A. Boechler. 1975. Mouse spleen cells in cultures. *In Vitro* 10:367.

Fogh, J., ed. 1973. *Contamination in Tissue Culture.* Academic Press, New York and London.

Foley, G. E. and B. P. Drolet. 1964. Loss of neoplastic properties *in vitro,* I. Observations with S-180 cell lines. Cancer Res. 24:1461-1467.

Foley, G. E., A. H. Handler, R. D. Adams and J. M. Craig. 1962. Assessment of potential malignancy of cultured cells: Further observations on the differentiation of "Normal" and

"Neoplastic" cells maintained *in vitro* by hetero-transplantation in Syrian hamsters. U.S. Nat. Canc. Inst. Monograph 7:173-204.

Freedman, V. H. and S. Shin. 1974. Cellular tumorigenicity in nude mice: Correlation with cell growth in semi-solid medium. Cell 3:355-359.

Gartler, S. M. 1967. Genetic markers as tracers in cell culture. U.S. Nat. Canc. Inst. Monograph 26:167-194.

Gey, G. O. 1956. Some aspects of the constitution and behavior of normal and malignant cells maintained in continuous culture. The Harvey Lectures, Series L, 1954-55. Academic Press, New York. pp. 154-229.

Gey, G. O., W. D. Coffman and M. T. Kubicek. 1952. Tissue culture studies of the proliferative capacity of cervical carcinoma and normal epithelium. Cancer Res. 12:264-265.

Giller, E. L., X. O. Breakefield, C. N. Christian, E. A. Neale and P. G. Nelson, 1975. Expression of neuronal characteristics in culture: some pros and cons of primary cultures and continuous cell lines. In: *Golgi Centennial Symposium*. M. Santini, ed. Raven Press, New York. pp. 603-623.

Grimley, P. M., L. J. Deftos, J. R. Weeks and A. S. Robson. 1969. Growth *in vitro* and ultrastructure of cells from a medullary carcinoma of the human thyroid gland: transformation by simian virus 40 and evidence of thyrocalcitonin and prostaglandins. J. Nat. Cancer Inst. 42:663-680.

Ham, R. G. 1972. Cloning of mammalian cells. In: *Methods in Cell Biology*. D. M. Prescott, ed. Academic Press, New York and London. pp. 37-73.

Hamerton, J. L. 1973. Long-term lymphocyte cultures-cytogenetics. In: *Long-term Lymphocyte Cultures in Human Genetics*. D. Bergsma, ed. Birth Defects: Original Article. Series 9. The National Foundation-March of Dimes. U.S.A. pp. 183–187.

Harrison, R. G. 1910. The outgrowth of the nerve fiber as a mode of protoplasm movement. J. Exper. Zool. 9:787–846.

Harrison, R. G. 1912. The cultivation of tissues in extraneous media as a method of morphologic study. Anat. Rec. 6:181–193.

Harrison, R. G. 1906–1907. Observations on the living developing nerve fiber. Proc. Soc. Exper. Biol. and Med. 4:140–143.

Harvey, A. M. 1975. Johns Hopkins. The birthplace of tissue culture: The story of Ross G. Harrison, Warren H. Lewis and George O. Gey. Johns Hopkins Med. J. 136:142–149.

Hayflick. L. 1961. The establishment of a line (WISH) of human amnion cells in continuous cultivation. Exptl. Cell. Res. 23:14–20.

Hayflick, L. 1965. The limited *in vitro* life time of human diploid cell strains. Exptl. Cell. Res. 37:614–636.

Hayflick, L. 1972. Cell senescence and cell differentiation *in vitro*. In: *Aging and Development*. H. Brendt and J. W. Rohen, eds. F. K. Schaffauer Verlag, Stuttgart-New York. pp. 1–15.

Hayflick, L. 1973. The biology of human aging. Am. J. Med. Sci. 265:432–435.

Hayflick, L. and P. S. Moorhead. 1961. The serial cultivation of human diploid cell strains. Exp. Cell Res. 25:585–621.

Heidelberger, C. 1973. Chemical oncogenesis in culture. Adv. Cancer Res. 18:317–366.

Heidelberger, C. 1975. Chemical carcinogenesis. Annu. Rev. Biochem. 44:79–121.

Hsu, T. C. 1961. Chromosomal evolution in cell populations. Int. Rev. Cytol. 12:69–161.

Hsu, T. C. 1963. Mammalian chromosomes *in vitro* XVI Analysis of chromosome breakages in cell population of the Chinese hamster. In: *Canad. Cancer Conf. Vol. 5*. R. W. Begg, C. P. Leblond, R. L. Noble, R. J. Rossiter, R. M. Taylor and A. C. Wallace, eds. Academic Press. New York and London. pp. 117-127.

Hsu, T. C. and O. Klatt. 1959. Mammalian chromosomes *in vitro*, X Heteroploid transformation in neoplastic cells. J. Nat. Cancer Inst. 22:313-339.

Hsu, T. C., and P. S. Moorhead. 1957. Mammalian chromosomes *in vitro*. VII Heteroploidy in human cell strains. J. Nat. Canc. Inst. 18:463-470.

Hsu, T. C., D. Brillen and A. Levan, 1961. Mammalian chromosomes *in vitro* XV Pattern of transformation. U.S.A. J. Nat. Canc. Inst. 27:515-519.

Hsu, S. H., B. Z. Schacter, N. L. Delaney, T. B. Miller, V. A. McKusick, R. H. Kennett, J. G.

Bodmer, D. Young and W. F. Bodmer. 1976. Genetic characteristics of the HeLa cell. Science 191:392-394.

Jones, H. W., V. A. McKusick, P. S. Harper and K. D. Wuu. 1971. George Otto Gey. The HeLa cell and a reappraisal of its origin. Obs. and Gynaecology 38:945-949.

Kadanka, Z. K., J. D. Sparkes, and H. G. Macmorine. 1973. A study of the cytogenetics of the human cell strain WI-38. *In Vitro* 8:353-361.

Kaplan, A. S. ed. 1974. *Virus Transformation and Endogenous Viruses*. Academic Press, New York.

Kelly, F. 1975. Chromosome analysis of a simian virus 40 transformed mouse cell line and two variant sublines that are resistant to cytochalasin B[1]. Cancer Res. 35:1210-1213.

Krooth, R. S., G. A. Darlington and A. A. Velazquez. 1968. The genetics of cultured mammalian cells. Ann. Rev. Genet. 2:141-164.

Kurstack, E. and K. Maramorosch, eds. 1974. *Virus, Evolution and Cancer*. Academic Press, New York.

Lavappa, K. S. 1977. ATCC stock of human cell lines with HeLa characteristics. *In Vitro* 13:199.

Levan, A. and J. J. Biesele. 1958. Role of chromosomes in cancerogenesis as studied in serial tissue culture of mammalian cells. Ann. N.Y. Acad. Sci. 71:1022-1053.

Lipetz, J. and V. J. Cristofalo. 1972. Ultrastrucutral changes accompanying the aging of human diploid cells in culture. J. Ultrastruct. Res. 39:43-56.

Macieira-Coelho, A. 1974. Are non-dividing cells present in ageing cell cultures? Nature (Lond.) 248:421-422.

Macpherson, I. 1970. The characteristics of animal cells transformed *in vitro*. In: *Advances in Cancer Research*. Vol. 13. G. Klein and S. Weinhouse, eds. Academic Press, New York, pp. 169-215.

Marbrook, J. 1967. Primary immune response in cultures of spleen cells. Lancet 2:1279-1281.

Martin, G. M. and C. A. Sprague. 1973. Life histories of hyperplastoid cell lines from aorta and skin. Exp. Mol. Path. 18:125-141.

Martin, G., C. Sprague and C. J. Epstein. 1970. Replicative life-span of cultured human cells. Effects of donor's age, tissue and genotype. Lab. Invest. 23:86-92.

Marsel, M., L. Ridder, M. DeBrabander and L. Vakaet. 1975. Characterization of spontaneous, chemical and viral transformation of a $C_3H/3T3$-type mouse cell line by transplantation into young chick blastoderms. J. Natl. Cancer Inst. 54:923-929.

Mattern, M. R., and P. A. Cerotti. 1975. Age-dependent excision repair of damaged thymine from gamma-irradiated DNA by isolated nuclei from human fibroblasts. Nature (Lond.) 254:450-452.

Merz, G. S. Jr., and J. D. Ross. 1969. Viability of human diploid cells as a function of *in vitro* age. J. Cell Physiol. 74:219-222.

Mishra, N. K. and G. DiMayorca. 1974. *In vitro* malignant transformation of cells by chemical carcinogens. Biochim. Biophys. Acta. 355:205-219.

Mitsui, Y. and E. L. Schneider. 1976. Increased nuclear sizes in senescent human diploid fibroblast cultures. Exp. Cell Res. 100:147-152.

Moore, G. E. 1975. Cell lines from humans with hematopoietic malignancies. In: *Human Tumor Cells in Vitro*. J. Fogh, ed. Plenum Press, New York, London. pp. 299-331.

Moore, J., J. Kieler and B. Biczowa. 1968. Comparative studies of a near-tetraploid and a near-diploid line of Ehrlich's ascites tumor produced *in vivo* and *in vitro*. II Cytology and transplantability. Europ. J. Cancer 4:81-95.

Moorhead, P. S. 1965. Human tumor cell line with a quasi-diploid karyotype (RPMI-2650). Exp. cell Res. 39:190-196.

Morgan, J. F., C. P. Eng, M. D. Heuchert and H. D. Kirk. 1970. Loss of transplantability and induction of immunoprotection by mouse asites tumor cells in tissue culture. Proc. Soc. Exp. Med. 134:305-308.

MRC (Canada) 1977 Guidelines for handling recombinant DNA molecules and animal viruses and cells. Minister of Supply and Services, Canada, Ottawa.

Murphy, J. B. and J. J. Morton. 1915. The lymphocyte in natural and induced resistance to transplanted cancer. II Studies on lymphoid activity. J. Exper. Med. 22:204-211.

Murray, M. R. and G. Kopech, eds. 1953. *A Bibliography of the Research in Tissue Culture, 1884-1950.* Academic Press, Inc., New York.

Nelson-Rees, W. A. and R. R. Flandermeyer. 1976. HeLa cultures defined. Science 191:96-98.

Nelson-Rees, W. A. and R. R. Flandermeyer. 1977. Inter-and intraspecies contamination of human breast tumor cell lines HBC and BrCa5 and other cell cultures. Science 195:1343-1344.

Nelson-Rees, W. A., R. R. Flandermeyer and P. K. Hawthorne. 1974. Banded marker chromosomes as indicators of intraspecies cellular contamination. Science 184:1093-1096.

Nichols, W. W., D. G. Murphy, V. J. Cristofalo, L. H. Toji, A. E. Greene and S. A. Dwight. 1977. Characterization of a new human diploid cell strain, IMR-90. Science 196:60-63.

O'Brien, S. J. O., G. Kleiner, R. Olson and J. E. Shannon. 1977. Enzyme polymorphisms as genetic signatures in human cell cultures. Science 195:1345-1348.

Orgel, L. E. 1973. Ageing of clones of mammalian cells. Nature (Lond.) 243:441-445.

Parker, R. C. 1932a. The functional characteristics of nine races of fibroblasts. Science 76:219-220.

Parker, R. C. 1932b. The stability of functionally distinct races of fibroblasts. Science 76:446-447.

Parker, R. C. 1933. The races that constitute the group of common fibroblasts. III. Differences determined by origin of explant and age of donor. J. Exp. Med. 58:401-414.

Parker, R. C. 1934. The races that constitute the group of common fibroblasts. Arch. Für exp. Zellforschung 15:99-105.

Parker, R. C. 1938. *Methods of Tissue Culture.* P. B. Hoeber Inc., New York.

Parker, R. C. 1954. Cultivation of tumor cells *in vitro.* In: *Canadian Cancer Conference,* Vol. I. R. W. Begg, ed. Academic Press, New York, pp. 42-54.

Parker, R. C. 1964, *Methods of Tissue Culture.* 3rd ed. Hoeber Medical Division. Harper and Row, New York. p. 4.

Parker, R. C., L. N. Castor and E. A. McCulloch. 1957. Altered cell strains in continuous culture: a general survey. Special Publication, New York Academy of Sciences 5:303-313.

Paterson, J. A., A. Privat, E. A. Ling and C. P. Leblond. 1973. Investigation of glial cells in semi-thin sections. III transformation of subependymal cells into glial cells, as shown by radioautography after $_3$H-thymidine injection into the lateral ventricle of the brain of young rats. J. Comp. Neurol. 149:83-102.

Pontén, J. 1973. In: *Symposium on molecular and cellular mechanisms of aging.* Paris Inst. Natl. Sante Recherche Med. 27:53.

Pontén, J. and E. H. Macintyre. 1968. Long term culture of normal and neoplastic human glia. Act. Pathol. Microbiol. Scand. 74:465-486.

Rapp, F. 1974. Herpes viruses and cancer. Adv. Cancer Res. 19:265-302.

Rapp, F. and S. C. St. Jeor. 1974. Transformation by Herpes simplex and cytomegalo virus. In: *Viral Transformation and Endogenous Viruses.* A. S. Kaplan, ed. Academic Press, New York, pp. 35-53.

Report of the Working Party on the Practice of Genetic Manipulation. 1976. Presented to Parliament by the Secretary of State for Education and Science by Command of Her Majesty, CMND. 6600, Her Majesty's Stationery Office, London. (The Williams Report).

Rothfels, K. H., A. A. Axelrad, L. Siminovitch, E. A. McCulloch and R. C. Parker. 1959. The origin of altered cell lines from mouse, monkey and man as indicated by chromosome and transplantation studies. In: *Canadian Cancer Conf.* Vol. 3. R. W. Begg, ed. Academic Press, New York and London. pp. 189-214.

Rothfels, K. H., E. B. Kupelwieser and R. C. Parker. 1963. Effects of X-irradiated feeder layers on mitotic activity and development of aneuploidy in mouse-embryo cells *in vitro.* In: *Canadian Cancer Conference* Vol. 5. R. W. Begg, C. P. Leblond, R. L. Noble, R. J. Rossiter, R. M. Taylor and A. C. Wallace, eds. Academic Press, New York and London. pp. 191-223.

Ruddle, F. H. 1961. Chromosome variation in cell populations derived from pig kidney. Cancer Res. 21:885-894.

Sanford, K. K. 1965. Malignant transformation of cells *in vitro*. Int. Rev. Cytol. 18:249-311.

Sanford, K. K. 1967. "Spontaneous" neoplastic transformation of cells *in vitro:* some facts and theories. U.S. Nat. Canc. Inst. Monograph 26:387-418.

Sanford, K. K. 1974. Biologic manifestations of oncogenesis *in vitro*. A. critique. J. Natl. Cancer Inst. 53:1481-1485.

Sanford, K. K., W. R. Earle and G. D. Likely. 1948. The growth *in vitro* of single isolated tissue cells. J. Nat. Cancer Inst. 9:229-246.

Sanford, K. K., W. R. Earle, E. Shelton, E. L. Schilling, E. M. Duchesne, G. D. Likely and M. M. Becker. 1950. Production of malignany *in vitro*. XII Further transformations of mouse fibroblasts to sarcomatous cells. J. Natl. Cancer Inst. 11:351-375.

Sanford, K. K., G. L. Hobbs and W. R. Earle. 1956. The tumor-producing capacity of strain L mouse cells after 10 years *in vitro*. Cancer Research 16:162-166.

Saksela, E. 1962. The effect of different human sera on the karyologic picture of HeLa cells. A contribution to the knowledge of environmental influences on cell populations. Acta Path. et microbiol. Scand. Supple. 153:1-73.

Saksela, E. and E. Saxén. 1962. Chromosome heterogeneity of cell lines *in vitro*. III The effect of individual fresh human sera on the mitosis of HeLa cells. Acta Pathol et Microbiol. Scand. 54:253-259.

Saksela, E., E. Saxén and K. Penttinen. 1961. Chromosome heterogeneity of cell lines *in vitro*. I Differences in the effect of individual human sera on the chromosome number of HeLa cells. Acta Pathol et Microbiol Scand. 51:127-131.

Sax, H. J. and K. N. Passano. 1961. Spontaneous chromosome aberrations in human tissue culture cells. Am. Naturalist 95:97-102.

Saxén, E. and K. Penttinen. 1956. Factors in serum affecting the growth of HeLa cells. Ann. Med. Exp. Biol. Fenn. 34:396-401.

Saxén, E. and K. Penttinen. 1965. Differences in the effects of individual human sera on cell cultures. J. Nat. Cancer Inst. 35:67-73.

Schneider, E. L. and Y. Mitsui. 1976. The relationship between *in vitro* cellular aging and *in vivo* human age. Proc. Natl. Acad. Sci., U.S.A. 73:3584-3588.

Schwarzacher, H. G. and W. Schnedl. 1965. Endoreduplication in human fibroblast cultures. Cytogenetics 4:1-18.

Sidman, R. L. 1970. Cell proliferation, migration and interaction in the developing mammalian central nervous system. In: *The Neurosciences: Second study program*. F. O. Schmitt, ed. Rockfeller Univ. Press, New York. pp. 100-107.

Smith, J. R. and L. Hayflick. 1974. Variation in the life-span of clones derived from human diploid cell strains. J. Cell. Biol. 62:48-53.

Stanley, J. F., D. Pye and A. MacGregor. 1975. Comparison of doubling numbers attained by cultured animal cells with lifespan of species. Nature (Lond.) 255:158-159.

Stanbridge, E. J. and F. T. Perkins. 1969. Tumor nodule formation as *in vivo* measure of the supression of cellular immune response by antilymphocytic serum. Nature (Lond.) 221:80-81.

Studzinski, G. P. 1977. Viral transformation of cultured mammalian cells as a multi-step process. In: *Growth, Nutrition, and Metabolism of Cells in Culture*. Vol. III. G. H. Rothblat and V. J. Cristofalo, eds. Academic Press, New York, San Francisco and London, pp. 355-419.

Taliaferro, W. H. 1957. Modification of the immune response by radiation and cortisone. 69:745-764.

Thompson, L. H. and R. M. Baker. 1973. Isolation of mutants of cultured mammalian cells. In: *Methods in Cell Biology*. Vol. VI. D. M. Prescott, ed. Academic Press, New York and London. pp. 209-281.

Thompson, K. V. and R. Holliday. 1973. Effect of temperature on the longevity of human fibroblasts in culture. Exp. Cell Res. 80:354-360.

Todaro, G. J. and H. Green. 1963. Quantitative studies of the growth of mouse embryo cells in culture and their development into established lines. J. Cell Biol. 17:299-313.

Todaro, G. J. and G. M. Martin. 1967. Increased susceptibility of Down's syndrome fibroblasts to transformation by SV40. Proc. Soc. exp. Biol. Med. 124:1232-1236.

U.S. National Institutes of Health. 1976. *Guidelines for research involving recombinant DNA molecules.* Federal Register 41:27902 and 41:38426.

Van Duuren, B. L., G. Witz, A. Sivak. 1974. Chemical carcinogenesis. In: *The Physiopathology of Cancer*, Vol. I. F. Homburger, ed. Karger, Basel, pp. 1-63.

Willmer, E. N. 1935. *Tissue Culture. The growth and differentiation of normal tissues in artificial media.* Methuen and Co. Ltd., London.

Willmer, E. N. 1960. In: *Cytology and Evolution.* Academic Press, New York and London.

White, P. C. *The cultivation of animal and plant cells.* Roland Press Co., New York. 1954. pp. 21 and 29.

GLIAL CELL LINES

S. E. PFEIFFER, B. BETSCHART, J. COOK*, P. MANCINI AND R. MORRIS

Department of Microbiology
University of Connecticut Health Center
Farmington, Connecticut

"Our experiments show that at least in rats, serially propagated brain cells are capable of performing at least one organ-specific function over a period of a year."

Benda, Lightbody, Sato, Levine and Sweet, 1968.

I. Introduction

The purpose of developing clonal lines of biochemically differentiated cells growing in cultures is to provide permanently established, functionally and genetically homogeneous, viable populations of normal specific cell types which can be produced in quantities sufficient for most biochemical studies (Table I). This means, in effect, lines which can be handled in many respects as microorganisms with the consequent advantages of availability and possibilities for easy exchange among investigators, quantification, genetic manipulation (e.g., cell hybridization and isolation of genetic mutants), cell synchronization and controlled mixing of clonal lines (with other clonal lines or with various primary cell preparations), for the study of cell interactions *in vitro.*

Given the tremendous potential advantages that clonal lines can offer they are highly rational, useful tools for the study of certain aspects of neurobiology and may be useful in detecting cellular constituents or processes unique to the nervous systems.

The extent to which the clonal lines, as well as some alternative preparations, satisfy the desired properties is shown in Table II. It can be seen that the clonal lines have not yet lived up to early expectations for their contributions. This is probably due as much to the unrealistic goals many of us set for them as to some inherent limitations present in both tumor cell lines and continuous cultures of supposedly normal cells which nevertheless have undergone some extent of "transformation" during culturing. Thus, relative to the amount of time and effort invested, and with some important

*Present Address: Roswell Park Memorial Institute
Buffalo, New York

exceptions, rather little truly original has been learned about the nervous system from studies using clonal tumor cell lines.

What has been established is that at least some glial lines can carry out some biochemical and physiological properties characteristic of the nervous system in an apparently normal fashion. However, characterization as to specific cell types is, at best, uncertain; normal intra- and inter-cellular regulation is doubtful and not really known, and it remains to be seen whether glial lines can ever be developed that carry out a whole concert of normal differentiated functions in normally regulated proportions.* In other words, many of the studies of clonal glial lines have been directed towards analyzing the capabilities of these lines, a sort of cataloguing effect, without the investigators necessarily being particularly concerned what such studies will at once tell us about nervous system function. Nevertheless, this period of testing has been important, and we can now continue to use glial cell lines in at least certain types of studies with a fair degree of confidence in the validity of the results.

In general, the most useful type of study will be the elucidation of specific biochemical events, already known to occur in the nervous system, in an environment optimally free of normal interactions among different cell types and organismic regulation, and optimally open to experimental control of the external environment. Thus, we are using systems in which we gain certain technical advantages at the expense of some of the most important aspects of cellular growth and function occurring at the organ level. In addition, the relative simplicity and microbiological nature of these cultures should also provide an important source of material for the detection of currently unrecognized brain specific molecules.

Against these rather ambivalent but overall hopeful feelings, the rest of the chapter will deal with some areas of glial cell line research that to us seem to support these propositions.

TABLE I

SOME ADVANTAGES OF CLONAL CELL LINES

Viability
Functional and genetic homogeneity
Quantities sufficient for biochemical studies
Availability; easy exchange among investigators
Possibilities for quantification
Possibilities for cell synchronization
Possibilities for genetic manipulation
Potential for controlled mixing experiments among cell types
Potential for use in detecting cellular constituents unique to the nervous system.

*Quite possibly the final answer to this question won't come until we understand the mechanism by which the nervous system operates so well that it will become quite academic.

TABLE II

SOME ALTERNATIVES TO CLONAL CELL LINES AND THE EXTENT TO WHICH THEY PROVIDE THE "DESIRED PROPERTIES" OF TABLE I

Alternatives	Viable Cells	Permanently Established Lines	Biochemically Differentiated Cells	Functionally and Genetically Homogeneous Populations	Quantities Sufficient for Biochemical Studies	Characterization with Respect to Specific Cell Type	Normal Regulation Intra- and Intercellularly
Clonal Lines	Yes	Yes	Yes	Yes	Yes	No, or at best uncertain	Doubtful and not really known
Primary Culture (explant and dissociated)	Yes	No; but could provide a starting point	Yes	Not generally, but methods are being developed in this direction.	Yes/No	Potentially	Potentially in some cases
Organ Culture (including reaggregation systems)	Yes	No	Yes	No	Yes/No	No	Potentially
Bulk Dissociation and Purification	Yes/No	No, but could provide a starting point	Yes	Possibly	Yes	Yes, ideally	No?
Hand Dissection of Specific Cells	Yes/No	No	Yes	Possibly	Only with great difficulty	Yes, ideally	No?
Localization *in situ* of specific cell types using specific staining or autoradiographic techniques.	Yes	No	Yes	"Yes" relative to the extent that the tagged function is cell type specific.	Cytochemistry	Potentially	Yes
Brain Slices	Yes/No	No	Yes	No	Yes	No	Yes?
Brain Homogenates	No	No	Doubtful	No	Yes	No	No

II. Glial Cell Function

In order to define problems of nervous system function to which clonal lines of glial cells might best be applied, as well as to measure the extent to which glial cell lines resemble their normal *in vivo* counterparts, let us consider briefly some of the suspected functions of glial cells in the central and peripheral nervous system. For the most part the reader is referred to other reviews for specific references (Kuffler and Nicholls, 1966, 1976; Bunge, 1968; Kuffler, 1967).

A. Classification

The neuroglia, or "nerve glue" as Virchow named these cells in the mid-19th century, are a group of satellite cells that surround most neurons, and outnumber them, perhaps as much as ten to one. They comprise about one-half of the bulk of the mammalian brain. A few main classes are currently recognized.

Astrocytes are either *fibrous* filament-containing cells found chiefly among bundles of myelinated nerve fibers, or *protoplasmic* with less fibrous material and located more abundantly in gray matter around nerve cell bodies, axons, and synapses. *Ependymal cells* line the inner surface of brain in the ventricles. *Oligodendrocytes* (central nervous system) and *Schwann cells* (peripheral nervous system) surround smaller axons, and elaborate to form myelin sheaths around larger axons, producing a high resistance covering which increases conduction velocity. Specialized close contacts occur between the membranes of axons and the myelin wrapping, apparently limiting the longitudinal spread of current.

B. Glial Properties

On the negative side, glial cells do not have axon-like processes, do not form synapses, are inexcitable by electrical current, and do not produce impulses.

On the positive side, they have high resting membrane potentials of up to -90 mV in vertebrates (higher than that of the neurons which they surround); are depolarized by potassium accumulation in their immediate environment (such as occurs when neighboring neurons discharge impulses); have a high internal potassium concentration; and are, at least in certain species, linked by gap junctions forming contiguous arrays of coupled cells with substantial continuity, thus allowing current flow among glia away from cells depolarized by increases in external potassium concentrations.

Glial cells are separated from neurons by an extracellular space in the form of a characteristic narrow cleft of about 20 nm containing a low resistance fluid. Glial-neuronal low resistance pathways are apparently absent. Although glial cells do not seem to be necessary for short-term survival of neurons or for impulse conduction in primary culture, the manner of impulse signalling is different in naked neurons upon excitation by a train of stimuli. This is apparently due to the absence of an accumulation of external potassium that would otherwise occur in the intercellular clefts between the axonal and glial membranes, affecting as a result the slower afterpotentials (Kuffler and Potter, 1964; Baylor and Nicholls, 1969).

C. Glial Functions

1. Structural. For many years glial cells were in general regarded as simply structural elements in the nervous system, a sort of packing material around the more delicate instrument of impulse conduction. In fact, we have already discussed two widespread examples of this sort, myelin formation and neuron-glial cleft formation in which the mere (although carefully controlled) physical presence of glial cells appears to be of major importance regardless of any subsequent glial metabolism.

In more recent times, speculations on the metabolic roles for glial cells have diversified in many directions.

2. Neurotransmitter metabolism. There is evidence that at least certain types of glial cells can take up GABA from the environment, and subsequently release it (Orkand and Kravitz, 1971; Schon and Kelly, 1974 a, b; Bowery and Brown, 1972; Iversen and Kelly, 1975). Dennis and Miledi (1974) have obtained evidence for acetylcholine release by Schwann cells. Thus glia may supply neurotransmitters and/or regulate their extracellular levels.

3. Nutritive roles, transfer of substances to neurons. In addition to possible withdrawal and supply of neurotransmitters, glial cells may supply a variety of as yet unknown nutritive and regulatory substances. For example, in the absence of satellite cells, neurons in dissociated nerve cells from sympathetic and dorsal root ganglia reversibly lose the ability to synthesize acetylcholine (Patterson and Chun, 1974). In addition, tumor cell lines of glia have been shown to produce several growth or neurite elongation factors (see below).

4. Developmental role. During the growth and development of the nervous system there are several examples in which neurons appear to migrate along glial processes. This has led to the hypothesis that glia may help to define neuronal organization (Rakic, 1971; Sidman and Rakic, 1973).

5. Repair and Regeneration. If glial cells do in fact guide or influence the positioning of neurons, they would be expected to have a role in regeneration after injury as well. In addition, the ability of glia to divide in the adult leads to another function. After loss of neurons due to aging or injury, glial cells proliferate and invade vacant spaces, forming a scar tissue. Phagocytic activity in glia cells may permit them to help clean up a site of injury.

D. Glial-Neuronal Signalling

If glial cells carry out functions such as supplying nutritive and regulatory factors to neurons, the existence of a signalling system between the two cell types would seem necessary. Indeed, there do appear to be possible mechanisms for this.

For example, many glial cells have β-receptors which permit them to respond to β-agonists such as norepinephrine with a resultant increase in intracellular cyclic AMP. The subsequent role for this cyclic AMP is still largely a matter of conjecture (see below), but activation of enzymes leading to the synthesis of specific glial factors remains a possibility.

Perhaps one of the most elegant possibilities for neuron-glial signalling, however, is via the glial sensitivity to potassium. The number and frequency of nerve impulses is reflected in the potassium concentrations in the intercellular clefts, which in turn determine glial membrane potentials. Since potassium is apparently not confined to special structures nor peculiar to either excitatory or inhibitory transmitters, this represents of course a rather generalized response. Nevertheless, it seems possible that potassium induced depolarization could be linked to intracellular metabolism such as enzyme induction, membrane transport, or secretion of specific substances. For example, increasing the concentration of external potassium to 15mM increases the uptake and metabolism of labeled glucose in glial cells (Salem *et al.*, 1975). In addition, raising the concentration of potassium to 20 mM or higher stimulates the release of GABA from satellite cells in sympathetic and dorsal root ganglia.

E. Conclusions

Thus for the purposes of this chapter, at a minimum, it will be assumed that glial cells can and do perform specific metabolic roles in the nervous system beyond their important but rather passive contributions through delineation of the neuronal space via myelination and cleft formation. The question is, then, to what extent *can* glial lines growing in culture contribute to our understanding of glial function, and to what extent have they actually done so.

III. Development of Clonal Lines

Following principles originally laid down by Sato and co-workers (Sato and Yasumura, 1966), the general approach to establishing cell lines which retain the ability to express organ-specific biochemical functions in tissue culture involves five principal steps (Pfeiffer, 1973) (Table III). Each of these steps will be considered briefly in the context of glial lines.

TABLE III

STEPS FOR DEVELOPING DIFFERENTIATED CLONAL LINES

1. Recognition of organ-specific biochemical functions expressed at the cellular level.
2. Development of relatively rapid and sensitive assays for these functions (usually biochemical or serological).
3. Development of an initial source (such as a transplantable tumor) of the organ or cell type under investigation.
4. Development of cell lines in culture.
5. Selection of clonal lines that retain the functions under consideration.

1. *Glial specific markers* (Step 1-2 in Table III). The difficulty in accurately identifying clonal lines with respect to cell type probably reflects two main factors. First, the cloned cells may be relatively immature or uncontrolled in their expression of differentiated properties, giving them an atypical set of properties that may even reflect mixed neuronal-glial characteristics. Second, there is a substantial controversy as to the specificity of most of the biochemical glial "markers" currently in use. It is even not generally agreed which cells in the brain express a given marker. Although this is one of the questions that one might have hoped could be answered using cloned glial lines, the circularity of the argument quickly becomes apparent.

Some of the more common biochemical markers for glial cells are summarized below.

a. *S100 protein* is an acidic nervous system specific protein described by Moore (1965, 1972) (see section V,*C* of the chapter for further discussion of its properties and for references). While it is thought to be a general glial marker, its location in neurons as determined by immunohistochemical methods has also been reported. Further, Schubert *et al.* (1974) detected S100 in clonal lines that on the basis of their electrical properties were considered to be neuronal*. Since its function is unknown, it is measured by serological methods such as complement fixation (e.g., Pfeiffer *et al.*, 1970).

b. *2′, 3′-cyclic nucleotide 3′-phosphohydrolase.* This enzyme, bound to membranes (both plasma and intracellular membranes; see Pfeiffer, 1973),

*This is a prime example where clonal lines do not help to answer a question about cellular characteristics in the normal brain, since we cannot know whether these "neuronal" clones make S100 because neurons do that sort of thing or because these "neuronal" tumor lines are poorly regulated and in fact display neuronal-glial (gleuronal?) characteristics.

has been a very useful marker for myelination. Nevertheless, it is present at relatively high specific activity in a number of non-myelinating glial lines (Zanetta *et al.*, 1972; Pfeiffer and Wechsler, 1972). Although it is often assayed by thin layer chromatographic demonstration of the 2′-AMP produced from 2′, 3′-cyclic AMP (see Glastris and Pfeiffer, 1974), a new assay based on the hydrolysis of 2′, 3′-cyclic NADP has significant advantages (Sogin, 1976; Guha and Moore, 1975).

c. *Glycerol phosphate dehydrogenase* (GPDH) is regulated by hydrocortisone in the brain and in clonal lines in culture (McGinnis and de Vellis, 1974; see also J. de Vellis, this volume). Recently P. J. Leveille and J. de Vellis (personal communication) have demonstrated immunohistochemically that it is located primarily in oligodendroglia in the rat central nervous system. Thus, this may be one of the more specific markers available. However, G. D. West and co-workers were recently surprised to find hydrocortisone inducible GPDH in three lines which on the basis of the presence of the veratridine-sensitive action potential ionophore were presumed to be neuronal (G. D. West, J. Uki, R. Stahn and H. R. Herschman, unpublished results). In view of some current reservations as to the specificity of the veratridine-stimulated sodium flux for neurons (Villegas, *et al.*, 1976), the source of the inconsistency must await further study. An assay for this enzyme has been described by de Vellis and Inglish (1973).

d. *Glial fibrillary acidic protein* (GFA) has been shown to be an astrocyte marker by virtue of immunofluorescent staining of both normal and tumor material (Eng *et al.*, 1971; Bignami *et al.*, 1972; Bignami and Dahl, 1974; Bissell *et al.*, 1974; Dahl and Bignami, 1976). It is probably one of the better glial markers available.

e. *β-receptors.* The response of many glial lines to catecholamines with an elevation of intracellular 3′, 5′-cyclic AMP (cAMP) has made this a useful marker for a specific class of surface receptors. A common form of the assay for β-agonist induced cAMP elevation is that described by Gilman and Nirenberg (1971); more recently Mukherjee *et al.* (1975) have used ^{3}H-alprenolol for β-receptor identification. This area is discussed further in this chapter in section V,*A*.

f. *Myelin basic protein* could be a particularly important marker for oligodendroglia and Schwann cells in view of its specificity for myelin. It can be determined either by the induction of experimental allergic encephalomyelitis (or neuritis), or more conveniently by radioimmunoassay (Sundarraj *et al.*, 1975; Sheffield and Kim, 1977).

g. *Specific lipids.* Among the wide variety of lipid components, galactosylceramide (cerebroside) and its 3-0-sulfate ester (sulfatide) are particularly useful as markers for glial lines in view of their presence in myelin. Identification of these lipids can be made by thin-layer chromatography.

h. *Other markers.* There are numerous other markers that could be used as well (see E. Bock, this volume). For example, carbonic anhydrase is considered to be primarily a glial enzyme (e.g., Cammer *et al.*, 1976) but thus far it has found relatively little use for the characterization of glial lines. A number of other components also appear to be unique to, or at least highly enriched in, the nervous system; these include the NSA-1 glycoprotein [found in astrocytes (Maunoury *et al.*, 1976)] and the NS-1 surface antigen (Schachner; 1974; see also Miani *et al.*, 1976 for further references).

2. *Development of an initial source of the desired cell type* (Step 3 in Table II). The primary culture from which the clonal line will eventually be isolated can originate either from normal material or from a tumor. Most of the cell lines established to date have been derived from tumors in order to take advantage of the loss of growth regulation in such cells, thus greatly improving the chances for unlimited growth *in vitro.* Tumors have been obtained for this purpose from four principle sources: (1) so-called "*spontaneous*" transformations *in vivo* (e.g., human brain tumors) or *in vitro* during repeated passage; (2) use of *general chemical carcinogens* such as methyl-cholanthrene (Zimmerman and Arnold, 1941) that will produce tumors generally in the treated animal; (3) use of *organotropic carcinogens* (Druckrey *et al.*,1969; Druckrey,1971; Wechsler *et al.*, 1973) which in the case of methyl- and ethylnitrosourea are organotropic for the nervous system, and which have been used for producing tumors from which many of the cell lines listed in Table III were derived (see below); (4) *viral transformation* either *in vivo* (e.g., Kumanishi *et al.*, 1973) or *in vitro* (e.g., Shein, 1968; de Vitry *et al.*, 1974; Benda *et al.*, 1977). It is expected that a combination of cell separation methods, primary culture techniques, and various forms of cellular transformation will lead to better defined clonal lines in the future.

3. *Clonal Cell Lines.* The techniques for establishing cells in culture and isolating clonal lines with interesting functions are varied, and clearly beyond the scope of this chapter (Buonassisi *et al.*, 1962: Pfeiffer and Pollack, 1971). The critical factors here are the eventual isolation of *clonal* lines with the consequent advantages discussed in the introduction, followed by adequate *karyotypic* and *biochemical characterization.*

IV. Glial Lines

Numerous glial cell lines described in the literature along with some of their properties are summarized in Table IV (see also Richelson, 1974, and Varon, 1975b). Some of these will be discussed in more detail.

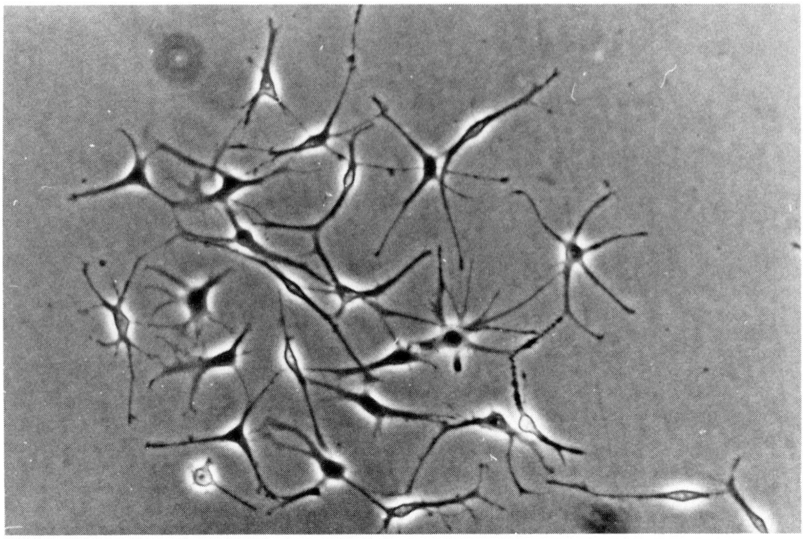

FIG. 1. Cell line C6-la grown in MEM supplemented with 15% fetal calf serum.

A. *Rat Glial Lines:*

C6: While most of the lines listed in Table IV are referred to in subsequent sections of the chapter in regard to specific aspects of glial research, rat line C6 has been far and away the most studied. As such it is a standard against which other lines can be compared. In addition to the original clone of C6 established by Benda *et al.* (1968, 1971), numerous subclones have been developed, including RG6 (Davidson and Benda, 1970); $C_6$2B (de Vellis and Brooker, 1973); C61a (Pfeiffer and Eagle, 1976; Fig. 1); C_6-BU-1 (bromodeoxyuridine resistant; Hamprecht *et al.*, 1976); C_6TK^- (bromodeoxyuridine resistant; Schimmer *et al.*, 1974); C_6-16 (EMS derived), G10, 9F, 11D, 9E, and 2B (de Vellis and Inglish, 1973); and Cl_2 (spontaneous 6-thio-guanine resistant; Skaper and Seigmiller, 1976). In addition to studies of cyclic AMP metabolism, neurotransmitter enzymes and transport, glial growth and differentiation factors, S100 protein synthesis, ion fluxes and electrical properties, and lipids (areas covered in more detail below), C6 cultures have been used to examine aspects of somatic genetics (Davidson and Benda, 1970; Benda and Davidson; 1971); acid mucopolysaccharide synthesis (Dorfman and Ho, 1970); nucleoside metabolism (Gibson *et al.*, 1974); a variety of enzyme activities [2′,3′-cyclic nucleoside 3′-phosphohydrolase (Zanetta *et al.*, 1972; Pfeiffer and Wechsler, 1972); aldolase C. (Kumanishi *et al.*, 1975); pyruvate kinase (Tolle *et al.*, 1976); 5′-nucleotidase (Stefanovic *et al.*, 1975, 1976c)]; glial fibrillary acidic protein (Bissell *et al.*, 1975); vitamin B12 deficiency (Barley *et al.*, 1972);

surface antigens (Pfeiffer *et al.*, 1971; Day and Bigner, 1973; Joseph and Oldstone; 1974); hormone induction of enzymatic induction and glucose uptake (Shinwari and de Vellis; 1972; see J. de Vellis; this volume); effects of sialic acid removal on antigenicity (Herschman *et al.*, 1972); tubulin polymerization (Wiche and Cole; 1976); and myosin synthesis (Miller and Kuehl, 1976).

33B and 21A: The lines mentioned above may have been developed from the Wistar-Furth strain, though the strain was not stated in the original description. A number of other rat lines (e.g., 33B and 21A) were induced with ethylnitrosurea in Wistar-Furth strain rats (Fields *et al.*, 1975). Their primary use has been in the study of nervous system specific and tumor antigens.

RN2 (and subclone RN22) (Pfeiffer and Wechsler, 1972; Wechsler *et al.*, 1977a): This rat line was derived from a peripheral neurinoma in BD-IX strain rats and has been described as possibly being an immature Schwann cell. It produces a basic membrane protein with some characteristics which originally suggested a similarity in size (SDS-acrylamide gel electrophoresis) and amino acid composition to myelin basic protein (MPB) from beef brain and to the larger component from rat brain. This protein gave weak positive reactions with antisera against beef brain MBP obtained from two different laboratories in subcutaneous anaphylaxis and microprecipitin tests (Pfeiffer and Wechler, 1972; Pfeiffer, 1973). Its characterization was subsequently pursued further (N. Sundarraj and S. E. Pfeiffer, unpublished results) using material from cells grown both *in vitro* and *in vivo* as clonal tumors. In extracting the protein from RN2 cells care was taken to disrupt the cells in such a way as to eliminate most of the nuclei, and thus the basic histones, early in the procedure. Using antisera prepared against MBP purified from beef brain, rat brain (both fast and slow components), or rabbit peripheral nerve (two sizes of MPB) (N. Sundarraj and S. E. Pfeiffer, unpublished results) together with a sensitive radioimmunoassay (Sundarraj *et al.*, 1975), we assayed for the presence of MBP in acid-extracts of the non-nuclear fractions, using brain and nerve samples as positive controls. Although the microprecipitin test gave occasional weak positives with RN2 basic protein, we were unable to observe any positive reaction with radioimmunoassay. Correspondingly, we were unable to elicit either experimental allergic encephalomyelitis or neuritis in Lewis rats with either tumor homogenate, acid extracts, or purified RN2 tumors. Although two out of five animals injected with acid extract did develop hind leg paralysis, subsequent histological examination did not reveal any accompanying demyelination (U. Greesbach, W. Wechsler, and S. E. Pfeiffer, unpublished results). We concluded that RN2 basic protein is not myelin basic protein. Tests by Dr. L. Greene (Harvard Medical School) on primary chick ganglia cultures indicate that it is also not the basic protein nerve growth factor, and

TABLE IV
GLIAL CELL LINES AND SOME OF THEIR PROPERTIES

Cell Line	Species	Strain	Source	Cell Type Description	Chromosome No.	Some Differentiated Properties	Cloned
C6[a]	Rat	Wistar-Furth? (Not stated in original description)	MNU-induced CNS tumor	Astrocyte	41	S100, beta receptors, CpNase, neurotransmitter transport, many others	Yes
$C2_1$[a]	Rat	Wistar-Furth?	MNU-induced CNS tumor	Astrocyte	—	Trace of S100	Yes
$C2_2$, $C2_3$, C3[a]	Rat	Wistar-Furth?	MNU-induced CNS tumor	—	—	—	Yes
33B[b]	Rat	Wistar-Furth	ENU-induced lumbar spinal cord and roots tumor	Oligodendroglioma	—	—	Yes?
21A[b]	Rat	Wistar-Furth	ENU-induced cerebral hemisphere tumor	Glial?	—	—	Yes?
RG-179[c]	Rat	Fisher	Multiple passages of five-day old rat brain culture	Glial	—	—	No?
RN-2[d]	Rat	BD-IX	ENU-induced peripheral neurinoma	Peripheral Neurinoma; Schwann cell?	43	S100, CpNase, unique collagen	Yes
B9, B11, B12, B15, B23, B38, B49, B82, B90, B92, B108 and others[e]	Rat	BD-IX	ENU-induced independent CNS tumors	Presumptive glial	Near diploid	S100, neurotransmitter uptake, beta-receptors, others	Yes

A15A5 A15A10[f]	Rat	BD-IX	ENU-induced CNS mixed glioma	Glial	—	Hydrocortisone induction of GPDH	Yes
R11, R113, R102, C143, Cx2 and others[g]	Rat	CDF	ENU-induced CNS tumor C172, oligodendroglioma	Glial	Near Diploid	S100, CpNase, beta-receptors, sulfatide (R11, Cx2)	Yes
A6, C1 and others[g]	Rat	CDF	ENU-induced CNS tumor C164, oligodendroglioma	Glial	Near diploid		
A, B, C, E, F, G-series[h]	Rat	?	ENU-induced tumors	Glial	—	Various, see text	Yes
CHB[i]	Rat**	—	?	Glial	—	S100	Yes
SATO[j]	Mouse	C59BL	Methylocholanthrene induced CNS glioma	Glioblastoma	—	S100, aldolase C	No?
G26-15, 19, G26-20, 24[k]	Mouse	C57BL/6	Methylcholanthrene-induced CNS tumor	Oligodendroglioma	60-62	S100, beta-receptors, NS-1, sulfatide (-20 and -24)	
G26-28, 29[k]	Mouse	C57BL/6		Glioma	Heteroploid		Yes
TR6B[l]	Mouse	C3H/Ho	Trigeminal nerve and brain	Schwannoma	—	—	Yes?
393VF[l]	Mouse	C3H	SV40 transformed hypothalamic primary culture	Glial	Subtetraploid	Hydrocortisone induction of GPDH, T-antigen	Yes
NN[m]	Hamster	Syrian	Spontaneous transformed newborn hamster brain primary culture.	Astrocyte	40% tetra- & octoploid	Fibrous cell processes	Yes

Table IV (Cont'd)

Cell Line	Species	Strain	Source	Cell Type Description	Chromosome No.	Some Differentiated Properties	Cloned
P5, P8[r]	Hamster	Syrian	Polyoma virus transformed fetal brain culture.	Astrocyte	—	Fibrous cell processes	?
THA15[q]	Hamster	Syrian	SV40 transformed fetal brain culture	Astrocyte	—	Astrocyte morphology, fibrous processes	?
138MG[p]	Human	—	Grade III astrocytoma-glioblastoma	Bipolar, glial, astrocytic?	Hypotriploid†	S100, GFA*	No
251MG[q] 343MG[r] 178MG[r]	Human	—	Malignant glioma Malignant glioma Grade III temporal astrocytic glioma	Glia-like	— 75† 45†	—	No†† No No
CG-series[s]	Human	—	Juvenile and adult normal brain biopsies	Astrocyte-like, glia-like	Diploid	—	No
1181N1 1181D1 1321N1[t, u]	Human	—	Cerebral glioma multiforme	Tumor astrocytes	Hyperploid, variable number	Stellate morphology, beta-receptors	Yes
EH118MG[p, v]	Human	—	Glioblastoma, Rous virus transformed 1181N1 in culture	Astrocyte	—	—	No
HFNG5[u]	Human	—	Three month old fetal brain.	Astrocyte	—	—	No?

TC526 TC593[w]	Human	—	Glioblastoma multiforme	Glioma	Aneuploid	Ganglioside pattern similar to brain.	No?
Various[x]	Human	—	Glioblastoma	Glioma	—	—	No?
SACF61[y]	Human	—	Human fetal primary culture	Astrocyte	—	NSA-1 antigen	No?
Glioma[z]	Human	—	Glioblastoma	Malignant glioma	—	GFA	No?
Various[aa]	Human	—	Glioblastoma	?	—	?	Yes

*Abbreviations: MNU: methylintrosourea; ENU: ethylnitrosourea; CNS: central nervous system; CpNase: 2',3'-cyclic nucleotide 3'-phosphohydrolase; GPDH: glycerol phsophate dehydrogenase; GFA: Glial fibrilary acidic protein.
**originally reported as human (see text)
†G-banding studies see Mark *et al.*, 1974a,b,c.
††now available in clonal form

[a]Benda *et al.*, 1968; [b]*Fields et al.*, 1975; [c]Wilson *et al.*, 1972; [d]Pfeiffer and Wechsler, 1972; [e]Schubert *et al.*, 1974; [f]Claisse and Roscoe, 1976; [g]S. E. Pfeiffer, W. Wechsler and N. Sundarraj, unpublished results; [h]G. J. West, J. Uki, R. Stahn and H. R. Herschman, unpublished results; [i]Lightbody *et al.*, 1970; Stoolmiller *et al.*, 1972; [j]Sato *et al.*, 1972; [k]Sundarraj *et al.*, 1975; [l]*Benda et al.*, 1977; [m]Shein *et al.*, 1970; [n] Shein, 1970; [o]Shein, 1967; [p]Ponten and MacIntyre, 1968 (other lines also described); [q]Westermark, 1973; [r]Westermark *et al.*, 1973 (43 different lines listed); [s]Lindgren *et al.*; 1975; Pontén *et al.* 1969; Brunk, 1973; [t]MacIntyre *et al.*, 1969b; [u]Perkins *et al.*, 1971; [v]MacIntyre *et al.*, 1969a; [w]Manuelidis, 1965 (see also Manuelidis, 1969); Manuelidis and Manuelidis, 1976); [x]Kornblith *et al.*, 1969; [y]Maunoury *et al.*, 1976; [z]Chen and Mealey, 1973; [aa]Bigner and unpublished results (personal communication from B. Westermark).

its identity remains undetermined. It may be significant that, to our knowledge, none of the clonal lines currently available produce true myelin basic protein (see, however, Volpe *et al.*, 1975, and comments in section *V,E* of this chapter). Possibly, myelin basic protein appears only in cells that are actively engaged in myelination of a fiber, an activity which has also not been described for clonal lines of glia.

RN2 has also been used for studies of collagen and procollagen synthesis (Church *et al.*, 1973), measles virus infection (Nakamura *et al.*, 1975); 5-hydroxytryptamine transport (Haber *et al.*, 1975); cyclic nucleotide regulation (Sheppard *et al.*, 1975); cell surface immunology (Joseph and Oldstone, 1964; Fields *et al.*, 1975) and the subcellular distribution of 5′-nucleotidase and 2′,3′-cyclic nucleotide-3′-phosphohydrolase (Pfeiffer, 1973). Line RN-1 carries the surface antigen Thy-1 (R. T. Acton and S. E. Pfeiffer, unpublished experiments) and synthesizes substantial amounts of prostaglandins E and F (L. Levine, unpublished results).

B9-B108 (Schubert *et al.*, 1974): This rat series, developed at the Salk Institute, constitutes a large, well-characterized series of nervous system cell lines. Neuronal, glial, and muscle cell types are included but the lines indicated in Table IV are all presumptive glial cells. They were induced in BD-IX rats using ethylnitrosourea. Of approximately 120 clonal lines isolated from independently arising tumors, 22 were originally characterized in detail (Schubert *et al.*, 1974). They are being studied extensively with regard to a wide variety of neurochemical phenomena (see ensuing sections), including neurotransmitter synthesis and uptake (Schubert, 1975; Schubert *et al.*, 1975), dibutyryl cAMP induced changes in morphology (Schubert, 1974; Klier *et al.*, 1975), surface antigens and veratridine-stimulated sodium flux (Joseph and Oldstone, 1974; Stallcup and Cohn, 1976; Fields *et al.*, 1975); cAMP metabolism (Schubert *et al.*, 1976), and protein secretion (Schubert, 1976).

Lines from CDF rat tumors (S. E. Pfeiffer, W. Wechsler and N. Sundarraj, unpublished results): In the development of several of the lines shown in Table IV, BD-IX rats were used because of the somewhat higher incidence of nervous system tumors induced in this strain. Subsequently, using ethylnitrosurea, (W. Wechsler, S. E. Pfeiffer and N. Sundarraj, unpublished results), we induced nervous system tumors in CDF rats since this strain is more readily available to laboratories around the world (there are very few BD-IX colonies in existence). In addition, we chose only two tumors, both from the central nervous system, with the intention of deriving as many distinct cell types as possible from each tumor. In fact our experience has been that even CNS tumors that appear fairly homogeneous by histological examination often yield a large number of apparently different cell types when introduced into cell culture. Since some of these appear only after several months of continuous culture with careful monitoring and selection procedures (to avoid overgrowth by the initially more numerous and/or vigorous cell types), it remains an interesting question whether each cell type existed already in the original tumor (although perhaps in very low numbers) or if they are differentiation products of stem cells present in the tumor (the latter would seem to remain a viable possibility since the chemical carcinogenesis occurs in fetal animals). Thus from one CNS tumor (C172) described histologically as an oligodendroglioma (Wechsler *et al.*, 1977), we derived fifty clonal lines which on the basis of morphological and biochemical criteria represent at least eight cell types. In contrast, 12 clonal lines derived from another CNS glioma (C164) all seem quite similar. In addition to containing S100 protein and β-receptors, and displaying CpNase activity, some of these clones produce interesting lipids (e.g., R21, R113; Dawson *et al.*, 1977), and Thy-1 antigen (R. Morris and S. E. Pfeiffer, unpublished results), and two presumptive neuronal lines synthesize immunoreactive thyrotropin releasing factor (Grimm-Jörgensen *et al.*, 1976).

A-G lines from Herschman's Laboratory (G. J. West, F. Uki, R. Stahn and H. R. Herschman, unpublished results): Thirty-five rat cell lines were established from 32 independent rat tumors induced by ethylnitrosourea. A number of these cell lines were classified as presumptive glial cells on the basis of presence of S100 protein or cortisone-induced GPDS activity. However, the authors point out the difficulty of authoritatively classifying tumor lines of this sort beyond "presumptive" glial or neuronal.

CHB (Lightbody *et al.*, 1970): This cell line was originally described as being of human origin and was used for studies of S100 protein synthesis (Herschman, 1971), neurotransmitter enzymes (Wilson *et al.*, 1972), ultrastructure (Weinstein and Kornblith, 1971), and lipids (Dawson *et al.*, 1971; Synder and Brody, 1970). Subsequently it was reported that evidenced by karyotype and isoenzyme patterns of three enzymes these cells are of rat origin (Stoolmiller *et al.*, 1972). Since they seem clearly not to be C6 glial cells, the origin and source of the confusion remains unclear. A review of the original chromosome studies has been reported to indicate a human karyotype as originally stated (J. Lightbody, personal communication), but apparently no new karyotypic examinations of freshly thawed vials of very early freezings of this line have been reported. Although its pedigree is thus in question, line CHB does possess some interesting differentiated characteristics of glial cells.

B. Mouse Glial Lines

G26-series (Sundarraj *et al.*, 1975) (Fig. 2, Table V): Six biochemically differentiated clonal lines were established from a transplantable glioma (G26) of the C57BL/6 inbred mouse strain (Zimmerman, 1955). These lines

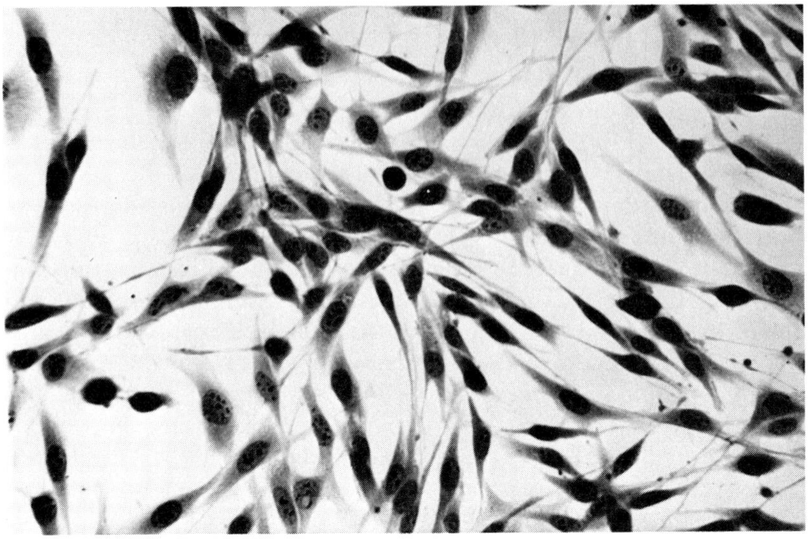

FIG. 2. Cell line G26-20.

produce S100 protein, have NS-1 surface antigen (Schachner, 1974), express CpNase activity, and have β-adrenergic receptors. The level of expression of these activities is sensitive to the environmental conditions under which they are grown (cell culture versus tumor; Table V), an occurrence which we have noted with several lines and which adds an interesting complication to characterizing cell lines (see Grimm-Jörgensen *et al.*, 1976). In addition, G26-20 and -24 synthesize the myelin lipids galactosyl ceramide and sulfatide (Dawson *et al.*, 1977).

393VF (Benda *et al.*, 1977): This mouse line was developed by SV40 transformation of primary cultures of hypothalamic cells derived from 14 day old fetuses. It responds to hydrocortisone treatment with an increase in the activity of GPDH. Tumors have been induced from this line only in athymic nude mice. It has a subtetraploid chromosome number (median 74) characteristic of most transformed lines.

C. Hamster Glial Lines

NN (Shein *et al.*, 1970): This hamster line was derived from a spontaneously transformed culture of newborn hamster brain. On the basis of morphology, including metallic impregnation techniques for observing fibrous cell processes (Shein, 1965) it is considered to be of astrocytic origin. NN cells are currently in use for studies of glial energy metabolism (Ciesielski-Treska *et al.*, 1976), lipids (Robert *et al.*, 1975, 1976; Eichberg *et*

TABLE V
DIFFERENTIATED BIOCHEMICAL FUNCTIONS OF G26 CLONAL LINES GROWN IN CULTURE AND OF CLONAL TUMORS GROWN SUBCUTANEOUSLY IN MICE

	S-100 protein*		Phosphohydrolase†		Intracellular cAMP levels‡		
					Untreated	Isopro-terenol-	Norepin-ephrine-
Tissue	In vitro	In vivo	In vitro	In vivo	control	treated	treated
Primary tumor	—	0.03	—	0.4	—	—	—
G26-15	0.2	<	<	0.1	33	68	—
					8	—	23
G26-19	0.2-0.5	0.01	<	0.4	20	—	486
					22	340	—
G26-20	0.2-0.9	<	<	0.3	14	1740	2000
G26-24	0.1-0.4	0.02	<	0.4	17	6420	—
					55	3160	—
G26-28	<	<	<	0.7	38§	113§	111§
G26-29	<	—	<	—	28§	216§	646§
C6 (rat astrocytoma)	0.1-0.3	—	0.9	—	33§	9150§	7420§
Mouse brain	—	0.65	—	4.5	—	—	—

*Expressed as percent of total water-soluble proteins in 700,000 *g*-min supernatant fraction; sensitivity of the assay is 0.01 μg of S-100 protein per ml.< indicates less than 0.01% S-100 protein of total soluble proteins.

†2′:3′-Cyclic nucleotide 3′-phosphohydrolase; μmol of substrate converted per min/mg of protein. Sensitivity of the assay is 7.5 nmol/min.< indicates no phosphohydrolase activity with samples of 1-3 mg of total protein per ml.

‡pmol/mg of total protein.

§Assayed directly from acid extracts without prior partial purification. The variations in the intracellular 3′:5′-cyclic AMP concentrations for the same cell line among experiments appear to be partly due to differences in cell density and growth phase at the time of catecholamine treatment. Except for G26-19 and G26-28, the approximate magnitude of the stimulations were reinforced for each clone by several additional earlier assays performed without prior purification. From Sundarraj *et al.*, 1975.

al., 1971), ethanol-induced modifications (Noble *et al.*, 1976), Na^+, K^+-ATPase (Embree *et al.*, 1971a,b; Mandel *et al.*, 1974), chloride uptake and Na^+ and K^+ content (Gill *et al.*, 1974; Lees and Shein, 1970), free amino acid content (Mokrasch, 1971), glial-neuronal interactions (Stefanovic *et al.*, 1976a; Ciesielski-Treska *et al.*, 1973), and ecto-ATPase (Mandel *et al.*, 1974, Stefanovic et al., 1976b).

D. Human Glial Lines

In general, less is known about the nervous system specific biochemical capacities of the available glial lines derived from human sources.

Nevertheless there are a substantial number of human glial lines of potential interest. Only a few are described further here.

138MG (Pontén and MacIntyre, 1968; Westermark *et al.*, 1973); Pontén, 1971, 1975). This bipolar astrocyte-like line was derived from a grade III astrocytoma-glioblastoma. It is perhaps the best characterized of the human lines, and produces both S100 and GFA protein (Edström *et al.*, 1973; Walum, 1975). It has found application in studies of glial glucose uptake (Edström *et al.*, 1975b, 1976; Walum and Edström, 1976), growth characteristics and oxygen consumption (Walum *et al.*, 1974; Lindgren *et al.*, 1975), tumor-specific membrane antigens (Wahlström *et al.*, 1974), and ATPase activity (Agren *et al.*, 1971a,b). Characteristics of other lines derived in a similar fashion are often also described in these reports (e.g., 87MG, 251MG, 343MG, 373MG, 787MG) (see also Westermark, 1973). All the MG (malignant glia) lines were derived from tumors. The corresponding CG (control glia) cells from non-transformed brain tissue have a limited life span (generally 15-30 passages) (Pontén and MacIntyre, 1968).

1181N1 and 1181D1 (Perkins *et al.*, 1971). These lines are two morphological variants of a cell line derived from a human glioblastoma multiforme. They have, for example, been used in studies of cyclic nucleotide metabolism (Perkins *et al.*, 1971).

Summary — We can now modify the conclusion of Benda *et al.* (1968), so hesitantly stated, to read "Experiments show that in rats, mice, humans, hamsters and presumably most or all mammals, serially propagated brain cells are capable of performing numerous organ-specific functions over a period of years."

V. Some Specific Areas of Glial Cell Metabolism Studied in Clonal Lines

A. Cyclic 3′,5′-Adenosine Monophosphate

Cyclic 3′,5′-adenosine monophosphate (cAMP) has been shown to be involved in the regulation of growth and proliferation (Sheppard and Bannai, 1974), morphological changes (Prasad and Kumar, 1974), macromolecular synthesis (Wicks, 1974), and hormonal control of cellular metabolism (Robison *et al.*, 1968) in a variety of systems (see also K. N. Prasad, this volume). In recent years an enormous body of evidence has accumulated which implicates cyclic nucleotides, and especially cAMP, in the regulation of nervous system function (for reviews see Rall and Gilman, 1970; Robison *et al.*, 1968; Gilman, 1972; Greengard, 1975, 1976), and suggests a role for cAMP as a possible mediator or "second messenger" in the control of glial function. In this section, we shall examine the relationship between levels of cAMP in glial cells and some of the proposed physiological roles of these cells in the nervous system, emphasizing the use of homogeneous populations of cells derived from cultured glial tumors as *in vitro* correlates of *in vivo* processes.

Most of the functions postulated for glia (see above) would necessitate some form of intercellular communication between neurons and glia. It has been proposed that cAMP may be involved in this process (Gilman and Nirenberg, 1971; Perkins *et al.*, 1971; Schimmer, 1971, 1973). Adenyl cyclase, phosphodiesterase (PDE) and cyclic nucleotide-dependent protein kinase activities were shown in early studies to be present in both the rat C6 and human 1181N1 astrocytoma-derived clonal lines (Jard *et al.*, 1972; Perkins *et al.*, 1971). Thus, these glial lines possess all the necessary enzymatic activities associated with cAMP metabolism. When C6 cells were tested for the effect of neurotransmitters on intracellular cAMP levels, it was found that they responded to norepinephrine and isoproterenol with approximately 200-fold increases in the intracellular content of cAMP (Gilman and Nirenberg, 1971) and actively secreted cAMP into the medium (Doore *et al.*, 1975; Penit *et al.*, 1974). This increase could be prevented by inhibitors of the β-receptor such as sotalol, but not by α-blockers, suggesting the presence of a functional β-receptor. Similarly, cAMP content can be increased in C6 cells by prostaglandin E_1 (Hamprecht and Shultz, 1973). The 1181N1 human astrocytoma line likewise responded to the β-adrenergic agonists norepinephrine and epinephrine as well as to adenosine (Clark and Perkins, 1971; Clark *et al.*, 1974). These responses in cultured tumor cell lines are similar to those observed in primary cultures of fetal rat brain (Gilman and Schrier, 1971) and in reaggregation cultures of fetal brain cells (Seeds, 1973). Other presumptive glial cell lines have since been shown to respond to catecholamines with elevations of intracellular cAMP contents (Sundarraj *et al.*, 1975; Schubert *et al.*, 1974; 1976, S. E. Pfeiffer, W. Wechter and N. Sundarraj, unpublished experiments, 1977), supporting the idea that the presence of catecholamine receptors may be a general phenomenon among glial cell types. These data lend support to the hypothesis that neurotransmitters released as a result of synaptic activity can stimulate β-receptors on the surface of surrounding glia, and that the resultant increase in cAMP mediates metabolic processes in glial cells which can modify neuronal-glial interactions. Although a mechanism for the proposed interaction is still under discussion, changes in cAMP level are associated with morphological changes, enzyme induction, expression of cell surface antigens, glucose metabolism, and altered β-receptor function in these cells, any or all of which may affect the response of glial cells to neuronal activity.

1. Morphological effects. One of the more obvious changes accompanying exposure of glial cells to agents which elevate cAMP levels is a morphological transformation of the cells from the flattened irregular shape characteristic of cultured glial lines to a bipolar or astrocyte-like morphology with extensive process formation similar to that seen in normal brain astrocytes. In several clonal glial lines (rat C6, rat RN22, human 138MG, rat B9, B15 and B82), stimulation with dibutyryl cAMP (Steinbach and Schubert, 1975; Hamprecht *et al.*, 1973; Sheppard *et al.*, 1975; MacIntyre *et al.*, 1972); catecholamines (Schlegel and Oey, 1975; Edström *et al.*, 1974, 1975a) or prostaglandin E_1 (PGE_1) (Edström *et al.*, 1974, 1975a) evokes a morphological response which can be potentiated by PDE-inhibitors such as theophylline (Schlegel and Oey, 1975), papaverine or RO-20-74 (Edström *et al.*, 1975a). The β-adrenergic response is inhibited by sotalol (Edström *et al.*, 1975a) or propanolol (Oey, 1975), while the morphological effect can be reversed by colchicine (Oey, 1975) or prolonged incubation with the agonist (Edström *et al.*, 1975a). The magnitude of the effect shows a positive correlation with the level of intracellular cAMP (Steinbach and Schubert, 1975; Oey, 1975). Interestingly, even cells which have been enucleated by cytocholasin B treatment, demonstrate the morphological effect when stimulated by norepinephrine (Oey, 1975). Thus, agents which exert their primary effect by elevating intracellular levels of cAMP may trigger a return to "normal" morphology in cultured glial cells in the absence of direct nuclear involvement, and the change may be maintained by high levels of intracellular cAMP acting in some way with intracellular microtubules. However, it should be emphasized that treatment with agents such as bromodeoxyuridine can lead to similar morphological changes in C6 cells in the absence of cAMP elevation (Schwartz *et al.*, 1973; Breckenridge, 1975). The morphological changes are thus more complicated than they might at first appear and there is probably no simple direct correlation between cAMP levels and morphological alterations in these cells.

2. *Enzyme activities.* Alterations of intracellular cAMP levels are known to affect protein synthesis in many experimental systems (Wicks, 1974), and changes have been demonstrated in the specific activity of several enzymes in glial cell cultures exposed to agents which elevate cAMP. In the RGC6 clone, for example (de Vellis and Brooker, 1973), treatment with isoproterenol or norepinephrine leads to a rapid, transient rise in cAMP levels, followed by an increase in lactic dehydrogenase (LDH) activity. This effect is prevented by propanolol (a β-blocker), but not by dibenzyline (an α-blocker), indicating that the effect is mediated via the β-adrenergic receptor in these cells. In addition, dibutyryl cAMP at millimolar concentrations (de Vellis and Brooker, 1972) can maximally induce LDH, while compounds such as histamine and dopamine have no effect on either cAMP levels or LDH induction. RNA and protein synthesis seem to be required for the induction since treatment with actinomycin D or acetoxycycloheximide will prevent the increase in LDH activity. To achieve LDH induction, maximal elevation of cAMP levels is necessary during the

initial period of RNA synthesis (up to 3.5 hours) but not during protein synthesis.

In contrast to LDH induction, hydrocortisone induction of GPDH in C6 cells does not seem to be mediated by cAMP, and agents such as dibutyryl cAMP or norepinephrine do not have any effect on the specific activity of this enzyme (deVellis and Brooker, 1973).

In C6BU-1 cells, the induction of ornithine decarboxylase is correlated with increased cAMP levels following exposure to norepinephrine, isoproterenol, or dibutyryl cAMP and theophylline (Bachrach, 1975).

In contrast, long term treatment of C6 cells with 1 mM theophylline apparently leads to decreases in the specific activities of enzymes involved in lipogenesis. However, this response may not be entirely due to increased cAMP levels as a result of the phosphodiesterase inhibition evoked by theophylline since dibutyryl cAMP treatment of the cells only partially reproduces the theophylline effect (Volpe and Marasa, 1976b).

The induction of both the low and high K_m forms of cAMP phosphodiesterase in C6 cells also appears to be correlated with intracellular cAMP levels, suggesting that cAMP can regulate the expression of enzymes involved in its own metabolism, thus allowing long term control of the cellular levels of this cyclic nucleotide (Schultz *et al.*, 1972; Browning, 1973; Weiss *et al.*, 1973; Browning *et al.*, 1974a,b; Schwartz and Passoneau, 1974). Norepinephrine or dibutyryl cAMP treatment leads to induction of PDE activities by 4 hours, and this increase can be prevented by actinomycin D or cycloheximide. It is not known whether protein synthesis is necessary for the increased synthesis of PDE itself, or of some other protein responsible for activation of the enzyme. It has been suggested that induction of PDE by elevated levels of cAMP might be a long term regulatory mechanism for controlling responsiveness of brain glia to catecholamines during periods of increased neuronal activity. Refractoriness to repeated hormonal stimulation may be due, in part, to induction of PDE activity over the long term (see below).

Little is known, in any of these systems, about the exact mechanism whereby enzyme induction or repression occurs, or indeed, whether increased cAMP levels are a cause of, or a concomitant result of, the expression of differentiated functions in these cells. It is known, for example, that increases in cAMP levels are also correlated with phosphorylation of specific proteins in the nervous system (see reviews by Greengard, 1975, 1976). However, the specificity of the phosphorylation, and the functions of the phosphorylated proteins, are not clear. For example, in synaptic membrane-enriched fractions from rat cerebral cortex, the phosphorylation of two proteins (molecular weights 86,000 and 48,000 daltons) seems to be cAMP-dependent (Johnson *et al.*, 1972; Ueda *et al.*, 1973), and the larger protein (protein I) may be involved in generation of the post-synaptic potential (Greengard, 1975, 1976). A cAMP-dependent phosphorylation of a 300,000 dalton microtubule-associated protein (MAP_2) of chick brain by an intrinsic protein kinase of microtubules has also been reported, suggesting an association of cAMP levels with microtubule function (Sloboda *et al.*, 1975). Activation of brain tyrosine hydroxylase, the rate-limiting enzyme in catecholamine biosynthesis appears to be regulated by cAMP via the action of a protein kinase (Morgenroth *et al.*, 1975). In glial cell lines cAMP-dependent protein kinase activity has been reported using histone as an exogeneous acceptor (Salem and de Vellis, 1976; Schlaeger and Köhler, 1976), and the phosphorylation of a 100,000 dalton protein present in nuclear and microsomal fractions of C6-2B cells seems to be related to norepinephrine stimulation (Salem and de Vellis, 1976). However, the relationship between phosphorylation of glial cell proteins and the physiological role of glial cells in brain function is still a mystery.

3. Surface antigens. In addition to its presumed role in the regulation of protein synthesis, cAMP may be involved in the expression of surface antigens in glial cell lines. A nervous system-specific cell surface antigen(s), NS-1 (Schachner, 1974), has been detected in several clonal lines derived from the G26 mouse glioma (Sundarraj *et al.*, 1975). These cell lines exhibit elevations of cAMP in response to treatment with isoproterenol or norepinephrine and the level of NS-1 seems to be inversely correlated with intracellular cAMP content (Sundarraj *et al.*, 1975). NS-1 expression is highest in clonal tumors, lower during *in vitro* growth of the cells, and further depressed by treatment of the cultured cells with dibutyryl cAMP or catecholamines. In contrast to NS-1 expression, H-2 antigen expression in these glial lines is unaffected by alteration of intracellular cAMP. Thus glial cell lines are potentially useful for studies of antigenic expression in nervous system cells and the possible mediation of this process on a molecular level by intracellular cAMP.

4. Neuronal nutrition. One of the possible functions of glial cells is thought to be as a nutrient or metabolite source for surrounding neurons. Thus, the control of glycogen metabolism by neurotransmitters has been investigated in cultured glial cell lines.

Opler and Makman (1972) have shown that the glycogen content of C6 cells is reduced upon stimulation by dibutyryl cAMP or catecholamines and that propanolol can block both glycogenolysis and the catecholamine-stimulated rise in cAMP. Norepinephrine has been shown to inhibit the uptake of glucose in these cells, and if cells which have been preloaded with isotopically labeled glucose, are stimulated by norepinephrine, there is an increased release of label into the surrounding medium (Newburgh and Rosenberg, 1972). Thus, neurotransmitter-induced glycogenolysis in glial cells has been suggested as a means of providing glucose or lactate to neighboring neuronal cells (Opler and Makman, 1972). The role of cAMP in mediating the conversion of phosphorylase *b* to phosphorylase *a* during glycogenolysis is known (Robinson *et al.*, 1968), and in C6 cells transformation from the inactive *b* to the active *a* form occurs during the first 30 seconds after norepinephrine treatment when cAMP levels are elevated approximately 10-fold above basal levels. The rate of glycogenolysis then declines and by 3 hours most of the phosphorylase activity is in the *b* form (Browning *et al.*, 1974a). Pretreatment of the cells with norepinephrine for 3 hours reduces the cAMP response to a subsequent norepinephrine challenge, but again conversion of phosphorylase *b* to the *a* form occurs and glycogenolysis is stimulated. Thus, desensitization of the β-receptor to repeated catecholamine stimulation does not seem to significantly affect glycogenolysis. This would suggest that control of glycogenolysis in C6 cells is either mediated by factors in addition to cAMP levels, which may be responsible for interconversion and maintenance of the two forms of phosphorylase (Browning *et al.*, 1974a), or that only minimal changes in cAMP levels are necessary for triggering the phosphorylase conversion.

It has recently been shown in C6 cells that regulation of glycogen metabolism is related not only to changes in cAMP level but also to the available energy supply (Passonneau and Crites, 1976). The activities of glycogen synthetase and phosphorylase can vary according to the glucose content of the medium. As glucose levels in the medium decrease, phosphorylase *a* activity rises and intracellular glycogen decreases; the reverse is seen when the medium glucose level is high. These changes occur in the absence of altered cAMP levels. However, in agreement with earlier findings, elevation of intracellular cAMP levels by treatment with norepinephrine or isobutyl methylxanthine (IBMX) also leads to increased phosphorylase *a* activity and glycogen breakdown. Thus it would seem that while cAMP-mediated activity is associated with altered glycogen metabolism under certain conditions in these cells, it may not be the only mechanism by which this can be achieved. The evidence does, however, lend support to the hypothesis that neurotransmitters, produced as a result of neuronal activity, may provide a form of neuronal-glial communication which activates subsequent glycogenolysis by glial cells and the release of glucose into the extracellular space.

5. Receptor-agonist Interactions. Perhaps one of the most fruitful areas of investigation in which glial cell lines have proven useful as a model system is in the study of β-adrenergic receptor-agonist interactions and the control of the response to catecholamines (Perkins *et al.*, 1975). Astrocytes surrounding the adrenergic synapses in brain are repeatedly exposed to catecholamine stimulation (Breckenridge, 1975; Perkins *et al.*, 1975), and it is known that over- or underexposure of the pineal gland to catecholamines can affect the sensitivity of the β-receptor (Perkins *et al.*, 1975; Deguchi and Axelrod, 1972, 1973). Several groups have investigated this catecholamine-induced refractoriness or subsensitization in an attempt to determine whether it is a result of decreased synthesis or increased degradation of intracellular cAMP.

If C6 or 1181N1 cells are continually exposed to norepinephrine or isoproterenol for periods of up to three hours, an initial rise in cAMP is observed (30-200 fold above basal level), this is followed by a gradual decrease in intracellular levels of this nucleotide with time (de Vellis and Brooker, 1974; Su and Perkins, 1974; Johnson and Perkins, 1975; Browning *et al.*, 1976; Su *et al.*, 1976a,b). On subsequent challenge with the agonist, there is a decreased response, and cAMP accumulates only slightly above basal levels. If the cells are washed free of the catecholamine, 60-70% responsiveness is recovered by 5-7 hours (Su *et al.*, 1976a). This effect on cAMP levels has been shown not to be due to increased excretion of cAMP into the culture medium or to degradation of the neurotransmitter (Browning *et al.*, 1974a; Schwartz and Passonneau, 1974). In some glial lines, treated cells have increased PDE activity (Browning *et al.*, 1976; Schwartz and Passonneau, 1974), and the rate of degradation of cAMP in these cells is faster (Browning, 1974; Browning *et al.*, 1976), but desensitization can be only partially reversed by the PDE inhibitor IBMX (Browning *et al.*, 1976; Leichtling *et al.*, 1976). In addition, the 2B subclone of C6, which does not show PDE induction by norepinephrine, nevertheless can be made refractory to subsequent challenge (deVellis and Brooker, 1974). This suggests that while PDE induction and degradation of cAMP may be involved in long term regulation of cAMP levels, it is probable not the sole mechanism responsible for the development of refractoriness.

Decreased activity of adenyl cyclase has also been demonstrated in catecholamine-treated cells (Brooker and deVellis, 1975; Browning *et al.*, 1976; Su *et al.*, 1976b), and desensitization is highly agonist-specific (Leichtling *et al.*, 1976; Su *et al.*, 1976a). In pulse labeling experiments, Su *et al.* (1976b) have shown that there is a decreased rate of incorporation of label into cAMP during exposure to norepinephrine; they concluded that decreased adenyl cyclase activity is involved in the response, but that the agonist specificity of the desensitization response precludes an exclusive effect on a common component of the system such as cAMP. They suggest that alteration of the specific receptor and/or its coupling to adenyl cyclase is more likely.

Recently, desensitization of the β-receptor of frog erythrocytes has been shown to involve a loss of specific β-adrenergic binding sites with no change in the apparent affinity of the remaining sites for the ligand (Mickey *et al.*, 1975; Mukherjee *et al.*, 1975; Caron and Lefkowitz, 1976). This effect has been demonstrated in several hormone-receptor systems (Lesniak and

Roth, 1976) and it has been suggested that the same mechanism may be applicable to the β-receptor in glial cells as well (Su *et al.*, 1976b). Subclones of C6 cells which do or do not express the adenyl cyclase response to catecholamines have been reported (Gilman and Minna, 1973), and the progeny resulting from hybridization of β^+ and β^- parental clones are β^-, indicating a negative form of genetic control over the expression of this response. Binding studies using [^{125}I]-iodohydroxybenzylpindolol (IHYP), a potent β-adrenergic antagonist, have confirmed that the decreased responsiveness to catecholamines seen in β^- clones is paralleled by a reduction in the number of specific binding sites for IHYP (Maguire *et al.*, 1976). In fact, it has recently been suggested that changes in cAMP levels *per se* may not regulate desensitization since C6 cells treated with concentrations of isoproterenol too low to cause appreciable changes in intracellular cAMP can still exhibit desensitization upon further challenge with the drug (Franklin and Twose, 1976). Desensitization could therefore involve agonist-regulated loss of specific surface receptors and concomitant loss of receptor-coupled adenyl cyclase activity in these cells, although the available data indicate that desensitization is a complex process and may be controlled at several levels (Lefkowitz, 1976).

It is evident from these studies that cAMP has been implicated in several aspects of glial cell metabolism and neuronal-glial communication. However, the cellular and molecular basis for its involvement in the regulation of these phenomena remains to be investigated. The use of clonal lines of glial cells as model systems should lead to better understanding of glial cell function and the role of cAMP in these processes.

B. Glial Growth and Differentiation Factors

Extensive communication from neurons to glia as well as from glia to neurons is likely to occur in the nervous system (Kuffler and Nicholls, 1966; Hydén, 1967; Gilman and Nirenberg, 1971). This could occur both electrically and ionically (Grossman and Seregin, 1977) and biochemically, and could in some cases involve direct cell-cell contact. Studies in primary culture have emphasized the importance such interactions could have in the regulation of nervous system biochemistry. For example, muscle cells (Giller *et al.*, 1973) or non-neuronal cells (Patterson and Chun, 1974) markedly stimulated choline acetyltransferase in cultured neurons from spinal cord and sympathetic ganglia, respectively.

Glial cell lines are currently being used to detect factors which influence neuronal growth and differentiation, endothelial growth during tumorigenesis, and the glial cells themselves. A summary of the different factors discussed is presented in Table VI.

TABLE VI
DIFFERENT FACTORS PRODUCED BY GLIAL CELL LINES OR AFFECTING GLIAL GROWTH AND DIFFERENTIATION

Factor	Origin	Properties	Reference
NGF NGF-like factor	male submaxillary gland, mesenchymal tumors, snake venom glioma, glial cells?	involved in neuronal growth, neurite attachment	Varon, 1975a Murphy *et al.*, 1977
EPF	C_6, NB-44, CNS tumor	endothelial cell growth	Kelly *et al.*, 1976
GLF	C_6 cell line rat brain	neuroblastoma C 1300 differentiation	Monard *et al.*, 1973, 1975 Schuerch-Ratgeb and and Monard 1976
GMF	pig brain extract	primary glial cell maturation	Lim *et al.*, 1973, 1975, 1977 cf. also Sensenbrenner *et al.*, 1972
EGF	submaxillary glands	enhancement of epidermal growth, stimulation of normal human astroglia growth	Westermark and Wasteson, 1975
FGF	bovine pituitary gland	stimulation of normal human astroglial growth	Westermark and Wasteson, 1975
OSF	osteosarcoma	stimulation of normal human astroglial growth	Westermark and Wasteson, 1975
Somato-medin B	human plasma	stimulation of normal human astroglial growth	Westermark and Wasteson, 1975

NGF = nerve growth factor, EPF = endothelial proliferation factor, GLF = glial factor, GMF = glial maturation factor, EGF = epidermal growth factor, FGF = fibroblast growth factor, OSF = osteosarcoma factor.

The C6 cell line was shown to release a factor (GLF in Table VI) into the medium which induces process formation in a neuroblastoma cell line, which normally has a rounded morphology (Monard *et al.*, 1973). The evoked effect is linearly dependent on the amount of factor used, but not correlated with an increase in intracellular cAMP. Since the factor was found in rat brain primary cultures independent of the developmental age, it must be present in normal tissue as well and thus is not a tumor-specific product (Schürch-Ratgeb and Monard, 1976). The failure to induce morphological differentiation in other nervous tissues suggests that the factor is either very specific for one cell type, or that the chosen cell types are unable to respond because they have already fully expressed their differentiation. Monard *et al.* (1975) have reported that the C6 factor is not related to nerve growth factor (NGF) because anti-NGF does not crossreact with the glial factor, the SDS gel electrophoresis patterns are different, and the glial factor induces differentiation only in neuroblastoma, whereas NGF has no effect on the neuroblastoma cells.

The involvement of glial cells in the maintenance of neurons has been reported by Varon (1975a). In normal dissociated ganglion cultures non-neuronal cells form a background layer. If these cells, identified as glial cells, are eliminated, exogenous NGF has to be added to support the neurite formation and attachment of the neurons. Supposing an involvement of the glia cells as a source of NGF, Varon *et al.* (1974a,b) treated them with anti-NGF before presenting them to neurons. This treatment impaired the neuron supporting activity of the glial cells, suggesting that they supply NGF-like molecules. NGF, which has been isolated from mouse submaxillary glands, mouse sarcomas and snake venom, acts on a variety of sympathetic neurons and on dorsal root ganglia (DRG) (containing sensory neurons) (Levi-Montalcini, 1966). In contrast, the NGF-like molecules produced by the glial cells in the experiments of Varon and his co-workers seem to have a much higher specificity since only glial cells co-cultured with neurons from the same DRG source can substitute for NFG. Thus, it seems that either NGF isolated from submaxillary glands has a broader specificity than the molecule produced by the glial cells, or that the glial cells are able to produce it only when the correct signal from the nerve cell is provided. Such a narrow specificity would be in agreement with the specificity of the glial factor found in C6 cells described above. Certainly, the demonstration of factors in different glial cell populations which show a high specificity towards homologous origin tissues would be of considerable interest, and may be an area in which clonal glial lines could play a role.

Longo and Penhoet. (1974) detected a protein in rat glioma crossreacting with antibody against 2.5 S NGF. The protein induced dorsal root ganglia to extend neurites, and influenced the morphological appearance of neuroblastoma cells. It is not clear exactly what kind of glioma produced the NGF-like molecule. Since mouse sarcoma and other mesenchymal tumors produce NGF, it has to be ruled out that the gliomas were of mesenchymal origin or at least contaminated with mesenchymal cells. Nevertheless, it seems probable that glial cells can supply NGF-like molecules, as seen above. Recently Murphy *et al.* (1977) have reported that C6 cells make NGF.

An endothelial-proliferative factor (EPF) has been reported to be produced by neural cell lines. Human endothelial cells *in culture* were treated with conditioned medium from central nervous system tumors and clonal cell lines (Suddith *et al.*, 1975b; Kelly *et al.*, 1976). Conditioned medium from all the tumors tested produced an increase in the percent of ^{3}H-labeled nuclei (with ^{3}H-thymidine in the medium) in the endothelial cell cultures. The highest stimulation was found with medium from an angioblastic meningioma and a neuroblastoma tumor. Three clones of the C6 cell lines were tested; two clones had a high capacity to condition the medium whereas one induced a three-fold diminished activity. Since no data were shown correlating stimulation with the growth of the tumor cells, the differences may represent differences in the capacity of the tumors to condition the medium due to varying growth capacities rather than to a property of a specific cell type. It remains to establish a relationship between the endothelial growth factor and the tumor angiogenesis factor (Folkman *et al.*, 1971). The use of the C6 cell line is expected to be helpful in isolating the factor since this cell provides a homogeneous cell population and has a high stimulating activity.

Knowledge about the differentiation and development of the glial cells is rather restricted (see chapters by A. Privat, S. Fedoroff and L. Hertz, this volume). It is therefore interesting that Sensenbrenner *et al.* (1972), M. Sensenbrenner (this volume) and Lim *et al.* (1973, 1975), have reported that a factor extracted from brain can induce morphological alterations in glial cells in primary cultures. The dissociated embryonic brain cells used generally have a flat epithelial-like morphology. Addition of the partially purified brain factor induces the cells to retract around the cell bodies and to extrude processes. Under the experimental conditions used by Lim and co-workers, the effect starts around 8 hours after addition and peaks at 20 hours. The cells were identified as glia on the basis of the presence of glial fibrillary acidic protein and rhythmic pulsation (Lim *et al.*, 1975, 1976, 1977). An increase in cAMP and S100 protein is found after a lag of at least 24 hours (Lim *et al.*, 1977). An increase of DNA synthesis during the first 24 hours suggests that the cells undergo a rather extensive mitotic cycle before the increase in cAMP and S100 can be detected. It might be postulated that the factor induces cell division in the glioblast population which leads to differentiated glial cells. Thus it would be interesting to study the effects of other treatments that induce cell division in growth controlled cell populations.

In another series of studies (Westermark *et al.*, 1975; Westermark and Wasteson, 1975, 1976; Lindgren *et al.*, 1975), glial cell cultures were developed from normal, adult human brain (specimens from operations for non-malignant diseases). These cultures could be passaged 15-30 times. The normal glial cells have a pronounced topoinhibition as shown by a low terminal cell density at which proliferaton is virtually suppressed and the cells arrested in G_1 phase (Pontén *et al.*, 1969). Using this cell system, a variety of growth promoting substances were tested (lower half of Table VI). The effect of the growth factors was scored by following the increase of tritiated thymidine incorporation into G_1-arrested cells. Fibroblast and epidermal growth factor, as well as somatomedin B, stimulated the glial cells to divide. A variety of tumor cells were tested for the ability to produce glial-growth factor using conditioned medium from such cultures as the stimulating agent. From four glioma lines, one glial line, one line of human embryonic lung fibroblasts, and one osteosarcoma line, only the last one showed a growth promoting activity. The produced factor is nondialyzable and relatively heat-stable. It is entirely destroyed by trypsin.

The possibility of initiating quantifiable glial-neuronal interaction using mixed cell cultures has been investigated using M1 neuroblastoma cells and the NN hamster line (Ciesielski-Treska *et al.*, 1973; Stefanovic *et al.*, 1976a; P. Mandel, this volume). When M1 and NN cells are co-cultured, M1 undergoes morphological changes, and the activity of Mg^{2+}- and Ca^{2+}-dependent ecto-ATPase activity increases up to five-fold. Even after reisolation of the two cell types, higher ATPase activity was observed through 48 passages or more. It would be interesting to test the generality of this phenomena, since the data could be interpreted as an irreversible induction of differentiation. One important control experiment would be to repeat the study using M1 neuroblasts, or NN astroblasts, co-cultured with a fibroblast (or other non-neuronal line). Pfeiffer *et al.* (1971) observed a very pronounced stimulation of the measurable serological activity of a C6 antigen upon co-culture of C6 glia and C3 cells (which are not thought to express nervous system properties).

The ultimate co-culture of neurons and glial cells might be, of course, the formation of glial-neuronal hybrids. Since this seems to be an unlikely mechanism of communication in the normal brain, it perhaps is an area of study more pertinent to somatic geneticists. Nevertheless, since glial-neuronal hybrid lines have been isolated that express nervous system properties not recognized in either parent, the possibility cannot be dismissed that hybrid formation in culture is in fact a technical tool for overcoming some of the deficiencies of clonal lines, and inducing neurons and glial cells to do intracellularly what they normally do *in vivo* via the transfer of soluble factors intercellularly. Thus, a hybrid was made between the C6 glial line and the N18 neuroblastoma line and named NG108 ($6 \times 18 = 108$) (Amano *et al.*, 1974; Hamprecht, 1974; B. Hamprecht, M. Nirenberg and T. Amano, unpublished experiments). This hybrid has a high activity of choline acetyltransferase comparable to brain levels while neither parent line has appreciable activity. It has morphine receptors, again in spite of their absence on either parental line, leading to its use in opiate and encephalin research (Klee, Nirenberg, 1974; Traber *et al.*, 1975a,b; Sharma *et al.*, 1975a,b). After prolonged treatment with opiates, effects resembling tolerance and withdrawal phenomena occur (Traber *et al.*, 1975a,b; Klee *et al.*, 1975; Sharma *et al.*, 1975a,b; Lampert *et al.*, 1976). Recently, synapse formation with cultured muscle cells has also been reported to occur with NG108 at relatively high frequency (Nelson *et al.*, 1976; P. G. Nelson, this volume; Puro and Nirenberg, 1976); R. N. Rosenberg *et al.* (this volume) have detected one hybrid protein not obviously present in either parent.

The cited examples of the use of glial cell lines suggests that there is an enormous potential of information available in these lines relevant to growth and differentiation factors. Isolation and purification of the different factors and a comparison of their effects could provide a basis for a better understanding of the role of glial cells in the nervous system.

C. Synthesis and Degradation of S100 Protein

Glial cell lines offer a system in which the accumulation, synthesis, and degradation of S100 protein can be studied in relationship to growth. Experiments on the synthesis and degradation of S100 protein in glial lines, rat and mouse brain, brain slices, and cell-free systems will be discussed. Problems associated with the measurement of S100 synthesis and degradation, which in some cases question the validity of the conclusions drawn in these experiments, are also considered.

S100 protein, named for its solubility in 100% saturated ammonium sulfate at neutral pH, is a small, acidic, nervous system-specific protein (Moore, 1965). Although it is found primarily in glial cells (Hydén and McEwen, 1966; Benda *et al.*, 1968; Moore, 1972) there is still controversy as to whether any S100, localized immunohistochemically, is located in neurons (e.g. Matus and Mughal, 1975; Hagflid *et al.*, 1974, 1975; Ludwin *et al.*, 1976). A membrane-bound fraction of S100 has been reported (Rusca *et al.*, 1973). Bovine S100 has a molecular weight of 19,500 to 21,000 and consists of 3 or 4 subunits (Dannies and Levine, 1969; Stewart, 1972a,b). Although the function of S100 is unknown, it has been implicated in learning (Hydén and Lange, 1970a,b), ion transport (Calissano *et al.*, 1974), regulation of genomic expression (Michetti *et al.*, 1974, 1976), and synaptic function (Haglid *et al.*, 1974).

Synthesis and degradation of S100 in vivo. Originally a half-life of 6 hours for S100 was determined *in vivo* in rat brain (McEwen and Hydén, 1966), while later experiments indicated a half-life of 16-days (Cicero and Moore, 1970).

The rate of S100 synthesis in developing mouse brain increases 4-fold during days 16 to 21 of postnatal development (Stewart and Urban, 1972). The calculated half-life ranged from 3 to 12 hours. The investigators obtained unusual results of increases in the rates of both synthesis and degradation during the period when S100 content increased and a decrease in these rates when the accumulation stopped. Experiments in which RNA synthesis was inhibited with actinomycin D indicated that an increase in the stability of S100 mRNA during accumulation contributed to the increase in the synthesis of S100 (Stewart, 1975).

S100 protein in clonal cell lines. For the C6 line, S100 constitutes about 0.2% of the soluble protein in stationary phase culture. S100 accumulates only during stationary phase under conditions of homologous cell contact (Pfeiffer *et al.*, 1970; 1971), and this accumulation can be markedly influenced by a variety of ethanesulfonic acids often used as buffers (Pfeiffer and Eagle, 1976). In contrast, in CHB cells the amount of S100 per culture accumulates during the rapid logarithmic phase of growth as well as after the cells reach confluency (Lightbody *et al.*, 1970). After adaptation to serum-free medium, both of these cell lines retain similar levels of S100 production (Donta, 1973, 1975), while treatment of C6 cells with aminopterin was reported to enhance S100 accumulation (Silbert and Goldstein, 1972).

Unlike the C6 cell line, in which neither S100 accumulation nor synthesis is increased by serum deprivation of log phase cultures (Pfeiffer *et al.*, 1970; Labourdette and Marks, (1975), a human glioma cell line (138 MG) appears to have a higher level of S100 protein after exposure to serum-free medium (Edström *et al.*, 1973).

Synthesis and degradation of S100 in clonal cell lines. Herschman *et al.* (1973) concluded that increased synthesis contributes to the accumulation of S100 in the C6 cell line during stationary phase. They calculated that the relative rate of S100 synthesis, defined as cpm in S100 protein divided by cpm in total soluble protein, was 10 times greater in stationary cultures than in logarithmic phase cultures, and that the rate of incorporation of label into S100 per cell was 5 times greater in stationary cells. Since the rate of total protein synthesis decreases in stationary phase, the actual increase in S100 protein synthesis will of course be less. The half-life of both S100 and total protein was 4 days during stationary phase; in contrast, S100 in CHB cells had a much longer apparent half-life of several weeks, while that for the total soluble protein was 3 days (Herschman, 1971).

Labourdette and Marks (1975), also concluded that the increased content of S100 in stationary C6 cells, is due primarily to increased S100 synthesis. The relative rate of S100 synthesis continually kept increasing during S100 accumulation, while the half-life was 3 days in both low-density and confluent monolayers.

S100 synthesis in cell-free systems. In a homologous cell-free system prepared from rat brain, S100 was synthesized primarily by free polyribosomes (Zomzely-Neurath *et al.*, 1973; Amaldi *et al.*, 1973). Polyadenylated polysomal RNA from rabbit brain directed the synthesis of S100 in cell-free extracts from wheat embryo (Mahony *et al.*, 1976).

Problems in the measurement of S100 synthesis and degradation. A major problem is that S100 is not a well-characterized protein, and is defined mainly by its interaction with antiserum against S100. Only one peak reacting with antiserum against S100 was found after purification of the protein by column chromatography under conditions preventing aggregation due to disulfide bond formation (Dannies and Levine, 1971); on the other hand, the reaction of antiserum against S100 with multiple peaks from DEAE-cellulose and Sephadex column chromatography and with multiple bands after electrophoresis has led to the suggestion that S100 is a class of proteins (Uyemura *et al.*, 1971; Gombos *et al.*, 1971; Stewart, 1972b; Abe *et al.*, 1974). Stewart (1972a) has suggested that the bovine S100 subunits, of which 6 were detected, can form different combinations which cause heterogeneity in the S100 proteins. Studies on the purification of bovine, porcine, and ovine S100 (Dannies and Levine, 1971; Uyemura *et al.*, 1971; Stewart, 1972a,b; Abe *et al.*, 1974) have led to contradictory results on its properties and subunit structure. Much less work has been done in S100 from the rat and mouse, which have been used for synthesis and degradation studies: Rat S100 does differ from bovine S100 in that it elutes at higher NaCl concentrations during DEAE-Sephadex chromatography and separates into only 1 instead of 2 different subunits after electrophoresis on an SDS-urea polyacrylamide gel or isoelectrofocusing urea polyacrylamide gel (J. L. Cook, unpublished results). More knowledge about the biochemistry of S100 and its heterogeneity is important for knowing which proteins are being isolated in particular assays and in understanding the synthesis of S100.

The S100 assays for the synthesis and degradation experiments done *in vivo* or with tissue slices involved acrylamide gel electrophoresis only (McEwen and Hydén, 1966) or ammonium sulfate fractionation and/or DEAE-cellulose chromatography in addition to electrophoresis (Cicero and Moore, 1970; Stewart and Urban, 1972; Wroński and van der Decken, 1975). The C6 and CHB cell lines were assayed by an immune precipitation followed by SDS acrylamide gel electrophoresis (Herschman, 1971; Herschman *et al.*, 1973; Labourdette and Marks, 1975). McEwen and Hydén (1966) isolated a class of S100 proteins, and it is likely that others did as well.

Purity of the isolated S100 is necessary for meaningful measurements of synthesis and degradation rates, and the final product was apparently in no case characterized by peptide mapping, for example, to confirm that only S100 was present. Contamination of the S100 isolated by an 11% acrylamide gel electrophoresis only (McEwen and Hydén, 1966) is likely. The S100 isolated by Cicero and Moore (1970) and Stewart and Urban (1972) was pure as judged by complement fixation or quantitative immune precipitation, although this method is not sensitive enough to detect the presence of a small amount of highly labeled contaminant. Another control, in which a tissue or cell line, not expected to synthesize S100, was assayed for incorporation of label into the S100 region of the final gel, showed impure S100 in the assay of Herschman *et al.* (1973), but suggested that the S100 was pure in the assays of Cicero and Moore (1970) and Labourdette and Marks (1975).

The yield of the isolated S100 has usually not been measured. An exception was the experiments on mouse development (Stewart and Urban, 1972), in which the yield was determined by measuring the specific activity of the protein in the S100 band and the total amount of S100 in the brain. Also, successive immune precipitations were used to show complete precipitation of the S100 from the cell lines (Herschman, 1971; Herschman *et al.*, 1973; Labourdette and Marks, 1975). However, loss of S100 was still possible in the assay containing a heat step at 70° for purification (Herschman, 1971; Herschman *et al.*, 1973).

The rate of incorporation of a labeled amino acid into S100 has been measured in order to determine the relative rate of S100 synthesis as a function of time during development (Stewart and Urban, 1972), nutrition (Wroński and von der Decken, 1976), or cell growth (Herschman *et al.*, 1973; Labourdette and Marks, 1975). It is important to know the relative specific activity of the amino acid precursor pool for protein synthesis under different conditions, which can be affected by the intensity of the amino acid transport and of amino acid metabolism. In the measurement of S100 synthesis during development, the specific activities of the free amino acid pools were used to calculate the rate of synthesis (Stewart and Urban, 1972); since protein restrictions did not affect the incorporation of labeled leucine into the free leucine pool, Wroński and von der Decken (1975, 1976) directly examined the rate of incorporation of label into proteins in cerebral tissue slices from rats fed a 20% or a 3% protein diet. By this method the rate of S100 synthesis appeared to be higher in rats fed the high protein diet. In the C6 cells no measurement of the specific activity of the pools was made. Comparison of the rate of S100 synthesis in logarithmic and stationary phase could only be made by measuring the absolute incorporation of radioactivity into S100 and assuming that the precursor pool was unaffected by growth rate (Herschman *et al.*, 1973), or by measuring the incorporation of radioactivity into S100 relative to total soluble protein (Herschman *et al.*, 1973; Labourdette and Marks, 1975). In this case the specific activity of the precursor could be assumed to be identical for S100 and total soluble protein, but the observed increase in relative S100 synthesis in stationary phase could be affected by decreased total protein synthesis as well as increased S100 synthesis. However, it is questionable at the present time whether even measuring the specific activity of the free amino acid pool is a valid method for correcting the rates of incorporation into protein, since it is unknown whether the free amino acid pool is actually the precursor pool (Goldberg and Dice, 1974). Experiments have led to conflicting models, in which the actual precursor pool has a specific activity either equal to (e.g., Li *et al.*, 1973), larger than (e.g., Hider *et al.*, 1969), or less than (Righetti *et al.*, 1971) the free amino acid pool.

The degradation rate of S100 has been measured by pulse-labeling rats or cultured cells with radioactive leucine, and measuring the rate of decrease in the amount of radioactive S100 following the initial labeling. Reutilization of labeled amino acids released by protein degradation for protein synthesis after the pulse-label can result in an apparent half-life that is too large. The half-life of 16 days that Cicero and Moore (1970) obtained for rat brain may have been overestimated because of reutilization. Since the specific activity of the amino acid pool in the mouse brain remained high after a pulse-label, Stewart and Urban (1972) avoided the problem of reutilization by calculating the rate of degradation from the rate of synthesis, and also by using a method in which the approach to maximum incorporation of label was measured. Both of these methods indicated much lower half-lives of 6 to 12 hours. In the C6 cells, Labourdette and Marks (1975) measured a half-life for S100 of 2 to 3 days in both the low density and confluent cultures. In cultured cells it should be possible to minimize the reutilization by adding large amounts of nonradioactive amino acid, although this has not been studied carefully (Goldberg and Dice, 1974). Also, the question arises whether the amount of reutilization changes with growth rate. One possible way of reducing reutilization is by using ^{14}C-1-glutamate (Goldberg and Dice, 1974). The ^{14}C is released into the bicarbonate pool after transamination followed by decarboxylation, or directly by glutamate decarboxylase present in the C6 cells (Schrier and Thompson, 1974).

Regulation of the level of S100 protein. An important question is the level at which the synthesis of S100 is regulated. It has already been mentioned that posttranscriptional regulation, in which the amount of S100 mRNA is increased by a decrease in mRNA degradation, probably contributes to S100 accumulation during development (Stewart, 1975). Another question concerns the biochemistry of S100 protein. If S100 is a class of proteins consisting of different combinations of chemically similar subunits, the subunits could either be synthesized independently or derived from a common precursor through proteolytic cleavage or modification. Knowledge of the biochemical relationship of the S100 subunits and proteins would be useful in studying possible processing and assembly of S100, and in determining whether different S100 proteins are regulated similarly at the transcriptional and translational levels.

The clonal cell lines provide a system in which the synthesis of S100 can be studied in one isolated cell type. In addition, the C6 cell line provides a system for examining the regulation of S100 accumulation as a function of cell growth and contact.

D. Neurotransmitter Metabolism

Glial cells are thought to be directly involved in inactivation of neurotransmitters and may also play a role in their production and release. Thus, although it is thought that a major mode of inactivation for most neurotransmitters may occur through re-uptake into presynaptic terminals (Axelrod, 1965; Iversen and Neal, 1968; Bloom and Iversen, 1971; Martin, 1973), subsequent data have revealed an important glial contribution to this process as well, perhaps especially in the case of the amino acid transmitters (Curtis and Johnston, 1970). For example, Henn and Hamberger (1971) measured the high efficiency of GABA uptake into semi-purified glial cells. Others, using autoradiographic techniques (Iversen and Schon, 1973) that have probably contributed most to our understanding of the phenomenon, have demonstrated GABA uptake into rat retina glial cells (Neal and Iversen, 1972), satellite glia in sensory ganglia (Schon and Kelly, 1974a,b) and Bergman glia (Hökfelt and Ljungdahl, 1971, 1972).

Data from similar studies using glial cell lines as well as glial cells in primary cultures (A. Schousboe, this volume) have been largely consistent with these ideas, and in addition have allowed the measurement of several kinetic parameters and the evaluation of the need for additional factors for optimal uptake and efflux. Evidence for uptake into glial lines of choline, GABA, glutamate, taurine, and other putative neurotransmitters has been obtained (Table VII). In addition, enzymatic activities related to both neurotransmitter synthesis and inactivation have also been observed (Table VIII).

It is expected that this is an area in which glial lines can continue to be useful in investigating various aspects of the uptake mechanisms (see also B. K. Schrier, this volume).

TABLE VII
SUMMARY OF SOME STUDIES OF PUTATIVE NEUROTRANSMITTER UPTAKE, SYNTHESIS, AND EXCRETION IN GLIAL LINES

Cell Line	Putative Neurotransmitter	Uptake Transport Properties	Intracellular Content or Rate of Synthesis	Ref.[a]
C6	Choline	Two uptake systems of high and low affinity. Km 12 μM. Stimulated by sodium.		16, 17
		K_T 2 μM, V_{max} 54 fmol/min/mg protein in prestarved cells.		5
		Km 0.7 μM, Vmax 0.1 nmoles/hr/mg protein; Km 24 μM, Vmax 1.4 nmoles/hr/mg protein.		10
	γ-amino-butyric acid	Km 0.2 μM, Vmax 1.4 /pmoles/min/mg protein. Sodium dependent, inhibited by amino-oxyacetic acid.		9 (see 4 also)
		K_T 32 μM, Vmax 23 /pmoles/min/mg protein for saturable component. Maximum concentration gradient of 4.5 fold.	Synthesis, 18 pmoles/min/mg protein. Also excreted into medium.	11
			1.6% of free amino acids.	15
	Glutamate	High affinity transport, Km 66 μM.		8
			Synthesis of 2 pmoles/min/mg protein from aspartate.	11
		High affinity uptake. K_T 10 μM, Vmax 385 nmol/min/mg protein.	30.9% of free amino acids.	15
	Taurine		20 pmol/min/mg protein from cystine.	11
			3.4% of free amino acids.	14
		High affinity uptake. Km 30 μM, Vmax 158 nmol/min/mg protein. Km 25 μM, Vmax 64 nmol/min/mg protein for subcellular fraction prepared in similar way as the synaptosomal fraction is prepared from brain tissue. Sodium and temperature dependent, sensitive to osmotic shock.		21

Table VII continued

C6, cont.	Trypotophan	High and low affinity uptake Km 7μM, Vmax 1.3 nmol/min/mg protein (for high), and Km 800 μM, Vmax 3.6 (low). Temperature dependent.		7
	β-alanine		0.3% of free amino acids.	15
	L-tyrosine	K_T 20 μM, Vmax 671 fmol/min/mg protein.		5
	L-phenylalanine	K_T 8 μM, Vmax 240 fmol/min/mg protein.		5
	Glycine		16.3% of free amino acids.	15
	Aspartate and asparagine		6.5% of free amino acids.	15
	Histamine		Negative (also negative in C2).	13
NN	Choline	High affinity uptake, inhibited by strychine, hexamethonium, and hemicholinium -3, Km 1.5 μM.		18 10
	Glutamate	Km 14 μM	+	1 23
	Aspartate	Temperature dependent, sodium requirement.	+	1 23
B11, 12, 15,23,49, 82, 90, 92, 108	γ-amino-butyric acid	High affinity uptake system is sodium dependent and temperature sensitive. Concentration intracellularly is calcium ion dependent.	0.1-0.6% of free amino acids.	14
	Glutamate	Uptake is sodium requiring and temperature sensitive.	8.4-27.4% of free amino acids.	15
	Taurine, beta-alanine, glycine, and aspartate	+	2.3-24.0% of free amino acids. 0.2-9.3 6.6-15.4 5.0-10.5	15
RN2	γ-amino-butyric acid	+	0.4% of free amino acids.	14 15
	Taurine, beta-alanine, glycine, aspartate		12.7% of free amino acids. 3.9 22.7 2.0	15
U251MG U787CG	α-amino-butyric acid	Km 1.5 mM and 1.4 mM, resp. Sodium ion dependent.		20

[a]For references see footnote to Table VIII.

TABLE VIII
SUMMARY OF SOME STUDIES OF ENZYMES INVOLVED IN NEUROTRANSMITTER SYNTHESIS AND MODIFICATION IN GLIAL CELL LINES

Cell Line	Enzyme	Specific Activity and Comments (pmol/min/mg protein)	Ref.[a]
C6	Acetylcholine-esterase	730; hydrolysis of acetylcholine.	3
		5000; hydrolysis of acetylthiocholine; estimated assuming 400 picograms protein per cell in log phase (see Pfeiffer *et al.*, 1970).	22
	Choline acetyl transferase	9	3
	Glutamic acid decarboxylase	8[b], 79[c]	3
		21[b]	6
	Monoamine oxidase	2600, with serotonin as substrate; inhibited by clorgyline; type A.	19
		Higher than in rat.	2
	Catechol-O-methytransferase	4, same as in rat brain.	13
		Higher than in rat brain.	2
	Histamine-N-methyltransferase	3, same as in rat brain.	13
$C2_1$	Acetylcholine-esterase	177	6
	Glutamic acid decarboxylase	36[b]	6
		5[b], 71[c]	3
		118[b], 348[c]	11
	Monoamine oxidase	Higher than in rat (also in C3)	2
	Catechol-O-methyltransferase	Higher than in rat.	2
		6, same as in rat.	13
	Histamine-N-methyltransferase	5, same as in rat.	13
RG-179	Acetylcholine esterase	280	3
		42	6
	Choline acetyl transferase	2	3
	Glutamic acid decarboxylase	3[b], 43[c]	3
NN THA15	Monoamine oxidase	More than in hamster brain;	2
P18	Catechol-O-methyltransferase	More than in hamster brain; (NN and P18); less than in hamster brain (THA 15)	2
B11, 12,	Acetylcholine-	3-5800 OD_{412}/min/mg prot.	12

15, 23	transferase		
49, 82, 90, 92,	Choline acetyl transferase	Less than 4-41	12
108	Glutamic acid decarboxylase	Less than 1[c]	12
	Tyrosine hydroxylase	Less than 0.2-1	12
CHB	Glutamic acid decarboxylase	1[b], 84[c] 6[b], 411[c]	3 11

[a]References: (1) Mokrasch, 1971; (2) Silberstein *et al.*, 1972; (3) Wilson *et al.*, 1972; (4) Haber *et al.*, 1973; (5) Richelson and Thompson, 1973; (6) Schrier and Shapiro, 1973; (7) Bauman *et al.*, 1974; (8) Henn *et al.*, 1974; (9) Hutchison *et al.*, 1974; (10) Massarelli *et al.*, 1974; (11) Schrier and Thompson, 1974; (12) Schubert *et al.*, 1974; (13) Garbarg *et al.*, 1975; (14) Schubert, 1975; (15) Schubert *et al.*, 1975; (16) Suddith *et al.*, 1975a; (17) Haber and Hutchison, 1975; (18) Massarelli *et al.*, 1976; (19) Murphy *et al.*, 1976; (20) Ronquist *et al.*, 1976; (21) Sieghart and Karobath, 1976; (22) Vernadakis *et al.*, 1976; (23) Balcar *et al.*, 1977.

[b]As measured by GABA production.

[c]As measured by Co_2 production.

E. Lipid Metabolism

A number of studies of lipid metabolism have been carried out using clonal glial lines, mostly C6 cells, but also lines NN, CHB, $C2_1$, TC526, and G26-20 and G26-24.

Dawson *et al.* (1971) and Robert *et al.* (1975) studied the glycosphingolipid composition of C6, CHB, and NN cells in culture. They found a relatively simple ganglioside pattern dominated by GM3, and essentially no evidence for glycosphyngolipids associated with myelin.

In contrast, Volpe *et al.* (1975) suggested that their analysis of C6 cells demonstrated some myelin-like components and that C6 cells had oligodendroglial properties. They detected 2′,3′-cyclic nucleotide 3′-phosphohydrolase (CpNase) activity, an enzyme having a high activity in myelin, and thus confirmed earlier reports by Zanetta *et al.*, (1972) and Pfeiffer and Weschler (1972). However, although CpNase activity is 20-40 fold higher in certain glial lines than non-neural lines, it is nevertheless much lower than in myelin (Pfeiffer and Wechsler, 1972). An alternative conclusion would seem to be to question the specificity of this enzyme for oligodendroglial and Schwann cells. In addition, Volpe *et al.* reported a marked localization of CpNase in the plasma membrane, a result that is in some disagreement with that of Pfeiffer (1973) who found it more generally distributed. Volpe *et al.* also examined the membrane protein pattern of these cells in polyacrylamide gel electrophoresis (PAGE), and found proteins that migrated similarly to myelin proteolipid protein (Folch *et al.*, 1957) and myelin basic protein. However, similarity in PAGE mobility is not sufficient to identify at least the myelin basic protein as discussed earlier in this chapter (see section IV, discussion of RN2 basic protein). Further, our own analysis of acid extracts of C6 cells with anti-myelin basic protein sera have been negative (N. Sundarraj and S. E. Pfeiffer, unpublished data). In agreement with other studies of C6 (Dawson *et al.*, 1971; Robert *et al.*, 1975; Dawson *et al.*, 1977), the lipid composition was suggestive of myelin.

Eichberg *et al.* (1971) have studied polyphosphoinositides in C6 and NN cells, finding this fraction to be metabolically very active. Subsequent studies by this group demonstrated marked differences in phospholipid metabolism between neuroblastoma and $C2_1$ cells, but no phospholipid response to either norepinephrine or acetylcholine as occurs in brain (Eichberg *et al.*, 1975).

Robert *et al.* (1976) have also studied the phospholipids, as well as neutral lipids, in C6 and NN cells. They reported a lipid composition for NN cells that resembles that found in glial enriched fractions from mammalian brain, although the increased phospholipid fatty acid saturation. In other studies, with hamster subcutaneous astrocyte nodules, virus-transformed hamster astroglia, and C6 and C2, subcutaneous tumors, Shein *et al.* (1970) and Embree *et al.*, (1972, 1973) concluded that astrocytes are very low or lacking in cerebrosides, and low in gangliosides.

Volpe and Marasa (1975, 1976a, b) have done some interesting studies

TABLE IX
GLYCOSPHINGOLIPID ANALYSIS OF TUMORS DERIVED FROM CELL LINES OF NEUROLOGICAL ORIGIN[a,b]

Lipid Fraction	G26-19	G26-20	Cell Line G26-24	C6	N18	Brain
Galactosyl ceramide	18	161	154	0	10	870
Glucosyl ceramide	18	52	33	23	14	10
Lactosyl ceramide	20	19	7	10	48	10
Sulfatide	7	109	104	Trace	Trace	292

[a] Adapted from Dawson, *et al.*, 1977.
[b] micromoles lipid per 100 g weight cells.

on lipogenic enzymes and their products in C6 cells, including fatty acid synthetase and acetyl-CoA carboxylase. They found that synthesis of these can be regulated by certain fatty acids and by hydrocortisone.

In contrast to the preponderance of GM3 in C6 cells (see above), Manuelidis *et al.* (1976) have stated that the ganglioside pattern of line TC526 is very similar to brain, and distinct from extraneural tissues. Thus, only a trace of GM3 was detected, while GM2, GM1, GD1a, GD2, GD1b, and GT1 were easily recognizable.

Recently Dawson *et al.* (1977) have analysed a number of cell lines derived from rat and mouse tumors for glycolipids. The myelin lipids galactosylceramide and its sulfated ester sulfatide were found in two mouse lines, G26-20 and G26-24 (Table IX). These lines, along with a few rat lines (R21, R113, R102), also incorporated high levels of $^{35}SO_4^{--}$ into sulfatide *in vitro.* These glial lines do not synthesize GM1 or GD1a, which seem generally to be characteristics of neuroblastoma-derived cell strains (Dawson and Stoolmiller, 1976; cf, however, also the results by Manuelidis *et al.*, mentioned above), but synthesize substantial amounts of GM3.

Thus there are cell lines available for studying certain aspects of the regulation of lipid metabolism. To our knowledge, however, a clonal cell line has not been developed that produces all the components of myelin and which actually myelinates a nerve fiber. Development of such a line would, of course, be a tremendous advance in myelin research.

F. Ion Fluxes and Membrane Potentials of Glial Lines

One of the principal communication links postulated to exist between neurons and glia involves the ability of glia to concentrate potassium ions (see Section II). There is an active uptake of K^+ in mammalian glial cells (Haljamae and Hamberger, 1971; Henn, *et al.*, 1972) which may be reflected by an increased oxygen uptake in postassium rich media (Hertz, 1966), and substantial levels of Na^+,K^+-ATPase activity have been demonstrated in glial cell fractions (Cummins and Hydén, 1962; Medzihradsky *et al.*, 1971). Since the ionic barriers and cellular heterogeneity in the brain make this an area of glial metabolism especially difficult to study in the intact brain, several laboratories have turned to glial lines. Again, the majority of these studies have used the C6 line.

The membrane resting potential in C6 cells is much lower than in brain glia (Dennis and Gerschenfeld, 1969; Hild *et al.*, 1958; Hild and Tasaki, 1962) ranging from -28 to -36 mV (Kukes *et al.*, 1973, 1976a; Amano *et al.*, 1974; Hamprecht *et al.*, 1976; de Vellis and Kukes, 1973); it is even lower, -10 to -27 mV, in several lines of human glioma cells (Kornblith *et al.*, 1969; Trachtenberg *et al.*, 1972; Manuelidis *et al.*, 1976). The human glioma cells had short time constants (on the order of milliseconds or less), and specific membrane resistances around 325 $\Omega \cdot cm^2$, higher than the even lower values found for C6 cells (34-72 $\Omega \cdot cm^2$) (Kukes *et al.*, 1976a). The resting potential is apparently determined by the ratio of external to internal K^+, but with a possible Na^+ involvement as well. Thus when Kukes *et al.* (1976a) depleted C6 cells of internal Na^+, K^+ influx was reduced, suggesting that part of the K^+ influx is linked to Na^+ efflux. Kukes *et al.* (1976a) have reported an internal concentration ratio K^+/Na^+ of 8 for C6, comparable to a ratio of 4.6 reported for NN cells (Gill *et al.*, 1974; see also Lees and Shein, 1970 and L. Hertz, this volume).

C6 cells have sufficient Na^+, K^+-ATPase activity to account for these ratios (Kimelberg, 1974; Kukes *et al.*, 1976b; Cotman *et al.*, 1971), as do other hamster and rat glial lines (Embree *et al.*, 1971a,b; Hess *et al.*, 1972). When this ATPase activity was inhibited in C6 cells with ouabain, there resulted a sixfold increase in intracellular Na^+, a two-thirds decrease in intracellular K^+, and a 12 mV decrease in the membrane resting potential (Kukes et al., 1976b).

Hamprecht and co-workers (Amano *et al.*, 1974; Hamprecht *et al.*, 1976) grew C6 cells in media containing dibutyryl cAMP and then applied acetylcholine iontophoretically. The cells underwent a slow hyperpolarization. This sensitivity could be strongly reduced by high concentrations of acetylcholine, and inhibited by atropine or 2-bungarotoxin but not tubocarine.

A variety of studies have suggested that since mammalian glia do not behave as ideal potassium electrodes (in contrast to glia in leech and *Necturus*) (Kuffler and Nicholls, 1976), there may be permeability to ions other than K^+ (Dennis and Gerschenfeld, 1969; Pape and Katzman, 1972; Ransom and Goldring, 1973; Wardell, 1966). Gill *et al.* (1974) found that the K^+/Na^+ ratio of 4.6 in NN cells was reduced to 0.3 when the cells were grown in chloride-free medium. Further, they were able to demonstrate a K^+-dependent uptake mechanism for chloride that showed saturation kinetics (Km 38.5 mM) and was temperature and energy dependent. Nevertheless, Manuelidis *et al.* (1976), using line TC526, found that the changes in membrane potential observed when the external potassium concentration was varied between 2.5 and 5.3 mM were close to those predicted by the Nernst equation.

Ciesielski-Treska *et al.* (1976) found that excess K^+ did not stimulate oxygen uptake into NN glial cells, although this has been observed in microdissected glial cells from the cortex (Hertz, 1966).

In summary, these data add evidence to the idea that the mammalian glial cells have the capacity to actively transport Na^+ and K^+ by an enzymatically catalyzed mechanism. Such a pump could conceivably be altered in response to K^+ released from neurons, and thus be a link in the communication system between neuronal activity and responsive glial metabolism.

Conclusions

In spite of some severe reservations stated in the introduction as to the long range value and originality of the contributions that can be made using clonal glial lines, it seems that maybe they have helped us to at least have some interesting ideas about the nervous system. Further, we are clearly in a position to exploit them in a variety of ways now, as indicated in section V.

However, there are several areas in which the field could use some improvement. Thus, while we are aware of over 100 reported mammalian glial lines, relatively few of these lines have been well-characterized with regard to their capacities for expressing differentiated biochemical parameters. In addition, most or all of these lines suffer to varying extents from uncertainties as to their exact cell type, often due to incomplete characterization with the existing tools, but also because of inherent difficulties with both the available assays and with the cells which may express mixed cell functions. Further, the "normality" of the cell lines relative to normal glia *in situ* is either clearly lacking or far from certain. Much more careful karyotype analyses, including chromosome banding pattern studies, could perhaps give us some clues in this regard.

It is suggested that any further work devoted to establishing new lines of glial cells has as one of its goals the development of clonal cell lines of *clearly defined specific cell types*, and of *normal diploid karyotype*, for the purpose of studying a particular glial property not presently approachable with existing glial lines. One of the most important cell types currently missing from the clonal arsenal is one capable of controlled myelination of neuronal processes. Following the criteria presented in this chapter, we are presently attempting to remedy this situation.

Acknowledgments

This work was supported by a grant from the National Institutes of Health No. NS 10861, and fellowships from the Public Health Service No. CA 00324-01 (JC), Jane Coffin Childs Foundation (RM), the Swiss National Foundation (BB), and the Josiah Macy, Jr. Foundation (SEP). One of us (SEP) wishes to thank Prof. Francois Jacob (Pasteur Institute, Paris) for his contributions of laboratory space and support during the period during which this chapter was written, and Dr. Bernd Hamprecht (Max Planck Institute, Munich) for his generous allocations of encouragement, library facilities, and photocopies during a critical week in the preparation of this manuscript.

References

Abe, T., K. Takahashi and T. Ando, 1974. Purification and properties of S-100 protein from porcine brain. J. Biochem. 75, 11–22.

Ågren, G., J. Pontén, G. Ronquist and B. Westermark. 1971a. Formation of extracellular adenosine triphosphate by normal and neoplastic human cells in culture. J. Cell. Physiol. 77:331-336.

Ågren, G., J. Pontén, G. Ronquist and B. Westermark. 1971b. Demonstration of an ATPase of the cell surface of intact normal and neoplastic human cells in culture J. Cell. Physiol. 78:171-176.

Amaldi, P., G. Rusca and P. Calissano. 1973. *In vitro* synthesis of a brain-specific protein (S-100) by free and membrane-bound polysomes. Biochim. Biophys. Acta 199, 634–641.

Amano. T., B. Hamprecht and W. Kemper. 1974. High activity of choline acetyltransferase induced in neuroblastoma × glioma hybrid cells. Exp. Cell Res. 85:399–408.

Axelrod, J. 1965. The metabolism, storage, and release of catecholamines. Recent Progr. Hormone Res. 21, 597–622.

Bachrach, U. 1975. Cyclic AMP-mediated induction of ornithine decarboxylase of glioma and neuroblastoma cells. Proc. Natl. Acad. Sci. U.S.A. 72:3087–3091.

Blacar, V. J., J. Borg and P. Mandel. 1977. High affinity uptake of L-glutamate and L-aspartate by glial cells. J. Neurochem. 28:87–93.

Barley, F. W., G. H. Sato and R. H. Abeles. 1972. An effect of vitamin B12 deficiency in tissue culture. J. Biol. Chem. 247:4270–4276.

Bauman, A., S. Bourgoin, P. Benda, J. Glowinski, and M. Hamon. 1974. Characteristics of tryptophan accumulation by glial cells. Brain Res. 66:253–263.

Baylor, D. A. and J. G. Nicholls. 1969. Changes in extracellular potassium concentration produced by neuronal activity in the central nervous system. P. Physiology (Lond.) 203:555–569.

Benda, P. and R. L. Davidson. 1971. Regulation of specific functions of glial cells in somatic hybrids. I. Control of S100 protein. J. Cell Physiol. 78:209–216.

Benda, P., J. Lightbody, G. Sato, L. Levine, and W. Sweet. 1968. Differentiated rat glial cell strain in tissue culture. Science 161:370–371.

Benda, P., K. Someda, J. Messer, and W. H. Sweet. 1971. Morphological and immunochemical studies on rat glial tumors and clonal strains propagated in culture. J. Neurosurg. 34:310–323.

Benda, P., D. Illinger, P. Poindron, M. Fiszman, and J.-C. Salomon. 1977. Passage through athymic nude mice of SV40 virus-transformed mouse fetal brain cells. Brain Res. 122:191–196.

Bignami, A., L. F. Eng, D. Dahl, and C. T. Uyeda. 1972. Localization of the glial fibrillary acidic protein in astrocytes by immunofluorescence. Brain Res. 43:429–435.

Bignami, A. and D. Dahl. 1974. Astrocyte-specific protein and neuroglial differentiation. An immunofluorescence study with antibodies to the glial fibrillary acidic protein. J. Comp. Neurol. 153:27–38.

Bissell, M. G., L. F. Eng, M. M. Herman, K. G. Bensch, and L. E. M. Miles. 1975. Quantitative increase of neuroglia-specific GFA protein in rat C-6 glioma cells *in vitro*. Nature (Lond.) 255:633–634.

Bissell, M. G., L. J. Rubinstein, A. Bignami, and M. M. Herman. 1974. Characteristics of the rat C-6 glioma maintained in organ culture systems. Production of glial fibrillary acidic protein in the absence of gliofibrillogenesis. Brain Res. 82:77–89.

Bloom, F. E. and L. L. Iversen. 1971. Localizing ^{3}H-GABA in nerve terminals of rat cerebral cortex by electron microscopic autoradiography. Nature (Lond.) 229:628-630.

Bowery, N. G. and D. A. Brown. 1972. Gamma-aminobutyric acid uptake by sympathetic ganglia. Nature (Lond.) New Biol. 238:89-91.

Breckenridge, B. McL. 1975. Regulation of cyclic nucleotide-related enzymes in an astrocytoma cell line. In: *Cyclic Nucleotides in Disease*. B. Weiss, ed. University Park Press, Baltimore. pp. 67-78.

Brooker, G. and J. de Vellis. 1975. Catecholamine refractoriness: evidence for inactivation of catecholamine-sensitive adenylate cyclase. Fed. Proc. 34:260a.

Browning, E. T. 1973. Metabolic effects of norepinephrine and papaverine in C6 astrocytoma cells. Fed. Proc. 32:679A.

Browning, E. T. 1974. Suppression of the norepinephrine (NE) effect on cAMP content of the C_6 astrocytoma cells: mode of expression. Fed. Proc. 33:507a.

Browning, E. T., J. P. Schwartz, and B. McL. Breckenridge. 1974a. Norepinephrine-sensitive properties of C_6 astrocytoma cells. Mol. Pharmacol. 10:162-174.

Browning, E. T., V. E. Groppi, Jr., and C. Kon. 1974b. Papaverine, a potent inhibitor of respiration in C_6 astrocytoma cells. Mol. Pharmacol. *10*:175-181.

Browning, E. T., C. O. Brostrom, and V. E. Groppi, Jr. 1976. Altered adenosine cyclic 3', 5' -monophosphate synthesis and degradation by C-6 astrocytoma cells following prolonged exposure to norepinephrine. Mol. Pharmacol 12:32-40.

Brunk, U. 1973. Distribution and shifts of ingested marker particles in residual bodies and other lysosomes. Exp. Cell Res. 79:15-27.

Bunge, R. P. 1968. Glial cells and the central myelin sheath. Physiol. Rev. 48:197-251.

Buonassisi, V., G. Sato, and A. J. Cohen. 1962. Hormone-producing cultures of adrenal and pituitary tumor origin. Proc. Natl. Acad. Sci. U.S.A. 48:1184.

Calissano, P., S. Alema, and P. Fasella. 1974. Interaction of S-100 protein with cations and liposomes. Biochem. 13, 4553-4560.

Cammer, W., T. Fredman, A. L. Rose, and W. T. Norton. 1976. Brain Carbonic Anhydrase: Activity in isolated myelin and the effect of hexachlorophene. J. Neurochem. 27:165-171.

Caron, M. C. and R. J. Lefkowitz. 1976. Solubilization and characterization of the β-adrenergic receptor binding sites of frog erythrocytes. J. Biol. Chem. 251:2374-2384.

Chen, T. T., and J. Mealey, Jr. 1973. Effect of corticosteroid on protein and nucleotide synthesis in human glial tumor cells. Cancer Res. 33:1721-1723.

Church, R. L., M. Tanzer, and S. E. Pfeiffer. 1973. Synthesis of collagen and procollagen by a clonal line of rat Schwann cells. Proc. Natl. Acad. Sci. U.S.A. 70:1943-1946.

Cicero, T. J. and B. W. Moore. 1970. Turnover of the brain specific protein, S-100. Science 169:1333-1334.

Ciesielski-Treska, J., J. Hermetet, and P. Mandel. 1973. Cultures associées de cellules gliales et d'un clone du neuroblastome C 1300. Compt. rend. Soc. Biol. 167:1971-1973.

Ciesielski-Treska, J., B. Roth-Schechter, D. Beya, P. Mandel. C. Nissen, and L. Hertz. 1976. Rate of oxygen uptake by the NN cell line of hamster astroglia: lack of effect by excess potassium. J. Neurochem. 26:197-199.

Claisse, P. J. and J. P. Roscoe. 1976. The inducibility of glycerol phosphate dehydrogenase in two rat glial clones. Brain Res. 109:423-425.

Clark, R. B. and J. P. Perkins. 1971. Regulation of adenosine 3', 5' -cyclic monophosphate concentration in cultured human astrocytoma cells by catecholamines and histamine. Proc. Natl. Acad. Sci. U.S.A. 68:2757-2760.

Clark, R. B., R. Gross, Y. -F. Su, and J. P. Perkins. 1974. Regulation of 3' :5' -monophosphate content in human astrocytoma cells by adenosine and the adenine nucleotides. J. Biol. Chem. 249:5296-5303.

Cotman, C., H. Herschman, and D. Taylor. 1971. Subcellular fractionation of cultured glial cells. J. Neurobiol. 2:169-180.

Cummins, J. and H. Hydén. 1962. Adenosine triphosphate levels and adenosine triphosphatases in neurons, glia and neuronal membranes of the vestibular nucleus. Biochim. Biophys. Acta 60:271-283.

Curtis, D. R. and G. A. R. Johnston. 1970. Amino Acid transmitters. In: *Handbook of Neurochemistry*, Vol. 4. A. Lajtha, ed. Plenum Press, New York. pp. 115-134.

Dahl, D. and A. Bignami. 1976. Immunogenic properties of the glial fibrillary acidic protein. Brain Res. 116:545-547.

Dannies, P. S. and L. Levine. 1969. Demonstration of subunits in beef brain acidic protein (S-100). Biochem. Biophys. Res. Comm. 37:587-592.

Dannies, P. S. and L. Levine. 1971. Structural properties of bovine S-100 protein. J. Biol. Chem. 246:6276-6283.

Davidson, R. and P. Benda. 1970. Regulation of specific functions of glial cells in somatic hybrids. II. Control of inducibility of glycerolphosphate dehydrogenase. Proc. Natl. Acad. Sci. U.S.A. 67:1870-1877.

Dawson, C. and A. C. Stoolmiller. 1976. Comparison of the ganglioside composition of established mouse neuroblastoma cell strains grown *in vivo* and in tissue culture. J. Neurochem. 26:225-226.

Dawson, G., S. F. Kemp, A. C. Stoolmiller, and A. Dorfman. 1971. Biosynthesis of glycosphingolipids by mouse neuroblastoma (NB41A), rat glia (RGG-6) and human glia (CHB-4) in cell culture. Biochem. Biophys. Res. Comm. 44:687-694.

Dawson, G., N. Sundarraj, and S. E. Pfeiffer. 1977. Synthesis of myelin glycosphingolipids (galactosylceramide and galactosyl (3-0-sulfate) ceramide (sulfatide) by cloned cell lines derived from mouse and rat neurotumors. J. Biol. Chem. 252:2777-2779.

Day, E. D. and D. D. Bigner. 1973. Specificity, cross-reactivity, and affinity of ^{125}I-labeled antiglioma antibodies of monolayers of cultured glioma cells. Cancer Res. 33:2362-2372.

Deguchi, T. and J. Axelrod. 1972. Induction and superinduction of serotonin N-acetyltransferase by adrenergic drugs and denervation in rat pineal organ. Proc. Natl. Acad. Sci. U.S.A. 69:2208-2211.

Deguchi, T. and J. Axelrod. 1973. Supersensitivity and subsensitivity of the β-adrenergic receptor in pineal gland regulated by catecholamine transmitter. Proc. Natl. Acad. Sci. U.S.A. 70:2411-2414.

Dennis, M. J. and H. M. Gerschenfeld. 1969. Some physiological properties of identified mammalian neuroglial cells. J. Physiol. (Lond.) 203:211-222.

Dennis, M. J. and R. Miledi. 1974. Electrically induced release of acetylcholine from denervated Schwann cells. J. Physiol. (Lond.) 237:431-452.

de Vellis, J. and G. Brooker. 1972. Effect of catecholamines on cultured glial cells: correlation between cyclic AMP levels and lactic dehydrogenase induction. Fed. Proc. 31:513a (#1690).

de Vellis, J. and G. Brooker. 1973. Induction of enzymes by glucocorticoids and catecholamines in a rat glial cell line. *In: Tissue Culture of the Nervous System.* G. Sato, ed. Plenum Press, New York pp. 231-246.

de Vellis, J. and G. Brooker. 1974. Reversal of catecholamine refractoriness by inhibitors of RNA and protein synthesis. Science 186:1221-1223.

de Vellis, J. and D. Inglish. 1973. Age-dependent changes in the regulation of glycerolphosphate dehydrogenase in the rat brain and in a glial cell line. *Progress in Brain Research.* Vol. 40. D. H. Ford, ed. pp. 321-330.

de Vellis, J. and G. Kukes. 1973. Regulation of glial cell functions by hormones and ions. Texas Reports on Biology and Medicine 31:271-293.

De Vitry, F., M. Camier, P. Czernichow, P. Benda, P. Cohen, and A. Tixier-Vidal. 1974. Establishment of a clone of mouse hypothalamic neurosecretory cells synthesizing neurophysin and vassopressin. Proc. Natl. Acad. Sci. U.S.A. 71:3575-3579.

Donta, S. T. 1973. The growth of functional rat glial cells in a serumless medium. Exp. Cell Res. 82:119-124.

Donta, S. T. 1975. Growth of functional glial cells in a serumless medium. Methods Cell Biol. 9:123-137.

Doore, D. J., M. M. Bashor, N. Spitzer, R. C. Mawe, and H. M. Saier, Jr. 1975. Regulation of adenosine 3' ,5' -monophosphate efflux from rat glioma cells in culture. J. Biol. Chem. 250:4371-4372.

Dorfman, A. and P. -L. Ho. 1970. Synthesis of acid mucopolysaccharides by glial tumor cells in tissue culture. Proc. Natl. Acad. Sci. U.S.A. 66:495-499.

Druckrey, H. 1971. Genotypes and phenotypes of ten inbred strains of BD-IX rats. Arzneim.-Forsch. 21:1274-1278.

Druckrey, H., R. Preussman, and S. Ivancovic. 1969. N-nitroso compounds in organotropic and transplacental carcinogenesis. Ann. N. Y. Acad. Sci. 163:676-696.

Edström, A., K. G. Haglid, M. Kanje, L. Rönnbäck and E. Walum. 1973. Morphological alterations and increase of S100 protein in cultured human glioma cells deprived of serum. Exp. Cell Res. 83:426-429.

Edström, A., M. Kanje and E. Walum. 1974. Effects of dibutyryl cyclic AMP and prostaglandin E1 on cultured human glioma cells. Exp. Cell. Res. 85:217-223.

Edström, A., M. Kanje, P. Löfgren, and E. Walum. 1975a. Drug-induced alterations in morphology and level of cAMP in cultured human glioma cells. Exp. Cell Res. 95:359-364.

Edström, A., M. Kanje, and E. Walum. 1975b. Uptake of 3-0-methyl D-glucose into cultured human glioma cells. J. Neurochem. 24:395-401.

Edström, A., M. Kanje, and E. Walum. 1976. Density-dependent inhibition of 2-deoxy-D-glucose uptake into glioma and neuroblastoma cells in culture. Exp. Cell Res. 97:6-14.

Eichberg, J., G. Hauser, and H. Shein. 1971. Polyphosphoinositides in normal and neoplastic rodent astrocytes. Biochem. Biophys. Res. Commun. 45:43-50.

Eichberg, J., H. M. Shein, and G. Hauser. 1975. Phospholipid metabolism in cultured neuroblastoma and glioma cells incubated with carbamylcholine and norephephrine. J. Neurochem. 24:67-70.

Embree, L. J., H. H. Hess, and H. M. Shein. 1971a. Sodium-potassium ATPase activity of normal and virally transformed hamster astroglia grown subcutaneously. Brain Res. 27:422-425.

Embree, L. J., H. H. Hess, and H. Shein. 1971b. Sodium-potassium adenosine triphosphate activity in N-nitrosomethylurea-induced rat astrocytoma cells. Exp. Neurol. 31:383-390.

Embree, L. J., H. H. Hess, and H. M. Shein. 1972. Biochemical structural components of cloned N-nitrosomethylurea-induced astrocytoma cells grown subcutaneously. Neurology (Minn.) 22:194-201.

Embree, L. J., H. H. Hess and H. M. Shein. 1973. Microchemical studies of lipids, proteins and nucleic acids in polyoma virus-transformed hamster astroglia. J. Neuropath. Exp. Neurol. 32:542-551.

Eng, L. F., J. J. Vanderhaeghen, A. Bignami, and B. Gerstl. 1971. An acidic protein isolated from fibrous astrocytes. Brain Res. 28:351-354.

Fields, K. L., C. Gosling, M. Megson, and P. L. Stern. 1975. New cell surface antigens in rat defined by tumors of the nervous system. Proc. Natl. Acad. Sci. U.S.A. 72:1296-1300.

Folch, J., M. Lees, and G. J. Sloane-Stanley. 1957. A simple method for the isolation and purification of total lipids from animal tissues. J. Biol. Chem. 226:497-509.

Folkman, Y., E. Merler, C. Abernathy, and C. Williams. 1971. Isolation of a tumor factor responsible for angiogenesis. J. Exp. Med. 133:275-288.

Franklin, T. J. and P. A. Twose. 1976. Desensitization of beta-adrenergic receptors of glioma cells: studies with intact and broken cell preparations. FEBS Lett. 66:225-229.

Garbarg, M., M. Baudry, P. Benda, and J. C. Schwartz. 1975. Simultaneous presence of histamine-N-methyltransferase and catechol-O-methyltransferase in neuronal and glial cells in culture. Brain Res. 83:538-541.

Gibson, D. A., S. Reichlin, and A Vernadakis. 1974. [^{3}H]Uridine uptake and incorporation into RNA in the C-6 glial cells following dibutyryl cyclic AMP treatment. Brain Res. 81:354-360.

Gill, T. H., O. M. Young, and D. B. Tower. 1974. The uptake of ^{36}Cl into astrocytes in tissue culture by a potassium-dependent saturable process. J. Neurochem. 23:1011-1018.

Giller, E. A., B. K. Schrier, B. K. Shainberg, H. R. Fisk, and P. G. Nelson. 1973. Choline acetyltransferase activity is increased in combined cultures of spinal cord and muscle cells from mice. Science 182:588-589.

Gilman, A. G. 1972. Regulation of cyclic AMP metabolism in cultured cells of the nervous system. Adv. Cyclic Natl. Res. 1:389-410.

Gilman, A. G. and J. D. Minna. 1973. Expression of genes for metabolism of cyclic adenosine 3' :5' -monophosphate in somatic cells. I. Responses to catecholamines in parental and hybrid cells. J. Biol. Chem. 248:6610-6617.

Gilman, A. G. and M. Nirenberg. 1971. Effect of catecholamines on the adenosine 3' 5' -cyclic monophosphate concentrations of clonal satellite cells of neurons. Proc. Natl. Acad. Sci. U.S.A. 68:2165-2168.

Gilman, A. G. and B. K. Schrier. 1971. Adenosine cyclic 3',5' -monophosphate in fetal rat brain cell cultures. I. Effect of catecholamines. Mol. Pharmacol. 8:410-416.

Glastris, B. and S. E. Pfeiffer. 1974. Mammalian membrane marker enzymes: sensitive assay for 5' -nucleotidase and assay for 2',3' -cyclic nucleotide-3' -phosphohydrolase. Meth. Enzymology-Biomembranes 32B:124-130.

Goldberg, A. L. and J. F. Dice. 1974. Intracellular protein degradation in mammalian and bacterial cells. Ann. Rev. Biochem. 43:835-869.

Gombos, G., J. P. Zanetta, P. Mandel, and G. Vincendon. 1971. Biochem. 53:645-655.

Greengard, P. 1975. Cyclic nucleotides, protein phosphorylation, and neuronal function. Adv. Cyclic Nucl. Res. 5:585-601.

Greengard, P. 1976. Possible role for cyclic nucleotides and phosphorylated membrane proteins in postsynaptic actions of neurotransmitters. Nature (Lond.) 260:101-108.

Grimm-Jörgensen, Y., S. E. Pfeiffer, and J. McKelvy. 1976. Metabolism of thyrotropin releasing factor in two clonal lines of nervous system origin. Biochem. Biophys. Res. Comm. 70:167-173.

Grossman, R. G. and A. Seregin. 1977. Glial-neural interaction demonstrated by the injection of Na^+ and Li^+ into cortical glia. Science 195:196-198.

Guha, A. and S. Moore. 1975. Solubilization of 2' ,3' -cyclic nucleotide 3' -phosphohydrolase from bovine brain without detergents. Brain Res. 89:279-286.

Haber, B. and H. T. Hutchison, 1975. Uptake of neurotransmitters and precursors by clonal cell lines of neural origin. In: *Symposium for Transport Phenomena in the Nervous System, Physiological and Pathological Aspects.* G. Levi, L. Battistin and A. Lajtha, eds. Vol. 69 of Advances in Experimental Medicine and Biology. Plenum Press, New York. pp. 179-198.

Haber, B., K. Werrbach, M. Risk and H. T. Hutchison. 1975. Abst. Commun. 4th Int. Meet. Int. Soc. Neurochem. Barcelona. p. 237.

Haglid, K., A. Hamberger, H. A. Hansson, H. Hydén, L. Persson, and L. Rönnbäck. 1974. S100 protein in synapses of the central nervous system. Nature (Lond.) 251:532-534.

Haglid, K. G., A. Hamberger, H. -A. Hansson, H. Hydén, L. Persson, and L. Rönnbäck. 1975. Immunohistochemical localization of S-100 protein in brain. Nature (Lond.) 258:748-749.

Haljamae, H. and A. Hamberger. 1971. Potassium uptake by bulk prepared neuronal and glial cells. J. Neurochem. 18:1903-1912.

Hamprecht, B. 1974. Cell cultures as model systems for studying the biochemistry of differentiated functions of nerve cells. In: *Biochemistry of Sensory Functions, 25th Mosbacher Colloquium der Gesellschaff für Biologische Chemie.* J. Jaenicke, ed. Springer-Verlag, New York. pp. 391-423.

Hamprecht, B., B. M. Jaffe, and G. W. Philpott. 1973. Prostaglandin production by neuroblastoma, glioma, and fibroblast cell lines: stimulation by N^6, O^2 -dibutyryl adenosine 3' :5' -cyclic monophosphate. FEBS Lett. 36:193-198.

Hamprecht, B., W. Kemper, and T. Amano. 1976. Electrical response of glioma cells to acetylcholine Brain Res. 101:129-135.

Hamprecht, B. and J. Schultz. 1973. Influence of noradrenalin, prostaglandin E_1 and inhibitors of phosphodiesterase activity on levels of the cyclic adenosine 2' :5' -monophosphate in somatic cell hybrids. Hoppe-Seyler's Z. Physiol. Chem. 354:1633-1641.

Henn, F. A. and A. Hamberger. 1971. Glial cell function: uptake of transmitter substances. Proc. Natl. Acad. Sci. U.S.A. 68:2686-2690.

Henn, F. A., H. Haljamae, and A. Hamberger. 1972. Glial cell function: active control of extracellular K^+ concentration. Brain Res. 43:437-443.

Henn, F. A., M. N. Goldstein, and A. Hamberger. 1974. Uptake of the neurotransmitter candidate glutamate by glia. Nature (Lond.) 249:663-664.

Herschman, H. R. 1971. Synthesis and degradation of a brain-specific protein (S100 protein) by clonal cultured human glial cells. J. Biol. Chem. 246:7569-7571.

Herschman, H. R., J. Breeding, and J. Nedrud. 1972. Sialic acid masked membrane antigens of clonal functional glial cells. J. Cell Physiol. 79:249-258.

Herschman, H. R., B. P. Grauling, and M. P. Lerner. 1973. Nervous system-specific proteins in cultured neural cells. *In: Current Topics in Neurobiology, Vol. 1. Tissue culture of the nervous system.* C. Sato, ed. Plenum Press, New York, pp. 187-202.

Hertz, L. 1966. Neuroglial localization of potassium and sodium effects on respiration in brain. J. Neurochem. 13:1373-1387.

Hess, H. H., L. J. Embree, and H. M. Shein. 1972. Enzymic control of sodium and potassium active transport in normal and neoplastic rodent astroglia. *In: Progress in experimental tumor research.* Vol. 17. Karger, Basel. pp. 308-317.

Hild, W. and I. Tasaki. 1962. Morphological and physiological properties of glial cells in tissue culture. J. Neurophysiol. 25:277-286.

Hild, W., I. Tasaki and J. J. Chang. 1958. Electrical responses from astrocyte glia in the mammalian central nervous system. Experientia 14:222-226.

Hider, R. C., E. B. Fern, and D. R. London. 1969. Relationship between intracellular amino acids and protein synthesis in the extensor digitorum longus muscle of rats. Biochem. J. 114:171-178.

Hökfelt, T. and A. Ljungdahl. 1971. Uptake of [^{3}H]noradrenaline and [^{3}H]α-aminobutyric acid in isolated tissues of rat: an autoradiographic and fluorescence microscopic study. In: *Histochemistry of Nervous Transmission. Progress in Brain Research,* Vol. 34. O. Fränkö, ed. Elsevier, Amsterdam. pp. 87-102.

Hökfelt, T. and A. Ljungdahl. 1972. Autoradiographic identification of cerebral and cellebellar cortical neurons accumulating labeled gammaminobutyric acid (^{3}H-GABA). Exp. Brain Res. 14:354-362.

Hutchison, H. T., K. Werrbach, C. Vance, and B. Haber. 1974. Uptake of neurotransmitters by clonal lines of astrocytoma and neuroblastoma in culture. I. Transport of γ-aminobutyric acid. Brain Res. *66:*265-274.

Hydén, H. 1967. *The Neuron.* Elsevier, Amsterdam.

Hydén, H. and B. McEwen. 1966. A glial protein specific for the nervous system. Proc. Natl. Acad. Sci. U.S.A. 55:354-358.

Hydén, H. and P. W. Lange. 1970a. Correlation of the S100 brain protein with behavior. Exp. Cell Res. 62:125-132.

Hydén, H. and P. W. Lange. 1970b. Brain-cell protein synthesis specifically related to learning. Proc. Natl. Acad. Sci. U.S.A. 65:898-904.

Iversen, L. L. and M. S. Neal. 1968. The uptake of [^{3}H]GABA by slices of rat cerebral cortex. J. Neurochem. 15:1141-1149.

Iversen, L. L. and F. E. Schon. 1973. The use of autoradiographic techniques for the identification and mapping of transmitter-specific neurons in CNS. In: *New Concepts in Neurotransmitter Regulation.* A. J. Mandell, ed. Plenum Press, New York. pp. 153-193.

Iversen, L. L. and J. S. Kelly. 1975. Uptake and metabolism of x-aminobutyric acid by neurones and glial cells. Biochem. Pharmacol. 24:933-938.

Jard, S., J. Premont, and P. Benda. 1972. Adenylate cyclase, phosphodiesterases and protein kinase of rat glial cells in culture. FEBS Letters 26:344-48.

Johnson, G. L. and J. P. Perkins. 1975. Desensitization of adenylate cyclase (AC) in human astrocytoma cells. Pharmacologist 17: N2 233.

Johnson, E. M., T. Ueda, H. Maeno, and P. Greengard. 1972. Adenosine 3', 5' -monophosphate-dependent phosphorylation of a specific protein in synaptic membrane fractions from rat cerebrum. J. Biol. Chem. 247:5650-5652.

Joseph, B. S. and M. B. A. Oldstone. 1974. Expression of selected antigens on the surface of cultured neural cells. Brain Res. 80:421-434.

Kelly, P. J., R. L. Suddith, H. T. Hutchison, K. Werrbach, and B. Haber. 1976. Endothelial proliferation factor present in tissue culture of CNS tumors. J. Neurosurg. 44:342-346.

Kimelberg, H. K. 1974. Active potassium transport and Na^+ -K^+ ATPase activity in cultured glioma and neuroblastoma cells. J. Neurochem. 22:967-971.

Klee, W. A. and M. Nirenberg. 1974. A neuroblastoma X glioma hybrid cell line with morphine receptors. Proc. Natl. Acad. Sci. U.S.A. 71:3474-3477.

Klee, W. A., S. K. Sharma, and N. Nirenberg. 1975. Life Sci. 16:1869.

Klier, G., D. Schubert, and S. Heineman. 1975. Ultrastructural changes accompanying the induced differentiation of clonal rat nerve and glia. Neurobiology 5:1-7.

Kornblith, P. L., A. Prieto, and D. A. Pollen. 1969. Alterations in glial and mesenchymal tumor cell membrane resistance with heteroimmune and autologous sera. Ann. N. Y. Acad. Sci. 159:585-590.

Kuffler, S. W. 1967. Neurological cells: Physiological properties and a potassium mediated effect of neuronal activity on the glial membrane potential. Proc. Roy. Soc. Lond. Ser. B 168:1-21.

Kuffler, S. W. and J. G. Nicholls. 1966. The physiology of neuroglial cells. Ergebn. Physiol. 57:1-80.

Kuffler, S. W. and J. G. Nicholls. 1976. *From Neuron to Brain.* Sinauer Associates, Inc. Sunderland, Mass. pp. 255-288.

Kuffler, S. W. and D. D. Potter. 1964. Glia in the leech central nervous system: Physiological properties and neuron-glia relationship. J. Neurophysiol. 27:290-320.

Kukes, G., J. de Vellis, and R. Elul. 1973. Linked active transport of Na^+ and K^+ in an astrocytic cell line. Trans. Amer. Soc. Neurochem. 4:141.

Kukes, G., R. Elul, and J. de Vellis. 1976a. The ionic basis of the membrane potential in a rat glial cell line. Brain Res. 104:71-92.

Kukes, G., J. de Vellis, and R. Elul. 1976b. A linked active transport system for Na^+ and K^+ in a glial cell line. Brain Res. 104:93-105.

Kumanishi, T., F. Ikuta and T. Yamamoto. 1973. Brain tumors induced by Rous sarcoma virus, Schmidt-Ruppin Strain. III. Morphology of brain tumors induced in adult mice. J. Natl. Cancer Inst. 50:95-109.

Kumanishi, T., F. Ikuta, T. Yamamoto, N. Maruyama, N. Nishida, K. Nishida, and K. Ueda. 1975. Aldolase c in cultured rat glioma. J. Neurochem. 24:1081-1082.

Labourdette, G. and A. Marks. 1975. Synthesis of S-100 protein in monolayer cultures of rat-glial cells. Eur. J. Biochem. 58:73-79.

Lampert, A., M. Nirenberg, and W. A. Klee. 1976. Tolerance and dependence evoked by an endogenous opiate peptide. Proc. Natl. Acad. Sci. U.S.A. 73:3165-3167.

Lees, M. B. and H. M. Shein. 1970. Sodium and potassium content of normal and neoplastic rodent astrocytes. Brain Res. 23:280-283.

Lefkowitz, R. J. 1976. The β-adrenergic receptor. Life Sci. 18:461-472.

Leichtling, B. H., A. M. Drotar, R. Ortmann, and J. P. Perkins. 1976. Growth of astrocytoma cells in the presence of prostaglandin E_1: Effect on the regulation of cyclic AMP metabolism. J. Cyclic Nucl. Res. 2:89-98.

Lesniak, M. A. and J. Roth. 1976. Regulation of receptor concentration by homologous hormone. Effect of human growth hormone on its receptor in IM-9 lymphocytes. J. Biol. Chem. 251:3720-3729.

Levi-Montalcini, R. 1966. The nerve growth factor: its mode of action on sensory and sympathetic nerve cells. Harvey Lect. 60:217-259.

Li, J. B., R. M. Fulks, and A. L. Goldberg. 1973. Evidence that the intracellular pool of tyrosine serves as precursor for protein synthesis in muscle. J. Biol. Chem. 248:7272-7275.

Lightbody, J., S. E. Pfieffer, P. L. Kornblith, and H. Herschman. 1970. Biochemically differentiated clonal human glial cells in tissue culture. J. Neurobiol. 1:411-417.

Lim, R. and K. Mitsunoba. 1975. Partial purification of a morphological transforming faction from pig brain. Biochem. Biophys. Acta 400:200-207.

Lim, R., K. Mitsunobu and W. K. P. Li. 1973. Maturation stimulating effect of brain cells in culture. Exp. Cell Res. 79:243-246.

Lim, R., D. E. Turriff, and S. S. Troy. 1976. Response of glioblasts to a morphological transforming factor: cinematographic and chemical correlations. Brain Res. 113:165-170.

Lim, R., D. E. Turriff, S. S. Troy, B. W. Moore, and L. F. Eng. 1977. Glia maturation factor: effect on chemical differentiation of glioblasts in culture. Science 195:195-196.

Lindgren, A., B. Westermark, and J. Pontén. 1975. Serum stimulation of stationary human glia and glioma cells in culture. Exp. Cell Res. 95:311-319.

Longo, A. M. and E. E. Penhoet. 1974. Nerve growth factor in rat glioma cells. Proc. Natl. Acad. Sci. U.S.A. 71:2347-2349.

Ludwin, S. K., J. C. Kosek, and L. F. Eng. 1976. The topographical distribution of S-100 and GFA proteins in the adult rat brain: an immunohistochemical study using horseradish peroxidase-labeled antibodies. J. Comp. Neur. 165:197-208.

MacIntyre, E. H., R. A. Grimes, and A. E. Vatter. 1969a. Cytology and growth characteristics of human astrocytes transformed by Rous sarcoma virus. J. Cell Sci. 5:583-602.

MacIntyre, E., J. Pontén, and A. E. Vatter. 1969b. Fine structure of human glial cells in tissue culture. Fed. Proc. 28:754.
MacIntyre, E. H., C. J. Wintersgill, J. P. Perkins, and A. E. Vatter. 1972. The responses in culture of human tumor astrocytes and neuroblasts to N^6, $O^{2'}$- dibutyryl adenosine 3′,5′-monophosphoric acid J. Cell Sci. 11:639-660.
Maguire, M. E., R. A. Wiklund, H. J. Anderson, and A. G. Gilman. 1976. Binding of [^{125}I] iodohydroxybenzylpindolol to putative β-adrenergic receptors of rat glioma cells and other cell clones. J. Biol. Chem. 251:1221-1231.
Mahony, J., I. Brown, G. Labourdette, and A. Marks. 1976. Synthesis of the brain-specific S-100 protein in a cell-free system from wheat embryo programmed with poly (A)-containing RNA from rabbit brain. Eur. J. Biochem. 67:203-208.
Mandel, P., V. Stefanovic, J. Ciesielski-Treska, and A. Ebel. 1974. ATPase activity at the cell surface of astroglia in culture. FEBS Lett. 45:337-339.
Manuelidis, E. E. 1965. Long term tissue culture lines of intracranial tumors. J. Neurosurg. 22:368-373.
Manuelidis, E. E. 1969. Experiments with tissue culture and heterologous transplantation of tumors. Ann. N. Y. Acad. Sci. 159:409-431.
Manuelidis, L. and E. E. Manuelidis. 1976. Cholera Toxin-Peroxidase: Changes in surface labeling of glioblastoma cells with increased time in tissue culture. Science 193:588-590.
Manuelidis, L., E. E. Manuelidis, and J. Prichard. 1976. Relationship between membrane potential and external potassium in human glioblastoma cells in tissue culture. J. Cell. Physiol. 87:179-188.
Mark, J., J. Pontén, and B. Westermark, 1974a. Origin of the marker chromosomes in an established hypotriploid glioma cell line studied with G-band technique. Acta Neuropath. 29:223-228.
Mark, J., J. Pontén, and B. Westermark. 1974b. G-band analyses of an established cell line of a human malignant glioma. Humangenetik 22:323-326.
Mark, J., J. Pontén, and B. Westermark. 1974c. Cytogenetical studies with G-band technique of established cell lines of human malignant gliomas. Hereditas 78:304-308.
Martin, D. L. 1973. Kinetics of the sodium-dependent transport of gamma-amino-butyric acid by synaptosomes. J. Neurochem. 21:345-356.
Massarelli, R., J. Ciesielski-Treska, A. Ebel, and P. Mandel. 1974. Choline uptake in glial cell cultures. Brain Res. 81:361-363.
Massarelli, R., V. Stefanovic, and P. Mandel. 1976. Cholinesterase activity and choline uptake in intact nerve cell cultures. Brain Res. 112:103-112.
Matus, A. and S. Mughal. 1975. Immunohistochemical localization of S-100 protein in brain. Nature (Lond.) 258:746-748.
Maunoury, R., A. Delpech, B. Delpech, M. N. Virard, and C. Vedrenne. 1976. Presence of neurospecific antigen NSA 1 in fetal human astrocytes in long-term culture. Brain Res. 112:383-387.
McEwen, B. S. and H. Hydén. 1966. A study of specific brain proteins on the semi-micro scale. J. Neurochem. 13:823-833.
McGinnis, J. F. and J. de Vellis. 1974. Cortisol induction of glycerol phosphate dehydrogenase in a rat brain tumor cell line. Nature (Lond.) 250:422-424.
Medzihradsky, F., P. S. Nandhashri, V. Inoyaga-Vargas, and O. Z. Sellinger. 1971. A comparison of ATPase activity of a glial cell fraction and the neuronal perikaryal fraction isolated in bulk from rat cerebral cortex. J. Neurochem. 18:1599-1603.
Miani, N., A. Caniglia, and V. Panetta. 1976. A brain-specific protein with affinity for DNA. J. Neurochem. 27, 145-150.
Michetti, F., N. Miani, G. De Renzis, A. Caniglia, and S. Correr. 1974. Nuclear localization of S-100 protein. J. Neurochem. 22:239-244.
Michetti, F., G. de Renzis, R. Donato, and N. Miani. 1976. Brain-specific effect of the S-100 protein on the RNA polymerase I activity in isolated nuclei. Brain Res. 105:372-375.
Mickey, J., R. Tate, and R. J. Lefkowitz. 1975. Subsensitivity of adenyl cyclase and decreased β-adrenergic receptor binding after chronic exposure to isoproterenol *in vitro*. J. Biol. Chem. 250:5727-5729.

Miller, C. and W. M. Kuehl. 1976. Isolation and characterization of myosin from cloned rat glioma and mouse neuroblastoma cells. Brain Res. 108:115-124.

Mokrasch, L. C. 1971. Free amino acid content of normal and neoplastic rodent astroglia. Brain Res. 25:672-676.

Monard, D., F. Soloman, M. Rentsch and R. Gysin. 1973. Glia-induced morphological differentiation in neuroblastoma cells. Proc. Natl. Acad. Sci. U.S.A. 70:1894-1897.

Monard, D., K. Stockel, R. Goodman, and H. Thoenen. 1975. Distinction between nerve growth factor and glial factor. Nature (Lond.) 258:444-445.

Moore, B. W. 1965. A soluble protein characteristic of the nervous system. Biochem. Biophys. Res. Comm. 19:739-744.

Moore, B. W. 1972. Chemistry and biology of two proteins, S-100 and 14-3-2, specific to the nervous system. Int. Rev. Neurobiol. 15:215-225.

Morgenroth, V. H. III, L. R. Hegstrand, R. H. Roth, and P. Greengard. 1975. Evidence for involvement of protein kinase in activation by cyclic AMP of brain tyrosine 3-monooxygenase. J. Biol. Chem. 250:1946-1948.

Mukherjee, C., M. G. Caron, and R. J. Lefkowitz. 1975. Catecholamine induced subsensitivity of adenylate cyclase associated with loss of β-adrenergic receptor binding sites. Proc. Natl. Acad. Sci. U.S.A. 72:1945-1949.

Murphy, D. L., C. H. Donnelly, and E. Richelson. 1976. Substrate and inhibitor-related characteristics of monoamine oxidase in C_6 rat glial cells. J. Neurochem. 26:1231-1235.

Murphy, R. A., J. Oger, J. D. Saide, M. H. Blanchard, B. G. W. Arnason, C. Hogan, N. J. Patazis, and M. Young. 1977. Secretion of nerve growth factor by central nervous system glioma cells in culture. J. Cell Biol. 72:769-773.

Nakamura, K., M. Homma, and N. Ishida. 1975. Growth of measles virus in cultures of rat glioma. Infection and Immunity 12:614-620.

Neal, M. J. and L. L. Iversen. 1972. Autoradiographic localization of ^{3}H-GABA in rat retina. Nature New Biology (Lond.) 235:217-218.

Nelson, P., C. Christian and M. Nirenberg. 1976. Synapse formation between clonal neuroblastoma X glioma hybrid cells and striated muscle. Proc. Natl. Acad. Sci. U.S.A. 73:123-127.

Newburgh, R. W. and R. N. Rosenberg. 1972. Effect of norepinephrine on glucose metabolism in glioblastoma and neuroblastoma cells in cell culture. Proc. Nat. Acad. Sci. U.S.A. 69:1677-1680.

Noble, E. P., P. J. Syapin, R. Vigran, and A. Rosenberg. 1976. Neuraminidase-releasable surface sialic adic of cultured astroblasts exposed to ethanol. J. Neurochem. 27:217-221.

Oey, J. 1975. Noradrenaline induces morphological alterations in nucleated and enucleated rat C_6 glioma cells. Nature 2257 (Lond.):317-319.

Opler, L. A. and M. H. Makman. 1972. Mediation by cyclic AMP of hormone-stimulated glycogenolysis in cultured rat astrocytoma cells. Biochem. Biophys. Res. Comm. 46:1140-1145.

Orkand, P. M. & E. A. Kravitz. 1971. Localization of the sites of α-amino-butyric acid (GABA) uptake in lobster nerve-muscle preparations. J. Cell Biol. 49:75-79.

Pape, L. and A. Katzman. 1972. Response of glia in cat sensorimotor cortex to increased extracellular potassium. Brain Res. 38:71-92.

Passonneau, J. V. and S. K. Crites. 1976. Regulation of glycogen metabolism in astrocytoma and neuroblastoma cells in culture. J. Biol. Chem. 251:2015-2022.

Patterson, P. H. and L. L. Y. Chun. 1974. The influence of non-neuronal cells on catecholamine and acetylcholine synthesis and accumulation in cultures of dissociated sympathetic neurons. Proc. Natl. Acad. Sci. U.S.A. 71:3607-3610.

Penit, J., S. Jard, and P. Benda. 1974. Probenecide sensitive 3' ,5' -cyclic AMP secretion by isoproterenol stimulated glial cells in culture. FEBS Letters 41:156-160.

Perkins, J., E. H. MacIntyre, W. D. Riley, and R. B. Clark. 1971. Adenyl cyclase, phosphodiesterase, and cyclic AMP-dependent protein kinase of malignant glial cells in culture. Life Sci. 10:1069-1080.

Perkins, J. P., M. M. Moore, A. Kalisher, and Y. -F. Su. 1975. Regulation of cyclic AMP content in normal and malignant brain cells. Adv. Cyclic Nucl. Res. 5:641-660.

Pfeiffer, S. E. 1973. Clonal lines of glial cells. *In: Tissue Culture of the Nervous System*. G. Sato, ed. Plenum Press, New York. pp. 203-210.

Pfeiffer, S. E. and H. Eagle. 1976. Effect of ethanesulfonic acid buffers and pH on the accumulation of a nervous system-specific protein (S-100) and a glial-enriched enzyme in a clonal line of rat astrocytes (C6). J. Biol. Chem. 251:5112-5114.

Pfeiffer, S. E. and R. E. Pollack, 1971. *Animal Cell Culture*. Cold Spring Harbor Laboratory, Cold Spring Harbor, N.Y.

Pfeiffer, S. E. and W. Wechsler. 1972. Biochemically differentiated neoplastic clone of Schwann cells. Proc. Natl. Acad. Sci. U.S.A. 69:2885-2889.

Pfieffer, S. E., H. R. Herschman, J. Lightbody and G. Sato. 1970. Synthesis by a clonal line of rat glial cells of a protein unique to the nervous system. J. Cell. Physiol. 75:329-339.

Pfeiffer, S. E., H. R. Herschman, J. E. Lightbody, G. Sato and L. Levine. 1971. Modification of cell surface antigenicity as a function of culture conditions. J. Cell. Physiol. 78:145-151.

Pontén, J. 1971. *Spontaneous and virus induced transformation in cell culture*. Virology Monographs No. 8. Springer Verlag, Wien-New York.

Pontén J. 1975. Neoplastic human glia cells in culture. *In: Human tumor cells in vitro*. J. Fogh, ed. Plenum Press, New York, pp. 175-206.

Pontén, J. and E. H. MacIntyre 1968. Long term culture of normal and neoplastic human glia. Acta Pathol. Microbiol. Scand. 74:465-486.

Pontén, J., B. Westermark and R. Hugosson. 1969. Regulation of proliferation and movement of human glia-like cells in culture. Exp. Cell Res. 58:393-400.

Prasad, K. N. and S. Kumar. 1974. Cyclic AMP and the differentiation of neuroblastoma cells in culture. *In: Control of Proliferation in Animal Cells*. B. C. Clarkson and R. Baserga eds. Cold Spring Harbor Laboratory. pp. 581-594.

Puro, D. G. and M. Nirenberg. 1976. On the specificity of synapse formation. Proc. Natl. Acad. Sci. U.S.A. 73:3544-3548.

Rakic, P. 1971. Neuron-glia relationship during granule cell migration in developing cerebellar cortex. A Golgi and electronmicroscopic study in Macacus Rhesus. J. Comp. Neurol. 141:283-312.

Rall, T. and A. G. Gilman. 1970. The role of cyclic AMP in the nervous system. Neurosciences Res. Progr. Bull. 8:221-323.

Ransom, B. R. and S. Goldring. 1973. Ionic determinants of membrane potential of cells presumed to be glia in cerebral cortex of rat. J. Neurophys. 36:855-866.

Richelson, E. 1974. The culture of established clones for neurobiologic investigation. In: *Metabolic compartmentation and neurotransmission* S. Berl, D. D. Clarke and D. Schneider, eds. Plenum Press, New York. pp. 305-326.

Rechelson, E. and E. J. Thompson. 1973. Transport of neurotransmitter precursor into cultured cells. Nature (Lond.) New Biol. 241:201-204.

Righetti, P., E. P. Little and G. Wolf. 1971. Reutilization of amino acids in protein synthesis in HeLa cells. J. Biol. Chem. 246:5724-5732.

Robert, J., L. Freysz, L. Sensenbrenner, P. Mandel and G. Rebel. 1975. Gangliosides of glial cells: A comparative study of normal astroblasts in tissue culture and glial cells isolated on sucrose-ficoll gradients. FEBS Lett. 50:144-146.

Robert, J., P. Mandel and G. Rebel. 1976. Neutral lipids and phospholipids from cultured astroblasts. J. Neurochem. 26:771-777.

Robison, G. A., R. W. Butcher and E. W. Sutherland. 1968. Cyclic AMP. Ann. Rev. Biochem. 37:149-174.

Ronquist, G., G. Ågren, J. Pontén and B. Westermark. 1976. α-Aminoisobutyric acid transport into human glia and glioma cells in culture. J. Cell. Physiol. 89:433-440.

Rusca, G., P. Calissano and S. Alema. 1973. Identification of a membrane bound fraction of the S-100 protein. Brain Res. 49:223-227.

Salem, R. and J. deVellis. 1976. Protein kinase (PK) activity and cyclic AMP-dependent protein phosphorylation in subcellular fractions after norepinephrine treatment of glial cells. Fed. Proc. 35:296a.

Salem, R. D., R. Hammerschlag, H. Bracho and K. Orkand. 1975. Influence of potassium ions on accumulation and metabolism of [^{14}C] glucose by glial cells. Brain Res. 86:499-503.

Sato, G. H. and Y. Yasumura. 1966. Retention of differentiated function in dispersed cell culture. Trans. N.Y. Acad. Sci. 28:1063-1072.

Sato, S., K. Shimizu, T. Sugimura, T. Takaoka and H. Katsuta. 1972. Aldolase C in cultured mouse glioblastoma cells. Cancer Res. 32:1290-1292.

Schachner, M. 1974. Identification of a nervous system specific cell surface antigen NS-1 on G26 glioma tumor. Proc. Natl. Acad. Sci. U.S.A. 71:1795-1799.

Schimmer, B. 1971. Effects of catecholamines and monovalent cations on adenylate cyclase activity in cultured glial tumor cells. Biochim. Biophys. Acta. 252:567-573.

Schimmer, B. 1973. Influence of lithium ions on epinephrine-stimulated adenylate cyclase activity in cultured glial tumor cells. Biochim. Biophys. Acta. 327:186-192.

Schimmer, B. P., L. F. Stevenson, C. T. Hofstede, N. H. Cheung and A. Marks. 1974. Glial tumor cells deficient in thymidine kinase: isolation and characterization. Exp. Cell Res. 86:425-428.

Schlaeger, E. -J. and G. Köhler. 1976. External cyclic AMP-dependent protein kinase activity in rat C-6 glioma cells. Nature (Lond.) 260:705-707.

Schlegel, B. and J. Oey. 1975. Catecholamine-induced cyclic AMP and morphologic alterations in glioma cells. Naturwissensch. 62:534-535.

Schon, F. and J. S. Kelly. 1974a. Autoradiographic localisation of [^{3}H]-GABA and [^{3}H]-glutamate over satellite glial cells. Brain Res. 66:275-288.

Schon, F. and J. S. Kelly. 1974b. The characterization of [^{3}H]-GABA uptake into the satellite glial cells of rat sensory ganglia. Brain Res. 66:289-300.

Schrier, B. K. and D. L. Shapiro. 1973. Effects of N^6 -monobutyryl cyclic AMP on glutamate decarboxylase activity in fetal rat brain cells and glial tumor cells in culture. Exp. Cell Res. 80:459-462.

Schrier, B. K. and E. J. Thompson. 1974. On the role of glial cells in the mammalian nervous system. Uptake, excretion and Metabolism of putative neurotransmitters by cultured glial tumor cells. J. Biol. Chem. 249:1769-1780.

Schubert, D. 1974. Induced differentiation of clonal rat nerve and glial cells. Neurobiology 4:376-387.

Schubert, D. 1975. The uptake of GABA by clonal nerve and glia. Brain Res. 84:87-98.

Schubert, D. 1976. Proteins secreted by clonal cell lines. Exp. Cell Res. 102:329-340.

Schubert, D., W. Carlisle and C. Look. 1975. Putative neurotransmitters in clonal cell lines. Nature (Lond.) 254:341-343.

Schubert, D., S. Heinemann, W. Carlisle, H. Tarikas, B. Kimes, J. Patrick, J. H. Steinbach, W. Culp and B. L. Brandt. 1974. Clonal lines from the rat nervous system. Nature (Lond.) 249:224-227.

Schubert, D., H. Tarikas and M. LaCorbiere. 1976. Neurotransmitter regulation of adenosine 3', 5' -monophosphate in clonal nerve, glia, and muscle cell lines. Science 192:471-472.

Schultz, J., B. Hamprecht and J. W. Daly. 1972. Accumulation of adenosine 3':5' -cyclic monophosphate in clonal glial cells: labeling of intra-cellular adenine nucleotides with radioactive adenine. Proc. Natl. Acad. Sci., U.S.A. 69:1266-1270.

Schürch-Ratgeb, Y. and D. Monard. 1976. Glial factor activity in rat brain primary cultures. Experientia 32:811.

Schwartz, J. P. and J. V. Passonneau. 1974. Cyclic AMP-mediated induction of the cyclic AMP phosphodiesterase of C-6 glioma cells. Proc. Natl. Acad. Sci., U.S.A. 71:3844-3848.

Schwartz, J., N. R. Morris and B. McL. Breckenridge. 1973. Adenosine 3' :5' -monophosphate in glial tumor cells. Alterations by 5-bromodeoxy-uridine. J. Biol. Chem. 248:2699-2704.

Seeds, N. 1973. Differentiation of Aggregating Brain Cell Cultures. In: *Tissue Culture of the Nervous System.* G. Sato, ed. New York Plenum Press. pp. 35-54.

Sensenbrenner, M., N. Springer, J. Booher and P. Mandel. 1972. Histochemical studies during the differentiation of dissociated nerve cells cultivated in the presence of brain extracts. Neurobiology 2:49-60.

Sharma, S. K., M. Nirenberg and W. A. Klee. 1975a. Morphine receptors as regulators of adenylate cyclase activity. Proc. Natl. Acad. Sci., U.S.A. 72:590-594.

Sharma, S. K., W. A. Klee and M. Nirenberg. 1975b. Dual regulation of adenylate cyclase accounts for narcotic dependence and tolerance. Proc. Natl. Acad. Sci. U.S.A. 72:3092-3096.

Sheffield, W. D. and S. U. Kim. 1977. Basic protein radioimmunoassay as a monitor of myelin in tissue culture. Brain Res. 120, 193-196.

Shein, H. M. 1965. Propagation of human fetal spongioblasts and astrocytes in dispersed cell cultures. Exp. Cell Res. 40:554-569.

Shein, H. M. 1967. Neoplastic transformation induced by simian virus 40 in Syrian hamster neuroglial and meningeal cell cultures. Arch. Ges. Virusforsch. 22:122-142.

Shein, H. M. 1968. Neoplastic transformation of hamster astrocytes *in vitro* by simian virus 40 and polyoma virus. Science 159: 1476-1477.

Shein, H. M. 1970. Neoplastic transformation of hamster astrocytes and choroid plexus cells *in vitro* by polyoma virus. J. Neuropath. Exp. Neurol. 29:70-88.

Shein, S. H., A. Britva, H. H. Hess and D. J. Selkoe. 1970. Isolation of hamster brain astroglia by *in vitro* cultivation and subcutaneous growth, and content of cerebroside, ganglioside, RNA and DNA. Brain Res. 19:497-501.

Sheppard, J. R. and S. Bannai. 1974. Cyclic AMP and cell proliferation. In: *Control of proliferation in animal cells.* B. C. Clarkson and R. Baserga, eds. Cold Spring Harbor Laboratory. pp. 571-580.

Sheppard, J., T. H. Hendson and J. R. Larson. 1975. Adenosine 3',5'-monophosphate analogs promote a circular morphology of cultured Schwannoma cells. Science 187:179-181.

Shinwari, M. A. and J. de Vellis. 1972. Effects of cortisol and norepinephrine on the uptake of amino acids and glucose into cultured glial cells. Trans. Am. Soc. Neurochem. 3:121.

Sidman, R. L. and P. Rakic. 1973. Neuronal migration, with special reference to developing human brain: a review. Brain Res. 62:1-35.

Silberstein, S. D., H. M. Shein, and K. R. Berv. 1972. Catechol-o-methyl transferase and monoamine oxidase activity in cultured rodent astrocytoma cells. Brain Res. 41:245-248.

Silbert, S. M. and M. M. Goldstein. 1972. Drug induced differentiation of a rat glioma in vitro. Cancer Res. 33:1422-1427.

Sieghart, W. and M. Karobath. 1976. Uptake of taurine into subcellular fractions of C6 glioma cells. J. Neurochem. 26:981-986.

Skaper, S. D. and J. E. Seegmiller. 1976. Purine metabolism in thioguanine-resistant glioma cells. Exp. Cell Res. 100:415-417.

Sloboda, R. D., S. A. Rudolph, J. L. Rosenbaum and P. Greengard. 1975. Cyclic AMP-dependent endoginous phosphorylation of a microtubule associated protein. Proc. Nat. Acad. Sci. U.S.A. 72:177-181.

Sogin, D. C. 1976. 2',3'-cyclic nucliotide 3'-phosphohydrolase. J. Neurochem. 27:1333-1337.

Stallcup, W. B. and M. Cohn. 1976. Correlation of surface antigens and cell type in cloned cell lines from the rat central nervous system. Exptl. Cell Res. 98, 285-294.

Stefanovic, V., P. Mandel and A. Rosenberg. 1975. Concanavalin A inhibition of ecto-5'-nucleotidase of intact cultured C6 glioma cells. J. Biol. Chem. 250:7081-7083.

Stefanovic, V., J. Ciesielski-Treska, A. Ebel and P. Mandel. 1976a. Neuroblast-glia interaction. The effect of co-cultivation upon ecto-ATPase activity of neuroblastoma and glioma cells. Exp. Cell Res. 98:191-203.

Stefanovic, V., M. Ledig and P. Mandel. 1976b Divalent cation activated ecto-nucleoside triphosphate activity of nervous system cells in tissue culture. J. Neurochem. 27:799-805.

Stefanovic, V., P. Mandel and A. Rosenberg. 1976c. Ecto-5'-nucleotidase of intact cultured C6 rat glioma cells. J. Biol. Chem. 251:3900-3905.

Steinbach, J. H. and D. Schubert. 1975. Multiple modes of dibutyryl cyclic AMP-induced process formation by clonal nerve and glial cells. Exp. Cell Res. 91:449-453.

Stewart, J. A. 1972a. Subunit composition of the tissue specific bovine brain S-100 protein. Bioch. Biophys. Res. Comm. 46:1405-1410.

Stewart, J. A. 1972b. Tissue specific brain S-100. A demonstration of multiple proteins. Biochim. Biophys. Acta 263:178-192.

Stewart, J. A. 1975. Contribution of a change in mRNA half-life to the accumulation of the tissue specific S-100 protein during postnatal development of the mouse brain. Devel. Biol. 29:372:384.

Stewart, J. A. and I. M. Urban. 1972. The postnatal accumulation of S-100 protein in mouse central nervous system. Modulation of protein synthesis and degradation. Devel. Biol. 29:372-384.

Stoolmiller, A. C., A. Dorfman and G. H. Sato. 1972. Rat origin of CHB cells. Science 178:1308.

Su, Y. -F. and J. P. Perkins. 1974. Effect of norepinephrine on cyclic AMP content of astrocytoma cells: biphasic changes during prolonged exposure. Fed. Proc. 33:493a.

Su, Y. -F., L. Cubeddu X. and J. P. Perkins. 1976a. Regulation of adenosine 3' :5' monophosphate content of human astrocytoma cells: desensitization to catecholamines and prostaglandins. J. Cyclic Nucl. Res. 2:257-270.

Su, Y. -F., G. L. Johnson, L. Cubeddu, X., B. H. Leichtling, R. Ortmann and J. P. Perkins. 1976b. Regulation of adenosine 3' :5' -monophosphate content of human astrocytoma cells: mechanism of agonist-specific desensitization. J. Cyclic Nucl. Res. 2:271-285.

Suddith, R. L., H. T. Hutchison and B. Haber. 1975a. Uptake of choline by clonal cell lines of neural origin: Neural-glial differences. Trans. Amer. Soc. Neurochem. 6:47.

Suddith, R. L., P. J. Kelly, H. T. Hutchison, E. A. Murray and B. Haber. 1975b. *In vitro* demonstration of an endothelial proliferative factor produced by neural cell lines. Science 190:682-684.

Sundarraj, N., M. Schachner and S. E. Pfeiffer. 1975. Biochemically differentiated mouse glial lines carrying a nervous system specific cell surface antigen (NS-1). Proc. Natl. Acad. Sci. U.S.A. 72:1927-1931.

Synder, R. A. and R. O. Brady. 1970. Ganglioside patterns of cultured human glioma cells. Neurology (Minneap.) 20:412a.

Tolle, S. W., R. D. Dyson, R. W. Newburgh and J. M. Cardenas. 1976. Pyruvate Kinase isozymes in neurons, glia, neuroblastoma, and glioblastoma. J. Neurochem. 27:1355-1360.

Traber, J., K. Fischer, S. Latzin, and B. Hamprecht. 1975a. Morphine antagonizes action of prostaglandin in neuroblastoma and neuroblastoma X glioma hybrid cells. Nature (Lond.) 253:120-122.

Traber, J., R. Gullis and B. Hamprecht. 1975b. Influence of opiates on the levels of adenosine 3' :5' -cyclic monophosphate in neuroblastoma X glioma hybrid cells. Life Sci. 16:1863-1868.

Trachtenberg, M., P. Kornblith and J. Häuptli. 1972. Biophysical properties of cultured human glial cells. Brian Res. 38:279-298.

Ueda, T., H. Maeno and P. Greengard. 1973. Regulation of endogenous phosphorylation of specific proteins in synaptic membrane fractions from rat brain by adenosine 3': 5' -monophosphate. J. Biol. Chem. 248:8295-8305.

Uyemura, K., G. Vincedon, G. Gombos and P. Mandel. 1971. Purification and properties of S-100 protein fractions from sheep and pig brains. J. Neurochem. 18:429-438.

Varon, S. 1975a. Nerve growth factor and its mode of action. Exp. Neurol. 48:75-92.

Varon, S. 1975b. Neurons and glia in neural cultures. Exp. Neurol. Part 2 48:93-134.

Varon, S., C. Raiborn and P. A. Burnham. 1974a. Comparative effects of NGF and ganglionic non-neuronal cells for the support of dissociated ganglion neurons in culture. J. Neurobiology 5:355-371.

Varon, S., C. Raiborn and S. C. Norr. 1974b. Association of antibodies to NGF with ganglionic non-neurons (glia) and consequent interference with their neuron supportive action. Exp. Cell Res. 88:247-256.

Vernadakis, A., R. Nidess, M. L. Timiras and R. Schlesinger. 1976. Responsiveness of acetylcholinesterase and butyrylcholinesterase activities in neural cells to age and cell density in culture. Exp. Cell Res. 97:453-457.

Villegas, J., C. Sevcik, F. V. Barnola and R. Villegas. 1976. Grayanotoxin, veratrine, and tetrodotoxin sensitive sodium pathways in the Schwann cell membrane of squid nerve fibers. J. Gen. Physiol. 67:369-380.

Volpe, J. J. and J. C. Marasa. 1975. Regulation of palmitic acid synthesis in cultured glial cells: effects of lipid on fatty acid synthetase, acetyl-CoA carboxylase, fatty acid and sterol synthesis. J. Neurochem. 25:333-340.

Volpe, J. J. and J. C. Marasa. 1976a. Regulation of palmitic acid synthesis in cultured glial cells: effects of glucocorticoid on fatty acid synthetase, acetyl-CoA carboxylase, fatty acid and sterol synthesis. J. Neurochem. 27:841-845.

Volpe, J. J. and J. C. Marasa. 1976b. Longterm regulation by theophylline of fatty acid synthetase, acetyl-CoA carboxylase and lipid synthesis in cultured glial cells. Biochim. Biophys. Acta 431:195-205.

Volpe, J. J., K. Fujimoto, J. C. Marasa and H. C. Agrawal. 1975. Relation of C-6 glial cells in culture to myelin. Biochem. J. 152:701-703.

Wahlström, T., E. Linder, E. Saksela and B. Westermark. 1974. Tumor-specific membrane antigens in established cell lines from gliomas. Cancer 34:274-279.

Walum, E. 1975. Glucose uptake into cultured tumour cells from the nervous system. Thesis. University of Göteborg, Sweden.

Walum, E. and A. Edström. 1976. Alterations in the characteristics of 2-deoxy-D-glucose transport in cultured human glioma cells during cell growth. Exp. Cell Res. 100:111-116.

Walum, E., C. Nissen, L. Hertz and A. J. Edström. 1974. Growth curve and rate of oxygen consumption by human glioma cells in culture. J. Neurochem. 23:881-883.

Wardell, W. 1966. Electrical and pharmacological properties of mammalian neuroglial cells in tissue culture. Proc. Roy. Soc. Ser. B. 165:326-361.

Wechsler, W., S. E. Pfeiffer, J. A. Swenberg and A. Koestner. 1973. S100 protein in methyl- and ethylnitrosourea-induced tumors of the rat nervous system. Acta Neuropath. 24:287-303.

Wechsler, W., M. A. Ramadan and S. E. Pfeiffer. 1977. Morphological and biochemical characteristics of transplantable neurogenic tumors induced with ethylnitrosourea in inbred BD-IX rats. J. Natl. Cancer Inst., in press.

Weinstein, R. S. and P. L. Kornblith. 1971. Ultrastructure of a cloned astrocytoma in tissue culture. Cancer 27:1174-1181.

Weiss, B., H. Shein and P. Uzunov. 1973. Induction by norepinephrine of a specific molecular form of cyclic 3',5' -AMP phosphodiesterase in cloned rat astrocytoma cells (C-2A). Fed. Proc. 32:679a.

Westermark, B. 1973. The deficient density-dependent growth control of human malignant glioma cells and virus-transformed glia-like cells in culture. Int. J. Cancer 12:438-451.

Westermark, B. and Å. Wasteson. 1975. The response of cultures of human normal glial cells to growth factors. Adv. Met. Disorder 8:85-100.

Westermark, B. and Å. Wasteson. 1976. A platelet factor stimulating human normal glial cells. Exp. Cell Res. 98:170-174.

Westermark, B., J. Pontén and R. Hugosson. 1973. Determinants for the establishment of permanent tissue culture lines from human gliomas. Acta Path. Scand. Sect. A 81:791-805.

Westermark, B., Å. Wasteson and K. Uthne. 1975. Initiation of DNA synthesis of stationary human glia-like cells by a polypeptide fraction from human plasma containing somatomedin activity. Exp. Cell Res. 96:58-62.

Wiche, G. and R. D. Cole. 1976. Reversible *in vitro* polymerization of tubulin from a cultured cell line (rat glial cell clone C_6). Proc. Natl. Acad. Sci. U.S.A. 73:1227-1231.

Wicks, W. D. 1974. Regulation of protein synthesis by cyclic AMP. Adv. Cyclic Nucl. Res. 4:335-438.

Wilson, S. H., B. K. Schrier, J. L. Farber, E. J. Thompson, R. N. Rosenberg, A. J. Blume and M. Nirenberg. 1972. Markers for gene expression in cultured cells from the nervous system. J. Biol. Chem. 247:3159-3169.

Wroński, A. and A. von der Decken. 1975. Synthesis of brain-specific acidic proteins in rat and mouse cerebral slices. Acta Physiol. Scand. 95:482-493.

Wroński, A. and A. von der Decken. 1976. Protein synthesis with special reference to S-100 protein in brain slices from rats receiving a restricted protein supply. Acta Physiol. Scand. 97:20-30.

Zanetta, J. P., P. Benda, G. Gombos and I. G. Morgan. 1972. The presence of 2', 3' -cyclic AMP 3' -phosphohydrolase in glial cells in tissue culture. J. Neurochem. 19:881-883.

Zimmerman, H. M. 1955. The nature of gliomas as revealed by animal experimentation. Amer. J. Pathol. 31:1-29.

Zimmerman, H. M. and H. Arnold. 1941. Experimental brain tumors. I. Tumors produced with methylcholanthrene. Cancer Res. 1:919-938.

Zomzely-Neurath, C., C. York and B. W. Moore. 1973. *In vitro* synthesis of two brain-specific proteins (S100 and 14-3-2) by polyribosomes from rat brain. I. Site of synthesis and programming by polysome-derived messenger RNA. Arch. Biochem. Biophys. 155:58-69.

NEURONAL CELL LINES

PHILLIP G. NELSON

Behavioral Biology Branch
National Institutes of Child Health and Human Development
National Institutes of Health
Bethesda, Maryland

I. Introduction

Neuronal cell lines are defined by two characteristics: 1) They are capable of indefinite cell division and 2) They are capable of exhibiting some interesting neuronal properties. The continued capability to express the neuronal phenotype in the face of unlimited mitotic activity implies a remarkable stability of the committed state. The mitotic activity of the cell lines allows a number of genetic manipulations to be performed, including cloning and somatic cell hybridization. This means that an essentially unlimited variety of genotypes and phenotypes can be generated. Large quantities of relatively homogeneous neuronal cells can be readily obtained and the process of development of the neurobiologic properties which the various clones are capable of expressing can be studied as the clones are caused to differentiate.

Two different orders of questions can be addressed with these kinds of preparations. On the one hand they can be used as a convenient material for studying any individual neurobiologic property of interest that they express. Thus, detailed characterization of various features of the voltage dependent sodium ionophore has been done on neuroblastoma cells using different toxins (Catterall, 1975a,b and see below). On the other hand, the relationship between various properties has been examined to see if there are patterns of expression — some "genetic program" for neurodifferentiation — such that coordinate or exclusionary regulatory mechanisms must be involved. Some evidence for a reciprocal expression of choline acetyltransferase and tyrosine hydroxylase has been obtained (Amano *et al.*, 1972; Minna *et al.*, 1972) implying that differentiation of a given clone into either cholinergic or catecholaminergic clones, but not both, may occur. Exceptions to this rule have been found, however (Prasad, 1975), and indeed recent experiments with primary cell cultures of sympathetic neurons show quite conclusively that interconversion between the cholinergic and catecholaminergic state can occur as well as joint expression of these two characteristics in the same neuron (Furshpan *et al.*, 1976; Landis, 1976). At present, it would appear that neurodifferentiation in neuroblastoma, while not entirely random in character, is not intrinsically highly organized.

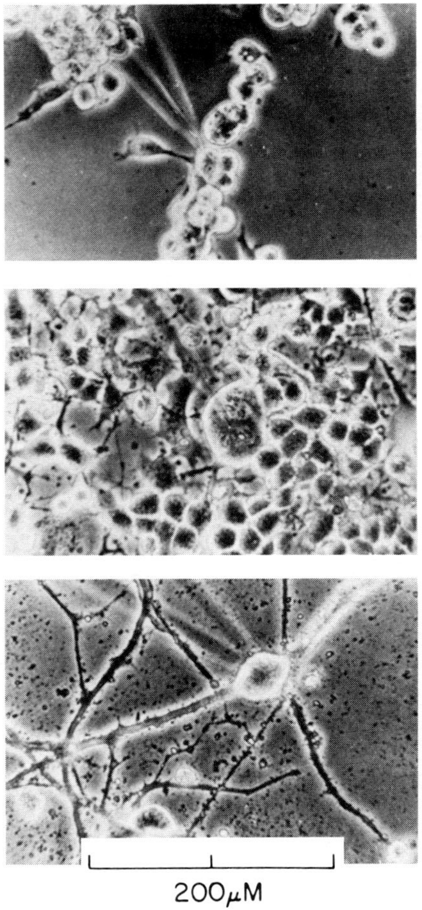
200μM

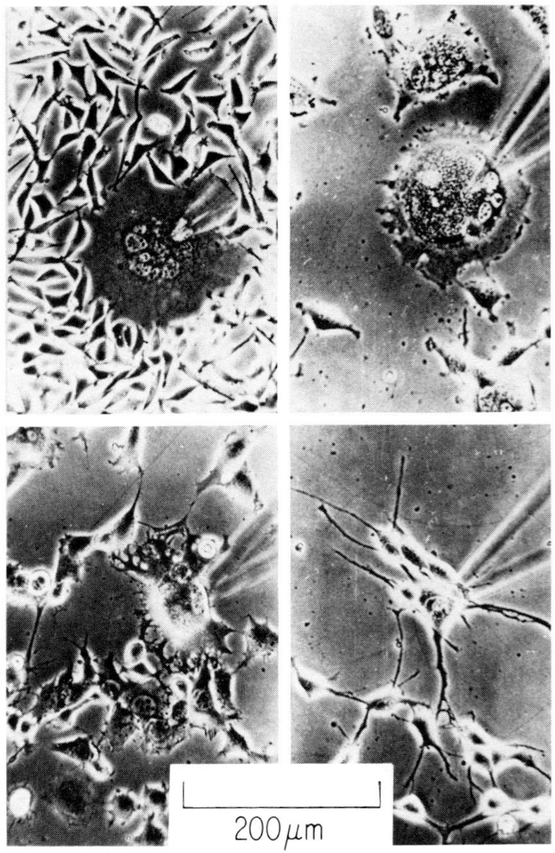
200μm

Even if a rather random process of phenotypic expression is the rule, however, it is clear that the cell lines are capable of a very substantial and complex degree of neurodifferentiation. A wide variety of neural transmitter-related enzymes and receptors have been identified in different clones. The apparatus required for the development of functioning synapses has been shown to develop in a completely clonal system (Christian *et al.*, 1977). It is an optimistic but not unreasonable assumption that each component of the molecular apparatus necessary for the initiation and development of specific synaptic networks can be isolated and studied in these systems. The present state of research is far removed from this, however, and the present review will deal only with a few examples of current work that illustrate advantages and limitations of the clonal approach.

II. Technology and General Properties of Clonal Material

Substantial reviews with reference to original material are available (Sato, 1973; McMorris *et al.*, 1973; Giller *et al.*, 1975; Fischbach and Nelson, 1977) and only a brief outline of clonal technology will be given here. Four steps can be defined in establishing a comprehensive *system* of cell lines for neurobiologic research.

A. Tumors must be identified that may represent material of neurobiologic interest.

B. The cells in the tumors must be established in tissue culture and clonal lines isolated.

C. Genotypic diversity must be broadened by genetic manipulations such as mutagenesis, selection (Breakefield and Nirenberg, 1974) and somatic cell hybridization.

D. Differentiated function must be elicited and characterized. Clonal and developmental differences in neuroblastoma cells shown in Fig. 1. illustrate the broad range of the neuronal phenotype that can be attained.

A. Neuronal Tumors

Neuronal cell lines have been derived from two kinds of sources. Most work has been done on spontaneously arising tumors of the nervous system, in particular a mouse neuroblastoma, the C-1300. This arose in 1940 and

FIG. 1. (Left) Phase contrast microphotographs of clone N-18 neuroblastoma cells under three culture conditions. Left upper: log phase culture 1 day after plating. Left middle: grown in medium with no serum for 4 days. Left lower: aminopterin selected cells 4 days after plating.
(Right) Neuroblastoma and L-cell parents with two daughter hybrid clones. Upper left: B82 L cell; Lower left: N4TG1 neuroblastoma; Upper right: NL-2 hybrid; Lower right: NL-1 hybrid. (From Peacock *et al.*, 1972 and Minna *et al.*, 1971)

was carried by serial passage in mice until Augusti-Tocco and Sato (1969) adapted it to tissue culture. Since then, this tumor has been extensively cloned by a variety of techniques in several laboratories. Human neuroblastomas have been similarly adapted and cloned. Nitroso-urea compounds have been found effective in inducing a number of central nervous system tumors and a large number of such tumors of neuronal, glial or mesenchymal origin have been established in culture (Schubert *et al.*, 1974; Stallcup and Cohen, 1976a, b; Pfeiffer, this volume).

B. Cloning

Entire lines can be derived from a single parent cell by a variety of isolation and propagation procedures. In principle, completely homogeneous subpopulations can be derived from a heterogeneous mixture of cell types.

It must be realized that problems of heterogeneity are not entirely resolved by the cloning methodology. The mouse neuroblastoma is not diploid for instance, and typically a rather wide range of chromosomal number is found when cloned lines are karyotyped. Some degree of instability in the karyotype is usually found with continued propagation of the clones. This reservation must be kept in mind in evaluating findings with the cloned material.

C. Somatic Cell Hybridization

Somatic cell hybridization is a powerful cell biologic technique useful both for generating cell lines with a variety of genotypes and for analyzing mechanisms regulating the expression of various phenotypes (Minna *et al.*, 1971, 1972; Amano, *et al.*, 1974; Hamprecht, 1974; McMorris and Ruddle, 1974; Greene *et al.*, 1975). In effect, the genetic material from two parental lines can be incorporated into a single genome and propagated as a new "hybrid" line. With subculturing, chromosomes are slowly lost from the hybrid cells so that a gradual deletion of genetic material, usually selective for one of the parental genomes, occurs with time.

D. Differentiation

A number of factors have been shown to affect the expression of the differentiated neuronal phenotype in neuroblastoma and a comprehensive review is available (Prasad, 1975). The role of cyclic nucleotides is discussed elsewhere in this volume (K. N. Prasad). The mitotic activity of the clonal material poses an obvious problem for the use of cloned cells in studying differentiated function. Even though differentiation and mitosis are not necessarily incompatible, it is clear that in a cell undergoing mitosis, the biochemical and morphologic processes involved in mitosis will greatly complicate study of differentiated functions. Most of the procedures developed to induce differentiation have consequently involved a complete or partial shutting off of mitosis. Indeed the neuroblastoma system has been extensively studied with regard to the question of control of mitosis (Prasad, 1975).

III. Experimental Studies

Neuroblastoma cells exhibit an obviously neuronal phenotype in that they are highly electrically excitable and develop processes which produce action potentials (Fig. 3 and Harris and Dennis, 1970). Process formation (Seeds *et al.*, 1970) and synthesis of transmitter related enzymes are two important neurobiologic functions which have been studied extensively in neuroblastoma cultures. They will not be dealt with at length here (see Fischbach and Nelson, 1977, for review).

A. Receptors

Electrophysiologic studies of the mouse C-1300 neuroblastoma had revealed early that these cells were responsive to acetylcholine and some of these studies will be described below. A large body of experimentation deals with biochemical measures of receptor activation, in particular intracellular concentrations of cyclic nucleotides. (Gilman and Nirenberg, 1971; Gilman and Minna, 1973; Matsuzawa and Nirenberg, 1975; Penit *et al.*, 1976). Much of this work has been reviewed elsewhere (Prasad, 1975) and in this volume (K. N. Prasad) and I will describe here only some studies on the opiate receptor in a neuroblastoma × glioma hybrid clone, the NG108-15.

A number of neuroblastoma and hybrid cell lines were examined for specific binding of narcotics and of these only one, the NG108-15 hybrid clone, was found to have large numbers (3×10^5/cell) of narcotic receptors. (Klee and Nirenberg, 1974, 1976). Non-specific binding occurred in all lines; the specific binding was characterized by its saturable nature and the 100-fold greater affinity of these saturable receptors for levorphanol as compared to dextrorphan. Morphine was found to inhibit adenylate cyclase activity in the hybrid and this effect was inhibited by the narcotic antagonist, naloxone (Sharma *et al.*, 1975b). The regulation of adenylate cyclase was found to be of a dual nature (Sharma *et al.*, 1975a; Lampert *et al.*, 1976). An initial inhibition by opiates of adenylate cyclase activity results in a lower than normal level of intracellular cAMP. In the continued presence of opiates total levels of adenylate cyclase rises so that intracellular cAMP levels return to normal. When the opiate is then withdrawn the release from opiate inhibition of the increased amount of adenylate cyclase activity results in a higher than normal level of intracellular cAMP. It was

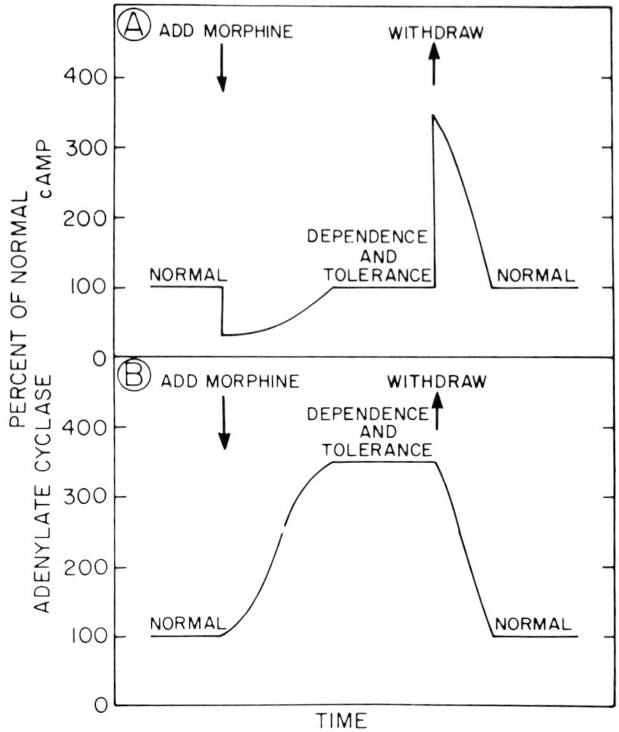

FIG. 2. Model of the role of adenylate cyclase regulation in the development of morphine tolerance and dependence. Part A shows the effect of morphine upon cAMP levels; and part B the effects of the opiate upon adenylate cyclase activity as a function of time. (From Sharma *et al.*, 1975a)

proposed that these effects might be involved in narcotic tolerance and dependence. The proposed mechanism is shown diagramatically in Fig. 2.

B. Voltage Dependent Na^+ Ionophores

When excitable cells are incubated with radioactive sodium, they take up measurable amounts of the label and an activator of the voltage-dependent sodium mechanism, veratridine, increases the amount of label taken up. The veratridine induced increase in Na^+ uptake is completely blocked by tetrodotoxin, which blocks voltage sensitive sodium channels in many excitable cells, including neuroblastoma. Catterall and Nirenberg (1973) have shown that this assay is a reliable and quantitative measure of electrical excitability in that inexcitable clones of neuroblastoma do not exhibit the increased uptake induced by veratridine and in excitable lines, differentiation is accompanied by an increase in the assayed activity.

Catterall has used this method in studying the mode of action of 4 neurotoxins (Catterall, 1975a,b). He identified 3 functionally separable components of the action potential mechanism. One regulatory site binds scorpion toxin and another regulatory site binds veratridine, batrachotoxin and aconitine; these two sites exhibit positive heterotopic cooperative interaction. These two sites in turn interact allosterically to control the activity of the ion transport site. Tetrodotoxin binds to and inactivates this latter site. This assay system has made possible detailed study of the ion dependence of the interaction of the different sites with the toxins. Routine screening of a variety of clones not favorable for electrophysiologic study is feasible and the techniques have been used in characterizing lines derived from the nervous system (Stallcup and Cohn, 1976a). It can also be used in studies of acetylcholine receptors in muscle cells (Catterall and Nirenberg, 1973; Catterall, 1975c). Both for the voltage dependent sodium conductance in neuroblastoma and for the ACh activated sodium conductance in muscle the rate of Na^+ flux is sufficiently high to require a channel mechanism rather than a carrier mechanism.

C. Electrophysiologic Studies

The mouse neuroblastoma system has proven technically favorable for the analysis of membrane electrical properties using intracellular microelectrode recording techniques. The cells are large and stable recordings can be obtained for long periods, up to several hours. Membrane potential and resistance and action potentials and responsiveness to iontophoretically applied neuroactive compounds can be monitored (Figs. 3 and 5). (Nelson *et al.*, 1969; Minna *et al.*, 1971; Peacock *et al.*, 1972, 1973; Chalazonitis and Greene, 1974; Hamprecht, 1974; Chalazonitis *et al.*, 1975).

1. The action potential. The neuroblastoma action potential involves both Na^+ and Ca^{2+} currents, and a variety of spike configurations occurs in different cells. Voltage clamp studies have been possible and results such as those shown in Fig. 4 indicate that the kinetics of Na^+, K^+ and Ca^{2+} currents are similar to those measured in *in vivo* systems (Moolenaar and Spector, 1977). The action potential mechanism is highly regulated and large differences in the degree of excitability occur between different clones and within a given clone under different conditions of differentiation.

Spontaneous action potentials are generated in some cells and slow oscillatory potentials are sometimes seen. Thus a variety of electrogenic mechanisms occur in the neuroblastoma cells and are available for study in this system. A large hyperpolarizing post-tetanic hyperpolarization occurs in some NG108-15 hybrid cells (P. G. Nelson and C. N. Christian, unpublished). This hyperpolarization in most instances has a reversal potential in the range of -70 to -80 mV and probably represents an increase in membrane permeability to K^+. The prolonged time course (several

A

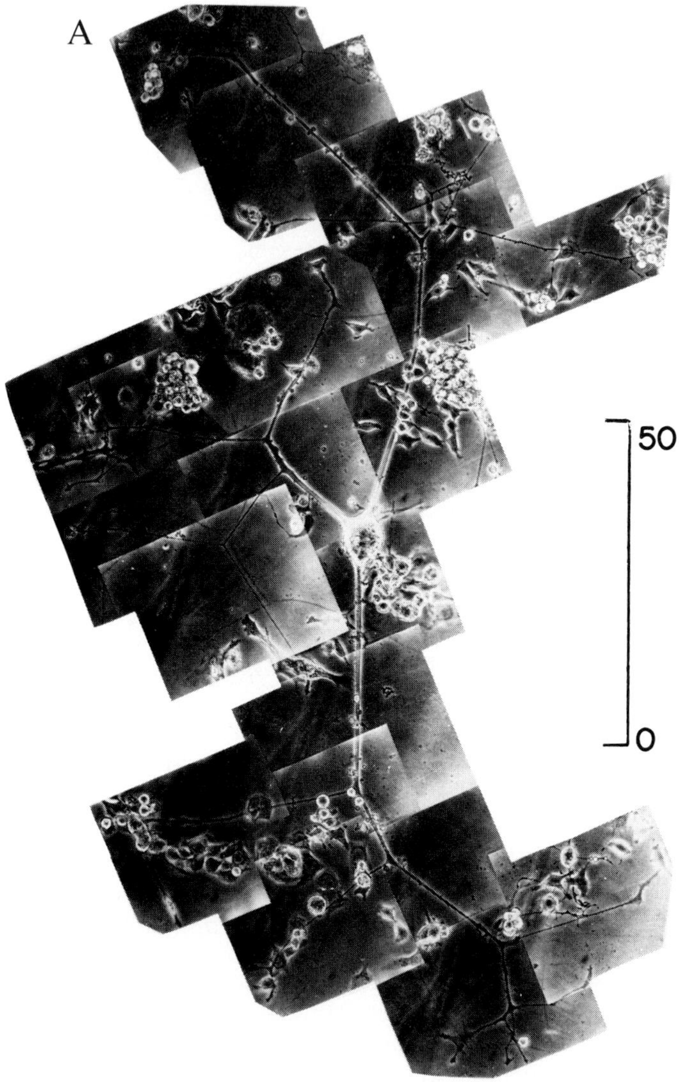

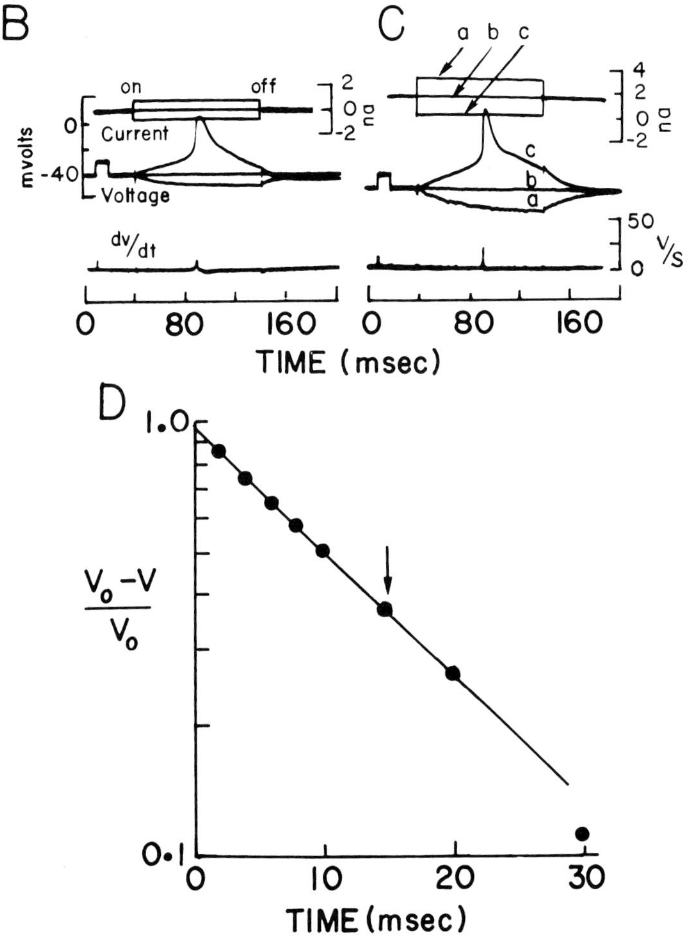

FIG. 3. A. Phase contrast photomicrograph of cell; scale as indicated. B. and C. Oscilloscope recordings obtained with an intracellular microelectrode from the cell shown in A. Upper traces represent the current passed across the cell membrane. Middle traces show the membrane potential changes produced by the current pulses. Lowest traces show the derivative of membrane potential with respect to time, dv/dt. D. A plot on semilogarithmic axes of the hyperpolarizing voltage transient (trace "a" in Fig. C above). Vo represents the voltage change at the end of the current pulse. The membrane time constant is the time required for $\frac{V_o - V}{V_o}$ to fall to a value 1/e of its initial value. (From Nelson *et al.*, 1971).

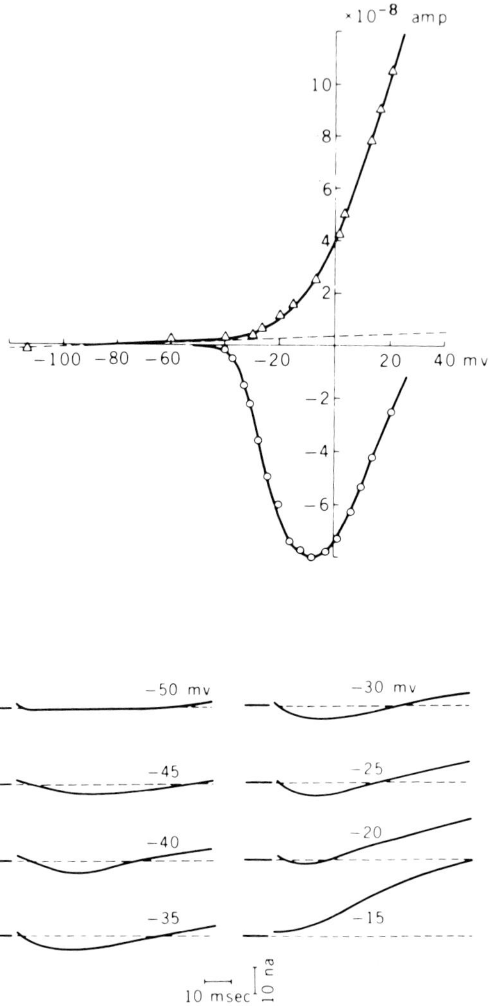

FIG. 4. (Upper) Current-voltage relations for peak values of net inward current (circles) and steady-state values of delayed outward current (triangles) in a neuroblastoma cell, clone N115. The holding potential was -85 mv. Steady-state values were measured at 50 msec after the onset of the clamp step. The dashed line represents leakage current.
(Lower) Slow membrane currents elicited at different levels of membrane potential in a voltage-clamped neuroblastoma cell bathed in sodium-free solution containing 10 mM calcium (sodium chloride was isosmotically replaced by tris-Cl). The holding potential was -80 mv. Delayed outward current can be seen with depolarizations beyond -25mv. (From Moolenaar and Spector, 1977)

seconds) and the fact that the action potential involves an ingress of Ca^{2+} ions suggests that the increased K permeability may be due to an intracellular accumulation of Ca^{2+} since such an increase in free intracellular Ca^{2+} has been shown to increase surface membrane K^+ permeability in a number of systems (Meech, 1974).

The neuroblastoma system is a favorable one for studying developmental and molecular aspects of electrical excitability. The possible modulating role that electrical activation may play on metabolic activity to neural tissue is also an attractive area for study. What might be the long term metabolic consequences, for instance, of the Ca^{2+} currents or Ca^{2+} mediated afterpotentials characteristic of neural spike activity? Despite the central role of calcium in a number of intracellular control processes, little information is available as to its possible mediation of a link between electrical and metabolic activity in neurons.

2. Chemosensitivity. A number of electrophysiologically defined receptors have been identified on neuronal tumor cells. Acetyl choline when iontophoresed onto neuroblastoma cells can produce either depolarizing excitatory responses or hyperpolarizing inhibitory response (Harris and Dennis, 1970; Peacock *et al.*, 1973; Peacock and Nelson, 1973). These can occur on the same cell in a topographically organized manner as shown in Fig. 5. Thus, not only can these cells synthesize receptors and associated specific ionophores but a mechanism for an orderly spatial distribution of these components must be present.

Both acetyl choline and serotonin receptors occur on some NG108-15 hybrids. Both compounds produce depolarizing responses, but separate receptors are involved since cross desensitization does not occur. Dopamine also produces a depolarizing response in these cells, but cross desensitization with the serotonin does occur so these two compounds apparently operate via a common receptor. Morphine in micromolar concentrations blocks the 5 HT response. This morphine block of the 5 HT response is not reversed by naloxone and is therefore apparently not mediated by an opiate type of receptor (C. N. Christian, P. G. Nelson, M. Nirenberg unpublished). A different dopamine response has been described in hybrid cells which was blocked by morphine, apparently via a narcotic receptor in that the morphine block was reversed by naloxone (Myers *et al.*, 1975).

The relationship between electrophysiologically defined chemosensitivity and biochemical measurement of receptor functions, especially those involving changes in cyclic nucleotides, requires comment. A priori, *no* relationship need exist between the two types of responses. Separate processes are involved in agonist binding to receptors and in the coupling between such receptors and either ionophore activation in the case of electrophysiologic responses, or cyclase or phosphodiesterase activity in the case of biochemically defined responses. Different membrane components

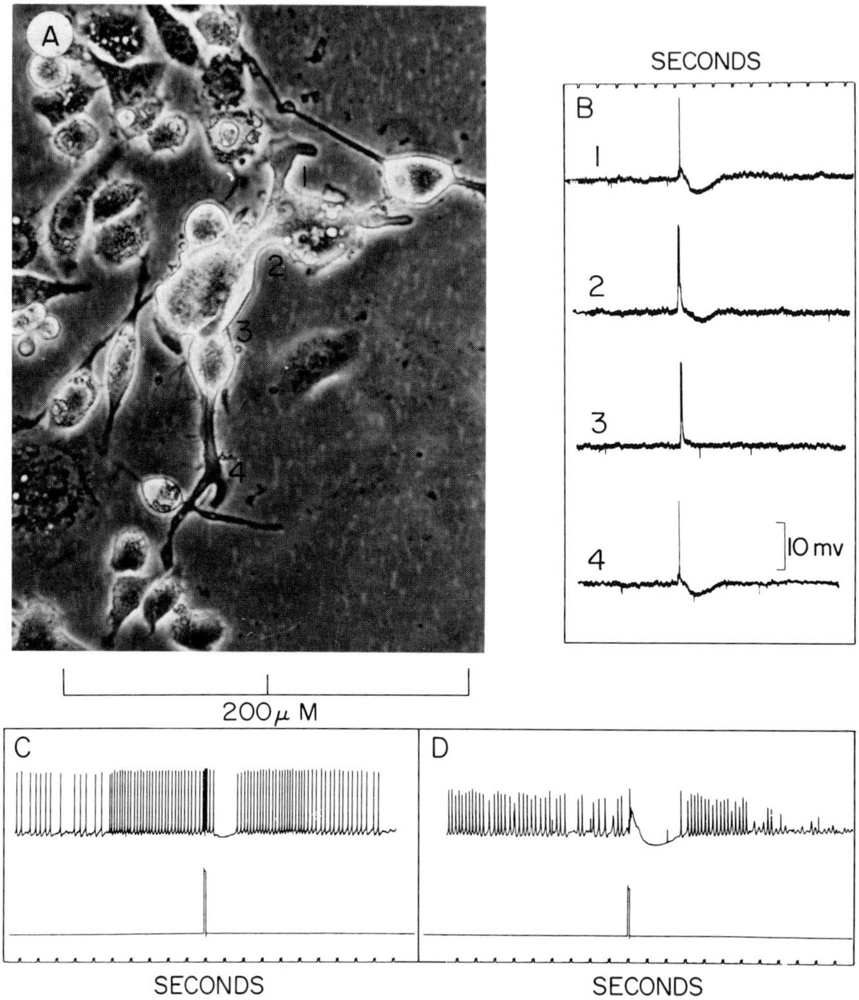

FIG. 5. Morphology and electrical response of mouse (NA) x human (fibroblast) hybrid NM VII-10. (A) Microphotograph of cell whose responses are shown in (B-D) (B 1-4). Responses to pulses of applied acetylcholine at positions on cell surface corresponding to the location of the traces. (C, D) Repetitive generation of action potentials interrupted by hyperpolarization (upper traces) produced by ACh pulse (ca. 30 nA) (lower traces). (D) Was recorded 9 h after (C). Time marks in B-D represent 1 sec. Voltage calibration in B applied to C, D. (From Peacock *et al.*, 1973)

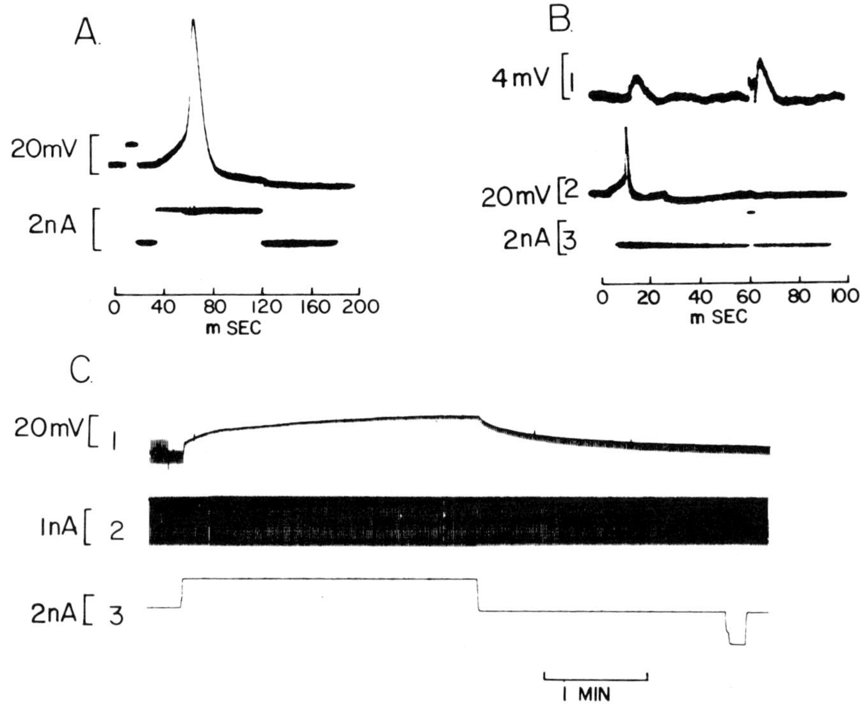

FIG. 6. Characteristics of the G-8 myotube and its innervation by the NG108-15 hybrid. All records are from tissue culture dishes of fused myotubes, co-cultured for at least 3 days with hybrid cells.

(A) An oscilloscope trace of an action potential (upper trace) in a myotube elicited by anodal break stimulation. A constant hyperpolarizing current (lower trace) was passed through a second electrode in the myotube, producing a membrane potential of -65 mV, then briefly terminated to produce an action potential.

(B) An oscilloscope trace demonstrating the similarity of a myotube depolarization produced by an action potential in a nearby hybrid and by direct focal application of acetylcholine from an iontophoretic pipette.

1) Intracellular recording from a G-8 myotube.

2) Intracellular record of contiguous hybrid cell caused to spike by depolarizing current passed through recording pipette.

3) Current passed through an acetylcholine iontophoretic pipette placed close to the myotube membrane.

(C) Chart record demonstrating a maintained myotube depolarization and conductance increase to a continued application of acetylcholine.

1) Myotube membrane potential; resting potential 48 mV.

2)Depolarizing pulses (1/sec) passed through a second intracellular pipette.

3) Current passed through an acetylcholine iontophoretic pipette. Baseline represents -2 nA holding current, which was turned off for approximately 3 minutes to release the drug. (From Christian *et al.*, 1977)

may be involved as in the case of morphine action where a narcotic receptor is involved in its action on adenylate cyclase while a non-narcotic receptor mediated process is involved in its action on the 5 HT response. Possible long term interaction between the two classes of response are clearly of great interest.

3. Synapse Formation. Functional cholinergic synapses can be found between clonal lines of nerve (Nelson *et al.*, 1976; Christian *et al.*, 1977) and muscle. Such synapses have been studied most extensively in a system involving primary mouse myotubes and a hybrid cell line formed between a mouse neuroblastoma and a rat glioma. This line was interesting because it exhibited high choline acetyltransferase (CAT) activity although neither parental line did so (Amano *et al.*, 1974). Synapses have been found between muscle cells and a non hybrid, high CAT neuroblastoma line so that the glial genome is not essential for synapse formation. Clonal muscle cells can also form synapses so that a biclonal pre- and postsynaptic system can be utilized for studies of synaptogenesis (Figs. 6, 7, 8). These synapses would seem to be of a rather primitive sort anatomically and with a low probability of transmitter release (Christian *et al.*, 1977). Nevertheless the existence of the synaptic connections show that the complete mechanism for transmitter synthesis storage and depolarization initiated release are functional in clonal cells. Membrane components responsible for the establishment of cell contacts are similarly present. No species specificity was found with regard to the establishment of synapses and it may be that the specificity is low during the initial relatively primitive stage of synapse formation (Puro and Nirenberg, 1976).

Various components of the synaptic machinery can be independently analyzed in the clonal systems.

1. Intercellular adhesion and its dependence on certain cell surface molecules and soluble agglutinins is being studied in a number of systems (Barbera *et al.*, 1973) and such techniques could be applied to the clonal preparations.

2. Choline uptake and choline acetyltransferase activity can be measured.

3. Acetylcholine accumulation within the neuroblastoma can be studied biochemically and the occurrence of vesicles of the sort usually associated with cholinergic terminals can be identified electron microscopically (Daniels and Hamprecht, 1974).

4. An assay has been developed for depolarization-dependent ACh release (McGee *et al.*, 1977).

5. Excellent biochemical, electrophysiologic and morphologic probes of receptor (especially ACh receptor) synthesis, degradation and distribution are available.

With these methodologic capabilities and with clonal material capable

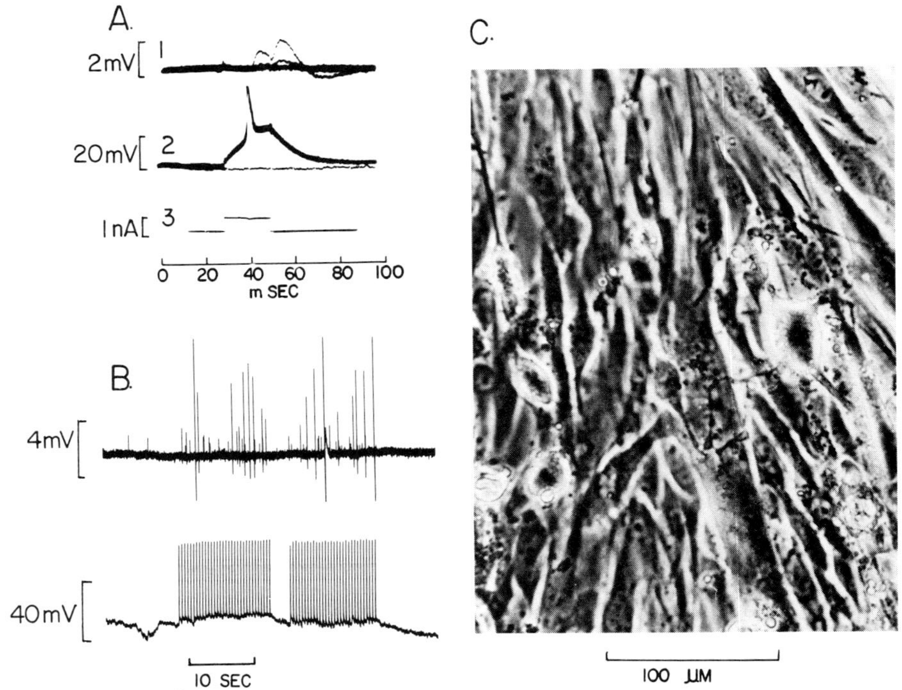

FIG. 7. (A) Multiple oscilloscope sweeps demonstrating a functional connection between the hybrid and myotube pictured in c.
1) Myotube recording; resting membrane potential is 45 mV.
2) Hybrid recording.
3) Current through hybrid recording electrode.
(B) Chart record of same connection showing dependence of myotube responses (upper trace) on spikes elicited in hybrid (lower trace) by depolarizing currents passed through hybrid recording pipette.
(C) Appearance of a co-culture plated 10 days previously with myotubes and 3 days previously with hybrid cells. (From Christian, *et al.* 1977).

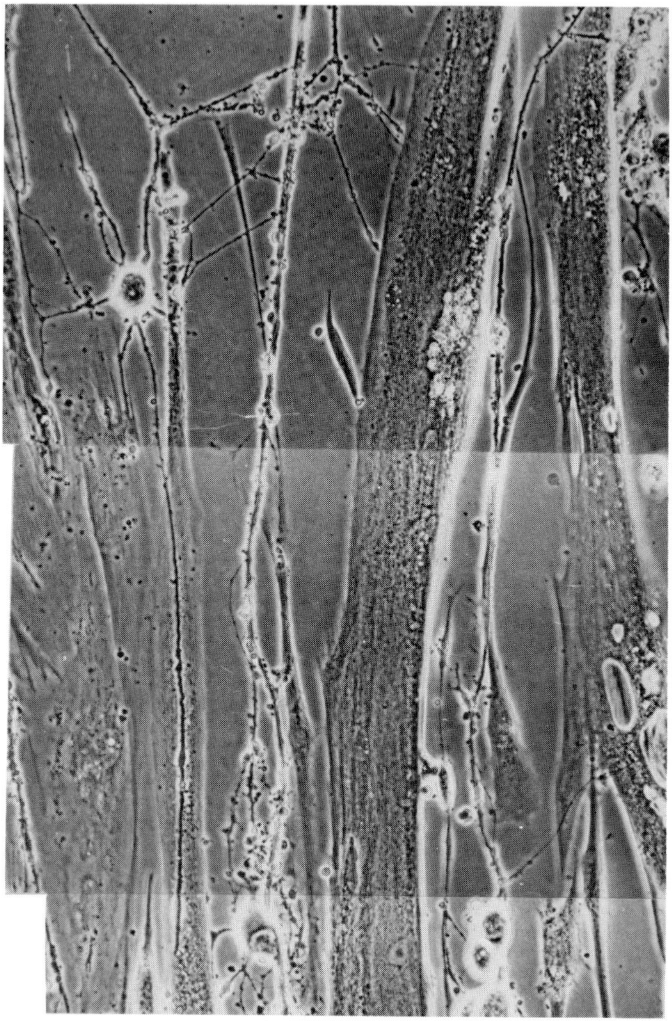

FIG. 8. Phase contrast microphotograph of co-cultured NG108-15 hybrid cells and G-8 clonal mouse muscle.

of exhibiting formation of at least a primitive form of synapses, a detailed study of each important component of the synaptogenesis process should be possible. Clones deficient at each stage can be developed. For instance, lines with high CAT activity have been identified which do not synthesize and store ACh, and lines that store, but do not exhibit depolarization dependent release of ACh, are also available (R. McGee, S. H. Wilson and M. W. Nirenberg, personal communication). Do clones exist that release ACh but which do not form synapses? Would these lack surface recognition molecules essential for synapse formation? Do some clones of neuroblastoma alter postsynaptic receptor distribution (Steinbach *et al.*, 1973) while others do not? The clonal lines, it is to be hoped, may be useful in demonstrating the molecular basis for each critical component of the synaptic machinery.

References

Amano, T., B. Hamprecht and M. Kemper. 1974. High activity of choline acetyltransferase induced in neuroblastoma × glia hybrid cells. Exp. Cell Res. 84:391-408.

Amano, T., E. Richelson and M. Nirenberg. 1972. Neurotransmitter synthesis by neuroblastoma clones. Proc. Nat. Acad. Sci. U.S.A. 69:258-263.

Augusti-Tosco G. and G. Sato. 1969. Establishment of functioning clonal lines of neurons from mouse neuroblastoma. Proc. Nat. Acad. Sci. U.S.A. 64:311-315.

Barbera, A. J., R. B. Marchase and S. Roth. 1973. Adehesive recognition and retinotectal specificity. Proc. Nat. Acad. Sci. U.S.A. 70:2482-2486.

Breakefield, X. O. and M. Nirenberg. 1974. Selection for neuroblastoma cells that synthesize certain transmitters. Proc. Nat. Acad. Sci. U.S.A. 71:2530-2533.

Catterall, W. A. 1975a. Activation of the action potential Na^+ ionophore of cultured neuroblastoma cells by veratridine and batrachotoxin. J. Biol. Chem. 250:4053-4059.

Catterall, W. A. 1975b. Cooperative activation of action potential Na^+ ionophore by neurotoxins. Proc. Nat. Acad. Sci. U.S.A. 72:1782-1786.

Catterall, W. A. 1975c. Sodium transport by the acetylcholine receptor of cultured muscle cells. J. Biol. Chem. 250:1776-1781.

Catterall, W. A. and M. Nirenberg. 1973. Sodium uptake with activation of action potential ionophores of cultured neuroblastoma and muscle cells. Proc. Nat. Acad. Sci. U.S.A. 70:3759-3763.

Chalazonitis, A. and L. A. Greene. 1974. Enhancement in excitability properties of mouse neuroblastoma cells cultured in the presence of dibutyryl cyclic AMP. Brain Res. 72:340-345.

Chalazonitis, A., L. A. Greene and W. Shain. Excitability and chemosensitivity properties of a somatic cell hybrid between mouse neuroblastoma and sympathetic ganglion cells. Exp. Cell Res. 96:225-238.

Christian, C. N., P. G. Nelson, J. Peacock and M. Nirenberg. 1977. Synapse formation between two clonal cell lines. Science 196:995-998

Daniels, M. P. and B. Hamprecht. 1974. The ultrastructure of neuroblastoma glioma somatic cell hybrids. J. Cell Biol. 63:691-699.

Fischbach, G. D. and P. G. Nelson. 1977. Cell cultures in neurobiology. In: *Handbook of Physiology*. Chapter 20. American Physiological Society, Bethesda, Md.

Furshpan, E. J., P. R. Macleish, P. H. O'Lague and D. D. Potter. 1976. Chemical transmission between rat sympathetic neurons and cardiac myocytes developing in microcultures: evidence for cholinergic, adrenergic and dual-function neurons. Proc. Nat. Acad. Sci. U.S.A. 73:4225-4229.

Giller, E. L., X. O. Breakefield, C. N. Christian, E. A. Neale and P. G. Nelson. 1975. Expression of neuronal characteristics in culture: some pros and cons of primary cultures and continuous cell lines. In: *Golgi Centennial Symposium Proceedings.* M. Santini, ed. Raven Press, N. Y. pp. 603-623.

Gilman, A. G. and J. D. Minna. 1973. Regulation of adenosine 3′, 5′ -cyclic monophosphate metabolism in hybrid somatic cells. In: *Frontiers in Catecholamine Res.* Pergamon Press, pp. 307-310.

Gilman, A. G. and M. Nirenberg. 1971. Regulation of adenosine 3′, 5′ -cyclic monophosphate metabolism in cultured neuroblastoma cells. Nature (Lond.). 234:356-358.

Greene, L. A., W. Shain, A. Chalazonitis, X. Breakefield, J. Minna, H. G. Coon and M. Nirenberg. 1975. Neuronal properties of hybrid neuroblastoma × sympathetic ganglion cells. Proc. Nat. Acad. Sci. U.S.A. 72:4923-4927.

Hamprecht, B. 1974. Cell cultures as model systems for studying the biochemistry of differentiated functions of nerve cells. In: *25 Colloquium der Gesellschaft für Biologische Chem.* L. Jaenicke, ed. Springer-Verlag, Berlin.

Harris, A. S. E. and M. J. Dennis. 1970. Acetylcholine sensitivity and distribution on mouse neuroblastoma. Science 167:1253-1255.

Klee, W. A. and M. Nirenberg. 1974. A neuroblastoma × glioma hybrid cell line with morphine receptors. Proc. Nat. Acad. Sci. U.S.A. 71:3474-3477.

Klee, W. A. and M. Nirenberg. 1976. Mode of action of endogenous opiate peptides. Nature (Lond.) 263:609-612.

Lampert, A., M. Nirenberg and W. A. Klee. 1976. Tolerance and dependence evoked by an endogenous opiate peptide. Proc. Nat. Acad. Sci. U.S.A. 73:3165-3167.

Landis, S. C. 1976. Rat sympathetic neurons and cardiac myocytes developing in microcultures: correlation of the fine structure of endings with neurotransmitter function in single neurons. Proc. Nat. Acad. Sci. U.S.A. 73:4220-4224.

Matsuzawa, H. and M. Nirenberg. 1975. Receptor-mediated shifts in cGMP and cAMP levels in neuroblastoma cells. Proc. Nat. Acad. Sci. U.S.A. 72:3472-3476.

McGee, R., P. Simpson, C. Christian, M. Mata, P. Nelson and M. Nirenberg. 1977. Regulation of acetyl choline storage and release from clonal hybrid cells. Trans. Am. Soc. Neurochem. 8:128.

McMorris, F. A. and F. H. Ruddle. 1974. Expression of neuronal phenotypes in neuroblastoma cell hybrids. Develop. Biol. 39:226-246.

McMorris, F. A., P. G. Nelson and F. H. Ruddle. 1973. Contributions of clonal system to neurobiology. Neurosci. Res. Prog. Bull. 11:412-536.

Meech, R. W. 1974. The sensitivity of Helix aspersa neurons to injected calcium ions. J. Physiol. (Lond.) 237:259-277.

Minna, J., D. Glazer and M. Nirenberg. 1972. Genetic dissection of neural properties using somatic cell hybrids. Nature (Lond.) New Biology 235:225-231.

Minna, J., P. Nelson, J. Peacock, D. Glazer and M. Nirenberg. 1971. Genes for neuronal properties expressed in neuroblastoma × L cell hybrids. Proc. Nat. Acad. Sci. U.S.A. 68:234-239.

Moolenaar, W. H. and I. Spector. 1977. Membrane currents examined under voltage clamp in cultured neuroblastoma cells. Science 196:331-333.

Myers, P. R., D. R. Livengood and W. Shaw. 1975. Effect of morphine on a depolarizing dopamine response. Nature (Lond.) 257:238-240.

Nelson, P. G. 1975. Nerve and muscle cells in culture. Physiol. Reviews 55:1-61.

Nelson, P., C. Christian and M. Nirenberg. 1976. Synapse formation between clonal neuroblastoma × glioma hybrid cells and striated muscle cells. Proc. Nat. Acad. Sci. U.S.A. 73:123-127.

Nelson, P. G., J. H. Peacock, T. Amano and J. Minna. 1971. Electrogenesis in mouse neuroblastoma cells *in vitro.* J. Cell Physiol. 77:337-352.

Nelson, P., W. Ruffner and M. Nirenberg. 1969. Neuronal tumor cells with excitable membranes grown *in vitro.* Proc. Nat. Acad. Sci. U.S.A. 64:1004-1010.

Peacock, J. H., F. A. McMorris and P. G. Nelson. 1973. Electrical excitability and chemosensitivity of mouse neuroblastoma × mouse or human fibroblast hybrids. Exp. Cell Res. 79:199-212.

Peacock, J., J. Minna, P. Nelson and M. Nirenberg. 1972. Use of aminopterin in selecting electrically active neuroblastoma cells. Exp. Cell Res. 73:367-377.

Peacock, J. H. and P. G. Nelson. 1973. Chemosensitivity of mouse neuroblastoma cells *in vitro*. J. Neurobiol. 4:363-374.

Penit, J., J. Huot, and S. Jard. 1976. Neuroblastoma cell adenylate cyclase: direct activation by adenosine and prostaglandins. J. Neurochem. 26:265-273.

Prasad, K. N. 1975. Differentiation of neuroblastoma cells in culture. Biol. Rev. 50:129-165.

Puro, D. G. and M. Nirenberg. 1976. On the specificity of synapse formation. Proc. Nat. Acad. Sci. U.S.A. 73:3544-3548.

Sato, G. (ed.) 1973. *Tissue culture of the nervous system. Current Topics in Neurobiology*, Vol. I, New York, Plenum Press.

Schubert, O., S. Heinemann, W. Carlisle, H. Tarikas, B. Kimes, J. Patrick, S. H. Steinbach, W. Culp and B. L. Brandt. 1974. Clonal cell lines from the rat central nervous system. Nature (Lond.) 249:224-227.

Seeds, N. W., A. G. Gilman, T. Amano and M. W. Nirenberg. 1970. Regulation of axon formation by clonal lines of a neural tumor. Proc. Nat. Acad. Sci. U.S.A. 66:160-167.

Sharma, S. K., W. A. Klee and M. Nirenberg. 1975a. Dual regulation of adenylate cyclase accounts for narcotic dependence and tolerance. Proc. Nat. Acad. Sci. U.S.A. 72:3092-3096.

Sharma, S. K., M. Nirenberg and W. A. Klee. 1975b. Morphine receptors as regulators of adenylate cyclase activity. Proc. Nat. Acad. Sci. U.S.A. 72:590-594.

Stallcup, W. B. and M. Cohn. 1976a. Electrical properties of a clonal cell line as determined by measurement of ion fluxes. Exp. Cell Res. 98:277-284.

Stallcup, W. B. and M. Cohn. 1976b. Correlation of surface antigens and cell type in cloned cell lines from the rat central nervous system. Exp. Cell Res. 98:285-297.

Steinbach, J. H., A. J. Harris, S. Patrick, D. Schubert and S. Heinemann. 1973. Nerve-muscle interaction *in vitro*. The role of acetylcholine, J. Gen. Physiol. 62:255-270.

Phenotypic Cell Expression

GENETIC CONTROL OF NEURAL DEVELOPMENT

R. K. HUNT

Jenkins Biophysical Laboratories
Johns Hopkins University
Baltimore, Maryland

I. Introduction

Developmental genetics of the nervous system is an infant science, whose best years lie far in the future and whose promise is rooted more in faith and expectation than in any objective measure of present performance. Yet its fascination, its capacity to provoke speculation and debate, cannot be questioned, linking audiences as broad as nature/nurture psychologists and molecular biologists concerned with cell differentiation and recognition. It would be pretentious to say that this paper surveys current knowledge on genetic neuroembryology, although that currently impossible task shares a kindred spirit with the present effort. A rather more accurate statement of what is to follow is that it attempts to examine a diverse body of work on neurogenesis, and to garner clues and seek out trends with which to construct some provisional notions of gene control in brain development. What emerges is a picture, sketched in the broadest of outlines and dabbed with bits of speculation, of how the genome *might* control neurogenesis and how the problem might be most strategically attacked.

II. Mutants Affecting Neurogenesis

Certainly the ultimate, if not the initial, explanation depends upon the isolation and characterization of single gene mutations affecting neural ontogeny. Over the past decade, a number of excellent laboratories have taken up this charge. Yet the work has proved slow and difficult; and none of the mutants analyzed to date offers a major comment on the questions distinguishing alternative modes of gene control (Britten and Davidson, 1969; Wolpert, 1971; Kaufman, 1973; Garcia-Bellido, 1975). Nevertheless, a brief overview of the kinds they have provided offers, I believe, a useful assessment of how far and how near we have come, and how future efforts can best accommodate the lessons of the past (Fig. 1).

The first and most fundamental sort of information to be derived from mutant analysis is the identification (and spatiotemporal localizaton) of wild-type gene products. Because mutation selects molecules on the basis of functional priorities rather than on the basis of quantitative superiorities, crucial proteins can be identified which are present in concentrations too low to be detected by the routine screening methods of analytical biochemistry. The minor proteins involved in T4 phage assembly (Wood and Edgar, 1967) for example, might never have been found without the instructive guidelines provided by their mutants. Short of actually identifying the protein, however, mutants can be used to obtain a phenomenological characterization of the protein, addressing the question: "what commands and in what language do single genes communicate with neural cells?". The second category of information to be gotten from mutants is a sense of gene integration, the topographic landscape over which genes work in the embryo, the additivity or hierarchical nature of their interaction. The studies of Lewis (1964) and Garcia-Bellido (1975) on the bithorax system of Drosophila and its role in imaginal disc development is one such example; the T locus studies in the mouse is another. Finally, mutants can provide unique experimental preparations for correlation analysis, the sorts of 'odd brains' that one could never produce with a knife or lesioning electrode.

At the moment, however, there are no neurogenetic mutants in the first category, and many of the currently used experimental species with favorable genetics (Drosophila, nematodes) are likely to prove stubborn in yielding the molecular identity of wild-type gene products. Likewise there is no information of the second kind (integrative action of genes) beyond the preliminary descriptions of exaggerated neurogenetic perturbations in *reeler/weaver* double homozygotes in the mouse (Yoon and Coffey, 1975).

MUTANTS AFFECTING NEUROGENESIS

1. Identification of wild-type gene products (molecules / commands)
2. Integrative gene control
3. Unique experimental preparations
   a) cell deletions
   b) cell-chemistry correlations
   c) 'secondary' failures
      granule cell degeneration in ***weaver*** mouse
      'en cascade' degeneration in *staggerer* mouse
   d) 'secondary' successes
      ectopic / misaligned cells survive
      ectopic / misaligned cells make normal connections
      ↓ migration does not reduce cell production
      (*reeler* mouse ; *spastic* axolotl)

FIG. 1. Review of information that can be derived from analysis of neurological mutants, as discussed in section II of text.

Information on the "commands" type is beginning to become available: individual mutants affect single axonal bifurcations on sets of six nematode neurons (Brenner, 1975), morphologic maturation of paired sets of chemosensory neurons (Lewis and Hodgkin, 1976), specific migratory capacities shared by several different neuroblast cell types (Sidman, 1973), and death or abortive development of two or more functionally unrelated neuron cell types (Van Deusen, 1973; Brenner, 1975; Mullen *et al.*, 1976; Lewis and Hodgkin, 1976). As yet, however, the information is too scanty to inform on major alternative classes of command networks, especially as they operate in early neurogenesis.

Thus, the overwhelming majority of existing studies of neurogenetic mutants fall into our final category: correlation analysis in odd brains. Some of these studies have provided dramatic evidence on what neurons can do when pressed, on the secondary 'successes' of neurons developing in the perturbed milieu of the mutant. In the *reeler* mouse, for example, a number of cell types migrate abnormally, and take up abnormal positions and alignments in the final brain neuropil (Sidman, 1973). Nevertheless, these cells survive and often manage to entertain near normal afferent connections, thus showing normal physiologic response properties (Drager, 1975). A somewhat more selective example is provided by the *spastic* axolotl. Ide and his co-workers (1976) have shown that this vestibular mutant contains a translocation, probably due to failed migration, of a single subpopulation of cerebellar Purkinje cells. Malpositioned, misaligned and abnormally packed into the junctional zone between the cerebellar auricle and *area acoustico-lateralis*, these cells nevertheless have normal response properties, and they survive into full adult life (Ide, 1977; Ide *et al.*, 1977). Finally, the Siamese cat (a temperature-sensitive mutant at the albinism locus) has provided two exquisite examples of the adaptive behavior of lateral geniculate fibers innervating visual cortex after miswiring of retinogeniculate connections (Hubel and Wiesel, 1971; Berman and Cynader, 1972; Guillery *et al.*, 1974).

If these correlation analyses definitively show that cells need not be in their proper position for afferent fibers to find them and make specific synaptic connections with them, other mutant studies emphasizing the "secondary failures" of neurons in mutant brains have been much less definitive. The *weaver* mouse, for example, shows a failure of granule cell migration that was ascribed to a primary failure of Bergmann glia fibers to develop normally in the mutant (Rakic and Sidman, 1973); breeding the *weaver* gene into a different genetic background, however, revealed the granule cell lesion in the absence of morphologic abnormalities in the glia (Sotello and Changeux, 1974b; Bignami and Dahl, 1974). Similarly, the 'encascade' degeneration of granule and olivary cells in *staggerer* mice was inferred to be secondary to Purkinje cell lesions and to be reflective to the normal dependencies of these interneurons on making connections with Purkinje cells (Sotello and Changeux, 1974a). In both cases, primary and secondary lesion sites are being ascribed in the absence of any information on the sites of wild-type gene product expression; in fact, such causal linkages can never be definitively inferred. However, recent successes in embryonic grafting of developing mammalian brain tissue (Das and Altman, 1971; Lund and Hauschka, 1976) and in production of tetraparental individuals with mosaic neural structures (Mullen and Lavail, 1976; West, 1976; Dewey *et al.*, 1976) offer the promise of chimera analysis (cf. Benzer, 1973) for sorting out such primary and secondary defects.

Finally, a very recent third strategy has been added to the odd-brain correlation analysis: the search for missing molecular specificities in brain tissue from which a single cell type has been deleted by mutation. To this end, Mallet *et al.* (1974) were able to identify a number of bands characteristically absent from the granular cerebellum of *weaver.* Likewise, Harris *et al.* (1976) have used spectrophotometric measurements on Drosophila retinal mutants to localize three discrete photopigments to the three classes of photoreceptors of the fly retina. Such experiments, especially when followed up by *in situ*-localization experiments in genotypically normal tissue (for example, with fluorescent antibodies to the missing molecular species), can provide important first steps to characterizing molecular specificities of individual neuronal cell types.

In summary, then, mutant analysis shows much promise, and has already begun to provide definitive information on genetic commands, on a variety of issues attending terminal cytodifferentiation of neurons and their subsequent morphogenesis and maturation (e.g., on the dispensibilities of cell position and alignment in synaptic specificity) and on specific molecular correlates of cell-type deletions. Nevertheless, the mutants are too few and their focus is directed away from the early determination events so central to the primary issue of gene control. For the moment, any interim views or any clues to the genetic control mechanisms in neurogenesis must be drawn from experiments on genotypically wild-stock.

III. Comparative Cytology of the Neuron

A major source of clues to possible modes of genetic control lies in the systematic comparison of neurons and the cataloguing of whatever morphological, physiological and biochemical parameters are available to us. Elsewhere, I have argued from such a comparative survey that neurons differ from one another on several levels, and that the diversity is stereotyped, systematic and related to cell position. Further, it appears to be built up out of unique combinations of (shared) neuronal properties in concert with rather small numbers of truly unique chemical properties (Hunt, 1975a).

These principles are evident in the amphibian retina, which contains six distinct neuronal cell-types: rods and cones (photoreceptors); amacrine, bipolar, and horizontal cells (interneurons); and ganglion cells. Each phenotype has a characteristic shape and neurite pattern and (so far as it has been studied) appears to synthesize a characteristic set of neurotransmitters and postynaptic receptors. The characteristic synaptic relationships among these neurons are locally confined, such that a cohort of adjacent photoreceptors connect with the few bipolar (and horizontal) cells immediately in front of them; these in turn converge onto the ganglion cell (and interact with amacrine cells) directly in front of them. The ganglion cell axons join to form the optic nerve and innervate the visual centers of the brain, including the midbrain *optic tectum;* there they form a point-to-point retinotopic pattern of connections with tectal neurons that reproduces a "map" of the retina across the tectal surface. In addition to the six basic cell phenotypes in the retina, several 'subclasses' exist with subtle but characteristic anatomic, physiologic, and biochemical differences (Hunt and Jacobson, 1974a). The comparative cytology of the amphibian retina, and some simple experiments on it, serve to illustrate three principles in the generation of neuronal diversity.

First, *different levels of specificity are seen when different neurons are compared.* Thus, the ganglion cell differs from the rod in its position, ability to conduct action potentials, broadly arborized dendrites, long axon innervating the brain, connectivity with amacrine cells, ability to transduce specific chemical signals into electrotonic potentials, as well as its inability to synthesize rhodopsin, elaborate outer segments or transduce light into electrotonic potentials. The three subclasses of ganglion cell in the frog (Maturana *et al.*, 1960) differ in their dendritic tree anatomy, physiologic response properties, the precise pattern of synaptic connections they entertain with retinal interneurons, and the depth at which they terminate beneath the tectal surface. Likewise, the several subclasses of photoreceptors differ in absorption spectra, microscopic anatomy and retinal wiring (Liebman and Entine, 1968; Reuter, 1969; Laties and Liebman, 1970). In contrast to these aspects of cell type specificity, individual ganglion cells at different positions in the retina differ in their affinities for different loci on the tectal surface, a property termed locus specificity (Hunt and Jacobson, 1972a). Thus the genetic decisions needed to generate the diversity of ganglion cells and rods are superimposed upon (but likely to be very different from) those which generated the qualitative differences between Type I and Type II ganglion cells, and between ganglion cells of the nasal vs. temporal region of the retina.

Second, both *locus specificity and cell-type specificity are related to cell position.* The development of locus specificity, and perhaps even its maintenance, are position-dependent processes (Hunt and Jacobson, 1974a;

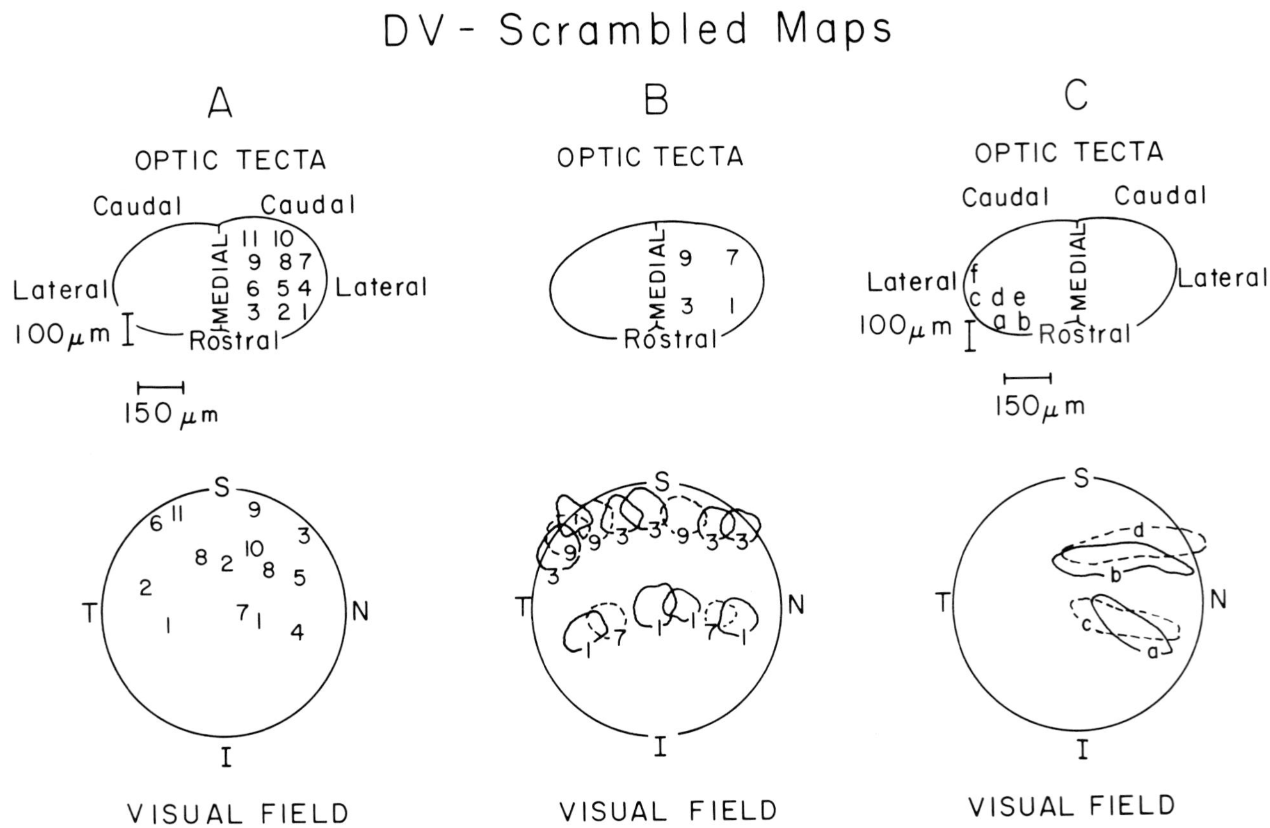
DV - Scrambled Maps
A
OPTIC TECTA
Caudal
Caudal
Lateral
Lateral
MEDIAL
Rostral
100 μm
150 μm
VISUAL FIELD
B
OPTIC TECTA
MEDIAL
Rostral
VISUAL FIELD
C
OPTIC TECTA
Caudal
Caudal
Lateral
Lateral
MEDIAL
Rostral
100 μm
150 μm
VISUAL FIELD

Yoon, 1973; see Sect. IV). The relation of cell phenotypic specificity to position is more complex. By subsystem, the differentiation of specific neuronal cell-types is roughly position-dependent; thus, ganglion cells and rods differentiate in the retina and nowhere else. Within subsystems, however, phenotypic differentiation is not position-dependent, but rather the ability to assume a particular final position is a property of the particular cell type. Thus, the six neuronal phenotypes of the frog retina are all generated from a single, monolayer neuroepithelium; and they will assume their normal proximodistal sequence of layers even if the presumptive retina is inverted in the mediolateral axis (Eakin, 1947; Hunt, 1977c; see also Goldberg, 1976).

Third, the specificity of the individual terminally differentiated neuron or neuronal cell-type is defined, to a large degree, by a *unique combination of neuronal properties*, drawn from a total set of shared neuronal properties, each of which is displayed two or more times across the nervous system. Thus, the ganglion cells may be the only cells in the retina which conduct true action potentials, exhibit long axons, and show synaptic affinity for tectal neurons; but each of these properties appears, in concert with others which the ganglion cells lack, in a variety of other neuronal phenotypes. This concept of information-sharing in neuronal diversity extends to position (rods and cones occupy a common layer) and orientation (the long axis of amocrines and horizontals is parallel to, that of all other retinal neurons is perpendicular to, the retinal layers), and to physiologic response (amocrines and ganglion cells show spike discharges, horizontal and photoreceptor cells show slow hyperpolarizations; cf. Werblin and Dowling, 1969).

The principle of combinatorial determination of neuron specificity extends to the synaptic specificities by which individual neurons recognize and associate selectively with their correct postsynaptic cells. Xenopus retinal ganglion cells, for example, form point-to-point connections that are precise and specific in the three dimensions of tectal space: across the tectal surface, they connect topographically, constructing a *map* of their cell body

FIG. 2. Demonstration that independent (combinatorial) *specifiers* mediate the discrimination and locus specificity involved in topographic mapping. In the visual field, numbers (A) or numbered receptive field patches (B,C) are used to show the area of visual field from which a shot of light evoked action potential recordings when a microelectrode was placed at the corresponding numbered position on the tectum. This figure shows the residual discrimination for one tectal axis (mediolateral) but not for the second tectal axis (rostro-caudal) by Xenopus retinal ganglion cells assembling a retinotectal map. The disorganized projection (A) from the right eye visual field to the left optic tectum (contrast this with the topographically ordered projection in Fig. 3B and C) does not show, at first glance, the accurate dorso-ventral specificity in the mapping behaviour of single optic fibers as assessed by single unit receptive fields of optic terminals (B) or of tectal output cells.

positions in the retina; beneath the correct surface locus, the three ganglion cell classes connect at different depths in the tectal neuropil. Recently we have used microsurgical procedures to 'jam' orientational signals between the embryo and early eye-bud, to produce various kinds of scrambled retinotectal maps. Analysis of the residual discriminations of optic fibers in the context of these scrambled maps indicates that at least three distinct recognition systems *(specifiers)* exist on each optic fiber tip to specify its connection in the tectum (Hunt, 1977a-c).

In one group of animals, the visuotectal projection topography was completely disorganized. Single-unit electrophysiologic analysis, on optic fibers and tectal output cells, showed that individual ganglion cells had made functional connections in the tectum, but had done so randomly with respect to their retinal position and to surface topography of the tectum.

Nevertheless, when the depth distribution of optic-fiber single units were analyzed, virtually all frogs showed a nearly normal segregation of the three ganglion cell subclasses to different depths in the tectal neuropil. These results demonstrate that an independent *phenotypic specifier* exists on the ganglion cell, and it can execute the depth discrimination even if the specifier(s) for map locus is/are absent or dysfunctional.

More recently we have applied similar analytical methods to frogs which show partial disorganization of the visuotectal projection (Fig. 2). Such cases arise as minority results after serial eye rotations or temporary fusion of eye fragments, at the early stages 24-33 of pattern programing in the eye cup. Although a variety of partial disorganizations have been recorded, undoubtedly the most common and most interesting are those in which the projection is orderly in the vertical dimension across the entire visual field, disorganized in the horizontal dimension of visual field, (Fig. 2) or the converse. Single unit analyses on the former reveal that retinotectal connectivity may be entirely normal in the dorsoventral axis of the retina (and mediolateral axis in the tectum), yet nearly random in the anteroposterior axis of the retina. Thus ganglion cells lying each horizontal line across the retina selected the appropriate mediolateral level in the tectum but randomly connected on any site along this mediolateral isobar. The reciprocal discrimination was spared in the vertically scrambled frogs.

These studies on scrambled retinotectal maps suggest that the information content of each ganglion cell is coded as a combinatorial set of at least *three* discrete specifiers that independently discriminate three different features of the target. One (Class X) specifier discriminates rostrocaudal level in the tectum; one (Class Y) discriminates mediolateral level; and one (Class Z, or *phenotypic)* specifier discriminates depth. Minimum estimates of the coded information content of the differentiated retina for wiring tectal connections (and based on the current limit of resolution of our electrophysiologic mapping technique: 10 cell diameters) is 30 specifiers of Class X, 30 of Class Y, and 3 of Class Z. Thirty specifiers in a class is well within the capacity of the genome to generate diversity and the capacity of cells to maintain and recognize uniqueness. Perhaps the most dramatic implication of such a code, however, is its applicability for neuronal wiring in general. By coupling the same array of 30 X and 30 Y specifiers with a *different* set of phenotypic (Z) specifiers in the auditory, leminscal, corticospinal and other pathways — all of the topographic projections of the CNS could be specified at only a modest increment in information over that used in the visual system alone.

The principle of combinatorial coding has another important implication. The diversity-generating mechanisms in neuronal differentiation must (1) work a standard set of genetic ingredients into the specific gene combinations that define the phenotypic specificity of Purkinje cells and rods; (2) avoid nonsense combinations, such as Purkinje cells which transduce light; (3) selectively discharge the Purkinje cell combination in the hindbrain area and the rod combination in the retina; and (4) superimpose individualized instruction for locus specificity upon single cells within an array of a single phenotype. Such a mechanism may squeeze a great deal of specificity out of a few structural genes. But this demands a complex system of regulatory genetics, which may require a long sequence of diverging intermediates to generate the many specific gene sets seen in the final state.

IV. Position-dependence of Locus Specificities

If the synaptic specificities of retinal ganglion cells are encoded combinatorially across the ganglion cell population, the *locus* specificities (Class X and Y specifiers) are also position-dependent; and these locus specificities provide the first and most convenient inroad into the general interrelation of cell position and developmental programing in the CNS. Beginning with the classic Salamander experiments of Stone (1944, 1960; see also Hunt and Piatt, 1974) a "critical period" has been recognized in the embryogenesis of visual patterns: before a critical embryonic stage, 180°-rotation of the eye-bud fails to prevent development of normal vision; the same operation performed a few stages later, gives rise to visuomotor responses that are orderly but misdirected upside-down and back-to-front. We now know that this critical period, which in *Xenopus* spans five hours at stages 28-31 and precedes the outgrowth of optic fibers from the eye (Jacobson, 1968a; Jacobson and Hunt, 1973), is the time of irreversible "locking-in" of the AP and DV retinal axes upon which retinal positional information (and ultimately the X and Y specifiers) is based (Hunt and Jacobson, 1972a).

During the earliest phases of optic cup formation, the stage 22-28

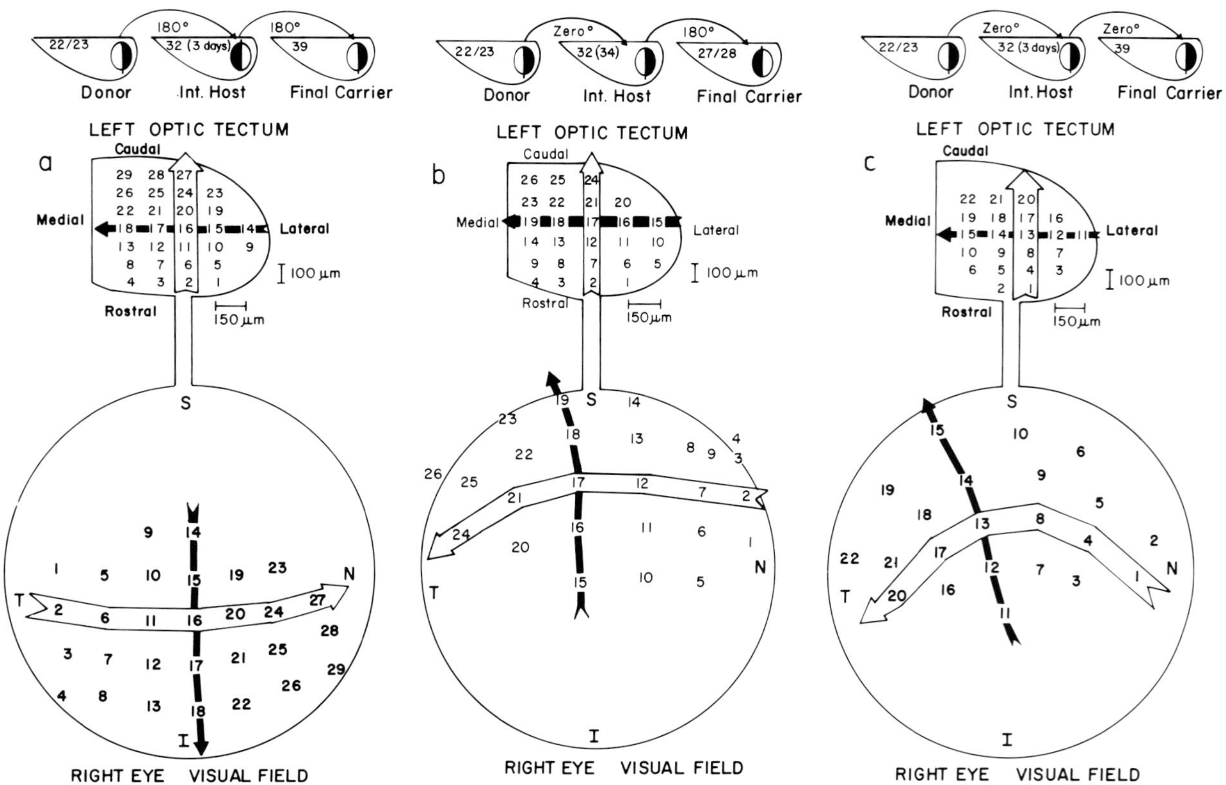
a
b
c
Donor
Int. Host
Final Carrier
22/23
32 (3 days)
32 (34)
27/28
39
180°
Zero°
LEFT OPTIC TECTUM
Caudal
Rostral
Medial
Lateral
100 µm
150 µm
S
I
T
N
RIGHT EYE VISUAL FIELD

eye-bud already possesses full information about its anteroposterior (AP) and dorsoventral (DV) orientation on the embryo, and the capacity to use this information to institute a complete program for patterned differentiation of the ganglion cells. An eye-bud cultured in saline for several days from stage 22/23 forms functional larval eyes *in vitro* and upon reintroduction into a host orbit, will assemble an orderly retinotectal map based on its original axial plan from the donor embryo (Hunt and Jacobson, 1973). However, the orientational information at stages 22-28, if sufficient and complete, is also 'replaceable': if the orientation of a stage 22-28 eye-bud is acutely changed by rotation or transplantation *in vivo*, the anlage interacts with the axial signals of the embryo and acquires new information appropriate to the new orientation after surgery (Fig. 3). Thus, even though the eyeball itself develops upside-down following an 180°-rotation at stage 28, the ganglion cells undergo their position-dependent differentiations in accordance with their new locations and assemble a normally-oriented retinotectal map (Jacobson, 1968a; Hunt and Jacobson, 1974b; Hunt, 1975b).

At stages 28-31, control mechanisms within the retina stabilize the existing information in the anlage, and establish a permanent developmental program for ganglion cell patterning. The stabilization process is called *axial specification* and takes place in two steps (AP before DV) over about five hours. After specification, the retinal field is refractory to the axial signals of the embryo; and even when rotated into a prestage 28 host, or serially passed through a sequence of four or more hosts, the retina follows its original program for position-dependent differentiation; and the orientation of the retinotectal map assembled in the final host deviates from the normal by the same angle as the retina, in its final disposition, deviates from what its orientation was at stages 28 to 31 (rev. Hunt, 1975b). Most importantly, the entire retinotectal map is consistent, even though only a few hundred retinal cells are present in the optic cup at stage 31(Jacobson, 1968b). The remainder of the 10^5 ganglion cells of the adult retina (and their attendant photoreceptors and interneurons) are added in annular 'rings' to

FIG. 3. Demonstration of "critical period" of axial establishment and persistence of axial cues in embryos beyond the "critical period". In the first experiment (A), the retina of a pre-stage 28 Xenopus eye bud is grafted with an inverted orientation into an intermediate host embryo (stage 28) and regrafted after a period of 3 days with the normal orientation into a post-stage 32 final host embryo. This experiment shows that axial cues persist in embryo beyond the "critical period" in which axis specification normally occurs (stages 28-32) as demonstrated by the inverted retinotectal map. Contrast this with the normally oriented maps in which the eye is present in a stage 32 intermediate host up to stage 34 only and, therefore, has not passed through the cirtical period before regrafting into a final carrier at stage 27/28 (B) or in which the eye has remained normally oriented throughout (C).

the retinal periphery over many weeks of larval life (Hollyfield, 1971; Straznicky and Gaze, 1971; Jacobson, 1976). The consistency of the rotated map indicates that all 10^5 ganglion cells added later underwent *their* position-dependent differentiation in accordance with the specified developmental program.

The mechanism by which positional information is conveyed to individual ganglion cells, as they are added to the growing larval retina, is less well understood. The fragmentary evidence, however, leaves no doubt that positional signalling is germane to the process and that this signalling shows great phenomenological similarities to the axial signalling between the embryo and a rotated pre-stage 28 eye-bud. While the specified state is very stable in whole eyes after stage 31, the intraretinal positional signals can be unmasked by surgical rearrangements of the post-stage 31 retina which produces dramatic perturbations in the axial program. For example, a mirror-image redundant pattern of ganglion cell differentiations can arise in a surgical recombinant (left-posterior/right-anterior) stage 32 eye. By fusing such eye fragments for 14 or 30 hours before separating them (and allowing the left-temporal fragment to round up to form a complete eye and map into the tectum alone) Frank and I could show that the departure of the left-temporal fragment from its original program involve a rapid, stepwise reprogramming of first its AP axis and then its DV axis. Following the interaction all ganglion cells added to the peripheral retina undergo position-dependent differentiation and deploy their locus specificities in accordance with these new positional signals. Control experiments confirm the dispensibility of body orientation and orientation of the non-neural parts of the eyeball, and confirm the intraretinal location of these signals (Hunt and Frank, 1975).

V. Lineage-dependent Events in Later Precursor Cells

As we stated earlier, cell-type or phenotypic specificity (at least in the late stages of neuron cytogenesis) does not arise from direct positional signals. Rather the ability to sort out or migrate to a specific position appears to be one of the traits with which the neuron is endowed. On the other hand, considerable evidence is accumulating which suggests that late neuronal precursor cells show a stereotyped mitotic schedule suggestive of lineage routines.

Clonal cell culture experiments have not yet been carried out on the precursor cells of the amphibian retina; nevertheless, this system shows a remarkable refractoriness to experimental "telescoping" of its divisional schedules. The first neurons are born at optic cup stages in all amphibia examined (stages 28-31 in Xenopus, about 1.5 days post-fertilization); their phenotypic properties are quickly evident, and they shortly connect to form a functional but miniature larval eye (by stage 39 in Xenopus, 2.5 days post-fertilization). In one of our earlier strategies, we showed that stage 22 eye primorida (primitive optic bulge) generate the same number of each neuronal phenotype of neuron, with roughly the same timecourse, when grafted to a stage 28-38 host as when grafted to a stage 22 (control) host, and when grafted onto the host flank vs. the host orbit. Moreover, half-eye primordia prepared at stage 25/26 generated about half the normal number of neurons of each phenotype, with roughly the normal timecourse, ruling out control of the time of neuron formation by "critical mass" (Berman and Hunt, 1975).

In another strategy, we attempted to block passage of neural precursor cells through the cell cycle, or to dissociate developmental programming from cell proliferation with special agents such as 5-bromodeoxyuridine. Treatment of stage 24±Xenopus with 5-bromodeoxyuridine (BrUdR) completely suppressed terminal cytodifferentiation of all neuronal cell types in the retina, yielding an abnormally large population of morphologically "undifferentiated" cells (Bergey *et al.*, 1973). A similar suppression of neuronal differentiation was seen in BrUdR-treated chick retina, chick brain and chick spinal cord (reviewed in Hunt, 1975a, 1977c). Fluorodeoxyuridine (FUdR), cytosine arabinoside, and KCl all blocked or slowed DNA synthesis in neural precursor cells of Xenopus optic cup, and definitive neuronal properties were never seen in the retina for the duration of the block (12-24 hrs.); in contrast to BrUdR, however, the DNA-synthetic blocking agents did not produce abnormally large numbers of neural precursor cells; in fact, the number of precursor cells (as expected) failed to rise normally over the duration of the block.

Recently we have deployed a direct cycle-counting method in such experiments (Fig. 4). Following a pulse of ^{3}H-thymidine (or ^{3}H-deoxyadenosine in the FUdR experiments), *in vivo,* and a "chase" *in vitro* with cold nucleoside, the two eye-buds (left and right) from the same embryo are reimplanted into normal hosts and allowed to complete cytodifferentiation under different conditions. Autoradiographs are prepared after fixation at stage 39; and by *varying the exposure time,* we reconstitute an optimal grain count histogram in all retinae. The results show that one to three grain dilutions occurs, for the various retinal cell types, between the mid-precursor cells present at eye-bud formation and the final birth of definitive neurons. Moreover, the results conclusively show that development on an older or younger host, *in vitro* or at an ectopic site outside the orbit, or with a temporary interruption of the program with FUdR, all have *no effect* on the scheduled number of divisions interposed between eye-bud formation and neuronal cytogenesis.

It appears, then, that the morphological features of neuronal phenotypic specificity do not appear when the lifetime of normal neural

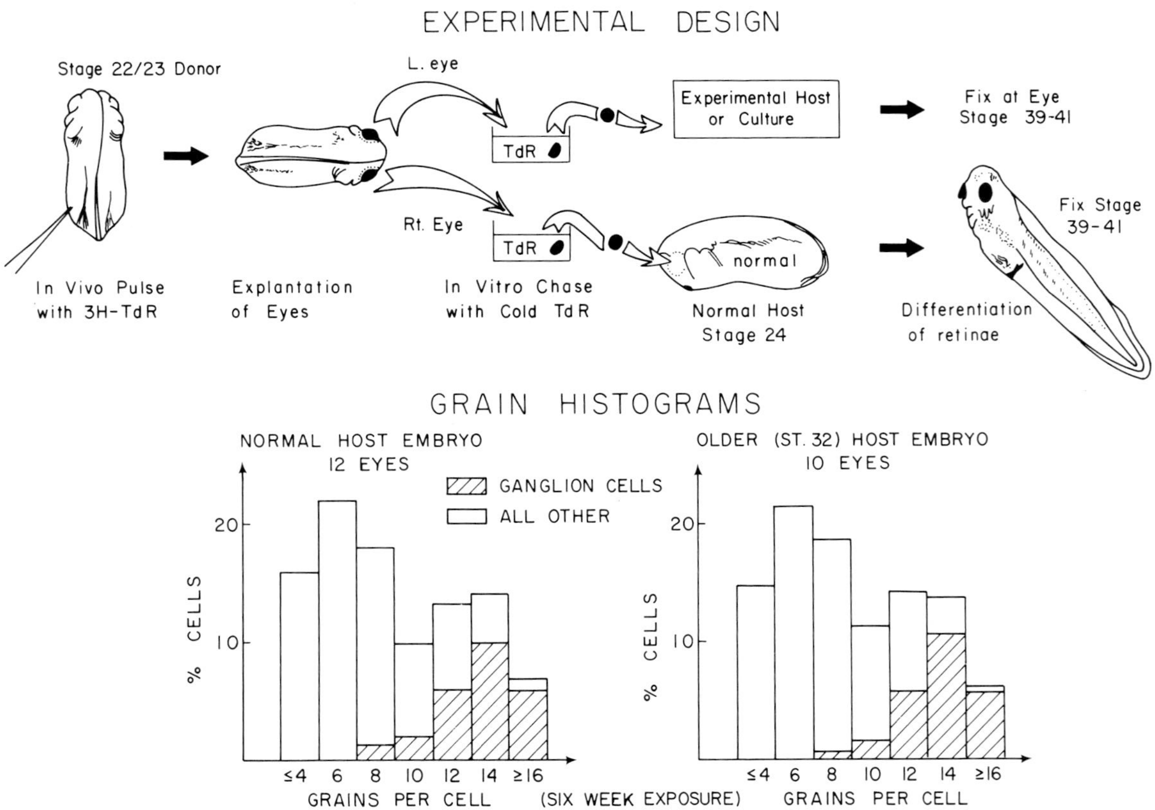

EXPERIMENTAL DESIGN
Stage 22/23 Donor
L. eye
Rt. Eye
TdR
TdR
Experimental Host or Culture
Fix at Eye Stage 39-41
Fix Stage 39-41
normal
In Vivo Pulse with 3H-TdR
Explantation of Eyes
In Vitro Chase with Cold TdR
Normal Host Stage 24
Differentiation of retinae
GRAIN HISTOGRAMS
NORMAL HOST EMBRYO 12 EYES
OLDER (ST. 32) HOST EMBRYO 10 EYES
GANGLION CELLS
ALL OTHER
% CELLS
20
10
≤4 6 8 10 12 14 ≥16
GRAINS PER CELL
(SIX WEEK EXPOSURE)

precursor cells is prolonged by DNA-synthetic blockade. Nor do neurons with "nonsense combinations" of shared neuronal properties appear in the presence of such a block. The results suggest a major role for cell cycle and cell lineage in the complex programming events needed to generate the stereotyped sets of shared and unshared phenotypic properties that mark the six neuronal cell types of frog retina. Unfortunately, little work of this sort has been done on other systems to extend or limit the generalizability of the Xenopus data. Yet two interesting experiments on invertebrate development offer some support and an important clue to the possible role of cell lineage in terminal cytodifferentiation. The first stresses the role of the cytoplasm in directing neuronal differentiation. Carlson (1952) rotated the mitotic spindle during the asymmetric division of the grasshopper neuroblast. When the chromosomes destined for one daughter cell (neuroblast) came to lie in cytoplasm destined for the other (ganglion mother cell) and vice versa, the respective cytoplasms determined the daughter cell types. In the second experiment, Whittaker (1973) blocked cytokinesis at the 16- or 32-cell stage in an ascidian embryo (using cytochalasin-B) and found that the blastomeres from which the two pigmented brain cells (otolith and ocellus) normally derive became pigmented at the normal time. Thus, it appears that some neuronal information can be activated in the absence of cytokinesis. But can specified combinations of shared and unshared neuronal information be mobilized in the appropriate cells under these conditions? It is of great interest that Whittaker's pigmented blastomeres never developed the other properties of the otolith and ocellus, suggesting that cytokinesis specifically (and cell lineages generally) may help to compartmentalize different cytoplasms for cytoplasmic-nuclear interactions.

VI. Coupling Functions

One juncture at which the positional-signalling system collides with the cycle-dependent cytogenesis program is in the coordination of axial specification and ganglion cell cytodifferentiation in stage 28-31 optic cup. Axial specification can occur *in vitro* (Hunt and Jacobson, 1973) and its timing is intrinsically controlled within the eye primordium. Thus the time of specification always correlates with the stage of the optic cup and, in eyes

FIG. 4. Shows constant-grain histograms after normal development or experimentally perturbed development of the retinal primordia, taken from a single donor embryo pulsed with ^{3}H-thymidine as shown. Following the *in vivo* pulse at stage 22/34 and the *in vitro* chase, one eye is grafted into an unlabeled host of its own stage, the other into an experimental host, and both eyes are allowed to differentiate (top). Autoradiograms of the two retinae are prepared and grain counts are made. The histograms from normal (left) and experimental (right, here an older stage 32 host) show identical grain distribution for both ganglion (shaded) and non-ganglion neurons, indicating that the divisional schedule of the retinal precursor cells was not altered by an older host microenvironment.

grafted between embryos, could be neither accelerated nor delayed by altering the stage of the host (Hunt and Jacobson, 1974b). Hunt *et al.*, (1977) exploited the effects of BrUdR on the developing optic cup, to show that the time of specification was a programmed component of ganglion cell differentiation. When BrUdR (from stage 24) blocked the onset of differentiation of all retinal neurons, the eyes were still able to undergo replacement of their retinal axes at stage 32. Not the prolonged time needed for these eyes to reach stage 32, nor the attainment of unusually large size, nor the appearance of terminally differentiated non-neural cell types was sufficient to trigger axial specification. However, when some of the ganglion cells were allowed to commence their terminal cytodifferentiation (after BrUdR from stage 27), axial specification occured on schedule. An obligatory role for the interneurons and photoreceptors was thus eliminated in the control of axial specification. Hunt *et al.*, (1977) inferred that the trigger for specification is localized to the gangliogenic precursor cells in the pre-stage 29 optic cup, and is activated at a predetermined point in their differentiation.

The importance of the precursor cell cycle is further underscored by observations on axial specification after treatment with FUdR (rev. Hunt, 1975a). FUdR was injected at concentrations which (1) completely arrest retinal cytodifferentiation and kill 100% of the retinal cells after two days without thymidine rescue, but which (2) permit recovery of functional eyes and histologically normal retinae if exogenous thymidine is provided within 18 hours or less. At stage 32, the treated eyes were challenged by grafting, in rotated orientation, into a normal stage 27/28 host orbit. Suprisingly, when a block was imposed from early stage 28 (arresting the last generation of gangliogenic precursors in G_1 or S), AP-specification always occurred and DV-specification almost always occurred by stage 32. Following the axial replacement challenge and rescue with thymidine, these eyes went on to generate completely rotated, or AP-rotated retinotectal maps. Similar results have been obtained using cytosine arabinoside and KC1.

These experiments permit two important inferences: First, the trigger for axial specification may be activated several hours in advance of the time at which the optic cup actually loses the capacity to undergo axial replacement — early in the lifetime of the immediate gangliogenic precursor cell or, perhaps, late in the lifetime of the penultimate precursor. Once activated, the events of specification can proceed to completion even if ganglion cell differentiation is halted. One important prediction of this model has recently been met in preliminary experiments: early FUdR treatment (from stage 20-24) did delay the time of axial specification by intervals approximating the duration of the block. Second, coupling axial specification to cell-cycle related cytogenesis events provides an elegant means for time-separation of an nth use of positional signals (setting up retinal axes in the first place) and an $(n + 1)^{th}$ use of the same signals (conveying specific address to single ganglion cells after stage 31).

VII. Earlier Phases of Neurogenesis

Moreover, if the role of positional signals can be traced forward from the phase of retinal axial patterning to that of intraretinal determination of cellular address, they can likewise be traced backward into the early formative events in the eye-bud and other CNS anlagen. Elsewhere I have reviewed existing data on the anatomic patterning of the eyeball (Hunt, 1975b; 1977c), and it shows a startling phenomenological similarity to the later events in the origin of retinal locus specificities. The comparison is limited somewhat by the lack of detailed anatomic asymmetries (pattern) along the AP axis of the eyeball; hence, concurrence with the AP-before DV rule cannot be assessed. However, a host of choroidal, iridial, vascular and pigmentation markers differentiate the DV axis of the eyeball. These markers are sufficient to demonstrate that a phase of "stable but replaceable" information (normal pattern develops *in vitro*, pattern alignment occurs after rotation *in vivo)* exists early on that gives way to the fixed information seen at the later stages (as when eye rotation, at stages 25-28 produces upside-down *anatomic* pattern but *normal* retinal axes and retinotectal map). In similar fashion, the tectum (although much less thoroughly studied) shows reiterative use of positional signals for anatomic pattern and for determining the reference axes for deploying retinotectal "target properties" (Hunt, 1977b). And while "stable but reversible" information states have not been detected because the tectum does not culture well *in vitro*, examination of both AP and DV (or mediolateral) axes *has* been possible in the tectum and both processes appear to obey the AP-before-DV rule (Corner, 1963; Chung and Cooke, 1975; M. Jacobson, cit. Hunt, 1977c).

Much more work needs to be completed to elucidate "coupling functions" which link lineage-dependent events and signalling events in the specification of anatomic pattern in eye and tectum. Yet, the process clearly involves a ubiquitous phenomenology in which positional signals come into reiterative play, shaping a *succession* of programming events in a developing cell population, even as (and perhaps because) the responsiveness of that cell population evolves through a succession of *category-states:* culminating first in the specification of anatomic pattern; later in the axial plan for patterning of locus specificities; finally in the transduction of cellular positional information by individual ganglion and tectal cells.

The final building block for our model derives from a reconsideration of the earliest phases of neurogenesis: the developmental programming events in the neural plate by which small groups of early embryonic cells become "founder cells" for specific neural structures and neuron populations and begin to move down particular pathways of neural differentiation. Classic approaches to regional determination of the neural plate have generally emphasized three methods: cell marking, transplantation and excision/explantation. The cell marking strategy provides no insight into the developmental programs or differentiated states of early embryonic cells, but simply details the size and positions of cell groups destined (in the undisturbed natural history of the embryo) to give rise to particular final structures. In the amphibia, such studies have demonstrated as early as gastrula stages (1) a large population of contiguous cells destined to give rise to CNS, made up of (2) smaller populations destined to give rise to individual neural structures (e.g., retina, hypothalamus, etc.) Individually, such cell groups remain contiguous throughout early neurogenesis, indicating that individual cells do not migrate about the neural plate or neural tube, or disseminate from a focal point in the plate to populate many different final structures. Collectively, these cell groups show a rough and ready topographic correspondence to the positions of their descendant structures along the neural axis; but the detailed spatial relationships and immediate proximity relationships in the neural plate are not related to final position or to functional or synaptic association (Jacobson, 1959; Corner, 1963, 1964). These observations alone establish the important point that the early determinative decisions in neurogenesis cannot be based on "master properties" (e.g., establishment of a primordial cell line for all cholinergic neurons) or on future functional association (e.g., establishment of a primordial cell line for all visual system cells): the geography of the system simply will not allow it (Hunt, 1975a).

In conjunction with these marking experiments, transplantation and excision/explantation experiments (Harrison, 1929; Spemann, 1938; Stefanelli, 1950; Corner, 1963, 1964; Jacobson, 1964) have shown that the regional diversification of neural plate passes through phases of labile and then fixed determination. Thus at the earliest stages, excision and explantation of the founder cells for retina are unable to uncover covert bias toward retinogenesis in this tissue. Slightly later, this bias is evinced by explantation into tissue culture, yet the rudiment is also capable of forming other neural structures if transplanted elsewhere in the neural plate (or to replenish another adjacent founder cell population which has been excised). This liability — the capacity of a cell group to reroute the differentiation of some or all of its cells in response to environmental cues — is gradually and progressively restricted with developmental time in the neural plate. Finally, a stage is reached where the developmental program of the retinal precursor cells is sufficiently fixed that it appears unmodified under the range of physiological changes in environment which induce modifications earlier.

The early neural plate is treacherous terrain for those seeking to apply new ideas to early neurogenesis. The data are old in most cases, and in all cases, were designed with other conceptual frameworks in mind. Thus, despite their intrinsic elegance, they inevitably provoke in the contemporary theorist the wince of the one perturbation left undone, the chagrin of the critical omitted control. Nevertheless, even accepting that the information is fragmentary and was not generated with contemporary issues in mind, the data on progressive determination in the early neural plate offer three persuasive comments on the role of positional signals in early neurogenesis. First, the phenomenology of labile determination giving way to progressive determination is a familiar feature to the position-dependent processes seen in later differentiation of retina and tectum. Second, and more significantly, systematic examination of the *intermediate states* of labile determination (e.g., during the progressive loss of the presumptive retina's ability to reroute its cellular differentiation to give rise to non-retinal structures), reveals a sequence based on positional rules. Thus, very early retinal founder cells can, as a population, give rise to diencephalon, hypothalamus, pigment retina and nearly any other forebrain (and some midbrain) structures (Spemann, 1938; Corner, 1963, 1964): later cells in the retinogenic lineage in the plate can no longer give rise to midbrain structures; still later cells can generate only hypothalamus and pigment retina; and the final stage in which some developmental plasticity persists, the remaining tissue which can be generated from presumptive neural retina, is pigment retina (rev. Hunt, 1977c). Similar results have been obtained with Mauthner's cell in the medulla (Stefanelli, 1950). What is most remarkable about these patterns of loss is that they bear absolutely no relationship to the functional or known biochemical properties of the terminally differentiated cells: retina and hypothalamus could scarcely be more different, and the pigment epithelium of the retina is even non-neuronal! Indeed, the only thing the three are known to have in common is that *spatially*, in the local geography of the plate, hypothalamus is the retina's nearest neural neighbor — separated only, in fact, by the presumptive pigment epithelium.

The final impetus for applying our positional signalling theme to early neural plate derives from the one series of experiments in which the labile determination problem was studied, not by excision or translocation paradigms, but by simply rotating pieces of plate to translocate the cell groups, or by grafting left-side plates into right-side host sites (Jacobson, 1964). Under these conditions, the progressive spatial restriction of "rerouting capability" drops out of the analysis as a covariate of graft size: normal patterns obtain after early large grafts, but at later and later stages, only after smaller and smaller grafts. Remarkably, the system shows an all-or-none tendency to make the original (graft-specific) structures and pattern or a completely normal array of structures. That is, it either repolarizes or does not; and the phenomenology of results becomes a near perfect approximation of the eyeball, retinal and intraretinal systems, featuring labile and fixed polarity, distinct AP and DV axes, and concurrence with the AP-before-DV rule.

VIII. Epilogue

What emerges is a model of neurogenesis in which regulatory genetics are complex and sequential and structural genetics are combinatorial and largely redundant. I propose that positional signals are used reiteratively in

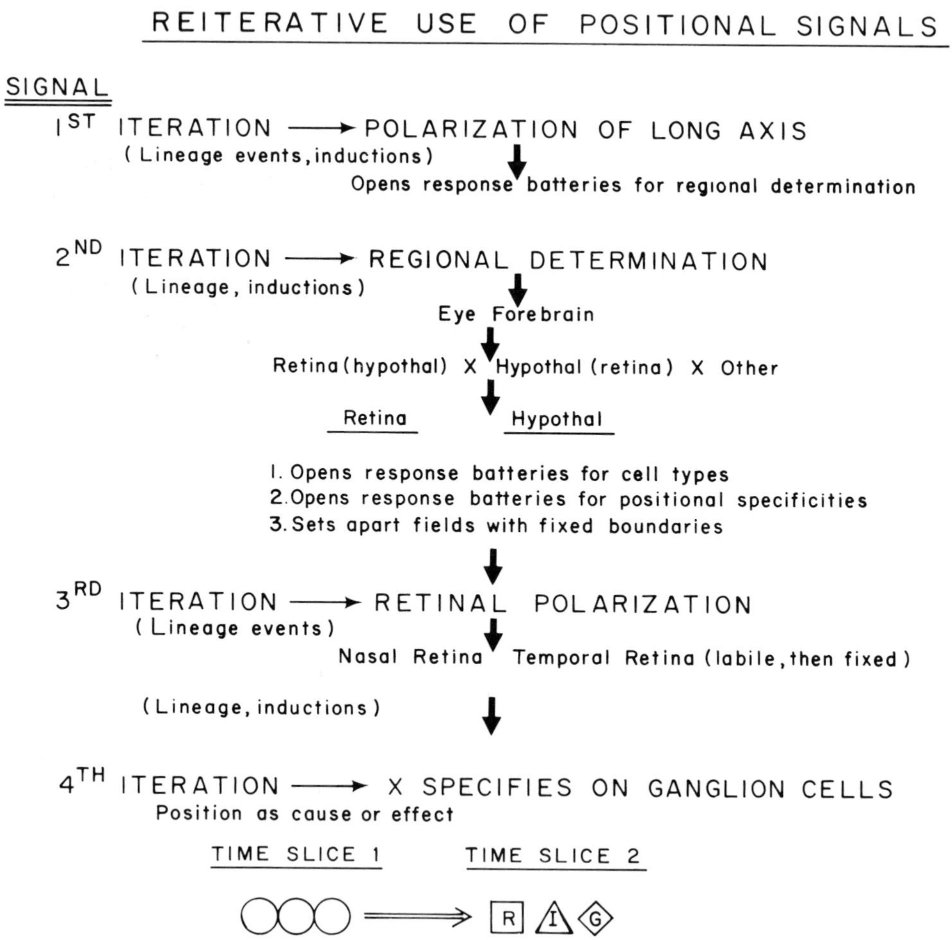

FIG. 5. To show a model of developmental programming in neurogenesis, using reiterative positional signals (horizontal) and changing category-states of response (vertical) in a cell population evolving through a neurogenic lineage. For explanation, see text.

time and redundantly in space (Fig. 5), interfacing at different timepoints in development with cell populations whose responsiveness evolves through a sequence of *category states*. The exact choice of response (e.g., to be nasal retina versus temporal retina, or to be retina/pigment versus hypothalamus) is a function of local parameters of the signal in space; the qualitative category of response (retina/pigment versus nasal retina) is determined by the maturational and differentiative histories of the cells. It remains to be determined whether lineage-dependent events provide the drive which cascades the cell groups through the sequence of *category states*, but the later events in retinal specification offer one clear example of this and provide a convenient model for coupling signal time and response time (see sec. VI above). It also remains to be determined whether the signal is in fact identical in all its iterations, but new data strongly points in that direction: stage 25-28 retina will undergo axial replacement on the body flank in the region of other developing fields, on very young neurula or very old hatched larvae, and in the orbits of species as distant as *Ambystoma maculatum* (Hunt, 1977c; Hunt and Piatt, 1977) — strongly arguing for the universality of the signalling mechanism. In addition, recent transplantations of dorsal lip organizers (and of neural plate cells with organizer properties) next to stage 25-28 eyes have succeeded in scrambling the retinal axes and the future retinotectal map (Cooke, 1977). In any case, the model may offer novel guidance in choice of future experiments: toward assessment of cross-reactivity of different developmental signals; toward examination of combinations of paired parameters of differentiation rather than single parameters (especially *in vitro)*: toward a supplementation of pure *in vitro* methods with *in vitro/in vivo* recombination methods, where some phases of differentiation can be controlled (eg., lineage progression) while the *in situ* signalling processes are also brought into play; and toward models of pattern formation and differentiation which are prized for the versatility in meeting the demands of many different phases of differentiation rather than their economy in accounting for the data in one phase. Finally, the model raises the spectre that our doleful lack of early neurogenic mutants may be no accident (Wolpert, 1971); perturbation of a master-signalling apparatus for any phase of development might well produce perturbations in every phase, leading to catastrophic early lethals whose isolation may require much more selective (e.g., temperature-sensitive) screening procedures than have been used previously in vertebrate genetic neurobiology.

Acknowledgments

The ideas in this paper have been immeasurably catalyzed by discussions with a number of embryological theorists, especially H. Holtzer,

L. Wolpert, S. Kaufman, A. Garcia-Bellido, and H. Meinhardt. Special thanks to Ms. Patricia Burck for preparation of the manuscript. The author's research is supported by NSF Grant BMS-75-18998 and NIH Grant NS-12606 and by a Biomedical Research Support Grant to The Johns Hopkins University.

References

Benzer, S. 1973. Genetic dissection of behaviour. Sci. Amer. 229. No. 6:24-37.

Bergey, G. K., R. K. Hunt and H. Holtzer. 1973. Selective effects of bromodeoxyuridine on developing *Xenopus laevis* retina. Anat. Rec. 175:271.

Berman, N. and M. Cynader. 1972. Comparison of the receptive field organizaton of the superior colliculus of Siamese and normal cats. J. Physiol. (Lond.) 224:363-389.

Berman, N. and R. K. Hunt. 1975. Visual projection to the optic tecta after partial extirpation of the embryonic eye. J. Comp. Neurol. 162:23-43.

Bignami, A. and D. Dahl. 1974. Development of Bergman glia in mice with cerebellar nerve formation: Reeler, Weaver and Staggerer. Immuno-fluorescence study with antibodies to the glial fibrillary acidic protein. J. Comp. Neurol. 155:219-234.

Brenner, S. 1975. Closing remarks: the genetic outlook. Ciba Found. Symp. 29:343-347.

Britten, R. J. and E. H. Davidson. 1969. Gene regulation for higher cells; a theory. Science 165:349-357.

Carlson, J. G. 1952. Micro dissection studies of the dividing neuroblast of the grasshopper, *Chortophaga viridifasciata* (DeGees). Chromosoma 5:199-220.

Chung, S. H. and J. Cooke. 1975. Polarity of structure and of ordered nerve connection in developing amphibian brain. Nature (Lond.) 258:126-132.

Cooke, J. 1977. Selective nerve connections: implanted organizer cells can disrupt retinal specification. Nature (Lond.) (in press).

Corner, M. A. 1963. Development of the brain of *Xenopus laevis* after removal of parts of the neural plate. J. Exp. Zool. 153:301-311.

Corner, M. A. 1964. Localizaton of capacities for functional development in the neural plate of *Xenopus laevis*. J. Comp. Neurol. 123:243-256.

Das, G. D. and J. Altman. 1971. Transplanted precursors of nerve cells: their fate in the cerebellum of young rats. Science 173:637-638.

Dewey, M. J., A. G. Gervais and B. Mintz. 1976. Brain and ganglion development from two genotypic classes of cells of allophenic mice. Develop. Biol. 50:68-81.

Drager, U. C. 1975. Receptive fields of single cells in mouse visual cortex. J. Comp. Neurol. 160:269-290.

Eakin, R. M. 1947. Determination and regulation of polarity in the retina of *Hyla regilla*. Univ. Calif. Publ. Zool. 51:245-288.

Garcia-Bellido, A. 1975. Genetic control of wing disc development in drosophila. Ciba Found. Symp. 29:161-182.

Goldberg, S. 1976. The invisible ocular map. Bull. New York Acad. Med. 52:201-217.

Guillery, R. W., V. A. Casagrande and M. D. Oberdorfer. 1974. Congenitally abnormal vision in Siamese cats. Nature (Lond.) 252:195-199.

Harris, W., W. Stark and J. Walker. 1976. Genetic dissection of photoreceptor systems in drosophila. J. Physiol. (Lond.) 256:415-439.

Harrison, R. G. 1929. Correlation in the development of the eye studied by means of heteroplastic transplantation. Roux Archiv. 120:1-55.

Hollyfield, J. G. 1971. Differential growth of the neuro-retina in *Xenopus laevis* larvae. Develop. Biol. 24:264-286.

Hubel, D. H. and T. N. Wiesel. 1971. Aberrant visual projection in the Siamese cat. J. Physiol. (Lond.) 218:33-62.

Hunt, R. K. 1975a. The cell cycle, cell lineage, and neuronal specificity. Results Problems Cell Diff. 7:43-62.

Hunt, R. K. 1975b. Developmental programming for retinotectal patterns. Ciba Found. Symp. 29:131-159.

Hunt, R. K. 1977a. Combinatorial specifiers on retinal ganglion cells for retinotectal map assembly in *Xenopus laevis*. Nature (Lond.) (in press).

Hunt, R. K. 1977b. Target properties in the optic tectum of *Xenopus laevis*. Nature (Lond.) (in press).

Hunt, R. K. 1977c. Sensory specificity in developing retina. In: *Handbook of Sensory Physiology: Development of Sensory Systems*. Springer-Verlag, New York, Berlin (in preparation).

Hunt, R. K. and E. Frank. 1975. Neuronal locus specificity: transpolarization of Xenopus retina after the time of specification. Science 189:563-565.

Hunt, R. K. and M. Jacobson. 1972a. Development and stability of neuronal specificity in Xenopus retinal ganglion cells. Proc. Nat. Acad. Sci. U.S.A. 69:2860-2864.

Hunt, R. K. and M. Jacobson. 1973c. Assays for analysis of the unspecified state. Proc. Nat. Acad. Sci. U.S.A. 70:507-511.

Hunt, R. K. and M. Jacobson. 1974a. Neuronal specificity revisited. Curr. Tops. Develop. Biol. 8:203-259.

Hunt, R. K. and M. Jacobson. 1974b. Rapid realignment of retinal axes in rotated Xenopus eyes. J. Physiol. (Lond.) 241:90-91.

Hunt, R. K. and J. Piatt. 1974. Axial specification in salamander embryonic retina. Anat. Rec. 178:515.

Hunt, R. K. and J. Piatt. 1977. Cross species axial signalling, with reversal of retinal polarity, in embryonic Xenopus eyes. Develop. Biol. (in press).

Hunt, R. K., G. K. Bergey and H. Holtzer. 1977. Bromodeoxyuridine localization of a developmental program in Xenopus optic cup. Develop. Biol. (in press).

Ide, C. F. 1977. Vestibular physiology of the spastic axolotl. Biophys. Journal 17:127a.

Ide, C. F., C. B. Kimmel, R. Tompkins, O. Elbert, E. Schabtach, N. Miszkowski and M. Duda. 1976. Analysis of spastic, a neurological mutant in the Mexican axolotl. In: *ICN-UCLA Symposia on Neurobiology* (in press).

Ide, C. F., R. Tompkins, and N. Miszkowski. 1977. Neuroanatomy of *spastic*, a behaviour mutant of the Mexican axolotl. Purkinje cell distribution in the adult cerebellum. J. Comp. Neurol. (in press).

Jacobson, C. O. 1959. The localization of the presumptive cerebral regions in the neural plate of axolotl larvae. J. Embryol. Exp. Morph. 7:1-21.

Jacobson, C. O. 1964. Motor nuclei cranial nerve roots, and fiber pattern in the medulla oblongata after reversal experiments on the neural plate of axolotl larvae. Zool. Bidr. Uppsala 36:73-160.

Jacobson, M. 1968a. Development of neuronal specificity in retinal ganglion cells of *Xenopus laevis*. Develop. Biol. 17:202-218.

Jacobson, M. 1968b. Cessation of DNA synthesis in retinal ganglion cells correlated with the time of specification of their central connections. Develop. Biol. 17:219-230.

Jacobson, M. 1976. Histogenesis of the retina in the clawed frog, with implications for the pattern of development of retinotectal connections. Brain Research 103:541-545.

Jacobson, M. and R. K. Hunt. 1973. The origins of nerve cell specificity. Sci. Amer. 228, No. 2:26-35.

Kauffman, S. A. 1973. Control circuits for determination and trans-determination. Science 181:310-317.

Laties, A. M. and P. A. Liebman. 1970. Cones in the living amphibian eye: selective staining. Science 168:1475-1477.

Lewis, E. B. 1964. Genetic control and regulation of developmental pathways. In: *Role of Chromosomes in Development*. M. Locke, ed. Academic Press, New York. pp. 232-251.

Lewis, J. A. and J. A. Hodgkin. 1976. Specific neuroanatomical changes in chemosensory mutants of the nematode. J. Comp. Neurol. 172:489-510.

Liebman, P. A. and G. Entine. 1968. Visual pigments of frog and tadpole (*Rana pipiens*). Vision Res. 3:1-18.

Lund, R. and S. Hauschka. 1976. Transplanted neural tissue develops connections with host rat brain. Science 193:582-584.

Mallett, J., M. Huchet, M. Shelanski and J. P. Changeux. 1974. Protein differences associated with the absence of granular cells in the cerebellum of mutant weaver mouse and X-irradiated rat. FEBS Letters 46:243-247.

Maturana, H. R., J. Y. Lettvin, W. S. McCulloch and W. H. Pitts. 1960. Anatomy and physiology of vision in the frog. J. Gen. Physiol. 43:129-175.

Mullen, R. J. and M. M. LaVail. 1976. Inherited retinal dystrophy: primary defect in pigment epithelium determined with experimental rat chimeras. Science 192:799-802.

Mullen, R. J., E. M. Eicher and R. L. Sidman. 1976. Purkinje cell degeneration: a new neurological mutation in the mouse. Proc. Nat. Acad. Sci. U.S.A. 73:208-212.

Piatt, J. 1949. Heterotopic transplantation of brachial spinal cord, followed by reimplantation into the orthotopic site. J. Exp. Zool. 111:1-25.

Rakic, P. and R. L. Sidman. 1973. Weaver mutant mouse cerebellum: Defective neuronal migration secondary to a specific abnormality of Bergman glia. Proc. Nat. Acad. Sci. U.S.A. 70:240-244.

Reuter, T. 1969. Visual pigments and ganglion cell activity in the retina of tadpoles and adult frogs. Acta Zool. Fenn. 122:1-67.

Sidman, R. L. 1973. Cell-cell recognition in the developing central nervous system. In: *The Neurosciences: Third Study Program.* F. O. Schmitt, ed. MIT Press, Cambridge. pp. 743-758.

Sotello, C. and J. P. Changeux. 1974a. Bergmann fibers and granule cell migration in the cerebellum of homozygous weaver mutant mice. Brain Research 77:484-491.

Sotello, C. and J. P. Changeux. 1974b. Trans-synaptic degeneration en cascade in the cerebellar cortex of staggerer mutant mice. Brain Research 67:519-526.

Spemann, H. 1938. *Embryonic Development and Induction.* Yale University Press, New Haven.

Sperry, R. W. 1950. Neuronal specificity. In *Genetic Neurology.* P. Weiss, ed. University of Chicago Press, Chicago, Ill. pp. 232-239.

Stefanelli, A. 1950. Studies on the development of Mauthner's cells. In: *Genetic Neurology.* P. Weiss. ed. University of Chicago Press, Chicago, Ill. pp. 161-165.

Stone, L. S. 1944. Functional polarization in retinal development and its re-establishment in regenerating retinae of rotated eyes. Proc. Soc. Exp. Biol. Med. 57:13-14.

Stone, L. S. 1960. Polarization of the retina and development of vision. J. Exp. Zool. 145:85-93.

Straznicky, K. and R. M. Gaze. 1971. Growth of the retina in *Xenopus laevis.* An autoradiographic study. J. Embryol. Exp. Morph. 26:67-79.

Van Deusen, E. 1973. Experimental studies on a mutant gene (e) preventing the differentiation of eye and normal hypothalamus in the axolotl. Develop. Biol. 34:135-156.

Werblin, F. S. and J. E. Dowling, 1969. Organization of the retina in the mudpuppy, *Necturus maculosus.* II intracellular recording. J. Neurophysiol. 32:339.

West, J. D. 1976. Clonal development of the retinal epithelium in mouse chimeras and X-inactivation mosaics. J. Embryol. Exp. Morph. 35:445-461.

Whittaker, J. R. 1973. Segregation during ascidian embryogenesis of egg cytoplasmic information for tissue specific enzyme development. Proc. Nat. Acad. Sci. U.S.A. 70:2096-3000.

Wolpert, L. 1971. Positional information and pattern formation. Curr. Topics Develop. Biol. 6:183-224.

Wood, W. B. and R. J. Edgar. 1967. Building a bacterial virus. Sci. Amer. 217:61-74.

Yoon, M. G. 1973. Retention of the original topographic polarity by the 180 degree-rotated tectal reimplant in young adult goldfish. J. Physiol. (Lond.) 233:575-588.

Yoon, C. H. and A. K. Coffey. 1975. Interaction of cerebellar mutant genes. Mice doubly affected by "Reeler" and "Weaver" conditions. Brain Research 96:219-232.

SPECIFIC PROTEIN SPECIES EXPRESSED BY DIFFERENTIATING NEUROBLASTOMA, GLIOMA, AND HYBRID CELLS IN CULTURE. EVIDENCE FOR NEUROBLAST-GLIOBLAST RECIPROCAL GENETIC REGULATION

ROGER N. ROSENBERG, CAROL K. VANCE, NAGINDRA PRASHAD AND FRED BASKIN

Laboratory of Cellular Neurobiology
Department of Neurology
The University of Texas Health Science Center
Southwestern Medical School
Dallas, Texas

I. Introduction

The use of clonal cell lines derived from tumors of the nervous system as models for both neurons and glia has become well-established (Rosenberg, 1973). Clonal cell lines derived from mouse neuroblastoma C1300 have been shown to possess many characteristics of differentiated neurons (Amano *et al.*, 1972, Anagnoste *et al.*, 1972, Augusti-Tocco and Sato, 1969, Benda *et al.*, 1968, Blume *et al.*, 1970, Hamprecht *et al.*, 1974, Harris *et al.*, 1971, Klebe and Ruddle, 1969, Murphy *et al.*, 1975, Nelson *et al.*, 1971a, b, Nelson *et al.*, 1969, Rosenberg *et al.*, 1971, Schubert *et al.*, 1969, Schubert *et al.*, 1971, Seeds *et al.*, 1970). These include action potentials in response to depolarizing stimuli (Nelson *et al.*, 1969/71b, Nelson *et al.*, 1969) and acetylcholine (Nelson *et al.*, 1971a); putative synapse formation with cultured muscle cells (Harris *et al.*, 1971); formation of neurites containing microtubules, neurofilaments and dense-core vesicles (Schubert *et al.*, 1969); the presence in various clones of cholineacetyltransferase (Amano *et al.*, 1972, Augusti-Tocco and Sato, 1969, Rosenberg *et al.*, 1971), tyrosine hydroxylase (Amano *et al.*, 1972, Augusti-Tocco and Sato, 1969), acetylcholinesterase (Blume *et al.*, 1970), catechol-O-methyl transferase (Blume *et al.*, 1970), dopamine-β-hydroxylase (Anagnoste *et al.*, 1972), and thymidylate synthetase (Rosenberg *et al.*, 1971), all at activities comparable to those in brain; catechol formation (Schubert *et al.*, 1969); presence of transport systems for various neurotransmitters (Hutchison *et al.*, 1974); and secretion of nerve growth factor (Murphy *et al.*, 1975). Clonal cell lines derived from a chemically-induced rat glioma (Benda *et al.*, 1968) possess properties characteristic of glial cells, such as the production of high levels of S100 (Pfeiffer *et al.*, 1970) and glial-specific hormonal induction of enzymes such as glycerol-3-phosphate dehydrogenase and lactate dehydrogenase (deVellis and Inglish, 1968).

Somatic cell hybrids have been utilized to discern many properties of parental cell lines (Minna *et al.*, 1971; Minna *et al.*, 1972). Genes for neuronal properties may be activated in neuronal-glial hybrids (Amano *et al.*, 1974). These hybrids resemble the parent neuroblastoma cells in many of their properties: acetylcholinesterase activity (Amano *et al.*, 1974), dopamine-β-hydroxylase activity (Hamprecht *et al.*, 1974), clear and dense-core vesicles (Hamprecht, 1974), ability to extend long neurites (Daniels and Hamprecht, 1974, Schubert *et al.*, 1971, Seeds *et al.*, 1970), and possession of electrically-excitable membranes (Hamprecht, 1974). Several neuronal properties are more strongly expressed in the somatic cell hybrid between N18 mouse neuroblastoma and C6 rat glioma cells. In contrast to the parents these hybrids possess high activity choline acetyltransferase (Amano *et al.*, 1974) and also express opiate receptors (Klee and Nirenberg, 1974).

It would seem from the biochemical and morphological properties reported for neuroblastoma cells that these cells resemble mature neurons *in vivo,* and therefore would be a good model for the study of differentiation of neurons in cell culture. Somatic cell hybridization has been used as an approach to obtain information on the mechanisms underlying cellular differentiation (Davidson, 1969). Differentiation of neurons has been previously studied through investigation of changes in protein spectra in neuroblastoma (Prashad and Rosenberg, 1977; Prashad *et al.*, 1977). Therefore, in this report, we will compare protein extracts from the somatic cell hybrid, NG105-15, and its two parents, N18TG2 and C6-BU-1, as run on high resolution, two dimensional polyacrylamide gel electrophoresis to determine whether length of time in culture involves any changes in the patterns of protein synthesis in the clones investigated and to see how the parental proteins are regulated in the hybrid lines.

II. Methods Used to Investigate Specific Proteins In The Present Study

A. Cell Culture

The clonal cell lines, designated NG105-15, are somatic cell hybrids from a Sendai virus-induced fusion of a 6-thioguanine resistant mutant, N18TG2, of mouse neuroblastoma line N18, and a bromodeoxyuridine-resistant mutant, C6-BU-1, of rat glioma line C6. All of these clones were obtained as generous gifts from Dr. Marshall Nirenberg, N.I.H., Bethesda, Maryland.

Cells were cultured in plastic petri dishes (Falcon) containing 10 ml Dulbecco's modified Eagle's medium (DMEM, GIBCO), supplemented with 10% fetal bovine serum (Reheis Chemical Co.) and incubated at 37°C in a water-saturated atmosphere containing 90% air-10% CO_2. Cells were subcultured via gentle trypsinization and plated at 1.5x10^6 cells/100 mm dish. Culture medium was changed every 1-3 days, dependant upon cellular density.

B. Determination of Growth Curve

Cells were plated in plastic dishes at 1.5x10^6 (100 mm in diameter) with 10 ml DMEM. Medium was changed daily. After aspiration of growth medium, cells were washed twice with isotonic saline and then scraped into 1 ml of phosphate buffered saline with a rubber policeman. Cells were counted on a Coulter Counter. All samples were prepared in duplicate.

C. Isolation and Separation of Proteins

Cells were plated for protein isolation in parallel with cells plated for growth rate determination. The extraction procedure is a modification of the O'Farrell (1975) procedure. Growth medium was completely aspirated, followed by two saline rinses. Liquid was then completely aspirated. Cells were harvested in the presence of 250 nmoles phenylmethylsulfonylfluoride (per 100 μl) by scraping into 100 microliters of sonication buffer (0.01 Tris-HCl, pH 7.4, 5 mM $MgCl_2$, 50 μg pancreatic RNase/ml). These extracts were sonicated through two 30-sec bursts on a Biosonik sonicator. The extracts were then digested via 50 μg/ml pancreatic DNase (stock: 1 mg/ml in 0.01 M Tris-HCl, pH 7.4, 1 mM $MgCl_2$) for 5 min at 0°C. Solid urea was then added to a concentration of 9M. An equal volume of lysis buffer [9.5 M urea, 2% (w/v) Nonidet P-40 (NP40), 2% Ampholines (BioRad) (comprised of 1.6% pH range 5-7, 4% pH range 3-10) and 5% (v/v) β-mercaptoethanol] was added to the extracts. The extracts were kept frozen at -20°C until use. Proteins were assayed by the TCA turbidity method of Comings and Tack (1972). It was demonstrated that freezing and thawing of samples did not change protein profiles.

The isoelectric focusing (IEF) electrophoresis was carried out on gels measuring 12.5 cm x 0.3 cm run for 17.5 hrs at 400 volts. The pH gradient ran from 5 to 7. The electrophoresis procedure is as outlined by O'Farrell (1975). IEF gels were stained with 0.1% (w/v) Coomassie Brilliant Blue in 40% (w/v) trichloroacetic acid. After 1 hr of equilibration in SDS sample buffer [10% (w/v) glycerol, 5% (v/v) β-mercaptoethanol, 2.3% (w/v) sodium dodecyl sulfate (SDS), and 0.0625 M Tris-HCl, pH 6.8], the first dimension gel is applied to the SDS slab gel (dimensions 13.7 cm x 10 cm) with 1% agarose in SDS sample buffer. The second-dimension electrophoresis is run for 600 volt-hours. Slabs were stained with 0.1% (w/v)

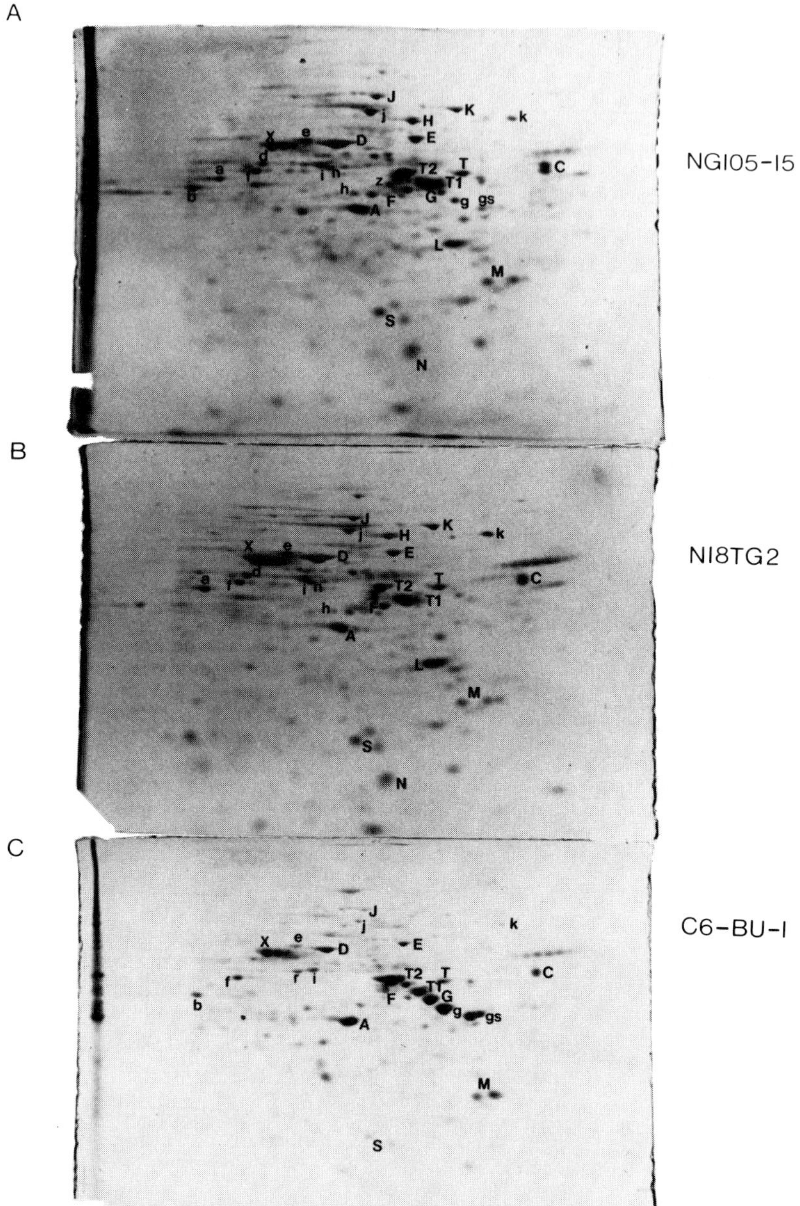

FIG. 1: Protein spectra of A) NG105-15; B) N18-TG2; and C) C6-BU-1 at Day 3.

Coomassie Brilliant Blue, 50% (v/v) methanol, and 7.5% (v/v) acetic acid. Destaining was with 7.5% (v/v) acetic acid.

III. Comparison of N18T62, C6BU-1, and NG105-15 Profiles

In these experiments we are addressing ourselves to two questions: 1) How are the proteins which have been derived from two types of neural tissues genetically regulated within a somatic cell hybrid derived from these two lines; and 2) are there temporal changes in the protein spectra of these cell clones, the parent neuroblastoma and glioma, and the hybrid, with differing times after plating.

Fig. 1 presents the visual protein profiles of a) NG105-15, b) N18T62, and c) C6-BU-1 at day 3 in culture, corresponding to log growth phase. On these gels are proteins with molecular weights ranging from 10,000 daltons (lower part of each gel) to 200,000 daltons (upper part of each gel). The most predominant proteins are represented in the 40,000 to 70,000 range.

One can observe many proteins in these three clones which migrate identically in regard both to their respective isoelectric points (Pi) and also their molecular weights (MW). Protein spot A of 43,000 MW, which has been identified as actin through comigration studies by Prashad *et al.* (1977), is found universally in each of the clones in seemingly the same concentration. Protein C, a doublet migrating with an acidic Pi (right side of gel) and MW of 58,000, also exists equally in all three clones. The protein cluster at M of a MW of 33,000 also is represented in all three clones. Further, the triplet protein X of molecular weight 68,000 can be seen in all clones. Protein f is more basic (left side of gel) than protein T, both having MW's of 55,000 and both existing in all cases. Another protein, D, MW 70,000, can also be seen universally in these clones. There are other less densely staining spots which also seem to be found equally in these three cell lines.

There are other proteins found in all three lines whose staining intensity varies with the particular line being investigated. Protein spot T_1 has been identified as the 53,000 MW subunit of tubulin by comigration studies (Prashad *et al.*, 1977); spot T_2 corresponds to the 55,000 MW subunit of tubulin. T_1 and T_2 are both densely stained in all three clones, tubulin being one of the predominant proteins of these particular clones. However, T_1 appears more heavily stained in the neuroblastoma and hybrid profiles while T_2 appears more heavily stained in the glioma. Spot E, MW 75,000, is of less intensity in the glioma than in either the neuroblastoma or the hybrid. This triplet is constant in both neuroblastoma and hybrid. The MW 26,000 and MW 28,000 subunits are likewise constantly expressed in the glioma. However, the 30,000 MW protein moiety has variable expression in the glioma from complete absence to very sparse staining. Proteins F, a cluster of proteins of MW 50,000 have been identified by Prashad *et al.* (1977) as neurofilament proteins. These are also mostly reduced in concentration or absent from glioma profiles while being nicely expressed in both neuroblastoma and hybrid cells. Among other proteins which have equal expression in both neuroblastoma and hybrid but diminished expresssion in glioma are protein e, MW 70,000; protein k, MW 95,000; protein i, MW 57,000; protein J, MW 100,000; and protein j, MW 95,000.

There are certain proteins which are expressed both in the glioma and hybrid clones but are essentially absent from the neuroblastoma clone. Proteins G, MW 49,000, and g, MW 47,000, are strikingly dominant proteins in the glial clone and are notably absent from the neuroblastoma clone; these two proteins are expressed to a lesser degree in the hybrid than in the glioma but are noticeably present. Protein gs, a 45,000 MW doublet, is completely absent in the neuroblastoma yet it presents a low intensity degree of expression in the hybrid. This is another dominant glioma moiety. Protein b, 50,000 MW, again is absent from the neuroblastoma, but similarly expressed in both the hybrid and the glioma. Protein r, slightly more basic than spot i (MW 56,000 and expressed in all lines), has a MW of 55,000 and is expressed only in the glioma lines.

There are a great many proteins present which are common to the neuroblastoma and hybrid clones but have no expression in the glioma. The protein triad at d, of MW 56-57,000, which forms a tetrad in the neuroblastoma and hybrid, is absent from the glioma. Likewise, protein N, a protein of MW 20,000 which resolves into a doublet, is absent from the glioma. Further, protein n, a spot more acidic than spot i and of MW 55,000, is totally absent in glioma. Among other proteins present in both neuroblastoma and hybrid but absent from the glioma are the following: protein K, MW 97,000; protein H, MW 90,000; protein a, MW 54,000; protein h, MW 50,000; and protein L, MW 38,000. Finally, a single protein, spot z of MW 53,000, is present only in the hybrid clone; whereas it was undetectable via visual staining in both the two parental clones.

IV. Comparison of Growth Profiles in each Clone

Whole cell proteins were analyzed by IEF and SDS slab gel electrophoresis from day 1 through day 6, therefore including the early log, midlog, and confluent phases of cell growth. In Fig. 2 can be seen a comparison of the growth rates of each of the cell lines. Confluency can be considered to have occurred by day 5; our confluent samples were taken from day 6. We have considered day 3 as constituting midlogarithmic cell growth and have used day 1 as the stationary phase of the growth curve. The hybrid clone is of larger cell size with more protein per unit cell; therefore a confluent plate possesses far fewer cells than does a confluent plate of either the glioma or neuroblastoma.

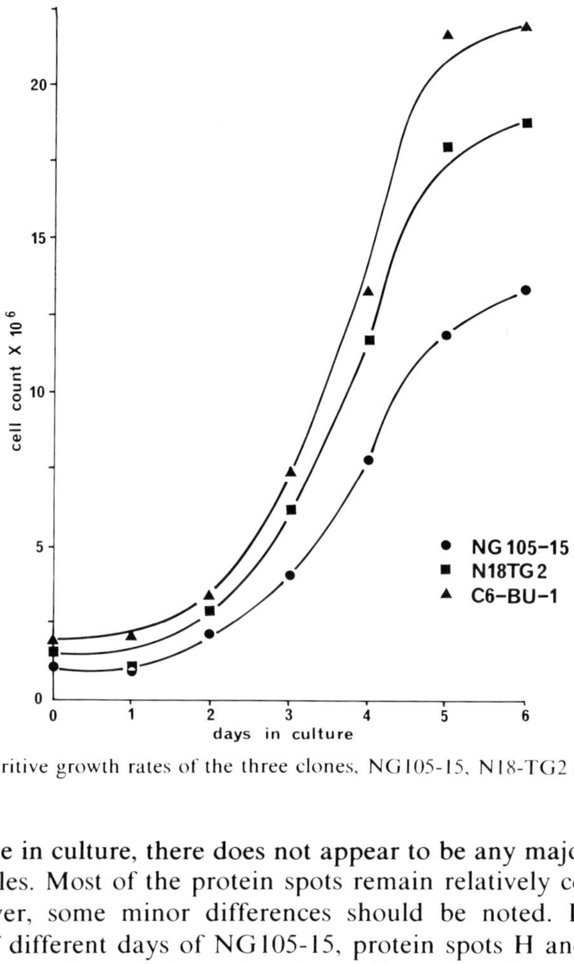

FIG. 2: Comparitive growth rates of the three clones, NG105-15, N18-TG2 and C6-BU-1.

With time in culture, there does not appear to be any major change in protein profiles. Most of the protein spots remain relatively constant with time. However, some minor differences should be noted. In Fig. 3, a composite of different days of NG105-15, protein spots H and K are less intense in the stationary phase than in midlog or confluent stages. The proteins G, g, and gs are most conspicuous by day 3, waning again at confluency. Protein z is very inconspicuous in the stationary phase, but intensifies with increasing time in culture.

In Fig. 4, one can observe the profiles of N18TG2 with differing times in culture. Again for the most part there is little change in the profile with time in culture beyond an intensification of certain proteins after the stationary phase. One can see that proteins T_1, T_2, L, n, f, and H all stain more darkly subsequent to day 1 in culture.

In Fig. 5, we can compare the protein profiles of C6-BU-1 with time after plating the cells. The most obvious change is the super-intensification of protein G and g at confluency. Protein b can be seen to be most intense

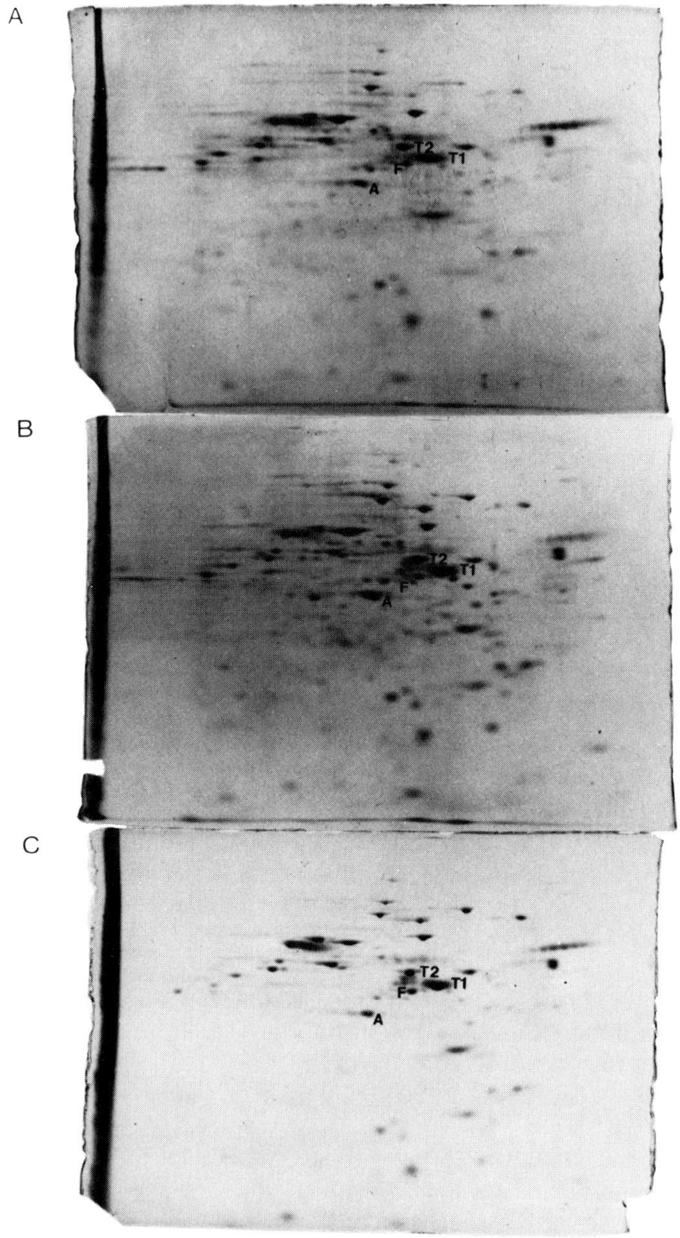

FIG. 3: Protein spectra of clonal somatic cell hybrid NG105-15: A) Day 1, B) Day 3, C) Day 6.

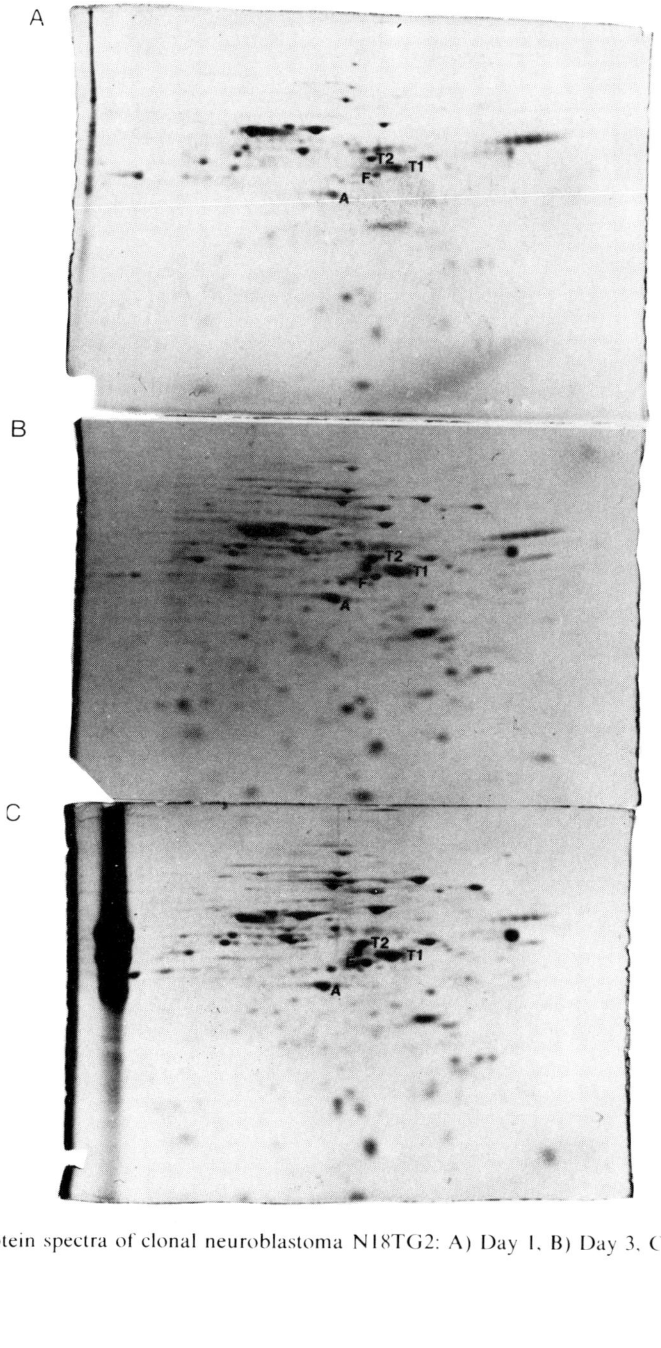

FIG. 4: Protein spectra of clonal neuroblastoma N18TG2: A) Day 1, B) Day 3, C) Day 5.

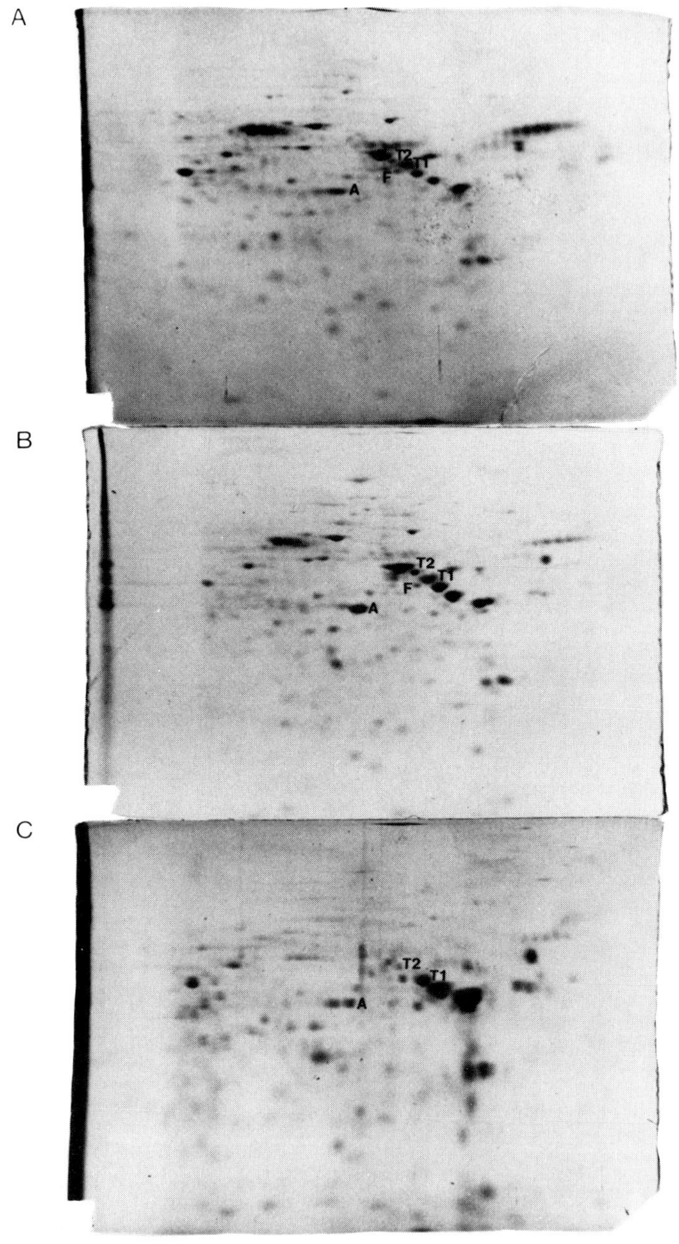

FIG. 5: Protein spectra of clonal glioma, C6-BU-1: A) Day 1, B) Day 3, C) Day 6.

on day 1. The high MW component of the Protein triad S decreases in intensity at day 3 and is almost absent at day 5. Proteins D and T_2 substantially decrease in staining intensity at confluency. Beyond these changes, the remainder of the protein profile seems to maintain a remarkable constancy.

V. Discussion

The molecular characterization of differentiating neuroblasts and glioblasts has previously emphasized enzyme markers for neurotransmitter synthesis and metabolism (Rosenberg, 1973). The present study has extended prior investigations by quantifying and qualifying the expression of protein species by differentiating neuroblastoma, glioma, and neuroblastoma-glioma hybrid clonal cell lines in monolayer cell culture. Proteins with known functional characteristics such as tubulin, filamentous protein and

TABLE I
PROTEINS EXPRESSED BY
NEUROBLASTOMA, GLIOMA, OR HYBRID CLONES

Constant to all Clones		Reduced in C6-BU-1	
D	70,000*	J	100,000
X	68,000	j	95,000
C	58,000	k	95,000
f	55,000	E	75,000
T	55,000	e	70,000
A	43,000	i	57,000
M	33,000	T_2	55,000
		T_1	53,000
		S	26-30,000

Specific to N18T62 and NG105-15		Specific to C6-BU-1 and NG105-15		Unique to NG105-15	
K	97,000	r	55,000	z	53,000
H	90,000	b	50,000		
d	56-57,000	G	49,000		
i	56,000	g	47,000		
n	55,000	gs	45,000		
a	54,000				
F	'50,000				
h	50,000				
L	38,000				
N	20,000				

*The molecular weight is indicated for each protein.

actin have also been identified to be expressed in culture and thus are now amenable to regulatory study.

Several important points should be emphasized. First, the two dimensional separation of cell proteins by the methodology employed here is highly reproducible and sensitive enough to identify the presence of small amounts of individual protein species from clone to clone and with time in cell culture. Second, as summarized in Table I, although some proteins were common to all clones, the segmentation of proteins expressed could be grouped into two major groups, i.e., a) the neuroblastoma and hybrid proteins, and b) the glioma and hybrid proteins. Proteins which were significantly expressed by neuroblastoma cells were also similarly expressed in the hybrid line. A smaller number of proteins which were dominant in gels of glioma cells were also expressed by the hybrid cells. However, the major neuroblastoma proteins (K, H, d, i, n, a, F, h, L, and N) were not expressed at all or only to a minor degree in glioma cell cultures, and conversely, the major glioma proteins (r, b, G, g, gs) were not expressed, or only minimally so, in neuroblastoma cell cultures. Thus cell lines derived from the two major cell types found in the mammalian brain, namely, neurons and glia, express strikingly different protein products during differentiation. This study points out for the first time, through the use of two dimensional separation of cell proteins, quantitative and qualitative differences in the genetic programs for neuroblast and glioblast differentiation in a cell culture system.

It appears from the protein spectra of the hybrid cell that its molecular and genetic expression is far more neuronal than glial. There appears to be a dominance for expression of the neuronal genome and inhibition of the glial genome within the hybrid (neuroblastoma-glioma) cell line (Fig. 1). Alternatively, the genetic dosage may be greater for neuroblastoma than rat glioma due to retention within the hybrid of more mouse than rat chromosomes, which is a point that requires further study. This observation may be important in considering the potential occurrence of 'trophic interaction' between neurons and glia in the intact brain. The dominance of neuroblast genetic expression in the hybrid cell suggests the possiblity of similar regulation and inhibition of glial functions by neurons *in vivo*. Further, neuron-glial interaction may be necessary for the genetic expression of differentiated functions which are absent or weakly expressed by either cell line alone. These would include the presence of opiate receptors and heightened choline acetyltransferase activity in hybrid cells (Amano *et al.*, 1974, Hamprecht, 1974) and our present demonstration of a protein species unique to hybrid cells, protein spot z.

These studies emphasize the point that the determination of an enzyme activity or protein concentration in a homogenate of brain may perhaps actually be the summated expression of both neurons and glia reciprocally monitoring and regulating the genetic expression of the opposite cell. By the use of clonal cell lines derived from neurons and glia, and comparing the protein spectra expressed with hybrid cells as shown here, it is clear that such modulation and reciprocal inhibition of the genetic material of the opposite cell type may be in operation and may be a central mechanism of differentiation within the mammalian brain.

Acknowledgments

Supported by Grants from the National Cancer Institute, NIH, U.S. Department of Health, Education, and Welfare Nos. 5-R01-CA16379-04, NIH 1-P01-CA17065-01, NO1-CM-53767, CA-21602-01, and Leland Fikes Foundation.

References

Amano, T., B. Hamprecht and W. Kemper. 1974. High activity of choline acetyltransferase induced in neuroblastoma × glia hybrid cells. Exp. Cell Res. 85:399-408.

Amano, T., E. Richelson and M. Nirenberg. 1972. Neurotransmitter synthesis by neuroblastoma clones. Proc. Natl. Acad. Sci. U.S.A. 69:258-263.

Anagnoste, B., L. S. Freedman, M. Goldstein, J. Broome and K. Fuxe. 1972. Dopamine-β-Hydroxylase activity in mouse neuroblastoma tumors and in cell cultures. Proc. Natl. Acad. Sci. U.S.A. 69:1883-1886.

Augusti-Tocco, G. and G. Sato. 1969. Establishment of functional clonal lines of neurons from mouse neuroblastoma. Proc. Natl. Acad. Sci. U.S.A. 64:311-315.

Benda, P., J. Lightbody, G. Sato, L. Levine and W. Sweet. 1968. Differentiated rat glial cell strain in tissue culture. Science 161:370-371.

Blume, A., F. Gilbert, S. Wilson, J. Farber, R. Rosenberg and M. Nirenberg. 1970. Regulation of acetylcholinesterase in neuroblastoma cells. Proc. Natl. Acad. Sci. U.S.A. 67:786-792.

Comings, D. E. and L. C. Tack. 1972. Similarities in the cytoplasmic proteins of different organs and species examined by SDS gel electrophoresis. Exp. Cell Res. 75:73-78.

Daniels, M. P. and B. Hamprecht. 1974. The heterostructure of neuroblastoma glioma somatic cell hybrids. J. Cell Biol. 63:691-699.

Davidson, R. L. 1969. Regulation of melanin synthesis in mammalian cells, as studied by somatic hybridization. Exp. Cell Res. 55:424-426.

deVellis, J. and D. Inglish. 1968. Hormonal control of glycerol-phosphate-dehydrogenase in the rat brain. J. Neurochem. 15:1061-1070.

Hamprecht, B., J. Traber and F. Lamprecht. 1974. Dopamine-β-hydroxylase activity in cholinergic neuroblastoma X glioma hybrid cells; increase of activity by N^6, ^{2}-dibutyryl adenosine 3′, 5′-cyclic monophosphate. FEBS Letters 42:221-226.

Hamprecht, B. 1974. Cell cultures as model systems for studying the biochemistry of differentiated functions of nerves. Hoppe-Seyler's Z. physiol. Chem. 355:109-110.

Harris, A. J., S. Heinemann, D. Schubert and H. Tarakis. 1971. Trophic interaction between cloned tissue culture lines of nerve and muscle. Nature (Lond.) 231:296-301.

Hutchison, H. T., K. Werrbach, C. Vance and B. Haber. 1974. Uptake of neurotransmitters by clonal lines of astrocytoma and neuroblastoma in culture: I. Transport of GABA. Brain Res. 66:265-274.

Klebe, R. J. and F. H. Ruddle. 1969. Neuroblastoma: Cell culture analysis of a differentiating skin cell system. J. Cell Biol. 43:69A.

Klee, W. A. and M. Nirenberg. 1974. A neuroblastoma X glioma hybrid cell line with morphine receptors. Proc. Natl. Acad. Sci. U.S.A. 71:3473-3477.

Minna, J., D. Glazer and M. Nirenberg. 1972. Genetic dissection of neural properties using somatic cell hybrids. Nature (Lond.) New Biol. 235:225-231.

Minna, J., P. Nelson, J. Peacock, D. Glazer and M. Nirenberg. 1971. Genes for neuronal properties expressed in neuroblastoma × L cell hybrids. Proc. Natl. Acad. Sci. U.S.A. 68:234-239.

Murphy, R. A., N. J. Pantazies, B. G. W. Arnason and M. Young. 1975. Secretion of a nerve growth factor by mouse neuroblastoma cells in culture. Proc. Natl. Acad. Sci. U.S.A. 72:1895-1898.

Nelson, P. G., J. Peacock and T. Amano. 1971a. Responses of neuroblastoma cells to iontophoretically applied acetylcholine. J. Cell. Physiol. 77:353-362.

Nelson, P. G., J. Peacock, T. Amano and J. Minna. 1971b. Electrogenesis in mouse neuroblastoma cells *in vitro*. J Cell. Physiol. 77:337-352.

Nelson, P. G., B. W. Ruffner and M. Nirenberg, 1969. Neuronal tumor cells with excitable membranes grown *in vitro*. Proc. Natl. Acad. Sci. U.S.A. 64:1004-1010.

O'Farrell, P. 1975. High resolution two-dimensional electrophoresis of proteins. J. Biol. Chem. 250:4007-4021.

Pfeiffer, S. E., H. R. Herschman, J. E. Lightbody and G. Sato. 1970. Synthesis by a clonal line of rat glial cells of a protein unique to the nervous system. J. Cell. Physiol. 75:329-339.

Prashad, N. and R. N. Rosenberg. 1977. Phosphorylation of proteins by dibutyryl cyclic AMP during differentiation of mouse neuroblastoma cells. Trans. Am. Soc. Neurochem. 8:86.

Prashad, N., B. Wischmeyer, C. Evetts, F. Baskin and R. Rosenberg. 1977. Dibutyryl cAMP-induced protein changes in differentiating mouse neuroblastoma cells. Cell Differentiation, in press.

Rosenberg, R. N. 1973. Regulation of neuronal enzymes in cell culture. In: Tissue Culture of the Nervous System (Current Topics in Neurobiology Vol. I). G. Sato, ed. Plenum Press, New York. pp. 107-134.

Rosenberg, R. N., L. Vandeventer, L. deFrancesco and M. E. Friedkin. 1971. Regulation of the synthesis of choline-O-acetyltransferase and thymidylate synthetase in mouse neuroblastoma in cell culture. Proc. Natl. Acad Sci. U.S.A. 68:1436-1440.

Schubert, D., S. Humphreys, C. Baroni and M. Cohn. 1969. *In vitro* differentiation of a mouse neuroblastoma. Proc. Natl. Acad. Sci. U.S.A. 64:316-323.

Schubert, D., S. Humphreys, F. De Vitry and F. Jacob. 1971. Induced differentiation of a neuroblastoma. Dev. Biol. 25:514-546.

Seeds, N. W., A. G. Gilman, T. Amano and M. Nirenberg. 1970. Regulation of axon formation by clonal lines of a neural tumor. Proc. Natl. Acad. Sci. U.S.A. 66:160-167.

IMMUNOCHEMICAL MARKERS IN PRIMARY CULTURES AND IN CELL LINES

ELISABETH BOCK

The Protein Laboratory
University of Copenhagen
Copenhagen, Denmark

I. Introduction

In the nervous system, specific functions are connected to the various cell types. Thus, nerve cell function includes the conduction of action potentials, synaptic transmission and the establishment of specific connections. Astrocytes have uptake mechanisms for neurotransmitters, and oligodendrocytes produce the closely packed plasma membranes of myelin. These specific functions are to a large extent mediated by proteins, soluble or membrane bound, which are unique to the nervous system.

The heterogeneity of the nervous system represents a serious obstacle for neurobiological studies. Culture techniques, with their ability to isolate organs, tissues, or cells from the rest of the organism, constitute a useful tool for the investigator, making it possible to study cellular interactions and biochemistry of a defined sample in the absence of homeostatic effects operative *in vivo*. The direct effect of various agents (drugs, hormones etc.) can be studied, and environmental variables can be manipulated in ways which are only possible *in vitro*. The choice of culture system depends on the scope of the investigation. Characterization of a culture system is necessary; the more data that can be accumulated on a culture system, the more reliable are the generalizations which can be made from the experiments. Characterization includes description of morphological, neurophysiological and biochemical parameters. For the proteinchemical characterization several approaches can be used to choose those proteins which probably are involved in functions specific for the various cell populations. One approach is to select proteins related to certain structures, which are important in the operation of the nervous system, such as the proteins of micro-tubules or myelin. Another is to make use of high affinity markers such as α-neurotoxins, which specifically bind to the nicotinic acetylcholine receptor. Determination of enzymes involved in specific functions such as metabolic pathways of neurotransmitters have also proven very useful. Finally, nervous system specific proteins, detected by immunochemical techniques, represent a group of valuable markers of various cells and cell structures. The study of proteins which, judged by immunochemical methods are specific to the nervous system, is based on the assumption that a protein, which is not required by any other tissue, has an important function, probably specific to the nervous system. Many of these proteins have been demonstrated in a wide range of species. Their conservation during evolution further supports this assumption. On the other hand, it was recently demonstrated that a neuronal cytoplasmic protein, the 14.3.2 protein, exhibits enolase activity (Bock and Dissing, 1975). This enzyme activity is common to virtually all living cells. However, this finding does not rule out the possibility that the 14.3.2 protein has other important functions.

In the following, a series of nervous system specific proteins found in normal mammalian brain will be described. Proteins, which are soluble or at least partly soluble at neutral pH in conventional buffer systems were the first to be isolated, and they are therefore the best characterized. Reports on these immunochemical markers in primary cell cultures and in cell lines will hereafter be briefly reviewed.

II. Nervous System Specific Proteins

A. Glial markers.

1. *The S100 protein.* This protein was first isolated by the pioneering work of Moore (1965) and Moore and McGregor (1965). It was named S100 because of its solubility in saturated ammonium sulphate at pH 7, a property which together with a very acidic isoelectric pH has been employed in the purification procedure (Moore, 1965, 1972). Purification procedures have also been described by Uyemura *et al.* (1971) and Dannies and Levine (1971). The small molecule (21,000 - 24,000 daltons) is of low immunogenicity. Levine and Moore (1965) first made an antiserum by complexing the antigen to methylated bovine serum albumin before immunization. The protein has been quantified by means of a complement fixation assay (Moore *et al.*, 1968), quantitative immunoelectrophoresis (Jacque *et al.*, 1974), and a radioimmunoassay (Uozomi and Ryan, 1973). S100 was found to be at least 10,000-fold more concentrated in brain than in other tissues and closely cross-reactive forms were found in all vertebrate species examined including mammals, birds, fish and reptiles. (Moore, 1972). S100 has been demonstrated both in the central and the peripheral nervous system, and generally the level in white matter is higher than in grey; particularly high levels were found in cerebellum (Moore, 1972). The protein has with certainty been localized in glial cells whereas a neuronal localization seems disputable. The glial localization was demonstrated by Hydén and McEwen (1966) by immunofluorescence staining and by microdissection of single neurons and glia which were subsequently assayed

for S100 content. Degeneration studies (Cicero *et al.*, 1970b, Perez *et al.*, 1970) and determination of S100 in bulk-prepared neurons and glial cells (Bock and Hamberger, 1976) also supported this view. Using the peroxidase-labeled antibody technique Ludwin *et al.* (1976) found S100 in astrocytes, both in the nucleus and the cytoplasm. It was also demonstrated in oligodendroglia, whereas no staining in any structure of the neurons were observed. Some authors have localized S100 in the nucleus (Michetti *et al.*, 1974) and specifically in the nuclei of nerve cells (Hydén and McEwen, 1966). S100 has also been demonstrated in isolated synaptosomes by Donato *et al.* (1975) and at the post-synaptic membranes by the peroxidase-labeled antibody technique (Haglid *et al.*, 1974). Presently neuronal localization of S100 seems open to question. The different findings may to some extent derive from differences in antibody specificity. Developmental studies have shown that S100 synthesis occurs relatively late in the development compared to histological maturity, the S100 level rising some time after cell division has ceased (Cicero *et al.*, 1970a). The function of S100 is unknown, but many chemical studies have been performed giving clues as to what to examine in the search for a possible function. The molecular weight is approximately 21,000-24,000 daltons (Moore, 1965). It is composed of three subunits of about 7,000 daltons which cannot be identical, since there is only one tryptophan residue per 21,000 daltons (Dannies and Levine, 1971). In the presence of calcium ions S100 has been shown to undergo a conformational change which exposes hydrophobic parts of the molecule and allows strong binding to artificial membranes, leading to a facilitation of the transport of monovalent cations (Calissano and Bangham, 1971). The 8-10 calcium binding sites have been shown to be of two types, and the high-affinity sites were inhibited in the presence of K^+ (Calissano, 1973). These effects occurred at physiological concentrations of both K^+ and Na^+ and Ca^{2+}. The appearance of hydrophobic sites in the presence of Ca^{2+} might play a role in determining the amount of membrane bound S100. Moore (1975) has shown that in the absence of Ca^{2+}, S100 did not bind measurably to synaptic membranes, myelin or red cell ghosts; whereas appreciable amounts were bound to all three types of membranes in the presence of Ca^{2+}.

2. *The glial fibrillary acidic protein (GFA).* GFA was first isolated by Eng *et al.* (1971) from tissues rich in fibrous astrocytes obtained from human brains with pathological conditions leading to fibrous gliosis, *e.g.*, multiple sclerosis plaques. The protein can also be isolated from normal brain tissue as described by Dahl and Bignami (1975). GFA isolated from multiple sclerosis plaques was very immunogenic, and antiserum could easily be prepared (Uyeda *et al.*, 1972). It has later been shown that GFA degrades very easily (Dahl, 1976) and that immunogenicity varies considerably depending on the degree of degradation (Dahl and Bignami,

1976a). Antisera with titer and avidity comparable to those prepared against protein isolated from multiple sclerosis plaques were obtained using GFA isolated from normal tissue after extensive autolysis. Non-degraded or partially degraded protein was only weakly immunogenic or not immunogenic at all. The protein has been determined by quantitative immunoelectrophoresis (Jacque *et al.*, 1974) and an immunoradiometric assay (Eng *et al.*, 1976). Uyeda *et al.* (1972) demonstrated the brain specificity of GFA. No GFA could be found in peripheral nerve or other organs. The level in white matter was higher than in grey matter. GFA could be demonstrated in a wide range of vertebrates including birds and fish (Dahl and Bignami, 1973). By immunofluorescence technique Bignami *et al.* (1972) found a selective staining of fibrous astrocytes and their processes. This was confirmed by Ludwin *et al.* (1976) using the peroxidase-labeled antibody technique. These authors also noted a somewhat weaker staining of protoplasmic astrocytes. During development, immunofluorescence first appeared in the period when bundles of glial filaments became visible with the electron microscope (Bignami and Dahl, 1973, 1974). In a quantitative study on brain specific proteins during ontogeny in mice (Jacque *et al.*, 1976) a peak of the specific concentration of GFA relative to total protein content was found between day 10-14 postnatally, corresponding to the outburst of astroglial differentiation at the time of myelination. The molecular weight of undegraded GFA determined by sodium dodecylsulphate polyacrylamide gel electrophoresis has been determined to be 54,000 daltons, but if the tissue is not rapidly frozen immediately after death lower molecular weight components in the range from 54,000-40,500 daltons appear (Dahl, 1976). Comparative studies of GFA and neurofilament proteins (Benitz *et al.*, 1976) and the isolation of a GFA-like protein from peripheral nerve where glial fibers are absent, indicate that there may be some similarity between the protein subunits of glial and nerve filaments (Dahl and Bignami, 1976b). Concerning the function of GFA, Dahl and Bignami (1973) hypothesized that one function of a fibrous astrocyte is to provide support for central nervous system axons, which are highly susceptible to tearing in trauma, and which are not surrounded, as in peripheral nerves, by the basal laminae of the Schwann cells and by the collagen fibrils of the endoneurium. The astroglial fiber, of which GFA presumably is a major constituent, may provide this support.

3. *Nervous system antigen-I* (NS-1) is a brain specific membrane protein (or proteins) defined by means of antiserum raised against a chemically induced glioblastoma in mice (Schachner, 1974). By the cytotoxicity test the protein was demonstrated on the cell surface. It was found in brains of other mammals. The content of NS-1 was higher in white than in grey matter and it was also present in peripheral nerve. Myelin

deficient mutant mice had low amounts of the antigen(s). These findings made Schachner suggest that it is located exclusively on glial membranes.

B. Neuronal Markers.

1. *The 14.3.2 protein.* Using the same general method of preparation as described for the S100 protein (Moore, 1965) it was also possible to isolate the 14.3.2 protein and prepare antibodies against it (Moore and Perez, 1968). Antigen-α isolated by Bennett and Edelman (1968) has been shown to be identical to 14.3.2 protein (Schneider, 1973). Suitable isolation procedures have been described by Bennett and Edelman (1968) and Moore (1972). The protein has been quantified by means of a complement fixation assay (Moore *et al.*, 1968), quantitative immunoelectrophoresis (Bock *et al.*, 1975) and a radioimmunoassay (Revoltella *et al.*, 1976). The protein is specific to the nervous system, and brain was shown by Moore and Perez (1968) to contain 200 times more 14.3.2 protein than any other organ tested. The protein was found in cross-reacting forms in all mammalian species and in birds. The amount of 14.3.2 protein was higher in grey matter than in white and it was present in peripheral nerve. The protein has been localized to the neuronal cytoplasm by degeneration studies (Perez *et al.*, 1970, Cicero *et al.*, 1970b). Microdissection of single neurons with subsequent quantitative measurement of the protein demonstrated its presence both in the cell bodies and in the surrounding neuropil (Moore, 1975). Developmental studies have shown that 14.3.2 synthesis occurs relatively late in the process of development of the nervous system (Cicero *et al.*, 1970a). Bennett and Edelman (1968) when analyzing brain extracts by immunoelectrophoresis, noted a heterogeneity in the electrophoretic mobility of 14.3.2, revealed as a double arc precipitate. They isolated the anodally migrating component; this component, which has a molecular weight of about 50,000 daltons (Moore, 1972), shows a tendency to aggregate. In 1975 Bock and Dissing demonstrated enolase activity connected with 14.3.2. They also showed that the two electrophoretically different forms of the protein were immunochemically partially identical. Furthermore, they were the first to demonstrate the presence of brain specific isoenzymes of enolase. This finding has been confirmed by Marangos *et al.* (1976). The isoenzymes which apparently are unique to the brain, have recently been shown to be two dimers, $\gamma\gamma$ and $\alpha\gamma$, composed of the brain specific γ subunit and the α subunit of wide tissue distribution (Bock *et al.*, 1977). The dimer $\gamma\gamma$ corresponds to the 14.3.2 protein as originally described (called 14.3.2 anodal component by Bock and Dissing, 1975). The dimer $\alpha\gamma$ (called 14.3.2 cathodal component by Bock and Dissing) cross-reacts with the $\gamma\gamma$ enolase. This is worth keeping in mind as the ratio of $\alpha\gamma/\gamma\gamma$ varies with age (E. Bock and W. Wechsler, unpublished observations). Crossed immunoelectro-

phoresis does discern between the two components whereas complement fixation assay or radioimmunoassay does not. This is the first brain specific protein to which a function has been attributed. The conversion of 2-phospho-glycerate to phosphoenol pyruvate is common to all cells. Since it has not been possible to detect any major kinetic differences between the several isoenzymes of enolase, specific functional differences for the neuronal form must be sought outside the active site.

2. *Synaptin.* By immunization with rat brain synaptic vesicles, Bock *et al.* (1974) raised antiserum against a vesicle membrane protein, synaptin (initially designated C1). This protein was specific to the nervous system and present in several mammalian species including mouse, ox, and man. By quantitative immunoelectrophoresis it was shown to be ten times more concentrated in the synaptic vesicle fraction than in whole brain homogenates (Bock and Jörgensen, 1975, Bock *et al.*, 1975). It was also demonstrated on synaptosomal plasma membranes, three times enriched compared to whole brain homogenate. In an ontogenic study in mice by Jacque *et al.* (1976) the level of synaptin expressed relative to total protein content, was shown to rise from day 8 to day 40 postnatally; at day 40 a plateau was reached. At days 1 and 5 an immunochemically partially identical form of synaptin was present. By means of an immunoabsorption technique synaptin has been demonstrated on the outside of the vesicle membrane and of the membrane of the chromaffine granule and on the inside of the synaptosomal plasma membrane (E. Bock and K. B. Helle, unpublished experiments). The protein contains a carbohydrate moiety, which reacts with various lectins; it has a molecular weight of 45,000 daltons and an isoelectric point 4.2. It behaves like an amphophilic protein and is thus regarded as an instrinsic membrane protein (E. Bock, B. Norrild and S. Bhakdi, unpublished observations). The function of synaptin is unknown, but the topographic localization combined with the presence of a carbohydrate moiety makes it a plausible hypothesis that it may be involved in the exocytotic process of synaptic vesicles during synaptic transmission.

3. *D1.* This brain specific membrane protein was identified by Jörgensen and Bock (1974) by means of an antiserum raised against rat synaptosomal plasma membranes. By quantitative immunoelectrophoresis synaptosomal plasma membranes were found to have four times as much D1 as whole brain homogenate (Bock and Jörgensen, 1975; Bock *et al.*, 1975). It has also been demonstrated in axonal preparations (O. S. Jörgensen, personal communication). The topographic localization in the synaptosomal membrane was investigated by an immunoabsorption technique (Jörgensen, 1976). The antigenic determinants were located mainly if not totally on the outside of the synaptosomal plasma membrane. In an ontogenic study on mice by Jacque *et al.* (1976) the level of D1, expressed relatively to the total protein content, rose steadily from day 12 to

day 40 where a plateau was reached. An immunochemically partially identical form of D1 was present at days 1, 5, and 8. The function of D1 is unknown.

4. *D2.* This brain specific membrane protein was identified in a similar manner to the D1 protein, using an anti-synaptosomal membrane antiserum (Jörgensen and Bock, 1974). By quantitative immunoelectrophoresis it was found to be three times concentrated in synaptosomal plasma membranes compared to whole brain homogenate (Bock and Jörgensen, 1975; Bock *et al.*, 1975). An immunochemically partially identical form of D2 has been demonstrated in human cerebrospinal fluid (Jörgensen and Bock, 1975). Topographically it has been localized on the outside of the synaptosomal plasma membrane (Jörgensen, 1976). Jacque *et al.* (1976) found that the level of D2 in mice, expressed relative to total protein content, decreased with age just after the brain growth spurt, and that the steady adult level reached at day 40 was only 50% of the level at day 12. The level of D2 thus paralleled the rate of synapse formation rather than the amount of formed synapses; this in connection with the topographic localization made Jörgensen (1976) hypothesize that D2 was involved in intercellular recognition processes during synaptogenesis.

5. *D3.* This brain specific membrane protein was demonstrated in the same investigations which led to the demonstration of the two previously mentioned membrane proteins (D1 and D2) (Jörgensen and Bock, 1974). By quantitative immunoelectrophoresis it was shown to be four times concentrated in synaptosomal plasma membranes compared to whole brain homogenate (Bock and Jörgensen, 1975; Bock *et al.*, 1975). Topographically it was localized on the inside of the synaptosomal plasma membrane (Jörgensen, 1976). Jacque *et al.* (1976) found that the level of D3 in mice, expressed relative to total protein, rose during ontogeny from day 12 to day 40 postnatally. At days 1, 5, and 8 an immunochemically partially identical form of D3 was shown. The function of D3 is unknown.

III. Studies on Nervous System Specific Proteins in Primary Cultures.

The difficulties in obtaining reagents (antisera and antigens) for determination of nervous system specific proteins have limited investigations in culture work. Moreover, working with primary cultures confronts the investigator with the problem of scarcity of material. Antanitus *et al.* (1975) identified astrocytes by immunofluorescence staining for GFA in primary tissue culture explants of human fetal forebrain (gestational ages 12-20 weeks). The cultures were examined after 8-39 days of cultivation. The monolayer outgrowth surrounding the explant revealed a population of small cells with small dark nuclei and scanty cytoplasm. These cells were argyrophilic with the Bodian protargol method, and did not stain for GFA;

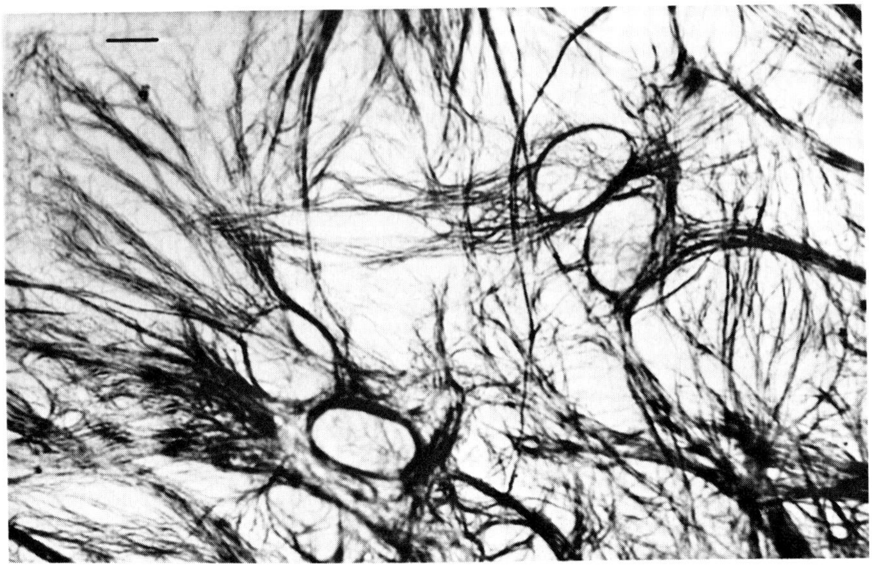

FIG. 1. Immunoperoxydase staining for GFA in a primary mouse astroglial cell culture (28 days of cultivation). (E. Bock, M. Möller, C. Nissen and M. Sensenbrenner, unpublished). The bar represents 10 μm. GFA is visualized in long fibers in the perikaryon and in the processes. There is no staining of the nuclei, although GFA-fibers below or above the nuclei are seen.

they were therefore interpreted as immature neurons. A second population consisted of larger cells with large nuclei and variable cytoplasmic morphology, some with quite long processes with frequent branching. These cells were not impregnated with silver by Bodian protargol method, but stained strongly for GFA and were regarded as immature astrocytes. Control cultures of fetal leptomeninges and WI 38 fibroblasts did not stain for GFA. Bock *et al.* (1975) determined the proteins 14.3.2, GFA, D1, D2, D3, and synaptin in a primary culture of hemispheres from newborn rats. The cells were cultivated for 21 days. At the time of harvesting no neuronal cells could be identified and the cells were supposed to be astroglial cells, contaminated with a few per cent of ependyma cells. The supposedly neuronal membrane proteins D1, D2, D3, and synaptin were all absent from the cultures. The anodal component of 14.3.2 was also absent whereas 14.3.2 cathodal component was present in an amount corresponding to 10% of adult whole brain homogenate, expressed relative to total protein. Whether this was due to a few contaminating neurons or to glial synthesis of the protein could not be determined. The level of GFA expressed relative to total protein was more than 40 times higher in the cultured cells than in whole adult brain. The authors considered the possibility that the amount of GFA in the cultured cells might be higher than in astroglial cells *in vivo* due to environmental factors during cultivation. Immunoperoxydase staining for GFA of a 28 day old primary astroglial cell culture is shown in Fig. 1 (E. Bock, M. Möller, C. Nissen and M. Sensenbrenner, unpublished experiments). Lim *et al.* (1977) investigated the effect of a glia maturation factor on S100 concentration in astrocytic cultures which had been subcultured once and were derived from dissociated brain cells from 17-day fetal rats. S100 was determined after various times of exposure to the factor. No change was seen after one day but considerable increases were observed after four and seven days exposure. Four to seven days cultivation in the presence of the glia maturation factor led to extension of processes in a criss-crossing network.

IV. Studies on Nervous System Specific Proteins in Cell Lines

In 1968 Benda *et al.* described the adaption to culture of cells derived from chemically induced rat glial tumors. One clonal cell line, the C-6 cell line, produced S100. It was shown that the ratio of S100 to soluble protein increased tenfold as the cells grew from a low cell density to confluency. When the cells reached confluency, the total amount of S100 remained constant. Two potentially separable phenomena, cessation of cell division and initial cellular contact, occur at the point in the growth curve where S100 accumulation begins. Pfeiffer *et al.* (1970) demonstrated the importance of homologous C-6 cell contact by arresting the division of exponentially growing cells with only limited cell contacts by addition of hydroxyurea and fluorodeoxyuridine or by withdrawal of serum. These treatments did not initiate S100 production, although they did not prevent the onset of S100 accumulation in stationary phase cultures. Moreover, S100 accumulation did not occur in stationary phase C-6 cells in suspension culture, and also mixed culture experiments with a cell line (C-3) which did not synthesize S100 suggested that S100 accumulation required homologous cell contact. This regulated initiation of accumulation of S100 in monolayer culture might be explained as an induction of the synthesis by cell contact. Alternatively, the pool of S100 could be continuously turned over at different rates in exponentially growing and in stationary cultures. Herschman *et al.* (1973) demonstrated that the absolute rate of synthesis of S100 protein increased by a factor of at least five to six, when C-6 cells entered the stationary phase, and they concluded that the observed increase of S100 in stationary phase was due to a density dependent, contact mediated induction of synthesis. The degradation rate of S100 in stationary phase C-6 cell cultures was approximately the same as the rate observed for total protein, and Herschman *et al.* (1973) concluded that this result was in agreement with the *in vivo* degradation data obtained by Cicero and Moore (1970), who found a half life of about 12-15 days. In contrast, McEwen and

Hydén (1966) found in rat brain that S100 had a half-life of several hours, while total protein had a half-life of several days. The differences may be due to different experimental approaches giving different degrees of contamination during isolation of S100. The methodology employed in these studies for determination of half-life, i.e., labeling the proteins with a radioactive precursor followed by isolation of the specific protein at various times after labeling, has, however, been shown to be of limited value (Poole, 1971). The presence of S100 has been demonstrated in several other cell lines. Pfeiffer and Wechsler (1972) demonstrated S100 in a Schwann cell tumor line RN-2, which also was able to synthesize myelin basic protein. Schubert *et al.* (1974) chemically induced tumors of neuronal and glial origin and adapted the resulting cells to clonal cell culture. S100 was demonstrated in both putative glial and putative neuronal clonal cell lines. Not all glial cell lines, however, contained S100. In the same study 14.3.2 was also measured, and this protein was similarly demonstrated in both glial and neuronal clonal cell lines. All the five neuronal clonal cell lines established by Schubert and coworkers contained 14.3.2; however, of two clones (S20 and C1A) derived from the C1300 mouse neuroblastoma only C1A contained measurable amounts of 14.3.2. Neither S100 nor 14.3.2 could be demonstrated in a clonal rat myoblast line (L6) or in a fibroblast cell line (3T3). Herschman and Lerner (1973) were unable to demonstrate 14.3.2 in two C1300 clones (N18 and NB41); whereas the human neuroblastoma line IMR-32 reacted strongly with antisera to human 14.3.2. In contrast to the results obtained for S100 protein in C-6 cells, 14.3.2 seemed to accumulate in growing IMR-32 cells. The amount of 14.3.2 and of total protein increased proportionally. Thus, the amount of 14.3.2 relative to total protein was the same in exponential and stationary phase IMR-32 cells, and it was concluded that cell division and production of 14.3.2 were not mutually exclusive phenomena in this cell line. Augusti-Tocco *et al.* (1973) were able to demonstrate 14.3.2 in the C1300 clone Nb41A3 and they found that the highest amount of 14.3.2 per cell was present at 24 hours after plating, which corresponded to the end of the lag phase. At the beginning of the logarithmic phase the content of 14.3.2 per cell decreased. Also, total protein per cell was highest at the end of the lag phase with subsequent decrease. Recalculation of the data shows that the amount of 14.3.2 relative to total protein is fairly constant; thus, these data are in agreement with the data of Herschman and Lerner (1973). The amount of 14.3.2 was the same for cells grown in monolayer and cells grown in suspension. These results indicate that contrary to the S100 production in C-6 cells, 14.3.2 production in neuroblastoma cells is independent of cell division and cell contact. 14.3.2 has also been demonstrated in the tumor Schwann cell line RN-2 (Braun *et al.*, 1976). The brain specific cell surface antigens NS-1 and S100 were determined by SundarRaj *et al.* (1976) in clonal cell lines and in clonal tumors grown *in vivo* and derived from the mouse glioma 26. NS-1 concentrations approximated the levels in the original tumor, when the cells were grown as clonal tumors *in vivo*, but were reduced significantly when the cells were grown *in vitro*. Conversely, the expression of S100 was high under *in vitro* conditions but negligible after one passage *in vivo*. NS-1 therefore also seemed to be subject to different

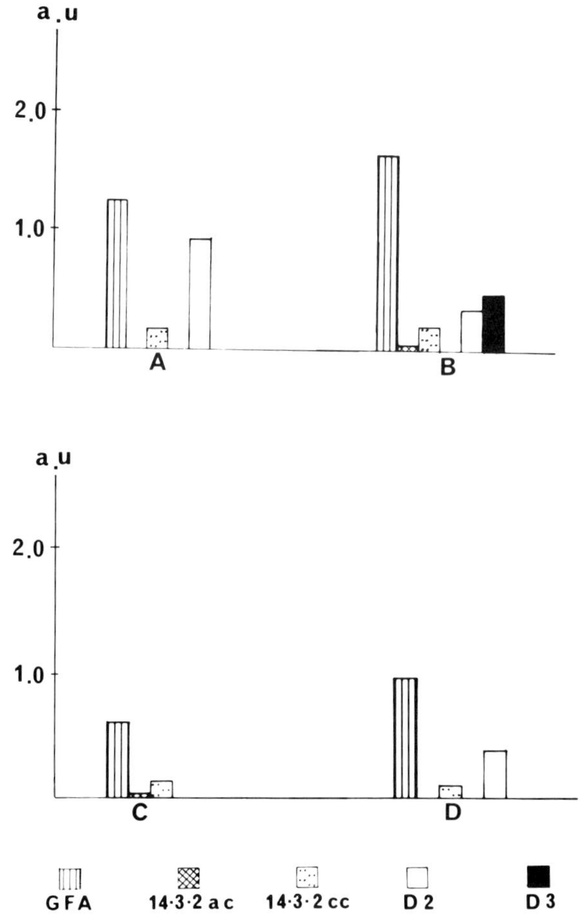

FIG. 2. Determination of GFA, 14.3.2 anodal and cathodal components (a.c. and c.c.), D1, D2, D3 and synaptin in A. Schwann cell tumor line RN-6, B. glioma cell line C-6, C. neuroblastoma cell line, subclone from B 35, D. neuroblastoma cell line subclone from B 103. The amount of the individual proteins is expressed in specific values relative to total protein. 1.0 arbitrary unit (a.u.) is the amount of a protein present per g total protein in a pool of forebrains from 35-day-old rats. The values represent the mean of three series of determinations on cultures harvested in stationary phase. (E. Bock and W. Wechsler, unpublished).

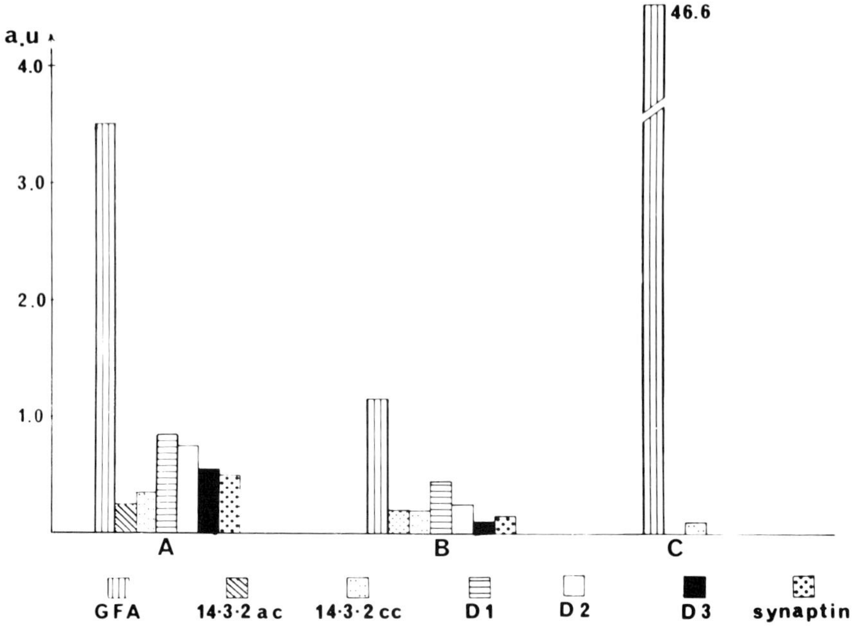

FIG. 3. Determination of GFA, 14.3.2 anodal component, 14.3.2 cathodal component, D1, D2, D3 and synaptin in bulk-prepared glial cells (A) and neuronal cells (B). The data are derived from Bock and Hamberger (1976). The amount of the individual proteins is expressed in specific values relative to total protein. One (1.0) arbitrary unit (a.u.) is the amount of a protein present per g total protein in the homogenate from which the cells were isolated. C is the amount of the same six proteins in a primary rat astroglial cell culture at 21 days of cultivation (Bock *et al.*, 1975).

regulatory mechanisms than S100. Incubation of the cells with dibuturyl 3′,5′-cyclic AMP or catecholamines led to decreased NS-1 concentrations. The authors suggested that this represented part of a mechanism of glial-neuronal interaction. The value of demonstration of a brain specific protein in a clonal cell line was investigated by E. Bock and W. Wechsler (unpublished studies). They determined six brain specific proteins: GFA, 14.3.2, synaptin, D1, D2, and D3, in five glioma lines (including the C-6 line and the RG-2 line), one Schwann cell tumor line (RN-6, subclone of RN-2), two neuroblastoma lines (subclones of B35 and B103), and two fibroblast lines, all derived from rat. In Fig. 2 the amount of the various proteins, expressed in specific values relative to the total protein content, is shown in the Schwann cell tumor line RN-6, the C-6 glioma cell line, neuroblastoma line subclone from B35 and neuroblastoma line subclone from B103. For comparison the amounts of the same proteins in bulk-prepared neuronal and glial cells (Bock and Hamberger, 1976) and in the previously mentioned primary astroglial cell culture (Bock *et al.*, 1975) are shown in Fig. 3. Synaptin and D1 were absent from all cell lines. The 14.3.2 protein was demonstrated in both glial and neuronal cell lines. The ratio of 14.3.2 cathodal component/14.3.2 anodal component was always above 1.0 in the cell lines. In brains from seven and fourteen day old rats, ratios of 2.0 and 1.3 were found, whereas the ratio in older animals was 1.0, indicating that ratios above 1.0 may be interpreted as a sign of immaturity. D2 was demonstrated in the RN-6 cells, the C-6 cells and in the subclone of B103. D3 was demonstrated only in the C-6 cells. GFA was present in all cell lines. The highest levels of GFA were found in the C-6 cells, the RN-6 cells and the two neuroblastoma cell lines. However, in none of the cell lines were amounts comparable to the amount demonstrated in the primary astroglial culture (Fig. 3C), not even to the amount of GFA demonstrated in bulk-prepared glial cells (Fig. 3A). Each cell line in stationary phase exhibited a fairly constant protein pattern in determinations repeated over a period of six months. Exponentially growing cells had a marker protein pattern similar to stationary phase cells.

Conclusion

Demonstration of nervous system specific proteins in a clonal cell line may be used as proof of a neuroectodermal origin, but the present data indicate that it is not possible to distinguish between neuronal and glial origin of a cell line by means of its actual content of immunochemical markers. The few investigations of primary cultures seem to indicate that these cultures are differently composed with regard to marker proteins, and in this respect probably more valid models of their *in vivo* counterparts than the clonal cell lines presently available. From an experimental point of view, however, clonal cell lines have many advantages and studies on clonal cells have, among other things, given useful information concerning regulatory mechanisms for S100 accumulation. The presence of a characteristic, reproducible pattern of marker proteins demonstrated in a series of clonal cell lines may prove valuable in the study of differentiation and cellular recognition mechanisms in cell lines, and in the evaluation of effects of various exogenous factors on cells derived from the nervous system.

References

Antanitus, D. S., B. H. Choi and L. W. Lapham. 1975. Immunofluorescence staining of astrocytes *in vitro* using antiserum to glial fibriallary acidic protein. Brain Res. 89:363-367.

Augusti-Tocco, G., L. Casola and A. Grasso. 1973. Neuroblastoma cells and 14.3.2, a brain specific protein. Cell Differentiation. 2:157-161.

Benda, P., J. Lightbody, G. Sato, L. Levine and W. Sweet. 1968. Differentiated rat glial cell strain in tissue culture. Science. 161:370.

Benitz, W. E., D. Dahl, K. W. Williams and A. Bignami. 1976. The protein composition of glial and nerve fibers. FEBS Letters. 66:285-289.

Bennett, G. S. and G. M. Edelman. 1968. Isolation of an acidic protein from rat brain. J. Biol. Chem. 243:6234-6241.

Bignami, A. and D. Dahl. 1973. Differentiation of astrocytes in the cerebellar cortex and the pyramidal tracts of the newborn rat. An immunofluorescence study with antibodies to a protein specific to astrocytes. Brain Res. 49:393-402.

Bignami, A. and D. Dahl. 1974. Heterogeneity of the glial fibrillary acidic protein in gliosed human brains. J. Neurol. Sci. 23:551-563.

Bignami, A., L. F. Eng, D. Dahl and C. T. Uyeda. 1972. Localization of the glial fibrillary acidic protein in astrocytes by immunofluorescence. Brain Res. 43:429-435.

Bock, E. and J. Dissing. 1975. Demonstration of enolase activity connected to the brain-specific protein 14.3.2. Scand. J. Immunol. 4, Suppl. 2:31-36.

Bock, E. and A. Hamberger. 1976. Immunoelectrophoretic determination of brain-specific antigens in bulk-prepared neuronal and glial cells. Brain Res. 112:329-335.

Bock, E. and O. S. Jörgensen. 1975. Rat brain synaptic vesicles and synaptic plasma membranes compared by cross immunoelectrophoresis. FEBS Letters, 52:37-39.

Bock, E., O. S. Jörgensen and S. J. Morris. 1974. Antigen-antibody crossed electrophoresis of rat brain synaptosomes and synaptic vesicles: Correlation to watersoluble antigens from rat brain. J. Neurochem. 22:1013-1017.

Bock, E., O. S. Jörgensen, L. Dittmann and L. F. Eng. 1975. Determination of brain specific antigens in short term cultivated rat astroglial cells and in rat synaptosomes. J. Neurochem. 25:867-870.

Bock, E., L. Fletcher, C. C. Rider and B. Taylor. 1977. The nature of the two proteins of brain specific antigen 14.3.2. J. Neurochem. (in press).

Braun, M., A. Grasso and W. Wechsler. 1976. 14.3.2 protein in rat primary and transplanted gliomas and neurinomas and in clonal cell lines. Exp. Brain Res. 25:93-97.

Calissano, P. 1973. Specific properties of the brain specific protein S-100. In: *Proteins of the Nervous System*. D. J. Schneider, ed. Raven Press, New York. pp 13-26.

Calissano, P. and A. D. Bangham. 1971. Effect of two brain specific proteins (S-100 and 14.3.2) on cation diffusion across artificial lipid membranes. Biochem. Biophys. Res. Commun. 43:504-509.

Cicero, T. J., W. M. Cowan and B. W. Moore. 1970a. Changes in the concentrations of the two brain specific proteins, S-100 and 14.3.2, during the development of the avian optic tectum. Brain Res. 24:1-10.

Cicero, T. J., W. M. Cowan, B. W. Moore and V. Suntzeff. 1970b. The cellular localization of the two brain specific proteins, S-100 and 14.3.2. Brain Res. 18:25-34.

Cicero, T. J. and B. W. Moore. 1970. Turnover of the brain specific protein, S-100. Science. 169:1333-1334.

Dahl, D. 1976. Glial fibrillary acidic protein from bovine and rat brain. Degradation in tissues and homogenates. Biochim. Biophys. Acta. 420:142-154.

Dahl, D. and A. Bignami. 1973. Immunochemical and immunofluorescence studies of the glial fibrillary acidic protein in vertebrates. Brain Res. 61:279-293.

Dahl, D. and A. Bignami. 1975. Glial fibrillary acidic protein from normal and gliosed human brain. Demonstration of multiple related polypeptides. Biochim. Biophys. Acta. 386:41-51.

Dahl, D. and A. Bignami. 1976a. Immunogenic properties of the glial fibrillary acidic protein. Brain Res. 116:150-157.

Dahl, D. and A. Bignami. 1976b. Isolation from peripheral nerve of a protein similar to the glial fibrillary acidic protein. FEBS Letters. 66:281-284.

Dannies, P. S. and L. Levine. 1971. Structural properties of bovine brain S-100 protein. J. Biol. Chem. 246:6276-6283.

Donato, R., F. Michetti and N. Miani. 1975. Soluble and membranebound S-100 protein in cerebral cortex synaptosomes. Properties of the S-100 receptor. Brain Res. 98:561-573.

Eng, L. F., J. J. Vanderhaeghen, A. Bignami and B. Gerstl. 1971. An acidic protein isolated from fibrous astrocytes. Brain Res. 28:351-354.

Eng, L. F., Y. L. Lee and L. E. M. Miles. 1976. Measurement of glial fibrillary acidic protein by a two-site immunoradiometric assay. Analyt. Biochem. 71:243-259.

Haglid, K., A. Hamberger, H. A. Hansson, H. Hydén, L. Persson and L. Rönnbäck. 1974. S-100 protein in synapses of the central nervous system. Nature (Lond.). 251:532-534.

Herschman, H. R., B. P. Grauling and M. P. Lerner. 1973. Nervous system-specific proteins in cultured neural cells. In: *Tissue culture of the nervous system.* G. Sato, ed. Plenum Press, New York, London. pp. 187-202.

Herschman, H. R. and M. P. Lerner. 1973. Production of a nervous-system-specific protein (14.3.2) by human neuroblastoma cells in culture. Nature (Lond.) New Biol. 241:242-244.

Hydén, H. and B. S. McEwen. 1966. A glial protein specific for the nervous system. Proc. Natl. Acad. Sci. U.S.A. 55:354-358.

Jacque, C. M., O. S. Jörgensen, N. A. Baumann and E. Bock. 1976. Brain-specific antigens in the quaking mouse during ontogeny. J. Neurochem. 27:905-909.

Jacque, C. M., O. S. Jörgensen and E. Bock. 1974. Quantitative studies of the brain specific antigens S-100, GFA, 14.3.2, D1, D2, D3 and C1 in quaking mouse. FEBS Letters. 49:264-266.

Jörgensen, O. S. 1976. Localization of the antigens D1, D2 and D3 in the rat brain synaptic membrane. J. Neurochem. 27:1223-1227.

Jörgensen, O. S. and E. Bock. 1974. Brain specific synaptosonal membrane proteins demonstrated by crossed immunoelectrophoresis. J. Neurochem. 23:879-880.

Jörgensen, O. S. and E. Bock, 1975. Synaptic plasma membrane antigen D2 measured in human cerebrospinal fluid by rocketline immunoelectrophoresis. Determination in psychiatric and neurological patients. Scand. J. Immunol. 4 Suppl 2:25-30.

Levine, L. and B. W. Moore. 1965. Structural relatedness of a vertebrate brain acidic protein as measured immunochemically. Neurosci. Res. Prog. Bull. 3:18-22.

Lim, R., D. E. Turriff, S. S. Troy, B. W. Moore and L. F. Eng. 1977. Glia maturation factor: Effect on chemical differentiation of glioblasts in culture. Science 195:195-196.

Ludwin, S. K., J. C. Kosek and L. F. Eng. 1976. The topographical distribution of S-100 and GFA proteins in the adult rat brain: An immunohistochemical study using horseradish peroxidase-labelled antibodies. J. Comp. Neurol. 165:197-208.

Marangos, P. J., C. Zomzely-Neurath and C. York. 1976. Determination and characterization of neuron specific protein (NSP) associated enolase activity. Biochem. Biophys. Res. Commun. 68:1309-1316.

Michetti, F., N. Miani, G. DeRenzis, A. Caniglia and S. Correr. 1974. Nuclear localization of S-100 protein. J. Neurochem. 22:239-244.

Moore, B. W. 1965. A soluble protein characteristic of the nervous system. Biochem. Biophys. Res. Commun. 19:739-744.

Moore, B. W. 1972. Chemistry and biology of two proteins, S100 and 14.3.2, specific to the nervous system. In: *International Review of Neurobiology*, vol. 15, Academic Press, New York. pp. 215-225.

Moore, B. W. 1975. Brain-specific proteins: S-100 protein, 14.3.2 protein, and glial fibrillary protein. In: *Advances in Neurochemistry.* B. W. Agranoff and M. H. Aprison, eds. vol. 1. Plenum Press, New York, London. pp. 137-155.

Moore, B. W. and D. McGregor. 1965. Chromatographic and electrophoretic fractionation of soluble proteins of brain and liver. J. Biol. Chem. 240:1647-1653.

Moore, B. W. and V. J. Perez. 1968. Specific acidic proteins of the nervous system. In: *Physiological and Biochemical Aspects of Nervous Integration.* F. D. Carlson, ed. Prentice-Hall, Englewood Cliffs, N. J. pp. 343-360.

Moore, B. W., V. J. Perez and M. Gehring. 1968. Assay and regional distribution of a soluble protein characteristic of the nervous system. J. Neurochem. 15:265-272.

Perez, V. J., J. W. Olney, T. J. Cicero, B. W. Moore and B. A. Bahn. 1970. Wallerian degeneration in rabbit optic nerve: Cellular localization in the central nervous system of the S-100 and 14.3.2 proteins. J. Neurochem. 17:511-519.

Pfeiffer, S. E., H. R. Herschman, J. Lightbody and G. Sato. 1970. Synthesis by a clonal line of rat glial cells of a protein unique to the nervous system. J. Cell. Physiol. 75:329-340.

Pfeiffer, S. E. and W. Wechsler. 1972. Biochemically differentiated neoplastic clone of Schwann cells. Proc. Nat. Acad. Sci. U.S.A. 69:2885-2889.

Poole, B. 1971. The kinetics of disappearance of labelled leucine from the free leucine pool of rat liver and its effects on the apparent turnover of catalase and other hepatic proteins. J. Biol. Chem. 246:6587-6591.

Revoltella, A., L. Bertolini, L. Diamond, E. Vignetti and A. Grasso. 1976. A radioimmunoassay for measuring 14.3.2 protein in cell extracts. J. Neurochem. 26:831-834.

Schachner, M. 1974. NS-1 (nervous system antigen-1), a glial-cell-specific antigenic component of the surface membrane. Proc. Nat. Acad. Sci. U.S.A. 71:1795-1799.

Schneider, D. J. 1973. Studies of nervous system proteins. In: *Proteins of the Nervous System*. D. J. Schneider ed. Raven Press, New York. pp. 67-94.

Schubert, D., S. Heinemann, W. Carlisle, H. Tarikas, B. Kimes, J. Patrick, J. H. Steinback, W. Culp and B. L. Brandt. 1974. Clonal cell lines from the rat nervous system. 249:224-227.

SundarRay, N., M. Schachner and S. E. Pfeiffer. 1975. Biochemically differentiated mouse glial lines: Modulation of nervous system specific cell surface antigen concentration by catecholamines. Proc. Natl. Acad. Sci. U.S.A. 72:4110-4115.

Uozumi, T. and R. J. Ryan. 1973. Isolation, amino acid composition and radioimmunoassay of human brain S-100 protein. Mayo Clin. Proc. 48:50-56.

Uyeda, C. T., L. F. Eng and A. Bignami. 1972. Immunological study of the glial fibrillary acidic protein. Brain Res. 37:81-89.

Uyemura, K., G. Vincendon, G. Gombos and P. Mandel. 1971. Purification and some properties of S-100 protein fractions from sheep and pig brains. J. Neurochem. 18:429-438.

TRANSMITTERS, PUTATIVE TRANSMITTERS AND TRANSMITTER-RELATED ENZYMES STUDIED IN CULTURED CELL SYSTEMS

BRUCE K. SCHRIER

Behavioral Biology Branch
National Institute of Child Health and Human Development
National Institutes of Health
Bethesda, Maryland

In this contribution I will first deal with the relationship between putative transmitters and glial cells in culture. I will attempt then to extend these findings so as to expand our concept of the role of glial cells in the central nervous system. Second, I will discuss our progress in the investigation of trophic effects on transmitter-related enzymes which may aid us in understanding the development of the neuromuscular junction.

I. Glial Control of Neuroactive Substances.

We first became interested in glial relationships to neurotransmitters when we found (Wilson *et al.*, 1972) minute amounts of activities of glutamic acid decarboxylase (GAD, E.C. 4.1.1.15) and choline acetyltransferase (CAT, E.C. 2.3.1.6) in the glioma cell line C6. We were not surprised then, when Henn and Hamberger (1971) demonstrated uptake of GABA by glial cells partially purified from adult rabbit brain. We decided to investigate glial handling of neurotransmitters in more detail and we chose to do this using the glioma cell lines C6 and C2. We looked for synthesis, catabolism, influx and efflux of two putative inhibitory neurotransmitters, GABA and taurine (Schrier and Thompson, 1974).

We found that both C6 and C2, glioma cell lines contained low activities of glutamate decarboxylase (GAD). This enzymatic activity was initially measured in cell homogenates, and it was in all respects found to be identical to the presumed "neuronal" GAD of brain: dependent upon pyridoxal-PO_4, inhibited by anions, and eliminated by 1 mM amino-oxyacetic acid (AOAA). These characteristics were only apparent when GABA production, rather than CO_2 production, was assayed (Table I). In the optimal reaction conditions the glioma cells formed 3 to 10 times more $^{14}CO_2$ than GABA from [U-^{14}C]glutamate, and different cell lines had different CO_2/GABA ratios. Homogenates of none of the cell lines formed any GABA from glutamate in the presence of 1 mM amino-oxyacetic acid,

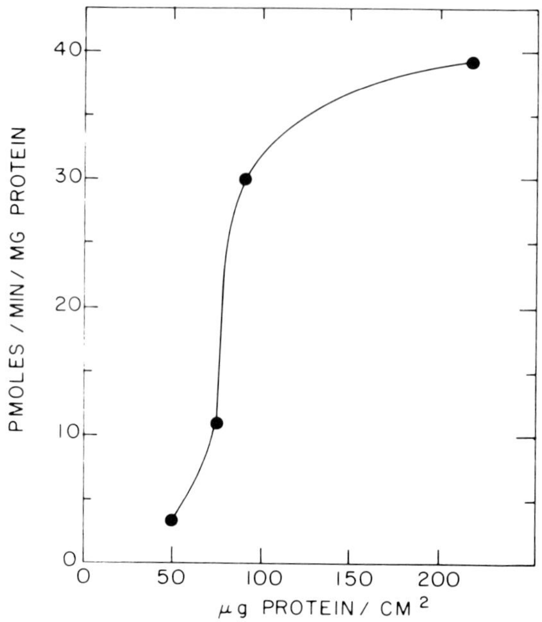

FIG. 1: Changes in specific activity of GAD with glioma cell density. Homogenates were made from C6 glioma cells at various times in a growth curve and assayed for GAD activity by GABA production. From Schrier and Thompson, 1974.

although $^{14}CO_2$ formation proceeded briskly in the presence of this inhibitor. In fact, although most tissues and cell lines will form $^{14}CO_2$ from [U^{14}C]glutamate in the presence or absence of AOAA, we have never found any homogenate which formed GABA in the presence of the inhibitor. Although we cannot distinguish between glial and neuronal GAD activities, and have not excluded the possibility that glial GAD might be a different polypetide from neuronal GAD, it is clear that the so-called "GAD II" of Haber *et al.* (1970) cannot be found in these sources, and that $^{14}CO_2$ evolution assays for GAD may give wildly spurious results. GAD activity in C6 cells also was regulated; that is, the specific activity of the cells depended on their density or rate of growth or length of time in culture, or some other factor with similar timing (Fig. 1). Although the ultimate specific activity was about 1/50 of that of a rat brain homogenate, its regulation in this manner is consistent with a conclusion that it has an important function for the glial cell.

TABLE I
EFFECTS OF GROWTH CONDITIONS (±BUCAMP) AND GAD INHIBITION (± AOAA) ON APPARENT (CO_2 ASSAY) AND REAL (GABA ASSAY) GAD ACTIVITIES[a] OF HOMOGENATES OF CULTURED GLIAL CELLS (C6, $C2_1$) OR SURFACE CULTURES OF FETAL RAT BRAIN CELLS (FRB).

Cells	BucAMP	AOAA	CO_2 Assay[b]	GABA Assay[c]
C6	—	—	365	21[d]; 38
—	+	—		20[d]
—	—	+	214	<0.7
$C2_1$	—	—	348	36[d]; 118
—	+	—		35[d]
—	—	+	254	<1.1
FRB	—	—	60[d]	750[d]; 791
—	+	—	68[d]	2000[d]
—	—	+		<5

[a]GAD activities given as pmoles of product formed/min/mg homogenate protein
[b]CO_2 production determined as described by Wilson *et al.*, 1972;
[c]GABA production assayed by high voltage electrophoresis as described by Schrier *et al.*, 1974.
[d]Data from Schrier and Shapiro 1973; all other data are from Wilson *et al.*, 1972.

We also found that living glioma cells formed GABA from radioactive glutamate placed in the growth medium (Table II); the apparent specific activity of this formation was of the same order of magnitude as that in homogenates. This showed that the cells were capable of uptake of glutamate; that uptake was not further investigated. At the end of the incubation, over 70% of the GABA formed had been released into the growth medium, but the cell/medium concentration ratio was still >48-fold. This also showed that there was little, if any GABA-T, in the glioma cells (cf. Schousboe, this volume). L cells did not make or excrete GABA under the same conditions (Table II). C6 glioma cells also formed taurine (Table III), another putative inhibitory neurotransmitter, from cystine in the medium, and this at a rate very similar to that of GABA formation. About ¼ of this taurine was further metabolized to isethionic acid, a compound without known function, which nevertheless is the major anion within the squid axon (Koechlin, 1955), and which may also be of importance to glial cells. Aspartic acid, although it entered C6 cells, was not converted appreciably to either β-alanine or GABA; therefore, some specificity with regard to synthesis and storage of these neuroactive substances was apparent. These findings led us to consider the possibility that glioma cells, as a model of glial cells *in situ*, might be capable of controlling the levels, in the extracellular space around neurons, of compounds which had excitatory and inhibitory effects on neuronal membranes.

TABLE II
SYNTHESIS OF GABA BY GLIOMA CELLS IN CULTURE

Cell Line	In Cells	In Medium	Total	Apparent Specific Activity
	pmoles/min/dish			pmoles/min/mg Protein
C6 Glioma	163	389	552	17.8
L-929 Fibroblast	<1.2	<19	<20	<1.0

From Schrier and Thompson, 1974.

Uptake of the amino acids glutamate, cystine, and aspartate, all of which affect the excitability of neurons, had been shown in a crude way above. Because of the difficulty in assessing whether such uptake was oriented toward protein synthesis, intermediary metabolism, or control of ECF levels of neuroactive substances, the ability of the cells to clear these substances from the surrounding extracellular fluid was noted, but not further investigated. We looked, instead at GABA and taurine influx and efflux.

There were two discernible uptake components (saturable and non-saturable) for both GABA and taurine in three glial cell lines (Table IV); cultured rat brain "fibroblasts" (RBF cells) had the same uptake characteristics for taurine, but lacked the saturable (high affinity) uptake component for GABA. Taurine uptake was more rapid than GABA uptake for all the cell lines, and maximum concentration gradients were greater for taurine by an order of magnitude. The uptake systems for the two substances were separate. Table V shows that in C6 but not in $C2_1$, picrotoxin inhibited GABA uptake and stimulated taurine uptake. Bicuculline lowered GABA uptake in $C2_1$ cells by an amount consistent with inhibition of the saturable component, but had variable effects on taurine

TABLE III
PUTATIVE NEUROTRANSMITTER SYNTHESIS BY GLIOMA CELLS IN CULTURE

Precursor	Product Formed	Rate of Formation pmol/min/mg Protein	% of Acid-soluble Radioactivity in Cells
[^{35}S]Cystine	Taurine	20.2	14.6
	Isethionic acid	3.8	4.3
[U-^{14}C]Aspartate	Glutamate	1.7	17
	Glutamine	2.5	25
	β-Alanine	0.1	1.2
	GABA	0.1	1.0

From Schrier and Thompson, 1974.

TABLE IV

GABA AND TAURINE UPTAKE SYSTEM CHARACTERISTICS IN CULTURED CELLS[a]

	Rate of Nonsaturable Component[b]		V_{max} Saturable Component[c]		K_t for Saturable Component μM		Max. Tissue/ Medium Gradient	
	GABA	Taurine	GABA	Taurine	GABA	Taurine	GABA	Taurine
Glial cells (C6; C2; NT-1)	1.0	0.7–2.0	23–24	385–443	13–32	10–17	3.4–4.6	48–65
Fibroblasts (RBF)	1.0	0.7	0	435		15	2.3	51

[a] Modified from Schrier and Thompson (1974), where values for each of the three glial cell lines are shown.
[b] Units in pmoles/min/mg protein/μM
[c] Units in pmoles/min/mg protein.

uptake. α-Aminoisobutyrate inhibited GABA uptake in both cell lines, and interfered, in C6, with both the non-saturable and the saturable components; it, however, had only minor effects on taurine uptake. In $C2_1$ cells, GABA uptake was double in the presence of D-α-amino-n-butyrate. The uptakes were not affected substantially by incubation with L-isoproterenol, which is known (Gilman and Nirenberg, 1971) to stimulate massively the accumulation in these cells of 3′, 5′-cyclic AMP. Taurine profoundly inhibited GABA uptake, but GABA did not decrease the influx of taurine. This showed that a major influx avenue for taurine was not shared with GABA. Interestingly, and consistent with recent findings from Kelly and co-workers (Schon and Kelly, 1974, 1975) β-alanine inhibited both components of GABA uptake, and at least the saturable component of taurine influx. There was profound inhibition also of both uptakes by low sodium media and low temperatures.

It was also determined by two different methods that neither GABA nor taurine influx was associated with "exchange diffusion", i.e., coupled exchange of radioactive GABA or taurine external to the cell with unlabeled GABA or taurine inside the cell. This was evaluated in two ways: first, C6 cells were preincubated with excess unlabelled GABA or taurine in order to load them with these substances (Table VI). The uptakes were then evaluated by adding tracers, and were compared to the uptakes in cells which had had only short pre-incubation. By a process of exchange diffusion, loaded cells should have shown accelerated uptake. These cells, on the other hand, showed decreased transport after preloading, indicating that exchange diffusion was not responsible for the uptakes of GABA and taurine. Second, exchange diffusion was ruled out by pre-unloading experiments. The cells were pre-incubated for various times in the uptake medium without GABA or taurine, in order to deplete their intracellular contents of these agents. A subsequent measurement of uptake should have shown a decrease of net influx with increasing pre-incubation time if the

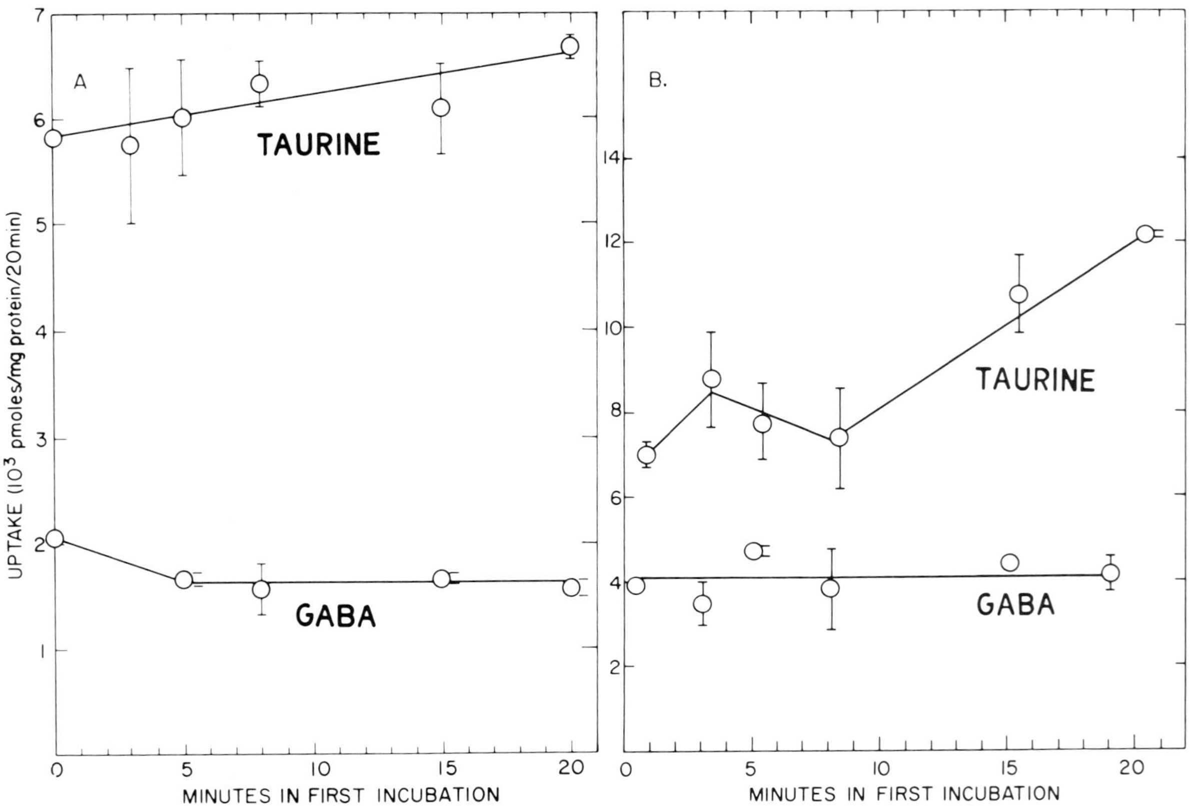

FIG. 2: Effect of pre-unloading of putative transmitters from glioma cells on subsequent uptake. After varying times of incubation without GABA or Taurine, C6 cells (panel *A*) or $C2_1$ cells (panel *B*) were incubated for 20 min with GABA or Taurine and uptakes were measured. From Schrier and Thompson, 1974.

TABLE V

EXTRACELLULAR CONTROL OF GABA AND TAURINE INFLUX BY GLIOMA CELLS

Extracellular Test Substance or Condition	C6 Uptake, % GABA	C6 Uptake, % Taurine	$C2_1$ Uptake, % GABA	$C2_1$ Uptake, % Taurine
Picrotoxin (100 μM)	69	172	90	80
Bicuculline (50 μM)	73	68	68	97
α-Aminoisobutyrate (2 mM)	27	119	53	73
β-Aminoisobutyrate (2 mM)	57	120	96	110
D-α-Amino-n-butyrate (2 mM)	77	88	208	—
L-Isoproterenol (20 μM)	89	93	113	146
Taurine (4 mM)	9		26	
GABA (3 mM)		116		121
β-Alanine (4 mM)	20	49	29	35
Low Temperature (0.7°)	10	7		
Low Na^+ (7 mM)	19	30	21	8

All uptakes are shown as per cent of control values (no added substance).
Data from Schrier and Thompson, 1974.

influx were due to exchange diffusion. Neither C6 nor $C2_1$ cells showed (Fig. 2) any change of uptake rate for GABA with pre-incubation. Since we know that the cells released GABA during exposure to a GABA free medium, if they contained any, it was obvious that they either contained none or that they had a fixed uptake rate regardless of the intracellular concentration. The result with taurine was different, in that the influx rate was increased with increasing pre-incubation. Since there was cysteine, the amino acid precursor of taurine, in the growth medium, it was likely that intracellular stores of this compound were being depleted during the pre-incubation, and the change in influx rate indicated some effort by the cells to recover these intracellular stores. Thus, exchange diffusion was ruled out, but an interesting possibility came to mind, that is, that glioma cells in culture might regulate, with some unknown degree of fidelity, the interior concentrations, and perhaps even the concentrations of GABA and taurine in the extracellular space.

TABLE VI

EFFECT OF PRE-LOADING OF C6 CELLS ON INFLUX OF NEUROACTIVE SUBSTANCES

Time	First Incubation Preload with:	Second Incubation Uptake pmoles/min/mg	Second Incubation Cell to Medium Ratio
21	GABA	14,200	0.3
0.5	GABA	24,000	0.5
21	Taurine	30,700	0.5
0.5	Taurine	39,900	0.7

From Schrier and Thompson, 1974

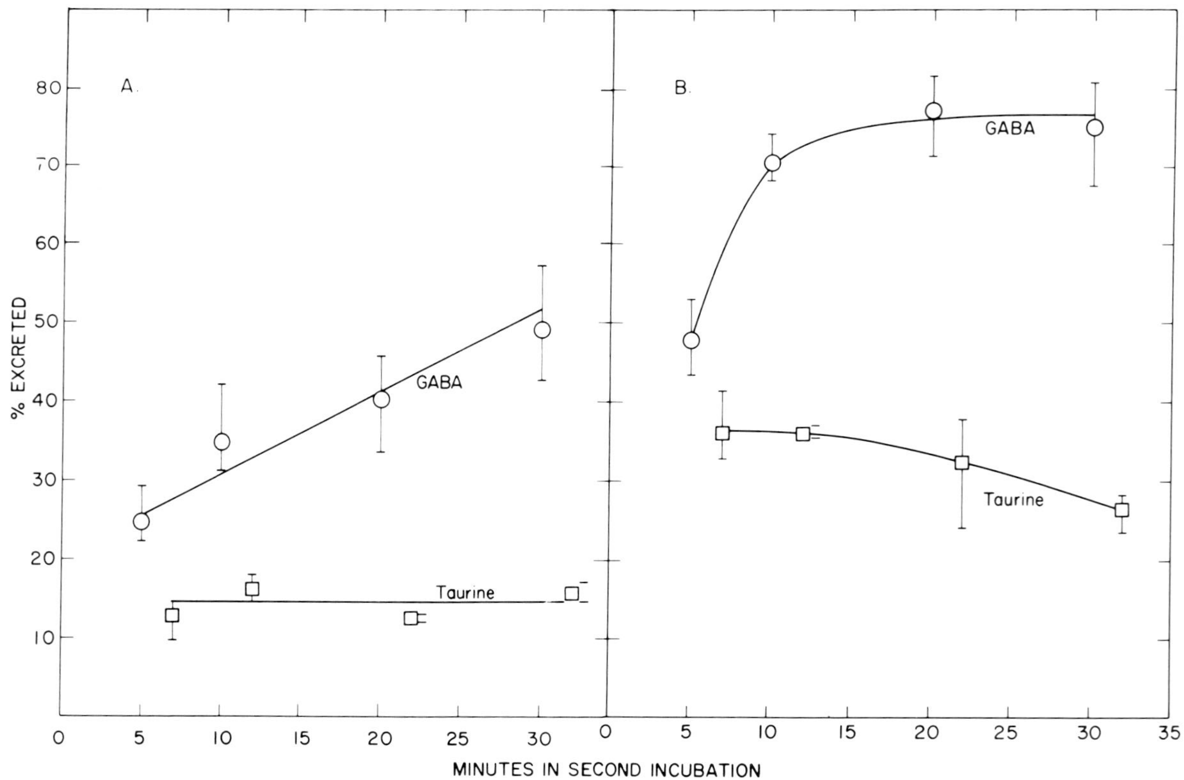

FIG. 3: Time courses of efflux of putative neurotransmitters by glioma cells in culture. Panel *A:* C6 cells; Panel *B:* C2 cells. From Schrier and Thompson, 1974.

We tested for such regulatory capacities by examining the rate of efflux from cells which had taken up labeled GABA or taurine (Fig. 3) when they were given medium without these compounds. With C6, taurine efflux seemed, after an initial release of 15% of the label, to be balanced by re-uptake, and the $C2_1$ cells seemed even to be involved in the net re-uptake after an initial release of one third of the radioactivity. Cell/medium concentration ratios of taurine were >1500-fold for C6, and >700 (and growing) for $C2_1$ after 32 minutes of efflux, indicating that the cells were actively maintaining very large transmembrane gradients. The highest recorded extracellular taurine concentration (at 7 min of efflux) was 0.7 μM, precisely the extracellular GABA concentration at which $C2_1$ cells were in influx-efflux equilibrium. After the $C2_1$ cells had excreted 75% of the GABA they had previously absorbed, a plateau was reached (Fig. 3); the transmembrane gradient on the plateau was still more than 80-fold. C6 cells had not reached equilibrium for GABA by 32 min of efflux, at which point the transmembrane gradient was still in excess of 300, so we don't know what level, if any, would have become a balance point. Thus, it seemed that, at least for three of the four cell line-substrate combinations, the cells were actively regulating the concentration of neuroactive substances in the extracellular space.

TABLE VII

EXTRACELLULAR CONTROL OF GABA AND TAURINE EFFLUX FROM GLIOMA CELLS

Conditions During Release	% of Radioactive Content Excreted after 24 min GABA	Taurine
Normal	32	20
Low Na^+ (7 mM)	38	7
High Substrate	57	20
High Substrate + L-Isoproterenol	42	25

Data from Schrier and Thompson, 1974

We attempted to determine whether the efflux could be altered by manipulations of the external environment. With $C2_1$ cells (Table VII), normal GABA excretion was little changed by low Na^+ concentration in the efflux medium, but taurine efflux was markedly inhibited; this last despite the fact that, in low Na^+ medium the reuptake of both substances would have been diminished considerably. The presence of high substrate concentrations in the efflux medium resulted in an increased GABA release (perhaps exchange diffusion under these conditions), but had little effect on taurine efflux. Treatment with L-isoproterenol was also without effect. Finally, we examined the effects of altered Ca^{2+} and K^+ concentrations on the uptake and efflux of GABA and taurine in C6 cells with the uptakes at 2 mM Ca^{2+} and 6 mM K^+ taken as 100% (Table VIII). In 2 mM Ca^{2+} there was little or no inhibition of GABA influx or efflux at any K^+ concentration tested, while both low and high K^+ concentration partially inhibited taurine influx without significant effect on its efflux (even though re-uptake in those conditions may have been inhibited). In normal K^+ concentration, both GABA and taurine uptakes were reduced in Ca^{2+}-free medium. In this medium, a slight stimulation of GABA efflux may have reflected this less efficient re-uptake, but taurine efflux was increased 5-fold in the absence of Ca^{2+}. Effects of higher Ca^{2+} concentrations were not significant except in the presence of increased K^+ concentrations. In a complex manner, high concentrations of both cations resulted in inhibition of both influx and efflux of GABA as well as in the uptake, but not the efflux, of taurine. As regards influx, it was clear that the combined effects of both ions were different from those of either ion separately. These results for GABA efflux are different from those obtained by Minchin and Iversen (1974) with glia of intact sensory ganglia. The differences may be consequent to the complexities of neuron-glia interactions which were absent from glioma cell cultures. Although it is not clear when *in situ*, glial cells might encounter zero Ca^{2+} or elevated concentrations of both Ca^{2+} and K^+, events of the nature outlined above could affect markedly the ability of glia to control the levels of GABA, taurine, and perhaps other neuroactive

TABLE VIII
INVOLVEMENT OF K^+ AND Ca^{2+} IONS IN GABA AND TAURINE INFLUX AND EFFLUX OF C6 GLIOMA CELLS[a]

Concentration in Medium		GABA		Taurine	
		Influx	Efflux	Influx	Efflux
Ca^{2+} mM	K^+ mM	% of Control	% of Control	% of Control	% of Control
2	6	100	100	100	100
2	0	128	118	70	117
2	12	111	105	89	117
2	24	85	96	59	112
0	6	70	123	75	504
5	6	102	84	95	88
10	6	85	90	94	114
5	12	76	70	63	80
5	24	61	69	60	92
10	24	64	74	51	104

[a]All values are given as per cent of those in a control medium with 2 mM Ca^{2+} and 6 mM K^+. In this medium GABA efflux corresponded to 18-21% of the influx and taurine efflux to 2.7-3.4% of the influx.

substances in the extracellular fluid of the nervous system. These data are thus consistent with the hypothesis that glial cells may affect the state of excitability of neurons by controlling the levels in the external milieu of the neuronal membranes, of inhibitory and excitatory neuroactive substances.

II. Trophic Effects of Muscle on Cholinergic Cells

When cells dissociated from spinal cords of 12-14 day fetal mice (sc) were inoculated onto cultures of mouse skeletal muscle myotubes (sc-m), we found (Giller *et al.*, 1973) a marked stimulation of the development of choline acetyltransferase (CAT) activity. The time course of development of this activity (Fig. 4, left panel), as well as the extent reached, were altered in co-cultures, the latter by 10-17 fold. This effect was not associated with an increase in the number of large neurons with many processes (perhaps the culture analogue of spinal motoneurons) in the co-cultures and, in fact, there appeared to be some decrease, due to co-cultures, in the number of medium-sized neurons with many processes. Usually during the course of CAT activity development, co-culture resulted (Fig.4, right panel) in a marked diminution in the activity of glutamate decarboxylase (GAD)

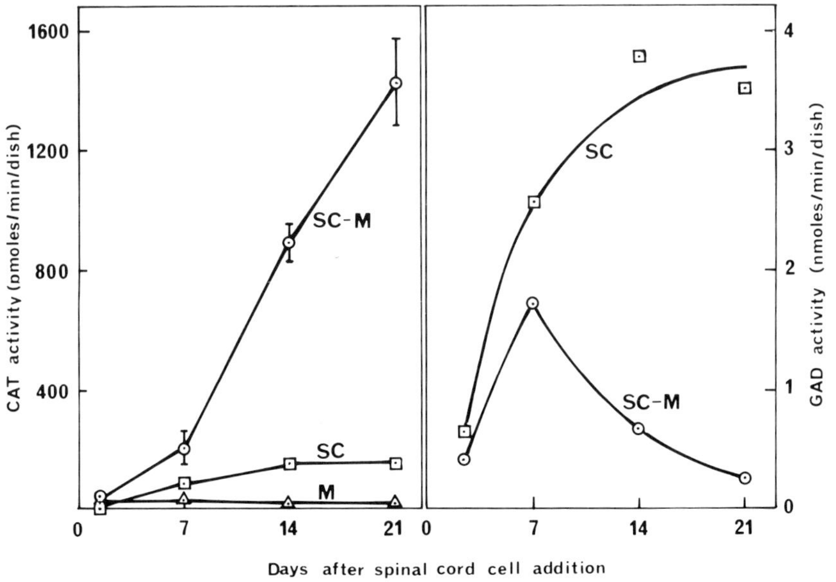

FIG. 4: Effects of co-culture with muscle cells on the development of SC cell CAT and GAD activities in culture. Left panel, activity of CAT. (From Giller *et al.*, 1973). Right panel, activity of GAD, M: muscle; SC: spinal cord; SC-M: spinal cord-muscle co-culture. From Giller *et al.*, 1977.

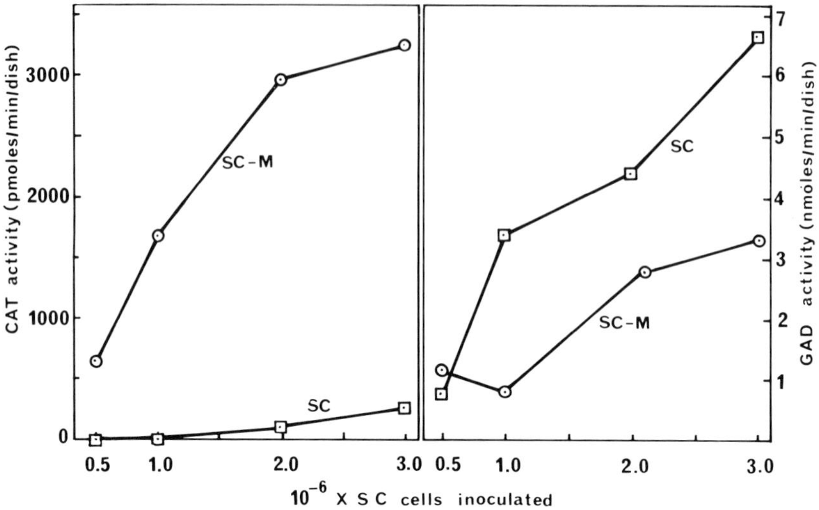

FIG. 5: Effects of SC cell inoculum size on development of CAT and GAD activities in cultures with (SC-M) and without (SC) muscle cells assayed at 21 days. From Giller *et al.*, 1977.

(Giller *et al.*, 1977). This loss of activity began at about 7 days of culture, the time at which a rapid acceleration in CAT activity development was detected. This suggested that the one effect was associated with the other. The marked CAT activity increase was specific (Giller *et al.*, 1973) among several muscle and nerve marker enzymes tested (CAT, ChE, CPK, MK, PH); thus it appeared that a highly selective effect of muscle cells on cholinergic neurons had been demonstrated.

The ability of a given myotube density to stimulate SC cell CAT activity was saturable by high inoclum densities of SC cells (Fig. 5, left panel). At the lowest SC cell dose, some of the phenomenon may have represented increased plating efficiency on myotubes; however, between inocula of 1×10^6 and 2×10^6, a doubling of stimulated CAT activity was observed. Thus, plating efficiency was probably not a factor in that range, and the saturation was clearly seen between inocula of 2×10^6 and 3×10^6 SC cells. The diminution of GAD activity by muscle cells in these same co-cultures depended on SC cell dose in a somewhat similar manner (Fig. 5, right panel). At the lowest inoculum, co-culture appeared to have stimulated GAD activity; this was probably an artifact of plating efficiency differences with and without muscle. At the other doses of SC cells, GAD activity was diminished by interaction with myotubes, but it appeared that a certain portion of this effect could be overcome by larger SC cell inocula. From these data it was also clear that there was enzymic specificity in the

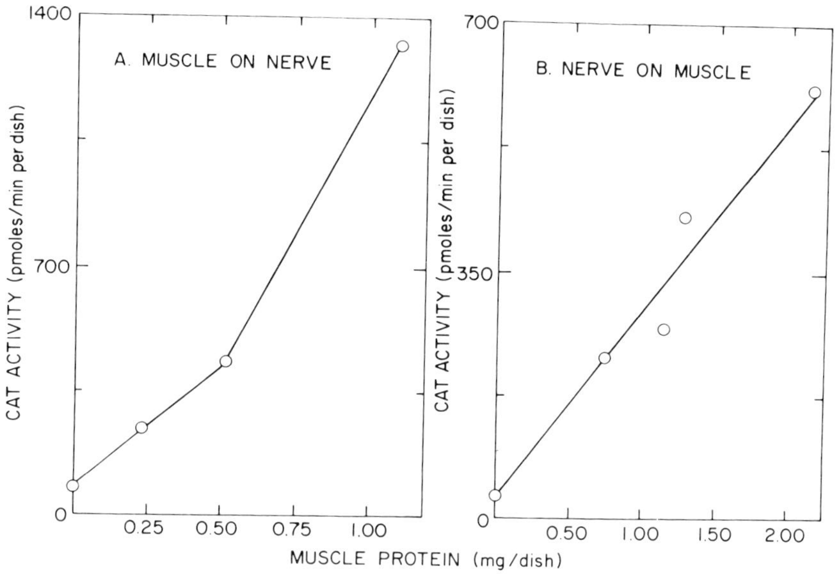

FIG. 6: Effects of muscle cell inoculum size on CAT activity development assayed after 14 days of culture. *A*. established myotubes inoculated onto 2 day cultures of SC cells. *B*. SC cells inoculated onto established muscle cells. From Giller *et al.*, 1977.

effect on CAT and that general increases in neuronal plating efficiency could not be invoked to explain the phenomenon of CAT stimulation.

Using replicate inocula of SC cells with varied amounts of muscle in the co-cultures, it was found that the degree of CAT stimulation also depended on the amount of muscle (Fig. 6). This was true whether the SC cells were inoculated onto established myotube cultures, as above, or intact myotubes from other cultures were added to SC cell cultures on day 2 of culture (see Giller *et al.*, 1977 for experimental details). The ability of the SC cells to respond by making additional CAT activity was not saturated at very high doses of muscle.

There was no necessity for physical contact between myotubes and SC cells in order to obtain the stimulation of CAT activity: growth medium, which had been conditioned by muscle cell cultures for 2-3 days, could be employed to stimulate CAT activity in SC cell cultures. The response after chronic treatments with such conditioned medium (CM) for 14 days was dependent on the dose of CM employed and rose almost linearly from about 50 pmol/min per dish in the absence of conditioned medium to about 450 pmol/min per dish in 100% muscle conditioned medium (Giller *et al.*, 1977). Conditioned medium treatments never resulted in a diminution of GAD activity; this suggested that that effect may have been associated with physical contact between SC and muscle cells. The magnitude of the effect on CAT activity was often as great as that obtained with co-culture (Table IX), and it seemed likely that the CM effect was at least a portion, if not all, of the co-culture effect. The finding that the CM and co-culture effects were additive (Table IX) did not dispel this theory, since it was likely (cf. above and Fig. 6) that neither effect had been saturated. It was however, clear that, with chronic poisoning of synaptic transmission (by α-bungarotoxin interaction with muscle ACh receptors), little if any decrease in co-culture effectiveness was found.

TABLE IX

EFFECTS OF CM, CO-CULTURE AND α-BUNGAROTOXIN (α-BTX) ON CAT ACTIVITY

Culture	CAT Activity (pmoles/min) Activity per Dish		Fold Increase per Dish		Activity per mg Protein	
	Expt.A	Expt.B	Expt.A	Expt.B	Expt.A	Expt.B
SC Cells alone	70	34	1	1	78	57
+ Muscle CM	282		4.0		390	
+ Muscle Cells	765	557	11	16	249	229
+ Muscle Cells + Muscle CM	1080		15		548	
+ Muscle Cells + α-BTX		515		15		238

From Giller *et al.*, 1977.

The degree of stimulation of CAT activity depended on the duration of treatment with CM or muscle, as well as on the time in culture of SC cells at which the treatment was begun (Table X). The effectiveness of CM (measured as increase per day of treatment) was substantially diminished when the inception of the treatment was delayed from day 3 to day 7. Similarly, muscle cells or CM added very late, but for long periods of time, had little, if any, effect. It was clear that SC cells, after a few days of culture, became considerably less responsive to the muscle factor(s).

Some properties of the CM factor have been determined (Giller *et al.*, 1977 and unpublished data of J. H. Neale and E. W. Godfrey). The factor(s) was stable to a variety of mild treatments, including dialysis (the factor was non-dialyzable), heating at 58°, rapid freeze-thaw, Millipore filtration, and storage at 4° for 3 days. Long-term storage or collection of large amounts in the freezer by freezing the CM was not successful. The factor appeared to be proteinaceous, as it was destroyed by trypsin but not by DNase or RNase. SC cell densities from 0.7 to 1.2 × 10^6 cells/35 mm dish (1.0 to 1.7 × 10^5 cells/cm^2 surface area) responded about equally well in terms of the

TABLE X
TIME OF INITIATION AND DURATION OF CM TREATMENT AND CO-CULTURE

Culture	Treatment of Co-Culture		CAT Activity	
	SC Age at Start of Treatment (days)	SC Age at Enzyme Assay (days)	Fold Increase	Increase per Day of Treatment (pmoles/min/dish)
Expt. A				
SC cells alone		13	1	
+ Muscle CM	3	10	3.5	20.7 ±3.2
+ Muscle CM	3	13	5.2	24.1±0.9
+ Muscle CM	5	13	3.3	16.4±3.2
+ Muscle CM	7	13	2.5	14.5±0.7
Expt. B				
SC cells alone		14	1	
+ Muscle Cells	2	14	4.9	49
+ Muscle Cells	15	29	0.8	0
+ Muscle Cells	19	33	1.2	3.1
+ Muscle CM	19	33	1.1	1.2

From Giller *et al.*, 1977.

fold-increase of CAT specific activity (pmols/min/μg DNA) attained. A molecular weight for the factor of >50,000 daltons was determined by its retention during pressure filtration through a membrane with that cutoff at that molecular weight. Muscle cultures produced a CAT-stimulating factor in the absence of horse serum in the medium, showing that the factor was not functioning simply to neutralize a toxic component in the serum NGF will not substitute for the CM factor(s).

The specificity of the factor may be examined from three different vantage points. Specificity as to the responding enzymes has been discussed above. Specificities of responsive cells and source, as determined to date, are shown in Table XI. Only preparations containing cholinergic neurons, other than myotubes themselves, which do not respond, have been tested as possibly responsive cells. The response of brain cell cultures is at least as vigorous as that of SC cells, but distinction between cortex (neurons of which never contact muscle in the animal) and subcortex needs to be accomplished before this can be interpreted as broad specificity. The response of the hybrid cells [NG108cc15 somatic cell hybrids (mouse neuroblastoma/rat glioma)] which have high CAT specific activity, allows a rapid and simplified bioassay for the factor during the process of purification (E. Godfrey, B. K. Schrier and P. G. Nelson, unpublished experiments). The stimulation of CAT activity produced by treatment with

TABLE XI

RESPONSIVE CELLS AND SOURCES FOR CAT STIMULATING FACTOR(S)

Sources		Responsive Cells		
	Species	SC Cells	Hybrid cells[b]	FMB[g]
Co-cultures				
Skeletal Muscle Cells	Mouse	+	+	+
Cardiac Muscle Cells[a]	Mouse	+		
Sk. Muscle Culture Fibroblasts[a]	Mouse	+		
Conditioned Media				
Skeletal Muscle Cells	Mouse	+	+	
Sk. Muscle Culture Fibroblasts[a]	Mouse	+		
Spinal Cord Cells[a]	Mouse	+		
G8 Skeletal Muscle Cell Line[b,c]	Mouse	0	+	
$BC3H_1$ Smooth Muscle Cell Line[b,e]	Mouse	±		
L8-S Skeletal Muscle Cell Line[b,d]	Rat	±		
Homogenates[b,f]				
Adult Skeletal Muscle	Mouse	+	+	
Fetal Skeletal Muscle	Mouse	+	+	
Fetal Liver	Mouse	±		
Fetal Brain	Mouse	+		

[a]Data from Giller *et al.*, 1977; [b]NG108 cc 15 mouse neuroblastoma/rat glioma; unpublished data of E. W. Godfrey; [c]Cell line obtained from J. Peacock, Stanford University; [d]Subclone "S" by D. A. Boenning of L8 line of D. Yaffe obtained from M. Nirenberg; [e]Cell line obtained from D. Schubert, Salk Institute; [f]unpublished data of J. Neale and P. Bullock; [g]fetal mouse brain cells, unpublished data of E. W. Godfrey.

dibutyryl-cyclic AMP in the hybrid cells has been shown to be enhanced by treatment with muscle CM. Cardiac muscle cells proved to be as potent a source of the factor as skeletal muscle cells (Table XI); fibroblasts derived from the latter by multiple culture passages were weakly positive as was their CM. SC cells conditioned their own medium to some degree, but it was not clear whether or not this represented the same type of stimulatory phenomenon; two established muscle cell lines, one of them from smooth muscle, made little, if any, assayable factor with SC cells as the possibly responsive cells. The hybrid cells, however, clearly responded to CM from the G8 cells; in all other regards tested thus far CAT stimulating factors for NG108cc15 cells and SC cells have appeared to have been the same. The only potential source examined from species other than mouse, the rat L8 muscle cell line (subclone S), proved to be provisionally positive. Clear evidence for a similarly acting factor in skeletal muscle homogenates from adult and fetal mice may provide a more convenient source for further studies than does CM. Fetal liver was probably without activity, but fetal brain homogenate was very effective (Table XI).

The studies of CAT stimulating factor, reported here, are proceeding in the areas of characterization, purification and specificities, and the biological events associated with production and function of the factor(s) are being studied. It is not without possibility that the factor(s) being investigated in Dr. Patterson's laboratory (Patterson, *et al.*, 1975; Furshpan, *et al.*, 1976) are similar to, or the same as, those discussed here.

References

Furshpan, E. J., P. R. Macleish, P. H. O'Lague and D. D. Potter. 1976. Chemical transmission between rat sympathetic neurons and cardiac myocytes developing in microcultures: Evidence for cholinergic, adrenergic and dual-function neurons. Proc. Nat. Acad. Sci., U.S.A. 73:4225-4229.

Giller, E. L., J. N. Neale, P. N. Bullock, B. K. Schrier and P. G. Nelson. 1977. Choline acetyltransferase activity of spinal cord cell cultures increased by co-culture with muscle and by muscle conditioned medium. J. Cell Biol. (in press).

Giller, E. L., B. K. Schrier, A. Shainberg, H. R. Fisk, and P. G. Nelson. 1973. Choline acetyltransferase activity is increased in combined cultures of spinal cord and muscle cells from mice. Science 182:588-589.

Gilman, A. G. and M. Nirenberg. 1971. Effect of catecholamines on the adenosine 3':5' cyclic monophosphate concentration of clonal satellite cells of neurons. Proc. Nat. Acad. Sci. U.S.A. 68:2165-2168.

Haber, B., K. Kuriyama, and E. Roberts. 1970. An anion stimulated L-glutamic acid decarboxylase in non-neural tissues. Occurrence and subcellular localization in mouse kidney and developing chick embryo. Biochem. Pharmacol. 19:1119-1136.

Henn, F. A. and A. Hamberger. 1971. Glial cell function: Uptake of transmitter substances. Proc. Nat. Acad. Sci. U.S. A. 68:2686-2690.

Koechlin, B. A. 1955. On the chemical composition of the axoplasm of squid giant nerve fibers with particular reference to its ion pattern. J. Biophys. Biochem. Cytol. 1:511-529.

Minchin, M. C. W. and L. L. Iversen. 1974. Release of (^{3}H) gamma-aminobutyric acid from glial cells in rat dorsal root ganglia. J. Neurochem. 23:533-540.

Patterson, P. H., L. F. Reichardt, and L. L. Y. Chun. 1975. Biochemical studies on the development of primary sympathetic neurons in cell culture. Cold Spring Harbor Symp. Quant. Biol. 40:389-397.

Schon, F. and J. S. Kelly. 1974. The characterization of (^{3}H)GABA uptake into the satellite glial cells of rat sensory ganglia. Brain Res. 66:289-300.

Schon, F. and J. S. Kelly. 1975. Selective uptake of (^{3}H)β-alanine by glia: association with the glial uptake system for GABA. Brain Res. 86:243-257.

Schrier, B. K. and D. L. Shapiro. 1973. Effects of N^6-monobutyryl-cyclic AMP on glutamate decarboxylase activity in fetal rat brain cells and glial tumor cells in culture. Exp. Cell Res. 80:459-462.

Schrier, B. K. and E. J. Thompson. 1974. On the role of glial cells in the mammalian nervous system. J. Biol. Chem. 249:1769-1780.

Schrier, B. K., S. H. Wilson and M. W. Nirenberg. 1974. Cultured cell systems and methods for neurobiology. Methods in Enzymology 32B:765-789.

Wilson, S. H., B. K. Schrier, J. L. Farber, E. J. Thompson, R. N. Rosenberg, A. J. Blume and M. W. Nirenberg. 1972. Markers for gene expression in cultured cells from the nervous system. J. Biol. Chem. 247:3159-3169.

DIFFERENCES BETWEEN ASTROCYTES IN PRIMARY CULTURES AND GLIAL CELL LINES IN UPTAKE AND METABOLISM OF PUTATIVE AMINO ACID TRANSMITTERS

ARNE SCHOUSBOE

Department of Biochemistry
Panum Institute
University of Copenhagen
Copenhagen, Denmark

It is intriguing that three of the most abundant amino acids in the brain, GABA, glutamate and taurine, seem to have functions other that taking a part in metabolic reactions. It is well established that GABA has a function as an inhibitory transmitter in the mammalian brain (for references, see Roberts, 1974; Schousboe *et al.*, 1977b) and there is strong evidence to support the concept that glutamate exerts excitatory effects on spinal cord (Curtis *et al.*, 1960) and cerebral neurons (Krnjevic and Phillis, 1963; Krnjevic, 1974). Furthermore, it has been suggested that taurine may act as an inhibitory transmitter both in brain and retina (Curtis and Watkins, 1965; Davison and Kaczmarek, 1971; Pasantes-Morales *et al.*, 1974). If these amino acids function as neurotransmitters there must be a process by which they are inactivated. Such an inactivation is in all probability brought about by a transport process (Curtis and Johnston, 1974) possibly an uptake into neurons or glia cells or both (for references, see Schousboe *et al.*, 1977b). In order to obtain information about glial transport systems for these putative transmitters, bulk-prepared glial cells (Henn and Hamberger, 1971) as well as neoplastic glial cell lines (Hutchison *et al.*, 1974; Schrier and Thompson, 1974; Faivre-Bauman *et al.*, 1974; Borg *et al.*, 1976; Balcar *et al.*, 1977) and primary cultures of glial cells from cerebrum (Hertz *et al.*, 1976; L. Hertz, A. Schousboe and S. Fedoroff, unpublished; Schousboe *et al.*, 1976b; 1977b) and cerebellum (Lasher, 1975) have been used.

The purpose of this paper is to discuss differences between results obtained using these different cell preparations on the basis of other biochemical parameters by which it may be possible to evaluate the usefulness of such preparations as models for glial cells *in vivo.*

Astrocytes were cultured from cerebral hemispheres as described by Booher and Sensenbrenner (1972) with the modifications introduced by Schousboe *et al.* (1975; 1976a, b). Briefly, mouse brain hemispheres were passed through a sterile nylon sieve (80 μm pore size), incubated and cultured for two weeks in a modified Eagle's minimum essential medium (Eagle, 1959) supplemented with 20% inactivated foetal calf serum and subsequently one week in an analogous serum-free medium to which 0.1 mM dibutyryl cyclic AMP had been added. The astrocytic character of the cells is indicated not only by the morphology (Schousboe *et al.*, 1976b; L. Hertz, A. Schousboe and S. Fedoroff, unpublished) but also by a high content (Bock *et al.*, 1975; L. Hertz and E. Bock, personal communication) of the GFA protein which is believed to be specific for astrocytes (Eng *et al.*, 1971).

The assessment of the present culture as a valid model for astrocytes *in vivo* is based both upon morphological and biochemical observations. Morphologically, the cultures were very much enriched in well differentiated astrocytic cells (Schousboe *et al.*, 1976b; L. Hertz, A. Schousboe and S. Fedoroff, unpublished) with abundant, radial, branching processes like their *in vivo* counterparts (Wolff, 1970). Furthermore, cells at this differentiated stage have been shown to have contents of ATP and glutamate (Schousboe *et al.*, 1975; 1976a), activities of such enzymes as the Na^+, K^+-ATPase (Schousboe *et al.*, 1976a), GABA-transaminase (Schousboe *et al.*, 1977b), glutamic-oxaloacetate-transaminase, glutamine synthetase and glutamate dehydrogenase (A. Schousboe and L. Hertz, unpublished), and an oxygen uptake (Mukerji and Hertz, 1976; Hertz, this volume) which are comparable to corresponding values for the brain *in vivo* (Awapara and Seale, 1952; Strominger and Lowry, 1955; Agrawal *et al.*,

TABLE I

ACTIVITIES OF ENZYMES INVOLVED IN METABOLISM OF GLUTAMATE AND GABA IN PRIMARY CULTURES OF ASTROCYTES, C-6 ASTROCYTOMA AND BULK-PREPARED GLIAL CELLS.

Enzyme Activity (nmol x min^{-1} x mg^{-1} Cell Protein)	Primary Cultures of Astrocytes	C-6 Cells	Bulk-prepared Glial Cells
GABA-T	1.6[1]	0.02[2] 0.008[5]	1[3]
GAD	<0.09[5]	0.04[6]	0.12[3]
GOT	188[4]	—	—
GLDH	13[4]	0.01[2]	—
GLU-S	23[4]	—	—

GABA-T: GABA-transaminase. GAD: L-glutamate decarboxylase. GOT: glutamic-oxaloacetate-transaminase. GLDH: glutamate dehydrogenase. GLU-S: glutamine synthetase. References: *1)* Schousboe *et al.* (1977b). *2)* Passonneau *et al.*(1976). *3)* Sellström *et al.* (1975). *4)* A. Schousboe and L. Hertz, unpublished. *5)* Wilson *et al.* (1972). *6)* Schrier and Thompson (1974).

1966; Berl, 1966; Medzihradsky *et al.*, 1971; Schousboe *et al.*, 1973; Hertz and Schousboe, 1975). It may accordingly be concluded that the present culture represents a valid model for astrocytes *in vivo*.

On the basis of this, the uptake of the above mentioned putative amino acid transmitters was studied together with activities of enzymes involved in GABA and glutamate metabolism in such cultures. Activities of GABA-transaminase, L-glutamate decarboxylase, glutamic-oxaloacetate-transaminase, glutamine synthetase and glutamate dehydrogenase in primary cultures of astrocytes, C-6 astrocytoma cells and bulk-prepared glial cells are given in Table I. Values in primary cultures seem to be comparable to those for bulk-prepared cells but with the exception of GAD which is probably neuronal in the brain, they are considerably higher than the corresponding activities found in C-6 cells indicating that cell lines may not always reflect the biochemical phenomena *in vivo*.

TABLE II

KINETIC PARAMETERS FOR UPTAKE OF PUTATIVE AMINO ACID TRANSMITTERS INTO PRIMARY CULTURES OF WELL DIFFERENTIATED ASTROCYTES, BRAIN SLICES AND GLIAL CELL LINES.

Amino Acid	Tissue or Cell Preparation	Temp.	K_m (μM)	V_{max} a (nmol $\times min^{-1} \times g^{-1}$ wet wt.)	V_{max} b (nmol $\times min^{-1} \times mg^{-1}$ prot.)	Ref.
GABA	brain slice	25°	11-31	34-167		*1*
GABA	astrocytes	37°	40	28	0.35	*2*
GABA	C-6 cells	37°	32		0.023	*3*
GABA	C-6 cells	37°	0.22		0.0014	*4*
Glutamate	brain slice	25°	20	240		*5*
Glutamate	brain slice	37°	30	65		*6*
Glutamate	brain slice	—	480	4000		*7*
Glutamate	astrocytes	37°	50	4500	60	*8*
Glutamate	C-6 cells	37°	15		3.8	*9*
Glutamate	NN cells	25°	14		0.7	*10*
Taurine	brain slice	37°	50-60	1-30		*11,12*
Taurine	astrocytes	37°	95	80	1.0	*13*
Taurine	C-6 cells	37°	10		0.39	*3*
Taurine	NN cells	37°	12		0.05	*14*

References: *1)* Martin (1976). *2)* Schousboe *et al.* (1977b). *3)* Schrier and Thompson (1974). *4)* Hutchison *et al.* (1974). *5)* Balcar and Johnston (1972). *6)* Benjamin and Quastel (1976). *7)* Cohen and Lajtha (1972). *8)* L. Hertz, A. Schousboe and S. Fedoroff, unpublished. *9)* Faivre-Bauman *et al.* (1974). *10)* Balcar *et al.* (1977). *11)* Kaczmarek and Davison (1972). *12)* Lähdesmäki and Oja (1973). *13)* Schousboe *et al.* (1976b). *14)* Borg *et al.* (1976).

The uptake of taurine, GABA and glutamate was studied in the cultured astrocytes and all three amino acids were taken up by the cells by very efficient, high affinity, sodium-dependent, carrier-mediated, concentrative mechanisms (Schousboe *et al.*, 1976b; 1977a, b; L. Hertz, A. Schousboe and S. Fedoroff, unpublished), values for K_m and V_{max} are given in Table II together with corresponding values for brain slices and other glial cell preparations.

It can be seen that the kinetic values determined for primary cultures of astrocytes are of orders of magnitude similar to the corresponding values for brain slices; whereas the V_{max} values for uptake into glial cells line are much lower. This reflects the results obtained for other biochemical parameters (cf. Table I and L. Hertz, this volume) and emphasizes the importance of comparison between different types of cultured cells. It furthermore lends support to the hypothesis that a transport of these amino acids into glial cells may be of quantitatively major importance for the inactivation process (cf. above). As far as glutamate is concerned, results obtained using a radioautographic technique have also recently indicated that glutamate uptake in brain cortex occurs mainly in glial cells (McLennan, 1976), and experiments on taurine uptake in brain slices of varying thickness led Riddall *et al.* (1976) to conclude that glial cells were the major site for the uptake of this amino acid. For GABA there is, however, strong evidence to support the concept that in addition, neuronal uptake is of quantitative importance (Bloom and Iversen, 1971; Iversen and Bloom, 1972).

The observed high activities of the enzymes involved in the metabolism of glutamate and GABA (cf. Table I) support the concept that glial cells may also be important for the subsequent metabolism of these amino acids (Benjamin and Quastel, 1972; Schousboe *et al.,* 1977b). Presently it can, however, not be deduced to what extent this metabolism involves a 'metabolic pool' or a'transmitter pool' of these amino acids.

Acknowledgments

The expert technical assistance by Miss Hanne Fosmark and the receipt of a travel grant (511-7230) from the Danish State Research Council are cordially acknowledged.

References

Agrawal, H. C., J. M. Davis and W. A. Himwich. 1966. Postnatal changes in free amino acid pool of rat brain. J. Neurochem. 13:607–615.

Awapara, J. and B. Seale. 1952. Distribution of transaminases in rat organs. J. Biol. Chem. 194:497–502.

Balcar, V. J., and G. A. R. Johnston. 1972. Glutamate uptake by brain slices and its relation to the depolarization of neurones by acidic amino acids. J. Neurobiol. 3:295–301.

Balcar, V. J. J. Borg and P. Mandel. 1977. High affinity uptake of L-glutamate and L-aspartate by glial cells. J. Neurochem. 28:87-93.

Benjamin, A. M. and J. H. Quastel. 1972. Locations of amino acids in brain slices from the rat. Biochem. J. 128:631-641.

Benjamin, A. M. and J. H. Quastel. 1976. Cerebral uptakes and exchange diffusion *in vitro* of L- and D-glutamates. J. Neurochem. 26:431-441.

Berl, S. 1966. Glutamine synthetase. Determination of its distribution in brain during development. Biochemistry 5:916-922.

Bloom, F. E. and L. L. Iversen. 1971. Localizing ^{3}H-GABA in nerve terminals of rat cerebral cortex by electron microscopic autoradiography. Nature (Lond.) 229:628-630.

Bock, E., O. S. Jörgensen, L. Dittmann and L. F. Eng. 1975. Determination of brain-specific antigens in short term cultivated rat astroglial cells and in rat synaptosomes. J. Neurochem. 25:867-870.

Booher, J. and M. Sensenbrenner. 1972. Growth and cultivation of dissociated neurons and glial cells from embryonic chick, rat and human brain in flask cultures. Neurobiology 2:97-105.

Borg, J., V. J. Balcar and P. Mandel. 1976. High affinity uptake of taurine in neuronal and glial cells. Brain Res. 118:514-516.

Cohen, S. R. and A. Lajtha. 1972. Amino acid transport. In: *Handbook of Neurochemistry*. Vol. 7. A. Lajtha, ed. Plenum Press, New York. pp. 543-572.

Curtis, D. R. and G. A. R. Johnston. 1974. Amino acid transmitters in the mammalian central nervous system. Ergebn. Physiol. 69:97-188.

Curtis, D. R. and J. C. Watkins. 1965. The pharmacology of amino acids related to gamma-aminobutyric acid. Pharmacol. Rev. 17:347-391.

Curtis, D. R., J. W. Phillis and J. C. Watkins. 1960. The chemical excitation of spinal neurones by certain acidic amino acids. J. Physiol. (Lond.). 150:656-682.

Davison, A. N. and L. K. Kaczmarek. 1971. Taurine — a possible neurotransmitter. Nature (Lond.) 234:107-108.

Eagle, H. 1959. Amino acid metabolism in mammalian cell cultures. Science 130:432-437.

Eng, L. F., J. J. Van der Haeghen, A. Bignami and B. Gerstl. 1971. An acidic protein isolated from fibrous astrocytes. Brain Res. 28:351-354.

Faivre-Bauman, A., J. Rossier and P. Benda. 1974. Glutamate accumulation by a clone of glial cells. Brain Res. 76:371-375.

Henn, F. A. and A. Hamberger. 1971. Glial cell function: Uptake of transmitter substances. Proc. Natl. Acad. Sci. U.S.A 68:2686-2690.

Hertz, L. and A. Schousboe. 1975. Ion and energy metabolism of the brain at the cellular level. Int. Rev. Neurobiol. 18:141-211.

Hertz, L., G. Svenneby and A. Schousboe. 1976. Glutamate and GABA uptake into cultures of normal glia cell. Trans. Amer. Soc. Neurochem. 7:192.

Hutchison, H. T., K. Werrbach, C. Vance and B. Haber. 1974. Uptake of neurotransmitters by clonal lines of astrocytes and neuroblastoma in culture. I. Transport of γ-aminobutyric acid. Brain. Res. 66:265-274.

Iversen, L. L. and F. E. Bloom. 1972. Studies on the uptake of ^{3}H-GABA and ^{3}H-glycine in slices and homogenates of rat brain and spinal cord by electron microscopic autoradiography. Brain Res. 41:131-143.

Kaczmarek, L. K. and A. N. Davison. 1972. Uptake and release of taurine from rat brain slices. J. Neurochem. 19:2355-2362.

Krnjevic, K. 1974. Chemical nature of synaptic transmission in vertebrates. Physiol. Rev. 54:418-540.

Krnjevic, K. and J. W. Phillis. 1963. Iontophoretic studies of neurones in the mammalian cerebral cortex. J. Physiol. (Lond.) 165:274-304.

Lasher, R. S. 1975. Uptake of GABA by neuronal and non-neuronal cells in dispersed cell cultures of postnatal rat cerebellum. J. Neurobiol. 6:597-608.

Lähdesmäki, P. and S. S. Oja. 1973. On the mechanism of taurine transport at brain cell membranes J. Neurochem. 20:1411-1417.

Martin, D. L. 1976. Carrier-mediated transport and removal of GABA from synaptic regions. In: *GABA in Nervous System Function* E. Roberts, T. N. Chase and D. B. Tower, eds. Raven Press, N. Y. pp. 347–386.

McLennan, H. 1976. The autoradiographic localization of L-^{3}H-glutamate in rat brain tissue. Brain Res. 115:139–144.

Medzihradsky, F., P. S. Nandhasri, V. Idoyaga-Vargas and O. Z. Sellinger, 1971. A comparison of the ATPase activity of the glial cell fraction and the neuronal perikaryal fraction isolated in bulk from rat cerebral cortex. J. Neurochem. 18:1599–1603.

Mukerji, S. and L. Hertz. 1976. Oxygen uptake by normal glia cells cultured under different conditions. Trans. Amer. Soc. Neurochem. 7:90.

Pasantes-Morales, H., J. Klethi, P. F. Urban and P. Mandel. 1974. The effect of electrical stimulation, light and amino acids on the efflux of ^{35}S-taurine from the retina of domestic fowl. Exp. Brain Res. 19:132–141.

Passonneau, J. V., W. D. Lust and S. K. Crites. 1976. GABA shunt metabolites and enzymes in glioma and neuroblastoma cell lines. Trans. Amer. Soc. Neurochem. 7:192.

Riddall, D. R., M. J. Leach and A. N. Davison. 1976. Neurotransmitter uptake into slices of rat cerebral cortex *in vitro:* Effect of slice size. J. Neurochem. 27:835–839.

Roberts, E. 1974. γ-Aminobutyric acid and nervous system function. A perspective. Biochem. Pharmacol. 23:2637–2649.

Schousboe, A., H. Fosmark and B. Formby. 1976a. Effect of serum withdrawal on Na^{+}-K^{+} ATPase activity in astrocytes cultured from dissociated brain hemispheres. J. Neurochem. 26:1053–1055.

Schousboe, A., H. Fosmark and L. Hertz. 1975. High content of glutamate and of ATP in astrocytes cultured from rat brain hemispheres: Effect of serum withdrawal and a cyclic AMP. J. Neurochem. 25:909–911.

Schousboe, A., H. Fosmark and G. Svenneby. 1976b. Taurine uptake in astrocytes cultured from dissociated mouse brain hemispheres. Brain Res. 116:158-164.

Schousboe, A., L. Hertz and G. Svenneby. 1977b. Uptake and metabolism of GABA in astrocytes cultured from dissociated mouse brain hemispheres. Neurochem. Res. 2:217-229.

Schousboe, A., J.-Y. Wu and E. Roberts. 1973. Purification and characterization of the 4-aminobutyrate-2-ketoglutarate transaminase from mouse brain. Biochemistry 12:2868-2873.

Schousboe, A., P. H. Wu, L. Hertz and S. Fedoroff. 1977a. GABA formation, uptake, release and metabolism in cultured normal astrocytes. Trans. Amer. Soc. Neurochem. 8:248.

Schrier, B. K. and E. J. Thompson. 1974. On the role of glial cells in the mammalian nervous system. J. Biol. Chem. 249:1769-1780.

Sellström, Å., L. -B. Sjöberg, and A. Hamberger. 1975. Neuronal and glial systems for γ-aminobutyric acid metabolism. J. Neurochem. 25:393-398.

Strominger, J. L. and O. H. Lowry. 1955. The quantitative histochemistry of brain. IV. Lactic, malic, and glutamic dehydrogenases. J. Biol. Chem. 213:635-646.

Wilson, S. H., B. K. Schrier, J. L. Farber, E. J. Thompson, R. N. Rosenberg, A. J. Blume and M. W. Nirenberg. 1972. Markers for gene expression in cultured cells from the nervous system. J. Biol. Chem. 247:3159-3169.

Wolff, J. R. 1970. The astrocyte as link between capillary and nerve cell. Triangle 9:153-164.

ROLE OF CYCLIC NUCLEOTIDE IN THE DIFFERENTIATION OF NERVE CELLS

KEDAR N. PRASAD

Department of Radiology
University of Colorado Medical Center
Denver, Colorado

I. Introduction

The process of neural differentiation involves many steps including induction, cell migration, increased expression of differentiated functions, and inhibition of cell division. Neuroblastoma cells are in many ways already differentiated, sharing several features of mature neurons, though these features are mostly expressed at low levels. Neuroblastoma cells are however undifferentiated compared to mature neurons in the sense that they rapidly divide and show a round cell morphology. Therefore, neuroblastoma cells in culture may be suitable primarily for studying the regulation of differentiated functions. Clonal lines of neuroblastoma cells obtained from transplanted mouse tumors, chemically induced central nervous system rat tumors and human tumors have been used to study the effect of cyclic AMP on the expression of differentiated functions. However, the mouse neuroblastoma cells are highly aneuploid; the chromosome number varies from 69-192, the diploid number being 40 (Amano *et al.*, 1972); whereas the human neuroblastoma cells (Biedler *et al.*, 1973) and rat central nervous system tumor cells (Schubert *et al.*, 1974) are nearly diploid. Therefore, one has to be cautious in comparing the effect of cyclic AMP on gene expression in various cell lines during differentiation. In addition, normal embryonic nervous tissue cultures have been used to study the effect of cyclic AMP on the expression of differentiated functions and it is now apparent that cyclic AMP may be involved in the regulation of certain differentiated functions in nerve cells. To study the involvement of cyclic AMP in neural induction, an experimental system allowing investigation of early embryogenesis must be used. Indeed, using explants of gastrulae of amphibia it has been shown (Wahn *et al.*, 1975) that cyclic AMP induces neural differentiation in undetermined presumptive epidermis.

The purpose of this paper is to evaluate the role of cyclic AMP in the induction and regulation of differentiated functions in nerve cells. The involvement of cyclic AMP in the regulation of a particular gene expression and morphological alteration is considered likely only when the following criteria are met: 1. The effect must be produced by at least one analog of

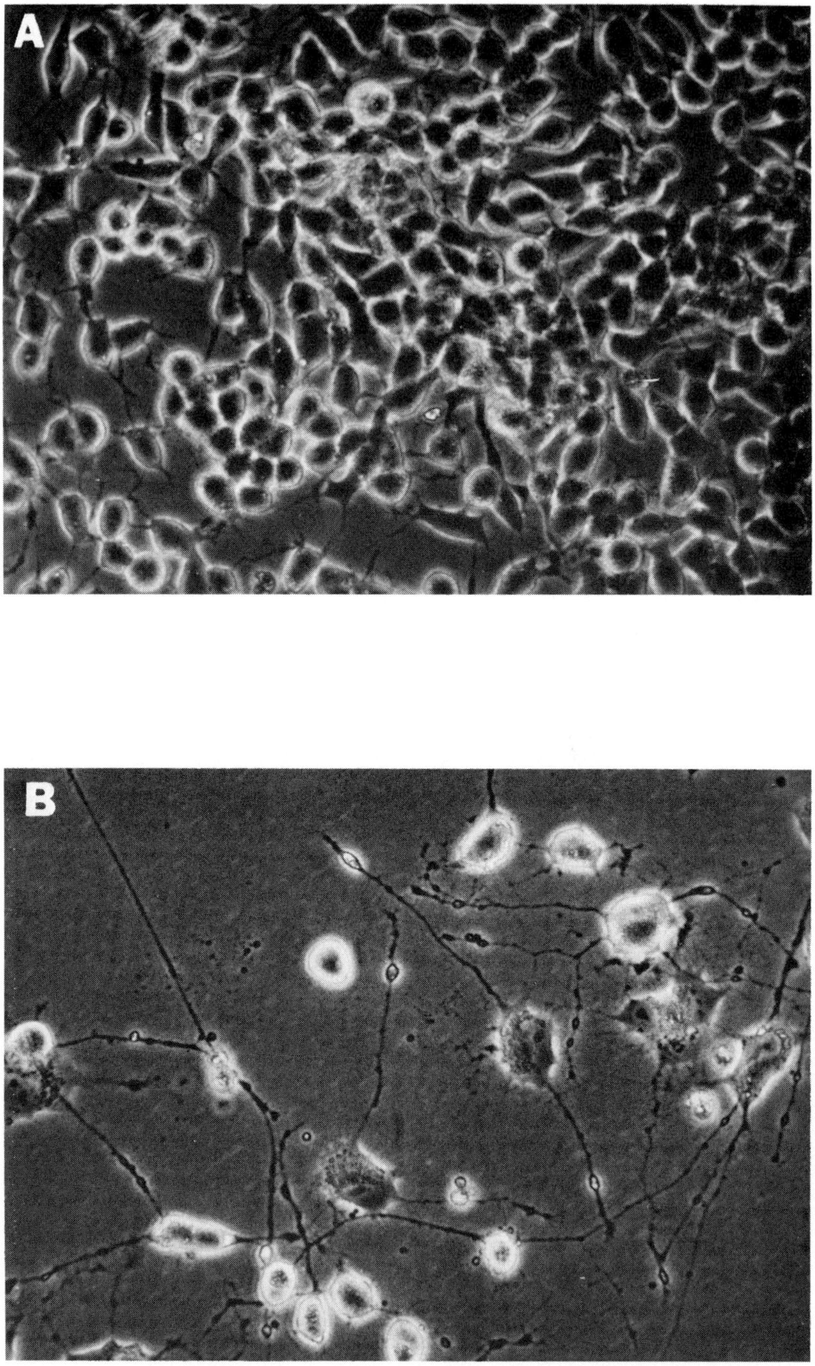
A
B

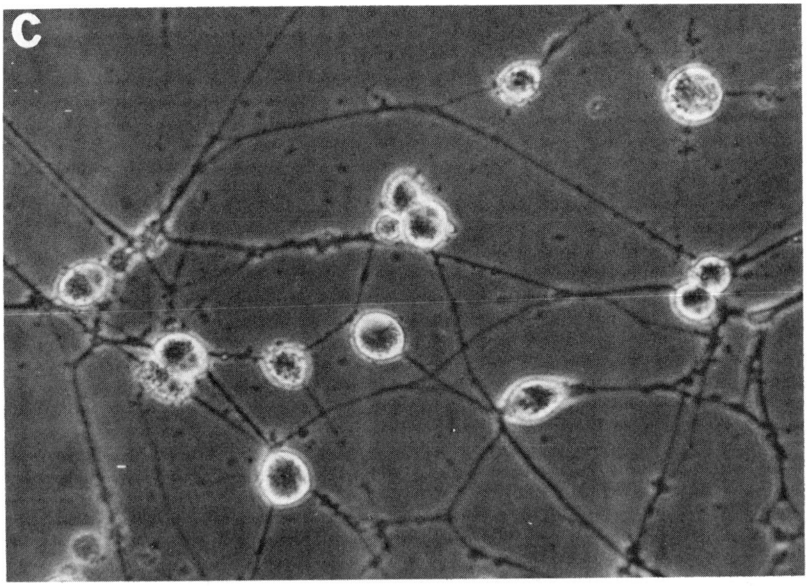

FIG. 1: Phase contrast micrographs of mouse neuroblastoma cells [NBA$_2$(1) clone] in culture. Cells (50,000) were plated in Falcon plastic dishes (60 mm) and prostaglandin E_1 (PGE_1) (10 μg/ml) and 4-(3 butoxy-4-methoxybenzyl)-2-imidazolidinone (R020-1724, 200 μg/ml), a specific inhibitor of cyclic AMP phosphodiesterase, were added separately 24 hrs. later. The medium and drug were changed at 3, 5, 8 and 11 days after treatment. Untreated malignant control culture (A) shows that cells grow in clumps and some of them have short cytoplasmic processes. PGE_1-treated culture 4 days after treatment (B) shows formation of long neurites. PGE_1-treated culture 14 days after treatment (C) shows that the remaining cells maintain their differentiated phenotype and show extensive network of thickened fibres. (From Prasad and Kumar, 1974.) (×92).

cyclic AMP, one stimulator of adenylate cyclase and one inhibitor of cyclic nucleotide phosphodiesterase. 2. Both the stimulator of adenylate cyclase and the inhibitor of phosphodiesterase must increase the intracellular level of cyclic AMP. 3. Changes in cyclic AMP-dependent phosphorylation must precede alteration in cellular functions. This is important because alteration of cyclic AMP-dependent phosphorylation activity can occur without any change in the cyclic AMP level if the phosphoprotein phosphatase activity is altered. However, it should be pointed out that variant clones which may have a defect in the sensitivity of adenylate cyclase to PGE_1 (prostaglandin E_1) or in the binding of cyclic AMP with proteins may not fulfill all the above-mentioned criteria. In addition, the possibility must be kept open that at least some effects of cyclic AMP may be mediated by mechanisms other than protein kinase(s).

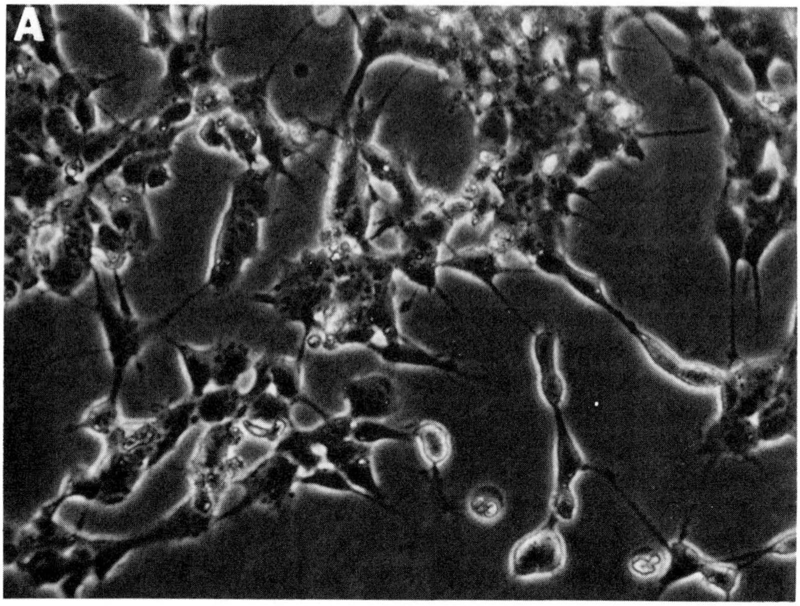

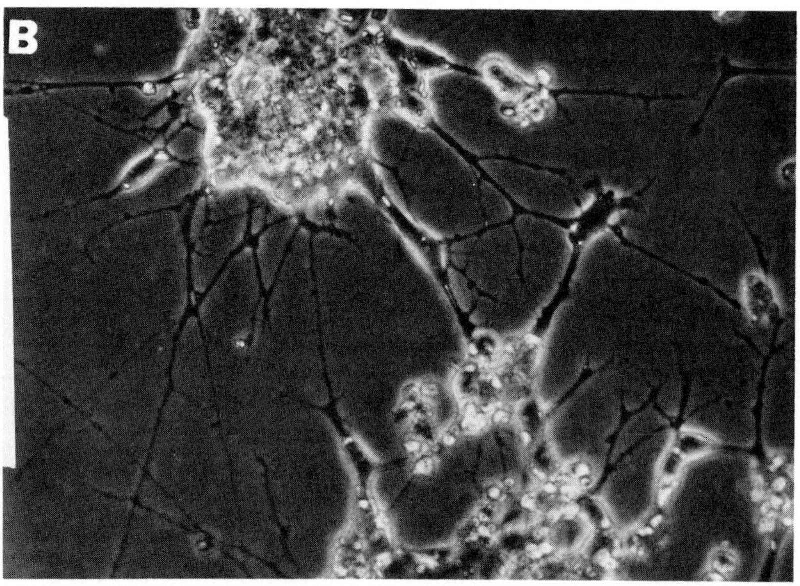

Legend page 452

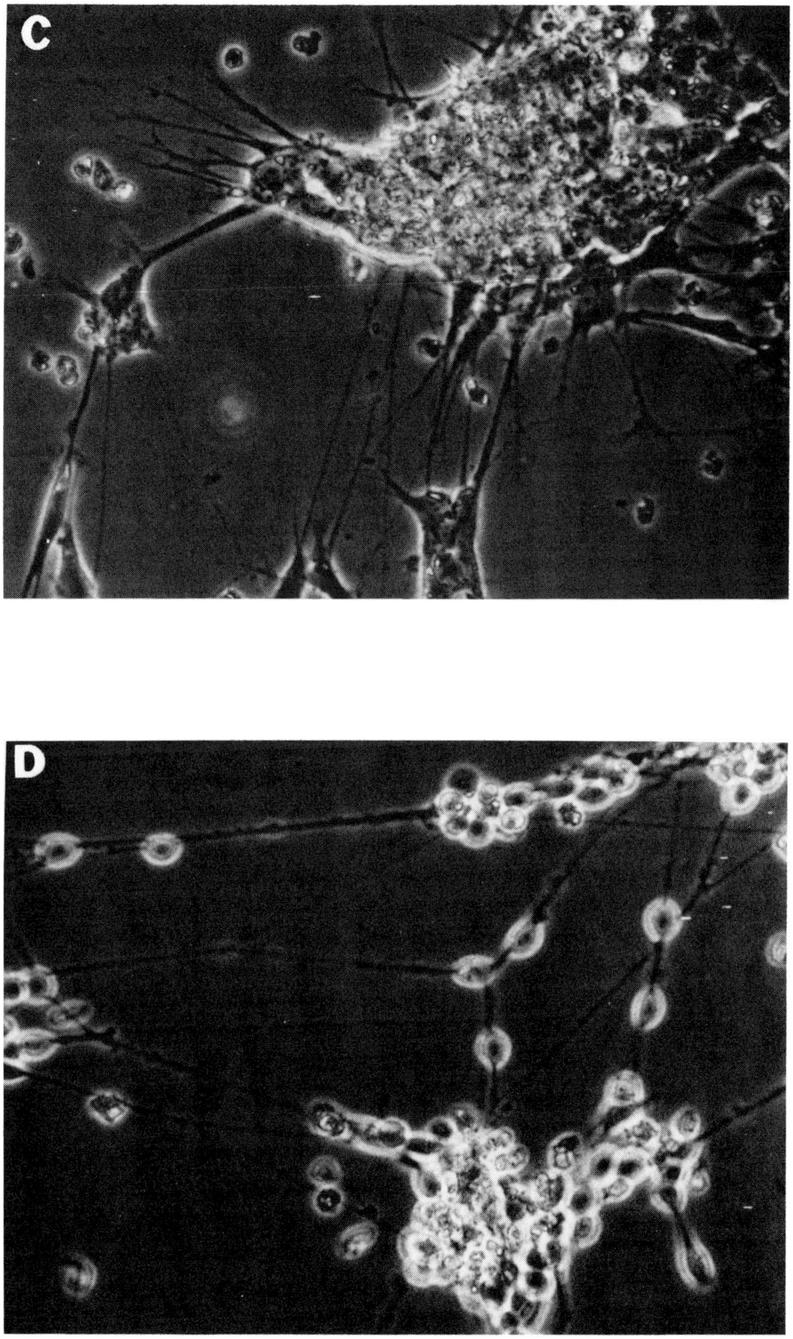

Legend page 452

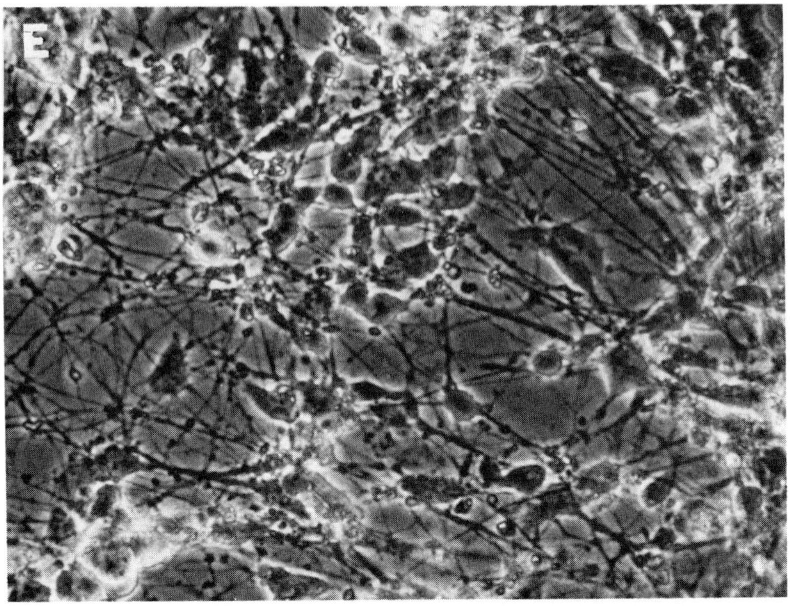

FIG. 2: Phase contrast micrographs of human neuroblastoma cells in culture (IMR-32 clone). Cells were plated in Falcon plastic dishes (60 mm), and papaverine (2.5 μg/ml), sodium butyrate (0.5 mM), serum-free medium (SFM) and 5-bromodeoxyuridine (5-BrdU, 2.5 μM) were added individually 4 days after plating. The drug and medium were changed every 2-3 days and the cultures were maintained for 10-13 days. Control (malignant) culture (A) shows that cells grow in clumps and exhibit no spontaneous morphological differentiation (i.e., cytoplasmic processes longer than 50 μm). Papaverine-treated culture, 10 days after treatment (B) shows extensive formation of neurites. Many cell deaths occurred during this period. Sodium butyrate-treated culture, 10 days after treatment (C) also shows the extensive formation of neurites. A lot of cell death occurred during this period. SFM-treated culture, 3 days after treatment (D) and 5-BrdU-treated cultures, 10 days after treatment (E) show an extensive neurite formation (From Prasad and Kumar, 1974 and 1975.) (×92).

II. Cyclic AMP and the Regulation of Differentiated Functions in Neuroblastoma Cells

To study the problem of differentiation in neuroblastoma cells in culture, the selection of a suitable clone is essential. After a given treatment, cells must not only express high levels of differentiated features, but they must also stop cell division. The neurites which are extended by neuroblastoma cells appear to be similar in appearance to the axons and dendrites of normal neurons (Ross *et al.*, 1975). Although these processes are electrically excitable and are capable of generating action potentials (Nelson *et al.*, 1969; Ross *et al.*, 1975), many of the biochemical differentiated functions may not be expressed in cells having such processes (Prasad, 1973; Prasad and Kumar, 1974). Agents such as dibutyryl cyclic

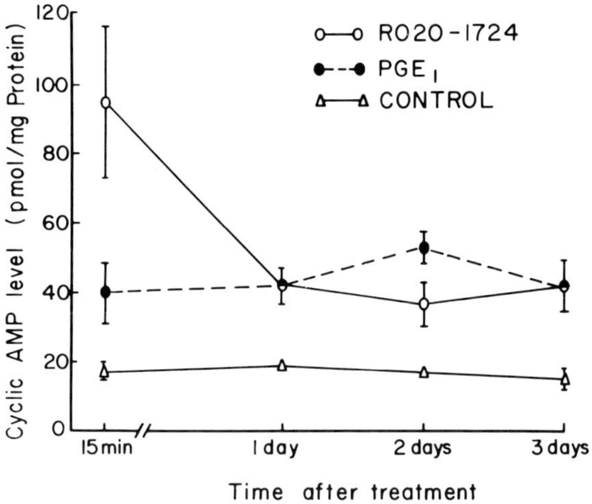

FIG. 3: Changes in the cyclic AMP level during morphological differentiation. Cells (0.5×10^6) of clone $NBA_{2(1)}$ were plated in large Falcon plastic flasks (75 cm^2) and PGE_1 (10 μg/ml dissolved in alcohol) and R020-1724 (200 μg/ml dissolved in alcohol) were added separately 24 hrs. later. The control cultures received an equivalent volume of alcohol. The drug and medium were changed 2 days after treatment and the level of cAMP was determined 3 days after treatment. Each value represents the average of 6-8 samples; the bar at each point shows the standard deviation. (From Prasad and Kumar, 1974.)

AMP, PGE_1, inhibitors of phosphodiesterase, serum-free medium, x-rays, 5-bromodeoxyuridine, tri-fluoro-methyl-2-deoxyuridine, 6-thioguanine, cytosine arabinoside, methotrexate, glial extract, hypertonic medium and nerve growth factor, increase the extent of neurite formation (for a review, see Prasad, 1975). Neurites are formed in both mouse (Figure 1A-C) and human neuroblastoma cells (Figure 2A-E) after treatment of cells with various agents. However, the rate of expression of differentiated phenotypes in human neuroblastoma cells is much slower than that in mouse cells (Prasad and Kumar, 1975). Figure 3 shows the changes in the levels of cyclic AMP as a function of time after treatment of mouse neuroblastoma cell cultures with PGE_1 and R020-1724 [4-(3-butoxy-4-methoxy-benzyl)-2-imidazolidinone], a specific inhibitor of cyclic AMP phosphodiesterase (Prasad, 1973). It should be noted that the level of cyclic AMP remained high for a period of at least 3 days. This is in contrast to nonneural cells in culture in which the increase in the level of cyclic AMP after treatment with cyclic AMP stimulating agents is transient. Some agents which do not increase the cyclic AMP level, nevertheless induce neurites, indicating that the expression of neurites in neuroblastoma cells can result from at least two modes of regulation, one of which involves changes in the cyclic AMP level. Agents such as x-irradiation and 6-thioguanine induce neurites in mouse cells, and 5-bromodeoxyrudine in human cells without changing the intracellular level of cyclic AMP (Prasad 1973, Prasad *et al.*, 1973a).

Promotion of the organization of microtubules and microfilaments must occur during neurite induction (Prasad and Hsie, 1971; Prasad, 1973), since the expression of differentiated phenotype is blocked by vinblastine sulfate (interferes with the assembly of microtubules) and cytochalasin B (interferes with the assembly of microfilaments). Cytochalasin B and vinblastine sulfate block the formation of R020-1724-induced neurites in mouse neuroblastoma cells; however, these agents do not decrease the R020-1724-stimulated increase in the cyclic AMP level (Prasad, 1973) indicating further the requirement of organization of microtubules and microfilaments for the expression of neurites. The expression of differentiated phenotype does not require the synthesis of new RNA (Prasad and Hsie, 1971), since cyclic AMP stimulating agents induce neurites in the presence of actinomycin D (inhibits RNA synthesis) and in enucleated neuroblastoma cells (Miller and Ruddle, 1974). However, there is some controversy regarding the requirement of synthesis of new protein. Neurites induced in serum-free medium do not require the synthesis of new protein (Seeds *et al.*, 1970); however, 5-bromodeoxyuridine and cyclic AMP-induced neurites require the synthesis of new protein (Schubert and Jacob, 1970; Prasad and Hsie, 1971). This difference may be due to differences in the experimental conditions. It has been suggested that the neurite formation can be initiated in the absence of protein synthesis but soon reaches a point beyond which *de novo* synthesis of protein is required (Schubert *et al.*, 1971). In some mouse neuroblastoma cells the neurite formation requires strong interaction between the cell and the surface of the culture dish (Schubert and Jacob, 1970), although this may not be equally true in another mouse neuroblastoma clone (Prasad, 1973).

The neurites which are formed after treatment of neuroblastoma cells with various agents may or may not be maintained after the removal of the drug. In a sensitive clone, cyclic AMP stimulating agents produce an irreversible effect on neurite formation. The irreversibility of neurite formation after treatment with agents which elevate the intracellular level of cyclic AMP has been confirmed in some clones by the following observations: When the 'differentiated' cells, 4 days after treatment with PGE_1 or R020-1724, are removed from dishes by means of Viokase solution and replated in separate dishes, cells attach and form long neurites within 24 hours even though no drug is present in the medium (Prasad and Kumar, 1974). The percentage of morphologically differentiated cells in the newly plated dishes is similar to that in dishes from which the drug is not removed. This indicates that the cellular factors which control the expression of differentiated phenotype remain functional after subculture. It is essential that the culture is maintained in the presence of drugs (PGE_1 and

R020-1724) at least for 4 days before the irreversible modulation can be observed. From the analysis of many studies it is now apparent that the difference in the cyclic AMP-effects observed in various laboratories may be due to one of the following reasons: a) A difference in clone may account for the difference in results. Most clones are sensitive to PGE_1 and R020-1724 in causing neurite formation. However, some clones, irrespective of their neuronal cell type, are sensitive to PGE_1 but not to phosphodiesterase inhibitor and vice versa (Prasad *et al.*, 1974). The clones which are insensitive to inhibitors of cyclic AMP phosphodiesterase are also unresponsive to dibutyryl cyclic AMP. b) Our preliminary data show that the expression of differentiated phenotype is markedly altered by different types of serum (Prasad, 1977). Table I shows that the expression of neurites in control cells (NBP_2 clone) varies from about 2% in newborn calf serum to 77% in agamma globulin calf serum. In addition, the effects of PGE_1 and R020-1724 on morphological differentiation vary markedly depending upon the type of serum. PGE_1 is least effective in fetal calf serum (14%) and it is most effective in heat inactivated fetal calf serum (81%). Similarly, R020-1724 is least effective in fetal calf serum (26%) and most effective in agamma globulin newborn calf serum (75%). It should be noted that the concentration requirement of each drug for a maximal effect on morphological differentiation also varies as a function of the type of serum used. It is not known whether the expression of biochemically differentiated functions and the growth rate are modified in different types of serum. It is also not known whether the alterations in the expression of differentiated functions in various types of serum are linked with changes in the cyclic AMP system.

Inhibition of cell division is not a prerequisite for the expression of neurites. On the contrary, neurites are formed in the dividing mouse neuroblastoma cells. Cyclic AMP (Prasad and Hsie, 1971) induces neurites prior to inhibition of cell division. Neurites are induced by 5-bromodeoxyuridine at a concentration which does not inhibit DNA synthesis (Schubert and Jacob, 1970). Several other differentiated functions such as the presence of neural specific enzymes (Augusti-Tocco *et al.*, 1970), synthesis of catecholamines (Helson and Biedler, 1973), presence of nervous system specific proteins (Herschmann and Lerner, 1973), increase in the intracellular level of cyclic AMP (Prasad, 1973), presence of cyclic AMP phosphodiesterase activity (Sinha and Prasad, 1977) and stimulation of adenylate cyclase activity by neurotransmitters (Prasad and Gilmer, 1974, Prasad *et al.*, 1975c) can be expressed in the dividing nerve cells. Some of these functions may however, be present at extremely low levels and only become fully expressed after cessation of cell division, at which time also other differentiated functions may appear. A neuroblastoma cell may permanently stop cell division without the expression of long neurites. The neurites may also be expressed in the absence of any increase in the expression of many biochemical differentiated functions (Prasad, 1975).

TABLE I
EFFECT OF DIFFERENT TYPES OF SERUM ON THE EXPRESSION OF DIFFERENTIATED PHENOTYPE IN MOUSE NEUROBLASTOMA CELLS IN CULTURE*

TYPES OF SERUM	MORPHOLOGICALLY DIFFERENTIATED CELLS (%)		
	CONTROL	PGE_1	R020-1724
Fetal calf (FC)	11 ± 2**	14 ± 2 (5 μg/ml)	26 ± 3 (100 μg/ml)
Heat inactivated FC	2 ± 0	81 ± 5 (15 μg/ml)	51 ± 4 (200 μg/ml)
Dialyzed FC	6 ± 1	72 ± 3 (2 μg/ml)	63 ± 4 (100 μg/ml)
Agamma globulin FC	highly toxic		
Newborn calf (NBC)	2 ± 0	20 ± 2 (10 μg/ml)	60 ± 4 (100 μg/ml)
Heat inactivated NBC	5 ± 0	34 ± 3 (10 μg/ml)	40 ± 3 (100 μg/ml)
Agamma globulin NBC	2 ± 0	41 ± 3 (10 μg/ml)	75 ± 4 (200 μg/ml)
Calf	highly toxic		
Heat inactivated calf	highly toxic		
Dialyzed calf	28 ± 3	56 ± 4 (10 μg/ml)	93 ± 3 (100 μg/ml)
Agamma globulin calf	77 ± 3	92 ± 2 (5 μg/ml)	89 ± 3 (100 μg/ml)

*Cells (25,000) were plated in Falcon plastic dishes (60 mm) and prostaglandin E_1 (PGE_1) and 4(-3-butoxy-4-methoxybenzyl)-2-imidazolidinone (R020-1724) was added separately 24 hrs. after plating. The drug and medium were changed at day 2 and 3, and the percentage of morphologically differentiated cells (processes >50 μm in length) were determined 4 days after treatment. The value in parentheses represents an optimal concentration. Each value represent an average of 9 samples. (These data are taken from K. N. Prasad, 1977).
**Standard deviation

A. Membrane Change

Mouse neuroblastoma cells show strong agglutination in the presence of concanavalin A (ConA) or wheat-germ agglutinin (WGA), whereas the 'differentiated' cells induced by inhibition of cyclic AMP phosphodiesterase with (R020-1725) show very little agglutination (Prasad, 1973). However, the 'differentiated' cells induced by dibutyryl cyclic AMP or PGE_1 show agglutination in the presence of ConA and WGA to an extent similar to that observed in malignant (control) neuroblastoma cells. Thus, changes in the agglutination sites are not necessarily linked with the 'differentiation' of neuroblastoma cells. The lack of agglutination in R020-1724-treated cells may be due to a unique effect of this agent on membranes. The transport of glucose does not significantly change in cAMP-induced differentiated cells (Prasad *et al.*, 1974).

B. Tyrosine Hydroxylase

A normal mature neuron expresses the activities of neural specific enzymes at high levels, but in neuroblastoma cells they are expressed at low levels. The activity of tyrosine hydroxylase (TH), a rate limiting enzyme in the biosynthesis of catecholamines, markedly increases in cyclic AMP-induced 'differentiated' mouse neuroblastoma cells (Waymire *et al.*, 1972; Richelson, 1973). This increase in enzyme activity can be blocked by cycloheximide but not by actinomycin D, indicating that the cAMP effect on TH activity is primarily at the translational level. Agents (phosphodiesterase inhibitors, PGE_1, analogs of cyclic AMP, sodium butyrate and 5-bromodeoxyuridine) which increase the cyclic AMP level also elevate tyrosine hydroxylase activity in mouse neuroblastoma cells (Prasad, 1975) and in human neuroblastoma cells (Prasad and Kumar, 1974). Agents (x-irradiation, 6-thioguanine, 5′-AMP and cytosine arabinoside) which do not elevate the cyclic AMP level, do not increase tyrosine hydroxylase activity. On the other hand, serum-free medium increases the cyclic AMP level in mouse neuroblastoma cells (Prasad, 1973), but does not increase tyrosine hydroxylase activity (Kates *et al.*, 1971, Waymire *et al.*, 1972), indicating that in addition to cyclic AMP, a second factor which is present in the serum is required to obtain an increase of the tyrosine hydroxylase activity. The tyrosine hydroxylase activity increases after treatment of human neuroblastoma cells with 5-bromodeoxyuridine (Prasad *et al.*, 1973a), and after treatment of mouse cells with 5-(3, 3-dimethyl-1-triazeno) imidazole-4-carboxamide (DTIC) (Culver *et al.*, 1977). These agents do not increase the cyclic AMP levels. Thus the activity of tyrosine hydroxylase is regulated by at least two modes, one of which involves changes in the level of cyclic AMP. X-irradiation and 6-thioguanine induce neurites without increasing TH activity, indicating that the morphological differentiation and the stimulation of TH activity are independently regulated (Prasad, 1975).

C. Choline Acetyltransferase and Acetylcholinesterase

The activities of choline acetyltransferase (CAT) and acetylcholinesterase (AchE) are markedly increased in cyclic AMP-induced 'differentiat-

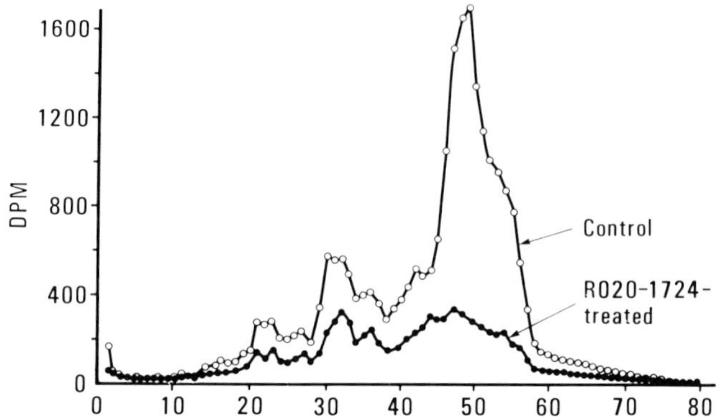

FIG. 4: Distribution of radiolabeled amino acids in SDS polyacrylamide gels of various histone fractions (0.25 N HCl fraction) from control (malignant) and cyclic AMP-induced 'differentiated' mouse neuroblastoma cells. Cells were treated with PGE_1 or R020-1724 for 3 days. Fresh growth medium was added one hr. before the addition of radioactive amino acids. Controls and treated cells were labeled with ^{14}C or ^{3}H-amino acids for 2 hrs. Total DPM ^{14}C:25,133; ^{3}H:8, 975. Gels were run for 17.5 hrs. at 6 ma/gel. (From Lazo *et al.*, 1976).

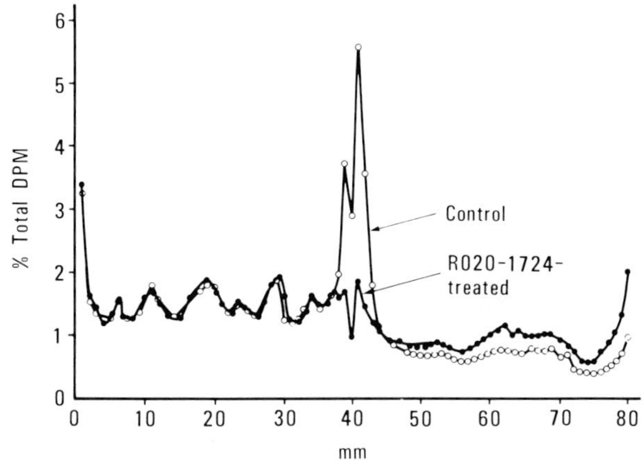

FIG. 5: Distribution of radiolabeled phosphate in SDS polyacrylamide gels of various histone fractions (0.25N HCl fraction) from control (malignant) and cyclic AMP-induced 'differentiated' mouse neuroblastoma cells. Cells were treated with PGE_1 or R020-1724 for 3 days. Fresh growth medium was added 1 hr. before the addition of radioactive phosphate. Controls and treated cells were labeled with ^{32}P or ^{33}P for 2 hrs. Total DPM ^{32}P:298, 079; ^{33}P:108, 670. Gels were run for 21 hrs at 6 ma/gel. (From Lazo *et al.*, 1976).

ed' cells as well as in cells treated with x-irradiation, 6-thioguanine, and 5′-AMP which do not increase the cyclic AMP level (Prasad, 1973). The increase in enzyme activity is blocked by cycloheximide but not by actinomycin D, indicating that the effects of cyclic AMP on CAT and AchE are primarily at the translational level. These data show that the activities of CAT and AchE are regulated by at least two modes one of which involves changes in the cyclic AMP level. The activities of CAT and AchE are inversely related to the growth rate. The increase in enzyme activities occurs without the formation of neurites.

D. *Catechol-O-Methyltransferase*

Catechol-O-methyltransferase (COMT) degrades catecholamines in mammalian cells. The enzyme activity is not increased in cyclic AMP-induced 'differentiated' mouse neuroblastoma cells, suggesting that cyclic AMP is not involved in the regulation of COMT activity; however, the COMT activity is increased after x-irradiation, or treatment with sodium butyrate or 5-bromodeoxyuridine (Prasad, 1973). In human neuroblastoma cells, the COMT activity is not detectable until the cells are treated with 5-bromodeoxyuridine (Prasad *et al.*, 1973a). The effect of 5-bromodeoxyuridine on COMT activity is similar in both mouse and human neuroblastoma cells in culture.

E. *Synthesis and Phosphorylation of Chromosomal and Nonchromosomal Proteins*

An elevation of the intracellular level of cyclic AMP in neuroblastoma cells markedly reduced the synthesis (Fig. 4) of histone and phosphorylation (Fig. 5) of H_1-histone (Lazo *et al.*, 1976). Since the above changes occur at the time of inhibition of cell division and DNA synthesis (Prasad *et al.*, 1972), they may be important biological signals for the dividing neuroblasts to turn off cell division. The occurrence of histone synthesis and phosphorylation of H_1-histone may be necessary biological events for the proliferating system (differentiated or embryonic); but for the embryonic nerve cells which must eventually stop cell division, a continuation of these events after a specified time during development may be indicative of malignant transformation.

We have reported that there was no prominent change in the synthesis or phosphorylation of nonhistone chromosomal proteins in cyclic AMP-induced 'differentiated' cells (Lazo *et al.*, 1976). Other investigators (Burdman, 1972; Fujitani and Holoubek, 1974) have been unable to demonstrate any significant changes in nonhistone chromosomal proteins during development of rat brains. However, we have observed a small decrease in the synthesis, and a small increase in phosphorylation of 40,000 dalton peptides in cyclic AMP-induced 'differentiated cells. The significance of this change in nonhistone proteins is unknown. It has been suggested (Elgin *et al.*, 1973) that a nonhistone protein in the 40,000-50,000 dalton range may be involved in DNA replication in mammalian cells. If this is true, the changes in the synthesis and phosphorylation of 40,000 dalton peptides in the 'differentiated' cells may be a reflection of the inhibition of DNA synthesis which occurs in these cells (Prasad *et al.*, 1972).

F. Reduction in Tumorigenicity of 'Differentiated' Cells

The tumorigenicity of 'differentiated' cells is markedly reduced or completely abolished depending upon the experimental condition. For example, in uncloned cell lines, analogs of cyclic AMP, R020-1724 (a phosphodiesterase inhibitor) and PGE_1 only partially reduced the tumorigenicity of cells, whereas the combination of PGE_1 with either analogs of cyclic AMP or a phosphodiesterase inhibitor completely abolished the tumorigenicity of the 'differentiated' cells. This is consistent with the observation that the tumor contains variant cells some of which are sensitive to PGE_1 but not to R020-1724, and vice versa (Prasad, 1972). We have repeated the above experiment using clonal lines (NBP_2). This clone is relatively more sensitive to R020-1724 than to PGE_1. In order to supress the tumorigenicity of differentiated cells, it was found essential to treat the cells with both PGE_1 and R020-1724 for 4 days.

III. Regulation of the Intracellular Level of Cyclic AMP

Since an elevation of the intracellular level of cyclic AMP in neuroblastoma cells both induces and increases the expression of many differentiated functions which are characteristic of mature neurons, it would be reasonable to assume that a defect in the cyclic AMP system may be responsible for an abnormal expression of differentiated functions. Since the intracellular level of cyclic AMP can be affected both by the relative activities of the adenylate cyclase and the cyclic AMP phosphodiesterase, the response of adenylate cyclase to neurotransmitters, and the levels of cyclic AMP binding proteins, we have investigated the regulation of the individual parameters of the cyclic AMP system.

A. Adenylate Cyclase Activity

The basal level of adenylate cyclase activity in homogenates and the sensitivity of adenylate cyclase to neurotransmitters vary from one clone of mouse neuroblastoma cells to another (Prasad and Gilmer 1974; Prasad and Kumar 1974; Prasad *et al.*, 1975d). In homogenates of cyclic AMP-induced 'differentiated' neuroblastoma cells, we have made the following observa-

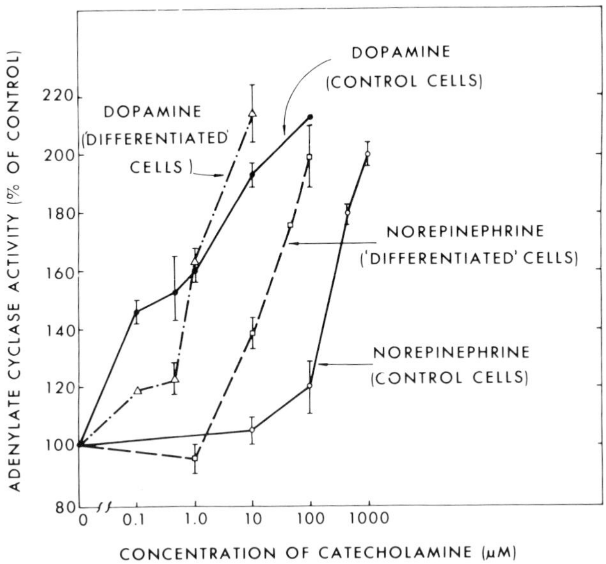

FIG. 6. Changes in adenylate cyclase activity in homogenates of control and 'differentiated mouse neuroblastoma cells $NBA_{2(1)}$ after treatment with dopamine and norepinephrine. R020-1724 was used to induce 'differentiation'. The basal activities of adenylate cyclase in control (15 ± 1.4 pmol/mg of protein/min) and 'differentiated' (21 ± 1pmol/mg of protein/min) cells were considered 100% control values; the adenylate cyclase values of treated cells were expressed as percentage of the value in the absence of the transmitters. Each value represents the average of 8-12 samples. The bar at each point shows the standard deviation. (From Prasad and Gilmer, 1974.)

tions: 1) the sensitivity of adenylate cyclase to dopamine, norepinephrine (Fig. 6) and acetylcholine increases; 2) the sensitivity of adenylate cyclase to PGE_1 and guanosine triphosphate (GTP) does not change; 3) GTP potentiates the PGE_1-stimulated adenylate cyclase activity (Fig. 7); and 4) high concentrations of calcium, magnesium and manganese inhibit adenylate cyclase activity. Thus an increase in the sensitivity of adenylate cyclase to neurotransmitters and ions, and an increase in the sensitivity of PGE_1-stimulated enzyme activity to GTP occurs during differentiation of neuroblastoma cells. Although dopamine and norepinephrine stimulate adenylate activity in homogenates, they do not increase the cyclic AMP level in intact cells until the activity of cyclic AMP phosphodiesterase is inhibited (Prasad *et al.*, 1974; Sahu and Prasad, 1975). PGE_1 stimulates the intracellular level of cyclic AMP to varying degrees in most of the mouse clones (Gilman and Nirenberg, 1971; Prasad, 1973; Hamprecht and Schultz, 1973; Blume *et al.*, 1973; Sahu and Prasad 1975), but in one clone (NBE^-), it does not increase the cyclic AMP level significantly until the phosphodiesterase activity is inhibited (Table II). PGE_1 in combination with a

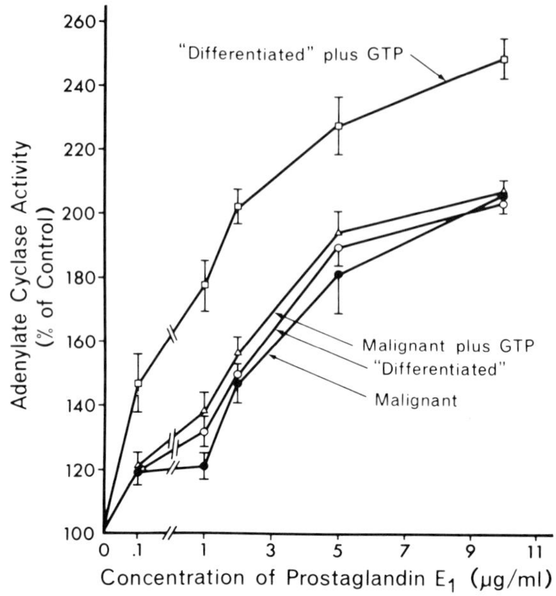

FIG. 7: Changes in adenylate cyclase activity in homogenates of 'differentiated' mouse neuroblastoma cells (NBP_2) after treatment with PGE_1 and guanosine triphosphate (10 μM). R020-1724 was used to induce differentiation. The basal activities of adenylate cyclase in (23 ± 1 pmol/mg of protein/min) and differentiated (25 ± 1 pmol/mg of protein/min) cells were considered 100% control values; the adenylate cyclase values of treated cells were expressed as percentage of control. Each value represents the average of 6 samples. The bar at each point shows the standard derivation. (From Prasad *et al.*, 1975c).

phosphodiesterase inhibitor increases the cAMP level more than the individual agent alone. A similar observation has been made with adenosine (Prasad, 1975). Thus, the rate limiting factor in the accumulation of cyclic AMP following the treatment of neuroblastoma cells with neurotransmitters, adenosine and PGE_1 appears to be the level of cyclic AMP phosphodiesterase activity. Therefore, we have suggested that an increase in cyclic AMP phosphodiesterase activity in the dividing neuroblasts may be one of the early lesions of malignancy of nerve cells (Prasad *et al.*, 1974).

B. Cyclic AMP Phosphodiesterase Activity

The cyclic AMP phosphodiesterase activity is markedly increased (Prasad, 1973; Kumar *et al.*, 1975) in cyclic AMP-induced differentiated mouse neuroblastoma cells in culture (Fig. 8). This increase in enzyme activity is blocked by cycloheximide but not by actinomycin D. The activity of cyclic AMP phosphodiesterase also increases during exponential growth and reaches a maximal value at confluency (Fig. 9). The increase in enzyme

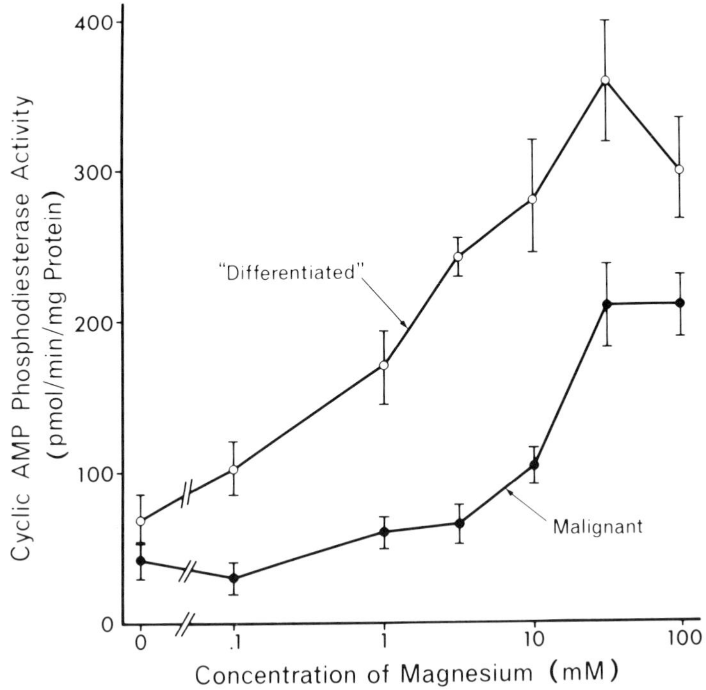

FIG. 8: Effect of magnesium on cyclic nucleotide phosphodiesterase activity in homogenates of malignant and differentiated mouse neuroblastoma cells. R020-1724 was used to induce differentiation. Each value represents the average of 8 to 12 samples. The bar at each point shows the standard deviation. (From Kumar and Prasad, 1975.)

TABLE II
EFFECT OF NEUROTRANSMITTERS AND PGE_1 ON CAMP LEVEL IN NEUROBLASTOMA CLONES*

Treatment	cAMP Levels in Various Clones, pmoles/mg Protein				
	NBA_2 (1)++	NBA_2 (1)	NBE⁻ (A)	NBDA⁻	NBP_2
Control, dopamine, or norepinephrine	25 ± 4**	14 ± 2	19 ± 2	19 ± 4	11 ± 2
R020-1724	41 ± 2	48 ± 7	107 ± 5	26 ± 4	65 ± 8
PGE_1	31 ± 4	20 ± 2	22 ± 2	56 ± 9	19 ± 2
R020-1724 + dopamine	117 ± 16	233 ± 14	139 ± 4	162 ± 49	66 ± 5
R020-1724 + norepinephrine	132 ± 6	191 ± 19	120 ± 11	122 ± 36	67 ± 4
R020-1724 + PGE	2,563 ± 121	2,803 ± 173	1,953 ± 118	2,426 ± 124	2,404 ± 212

*NBA^2(1)++ contains tyrosine hydroxylase (TH), but no choline acetyltransferase (ChA), and has been in culture less time than NBA_2(1); NBA_2(1) contains TH, but no ChA; NBE (A) contains ChA, but no TH; NBDB contains neither TH nor ChA; NBP_2 contans both ChA and TH. (These data are taken from A. Sahu and K. N. Prasad, 1975).

**Standard deviation

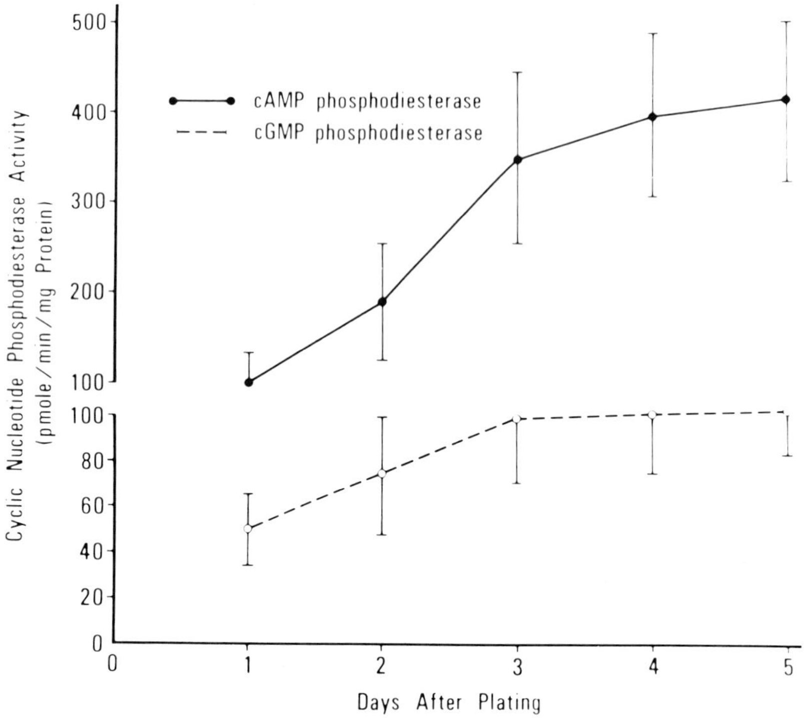

FIG. 9: Cyclic nucleotide phosphodiesterase activity in homogenates of mouse neuroblastoma cells as a function of days after plating. The reaction mixture (60 μl) contained 3 μg protein and 3.33 μM of ^{3}H-cAMP (11,000 CPM) or 3.33 μM of ^{3}H-cGMP (10,000 CPM). Each value represents the average of 12-15 samples. The bar at each point shows the standard deviation. (From Sinha and Prasad, 1977.)

activity during the growth period occurs without any change in the intracellular level of cyclic AMP, and it is blocked by both cycloheximide and actinomycin D (Sinha and Prasad, 1977). Thus the activity of cyclic AMP phosphodiesterase in neuroblastoma cells is regulated by at least two modes, namely, cyclic AMP and growth rate.

C. Binding of Cyclic AMP with Proteins

Since the intracellular level of both cyclic AMP and cyclic AMP phosphodiesterase activity increase in 'differentiated' neuroblastoma cells, the cells must develop the mechanism of protecting cyclic AMP from the enzymatic hydrolysis. The protein bound cyclic AMP is not hydrolyzed by phosphodiesterase; therefore, we have measured the binding of cyclic AMP with proteins of malignant (control) and 'differentiated' neuroblastoma cells. Indeed, the binding of cyclic AMP with soluble (100,000 × g supernatant) and non-soluble proteins becomes approximately doubled in

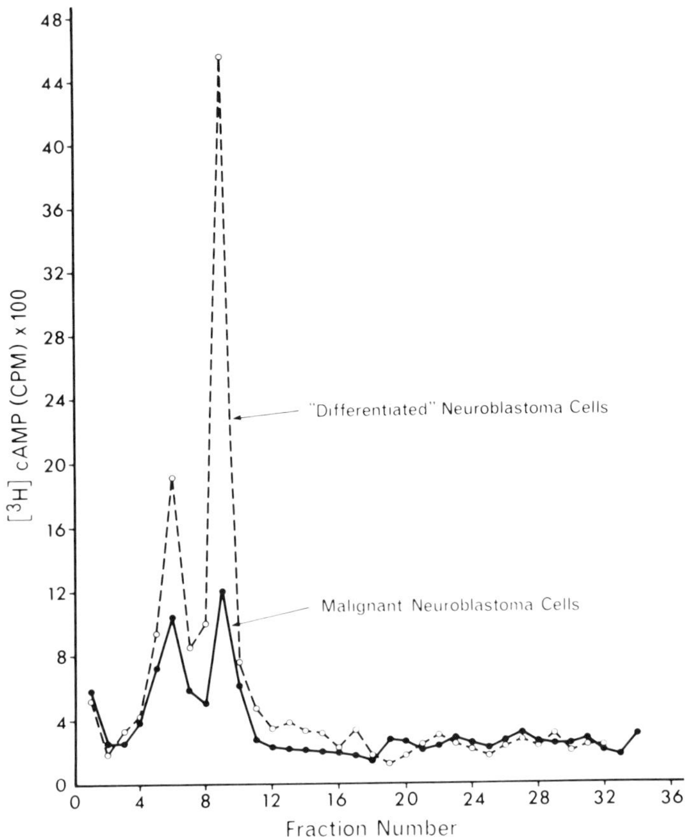

FIG. 10: Binding of cyclic (^{3}H) AMP with protein fractions obtained after polyacrylamide gel electrophoresis. R020-1724 was used to induce differentiation. The soluble proteins of control and 'differentiated' neuroblastoma cells were applied on the gel. The experiments were repeated twice. (From Prasad *et al.*, 1975e.)

cyclic AMP-induced 'differentiated' neuroblastoma cells (Prasad *et al.*, 1975e). In malignant (control) and 'differentiated' mouse neuroblastoma cells, two peaks of binding proteins are observed, and the magnitude of each of these peaks increases in 'differentiated' cells (Fig. 10). An increase in cyclic AMP binding occurs 24 hours after treatment of neuroblastoma cells with PGE_1 (Fig. 11) and is completely blocked by cycloheximide but not by actinomycin D (Prasad *et al.*, 1976b). The cyclic AMP binding is heat labile and is sensitive to the action of protease. These findings suggest than an increase in cyclic AMP binding in 'differentiated' cells is due to an increase in the levels of binding proteins. X-irradiation and 6-thioguanine, which increase the amount of total protein per cell, inhibit cell division, and induce some differentiated functions (Prasad, 1973) in neuroblastoma cells

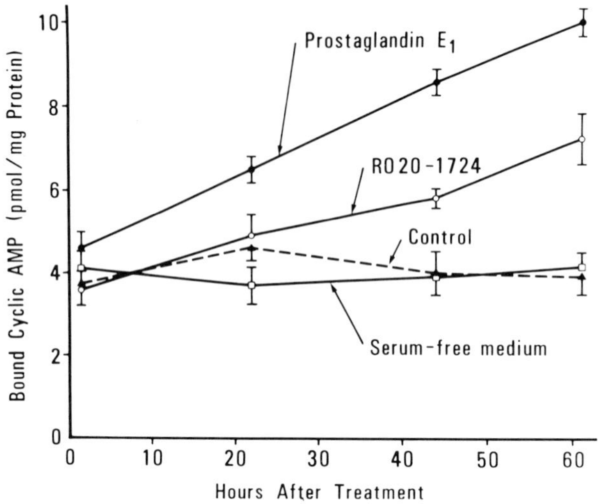

FIG. 11: Binding of cyclic AMP with soluble proteins of neuroblastoma cells as a function of time of treatment. PGE_1, R020-1724, and serum-free medium were added separately to culture, and the binding assay was performed 3 hrs., 1 day, 2 days and 3 days after treatment. The incubation mixture contained 100 μg protein. Each value represents the average of 6 samples. The bar at each point shows the standard deviation. From Prasad *et al.*, 1976c.)

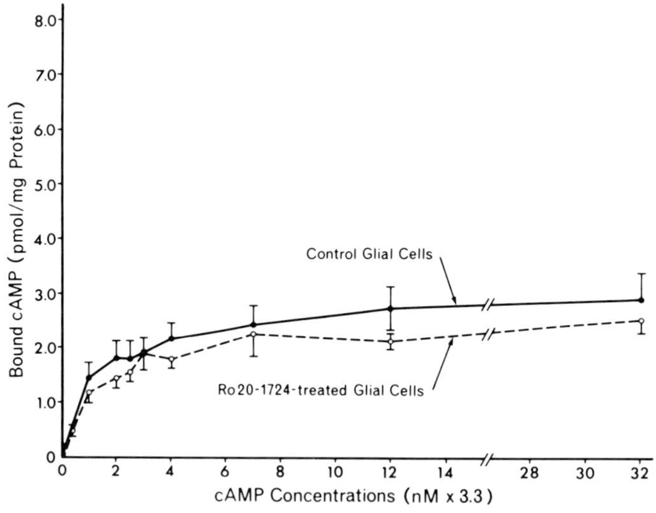

FIG. 12: Binding of cyclic AMP with soluble protein of rat glioma cells (C6). R020-1724 was used to induce differentiation. The incubation mixture contained 100 μg of protein. Each value represents the average of 6 samples. The bar at each point shows the standard deviation. (From Prasad *et al.*, 1976c.)

without changing the cellular cyclic AMP, do not elevate the level of cyclic AMP binding proteins. This indicates that neither the inhibition of cell division nor the increase in total protein is sufficient to increase the levels of binding proteins.

The increased amount of binding proteins in 'differentiated' cells may in part account for the existence of a high intracellular level of cyclic AMP even in the presence of elevated cyclic AMP phosphodiesterase activity. Thus, one of the mechanisms by which neuroblastoma cells could maintain a high intracellular level of cyclic AMP is by an increase in the amount of binding proteins as a result of differentiation. Glial and L-cells fail to develop such a mechanism after treatment with PGE_1 or R020-1724 (Fig. 12). This may in part explain the reversibility of cyclic AMP effects on nonneuronal tumor cells soon after the removal of the drug. Thus there is a defect in the regulation of cyclic AMP binding proteins in rat glioma cells in culture or the level of binding proteins in rat glioma cells is not regulated by cyclic AMP (Prasad, 1975e). The increase in the levels of binding proteins in 'differentiated' neuroblastoma cells is associated with the irreversible effect of cyclic AMP, but this is not the cause of irreversibility, since the increase in binding proteins occurs in dividing nerve cells 24 hrs. after treatment with PGE_1, at which time the effect is still reversible. However, this may be one of the important events which is essential for maintaining the irreversible neural differentiation.

D. Cyclic AMP-Dependent-Phosphorylation

The binding of cyclic AMP with the regulatory subunits of the protein kinase(s) results in dissociation of the regulatory proteins from the catalytic units with subsequent activation of the latter (Gill and Garren 1970, Tao *et al.*, 1970, Reimann *et al.*, 1971) which then phosphorylate various proteins. If this is true during differentiation of neuroblastoma cells, the cyclic AMP-dependent phosphorylation activity should increase in 'differentiated' cells. Therefore, we have first investigated this by measuring the activity of the phosphorylation of added histones by both the supernatant and the pellet fraction obtained after centrifugation at 100,000 × g for 90 min. We obtained the following results: (a) Cyclic AMP-dependent phosphorylation of H_1-histone did not change significantly in either the supernatant or the pellet fractions of 'differentiated' neuroblastoma cells (Prasad *et al.*, 1976a). (b) Cyclic AMP-dependent phosphorylation of H_2b histone decreased to about one-half in the supernatant fraction of 'differentiated' neuroblastoma cells (K. N. Prasad, unpublished observation). These data suggest that the level of cyclic AMP-dependent phosphorylation activity in the supernatant fraction either decreases or shows no change at the time when the level of cyclic AMP binding proteins increases. However, the phosphorylation activity of exogenous histones may not be a true reflection of endogenous

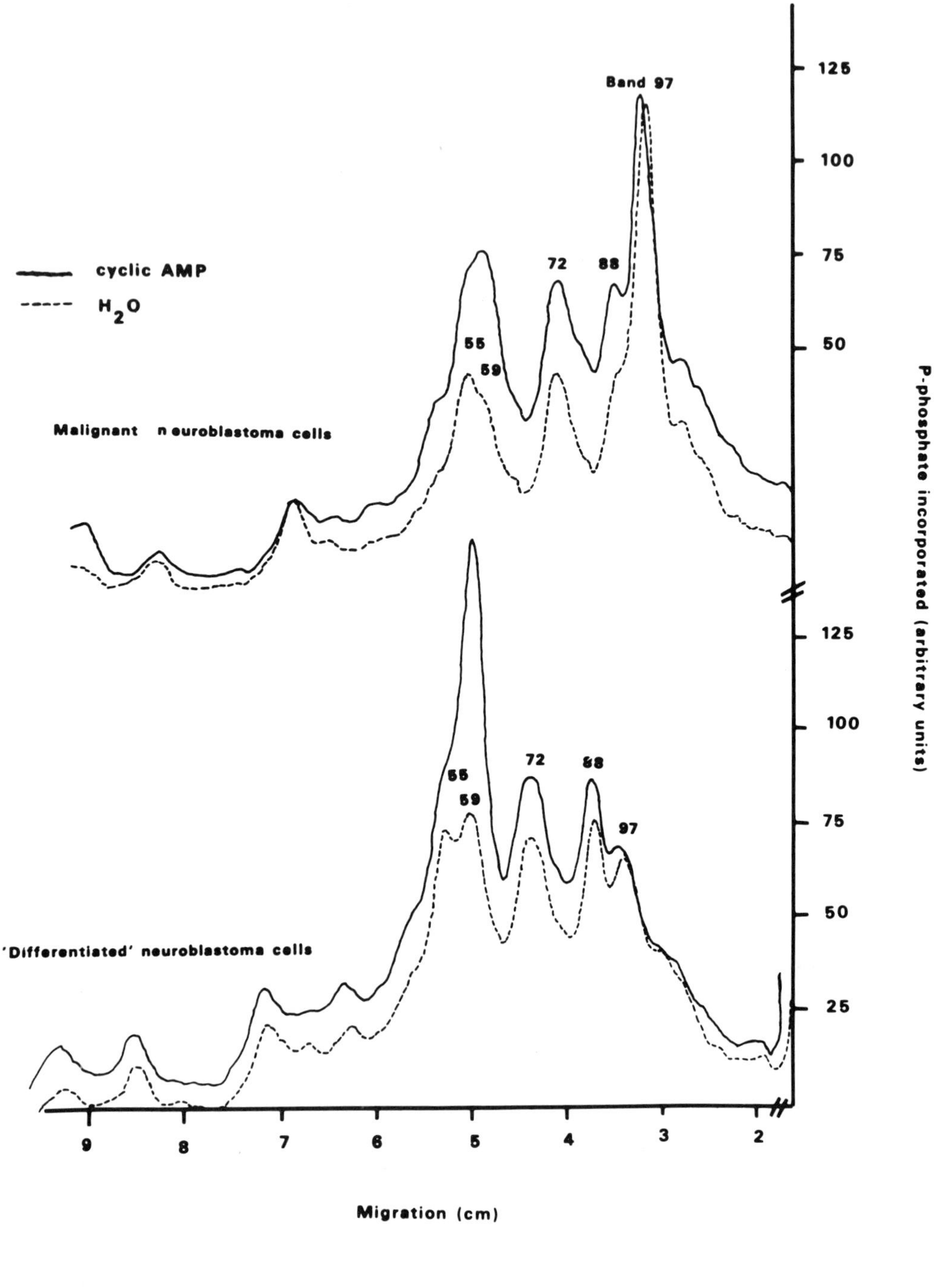
cyclic AMP
H_2O
Malignant neuroblastoma cells
'Differentiated' neuroblastoma cells
Band 97
55
59
72
88
97
125
100
75
50
25
9
8
7
6
5
4
3
2
Migration (cm)
^{32}P-phosphate incorporated (arbitrary units)

cyclic AMP-dependent phosphorylation activity. Indeed, we have observed (Ehrlich *et al.*, 1977) that the cyclic AMP-dependent-phosphorylation activity of cytosol proteins markedly increases, whereas, cyclic AMP-independent phosphorylation of another protein decreases by about 2-fold in 'differentiated' cells (Fig. 13). There is also an increase in cyclic AMP-independent phosphorylation activity of two specific proteins in the crude nuclear fractions of 'differentiated' cells (Fig. 14).

IV. Poly-A Containing Cytoplasmic RNA

The cytoplasm of cyclic AMP induced 'differentiated' neuroblastoma cells had more poly A-containing RNA (Bondy *et al.*, 1974), presumed to be messenger RNA, than that of the malignant cells (Fig. 15). This increase in poly A-containing cytoplasmic RNA was in part due to an increase in the rate of transport of mRNA from nucleus to cytoplasm (Prasad *et al.*, 1975b). The increase in messenger RNA contents in 'differentiated' cells was consistent with the observation that the activities of many enzymes increase in differentiated cells. However, the quantitative amount of genetic material available for the transcription might remain relatively constant during differentiation. This was substantiated by the similarity in binding of actinomycin with nuclei of malignant and 'differentiated' cells. The binding of actinomycin D has been used to estimate the amount of DNA within nuclei available for the template activity (Kernell *et al.*, 1971; Watson 1974). The lack of any quantitative change in template activity during cAMP-induced 'differentiation' of neuroblastoma cells could be due to the fact that some genetic codes are activated while others are suppressed during differentiation. Indeed, the activities of several enzymes in 'differentiated' neuroblastoma cells were elevated but the synthesis of histone, phosphorylation of H_1histone and cyclic AMP-independent phosphorylation activity of cytosol proteins were considerably reduced (Prasad 1975; Lazo *et al.*, 1976; Ehrlich *et al.*, 1977.)

V. Role of Cyclic GMP in Differentiation

Mouse neuroblastoma cells also have cyclic GMP (Prasad *et al.*, 1976b), the concentration of which is 100-fold less than that of cyclic AMP (Table III). In addition to the presence of cyclic GMP phosphodiesterase

FIG. 13: Autoradiogram showing, after electrophoretic separation, phosphorylated proteins in the cytosol of malignant and cyclic AMP-induced 'differentiated' neuroblastoma cells in culture. It should be noted that the incorporation of ^{32}P into proteins from gamma ^{32}P-ATP is a net result of the action of protein kinase and phosphoprotein phosphatase activities. Therefore, the term phosphorylation refers to changes in the amount of radioactive phosphate incorporation into specific bands under standard assay conditions, and does not refer to the capacity of specific enzymes. (From Ehrlich *et al.*, 1977.)

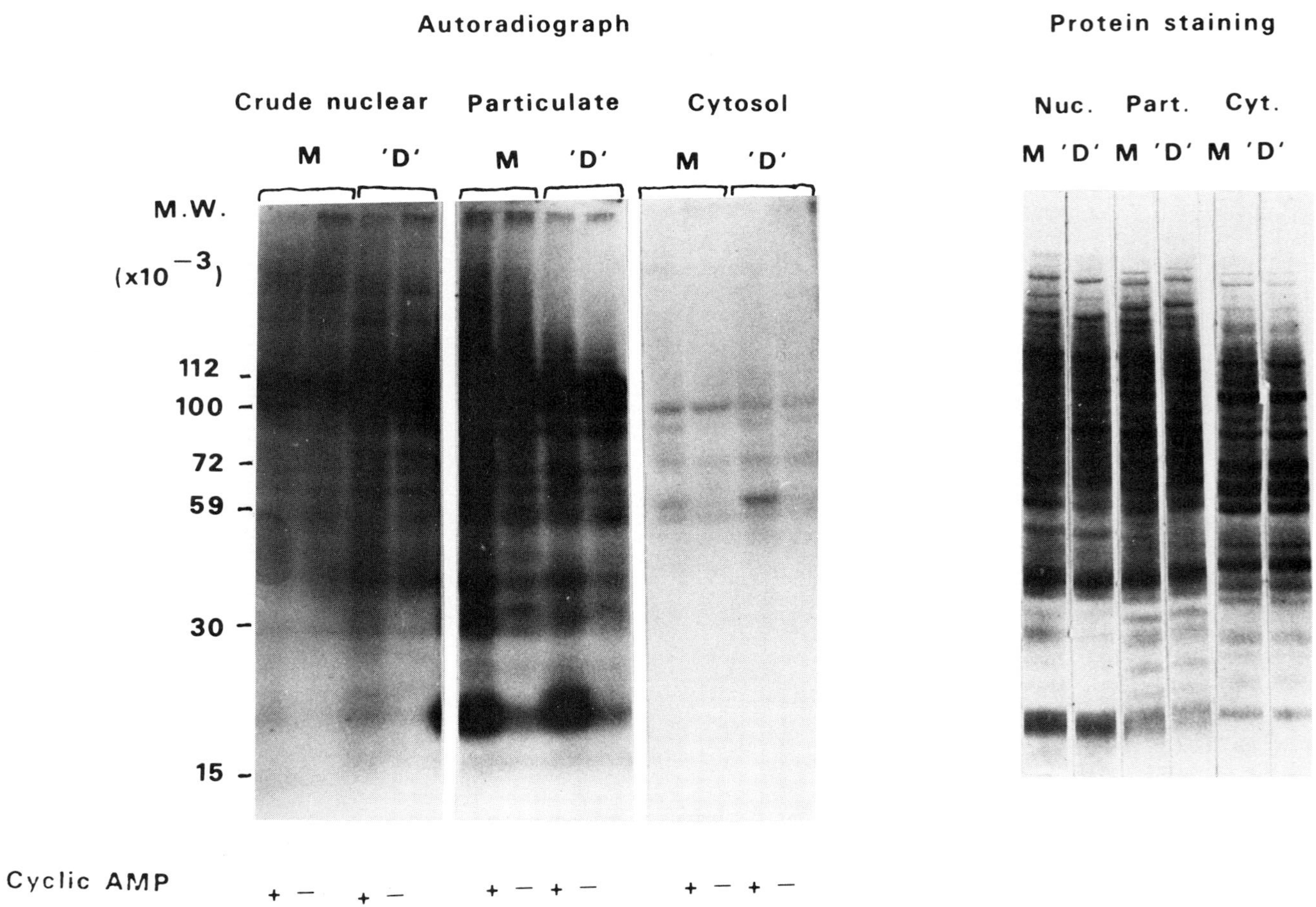
Autoradiograph
Protein staining
Crude nuclear
Particulate
Cytosol
Nuc.
Part.
Cyt.
M
'D'
M.W.
(x10^-3)
112
100
72
59
30
15
Cyclic AMP
+ − + − + − + − + − + −

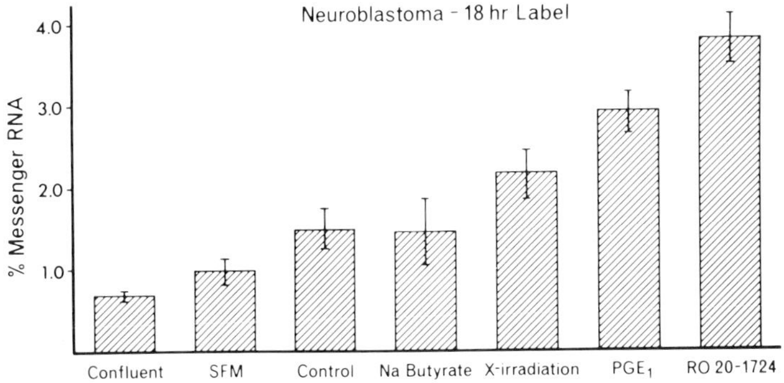

FIG. 15: Proportion of labeled cytoplasmic RNA that is messenger in mouse neuroblastoma cells after various treatments. Neuroblastoma cells were plated in Falcon plastic flasks (75 cm^2) and serum-free medium (SFM), sodium butyrate (Na butyrate), prostaglandin E_1 (PGE_1), R020-1724 or x-irradiation was given 24 hrs. later. The fresh growth medium and drug were changed every day. ^{3}H-adenosine (2 μCi/ml) and fresh drug were added 54 hrs. after treatment, and cells were further incubated for 18 hrs. The polyadenylic acid containing cytoplasmic RNA in control and treated cells was determined. Each value represents the average of at least 6 samples ± standard error of mean. (From Prasad *et al.*, 1975b.)

which is distinct from cyclic AMP phosphodiesterase (Prasad *et al.*, 1975a), the extremely low binding affinity of cyclic GMP with proteins (10 times less than that of cyclic AMP) may in part account for the low intracellular level of cyclic GMP. A preliminary study indicated that PGE_1 or R020-1724 does not increase the intracellular level of cyclic GMP 15 min. after treatment; these agents under similar experimental conditions are known to increase the cyclic AMP level. Acetylcholine approximately doubles the intracellular level of cyclic GMP; whereas it does not significantly change the level of cyclic AMP. Acetylcholine does not antagonize the effect of PGE_1 or R020-1724 on differentiation, and does not affect growth rate of differentiation in mouse neuroblastoma cells in culture (Prasad *et al.*, 1975d). The addition of either exogenous cyclic GMP or N^2-2^1-0-dibutyryl cyclic GMP to cultures of mouse or human neuroblastoma cells inhibits cell division without causing differentiation, and when added with PGE_1 or R020-1724, fails to antagonize the expression of differentiation (Prasad, 1975). These data suggest that cyclic GMP neither has any role in regulating the expression of differentiated functions in neuroblastoma cells nor

FIG. 14: Autoradiogram of phosphorylated proteins of various subcellular fractions obtained from malignant and cyclic AMP-induced 'differentiated' neuroblastoma cells in culture. It should be noted that the incorporation of ^{32}P into proteins from gamma ^{32}P-ATP is a net result of the action of protein kinase and phosphoprotein phosphatase activities. Therefore, the term phosphorylation refers to changes in the amount of radioactive phosphate incorporation into specific bands under standard assay conditions, and does not refer to the capacity of specific enzymes. (From Y. A. Ehrlich and P. K. Sinha, K. N. Prasad, L. G. David and E. G. Brunngraber, unpublished observations.)

antagonizes the effect of cyclic AMP. Possibly it has no role in the expression or maintenance of malignancy. Further work is needed to substantiate this.

TABLE III

EFFECT OF VARIOUS AGENTS ON THE INTRACELLULAR LEVEL OF CYCLIC NUCLEOTIDES IN NEUROBLASTOMA CELLS

Treatment	Dose μg/ml	cAMP Level pmol/g Protein	cGMP Level, pmol/g Protein
Control		5,000 ± 400*	50 ± 7
Acetylcholine	100	5,300 ± 450	109 ± 12
R020-1724	200	29,000 ± 400	39 ± 6
Papaverine	25	32,000 ± 600	84 ± 9
PGE_1	10	8,500 ± 500	41 ± 8

*Standard deviation
(The values are taken from Prasad *et al.*, 1976b.)

VI. A Working Hypothesis for the Differentiation of Nerve Cells

The differentiation of normal nerve cells may involve steps which are opposite to those involved in the expression of malignancy. Figure 16 shows a diagrammatic model to explain the malignancy of nerve cells (Prasad, 1974). This model suggests that a mutation in the regulatory gene for cyclic AMP phosphodiesterase within a single dividing neuroblast or a group of dividing neuroblasts may result from viruses, chemical carcinogens, ionizing radiation or any combination of these agents, and this mutational change may increase phosphodiesterase activity in the mutated cells. The high cyclic AMP phosphodiesterase activity could lead to low levels of cyclic AMP, and could be responsible for low levels of adenylate cyclase activity, and insensitivity of adenylate cyclase to neurotransmitters. These changes might then prevent the expression of neuronal differentiated functions in the mutated neuroblast, causing it to become a cancer cell. The regulatory genes for other differentiated functions in the daughter cancer cells may consequently become more susceptible to mutagenic changes which then may produce further molecular lesions. This may account for the fact that the neuroblastoma cells obtained from a tumor differ quantitatively and qualitatively from one another with respect to expression of cellular properties and to sensitivity towards different drugs. The mutated neuroblast appears to maintain the capacity to differentiate into various forms of nerve cells. This is supported by the fact that neuroblastoma contain four major types of nerve cell (Amano *et al.*, 1972; Prasad *et*

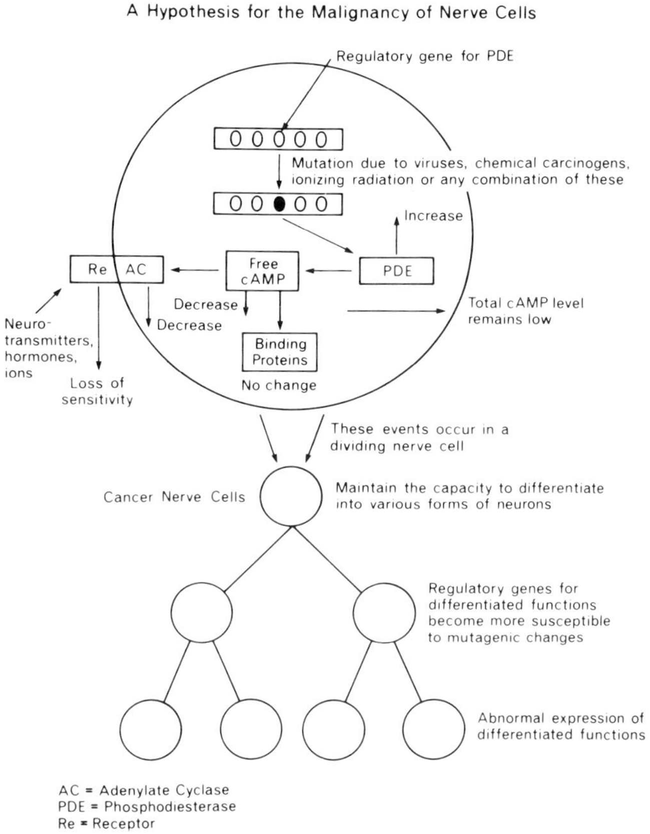

FIG. 16: A diagramatic model to explain the postulated mechanism for malignant transformation of nerve cells. Adenosine 3′ :5′-cyclic monophosphate=cAMP; phosphodiesterase=PDE; adenylate cyclase=AC; receptor=Re; dopamine=DA; norepinephrine=NE. (From Prasad, 1974.)

al., 1973b; Knapp and Mandel 1974); a) adrenergic cells; b) cholinergic: cells; c) sensory-like cells; and d) serontonergic cells. However, the differentiated functions of these nerve cells are not adequately expressed and therefore they continue to divide. It may be possible that the existence of more than one neural cell type in a neuroblastoma tumor may be due to the fact that these cells are highly aneuploid; and therefore, the expression of a neural specific enzyme in a given neuron may be due to the genetic abnormality rather than due to adrenergic or cholinergic differentiation. This may be unlikely because human neuroblastoma cells are diploid, but still contain more than one neural cell type. Whether or not the first cancer nerve cell will lead to the formation of a detectable neoplasm depends upon the host's immunological environment. If the host's immunological environment is normal, the mutated nerve cell may be rejected and no malignant lesion will ever appear. On the other hand, if the immunological environment of the host is abnormal, the malignant neoplasm may become detectable in a few months or years after the appearance of the first cancer cell. The following experimental data support the proposed hypothesis: a) a raised intracellular level of cyclic AMP in neuroblastoma cells irreversibly induces several differentiated functions which are characteristic of mature neurons (Prasad, 1975); b) the inhibition of cyclic AMP phosphodiesterase activity reduces the oncogenicity of 'differentiated' cells (Prasad, 1972); c) although dopamine and norepinephrine-sensitive adenylate cyclases are demonstrable *in vitro* (Prasad and Gilmer, 1974), these agents do not increase the cyclic AMP level until the phosphodiesterase activity is inhibited (Prasad *et al.*, 1974; Sahu and Prasad, 1975); d) a relatively high phosphodiesterase activity in neuroblastoma cells is associated with low activity of adenylate cyclase and a low level of cyclic AMP (Prasad *et al.*, 1975d). Although the basal level of cyclic AMP phosphodiesterase activity in 'differentiated' mouse neuroblastoma cells increases (Prasad, 1973; Kumar *et al.*, 1975), the intracellular level of cyclic AMP also increases (Prasad and Kumar, 1974). The association of a high level of cyclic AMP with the high level of cyclic AMP phosphodiesterase activity may be due to an increased level of cyclic AMP binding proteins (Prasad *et al.*, 1975e). The protein-bound cyclic AMP is relatively less susceptible to enzymatic hydrolysis.

Our evidence does not suggest that a defect in the adenylate cyclase regulation is one of the early lesions in the malignancy of nerve cells. For example, the adenylate cyclase activity in neuroblastoma homogenates can be stimulated by dopamine, norepinephrine, PGE_1 and acetylcholine (Prasad, 1975); however, only PGE_1 increases the intracellular level of cyclic AMP without the inhibition of AMP phosphodiesterase activity (Prasad, 1973). In one clone of neuroblastoma cells, even PGE_1 does not increase the cyclic AMP level until the phosphodiesterase activity is inhibited (Sahu and Prasad, 1975). Thus the rate limiting factor in the accumulation of cyclic AMP after treatment of cells with neurotransmitters is the high activity of phosphodiesterase.

The present model assumes that the malignant transformation occurs in either a single dividing nerve cell or in a group of dividing nerve cells. There is some indirect evidence to support this. For example, these tumors generally occur in children under five years of age. In addition, the neuroblastomas contain more than one type of nerve cell, indicating that the malignant transformation occurs in nerve cells which are not yet committed to form any one neuronal cell type. There is no evidence that the mature neurons ever undergo malignant transformation, although the glial elements of adult nervous tissue do so.

VII. Cyclic AMP and Regulation of Differentiated Functions in Normal Embryonic Nerve Cells

Cultures of dorsal root ganglia from chick embryos (Roisen *et al.*, 1972), fetal rat brain (Shapiro, 1973) or mouse sensory ganglia (Hass *et al.*, 1972) show neurite formation after treatment with dibutyryl cyclic AMP. In addition, dibutyryl cyclic AMP increases tyrosine hydroxylase activity in cultures of sympathetic ganglia (Mackay and Iversen, 1972) and dopamine-β-hydroxylase activity in isolated superior cervical ganglia (Keen and McLean, 1972). The adenylate cyclase in the brain undergoes an age-dependent increase in its sensitivity to norepinephrine (Schmidt and Robison, 1970). The activities of adenylate cyclase and cyclic AMP phosphodiesterase increase during the differentiation of mammalian brain (Weiss *et al.*, 1971).

Many effects of nerve growth factor (NGF) in chick embryo dorsal root ganglia are similar to those produced by dibutyryl cyclic AMP. Therefore, it has been suggested (Roisen *et al.*, 1972; Roisen and Murphy, 1973) that NGF-effect is mediated by cyclic AMP. This has been disputed by another group of investigators (Hier *et al.*, 1972; Hier *et al.*, 1973). A recent study Nikodijevic *et al.*, 1975) has shown that NGF (2.5s) produces a several fold increase in cyclic AMP concentration in rat superior cervical ganglia in organ culture within 5 min, and a return to the basal level in about 10 min. Thus, the effect of NGF may be mediated by cyclic AMP; however, one has to demonstrate an increase in cyclic AMP-dependent phosphorylation activity before the involvement of cyclic AMP in the mechanism of NGF-effect is considered. However, the possibility that all effects of cyclic AMP may not be mediated via cyclia AMP-dependent phosphorylation should remain open.

VIII. Cyclic AMP and Induction of Neural Tissue

Many substances have been shown to induce nervous tissue in ectoblasts (Brachet, 1974). These include organizer (dorsal lip), dead tissues, ribonuceloproteins, ions, changes in pH, etc. There is a difference in the organization of nervous tissue induced by living organizers and dead tissues. The living organizer stimulates the normal morphogenesis of nervous system which exhibits regional characteristics. However, the dead tissue induces nervous tissues which are abnormal and complex. Much evidence suggests that the primary inductors under physiological condition may be protein (Tiedeman, 1967). The role of cyclic AMP in the induction mechanism became a possibility when it was discovered that an elevation of cyclic AMP in neuroblastoma cells induces as well as increases the expression of many differentiated functions which are characteristic of mature neurons (Prasad and Hsie, 1971; Prasad, 1975). Indeed, it has been shown (Wahn *et al.*, 1975) that dibutyryl cyclic AMP, 8-bromo cyclic AMP, and cyclic AMP with theophylline can induce undetermined cells to differentiate into derivatives of neural ectoderm and neural crest (neurons, melanophore and glial cells). Theophylline alone induced limited neural differentiation only in case of *P. waltilii;* cyclic AMP alone gave variable results usually with limited neural differentiation. This is not unexpected, since theophylline has failed to increase cyclic AMP in intact cells (Prasad, 1973), although it inhibits the cyclic nucleotide phosphodiesterase activity *in vitro*. Cyclic AMP is known to penetrate poorly into intact cells. It has been reported (Wahn *et al.*, 1975) that two different analogs of cyclic AMP produces quantitatively different results. For example, in *S. mexicanum*, 8-bromo cyclic AMP causes the formation of a large number of astrocytes with few simple motor neurons and practically no central nervous system neurons, whereas dibutyryl cyclic AMP causes the formation of both types of neurons as well as astrocytes.

Dibutyryl cyclic AMP induces neural differentiation in small explants of presumptive epidermis, but fails to do so in large explants or in whole embryos (Wahn *et al.*, 1976). Cellular movements are known to occur in the presumptive neural ectoderm *in vivo;* in small explant cultures, this is achieved by the migration of cells on the surface of the culture dish, whereas such movements cannot occur in larger explants. To test the correlation between differentiation and cellular migration, a simple experiment was performed (Wahn *et al.*, 1976). Large explants of presumptive epidermis were first incubated in the presence of dibutyryl cyclic AMP for 12 or 24 hours, and were then washed. The treated large explants were then dissociated into small pieces and were cultured in the absence of dibutyryl cyclic AMP. It was found that the cells migrated out from the explant and underwent neural differentiation. These results show that the neural induction had taken place during the period of incubation of explants with dibutyryl cyclic AMP. Once the induction had taken place, external cyclic AMP is no longer required for the cytodifferentiation of the induced cells (Wahn *et al.*, 1976). It was also observed that cells from small explants that were not treated with dibutyryl cyclic AMP do also migrate away from each other. However, no neural differentiation occurs. Thus, the cell migration *per se* is not sufficient for neural differentiation. It appears that a combination of induction by cyclic AMP followed by cell migration in culture is essential for neural differentiation. It may also explain why Yamada (1961) has been able to achieve much more efficient induction of neural differentiation by keeping the large explants flat with a nylon mesh, creating a different typology for cell migration as compared to a ball of cells.

It has been admitted (Wahn *et al.*, 1975) that these results do not *per se* prove that cyclic AMP is the normal agent mediating the effect of primary inductors. Before the involvement of cyclic AMP in the mechanism of induction is further considered, at least one stimulator of adenylate cyclase must be shown to induce neural induction; as seen in neural tissue after normal induction, there must occur an increase in the intracellular level of cyclic AMP and in cyclic AMP-dependent phosphorylation activity.

IX. Effect of Ions

It has been reported (Barth and Barth, 1959) that transient treatment of monolayer cell cultures derived from presumptive epidermis of *Rana pipiens* with high concentrations of cations resulted in the sequential appearance of cells characteristic of neural fold origin. It has been shown (Barth and Barth, 1972) that the cations Li^+ or Ca^{2+} induce differentiation of neurons from presumptive epidermis. These neurons can be further induced to differentiate into pigment cells by later treatment with the same ions. Three types of neurons or pigment cells are produced by sequential inductions. The ectodermal cells competent for these inductions normally occur in the embryo. The result of treatment of cells with ionic inducers changes similarly with time, whether the cells are left in the embryo or cultured *in vitro*. Ectodermal cells cultured *in vitro* must be treated with an ion which can trigger them to become neurons before they are competent to be induced to become pigment cells, although the same ion inducer and lithium affect both inductions. Therefore, it has been suggested (McMahon, 1974) that changes in ionic flux normally accompany the induction of nervous tissue *in vivo*. These changes do not occur in an interspecies hybrid which does not gastrulate normally and in which, therefore, the ectoderm is not subjected to the natural inducer. The neural plate and its derivative cells are not induced, and the normal changes in permeability to ions do not occur. Cells from these abnormal gastrulae can be induced to form neurons and pigment cells by treating them with the same ions which induce neural differentiation in the cells of a normal embryo. This shows that the hybrid cells contain the information to make neurons and pigment cells and can express this if triggered. These experiments further suggest that the ionic changes which accompany the induction *in vivo* result from the process of induction and are sufficient for the induction of neurons and pigment cells.

A. Relationships of Cations with Cyclic Nucleotide

Lithium is known to inhibit adenylate cyclase activity of a variety of organs including brain *in vitro* (McMahon, 1974; Forn and Valdecasas, 1971; Smith *et al.*, 1972), but it is not known if it affects the intracellular level of cyclic AMP. Lithium ions may also inhibit the effect of cyclic AMP. A small amount of bound calcium is required for the activity of adenylate cyclase from brain and at least for one protein kinase. However, an increase of the calcium concentration to $>10^{-4}$M inhibits adenylate cyclase and protein kinase from brain and other tissues, and even causes cyclic AMP to inhibit cyclic AMP-dependent protein kinase. Cyclic AMP phosphodiesterase activity is also inhibited by calcium ($>10^{-4}$M), particularly in the absence of magnesium although smaller amounts of calcium ($<10^{-4}$M) may stimulate it. Calcium also inhibits the activity of phosphoprotein phosphatase (McMahon, 1974). Magnesium and manganese ions are required for the activity of adenylate cyclase, phosphodiesterase and protein kinases. Magnesium also affects the modulation of protein kinase activity by the cyclic AMP-dependent protein kinase modulator, but it is not required for activity of at least one phosphoprotein phosphatase (McMahon, 1974). Manganese stimulates the activity of brain protein phosphatase. Zinc can inhibit adenylate cyclase, phosphodiesterase, and protein phosphatase, but it neither increases the activity of protein kinases nor appears to inhibit them (McMahon, 1974). Sodium and potassium ions have slight effect on adenylate cyclase activity. A high concentration of either sodium or potassium inhibits the ability of protein kinases to phosphorylate their substrates (McMahon, 1974). For example, 0.1 M NaCl increases the Michaelis constant for casein substrate by six times; however, a similar concentration may stimulate phosphorylation of histone substrate. A high concentration of potassium also increases the cyclic AMP level in brain cortex (Schimizu *et al.*, 1972). Thus the relationship between cations and cyclic nucleotide is complex. These studies have been performed at abnormally high ion concentrations, and may not therefore, have any relevance to physiological conditions. It is certain that an elevation of cyclic AMP causes changes in the levels of intracellular ion concentrations. Hence one of the consequences of changes in the level of cyclic nucleotide would be to change the ionic environment which in turn may alter the expression of a variety of genes.

X. Therapeutic Relevance of Neuroblastoma Study

We have recently observed (Prasad *et al.*, 1977) that sodium ascorbate (vitamin C) potentiates the effect of several therapeutic agents such as x-irradiation (30-fold) vinblastine sulfate (17-fold), cyclophosphamide (18-fold), adriamycin (60-fold), R020-1724 (75-fold), and sodium butyrate (3-fold) on mouse neuroblastoma cells in culture; however, it produces no such effects on mouse fibroblasts (L-cells) and rat glioma cells.

Based on the current data on neuroblastoma cells in culture, the previous suggestion for the therapy (Prasad *et al.*, 1974) of neuroblastoma tumor should be modified as follows: (1) Sodium ascorbate and sodium butyrate are given daily for the entire period of treatment. Sodium ascorbate is a relatively nontoxic compound. The oral administration of sodium butyrate (6-10 g/day) for four months produced no detectable toxic effects in a child with neuroblastoma (L. Furman, personal communication); (2) After 5 days of this treatment cyclophosphamide, vincristine, adriamycin, and 5-fluorouracil are administered in a sequence, and with dosages and time intervals which are currently used. It is hoped that after the completion of the second phase, most of the tumor cells would be killed; (3) After the second phase, differentiating agents such as cyclic AMP stimulating agents (inhibitors of phosphodiesterase like papaverine and R020-1724, and stimulator of adenylate cyclase like prostaglandin E_1) are continuously infused for at least 4 days. The administration of PGE_1 in the presence of an inhibitor of cyclic AMP phosphodiesterase is suggested because the neuroblastoma tumor contains cells whose sensitivity to PGE_1 and phosphodiesterase inhibitor markedly vary. However, the combination of PGE_1 and phosphodiesterase inhibitor increase the intracellular level of cyclic AMP in all clones (Sahu and Prasad 1975). It is hoped that the administration of differentiating agents would produce differentiation in many of the remaining cells. The differentiated cells may evoke an immune response which would kill many of the drug-resistant tumor cells; (4) After the administration of differentiating agents, nonspecific immune stimulants such as BCG should be administered. It is hoped that a further stimulation of the host's immune system would reject the residual, drug resistant tumor cells. The phases 1, 3 and 4 should be repeated at least twice after the clinical remission has been achieved. Because of the carcinogenic effect of ionizing radiation, the use of radiation for the treatment of neuroblastoma is not recommended until an extreme emergency condition exists. The proposed concept is being tested in other neural crest tumor cell lines such as melanoma, and Wilm's tumors.

Dr. L. Helson of Memorial Hospital, New York has been using a combination of cytotoxic drugs such as cyclophosphamide and vincristine in combination with differentiating agents such as papaverine and tri-fluoromethyl-2-deoxyuridine in metastatic neuroblastoma patients who were unresponsive to previous treatments or who were untreated. Although a marked regression of tumor was observed in all patients, and the conversion from neuroblastoma to ganglioneuroma was observed in cases where biopsies were taken and examined (Helson, 1975), the response of the untreated patients appears to be the best. Dr. Helson has indicated (personal communication) that 17 out of 19 untreated patients remain free of disease between 12-24 months. Thus, a pharmacological treatment which largely has been developed on the basis of tissue culture experiments is proving to be of considerable clinical value in the treatment of this form of cancer.

Conclusions

The following conclusions can be made: 1) Cyclic AMP appears to be one of the important factors in induction as well as in regulation of several differentiating functions in mammalian nerve cells. 2) The increased level of proteins that bind cyclic AMP provides the important intracellular mechanisms of maintaining a high level of cyclic AMP during differentiation of neuroblastoma cells. 3) The reduction in histone synthesis and in H_1-histone phosphorylation may be an important biological signal for the dividing neuroblasts to turn off cell division. If these events do not occur, the neuroblasts would continue to divide, and the differentiated functions would not be able to express themselves at maximal levels.

References

Amano, A., E. Richelson and M. Nirenberg. 1972. Neurotransmitter synthesis by neuroblastoma clones. Proc. Nat. Acad. Sci. U.S.A. 69:258-263.

Augusti-Tocco, G., G. Sato, P. Calude and D. Potter. 1970. Clonal cell lines of neurons in: *Control mechanisms in the Expression of Cellular Phenotypes*. H. A. Padykula, ed. Academic Press, New York. pp. 109-120.

Barth, L. G. and L. J. Barth. 1959. Differentiation of cells of the *Rana pipiens* gastrula in unconditioned medium. J. Embryol. Exptl. Morphol. 7:210-222.

Barth, L. G. and L. J. Barth. 1972. Sodium and calcium uptake during embryonic induction in *Rana pipiens*. Dev. Biol. 28:18-34.

Biedler, J. L., L. Helson, and B. A. Spengler. 1973. Morphology and growth, tumorigenicity, and cytogenetics of human neuroblastoma cells in continuous culture. Cancer Res. 33:2643-2652.

Blume, A., C. Dalton, and H. Sheppard. 1973. Adenosine and mediated elevation of cyclic 3′, 5′-adenosine monophosphate concentrations in cultured mouse neuroblastoma cells. Proc. Natl. Acad. Sci. U.S.A. 70:3099-3102.

Bondy, S. C., K. N. Prasad, and J. L. Purdy. 1974. Neuroblastoma: Drug-induced differentiation increases proportion of cytoplasmic RNA that contains polyadenylic acid. Science 186:359-361.

Brachet, J. 1974. *Introduction to Molecular Embryology*. Springer-Verlag, New York.

Burdman, J. A. 1972. The relationship between DNA synthesis and the synthesis of nuclear proteins in rat brain during development. J. Neurochem. 19:1459-1469.

Ehrlich, Y. H., E. G. Brunngraber, P. K. Sinha, and K. N. Prasad. 1977. Specific alterations in phosphorylation of endogenous proteins of differentiated neuroblastoma cells in culture. Nature (Lond.) 265:238-241.

Elgin, S. C. R., J. B. Boyd, L. E. Hood, W. Wray, and F. C. Wu. 1973. A prologue to the study of the nonhistone chromosomal proteins. Cold Spring Harbor Symp. Quant. Biol. 38:821-833.

Forn, J. and F. G. Valdecasas. 1971. Effect of lithium on brain adenyl cyclase activity. Biochem. Pharmacol. 20:2773-2779.

Fujitani, H. and V. Holoubek. 1974. Nonhistone nuclear proteins of rat brain. J. Neurochem. 23:1215-1224.

Gill, G. N. and L. D. Garren. 1970. A cyclic-3′, 5′-adenosine monophosphate dependent protein kinase from the adrenal cortex: Comparison with a cyclic AMP binding protein. Biochem. Biophys. Res. Commun. 39:335-343.

Gilman, A. and M. W. Nirenberg. 1971. Regulation of adenosine 3′, 5′-cyclic monophosphate metabolism in cultured neuroblastoma cells. Nature (Lond.) 234:356-357.

Hamprecht, B. and J. Schultz. 1973. Stimulation by prostaglandin E_1 of adenosine 3′, 5′-cyclic monophosphate formation in neuroblastoma cells in the presence of phosphodiesterase inhibitors. FEBS Letters 34:85-89.

Hass, D. C., D. B. Hier, B. G. W. Aranson and M. Young. 1972. On a possible relationship of cyclic AMP to the mechanism of action of nerve growth factor. Proc. Soc. Exp. Biol. Med. 140:45-47.

Helson, L. and J. L. Biedler. 1973. Catecholamines in neuroblastoma cells from human bone marrow, tissue culture and murine C-1300 tumor. Cancer 31:1087-1091.

Helson, L. 1975. Management of dissemination neuroblastoma. CA 25:264-268.

Herschman, J. R. and M. P. Lerner. 1973. Production of a nervous system specific protein (14-3-2) by human neuroblastoma cells in culture. Nature (Lond.) New Biol. 241:242-244.

Hier, D. B., B. G. W. Aranson, and M. Young. 1972. Studies on the mechanism of action of nerve growth factor. Proc. Natl. Acad. Sci. U.S.A. 69:2268-2272.

Hier, D. B., B. G. W. Aranson, and M. Young. 1973. Nerve growth factor: Relationship to the cyclic AMP system of sensory ganglia. Science 182:78-81.

Kates, J. R., R. Winterton, and K. Schlesinger. 1971. Induction of acetylcholinesterase activity in mouse neuroblastoma tissue culture cells. Nature (Lond.) 224:345-346.

Keen, P. and W. G. McLean. 1972. Effect of dibutyryl cyclic AMP on levels of dopamine β-hydroxylase in isolated superior cervical ganglia. Arch. Pharmacol. 27:465-469.

Kernell, A. M., L. Bolound and N. R. Ringertz. 1971. Chromatin changes during erythropoiesis. Exp. Cell Res. 65:1-6.

Knapp, S. and A. J. Mandel. 1974. Serotonin biosynthetic capacity of mouse C-1300 neuroblastoma cells in culture. Brain Res. 66:547-551.

Kumar, S., G. Becker and K. N. Prasad. 1975. Cyclic AMP phosphodiesterase activity in malignant and cyclic AMP-induced 'differentiated' neuroblastoma cells. Cancer Res. 35:82-87.

Lazo, J. S., K. N. Prasad and R. W. Ruddon. 1976. Synthesis and phosphorylation of chromatin associated proteins in cAMP-induced 'differentiated' neuroblastoma cells in culture. Exp. Cell Res. 100:41-46.

MacKay, A. V. P. and L. L. Iversen. 1972. Increased tyrosine hydroxylase activity of sympathetic ganglia cultured in the presence of dibutyryl cyclic AMP. Brain Res. 48:424-426.

McMahon, D. 1974. Chemical messenger in development: A hypothesis. Science 185:1012-1021.

Miller, R. A. and F. H. Ruddel. 1974. Enucleated neuroblastoma cells form neurites when treated with dibutyryl cyclic AMP. Exp. Cell Res. 63:295-299.

Nelson, P., W. Ruffner and M. Nirenberg. 1969. Neuronal tumor cell with excitable membrane grown *in vitro*. Proc. Natl. Acad. Sci. U.S.A. 64:1004-1010.

Nikodijevic B., O. Nikodijevic, M. W. Yu, H. Pollard, and G. Guroff. 1975. The effect of nerve growth factor on cyclic AMP levels in superior cervical ganglia of the rat. Proc. Natl. Acad. Sci. U.S.A. 72:4769-4771.

Prasad, K. N. 1972. Cyclic AMP-induced differentiated mouse neuroblastoma cells lose tumorgenic characteristics. Cytobios 6:163-166.

Prasad, K. N. 1973. Role of cyclic AMP in the differentition of neuroblastoma cell culture in: *The Role of Cyclic Nucleotides in Carcinogenesis,* J. Schultz and H. G. Gratzner, eds. Academic Press, New York. pp. 207-237.

Prasad, K. N. 1974. Abnormal regulation of cyclic AMP phosphodiesterase: A hypothesis for the development of cancer of nerve cells. Differentiation 2:367-369.

Prasad, K. N. 1975. Differentiation of neuroblastoma cells in culture. Bio. Rev. 50:129-165.

Prasad, K. N. 1977. Differentiation and growth of neuroblastoma cells and serum types. Trans. Am. Soc. Neurochem.

Prasad, K. N. and K. N. Gilmer. 1974. Demonstration of dopamine-sensitive adenylate cyclase in malignant neuroblastoma cells and change in sensitivity of adenylate cyclase to catecholamines in "differentiated" cells. Proc. Natl. Acad. Sci. U.S.A. 71:2525-2529.

Prasad, K. N. and A. W. Hsie. 1971. Morphological differentiation of mouse neuroblastoma cells induced in vitro by dibutyryl adenosine 3′:5′-cyclic monophosphate. Nature New Biol. (Lond.) 233:141-142.

Prasad, K. N. and S. Kumar. 1974. Cyclic AMP and the differentiation of neuroblastoma cells. In: *Control of Proliferation of Animal Cells*. B. Clarkson and R. Baserga, eds. Cold Spring Laboratory, Cold Spring Harbor. pp. 581-594.

Prasad, K. N. and S. Kumar. 1975. Role of cyclic AMP in differentiation of human neuroblastoma cells in culture. Cancer 36:1338-1343.

Prasad, K. N., J. C. Waymire and N. Weiner. 1972. A further study on the morphology and biochemistry of X-ray and dibutyryl cyclic AMP-induced differentiated neuroblastoma cells in culture. Exp. Cell Rev. 74:110-114.

Prasad, K. N., B. Mandal and S. Kumar. 1973a. Human neuroblastoma cell culture: Effect of 5-bromodeoxyuridine on morphological differentiation and levels of neural enzymes. Proc. Soc. Exp. Biol. Med. 144:38-42.

Prasad, K. N., B. Mandal J. C. Waymire, G. J. Less, A. Vernadakis and N. Weiner. 1973b. Basal levels of neurotransmitter synthesizing enzymes and effect of cyclic AMP agents on the morphological differentiation of isolated neuroblastoma clones. Nature (Lond.) New Biol. 241:117-119.

Prasad, K. N., S. K. Sahu and S. Kumar. 1974. Relationship between cyclic AMP and differentiation of neuroblastoma cells in culture. In: *Differentiation and Control of Malignancy of Tumor Cells*, W. Nakahara, T. Ono, T. Sugimura, and H. Sugano. eds. University of Tokyo Press, Tokyo. pp. 287-309.

Prasad, K. N., G. Becker, and K. Tripathy. 1975a. Differences and similarities between guanosine 3', 5'-cyclic monophosphate phosphodiesterase and adenosine 3', 5'-cyclic monophosphate phosphodiesterase activities in neuroblastoma cells in culture. Proc. Soc. Exp. Biol. Med. 149:757-762.

Prasad, K. N., S. C. Bondy, and J. L. Purdy. 1975b. Polyadenylic acid containing cytoplasmic RNA increases in adenosine 3' :5'-cyclic monophosphate-induced 'differentiated' neuroblastoma cells in culture. Exp. Cell Res. 94:88-94.

Prasad, K. N., K. N. Gilmer, S. K. Sahu and G. Becker. 1975c. Regulation of adenylate cyclase activity in malignant and cyclic AMP-induced differentiated neuroblastoma cells: Effect of neurotransmitters, guanosine triphosphate and divalent ions. Cancer Res. 35:77-88.

Prasad, K. N., S. Kumar, G. Becker and S. K. Sahu. 1975d. Role of cyclic nulceotides in differentiation of neuroblastoma cells in culture. In: *Cyclic Nucleotides in Diseases*. B. Weiss, ed. University Park Press, Baltimore. pp. 45-66.

Prasad, K. N., P. K. Sinha, S. K. Sahu and J. L. Brown. 1975e. Binding of cyclic nucleotides with soluble proteins increases in 'differentiated' neuroblastoma cells in culture. Biochem. Biophys. Res. Commun. 66:131-138.

Prasad, K. N., D. Fogleman, M. Gaschler, P. K. Sinha and J. L. Brown. 1976a. Cyclic nucleotide-dependent protein kinase activity in malignant and cyclic AMP-induced differentiated neuroblastoma cells in culture. Biochem. Biophys. Res. Commun. 68:1248-1255.

Prasad, K. N., S. K. Sahu and P. K. Sinha. 1976b. Cyclic nucleotides in the regulation of expression of differentiated functions in neuroblastoma cells. J. Natl. Cancer Inst. 57:619-631.

Prasad, K. N., P. K. Sinha, S. K. Sahu and J. L. Brown. 1976c. Binding of cyclic nucleotides with proteins in malignant and cyclic AMP-induced 'differentiated' neuroblastoma cells in culture. Cancer Res. 36:2290-2296.

Reimann, E. M., C. O. Brostrom, J. D. Corbin, C. A. King and E. G. Krebs. 1971. Separation of regulatory and catalytic subunits of the the cyclic 3' :5'-adenosine monophosphate-dependent protein kinase(s) of rabbit skeletal muscle. Biochem. Biophys. Res. Commun. 42:187-194.

Richelson, E. 1973. Stimulation of tyrosine hydroxylase activity in an adrenergic clone of mouse neuroblastoma by dibutyryl cyclic AMP. Nature (Lond.) New Biol. 242:175-177.

Roisen, F. J., R. A. Murphy, M. E. Pichichero and W. G. Braden. 1972. Cyclic adenosine monophosphate stimulation of axonal elongation. Science 175:73-74.

Roisen, F. J. and R. A. Murphy. 1973. Neurite development in vitro. II. The role of microfilaments and microtubules in dibutyryl adenosine 3' :5'-cyclic monophosphate and nerve growth factor stimulation maturation. J. Neurobiol. 4:397-412.

Ross, J. R., J. B. Olmsted and J. L. Rosenbaum. 1975. The ultrastructure of mouse neuroblastoma cells in tissue culture. Tissue and Cell 1:106-136.

Sahu, S. K. and K. N. Prasad. 1975. Effect of neurotransmitters and prostaglandin E_1 on cyclic AMP levels in various clones of neuroblastoma cells in culture. J. Neurochem. 24:1267-1269.

Schimizu, P. N., C. R. Creveling and J. W. Daly. 1972. Effect of membrane depolarization and biogenic amines on formation of cyclic AMP in incubated brain slices. In: *Role of Cyclic AMP in Cell Function.* P. Greengard and E. Costa, eds. Raven Press, New York. pp. 135-154.

Schmidt, M. J. and G. A. Robison. 1970. Cyclic AMP adenyl cyclase and effect on norepinephrine in the developing rat brain. Fed. Proc. 29:479a.

Schubert, D. and F. Jacob. 1970. 5-bromodeoxyuridine-induced differentiation of a neuroblastoma. Proc. Natl. Acad. Sci. U.S.A. 67:247-254.

Schubert, D., S. Humphreys, F. Vitry and F. Jacob. 1971. Induced differentiation of a neuroblastoma. Dev. Biol. 52:514-546.

Schubert, D., S. Heinemann, W. Carlisle, H. Tarikas, B. Kimes, J. Patrick, J. H. Steinbach, W. Culp and B. L. Brandt. 1974. Clonal cell lines from rat central nervous system. Nature (Lond.) 249:224-227.

Seeds, N. W., A. G. Gilman, T. Amano and M. W. Nirenberg. 1970. Regulation of axon formation by clonal lines of neural tumor. Proc. Natl. Acad. Sci. U.S.A. 66:160-167.

Shapiro, D. L. 1973. Morphological and biochemical alteration in fetal rat brain cells cultured in the presence of monobutyryl cyclic AMP. Nature (Lond.) 241:203-204.

Sinha, P. K. and K. N. Prasad, 1977. A further study on the regulation of cyclic nucleotide phosphodiesterase activity in neuroblastoma cells. Effect of growth. In Vitro (in press)

Smith, B. M., C. A. Harris and P. W. Major. 1972. The effect of lithium ions on the activation of ovarian adenyl cyclase. In: *Advances in Cyclic Nucleotide Research.* P. Greengard, G. A. Robison and R. Paoletti, eds. Raven Press, New York. Vol. I, p. 588.

Tao, M., M. L. Salas and F. Lipmann. 1970. Mechanism of activation by adenosine 3′:5′-cyclic monophosphate of a protein phosphokinase from rabbit reticulocytes. Proc. Natl. Acad. Sci. U.S.A. 67:408-414.

Tiedeman, H. 1967. Biochemical aspects of primary induction and determination. In: *Biochemistry of Animal Development*, R. Weber, ed. Academic Press, New York. Vol. 2, pp. 4-55.

Wahn, H. L., L. E. Lightbody and T. T. Techen. 1975. Induction of differentiation in cultures of amphibian undetermined presumptive epidermis by cyclic AMP derivatives. Science 188:366-368.

Wahn, H. L., L. T. Lightbody, T. T. Techen and J. D. Taylor. 1976. Adenosine 3′:5′-monophosphate, morphogenetic movements and embryonic neural differentiation in *Pleurodeles waltilii.* J. Exp. Zool. 196:125-130.

Watson, W. E. 1974. The binding of actinomycin D to the nuclei of exotomized neurons. Brain Res. 65:317-322.

Waymire, J. C., N. Weiner and K. N. Prasad. 1972. Regulation of tyrosine hydroxylase activity in cultured mouse neuroblastoma cells. Elevations induced by analogs of adenosine 3′:5′-cyclic monophosphate. Proc. Natl. Acad. Sci. U.S.A. 69:2241-2245.

Weiss, B., H. M. Shein and R. Snyder. 1971. Adenylate cyclase and phosphodiesterase activity of normal and SV_{40} virus transformed hamster astrocytes in cell culture. Life Sci. 10:1253-1260.

Yamada, T. 1961. A chemical approach to the problem of the organizer. In: *Advances in Morphogenesis.* M. Abercrombie and J. Brachet, eds. Academic Press, New York. Vol. 1, pp. 1-54.

HORMONAL EFFECTS ON DIFFERENTIATION IN NEURAL CULTURES

J. de VELLIS, J. F. McGINNIS, G. A. M. BREEN, P. LEVEILLE, K. BENNETT AND K. McCARTHY

Laboratory of Nuclear Medicine and Radiation Biology
Department of Anatomy and the Mental Retardation Research Center
University of California
Los Angeles, California

I. Introduction

The importance of hormones in the development and differentiation of the nervous system was suspected a long time ago when it was demonstrated that hypothyroidism is associated with mental retardation. Later it was shown that thyroid hormones are essential for normal arborization of dendrites and synaptogenesis (for review and references, see de Vellis and Clemente, 1970; Balasz, 1974). Like thyroid hormones, glucocorticoids appear to affect most regions of the nervous system. They regulate the level of glycerol phosphate dehydrogenase activity (de Vellis and Inglish, 1968) and affect cellular proliferation (for review see Balasz, 1974). The role of steroid hormones in the sexual differentiation of the brain implies that they have an organizing effect on the brain. The presence of receptors for sex steroids in certain brain regions and other evidence suggests that these hormones selectively affect gene expression in those tissues (for review, see McEwen *et al.*, 1973; Balasz, 1974). Thus the nervous system can be considered as made up of several hormone target organs, each of which can be responsive to the same or different hormones. Hence the effect of certain hormones can only be observed in a detailed regional study of the nervous system.

The study of hormonal effects in cell cultures of the nervous system started about ten years ago. Hamburgh (1966) reported that thyroid hormones increase the formation of myelin in brain explant cultures. Vernadakis (1971) reported a growth effect of steroids on glial cells in primary neural cultures. More recently, there has been a growing interest in hypothalamic cultures as a model system for studying the effect of sex steroid hormones on neural differentiation. Testosterone and estradiol have now been shown to increase markedly neurite outgrowth of hypothalamic explant cultures (Toran-Allerand, 1976), suggesting that differential axonal growth may be important in the sexual differentiation of the brain. Canick *et al.* (1977) reported that dissociated hypothalamic cultures retain the *in vivo* differentiated function of aromatization of androgens to estrogens. Thus, the effect of testosterone on hypothalamic cultures appears to be due to its conversion to estradiol, as *in vivo*. This type of observation underlies the importance of considering the numerous factors influencing the expression of differentiated functions in cell culture. Most hormones lose biological activity when they are metabolized, whereas some have to be converted from an inactive to an active form (cf. I. S. Mendelson and R. Walter, this volume). Serum, used in culture, contains enzymes which may degrade some hormones even to the extent of preventing an effect. However, serum also may contain hormones at a sufficiently high concentration to produce an effect in the absence of exogenously added hormones. Such a possibility is even more disturbing in light of the fact that hormone levels in sera have been found to vary over a 23-fold range (Esber *et al.*, 1973). With sera from which thyroid hormones had been removed by charcoal absorption, a decrease in the number of synapses formed in cerebral reaggregate cultures was demonstrated. Supplementation of the media with thyroid hormones corrected this deficiency (Lu *et al.*, 1977a, b). These results confirm the *in vivo* observation that thyroid hormones are required for the formation of the normal number of synapses and validate the use of reaggregate cell culture as a model to study the regulation of synaptogenesis.

If cell differentiation is defined as the process by which a cell can modify its protein composition (for review see de Vellis, 1970), then enzyme induction by hormones is an appropriate model and endpoint to study the mechanisms by which hormones regulate neural cell differentiation. We chose to discuss the hydrocortisone induction of glycerol phosphate dehydrogenase because it is presently the only well-characterized hormonal effect in neural cell cultures and it has been demonstrated *in vivo* to be specific to the nervous system (de Vellis and Inglish, 1968).

II. Hydrocortisone Regulation of Glycerol Phosphate Dehydrogenase

A. General Considerations

In the brain of adult rats, the level of glycerol phosphate dehydrogenase (EC 1.1.1.8; GPDH) is regulated by the adrenal and pituitary glands. Following hypophysectomy or adrenalectomy, the specific activity of GPDH decreases by 60%. Injection of adrenocorticotrophic hormone or hydrocortisone into hypophysectomized rats, or hydrocortisone into adrenalectomized rats restores the enzyme to its normal adult level (de Vellis and Inglish, 1968). Since GPDH induction by hydrocortisone among the several tissues examined thus far, has only been found in CNS and PNS tissues, it has been considered a specific biochemical marker of nervous tissue (de Vellis and Kukes, 1973). The development of GPDH coincides with myelination, rising rapidly between 10 and 30 days in the rat brain (de Vellis *et al.*, 1967). The development of the enzyme in the postnatal rat brain is also regulated by hydrocortisone (de Vellis and Inglish, 1968, 1973). Injection of hydrocortisone to rat pups in the second postnatal week, when the pituitary-adrenal axis is essentially inoperative, results in a precocious appearance of GPDH (de Vellis and Inglish, 1973). Thus the expression and the maintenance of brain GPDH at its maximum level requires the presence of glucocorticoids. The lack of time restriction of the hydrocortisone action on GPDH differs from the effect of some other hormones on the CNS, such as thyroid hormones, which operate during a narrow window of development (see de Vellis and Clemente, 1970).

The finding of GPDH inducibility has raised questions that can be grouped into three areas: (a) the biochemical characterization of hydrocortisone action on nervous tissue, (b) its cellular localization, (c) its physiologic significance. We probed the first two points and realized that the mode of action of hydrocortisone would be best studied in cell culture systems because of the problems inherent in the *in vivo* situation, such as the difficulty in delivery and removal of compounds to brain cells and the heterogeneity of cell types. In 1968, listening to a seminar by Dr. Gordon Sato on differentiated clonal cell lines led us to test successfully the C6 rat glioma cell line (generously provided by Dr. Sato) for GPDH inducibility, (de Vellis and Inglish, 1969). More recently, GPDH induction has been found in several other rat clonal cell lines derived from chemically-induced tumors (West *et al.*, 1977; Claisse and Roscoe, 1976) and in mouse cell lines established by virus transformation of embryonic dissociated brain cultures (Benda *et al.*, 1977). Although clonal cells represent the best system to dissect the mechanisms of hormonal action and allow meaningful correlations of biochemical parameters to be made, a doubt remains that some biochemical effects of the hormone are perhaps altered by their tumorigenic state. For instance, certain cell-to-cell contacts and interactions may be lacking; certainly, heterotypic cell contacts are missing. Thus it is desirable to compare the characteristics of the induction in clonal cell lines with that observed in primary cultures. Toward this end, we have tested for the induction of GPDH in three types of primary culture: explant, dissociated and reaggregated. Each of these systems offers distinct advantages to answer neurobiological questions (Herschman, 1973; de Vellis *et al.*, 1975). Explants largely retain the original tridimensional tissue organization. Dissociated cell cultures lose the tissue organization and form a new, largely two-dimensional network while reaggregates recreate a histotypic organization. The importance of tissue organization in enzyme induction has been illustrated by the work of Morris and Moscona (1971) on the hydrocortisone induction of glutamine synthetase in the retina. Glutamine synthetase is inducible in organ and aggregate cultures but not in dissociated retinal cell cultures. Another advantage of primary cultures over clonal lines is the feasibility of performing regional studies of the nervous system. Indeed, it is hard to know specifically the regional cellular origin of tumors chemically induced *in vivo*. The proven alternative is the *in vitro* transformation of a primary culture of the desired nervous region (Shein, 1968; de Vitry *et al.*, 1974).

Bearing in mind the advantages and limitations of each of the four types of cultures, we selected one or more systems to answer each of the following questions:

1. Is an activator or inhibitor of the enzyme involved?
2. Are DNA, RNA and/or protein synthesis required?
3. Are different isoenzymes involved in the basal and induced states of the enzyme?
4. Does the induction represent an increase in the number of enzyme molecules or an increase in their catalytic efficiency?
5. Is the induction the result of an increase in the rate of synthesis or a decrease in the rate of degradation of the enzyme?
6. Are high affinity hormone receptors involved? Is the original glucocorticoid or one of its metabolites bound to the receptor(s)?
7. Are glucocorticoids (e.g., corticosterone, hydrocortisone and synthetic dexamethasone) the only hormones active? Is serum required?
8. Do antiglucocorticoid substances block the induction?
9. Is the continuous presence of the hormone required for maintenance of the induced enzyme?
10. Are cyclic nucleotides second messengers of glucocorticoid action?
11. Can other hormones, although not inducers of GPDH, modify the regulations?
12. Are the kinetics of induction and deinduction similar in glial cell lines and in explant, dissociated and reaggregated primary brain cultures?
13. Is there a relation between the basal level of the enzyme and its inducibility?
14. Does the maturation of the glucocorticoid response in culture match the *in vivo* developmental profile of the enzyme?
15. Is the regulation influenced by the spatial and/or preexisting relationship of the cells?
16. Which developmental age gives the most responsive primary cultures?
17. Is GPDH and its induction operative in one or more than one cell type of the nervous system?

B. Cell Culture Results

Addition of hydrocortisone in the physiological range, 10^{-8} to 10^{-7}M, to monolayers of confluent C6 cells results in a dramatic rise in GPDH activity (Fig. 1). Simple explanations for this increased activity — a direct effect of

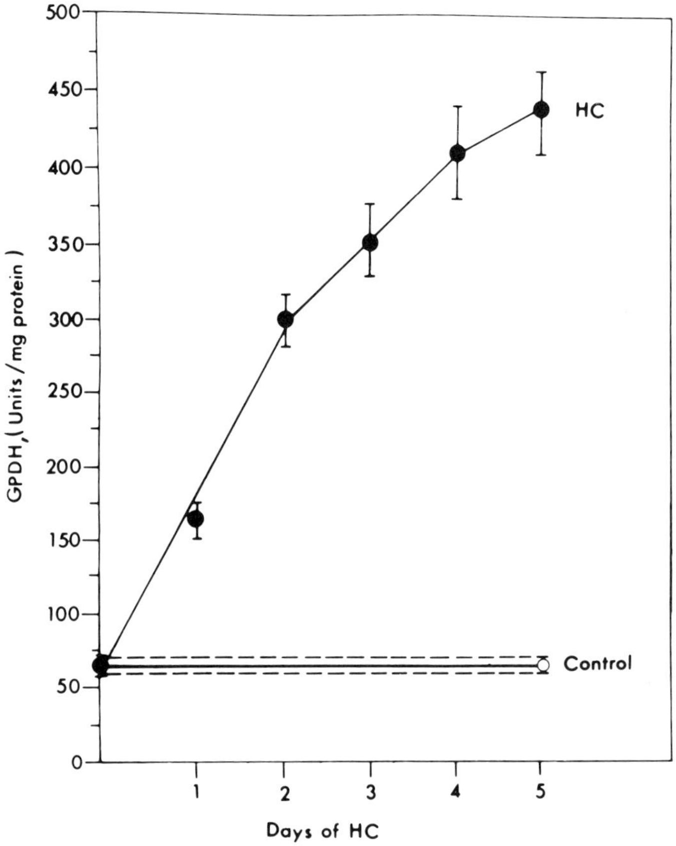

FIG. 1. Time course of GPDH induction by hydrocortisone (HC) in C6 cells. C6 cells were grown for 10 days as monolayers in T30 Falcon plastic dishes at which point they were in stationary phase. Then hydrocortisone-treated cultures received the hormone daily in fresh culture medium (Ham's F10 + 10% fetal calf serum) at a final concentration of 5×10^{-7}M for the number of days indicated. Each point represents the average specific activity of GPDH in 4 cultures ± SD. One unit of enzyme activity is that amount of enzyme which oxidizes one nanomole of NADH per min at 30° C at pH 7.3. (From de Vellis and Inglish, 1973).

hydrocortisone in the assay, increased overall protein synthesis and the presence of diffusible inhibitors or activators — have been ruled out (de Vellis *et al.*, 1971a). To initially investigate the role of transcription and translation, selective inhibitors of DNA, RNA and protein syntheses were used on C6 cells. Cytosine arabinoside, an inhibitor of DNA synthesis, did not affect the induction. Cordycepin, an inhibitor of mRNA processing, α-amanitin, a very selective inhibitor of RNA polymerase II, and actinomycin D blocked the induction. Acetoxycycloheximide, cycloheximide and puromycin totally inhibited protein synthesis and GPDH induction (de Vellis *et al.*, 1971a; de Vellis, 1973; de Vellis and Brooker, 1973). Entirely similar results were observed using explant and dissociated cell cultures (Breen and de Vellis, 1974, 1975). These data suggested that hydrocortisone induces GPDH by *de novo* synthesis of the enzyme with a requirement for RNA but not for DNA synthesis. To approach this hypothesis by more rigorous and direct methods, we purified rat brain GPDH first by classical methods (McGinnis and de Vellis, 1974c), and then by an affinity column using 6-phosphogluconate as the ligand (McGinnis and de Vellis, 1974a). Using rabbit anti-rat brain GPDH, we titrated GPDH in control and induced C6 cultures to resolve the following alternatives. If the induction is due to an increased number of molecules, then equal amounts of antiserum will be needed to inactivate the same number of GPDH units from either induced or control cells. Alternatively, if the induction represents an increased activity per molecule, then the amount of antiserum necessary to inactivate the same number of enzyme units from each source will be different. Fig. 2 shows that equal amounts of enzyme

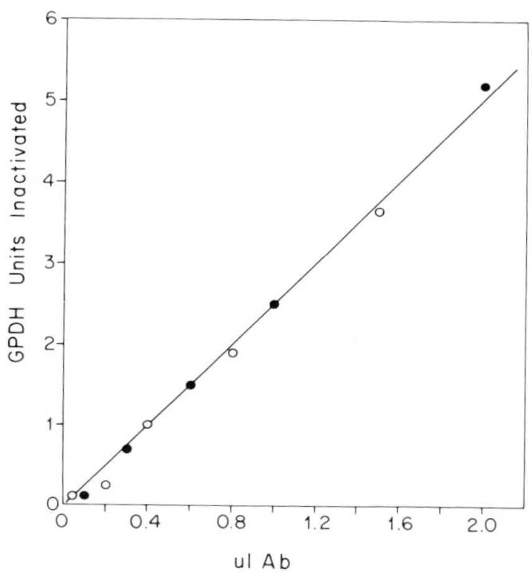

FIG. 2. Immunological titration of GPDH from control and hydrocortisone-induced C6 cells. Constant amounts of GPDH from induced (closed circles) and uninduced cells (open circles) were reacted with increasing amounts of GPDH antisera. The antibody-GPDH complex was enzymatically inactive. The amount of enzyme remaining after incubation with antiserum was substracted from that in the absence of antiserum and the results were plotted as the number of units of enzyme removed against the microliters of antiserum present (from McGinnis and de Vellis, 1974 b).

units are inactivated in each case by a given amount of antiserum. Therefore, we concluded that GPDH induction by hydrocortisone is the result of an increase in the number of enzyme molecules (McGinnis and de Vellis, 1974b). The identity of the induced and basal enzyme activity was confirmed by two other methods. First, GPDH activity in induced cells was electrophoretically identical to that in control cells (Fig. 3). Secondly, the time course of GPDH inactivation at 50°C showed that there is only one heat labile component in control and induced cells and that the two are identical (Fig. 4). Next, we asked if the increased number of molecules represented an increased rate of synthesis or decreased rate of degradation or a combination of both. Fig. 5 shows that 3 hours after hydrocortisone addition the rate of synthesis has tripled and that by 24 hours it has increased more than 10-fold. Since, as we will see later, the half-life of GPDH remains the same during induction and deinduction, we can conclude that hydrocortisone induction of GPDH is brought about solely by an increase in its rate of synthesis (McGinnis and de Vellis, 1976b). Using labeled dexamethasone, we detected a high-affinity glucocorticoid receptor in C6 cytosol (Fig. 6). The concentration of this receptor protein and the binding of the hormone receptor complex in the cell nucleus correlate well with GPDH inducibility (de Vellis *et al.*, 1974). It appears, then, that hydrocortisone induces GPDH in C6 cells according to the currently accepted model of steroid action (Fig. 7). Hydrocortisone diffuses into the cell, binds to a specific high-affinity soluble receptor which is then activated by a temperature-dependent process and translocated to the nucleus. The hormone-receptor complex binds to nuclear sites and by yet-undefined mechanisms, increases the concentration of mRNA or possibly increases the efficiency of its translation.

The inhibitory action of cytochalasin B on GPDH induction lends further credence to the role of glucocorticoid receptors in this model (Bennett *et al.*, 1977; Bennett and de Vellis, 1974). Cytochalasin B inhibited the accumulation of induced GPDH activity, induced GPDH synthesis (Fig. 8) and nuclear binding of ^{3}H-hydrocortisone (Fig. 9) — all three parameters by about 60%, whereas total protein synthesis was unaffected (Fig. 8). When cytochalasin B was removed from the medium, C6 cells recovered the ability to bind the hormone and induce the enzyme. This selective and reversible effect of cytochalasin B makes it an important tool to study the events of the hormone-receptor interaction with the cell.

Only glucocorticoids such as corticosterone, hydrocortisone and synthetic dexamethasone are inducers (de Vellis *et al.*, 1971a). Progesterone, an antiglucorticoid, which prevents glucocorticoids from binding to their cytoplasmic receptors, also inhibits the induction. C6 cells metabolize corticosterone and hydrocortisone, but it is only the unmetabolized form which binds to the nucleus (de Vellis *et al.*, 1974). As *in vivo*, the continuous

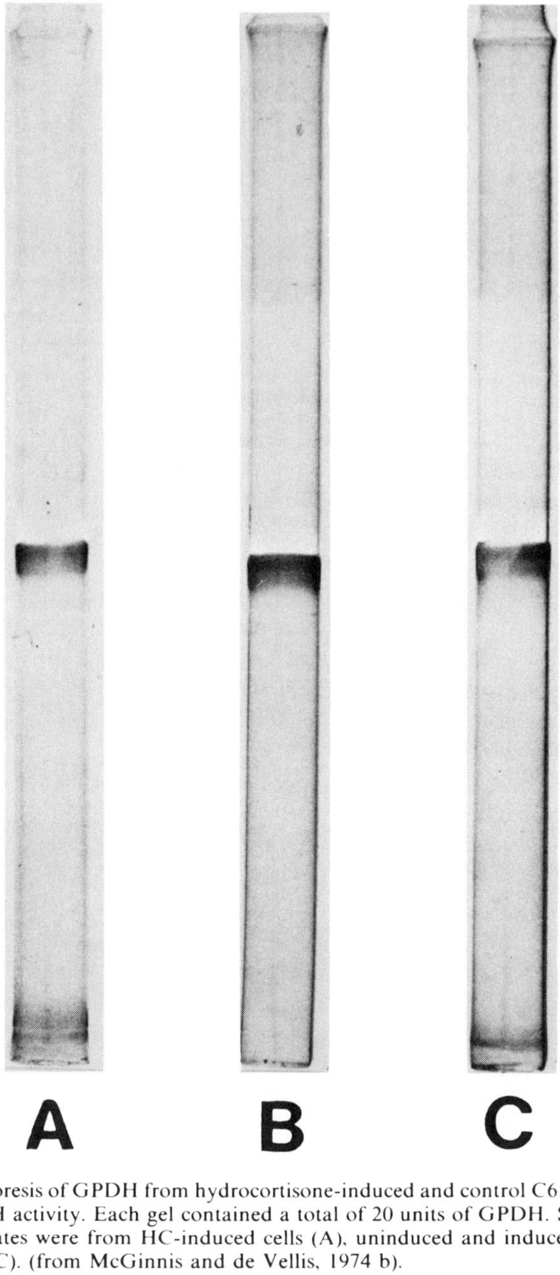

FIG. 3. Electrophoresis of GPDH from hydrocortisone-induced and control C6 cells. Gels were stained for GPDH activity. Each gel contained a total of 20 units of GPDH. Supernatants of C6 cell homogenates were from HC-induced cells (A), uninduced and induced cells (B) and uninduced cells (C). (from McGinnis and de Vellis, 1974 b).

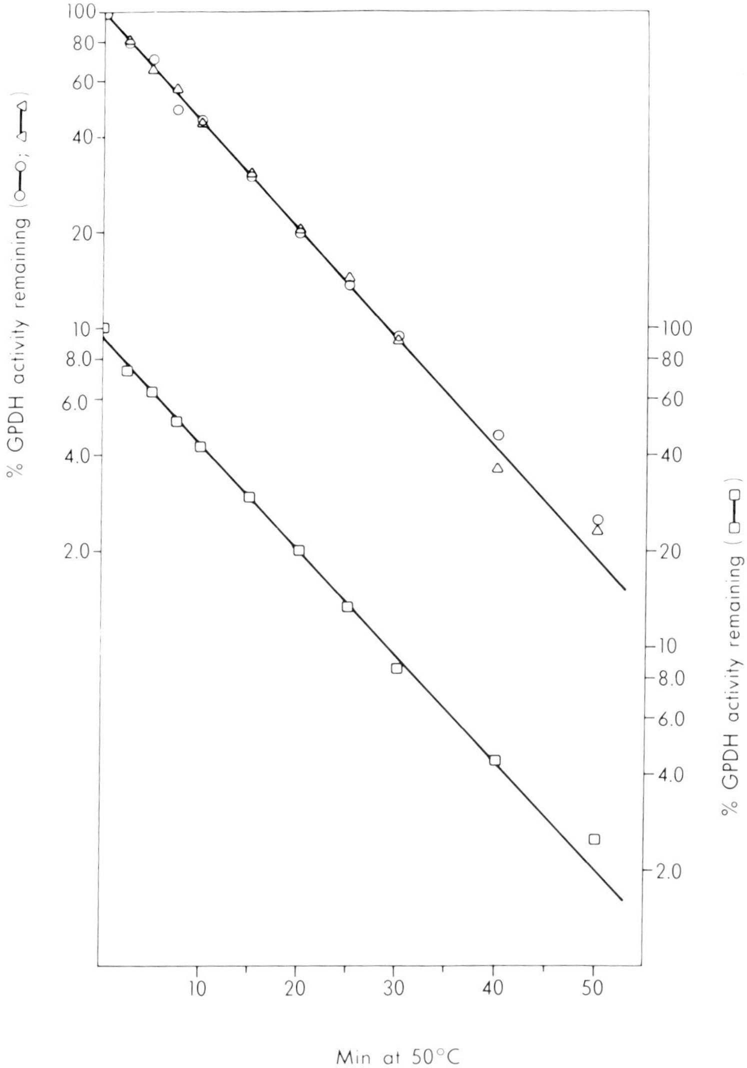

FIG. 4. Heat inactivation of GPDH from hydrocortisone-induced and control C6 cells. Before concentration by ultrafiltration, the specific activities of GPDH in the high speed supernatants of cell homogenates were 45 units per mg of protein and 188 units per mg of protein, for uninduced and induced cells respectively. The concentrations of GPDH at the zero time points (100% activity) were 207 units/ml for the induced (circles), 206 units/ml for the uninduced (triangles) and 214 units/ml for the mixture (squares) containing equal amounts of induced and uninduced enzyme activity. For clarity the data from the mixture are plotted using the ordinate on the right. Three sets of tubes (in duplicate) containing 200 μl of each of the enzyme solutions were incubated in a circulating water bath at 50.0 ± 0.05°C for the indicated times. Then GPDH was assayed (McGinnis and de Vellis, 1974b).

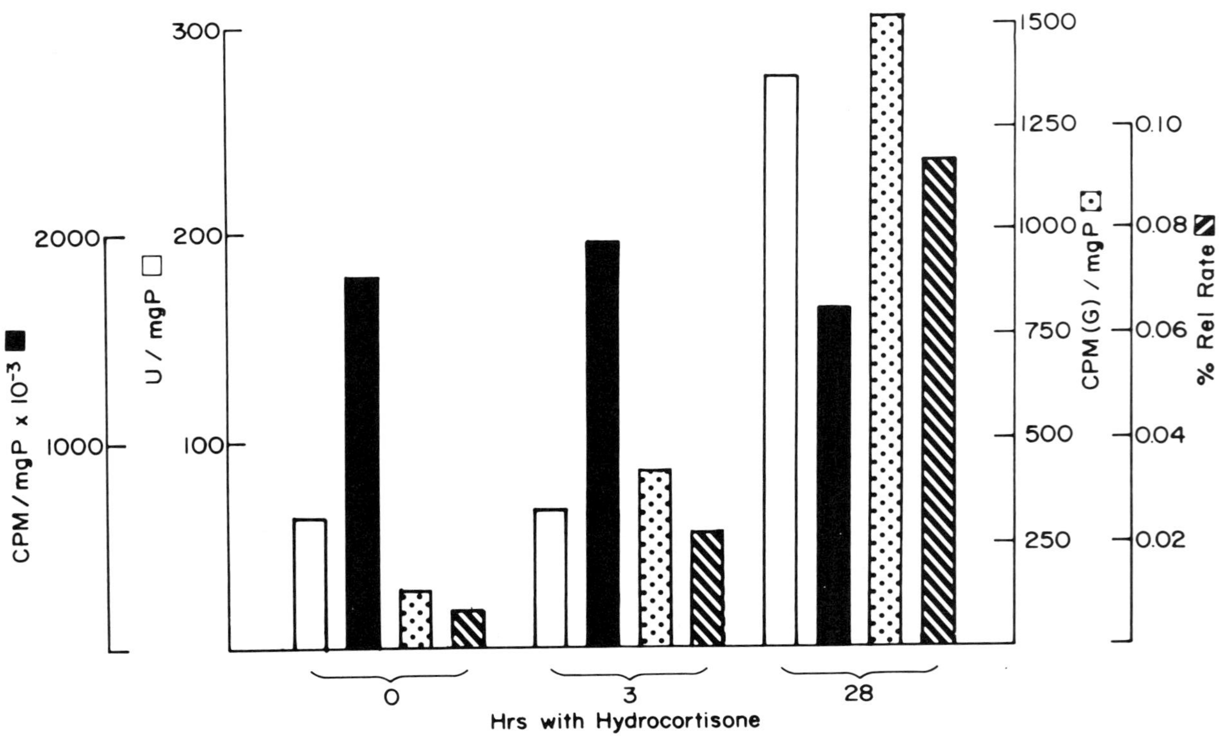

CPM / mgP x 10^-3
2000
1000
U / mgP
300
200
100
0
3
28
Hrs with Hydrocortisone
CPM (G) / mgP
1500
1250
1000
750
500
250
% Rel Rate
0.10
0.08
0.06
0.04
0.02

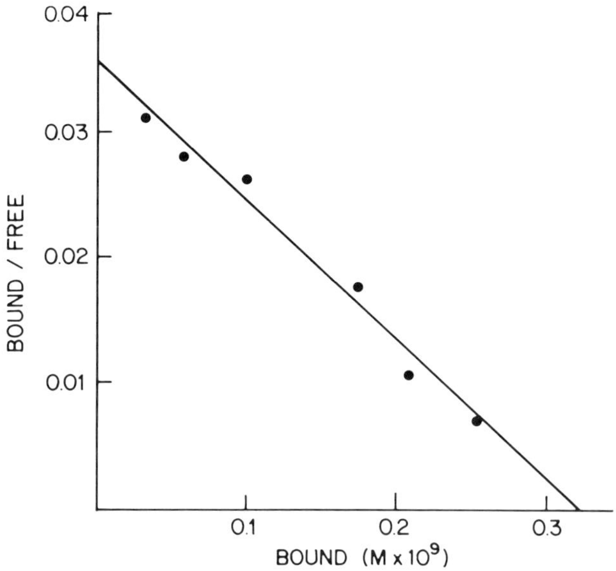

FIG. 6. Scatchard plot of specific binding of ^{3}H-dexamethasone to cytoplasmic receptor(s) of C6 cells. Aliquots of C6 cytosol were incubated for 2 hrs at 0° C with increasing concentrations of ^{3}H-dexamethasone alone or together with an 100-fold concentration of non-labeled dexamethasone. ^{3}H-dexamethasone binding was measured by a charcoal assay. Specific binding of ^{3}H-dexamethasone was calculated by substracting the amount bound in the presence of 100-fold excess of non-labeled steroid (non specific binding) from the total binding (in the absence of non-labeled steroid).

FIG. 5. Rate of synthesis of GPDH in hydrocortisone-induced and control of C6 cells. Cells (10 day cultures) were either uninduced, or induced with HC for either 3 hr. or 28 hr. During the last 30 min of the induction, the cells were pulse labeled with ^{3}H-leucine and the number of counts incorporated into GPDH molecules was determined. Uninduced cultures were similarly labeled. The radioactivity in the newly synthesized GPDH molecules was quantitated by partial purification of GPDH on an affinity column, quantitative immunoprecipitation with a monospecific antiserum, SDS-polyacrylamide gel electrophoresis and subsequent analysis of the number of counts in the isolated band of GPDH (squares with centre dot). The relative rate of synthesis (diagonals) was determined by dividing the number of counts in GPDH (squares with centre dot) by the number of counts incorporated into total cellular protein (solid squares). GPDH enzymic activity (squares) represents units/mg protein (McGinnis and de Vellis, 1976a).

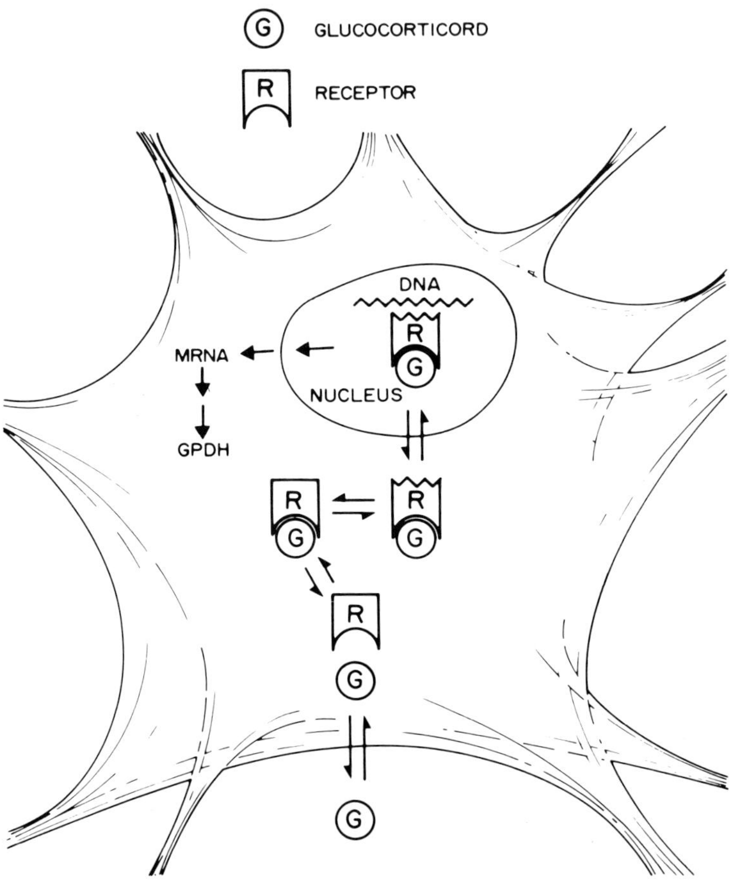

FIG. 7. Diagramatic model of the mechanism of action of glucocorticoids in C6 cells (from McGinnis and de Vellis, 1976b).

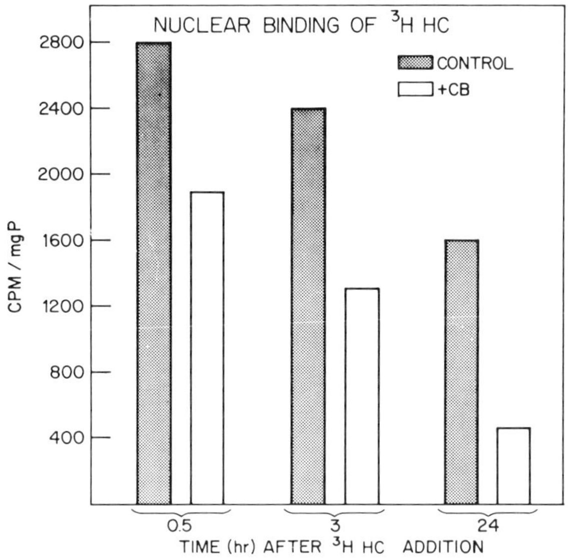

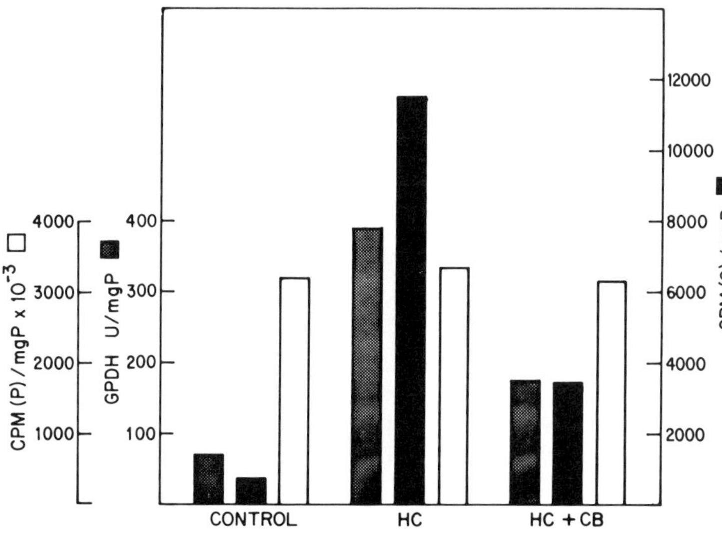

FIG. 8. Cytochalasin B (CB) inhibition of the specific rate of synthesis of GPDH in induced C6 cells. Stationary phase cells (10 days) were either uninduced, induced with HC alone or with HC + CB (25 μg/ml) for 24 hours. The cultures were then washed, pulse labeled with ^{3}H-leucine for 30 min and the incorporation of radioactivity into total cellular protein (squares) and into GPDH (solid squares) was determined as described in the legend to Fig. 5. Squares with criss-crosses represent GPDH enzymic activity. (From Bennett *et al.*, 1977).

FIG. 9. The effect of cytochalasin B on the nuclear binding of ^{3}H-hydrocortisone (^{3}H-HC) in C6 cells. Stationary phase cells (10 days) were incubated in complete culture medium for times indicated with 2×10^{-7}M ^{3}H-HC alone or together with 25 μg/ml cytochalasin B (CB). Cells were harvested and the radioactivity in the purified nuclear pellet determined. (K. Bennett, J. F. McGinnis and J. de Vellis, unpublished experiments).

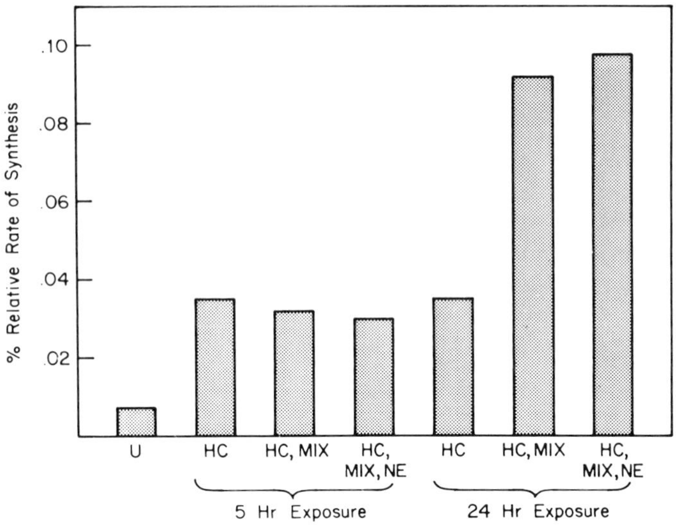

FIG. 10. Relative rate of synthesis of GPDH in C6 cells. C6 cells, 3 days in culture (log phase) were incubated with either medium alone (uninduced controls, U) or medium plus HC. Twenty-four hours later induced cultures received medium with either HC or HC + MIX (methyl isobutylxanthine) or HC + MIX + NE (norepinephrine). At either 5 hours or 24 hours after these additions, the cells were pulse labeled for 30 min with ^{3}H-leucine and harvested. The relative rate of GPDH synthesis (%) was determined as described in Fig. 5. (From Breen *et al.*, 1977).

presence of the hormone is required to maintain the induced level of GPDH. The fact that the induction in primary cultures of C6 cells does not require serum, rules out the necessity of a co-inducer, hormone, growth factor or any other substance present in sera. However, catecholamines (or dibutyryl cyclic AMP), although by themselves not inducers of GPDH (de Vellis *et al.*, 1971a; de Vellis, 1973) increase two-fold the rate of GPDH synthesis when administered with hydrocortisone (Fig. 10). Similarly, enzymic activity is also doubled. This effect occurs after a lag of approximately five hours and appears to be mediated by cyclic AMP at the transcriptional level (G. A. M. Breen, J. F. McGinnis and J. de Villis, unpublished results).

Since C6 is a tumor cell line, one wonders if the characteristics of the induction are different in primary cultures. It was rather astonishing to find that the time course of induction in explant cultures, dissociated cell cultures (Figs. 11A and B) and reaggregates (Breen *et al.*, 1973) was very similar to those obtained with C6. Furthermore, as mentioned earlier, inhibitors of DNA, RNA and protein synthesis gave identical results in all

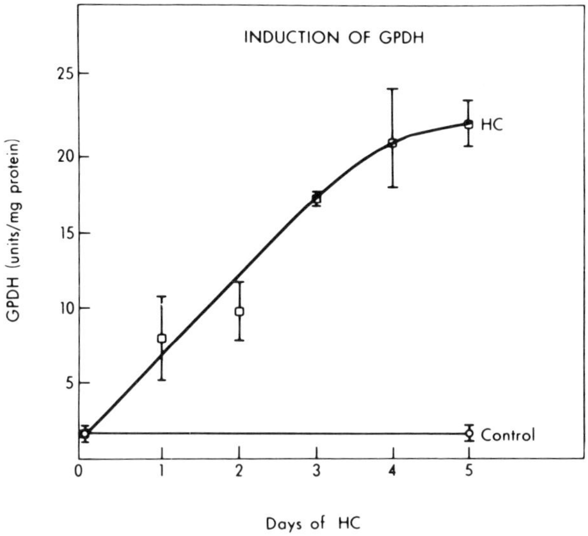

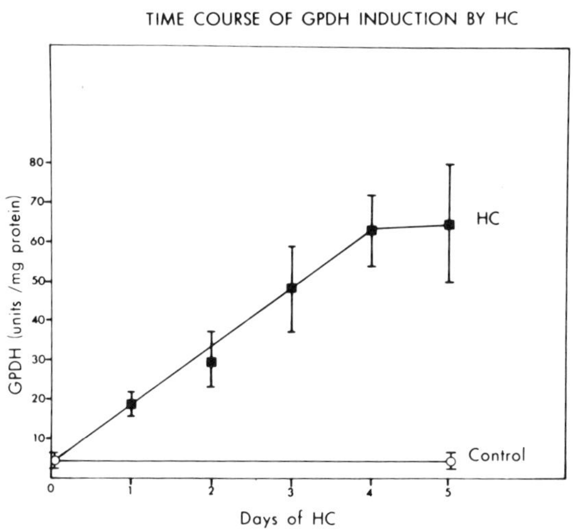

FIG. 11a. Time course of GPDH induction in rat cerebral explants. Cultures were established from the cerebral hemispheres of 20-21 day rat fetuses. Hydrocortisone was added to the culture media at 1.38×10^{-6}M concentration for the number of days indicated. All cultures were harvested at 12 days and assayed for GPDH. Each point represents the average specific activity of GPDH of 4 cultures ± S.D. (From Breen and de Vellis, 1975).

FIG. 11b. Time course of GPDH induction in rat cerebral dissociated cell cultures. Same conditions as in Fig. 11a except that the cerebral cells were dissociated before culturing. (From Breen and de Vellis, 1974).

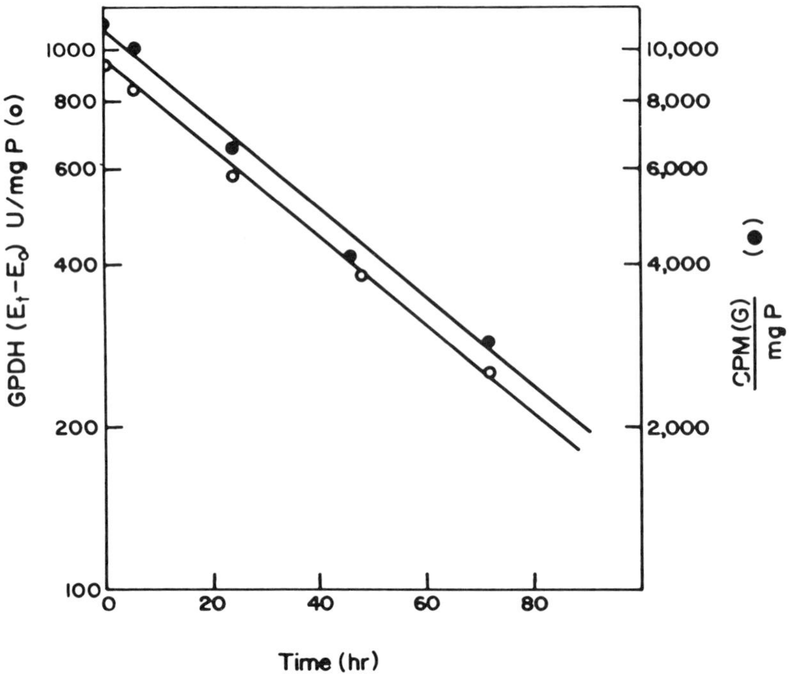

FIG. 12. Simultaneous loss of enzymatic activity and immunological reactivity of GPDH in C6 cells. C6 cells were induced for three days with HC and labeled for the last 24 hr with ^{3}H-leucine. At 0 time the cells were washed and fresh medium without hormone or ^{3}H-leucine was added. At the times indicated, cells were collected (5 flasks- portion) and the specific activity (open circles) and the amount of radioactivity in GPDH molecules (closed circles) were determined as described in the legend to Fig. 5, i.e., including immunoprecipitation with a monospecific antiserum. (From McGinnis and de Vellis, 1976a).

four systems (Breen and de Vellis, 1974 and 1975). Therefore, C6 is an adequate cell culture model, although it is a tumorigenic cell line.

The half life of the enzyme calculated from the kinetics of induction in C6 cells is approximately 40 hours. A similar half life was calculated from the loss of enzyme activity after removal of the hormone (deinduction) and from the disappearance of immunochemically purified ^{3}H-leucine-labeled GPDH (Fig. 12). These data suggest that the rate of degradation is the same during induction as it is during deinduction and that immunological identity is lost simultaneously with enzyme activity.

An increase in GPDH inducibility with age of the cultures from 20 day rat fetuses matches the *in vivo* developmental profile of the enzyme, both for dissociated cells (Fig. 13) and for explants (Breen and de Vellis, 1975), thus validating these models for developmental studies.The inducibility in C6 also increases until the cultures reach confluency (Davidson and Benda,

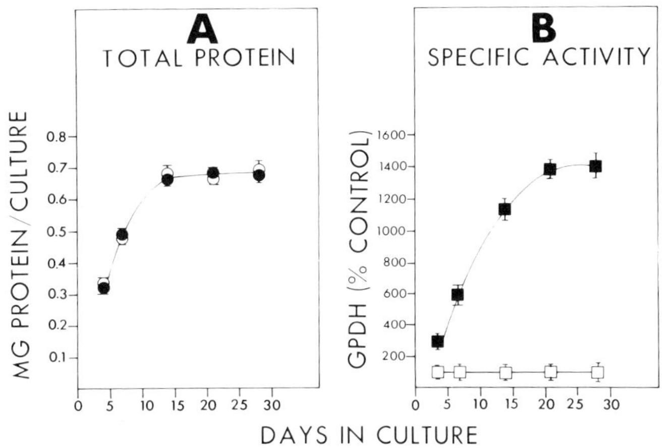

FIG. 13. GPDH induction and protein content of dissociated cerebral cell cultures as a function of the length of the culturing period. Dissociated cerebral cell cultures of pooled littermate animals were either uninduced or induced with 1.38×10^{-6}M HC for 48 hrs. at the times indicated on the graph. GPDH activity and total protein were then assayed. A: The total protein content of uninduced (open circles) and induced (closed circles) cultures is indicated. B: GPDH specific activities of induced cultures (closed squares) are expressed as percent of uninduced specific activities (open squares). GPDH specific activities of uninduced cultures at various days were as follows: Day 4, 5.0; Day 7, 5.6; Day 14, 5.4; Day 21, 5.5; Day 28, 5.6. Each point is the average of at least four cultures ± SD. (From Breen and de Vellis, 1974).

1970; de Vellis and Inglish, 1973), but it declines as the cultures age beyond 12 days. These changes in C6 cells appear to correlate specifically with receptor concentration and nuclear binding of the hormone and not with general cell parameters (de Vellis *et al.*, 1974).

Since GPDH induction occurs in clonal C6 cells, it obviously can occur in the absence of heterotypic cell contacts. The similarity in response among explants, dissociated cell and reaggregate cultures makes it unlikely that a histotypic organization is required. The pre-existing contacts maintained in explants are not necessary for the induction, nor do they appear to enhance its magnitude (Breen and de Vellis, 1975). Like most inducible enzymes in mammalian cells (de Vellis, 1970), GPDH has a basal level of activity. It is high in C6 (Fig. 1), although it varies considerably in C6 subclones (de Vellis and Inglish, 1973), and in hybrids (Davidson and Benda, 1970); however, its activity is low in primary cultures (Fig. 11). Since the inducibility in C6 subclones does not relate to the basal levels of the enzyme in the respective subclone, it suggests that basal and induced levels are regulated independently.

Since GPDH appears to be regulated by hydrocortisone in all brain regions, one would expect to have equal chances of establishing responsive

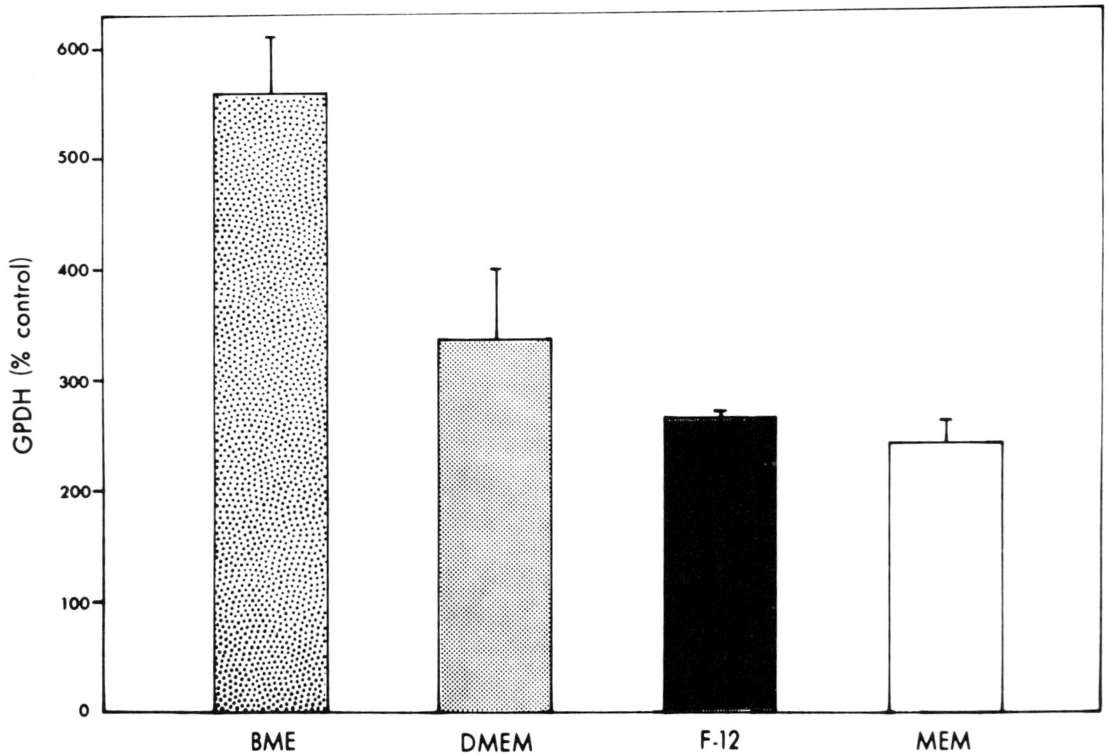
GPDH (% control)
600
500
400
300
200
100
0
BME
DMEM
F-12
MEM

primary cultures from different brain areas. This is indeed the case, although the magnitude of the inducibility varies moderately between areas (Breen and de Vellis, 1974, 1975). It remains to be determined if this variation reflects a difference in the proportion of GPDH-inducible cells in the culture or an heterogeneity among GPDH-inducible cells. These alternatives can now be investigated using the immunoperoxidase technique described below. The inducibility in primary cultures is influenced by other factors such as the choice of nutrient medium and the age of the starting tissue. Figure 14 shows that cerebral cortex cells are more inducible when cultured in BME than in several other media, suggesting that the composition of the medium influences plating, survival, growth or differentiation of GPDH-inducible cells. The age of the animal from which the tissue is taken is an even more important factor in obtaining cultures which express the desired function. GPDH induction, probably a glial characteristic (cf. below) is highest in cultures derived from postnatal rat cerebral cortex whereas veratridine-stimulated sodium uptake, a measure of neuronal function (Catterall and Nirenberg, 1973) is highest in cultures derived from embryonic tissues (Fig. 15).

C. Cellular Localization of GPDH in the Nervous System

Although GPDH is found in other tissues, and is immunologically similar in brain, liver and skeletal muscle (McGinnis and de Vellis, 1977), its induction is specific to nervous tissue. It was, therefore, concluded that the induction but not the enzyme itself should be considered a specific marker of nervous tissue. Furthermore, indirect evidence led us to speculate on a glial localization for GPDH induction (de Vellis and Kukes, 1973). In an attempt to verify this speculation, the localization of GPDH in the nervous system *in situ* was attempted by the immunoperoxidase method (Leveille, 1976). Since the electron microscopic characterization of cells is the only unequivocal method for distinguishing glial and neuronal cell types, the procedures for the immunoperoxidase technique were standardized for use both at the electron and light microscopic level. The strategy of this approach was to use rabbit monospecific antibodies generated against a homogenous preparation of purified rat GPDH. The localization

FIG. 14. Effect of different culture media on the induction of GPDH by HC. Dissociated cells from the cerebral hemispheres of pooled littermate animals were grown for 10 days in either Basal Medium Eagle's (BME), Dulbecco-Vogt modification of Eagle's medium (DMEM), Ham's F-12 (F-12) or Modified Eagle's Medium (MEM). Cultures were then induced for 48 hr. with 1.38×10^{-6}M HC and GPDH assayed. The specific activities of induced cultures are expressed as percent of uninduced cultures. Control specific activities of GPDH for the various media were as follows: BME, 4.3 ± 0.05; DMEM, 3.2 ± 0.30; F-12, 3.5 ± 0.25; and MEM, 4.0 ± 0.30. Each control or induced value represents the average of at least four cultures ± SD. (from Breen and de Vellis, 1974).

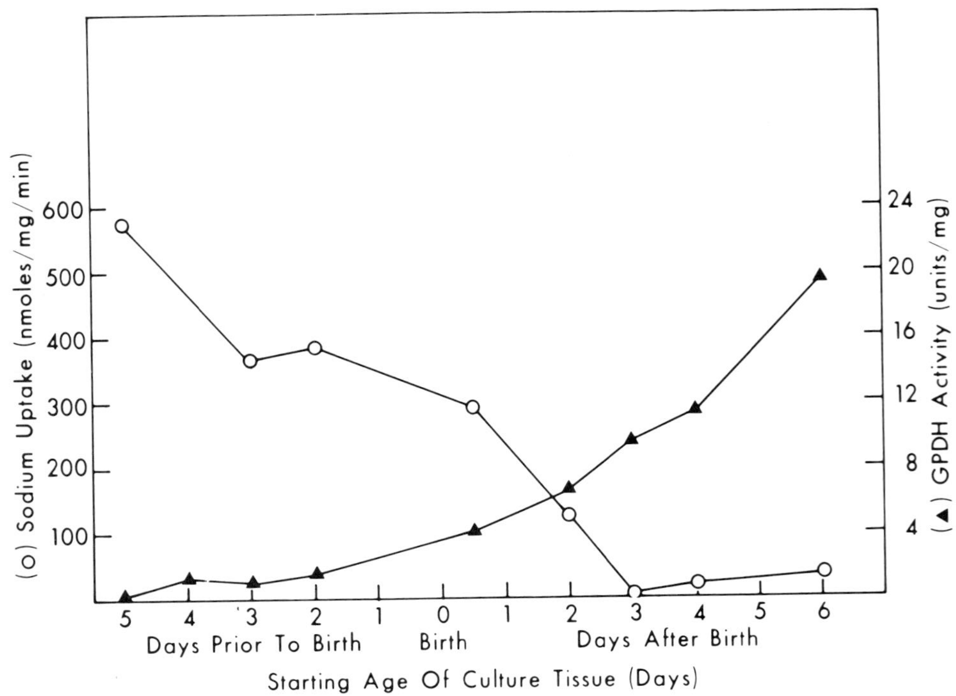

FIG. 15. Effect of age of the starting tissue on neuronal and glial cell markers in primary cultures of rat cerebral cortex. Veratridine-dependent sodium uptake was measured according to the procedure of Catterall and Nirenberg (1973) utilizing a two-minute incubation period. GPDH was induced by incubating the cultures with medium containing 1.38×10^{-6}M hydrocortisone (HC) for 48 hours prior to harvesting the cultures. Results represent stimulated minus control levels. All cultures were grown *in vitro* for seven days prior to harvest. (From McCarthy and de Vellis, 1977).

of GPDH *in vivo* by immunoperoxidase methods modified from Nakane and Pierce (1966) and Sternberger *et al.* (1970) would serve to relate histochemical function to cellular and subcellular structure and possibly provide an added dimension to the physiological characterization of morphological cell types.

GPDH-positive cells in the cerebellum are characterized by an intense staining reaction observed in both histological and ultrastructural preparations. By light microscopy, each cell was distinguished by its dark brown to black rim of cytoplasm from which a few delicate processes extended in various directions. In contrast, the small round to oval nucleus was unstained. Taking the plane of section into account, little variation in shape is seen between the positive cells of white and grey matter. Because tissue sections were exposed to the same experimental protocol, it has been possible to correlate histological observation of reaction product directly with ultrastructural findings. The arrangement of these stained cells between bundles of myelinated axons identify them as interfascicular oligodendrocytes (Figs. 16 and 17). Electron dense reaction product is found in the cytoplasm and processes extending from the soma to the surrounding myelinated axons (Fig. 17). Mitochondria, RER, Golgi apparatus and nuclei are unreactive (Figs. 16 and 17). Neuronal somas, synaptic profiles, dendrites, axons and astrocytes were GPDH negative. GPDH positive cells were most numerous in white matter with fewer cells present in the granular and lower molecular layers. GPDH positive cells have nuclear characteristics distinctive for oligodendrocytes (regular shape, frequently eccentric position, homogeneous nucleoplasm varying from pale to dense. (See Privat's chapter in this book).

Conclusions and Perspectives

The data presented clearly demonstrate that the induction of GPDH by hydrocortisone in C6 cells is due to an increased rate of synthesis without alteration of the rate of degradation of the enzyme, thus resulting in a greater number of molecules. Similarly, we have shown by immunotitration, Ouchterlony double diffusion, gel permeation chromatography, pH optimum, heat lability and polyacrylamide gel electrophoresis that GPDH in the brains of normal (induced) and hypophysectomized (uninduced) rats is identical (McGinnis and de Vellis, 1977). Thus *in vivo* as in cell culture, the hormonal regulation of GPDH activity is brought about by a change in the number of molecules, not in their catalytic efficiency. Similar questions now arise about the mRNA for GPDH, i.e., the influence of hydrocortisone on its rate of synthesis and degradation and on the efficiency of its translation. Since brain GPDH is identical to the enzyme present in muscle and liver, the mechanism for the specificity of the induction is an intrinsic

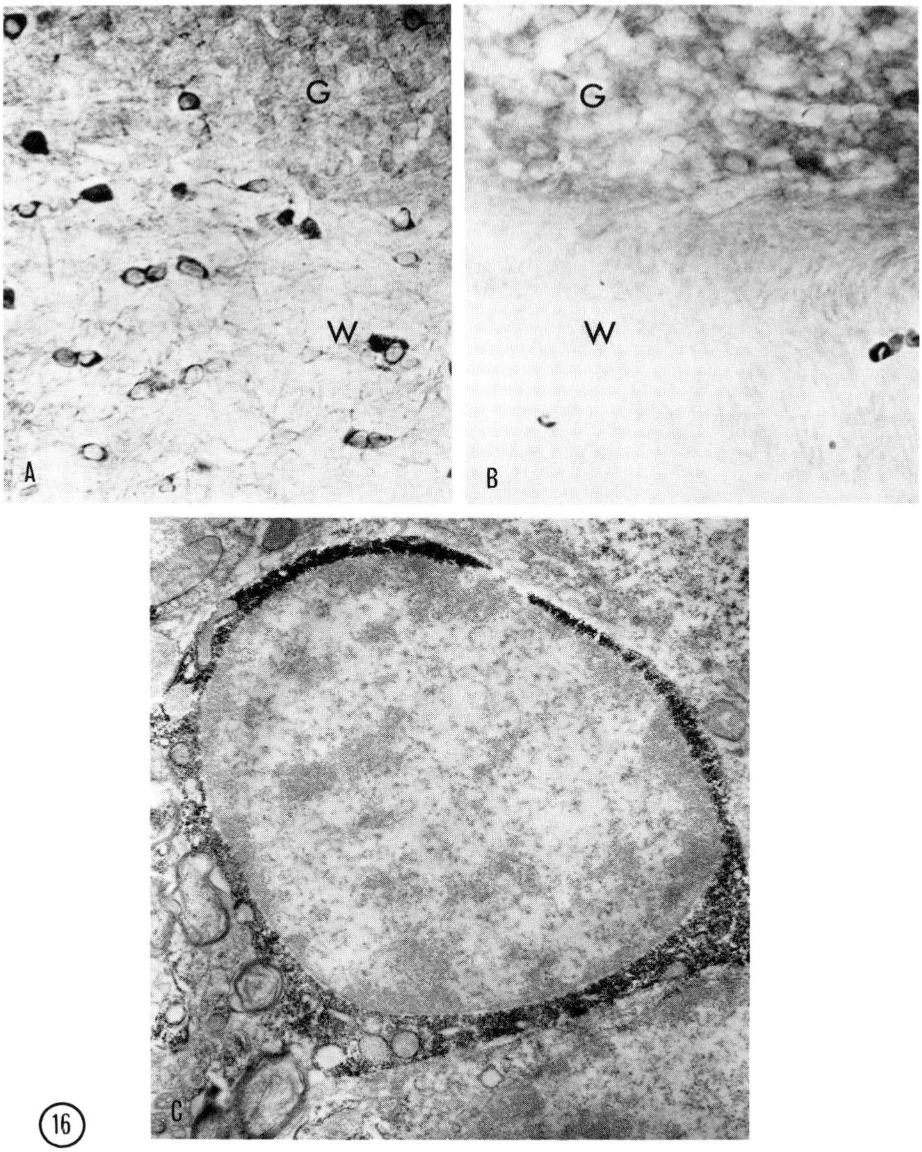

16

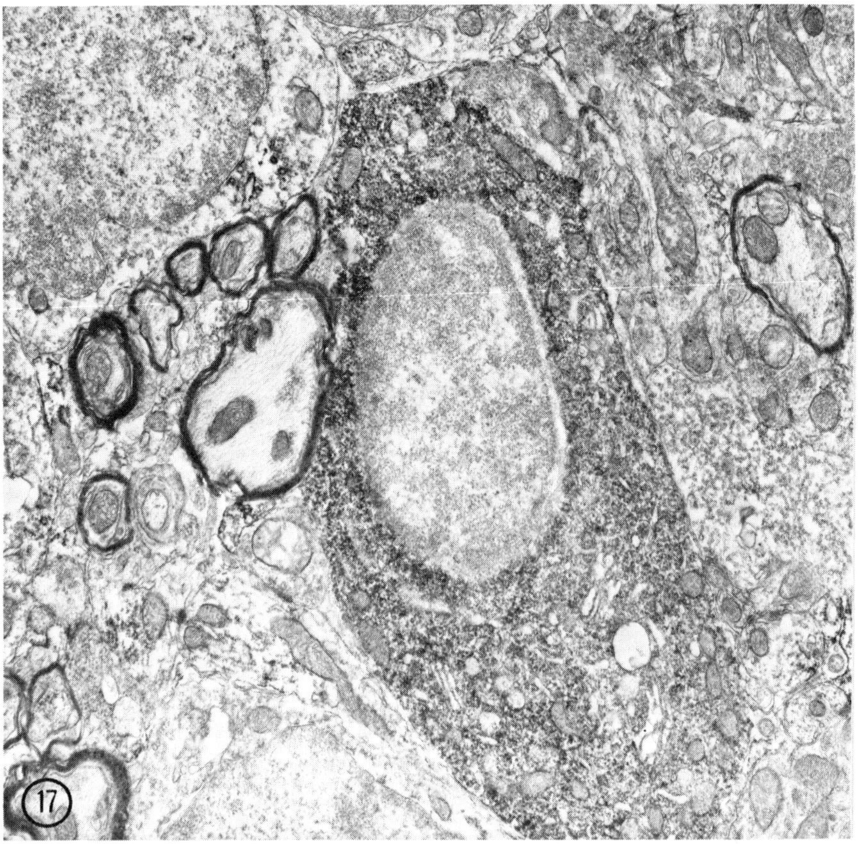

FIG. 16. Combined light and electron microscopy of GPDH in rat cerebellum. A. Experimental whole-mount section: rabbit anti-GPDH gamma globulins followed by peroxidase-labeled goat anti-rabbit gamma globulins (HRP-GARGG conjugate method of Nakane and Pierce, 1966). GPDH-positive oligodendrocytes, recognized by intense cytoplasmic staining, appear predominantly in white matter (W) and granule cell layer (G). Compare with Fig. 16B. × 150. B. Control whole-mount section: rabbit anti-GPDH followed by unlabeled GARGG, then reacted with HRP-GARGG and final DAB-H_2O_2 solution. Dense bodies observed in micrograph are red blood cells exhibiting endogenous reactivity to histochemical substrates. C. Electron micrograph of GPDH-positive cell stained by protocol as described for Fig. 16A prior to preparation for electron microscopy. No routine lead citrate nor uranyl acetate stains were used to enhance density of GPDH-positive DAB-osmium complex. Specific reaction product fills the cytoplasmic matrix, while nucleus is unstained. (From Leveille, 1976).

FIG. 17. Electron microscopy of GPDH-positive oligodendrocyte in granule cell layer of rat cerebellum. Experimental treatment pre-embedding: rabbit anti-GPDH gamma globulins; PAP-Fab complex method of Saito (modified from Sternberger *et al.*, 1970). Cytoplasmic localization of GPDH at the ultrastructural level; eccentric nucleus is unstained; numerous mitochondria and cisterns of Golgi apparatus and endoplasmic reticulum are apparent against the contrast of specific reaction product. Neuronal axoplasm (left) and mossy fiber terminal with synaptic complexes (right) are unreactive. Density of membrane profiles were enhanced by use of a routine lead citrate stain. × 9500 . (From Leveille, 1976).

property of brain tissue rather than related to the enzyme itself such as the coding by different structural genes (McGinnis and de Vellis, 1977).

Having employed a monospecific antibody probe to visualize GPDH present in brain as a single antigen, the immunoperoxidase methods used demonstrate that only one type of nervous cell is GPDH positive. This type is clearly defined by light and electron microscopic criteria as oligodendrocytic. This finding supports two novel conclusions: a) in nervous tissue, GPDH is a biochemical marker for oligodendrocytes and b) oligodendrocytic are target cells for glucocorticoids. Since the identification of oligodendrocytes has rested until now on morphological criteria, the present immunocytochemical method will be particularly useful in situations, such as development, non-myelinating primary cell cultures and pathological conditions of nervous tissue, where morphological criteria are usually lacking. For instance, the ability to identify and count GPDH positive cells in primary cultures would help answer the following questions. The variation of GPDH inducibility in primary cultures of nervous tissue with age of the culture (Fig. 13), choice of the nutrient medium (Fig. 14) and age of the animal from which the culture is established (Fig. 15) must result from either a change in the number of GPDH cells or the ability of each cell to accumulate GPDH, or possibly a combination of both. Since oligodendrocytes proliferate in the rat postnatally after the bulk of other nervous cells, it is probable that postnatal brain tissue gives rise to better inducible cultures than embryonic tissue (Fig. 15) simply because they contain more oligodendrocytes. The resulting cultures from which neurons are apparently absent (Fig. 15) may provide a good starting point to try various manipulations to obtain cultures enriched in GPDH cells, i.e., putative oligodendrocytes.

The similarity in the characteristics of GPDH induction in C6 clonal cells, explant, dissociated and reaggregate cultures show that they are all equally valid models to analyze the mechanisms of induction operating *in vivo.* Hydrocortisone has a pleiotropic effect on C6 cells. It transiently inhibits cell division (Grasso, 1976). The uptake of aminoacids and sugars is decreased (Shinwari and de Vellis, 1972) and the morphology of the cell is altered (de Vellis *et al.* 1971a and b). Particularly the cells lose microvilli, but increase their microfilament content (Berliner *et al.*, 1976). A transient inhibition of postnatal cell proliferation by hydrocortisone has been demonstrated in the rat brain (Balasz, 1974), but the other effects need to be validated *in vivo.*

Little is yet known about the regulation of GPDH in other species. The finding of GPDH induction in mouse brain cell lines (Benda *et al.*, 1977) is in apparent opposition to the reported lack of GPDH induction in mouse brain reaggregating cell cultures (Kozak, 1977). Kozak cultured the aggregates in a medium containing 15% horse serum. Since the activity of

GPDH was very high, it means that either GPDH is not inducible or that the horse serum contained a sufficient amount of glucocorticoids to fully induce the enzyme. However, a similarity with the rat (Fig. 16) is that the GPDH content of the cultures increases with the age of the animal, 6 day postnatal animals giving the cultures with highest GPDH activities. Mouse brain GPDH exists as embryonic and adult forms, each coded by a separate gene (Kozak, 1974).

GPDH is usually thought to play a role in energy metabolism and myelin synthesis (see de Vellis and Kukes, 1973). However, its precise role in the brain remains to be determined, but its exclusive localization in oligodendrocytes of the adult rat brain suggests an involvement in myelin and/or a supportive metabolic role for astrocytes and neurons which are lacking this enzyme.

A model of hormonal action different from the one of steroid hormones is represented by hormones which act by activating adenylcyclase and increasing intracellular cyclic AMP concentrations. Neurotransmitters, such as catecholamines, act in this fashion to induce enzymes in the brain and in cell cultures (de Vellis and Brooker, 1973, 1974; Klein, 1973; Bachrach, 1975; see also K. N. Prasad's chapter in this volume). Thus the two main types of hormonal action can be studied in cell cultures.

Acknowledgments

The work of the authors was supported by ERDA contract EY-76-C-03-0012 and USPHS grants HD-04612, HD-05615 and HL18567.

References

Bachrach, U. 1975. Cyclic AMP-mediated induction of ornithine decarboxylase of glioma and neuroblastoma cells. Proc. Nat. Acad. Sci. U.S.A. 72:3087-3091.

Balasz, R. 1974. Influence of metabolic factor on brain development. Brit. Med. Bull. 30:126-134.

Benda, P., D. Illinger, P. Poindron, M. Fiszman and J. C. Salimon. 1977. Passage through athymic nude mice of SV 40 virus-transformed mouse fetal brain cells. Brain Res. 122:191-196.

Bennett, K. and J. de Vellis. 1974. Inhibitory effect of cytochalasin B on glucocorticoid and catecholamine regulation of enzymes in a rat glial cell line. J. Cell Biol. 63:20a.

Bennett, K., J. F. McGinnis and J. de Vellis. 1977. Reversible inhibition of the hydrocortisone induction of glycerol phosphate dehydrogenase by cytochalasin B in rat glial C6 cells. J. Cell Physiol. (in press).

Berliner, J. A., K. Bennett, L. E. Gerchenson and J. de Vellis. 1976. The role of microfilaments in glucocorticoid action. J. Cell Biol. 70:116a.

Breen, G. A. M. and J. de Vellis. 1974. Regulation of glycerol phosphate dehydrogenase by hydrocortisone in dissociated rat cerebral cell cultures. Develop. Biol. 41:255-266.

Breen, G. A. M. and J. de Vellis. 1975. Regulation of glycerolphosphate dehydrogenase by hydrocortisone in rat brain explants. Exp. Cell Res. 91:159-169.

Breen, G., J. de Vellis and R. Cole. 1973. Induction of glycerolphosphate dehydrogenase by hydrocortisone in primary rat brain cultures. Trans. Soc. Neuroscience 3:427.

Canick, J. A., D. E. Vaccaro, K. J. Ryan and S. E. Leeman. 1977. The aromatization of androgens by primary monolayer cultures of fetal rat hypothalamus. Endocrinology 100:250-253.

Catterall, W. A. and M. Nirenberg. 1973. Sodium uptake with activation of cultured neuroblastoma and muscle cells. Proc. Nat. Acad. Sci. U.S.A. 70:3759-3763.

Claisse, P. J. and J. P. Roscoe. 1976. The inducibility of glycerol phosphate dehydrogenase in two rat glial clones. Brain Res. 109:423-425.

Davidson, R. L. and P. Benda. 1970. Regulation of specific functions of glial cells in somatic hybrids. II Control of inducibility of glycerol-phosphate dehydrogenase. Proc. Nat. Acad. Sci. U.S.A. 67:1870-1877.

de Vellis, J. 1970. Enzyme regulation during cell differentiation. In: *Cell Differentiation.* O. A. Schjeide and J. de Vellis, eds. Van Nostrand-Reinhold Co., New York. pp. 529-573.

de Vellis, J. 1973. Mechanisms of enzymic differentiation in the brain and in cultured cells. In: *Developing and Aging in the Nervous System.* M. Rockstein, ed. Academic Press, New York. pp. 171-198.

de Vellis, J. and G. Brooker. 1973. Induction of enzymes by glucocorticoids and catecholamines in a rat glial cell line. In: *Tissue Culture of the Nervous System.* G. Sato, ed. Plenum Press, New York. pp. 231-245.

de Vellis, J. and G. Brooker. 1974. Reversal of catecholamine refractoriness by inhibitors of RNA and protein synthesis. Science 186:1221-1223.

de Vellis, J. and C. D. Clemente. 1970. Neural cell differentiation. In: *Cell Differentiation.* O. A. Schjeide and J. de Vellis, eds. Van Nostrand Reinhold Company, New York. pp. 529-574.

de Vellis, J. and D. Inglish. 1968. Hormonal control of glycerolphosphate dehydrogenase in the rat brain. J. Neurochem. 15:1061-1070.

de Vellis, J. and D. Inglish. 1969. Effect of cortisol and epinephrine on the biochemical differentiation of cloned glial cells in culture and of the developing rat brain. Trans. 2nd Intern. Meet. Soc. Neurochem. pp. 151-152. Tamburini, Milan.

de Vellis, J. and D. Inglish. 1973. Age-dependent changes in the regulation of glycerolphosphate dehydrogenase in the rat brain and in a glial cell line. Prog. Brain Res. 40:321-330.

de Vellis, J. and G. Kukes. 1973. Regulation of glial cell functions by hormones and ions: a review. Texas Rep. Biol. and Med. 31:271-293.

de Vellis, J., O. A. Schjeide and C. D. Clemente. 1967. Protein synthesis and enzymic patterns in the developing brain following head X-irradiation of newborn rats. J. Neurochem. 14:499-511.

de Vellis, J., D. Inglish, R. Cole and J. Molson. 1971a. Effects of hormones on the differentiation of cloned lines of neurons and glial cells. In: *Influence of Hormones on the Central Nervous System.* D. Ford, ed. Karger, Basel. pp. 25-39.

de Vellis, J., D. Inglish and F. Galey. 1971b. Effects of cortisol and epinephrine on glial cells in culture. In: *Cellular Aspects of Growth and Differentiation in Nervous Tissue.* Daniel Pease, ed. UCLA Forum on Medical Sciences, No. 14. pp. 23-32.

de Vellis, J., B. S. McEwen, R. Cole and D. Inglish. 1974. Relations between glucocorticoid nuclear binding, cytosol receptor and enzyme induction in a rat glial cell line. J. Steroid Biochem. 5:392-393.

de Vellis, J., G. A. M. Breen and J. F. McGinnis. 1975. Biochemical studies in various culture systems of neural tissues. In: *Brain Mechanisms in Mental Retardation.* N. A. Buchwald and M. A. B. Brazier, eds. Academic Press, New York. pp. 115-122.

de Vitry, F., M. Carnier, P. Czernichow, P. Benda, P. Cohen and A. Tixier-Vidal. 1974. Establishment of a clone of mouse hypothalamic neurosecretory cells synthesizing neurophysins and vasopressins. Proc. Nat. Acad. Sci. U.S.A. 159:1476-1477.

Esber, H. J., I. J. Payne and A. E. Bogden. 1973. Brief communication: variability of hormone concentrations and ratios in commercial sera used for tissue culture. J. Natl. Cancer Inst. 50:559-562.

Grasso, R. J. 1976. Transient inhibition of cell proliferation in rat glioma monolayer cultures by cortisol. Cancer Res. 36:2408-2414.

Hamburgh, M. 1966. Evidence for a direct effect of temperature and thyroid hormone on myelinogenesis *in vitro.* Develop. Biol. 13:1253-1255.

Herschman, H. R. 1973. Tissue and cell culture as a tool in neurochemistry. In: *Proteins of the Nervous System.* D. J. Schneider, ed. Raven Press, New York. pp. 95-115.

Klein, D. C. 1973. Circadian rhythms in indole metabolism in the rat pineal gland. In: *The Neuroscience Third Study Program*. F. O. Schmitt and E. G. Worden, eds. The MIT Press, Cambridge. pp. 508-515.

Kozak, L. P. 1974. Genetic and developmental control of multiple forms of glycerol phosphate dehydrogenase. J. Biol. Chem. 249:7775-7781.

Kozak, L. P. 1977. The transition from embryonic to adult isozyme expression in reaggregating cell cultures of mouse brain. Develop. Biol. 55:160-169.

Leveille, P. J. 1976. The localization of glycerolphosphate dehydrogenase in adult rat cerebellum by immunoperoxidase methods: a light and electron microscopic study. Doctoral Dissertation, Department of Anatomy, UCLA.

Lu, E. J., W. J. Brown and J. de Vellis. 1977a. Effects of thyroxine on the developing cerebral cortex in a hormone free culture. Western Regional Society of Electron Microscopists 8:8.

Lu, E. J., W. J. Brown and J. de Vellis. 1977b. Ultrastructure of developing cerebral cortex in cultured cell aggregates. Western Regional Society of Electron Microscopists 8:7.

McCarthy, K. D. and J. de Vellis. 1977. Age dependent changes in neuronal and glial cell markers in cultures from rat cerebral cortex. Trans. Amer. Soc. Neurochem. 8:88.

McEwen, B. S., C. J. Denef, J. L. Gerlach and L. Plapinger. 1973. Chemical studies of the brain as a steroid hormone target tissue. In: *The Neurosciences Third Study Program*. F. O. Schmitt and E. G. Worden, eds. The MIT Press, Cambridge. pp. 599-620.

McGinnis, J. F. and J. de Vellis. 1974a. A novel affinity column for the purification of α-glycerol phosphate dehydrogenase. Biochem. Biophys. Res. Com. 60:186-195.

McGinnis, J. F. and J. de Vellis. 1974b. Cortisol induction of glycerol phosphate dehydrogenase in a rat brain tumour cell line. Nature (Lond.) 250:422:424.

McGinnis, J. F. and J. de Vellis. 1974c. Purification and characterization of rat brain glycerol phosphate dehydrogenase. Biochim. Biophys. Acta. 364:17-27.

McGinnis, J. F. and J. de Vellis. 1976a. Glucocorticoid regulation of the concentration of glycerol phosphate dehydrogenase in a rat glioma cell line. Fed. Proc. 35:1636.

McGinnis, J. F. and J. de Vellis. 1976b. Hormonal regulation of gene expression in a rat glioma cell line. In: Symposium "Uses and abuses of clonal system." Trans. Amer. Soc. Neurochem. 7:146.

McGinnis, J. F. and J. de Vellis. 1977. Differential hormonal regulation of L-glycerol-3-phosphate dehydrogenase in rat brain and skeletal muscle. Arch. Biochem. Biophys. 179:682-691.

Morris, J. E. and Moscona, A. A. 1971. The induction of glutamine synthetase in cell aggregates of embryonic neural retina: correlations with differentiation and multi-cellular organization. Develop. Biol. 25:420-444.

Nakane, P. K. and G. B. Pierce. 1966. Enzyme-labeled antibodies: preparation and application for localization of antigens. J. Histochem. Cytochem. 14:929-931.

Shein, H. M. 1968. Neoplastic transformation of hamster astrocytes *in vitro* by Simian virus 40 and polyoma virus. Science 159:1476-1477.

Shinwari, M. A. and J. de Vellis. 1972. Effects of cortisol and norepinephrine on the uptake of analogs of aminoacids and glucose into cultured glial cells. Trans. Amer. Soc. Neurochem. 3:121.

Sternberger, L. A., P. H. Hardy Jr., J. J. Cuculis and H. G. Meyer. 1970. The unlabeled antibody enzyme method of immunohistochemistry: preparation and properties of soluble antigen-antibody complex (horseradish peroxidase-antiperoxidase) and its use in identification of spirochetes. J. Histochem. Cytochem. 18:315-333.

Toran-Allerand, C. D. 1976. Sex steroid and the development of the newborn mouse hypothalamus and preoptic area *in vitro*: implications for sexual differentiation. Brain Res. 106:407-412.

Vernadakis, A. 1971. Hormonal factors in the proliferation of glial cells in culture. In: *Influence of Hormones on the Nervous System*. D. H. Ford, ed. S. Karger, Basel. pp. 42-55.

West, G. J., J. Uki, R. Stahn and H. R. Herschman. 1977. Neurochemical properties of cell lines from N-ethyl-N-nitrosourea induced rat tumors. Brain Res. (in press).

Cell Interactions

CELL AGGREGATION AND RECOGNITION IN THE SELF-ASSEMBLY OF BRAIN TISSUES

BEATRICE B. GARBER

Departments of Biology and Anatomy
University of Chicago
Chicago, Illinois

The vertebrate central nervous system arises from a discrete epithelial primordium by highly ordered sequences of cell proliferation, migration, selective association, and differentiation (Ramon y Cajal, 1929; Sidman, 1970a). The histogenetic schedule for each regionally differentiated area is specified very early in ontogenesis and quickly reaches a degree of complexity that tends to obstruct the resolution of fundamental problems concerning developmental mechanisms and regulatory controls. In attempts to overcome this difficulty, the use of cell dissociation and reaggregation techniques has provided a powerful analytical tool for examining critical cell interactions involved in the differentiation of the nervous system. The rationale for this experimental approach is essentially the reduction of developing tissues from their complex organized form *in situ* to simplified elemental *in vitro* systems comprised of individual cell components, randomly dispersed in suspension. Then as these dissociated cells are permitted to re-establish contact with one another under carefully defined experimental conditions, their morphogenetic interaction and differentiative potential can be monitored directly as they reorganize into tissues.

This approach has been applied effectively to a wide spectrum of developing tissues, revealing a number of fundamental histogenetic principles that have special significance for the elusive mechanisms underlying neural organization. Perhaps the most striking discovery is the basic tenet that completely dissociated embryonic cells retain the capacity to self-assemble and reconstruct organized tissue characteristic of their tissue of origin — that is, individual cells carry intrinsic developmental information which they are able to communicate to one another and utilize for the continuation of morphogenetic programs (Wilson, 1907; Townes and Holtfreter, 1955; Moscona, 1961). Cells have been demonstrated to recognize like or complimentary phenotypes, forming stable mutual associations and reconstituting coherent tissue fabrics, e.g., the histogenetic reassembly of dissociated liver cells into a characteristic liver parenchyma (Moscona, 1961) or the reconstruction of neural tissue from dissociated brain cells (Ishii, 1966; DeLong, 1970; Garber and Moscona, 1972a); (see

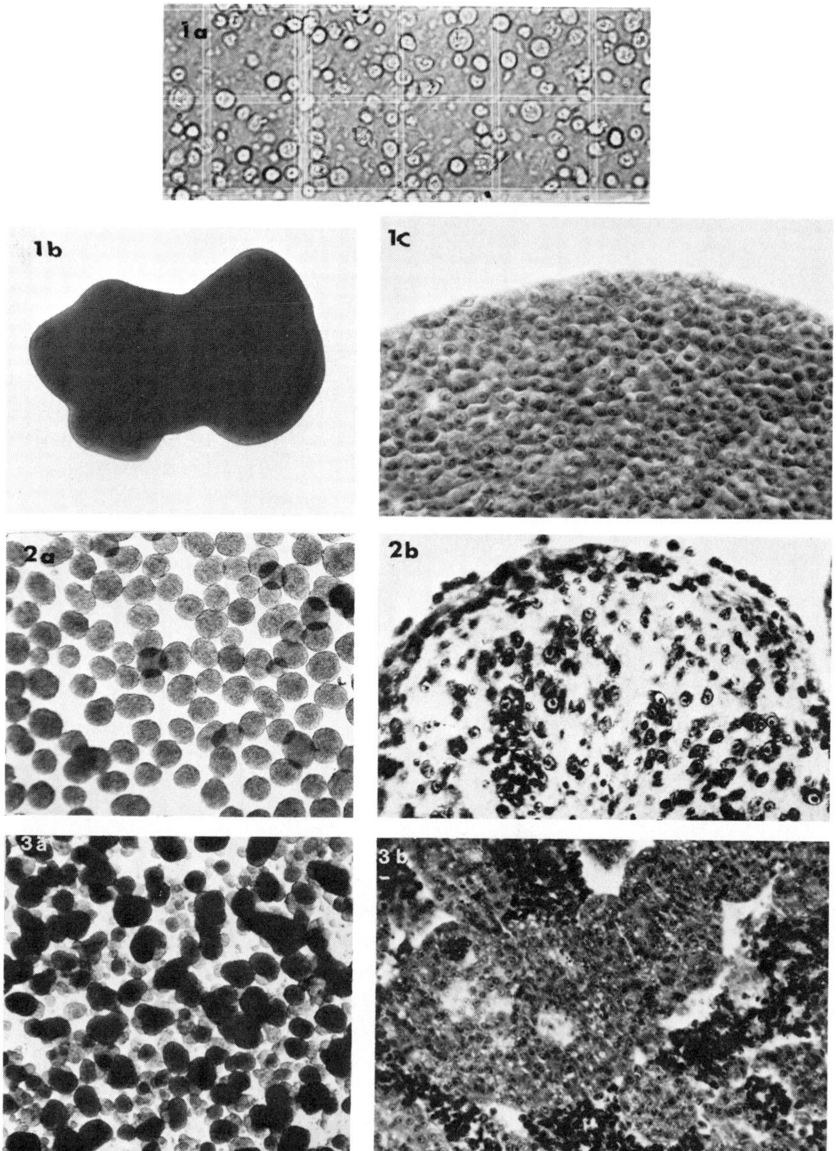
1a
1b
1c
2a
2b
3a
3b

Figs. 1,2). Cells also recognize unlike or incompatible phenotypes, separating out and failing to associate cooperatively, e.g., the sorting out of dissociated brain and liver cells commingled in suspension (Garber and Moscona, 1972a; see Fig. 3).

In certain interacting cell combinations mutual cell recognition induces a train of differentiative events that culminate in complex patterns of tissue and organ morphogenesis, e.g., completely dissociated chick epidermal and dermal papilla cells from young embryonic skin interact to reconstruct normal feather structures *de novo* (Garber, 1967). In other situations, cell recognition results in the suppression of further developmental sequelae, serving to stabilize and maintain a particular mode of differentiation, e.g., older chick skin cells combined in dissociated cell suspensions with young feather-forming chick skin cells completely suppress the formation of feathers and only mature keratinizing skin sheets are produced in the resulting integumental tissue (Garber and Moscona, 1967). Thus, it has been postulated that individual cells carry a communication code at their surfaces that specifies their phenotypic identity and regulates their developmental potential.

These guidelines were formulated largely on the basis of experimental histogenetic studies examining rather simply organized tissues such as liver, skin, kidney, or cartilage (Moscona, 1962; Garber 1976). However, in early investigations of more complex tissues, cells dissociated from developing muscle and nervous tissues were also found to reconstitute appropriate intercellular relationships after reaggregation and to attain advanced states of differentiation, as evidenced by normal myotube assembly, cytoarchitectural organization, synapse formation and neurotransmitter synthesis (Fischman, 1970; Sheffield and Moscona, 1970; Shimada *et al.*, 1969; Seeds, 1971). Therefore, it is timely and appropriate to focus these analytical techniques of cell dissociation and reaggregation on histogenetic investigations of brain tissue, where the extensive array of cellular phenotypes and the strict specificities ordering their complex associations and connections

FIGS. 1,2,3. Reconstruction of tissues from dissociated embryonic cell suspensions.

FIG. 1. a) Dissociated cell suspension. ×300. b) 24-hr. aggregate reconstituted by 7-day embryonic chick liver cells. ×20. c) Histological section of 1b showing internal organization of liver parenchyma. ×150.

FIG. 2. a) 24-hr. aggregates reconstituted by 7-day embryonic chick cerebrum cells. ×20. b) Histological section of an aggregate in 2a after 2 weeks of culture in rotary suspension. Its internal organization reveals differentiated neuronal perikarya, extensive neuropil, and a peripheral cortical arrangement. ×150.

FIG. 3. a) 24-hr co-aggregates formed by commingled embryonic mouse cerebrum cells (14-day) and chick liver cells (10-day). ×20. b) Histological section of aggregates in 3a. Note marked sorting out of the two cell types into discrete areas of dark staining mouse cerebrum cells and lighter staining chick liver cells. ×150.

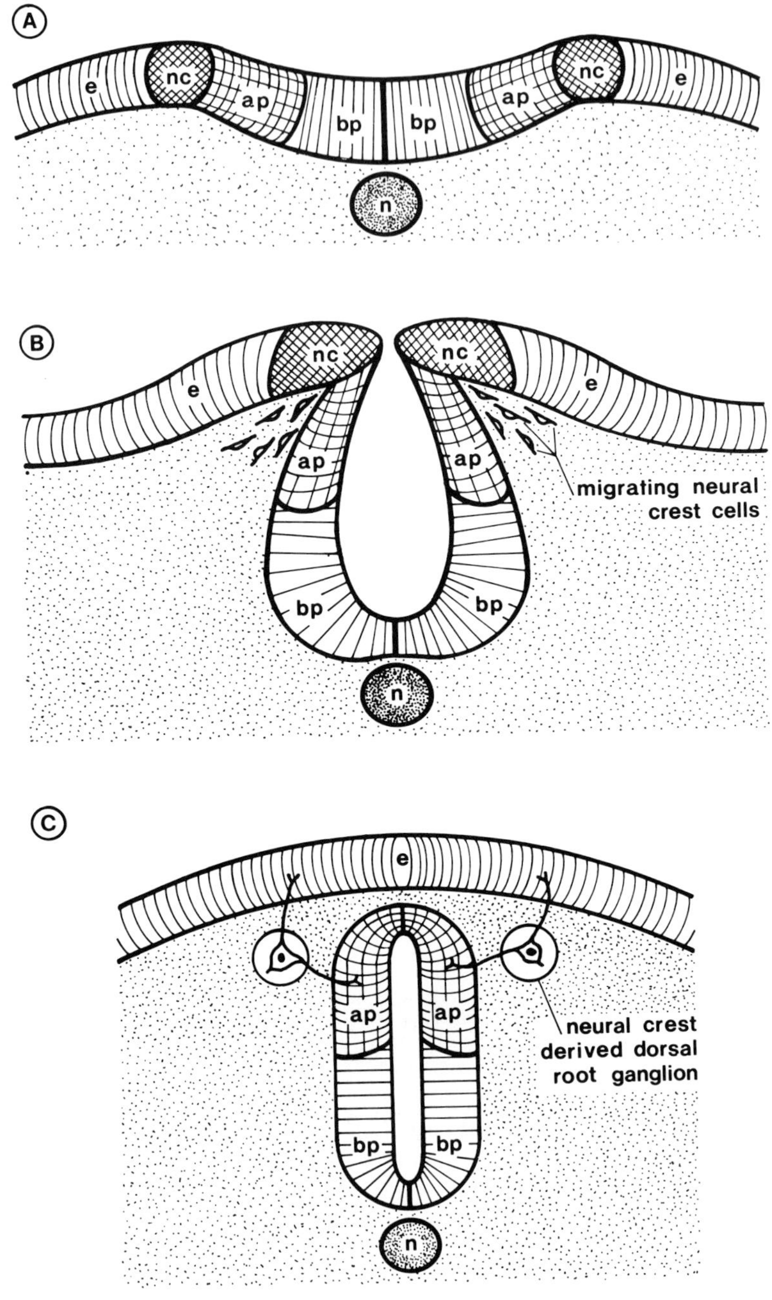
A
e
nc
ap
bp
bp
ap
nc
e
n
B
nc
nc
e
e
ap
ap
migrating neural crest cells
bp
bp
n
C
e
ap
ap
neural crest derived dorsal root ganglion
bp
bp
n

clearly suggest refined mechanisms of cell recognition and developmental interaction.

The specific regional patterns of brain histogenesis *in situ* have been carefully elaborated by the elegant birthdating and mapping procedures introduced by Sidman (1970b). By this technique specific generations of post-mitotic cells, permanently labeled with tritiated thymidine during their last DNA synthesis, can be traced from their origins in the germinal ventricular epithelium or subventricular depots throughout their extensive migratory routes in the mantle to their adult terminal sites (Sidman, 1961; Miale and Sidman, 1961; Angevine and Sidman, 1961; Fujita, 1964; Angevine, 1965, 1970; Altman and Das, 1966; Hinds, 1968a,b; LaVail and Cowan, 1971; Hanaway *et al.*, 1972; Rakic, 1974; Taber Pierce, 1973; Nornes and Das, 1974; Kahn, 1974; Lauder and Bloom, 1974; Tsai, 1977). The interplay of cell dissociations, reaggregations, and specific affinities observed during neurogenesis in the embryo appears to be regulated by a system of recognition cues. A salient example of this morphogenetic control through cell recognition occurs at the time of neural tube closure. There is a mass dissociation of neural crest cells from their neighbors in the ectodermal epithelium, disengaging from the lateral neural plate on one side and the presumptive epidermis on the other side (Fig. 4a,b). These neural crest populations migrate as single cells for various distances through the peripheral tissues of the embryo and later reaggregate in specified sites to form cranial, spinal, or autonomic ganglia as well as other distinct phenotypes (Fig. 4c).

For their part, the neural tube cells, now brought into direct contact by the departure of the neural crest populations, recognize one another as like cell types and fuse into a cohesive epithelium (Fig. 4c); moreover, the neural tube cells also recognize the cells of the approaching epidermal epithelium as unlike phenotypes, do not associate with them and become internalized.

FIG. 4. Cell recognition events in sequential stages of normal development of the nervous system: epidermal-neural crest-spinal cord cell interactions. a) "Nearest neighbor" relationships between ectodermal compartments in the primordial epithelium. Presumptive neural crest cells (nc) are adjacent to the presumptive alar plate cell (ap) of the spinal cord on one side and the presumptive epidermal epithelium (e) on the other side. b) Dissociation and migration of neural crest cells (nc) from neighboring alar plate (ap) and epidermal (e) cells as differentiation advances. c) Selective association of like and complimentary cells during normal histogenesis. Note the fusion of epidermal cells (e) to form the external skin covering and the fusion of alar plate cells (ap) to close the neural tube. This dissociation of spinal cord cells from epidermal cells effects the internalization of the neural tube. A migrating neural crest cell population aggregates to form a series of spinal ganglia, whose cells send out processes which reassociate with their earlier "nearest neighbors": alar plate cells (ap) of the dorsal spinal cord on one side and epidermal (e) cells of the skin on the other side.
Abbreviations: e (epidermal epithelium), ap (alar plate), bp (basal plate), nc (neural crest), n (notocord).

Similarly, the ectodermal cells formerly separated by both the neural crest and neural tube populations now come into approximation and, recognizing one another as like phenotypes, fuse to form an outer epithelial covering for the embryo (Fig. 4c). The rapid organizaton of proliferating neural cell populations within the neural tube into structural and functional compartments suggests that similar crucial recognition events continue to occur with ongoing neural differentiation, but their identification becomes obscured by the spatial and temporal complications of concomitant histogenetic processes in the CNS. These are the kinds of cell recognition events that can be addressed in the simplified histogenetic models afforded by *in vitro* cell aggregation systems.

In order to design effective experimental probes it is necessary to take stock of the basic developmental questions that remain unanswered. First, with respect to the proliferative compartment, it is still unknown when each of the many diverse neural phenotypes is first determined, whether a cell becomes irreversibly committed to a given phenotype within the germinal epithelium, and, if so, how and when each cell type is specified. It has been speculated that the number of mitotic divisions a cell undergoes within the proliferative compartment may influence its phenotype, since it has been widely observed that the earliest neurons to be released into the mantle layer differentiate as large neurons, the next generations become neurons of intermediate size, and the last neurons to be released from the proliferative phase form very small associative interneurons (Sidman, 1961; Jacobson, 1970). Glial cells continue to be proliferated after all neuronal generation has ceased.

Taking into account the presumptive affinities of various neuronal classes and their time and place of origin, another hypothesis can be offered. Since the recognition cues whereby neurons selectively associate or make synaptic connections are communicated via the cell surface, neurons of like type or complimentary fit may acquire matching cues by virtue of conserved portions of the cell membrane previously differentiated and shared in the germinal neuroepithelium. In the example of the neural crest-derived cells cited above, it could be argued that the spinal ganglionic neurons re-establish selective connections with the dorsal horn neurons of the spinal cord on one side and with sensory placodal-type cells in the skin epithelium on the other side, on the basis of cues acquired when they were "nearest neighbors" in the primordial ectoderm of the neurula (Fig. 4). Similar complimentarities could be suggested for the synaptic fit between Purkinje cells and inferior olive cells which have shared a common neuroepithelial site and generation time (Miale and Sidman, 1961; Taber Pierce, 1973). In a somewhat modified form this concept could be extended to both the horizontal lamination of cells of like type in each of the ganglion, bipolar, and receptor layers of the retina, and to the synaptic affinities between each of these three layers displayed in their vertical alignments, as suggested by Sidman (1961). Recent autoradiographic studies of developing visual areas in the avian telencephalon (Tsai, 1977) and comparative evolutionary studies of adult visual projections in the central nervous system (Ebbesson, 1972; Karten, 1969) provide cogent evidence that a primordial neuron pool gives rise to functionally related structures, i.e., brain structures that will later establish related neural circuits originate from a common neuroepithelial compartment. These suggestions are open to testing by dissociation and aggregation studies in which selected populations of the germinal epithelium can be birthdated with tritiated thymidine, then isolated at various developmental ages, combined with putative interacting populations, and followed in experimental tissues fabricated *in vitro* to determine their selective recognition properties and their differentiative fate.

Second, many unanswered questions abound with respect to the migratory phase of neurogenesis. Upon release from the proliferative compartment in the neural tube, most cells migrate away actively and begin to exhibit overt differentiative changes. There is still scanty understanding of the mechanisms permitting the dissociation of cells from the stem cell epithelium or of the cues directing neurons to specific destinations. The concept of contact guidance (Harrison, 1910; Weiss, 1934; Weiss and Garber, 1952) has long been accepted as an important factor in nerve outgrowth and stresses the requirement for an oriented physical substrate. However, except for the recent attention called to radial glial or neuroepithelial cells and fasciculated fiber tracts acting as migrational guides within the central nervous system (Simpson, 1970; Rakic, 1971, 1972; Nornes and Das, 1972), the basis for the specific migratory routes taken by different cell classes remains unexplored. Specific recognition cues may be necessary to account for the vectorial decisions of moving cells and the precise point at which they cease their movement and establish stable cell associations (Sperry, 1965). Here again, *in vitro* reaggregating cell systems can be used to address these histogenetic problems of neuronal migration and selective cell association involved in the formation of specific nuclear condensations and ordered cortical layers. Experimental confrontation between selected migratory cell populations and various types of neuronal or glial cells with which interaction during normal histogenetic patterning *in situ* is postulated, can be designed to test cell recognition with respect to directed cell movements and establishment of stable architectural configurations. In addition, recent preparations of extracellular recognition factors from brain cell structures (Garber and Moscona, 1972b; Spiegel and Garber, 1974; Balsamo and Lilien, 1974a,b) permits investigation of the informational role of these moieties in neural patterning.

Finally, with respect to the third aspect of neurogenesis, i.e., neural differentiation and maturation, the obligate cell interactions leading to expression of the adult phenotype are almost completely undefined. The question remains unanswered whether a particular neuron self-differentiates autonomously after being released from the proliferative compartment, or if it must migrate to specific sites in order to associate with other cell classes which then induce its definitive structural and functional maturation. For example, is contact with afferent systems prerequisite for neuronal maturation, as certain evidence from the peripheral nervous system suggests (e.g., Black *et al.*, 1971)? Is it a special case for certain central neurons, or a general rule, that failure to establish appropriate efferent connections results in the loss of phenotypic expression and even cell death (e.g., Hamburger, 1934)?

There are numerous advantages in using dissociation and aggregation procedures to produce tissue models for such histogenetic analysis. The first important advantage, shared in common with other *in vitro* approaches is, of course, that the complexities involved in studying the total embryo are substantially reduced and isolated tissue systems can be observed directly. But beyond this there are certain unique features. While the explantation of tissue fragments from the embryo to the culture dish provides an opportunity to isolate the normally organized cytoarchitecture in a controlled environment, this organization changes and may deteriorate to a considerable extent in culture as a peripheral "monolayer" zone of extensive cell outgrowth builds up. Essentially 2-dimensional monolayers are also produced when dissociated cells are plated onto stationary culture dishes, although neuronal cells tend to cluster into ganglion-like aggregates. While these planar cell sheets allow close inspection of individual living cells in culture, they represent an "open cell system" that characteristically adapts to the conditions of culture and tends to lose the "closed system" attributes of tissue *in vivo.*

In contrast, suspension cultures of dissociated cells maintained in rotating culture vessels reconstitute closed tissue systems, forming 3-dimensional masses whose size can be controlled by the speed of rotation. The cytoarchitecture of these aggregates is not dependent on a planar substrate presented by the culture vessel surface, but is reconstructed on the basis of cell-cell contacts, mutual cell affinities, and extracellular substrates produced by the cells themselves, and thereby serves as a measure of cell recognition and intercellular communication. Since the dissociation process essentially reduces tissues to their building blocks, the investigator can design and fabricate experimental tissues by commingling cell suspensions from different areas of the CNS, from different species and from different stages of development. Thus, specific questions can be asked regarding the interaction of designated cell types. The possibilities range even further: mutant cells can be combined histogenetically with wild type cells; neoplastic cells can similarly be combined with normal cells, exogeneous cell-free products can be tested for their abilities to substitute for the intact cells from which they are derived.

The obvious limitation of such a test system starting with a scrambled cell population is the availability of unambiguous cell markers to act as tracers for the assorted cast of participants during the histogenetic processes that follow, some of which may be totally unexpected and unpredictable. Fortunately, many such cell markers are well known, e.g., the distinctive nucleolar marker of quail cells (Le Douarin, 1973); the condensed chromatin patterns and large nuclear size of mouse cells which can be distinguished from the small nuclei of chick cells by hematoxylin staining (Wolff, 1954); the endogenous cell-specific histofluorescence which can be used to visualize catecholamine neurons (Falck *et al,* 1962); or the experimental labeling of selected cell populations with appropriate radioisotopic markers (Sidman, 1970b). As in any other *in vitro* situation the fidelity of cytodifferentiation and tissue reconstruction is dependent upon the physiological conditions presented by the culture system. While our knowledge of environmental requirements for full organotypic differentiation is incomplete, especially in the case of CNS tissue, this limitation, in a sense, favors the choice of tissue reconstruction in aggregates as a histogenetic assay system since defined factors can be sought using the restoration of full phenotypic differentiation as an endpoint. With these general considerations in mind with regard to the kinds of problems of neural differentiation amenable to analysis by dissociation and reaggregation assays of cell interaction, it is fruitful at this point to review and assess some of the specific findings revealed through this experimental approach to brain histogenesis.

In preparing brain cell aggregates for *in vitro* studies, mechanically dissociated cells can be obtained merely by pressing the delicate embryonic tissues through a fine Nitex filter. However, there is far less damage, complete dissociation, and, therefore, greater viability and more consistent results if gentle trypsinization procedures are used to achieve a chemical cell dissociation. A basic procedure (Garber and Moscona, 1972a) can be described briefly. The tissues selected for dissociation are dissected out using sterile techniques, washed free of cell exudates and debris in Tyrode's solution, then incubated in calcium-magnesium free Tyrode's solution (CMF) at 37°C for 15 min to remove cations that stabilize intercellular bonds. The tissues are then incubated for 30 min at 37°C in a mild trypsin solution (15,000 units/ml CMF;pH 7.4). After careful removal of the trypsin by repeated washings in CMF containing trypsin inhibitor, the culture medium is added (Eagle's basal medium plus 1% L-glutamine, 1% penicillin-streptomycin, 10% serum, and DNAase, 25 μg/ml). The tissues can now be mechanically dispersed by flushing them 20-30 times through a fine bore Pasteur pipette. The cell suspensions should be checked routinely to insure complete dissociation, particularly if they are to be used for assays of tissue reconstruction from single cell populations. Aliquots of the cell suspension are dropped into 25 ml Erlenmeyer flasks containing 3 ml of culture medium and the cultures are placed under an atmosphere of 5% CO_2-95% air in a gyratory incubator shaker at 70 rpm, 37°C. The aggregation of the dissociated cells begins immediately; the cell surfaces are repaired by 2-4 hr; and the size of the individual aggregates stabilizes at about 20 hr. For long term culture, the medium is changed every 2 to 3 days.

One of the first determinations to be made was whether the regionally

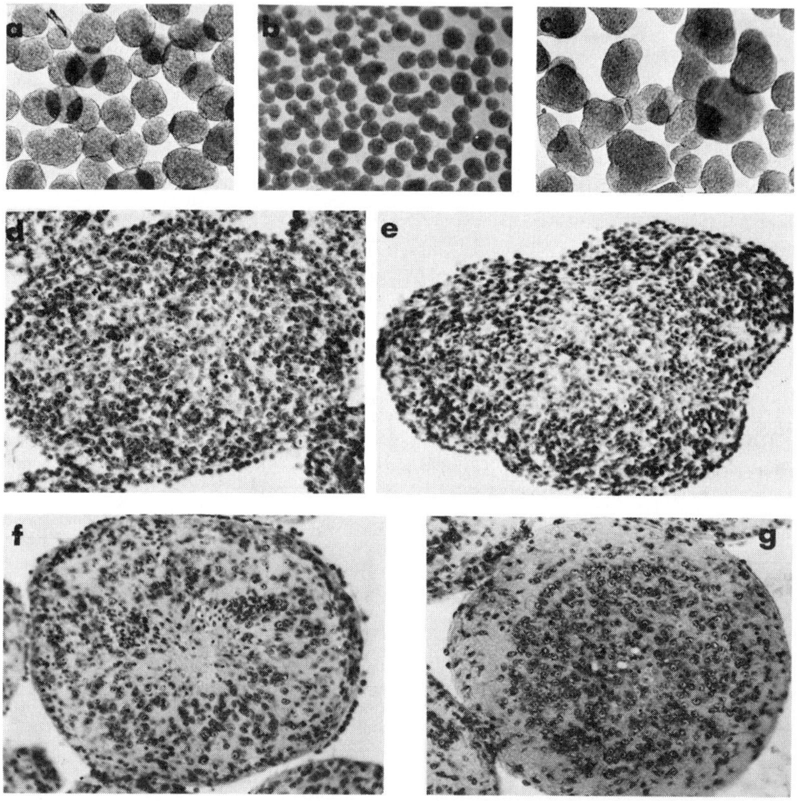

FIG. 5. Aggregates formed by cells dissociated from different areas of the embryonic chick brain. The external appearance of 24-hr aggregates formed by 7-day a) telencephalic cells (cerebrum) b) diencephalic cells, and c) mesencephalic cells (optic, tectum) shows differences in size and shape (×25). Their internal cellular organization appears relatively undifferentiated after 24 hrs of culture, as illustrated by d) cerebral and e) optic tectal aggregates (×150). However, after 2 weeks of culture f) cerebral and g) tectal aggregates develop distinctive patterns of cytoarchitecture (X150). (After Garber and Moscona, 1972a).

differentiated cell populations dissociated from various subdivisions of the developing brain had distinctive patterns of reassembly in aggregate cultures. In other words, would the cells from each specific brain area, such as cerebrum, optic tectum, or cerebellum reaggregate and reorganize according to a particular histogenetic pattern uniquely characteristic of their site of origin, or would all brain cells, regardless of their origin, display a common "brain-type" of histogenesis during reaggregation and establish identical morphological endpoints? It was found that cell populations from each of the various brain regions tested did show a distinctive aggregative behavior, just as liver, kidney, or cartilage cells had done in the classical experiments cited earlier. This was manifest in the gross external features of aggregate size and shape and, even more significantly, in the internal organization and cytodifferentiation of the reconstructed aggregate tissues after extended periods of culture (Garber and Moscona, 1972a; DeLong, 1970; Spiegel, 1977; see Fig. 5).

It was also found that the developmental age of the brain tissue at the time of dissociation was a critical factor affecting the capacity for cell aggregation and subsequent cytoarchitectural organization. Cell aggregability was greatest during the phases of active cell proliferation and migration and decreased progressively with advancing differentiation (Garber and Moscona, 1972a). Since the proliferative and migratory activities of cells appear to be closely regulated by recognition cues, this result encouraged the view that aggregate studies would prove to be incisive probes for the cell recognition events that influence the course of brain histogenesis.

In designing cell aggregation systems to analyze the significance of cell interactions within the proliferative compartment for the eventual pattern of neurodifferentiation, several questions have to be considered. First, is it possible to reconstruct the germinal neuroepithelium in brain cell aggregates? While this issue has not been addressed systematically, neuroepithelial rosettes have been observed in aggregates formed from 3-day embryonic chick optic tectum cells, but not in those formed from 6-day tectum (Adler, 1970). Similarly, large neuroepithelial vesicles have been recovered in aggregates reconstructed from 5-day chick cerebrum, but these structures are not characteristic of cerebral aggregates formed by older cerebral cells (Garber, unpublished observations). Thus, it appears that the cell recognition cues required for reassembly of a germinal epithelium *in vitro* may wane or be blocked long before the proliferative activity of this compartment *in situ* has terminated (see LaVail and Cowan, 1971; Tsai, 1977). Following dissociation, mitotic activity in aggregates is depressed for the first 12-24 hr, but is later found to resume in aggregates formed by young embryonic cells. Mitosis occurs even in the absence of a conspicuous neuroepithelial structure. Glial proliferation becomes increasingly active with ongoing culture and often threatens the survival of neurons in

6

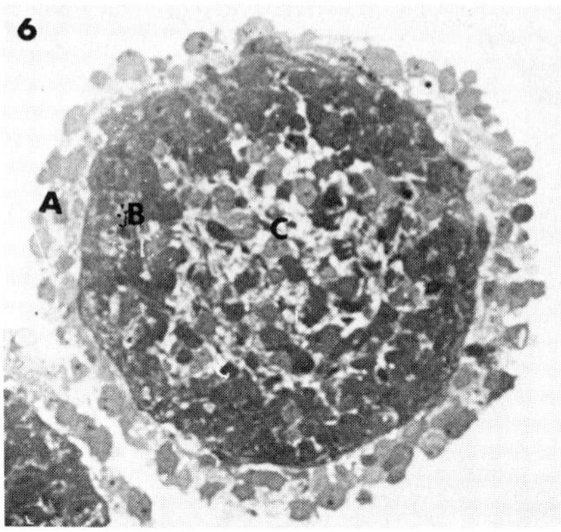

7a

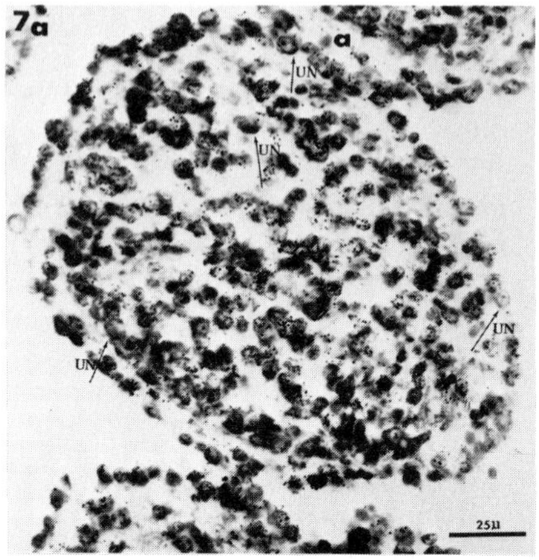

8

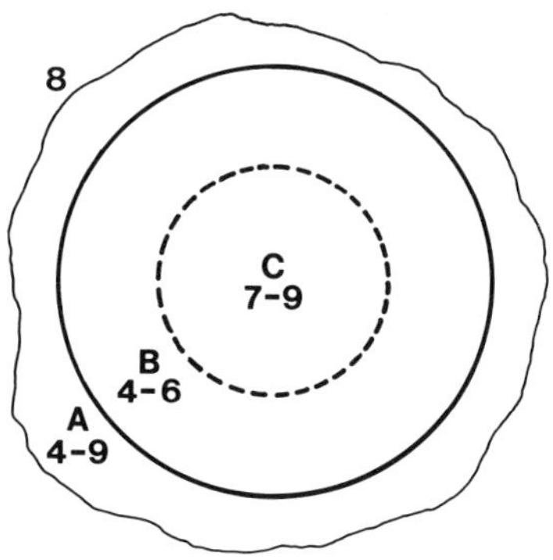

7b

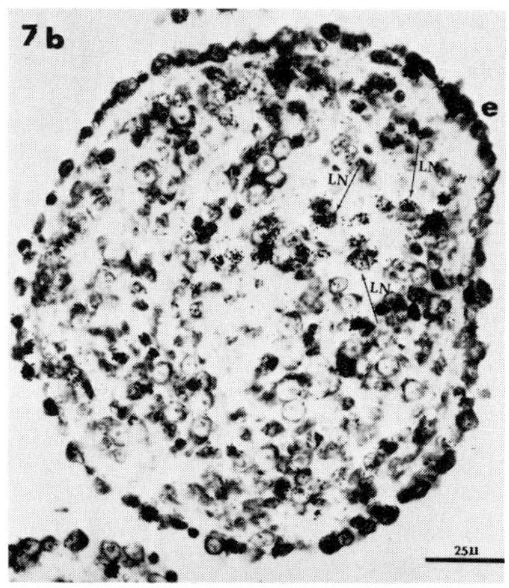

FIGS. 6,7,8. Autoradiographic analysis of chick brain cell aggregates. (From Tsai, 1977).
FIG. 6. One micron plastic section through the center of a 10-6 chick cerebral aggregate stained with methylene blue. Note 3 concentric areas: A) an external crown layer, B) a peripheral cortical-like zone, and C) a core region. (10-6: dissociated from 10-day embryonic tissue, cultured for 6 days in rotary suspension).
FIG. 7. Autoradiographs, counterstained with cresyl violet, of a) 5*-10-6 cerebral aggregate (labeled *in ovo* on embryonic day 5), showing the presence of unlabeled neurons (UN), with birthdates earlier than day 5, in the periphery of the aggregate and b) a 9*-10-6 cerebral aggregate (labeled *in ovo* on embryonic day 9), showing labeled neurons (LN), with birthdates on day 9, preferentially in the core region of the aggregate.
FIG. 8. Diagram of a 10-6 cerebral aggregate mapping the birthdates of the component neurons.

aggregates maintained longer than 2 weeks. These interesting observations bearing on the restrictions of the proliferative compartment *in situ* merit further investigation.

Another critical problem in this regard is the identity of cells that participate in the reassembly process during the formation and reorganization of aggregates. Since it has been found that the aggregability of a particular brain cell population is correlated with its proliferative activity *in situ*, certain questions arise: is it essentially the proliferating set of cells in the neuroepithelium that reconstructs the aggregate; do newly-born postmitotic cells retain the capacity to aggregate and if so, for how long; do older differentiating post-mitotic cells fail to be incorporated into the aggregates? These questions are being addressed systemically and some provocative answers have emerged. On the basis of autoradiographic birthdating, chick telencephalic aggregates have been analyzed to determine the time of origin for each of their cellular components, also, the histogenetic patterning and differentiative fate with respect to the time of cell generation has been investigated (Tsai, 1976, 1977). These studies exploit the fact that tritiated thymidine incorporated into neural cells undergoing a final round of DNA synthesis marks post-mitotic neurons with a permanent radioisotopic label that can be visualized by autoradiographic procedures. By designing a timed series of injections *in ovo*, it is possible to determine the precise birthdates for individual neurons (Sidman, 1970b; LaVail and Cowan, 1971; Tsai, 1977). In the chick egg labeling is cumulative because the tritiated thymidine is not excreted to the outside, and subtractive methods were used to define neuronal birthdates. Since the chick telencephalon has generated all of its neurons by day 10, a series of embryos were labeled *in ovo* on successive days of development, days 3-9, and telencephalic aggregates were prepared on day 10, then cultured for 6 days *in vitro* (these aggregates are designated as 10-6 aggregates). Autoradiographs of aggregates labeled on day 4 (4*-10-6; *designates the day of isotope injection) showed that *all* cells were labeled and indicated that the reaggregated neurons were derived from precursor cells actively dividing up to at least the fourth day of development. Autoradiographs of 5*-10-6 aggregates showed the first *unlabeled* cells in the outermost regions of the differentiated aggregate. Since these cells did not incorporate tritiated thymidine on day 5, they had already been generated as post-mitotic cells on day 4. Autoradiographs of 6*-10-6 aggregates showed an increase in the number of unlabeled cells in the outer region of the aggregate. The difference between the number of labeled cells seen in 5*-10-6 and 6*-10-6 aggregates represents the number of cells generated on day 5. By carrying this kind of analysis through a comprehensive series of labeling schedules, the following conclusions were drawn (see Fig. 6-8). First, neurons of all birthdates, days 4-9, were found in chick telencephalic aggregates in the same proportion as they occurred *in situ.* There was no evidence of selection or preferential survival for particular generations of neurons after dissociation, reaggregation and culture *in vitro.* Second, the histogenetic patterning within the aggregate, based on the times of neuronal origin, was basically an "outside-in" pattern, i.e., the earliest-generated neurons (birthdates 4-6) were preferentially localized in the outer zone of the aggregate, while the latest-generated neurons (birthdates 7-9) were concentrated in the innermost region. The significance of this result is that an "outside-in" pattern is also characteristic of the developing chick telencephalon *in situ* (Tsai, 1976, 1977). These findings strongly suggest that in the absence of pre-organized proliferative epithelium as a starting gate for outward migration of cells, the telencephalic neurons are able to recognize one another on the basis of cell type and reorganize by migrating within the aggregate to appropriate positions characteristic of their histotypic pattern *in situ.* Further autoradiographic studies have shown that there is no selective loss of specific neuronal cell generations with prolonged culture *in vitro* up to 16 days, and that the differentiative status of the cells at the time of dissociation — whether neuronal precursors still capable of mitosis, undifferentiated young post-mitotic neurons, or older differentiated neurons — does not limit their potential for expressing advanced states of neuronal differentiation, judged by the conventional staining procedures compatible with autoradiographic processing.

The sorting out of cell types within aggregates assumes special significance in the context of brain development because of the critical role this selective cell association plays in the establishment of functional neural circuitry. Cell sorting within a random assembly of reaggregated cells requires recognition mechanisms not only for preferential adhesion between certain cells, but also for the movements of cells from their initial chance site in the aggregate to their final position in the internally reorganized tissue. The "outside-in" pattern of chick telencephalic aggregates described above presents one example of brain cell sorting. In an effort to elucidate mechanisms underlying cell recognition and directed cell migration, another aggregating brain cell system has been systemically analyzed. This is the coaggregate system of commingled cerebral (telencephalic) and optic tectum cells using heterospecific cell combinations as a cell marking device (Garber and Moscona, 1972a; Spiegel and Garber, 1974; Spiegel, 1977). The discovery and preparation of tissue-specific aggregation factors from cell-free supernatants of plated brain cell cultures (Garber and Moscona, 1969, 1972b) invited the question of whether these neural cell recognition factors, which selectively promoted the aggregation of homologous cell types, also played a role in the internal organization of the reconstituted brain tissue. Therefore, a series of experiments was designed to provide a simple assay for cell sorting. Experimental coaggregates of dissociated

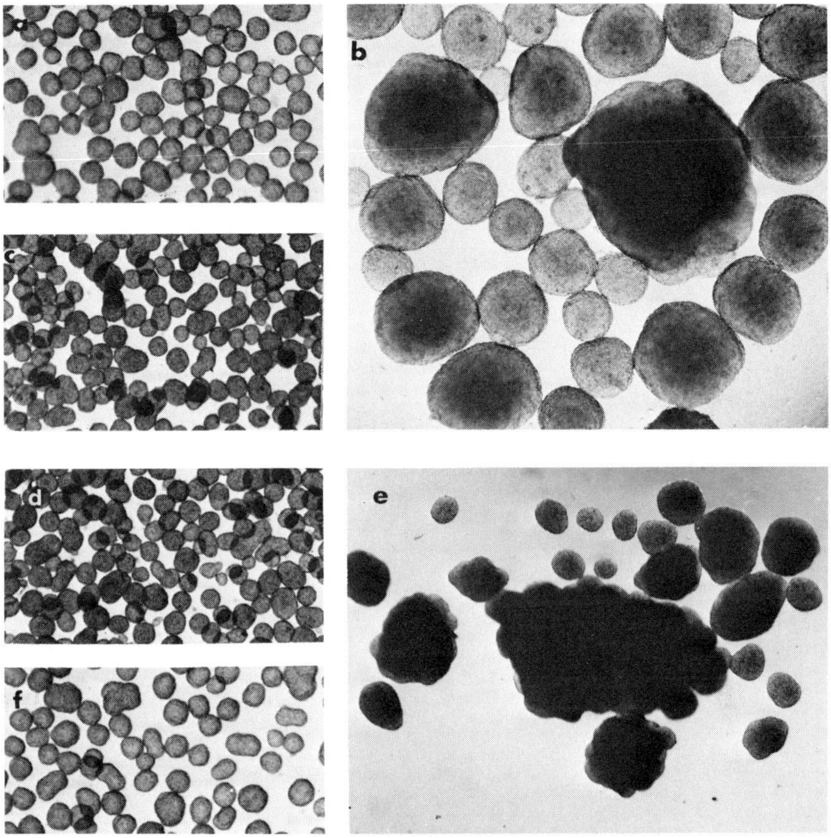

FIG. 9. Type-specific enhancement of brain cell aggregation by CRF (cerebrum cell recognition factor) and OTF (optic tectum cell recognition factor). 24-hr aggregates. ×20. a) cerebrum cell aggregates formed in control medium (serum-free). b) cerebrum cell aggregates formed in serum-free medium + CRF. c) optic tectum cell aggregates formed in serum-free medium + CRF. d) optic tectum cell aggregates formed in control serum-free medium. e) optic tectum cell aggregates formed in serum-free medium + OTF. f) cerebrum cell aggregates formed in serum-free medium + OTF. (From Spiegel, 1977).

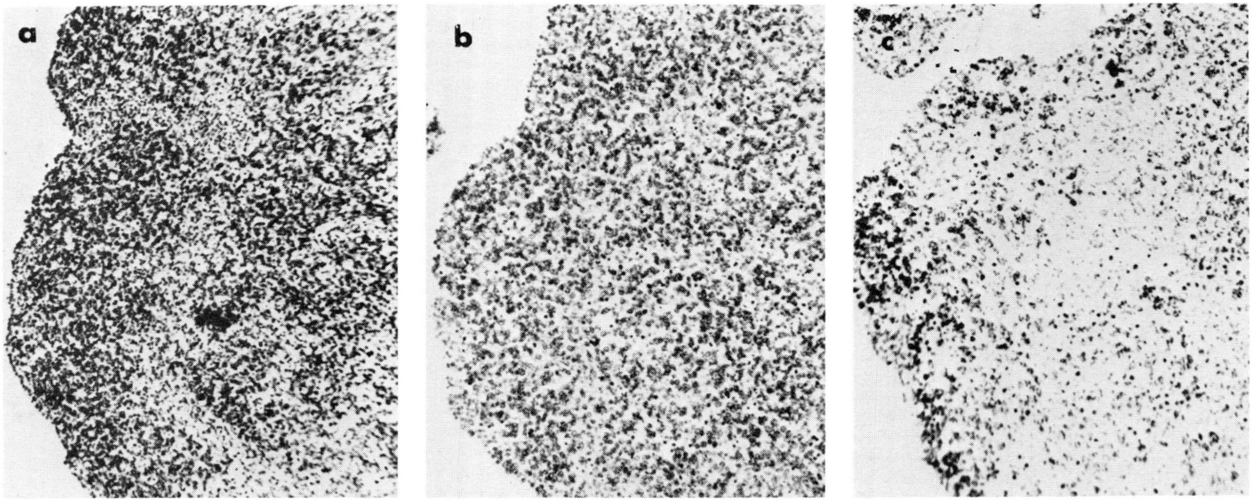

FIG. 10. Sorting out of brain cells in response to type-specific cell recognition factors. Histological sections of 24-hr co-aggregates formed by commingled suspension of dissociated 14-day embryonic mouse cerebrum cells (dark staining) and 10-day embryonic chick optic tectum cells (light staining) ×100. (From Spiegel, 1977). a) In serum-free medium + CRF. Note the marked selective association of dark cerebrum cells. b) In control serum-free medium. Note the homogeneous intermixing cerebrum and tectum cells. c) In serum-free medium + OTF. Note the marked selective association of light tectum cells.

14-day embryonic mouse cerebrum cells and 8-day chick optic tectum cells were fabricated either in the absence or presence of 1) a cerebral cell recognition factor which specifically promoted the aggregation of cerebral cells only, or 2) a tectal cell recognition factor which specifically promoted the aggregation of tectal cells only (Fig. 9). In the absence of serum, cerebral and tectal cells were found to be homogeneously intermingled in a 24 hr. aggregate (Fig. 10a). However, the addition of either neural recognition factor to the coaggregating cell suspension within the first 5-6 hr. of culture resulted in a distinctive and characteristic type of histogenetic patterning. When the cerebral cell recognition factor was added to this commingled cell suspension of cerebral and tectal cells, the mouse cerebral cells associated selectively and formed massive cell groups in a consistent pattern, illustrated in Fig. 10b. Conversely, if the optic tectum cell recognition factor was added to an identical dissociated cell mixture, an entirely different histogenetic pattern resulted from the selective association of chick tectal cells into massive groups (Fig. 10c). Long term cultures of these coaggregates indicated that the migratory propensities of the 8-day tectal cell population further influences the asymmetry of the sorted cell patterns. A quantitative analysis of this sorting out behavior has verified these conclusions based on histological assessments and opens new opportunities for rapid computer analysis of histogenetic assays for cell recognition as an obvious direction for the future.

The study of brain cell differentiation in cell aggregates developing *in vitro* offers other intriguing possibilities for revealing the essential cell interactions and environmental requirements that regulate the expression of specific neural phenotypes. The determination of advanced neural differentiation and the classification of brain cell types in reconstituted aggregates was severely limited until recently by dependence on conventional histological techniques. The successful application of Palacios's (1970) methods for Golgi impregnation to cerebral cell aggregates (Fig. 11) revealed the remarkable degree of differentiation obtained in these reconstructed brain tissues (Palacios *et al.*, 1974, 1975, 1976). Subsequent correlation with the Golgi cytoarchitectural atlas of the chick telencephalon *in situ* completed recently by Palacios (1976) has permitted reliable morphological identification of cell classes on the basis of their full silhouette, including extensive dendritic arborizations and axonal outgrowth, and a more rigorous analysis of the cytoarchitectural organization of the 10-6 aggregate. It has been found that multipolar corpus striatal cell types occupy the interior of the aggregate, while pyramidal and granule cells, typical of the dorsolateral cortex, are localized in the outer cortical-like layer of the aggregate. This organization, moreover, correlates very well with the audioradiographic data for cell origins cited earlier (Tsai, 1977), and experiments are now in progress to refine our understanding of

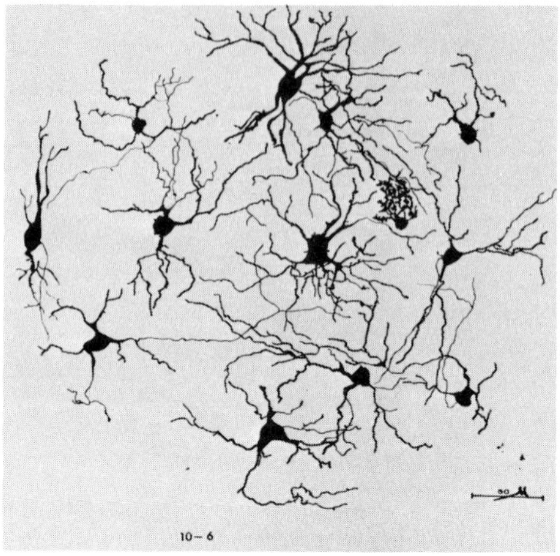

FIG. 11. Cytoarchitectural differentiation within a 10-6 chick cerebral cell aggregate (formed from dissociated 10-day embryonic chick cerebrum tissue and cultured for 6 days in rotary suspension). This drawing is reconstructed from Golgi impregnated aggregates to illustrate the various cell types characteristically found in these aggregates, their localization and organization. (From Palacios, Garber, and Larramendi, 1974).

these cellular relationships. Electron microscopic studies of developing aggregates has shown the presence of unusual elongated wide-gap appositions containing dense intercellular material and the rapid appearance of abundant synaptic connections. These cell surface specializations are being investigated carefully to determine if they are involved in self-assembly processes and histogenetic patterning (Larramendi and Garber, 1976).

The biochemical differentiation of brain cell aggregates has also been explored, particularly with respect to neurotransmitter substances and their relevant enzymes. Aggregates reconstituted from dissociated whole embryonic mouse brain were shown to develop a number of these specialized nerve cell products in amounts comparable to those found in adult brain *in situ* (Seeds, 1971; Seeds and Gilman, 1971, Honneger and Richelson, 1976). In addition, experimental interaction between cells from various chick brain cell regions resulted in significant alterations in levels of neurotransmitter and enzyme synthesis (Adler and Teitelman, 1974; Adler *et al.*, 1976).

In order to obtain a direct morphological and biochemical correlation of a specific brain cell class, our attention has turned recently to aggregate cell systems of CNS catecholamine neurons. These cells are unique in their advantages for neurogenetic analysis. Individual neurons can be identified and visualized morphologically by virtue of their catecholamine content, using Falck-Hillarp histofluorescence techniques (Falck and Owman, 1965), and are also amenable to rigorous biochemical analysis of their neurotransmitter content by very sensitive radiochemical assays (Coyle and

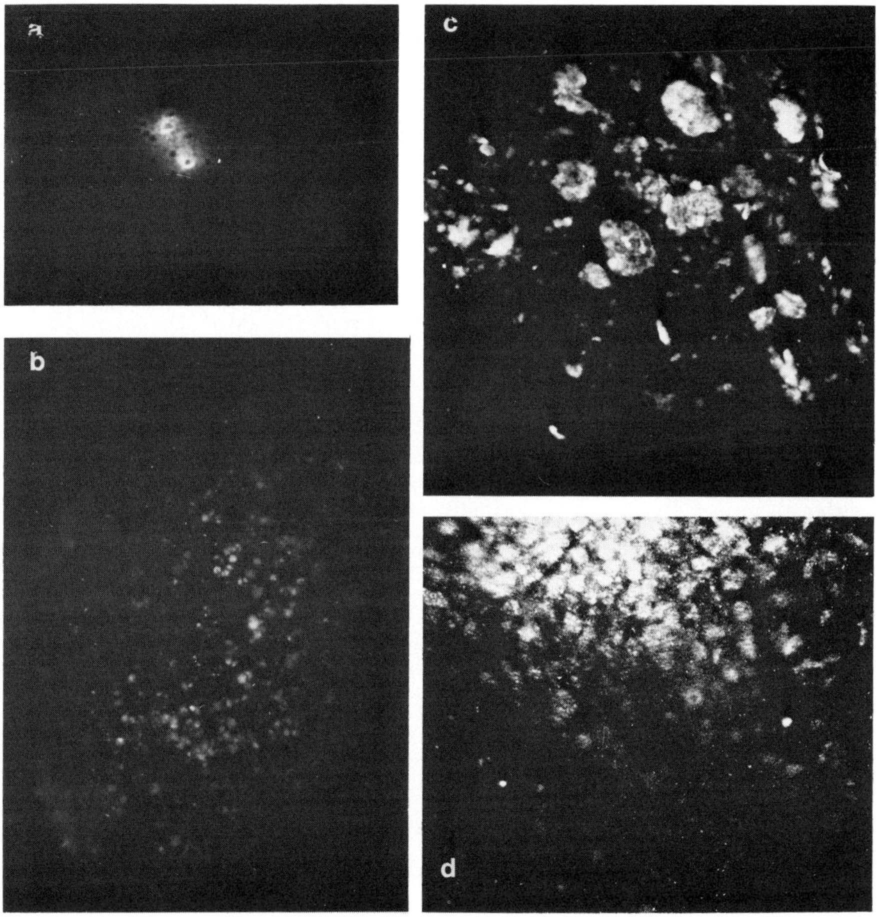

FIG. 12. Selective association of catecholamine-containing neurons (CA) in aggregates formed by cells dissociated from 15-day embryonic mouse midbrain stem. a) Fluorescent cells in suspension immediately after dissociation. ×200. b) 24-hr aggregate, showing dispersed dimly fluorescent CA neurons. ×100. c) 48-hr aggregate, showing clusters of highly fluorescent CA neurons. ×200. d) 96-hr aggregate, showing association of intensely fluorescent cells into a single dense band. ×200.

Henry, 1973). In a recent study of aggregates formed from dissociated mouse midbrain stem tissue containing the distinctive dopamine neurons of the substantia nigra (Levitt *et al.*, 1976), it was found that these neurons retained their histo-fluorescent properties immediately after dissociation and could be traced throughout the aggregation process. They were found to be randomly distributed in aggregates after 24 hr of culture. However, after 48 hr the dopamine neurons had formed numerous cell clusters with greatly increased fluorescence and by 4-5 days had selectively associated into elongated bands, resembling the substantia nigra structure at a comparable embryonic age *in situ* (Fig. 12). In addition, these intensely fluorescent cells developed abundant varicosities at a distance from the cell bodies, suggesting extensive terminal aborization. Biochemical analyses of these aggregates showed a large increase in dopamine levels with ongoing culture *in vitro* up to 8 days. Since present evidence indicates that these dopamine neurons are post-mitotic and overtly differentiated at the time of dissociation (Olson and Seiger, 1972; Hanaway *et al.*, 1972; Golden, 1973; Lauder and Bloom, 1974), these findings of selective cell grouping and increased neurotransmitter synthesis cannot reasonably be attributed to cell proliferation or *de novo* expression of catecholamine synthesis, but rather indicate self-assembly of pre-existing dopamine neurons. It seems likely that this mutal cell recognition displayed by the dopamine neurons in cell aggregates may be linked to catecholamines or associated moieties acting at cell surfaces during the developmental period predating their neurotransmitter function. This hypothesis is being tested in our laboratories using pharmacological agents that affect catecholamine synthesis, uptake, storage, release and receptor mechanisms.

The specific experimental studies described above serve to illustrate the analytical potential of reaggregated neural cell systems *in vitro*. Although this approach is just at the beginning of its development as a research tool in neurobiology, its significant contributions to the understanding of cell interaction and recognition during all phases of neural differentiation are stimulating increasing interest and encouraging optimism that some of the classical enigmas concerning ontogenetic mechanisms related to nerve cell proliferation, directed cell migration, neural specificities, and phenotypic expression may soon prove susceptible to these experimental strategies.

Acknowledgments

Portions of this research were supported by PHS grants NS 10714 and CA 19265.

References

Adler, R. 1970. Changes in reaggregation pattern during neural tube differentiation. Develop. Biol. 21:403–423.

Adler, R. and G. Teitelman. 1974. Aggregates formed by mixtures of embryonic neural cells: Activity of enzymes of the cholinergic system. Develop. Biol. 39:317–321.

Adler, R., G. Teitelman and A. Suburo. 1976. Cell interactions and the regulation of cholinergic enzymes during neural differentiation *in vitro*. Develop. Biol. 50:48–57.

Altman, J. and G. D. Das. 1966. Autoradiographic and histological studies of postnatal neurogenesis. I. A longitudinal investigation of the kinetics, migration and transformation of cells incorporating tritiated thymidine in neonate rats, with special reference to postnatal neurogenesis in some brain regions. J. Comp. Neurol. 126:337–390.

Angevine, J. B. 1965. Time of neuron origin in the hippocampal region: an autoradiographic study in the mouse. Exp. Neurol. 13, Suppl. 2:1–70.

Angevine, J. B. 1970. Time of neuron origin in the diencephalon of the mouse. An autoradiographic study. J. Comp. Neurol. 139:129–188.

Angevine, J. B. and R. L. Sidman. 1961. Autoradiographic study of cell migration during histogenesis of cerebral cortex in the mouse. Nature (Lond.) 192:766–768.

Balsamo, J. and J. Lilien. 1974a. Embryonic cell aggregation: Kinetics and specificity of binding of enhancing factors. Proc. Nat. Acad. Sci. U.S.A. 71:727–731.

Balsamo, J. and J. Lilien. 1974b. Functional identification of three components which mediate tissue-type specific embryonic cell adhesion. Nature (Lond.) 251:522–524.

Black, I. B., I. A. Hendry and L. L. Iversen, 1971. Trans-synaptic regulation of growth and development of adrenergic neurons in a mouse sympathetic ganglion. Brain. Res. 34:229–240.

Coyle, J. T. Jr. and D. Henry. 1973. Catecholamines in fetal and newborn rat brain. J. Neurochem. 21:61–68.

DeLong, G. R. 1970. Histogenesis of fetal mouse isocortex and hippocampus in reaggregating cell cultures. Develop. Biol. 22:563–583.

Ebbeson, S. O. E. 1972. A proposal for a common nomenclature for some optic nuclei in vertebrates and the evidence for a common origin of two such cell groups. Brain, Behav. Evol. 6:75–91.

Falck, B. and C. Owman. 1965. A detailed morphological description of the fluorescence method for the cellular demonstration of biogenic amines. Acta Univ. Lund II 7:1–23.

Falck, B., N. A. Hillarp, G. Thieme and A. Thorp. 1962. Fluorescence of catecholamines and related compounds condensed with formaldehyde. J. Histochem. Cytochem. 10:348–354.

Fischman, D. A. 1970. The synthesis and assembly of myofibrils in embryonic muscle. Curr. Topics Develop. Biol. 5:235-280.

Fujita, S. 1964. Analysis of neuron differentiation in the central nervous system by tritiated thymidine autoradiography. J. Comp. Neurol. 122:311-328.

Garber, B. 1967. Aggregation *in vivo* of dissociated cells. II. Role of developmental age in tissue reconstruction. J. Exp. Zool. 164:339-350.

Garber, B. 1976. Control of epithelial development. Curr. Prob. Dermatol. 6:154-190.

Garber, B. and A. A. Moscona. 1967. Suppression of feather morphogenesis in coaggregates of skin cells from embryos of different ages. J. Exp. Zool. 164:351-362.

Garber, B. B. and A. A. Moscona. 1969. Enhancement of aggregation of embryonic brain cells by extracellular materials from cultures of brain cells. J. Cell. Biol. 43:41a.

Garber, B. and A. A. Moscona. 1972a. Reconstruction of brain tissue from cell suspensions. I. Aggregation patterns of cells dissociated from different regions of the developing brain. Develop. Biol. 27:217-234.

Garber, B. B. and A. A. Moscona. 1972b. Reconstruction of brain tissue from cell suspensions. II. Specific enhancement of aggregation of embryonic cerebral cells by supernatant from homologous cell cultures. Develop. Biol. 27:235-271.

Golden, G. 1973. Prenatal development of the biogenic amine systems of the mouse brain. Develop. Biol. 33:300-311.

Hamburger, V. 1934. The effects of wing bud extirpation on the development of the central nervous system in chick embryos. J. Exp. Zool. 68:449-494.

Hanaway, J., J. A. McConnell and M. G. Netsky. 1972. Histogenesis of the substantia nigra, ventral tegmentum area of Tsai and interpeduncular nucleus. An autoradiographic study of the mesencephalon in the rat. J. Comp. Neurol. 142:59-74.

Harrison, R. G. 1910. The outgrowth of the nerve fiber as a mode of protoplasmic movement. J. Exp. Zool. 9:787-846.

Hinds, J. W. 1968a. Autoradiographic study of histogenesis in the mouse olfactory bulb. I. Time of origins of neurons and neuroglia. J. Comp. Neurol. 134:287-304.

Hinds, J. W. 1968b. Autoradiographic study of histogenesis in the mouse olfactory bulb. II. Cell proliferation and migraton. J. Comp. Neurol. 134:305-322.

Honneger, P. and E. Richelson. 1976. Biochemical differentiation of mechanically dissociated mammalian brain in aggregating cell culture. Brain Res. 109:335-354.

Ishii, K. 1966. Reconstruction of dissociated chick brain cells in rotation mediated culture. Cytologia 31:89-98.

Jacobson, M. 1970. *Developmental Neurobiology*. Holt, Rinehart and Winston, New York.

Kahn, A. J. 1974. An autoradiographic analysis of the time of appearance of neurons in the developing chick neural retina. Develop. Biol. 38:30-40.

Karten, H. J. 1969. The organization of the avian telencephalon and some speculations on the phylogeny of the amniote telencephalon. Ann. N.Y. Acad. Sci. 167:164-179.

Larramendi, L. M. H. and B. Garber. 1976. Electron-dense intercellular material in wide-gap appositions between neurons in chick embryo telencephalic aggregates. Neuroscience Abstr. 2:198.

Lauder, J. M. and F. E. Bloom. 1974. Ontogeny of monamine neurons in the locus coeruleus, raphe nuclei and substantia nigra of the rat. I. Cell differentiation. J. Comp. Neurol. 155:469-482.

LaVail, J. and M. Cowan. 1971. The development of the chick optic tectum II. Autoradiographic studies. Brain Res. 28:421-441.

LeDouarin, N. 1973. A biological cell labeling technique and its use in experimental embryology. Develop. Biol. 30:217-222.

Levitt, P., R. Y. Moore and B. Garber. 1976. Selective cell association of catecholamine neurons in brain aggregates *in vitro*. Brain Res. 111:311-320.

Miale, I. L. and R. L. Sidman. 1961. An autoradiographic analysis of histogenesis in the mouse cerebellum. Exp. Neurol. 4:277-296.

Moscona, A. A. 1961. Rotation-mediated histogenetic aggregation of dissociated cells: a quantifiable approach to cell interactions *in vitro*. Exp. Cell Res. 22:455-475.

Moscona, A. A. 1962. Cellular interactions in experimental morphogenesis. Int. Rev. Pathol. 1:371-428.

Nornes, H. O. and G. D. Das. 1972. Temporal pattern of neurogenesis in spinal cord: cytoarchitecture and directed growth of axons. Proc. Nat. Acad. Sci. U.S.A. 69:1962-1966.

Nornes, H. O. and G. D. Das. 1974. Temporal pattern of neurogenesis in spinal cord of rat I. An autoradiographic study — time and sites of origin and migration and settling patterns of neuroblasts. Brain Res. 73:121-138.

Olson, L. and A. Seiger. 1972. Early prenatal ontogeny of central monoamine neurons in the rat: fluorescence biochemical observations. Z. Anat. Entwickl.-Gesch. 137:301-316.

Palacios, E. 1970. Two useful variations of the Golgi silver-chromate method. Acta Cient. Venez. 21:105-106.

Palacios, E. 1976. Golgi studies of the telencephalon of the domestic chicken (Gallus domesticus). Anat. Rec. 184:495.

Palacios, E., B. Garber and L. M. H. Larramendi. 1974. Silver chromate impregnation of chick embryo brain cell aggregates. Brain Res. 66:173-178.

Palacios, E., B. Garber and L. M. H. Larramendi. 1975. The "Reazione Nera": A tool for the analysis of neuronal aggregates reconstituted *in vitro* from dissociated embryonic brain cells. In: *Golgi Centennial Symposium: Perspectives in Neurobiology*. M. Santini, ed. Raven Press, New York. pp. 593-601.

Palacios, E., B. Garber and L. M. H. Larramendi. 1976. Neuronal organization of aggregates prepared from chick embryo striatum and entorhinal cortex. Proc. Cajal Club.

Rakic, P. 1971. Neuron-glia relationship during granule cell migration in developing cerebellar cortex: A Golgi and electronmicroscopic study in Macacus rhesus. J. Comp. Neurol. 141:283-312.

Rakic, P. 1972. Mode of cell migration to the superficial layers of fetal monkey neocortex. J. Comp. Neurol. 145:61-84.

Rakic, P. 1974. Neurons in rhesus monkey visual cortex: Systemic relation between time of origin and eventual disposition. Science 183:425-427.

Ramon y Cajal. S. 1929. *Studies on Vertebrate Neurogenesis.* (Translated by L. Guth, 1960). Thomas, Springfield, Illinois.

Seeds, N. W. 1971. Biochemical differentiation in reaggregating brain cell cultures. Proc. Nat. Acad. Sci. U.S.A. 68:1858-1861.

Seeds, N. W. and A. G. Gilman. 1971. Norepinephrine stimulated increase of cyclic AMP levels in developing mouse brain cultures. Science 174:292.

Sheffield, J. B. and A. A. Moscona. 1970. Electron microscopic analysis of aggregation of embryonic cells: the structure and differentiation of aggregates of neural retina cells. Develop. Biol. 23:36-61.

Shimada, Y., D. A. Fischman and A. A. Moscona. 1969. Formation of neuromuscular junctions in embryonic cell cultures. Proc. Nat. Acad. Sci. U.S.A. 62:715-721.

Sidman, R. L. 1961. Histogenesis of the mouse retina studied with tritiated thymidine. In: *The Structure of the Eye.* G. K. Smelser, ed. Academic Press, New York. pp. 487-505.

Sidman, R. L. 1970a. Cell proliferation, migration, and interaction in the developing mammalian central nervous system. In: *The Neurosciences: Second Study Program.* F. O. Schmitt, ed. MIT Press, Cambridge, Mass. pp. 100-116.

Sidman, R. L. 1970b. Autoradiographic methods and principles for study of the nervous system with thymidine-H^3. In: *Contemporary Research Methods in Neuroanatomy.* W. J. H. Nauta and S. O. E. Ebbesson, eds. Springer-Verlag, New York, pp. 252-274.

Simpson, S. B. Jr. 1970. Studies on regeneration of the lizard's tail Am. Zool. 10:157-165.

Sperry, R. W. 1965. Embryogenesis of behavioral nerve nets. In: *Organogenesis.* R. L. DeHaan and H. Ursprung, eds. Holt, Rinehart and Winston, New York. pp. 161-186.

Spiegel, J. 1977. Enhancement of cell sorting in brain cell co-aggregates by nerve cell recognition factors. Thesis, University of Chicago.

Spiegel, J. and B. Garber, 1974. Enhancement of cell sorting in brain cell coaggregates by nerve cell recognition factors. Anat. Rec. 178:470.

Taber Pierce, E. 1973. The time of origin of neurons in the brain stem of the mouse. Prog. Brain Res. 40:53-65.

Townes, P. and J. Holtfreter. 1955. Directed movements and selective adhesion of embryonic amphibian cells. J. Exp. Zool. 128:53-120.

Tsai, H. M. 1976. Autoradiographic studies of chick telencephalic histogenesis *in vivo* and *in vitro.* Anat. Rec. 184:550.

Tsai, H. M. 1977. An autoradiographic analysis of histogenesis with H^3-thymidine in the developing chick telencephalon *in situ* and *in vitro.* Thesis. University of Chicago.

Weiss, P. 1934. *In vitro* experiments on the factors determining the course of the outgrowing nerve fiber. J. Exp. Zool. 68:393-448.

Weiss, P. and B. Garber. 1952. Shape and movement of mesenchyme cells as functions of the physical structure of the medium. Contributions to a quantitative morphology. Proc. Nat. Acad. Sci. U.S.A. 38:264-280.

Wilson, H. V. 1907. On some phenomena of coalescence and regeneration in sponges. J. Exp. Zool. 5:245-258;

Wolff, E. 1954. Potentialités et affinités des tissus reveleés par la culture in vitro d'organes en associations héterogènes et xénoplastiques. Bull. Soc. Zool. Fr. 79:357-367.

ORGANIZATON OF PATHWAYS IN CULTURES OF MOUSE CEREBELLUM

W. J. HENDELMAN, K. C. MARSHALL, A. S. AGGERWAL AND J. M. WOJTOWICZ

Departments of Anatomy and Physiology
University of Ottawa
Faculty of Medicine
Ottawa, Ontario
Canada

I. Introduction

Studies of neurons in tissue culture should provide data concerning the structure and function of the neuron. This would establish the integrity of tissue culture as a model system in the study of neurobiology and neuroanatomy. In this chapter we have reported results of the development of neurons and synaptic specificity in cultures of mouse cerebellum. Experiments have been conducted to correlate the morphology (synaptology) with function.

The cerebellum has been intensively investigated in organotypic culture (Maximow type), and by means of the light microscope (Wolf, 1964; Allerand, 1971; Seil, 1972) and the electron microscope (Hendelman, 1967; Wolf and Dubois-Dalcq, 1970; Seil and Herndon, 1970; Kim, 1970; Privat and Drian, 1975). These investigations have revealed characteristic cerebellar cells and synapses in tissue cultures. The cultures have been studied under experimental conditions and changes have been found in their growth and synaptology (Wolf, 1970; Privat and Drian, 1976).

The reports of others have not focused on the detailed synaptic organizaton of cerebellar cultures. Since the process of synaptogenesis takes place almost totally *in vitro* and in the absence of extracerebellar afferents and targets, the first question concerns the specificity of synaptic connections in such cultures. Electrophysiological activity has been recorded in the cerebellar cultures (Leiman and Seil, 1973; Calvet *et al.*, 1974; Calvet and Lepault, 1975), showing functional connections established *in vitro*, particularly in the cortical region. Using another type of culture system other investigators have studied the activity of the Purkinje neuron (Schlapfer *et al.*, 1972) and its response to neuroactive drugs (Gähwiler, 1975).

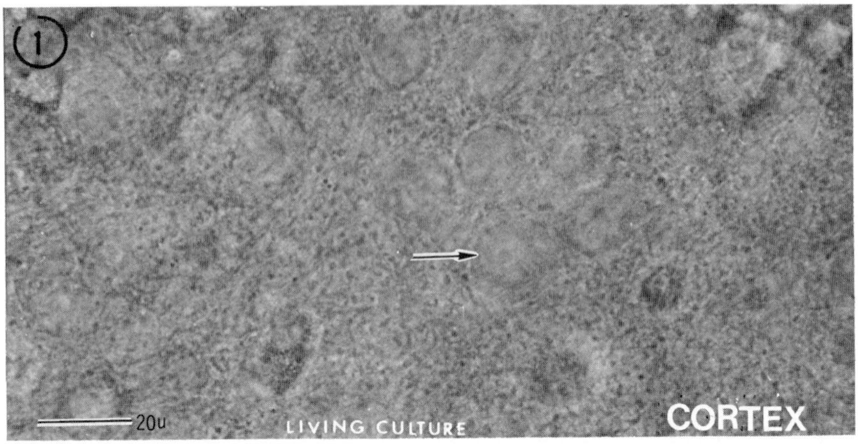
1
20u
LIVING CULTURE
CORTEX

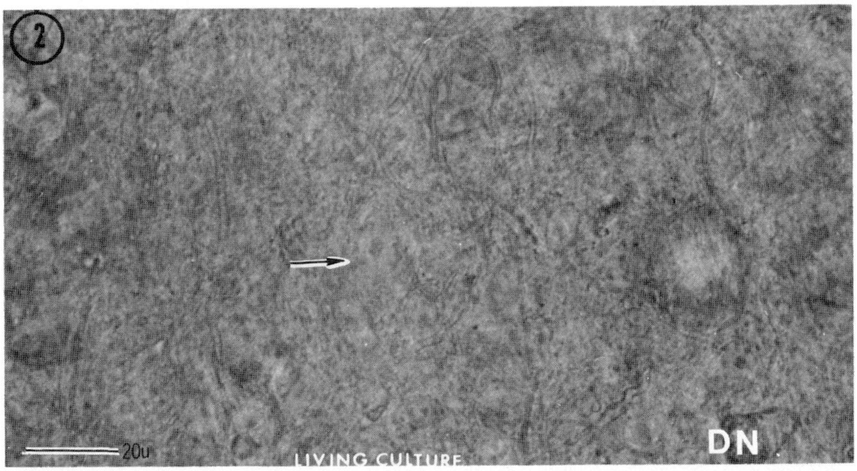
2
20u
LIVING CULTURE
DN

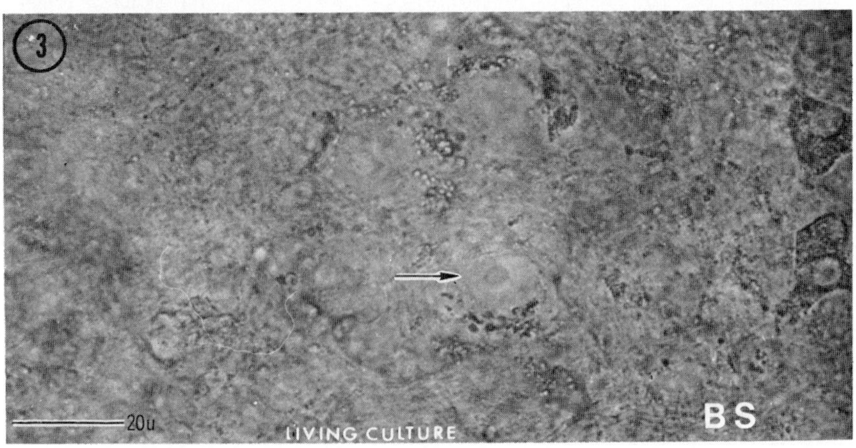
3
20u
LIVING CULTURE
BS

Neuronal pathways however, have not been studied in cerebellar cultures. Bunge and his co-workers have investigated synapse formation between the superior cervical ganglion and the spinal cord (Olson and Bunge, 1973; Bunge *et al.*, 1974; Ko *et al.*, 1976). Crain and his co-workers have investigated connections formed between dorsal root ganglion (DRG) neurons and other CNS explants (Crain and Peterson, 1974). It is now known that the DRG axons establish connections preferentially with certain appropriate regions of the spinal cord and the medulla (Crain, 1976).

II. Living Cultures

The cultures of newborn mouse cerebellum were explanted into the Maximow double coverslip assembly according to established practices (Wolf, 1964; Allerand, 1971). The medium consisted of 50% Eagle's MEM, 25% chick embryo extract (EE-50) and 25% human cord serum, supplemented with glucose to a final concentration of about 1000 mg %. Such cultures mature in 20 days and are suitable for experimental purposes until 35 days *in vitro* (DIV).

The cultures were explanted so as to retain their topographic features and this topographic organization has been the basis for the analytic work, both the electron microscopy and the electrophysiology. Three regions are distinctive. The cortical region is characterized by Purkinje cells (Fig. 1). Their number and random distribution and uniformity permit the cortical region to be identified. The surrounding neuropil presents a unique punctate appearance. Myelinated fibers originate from this region and converge upon another group of neurons near the ciliated ependyma. These deep nuclear neurons (DN) are not easily visualized and the region becomes recognizable by the presence of myelinated axons weaving through the area (Fig. 2), (Hendelman and Aggerwal, 1974). The region is considerably smaller than the cortical region. Explants from the peduncular region also contain a distinctive cluster of neurons, characterized to a certain extent by their refractile granules and their large size. Since they originate from the brain stem, they have been classified for the time being as brain stem neurons (BS) (Fig. 3). Until their identity is definitively established, we have not accepted their classification by other authors as vestibular neurons (Wolf, 1964; Wolf and Dubois-Dalcq, 1970; Allerand, 1971).

FIGS. 1, 2, 3: Neurons characteristics of the 3 regions of the culture are illustrated in living cultures photographed in the Maximow chamber. Fig. 1-Purkinje neuron (arrow) of the cortex region; Fig. 2-neuron of the deep cerebellar nucleus (DN) (arrow); Fig. 3-brain stem neuron (BS) (arrow) with characteristic granules in the cytoplasm.

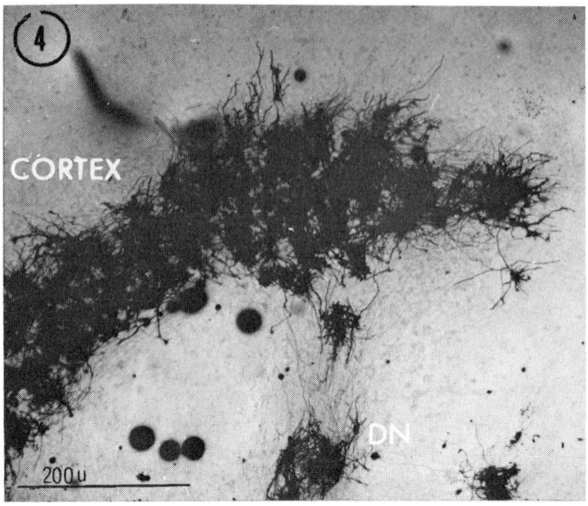
4
CORTEX
DN
200u

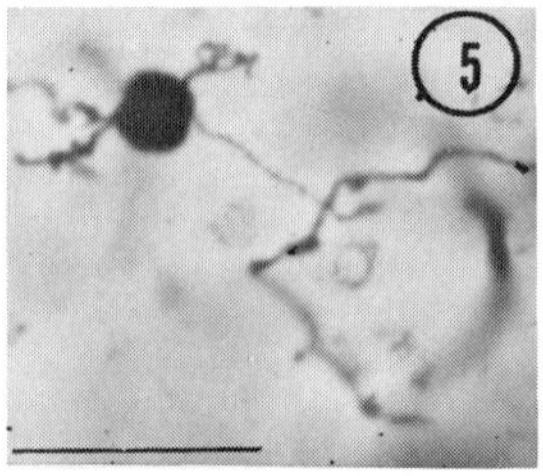
5

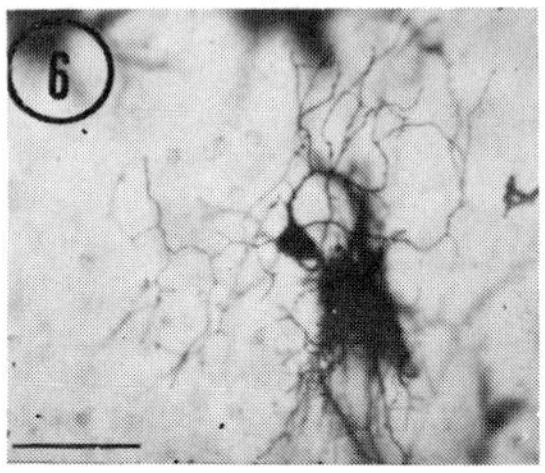
6

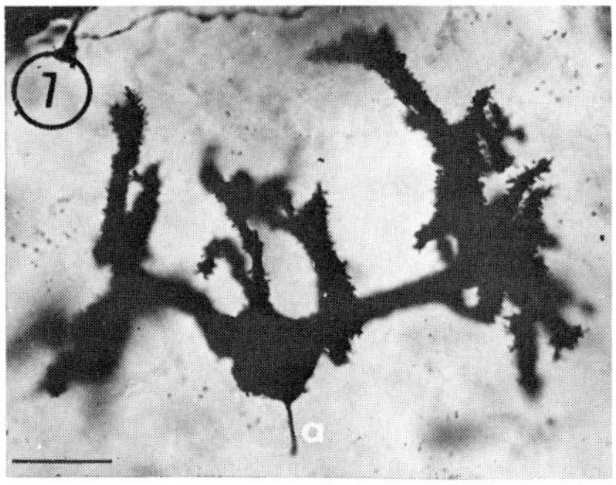
7
a

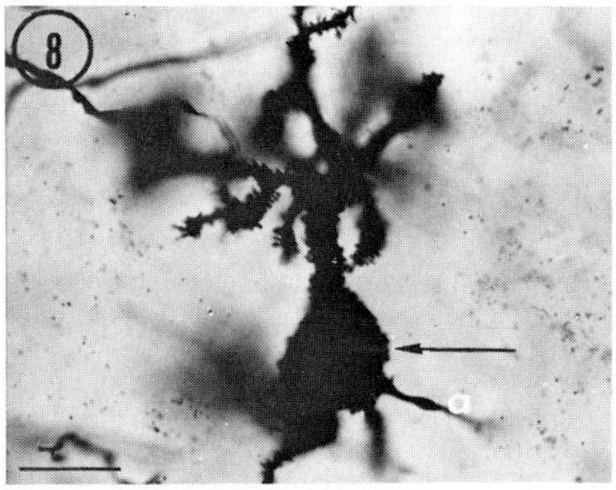
8
a

9

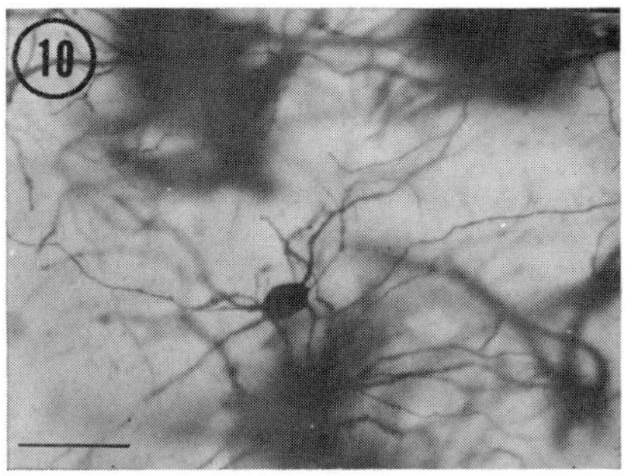
10

III. Golgi Studies

Cultures were fixed and stained as whole mounts with a modification of the Golgi-Cox technique. (Details of this method will be published elsewhere). In select cultures, both the cortical region and the DN region were stained, with enough separation between them to allow for their correct identification (Fig. 4). In the cortex region certain typical neurons were identifiable, namely granule cells (Fig. 5) with a characteristic claw, and the Golgi cell with its axonal arborization (Fig. 6).

In the cerebellar cortex region, the characteristic cell is the Purkinje neuron (Figs. 7, 8, 9). Its most recognizable feature is the presence of dendritic spines. The Purkinje neuron differs from its counterpart in the intact animal by the presence of several dendrites emerging from its soma region, and by the presence of spines on the soma. These are known as perisomatic dendrites and perisomatic spines, respectively. These differences are accounted for in the developmental history of the Purkinje neuron (Aggerwal, 1974; Hendelman and Rouf, 1974). The DN neurons are completely different (Fig. 10). They are multipolar neurons with long dendrites and few if any spines (Hendelman and Aggerwal, 1974). The BS neurons did not stain with the Golgi method.

IV. Electrophysiology

Select cultures were placed in a specially designed chamber mounted on the stage of an inverted microscope. This allowed stimulating and recording electrodes to be introduced from above, while neurons and electrode tips could be seen from below. The system was designed to control the pH (7.3 - 7.4), the temperature (35° - 37°C) and the rate of flow of fluid (modified Earle's BSS) over the culture. Recordings were done with single or multibarrelled glass micropipettes. The recording electrode was filled with 3N NaCl for extracellular recording and its resistence measured 5 - 10 Megohms. Other barrels were filled with test solutions for iontophoretic application. The stimulating electrode was made of platinum or tungsten needles, insulated to within 50 μm from the tip.

The spontaneous spiking activity of the different neurons was recorded (Fig. 11). Approximately one-half of the cortical neurons, presumably Purkinje cells, fire spontaneously in almost all cultures. The most characteristic pattern is an irregular firing interrupted by short bursts (upper record Fig. 11), although other patterns of firing are also seen. In most cultures there was no spontaneous activity, of either DN or BS

FIGS. 4-10: Golgi preparations of cerebellar cultures: Fig. 4, cortex-deep nuclear neurons(DN) Fig. 5, granule cell; Fig. 6. Golgi cell; Figs. 7, 8, 9, Purkinje neurons; Fig. 10, Deep nuclear neuron (DN). Bar in Fig. 4=200 μm and in all others 20 μ.

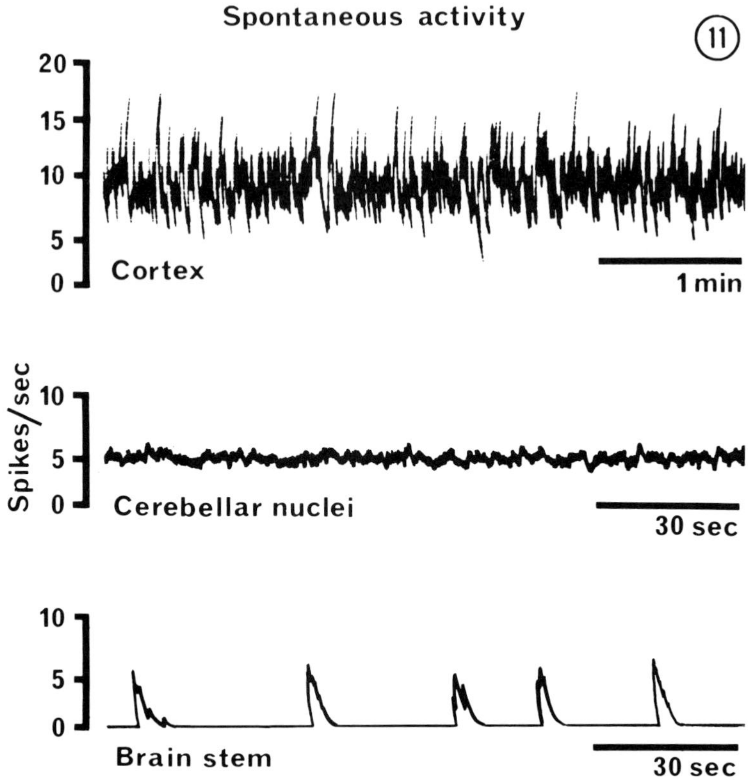

FIG. 11: Ratemeter records of spontaneous activity. The height of the record reflects the frequency of firing of the neurons indicated.

neurons. In those cultures which showed spontaneous activity, the pattern of firing of DN neurons was usually either regular (middle record, Fig. 11) or bursting (not shown). BS neurons sometimes had a phasic pattern of firing (lower record, Fig. 11) not seen in other neuron types. Electrophysiological interactions were studied between different regions of the cultures.

A. Pathways from Cerebellar Cortex to Deep Nuclear Neurons

Stimulation of the cortex region produces a depression of activity of DN neurons (Wojtowicz *et al.*, 1976). This depression lasts for 20-300 msec, and is frequently followed by a rebound excitation. The depression is blocked by a specific gamma-aminobutyric acid (GABA) blocker, bicuculline (4-12 μM), introduced into the perfusing solution. The depression of

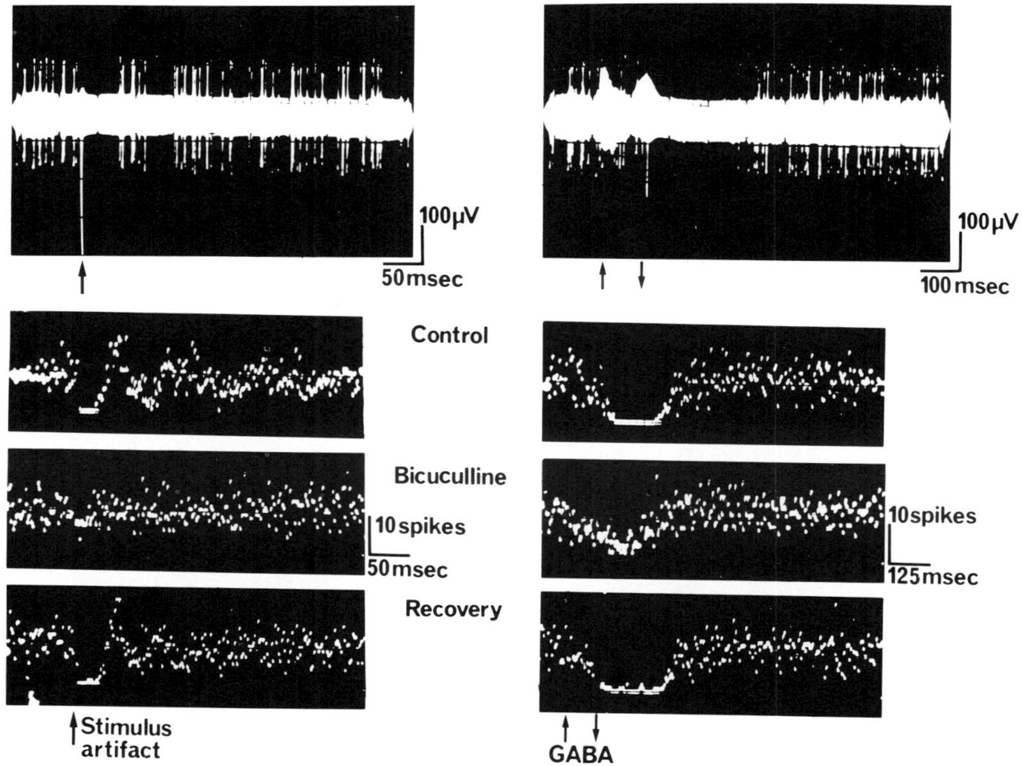

FIG. 12: Left Column: Electrical stimulation with single pulses in the cortical region results in inhibition of a DN neuron (top photograph, 10 superimposed oscilloscope sweeps). The lower 3 records are poststimulus histograms constructed from 200 superimposed sweeps (bin width 2 msec). The depression (upper of the records) was blocked by bicuculline (11.0 μM) (middle record) and recovery was complete (lower record).

Right Column: GABA iontophoresis to the DN neuron inhibits its activity.

Top Photograph: arrow up starts, and arrow down ends GABA application. The poststimulus histograms (200 oscilloscope sweeps, bin width 5 msec) show the inhibition (uppermost record), the bicuculline effect (middle record), and recovery (lower record).

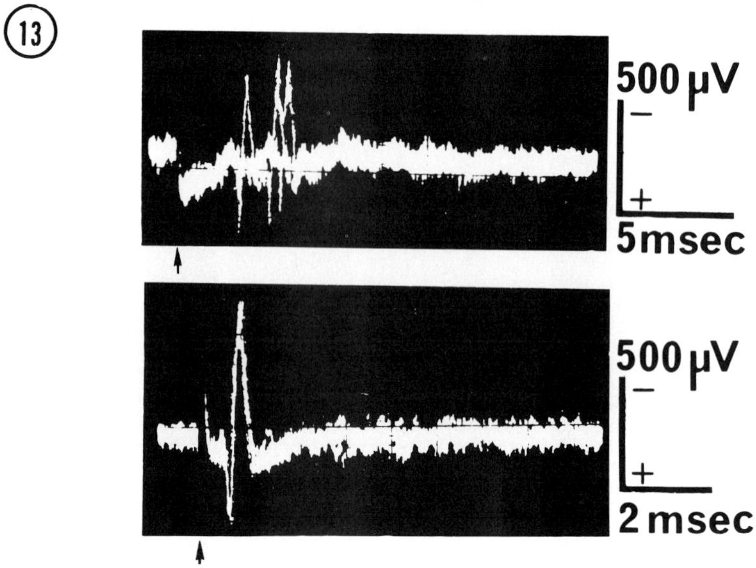

FIG. 13: The upper tracing shows evoked apparently synaptic responses in cortical neuron in culture after stimulation of the explant in the DN area, 500 μm away (5 superimposed sweeps). Note that 3 of the 5 stimuli resulted in an evoked spike. The lower record shows the antidromic responses in a DN neuron in the culture after stimulation in the cortical region of the explant, 500 μm away (5 superimposed sweeps). The arrow indicates the time of stimulus application. Spikes were retouched for clarity.

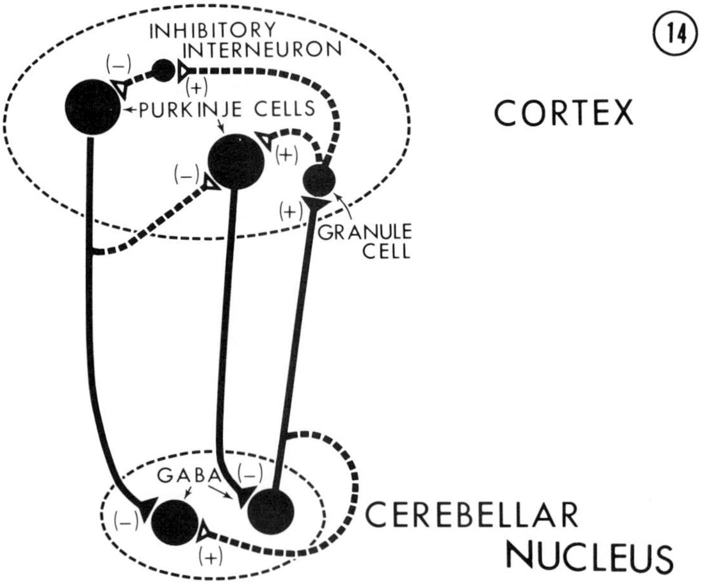

FIG. 14: Schematic representation of synaptic connections in cultures of mouse cerebellum. Cortex = cerebellar cortex; Cerebellar nucleus = deep cerebellar nucleus (DN). Solid lines indicate pathways based upon experimental data; interrupted lines are proposed pathways.

DN neuron activity could be reproduced by applying GABA iontophoretically onto the same neuron and the depression could be blocked by bicuculline (Fig. 12). The results are consistent with the known projection of the cerebellar cortex to DN and lateral vestibular neurons, and the proposed GABA neurotransmission (e.g., Obata *et al.*, 1970). These studies indicate that a specific pathway has been reproduced in CNS culture and that neurotransmission is mediated by the same neuro transmitter (GABA) as *in vivo*.

B. Pathways From Deep Nuclear Neurons to Cerebellar Cortex

The existence of a synaptic pathway from the DN to the cerebellar cortex has been disputed in the literature. The existence of such a pathway was first observed in our cultures (Wojtowicz *et al.*, 1975). With the stimulating electrode in the cortex and recording from the DN area, antidromic activation was frequently recorded (Fig. 13, lower record). Antidromic activation was identified by the collision technique (Wojtowicz *et al.*, 1977). With the stimulating electrode in the DN, evoked activity, apparently synaptic, was recorded in the cortical region (Fig. 13, upper record). Recently, the pathway from DN to cortex has been demonstrated in the intact animal by several neuroanatomical and neurophysiological techniques (Tolbert *et al.*, 1976; Bloedel *et al.*, 1976; Gould and Graybiel, 1976).

The results of the studies conducted on these pathways in culture can be summarized in a diagram (Fig. 14). The synaptic interactions depicted incorporate both the known electron microscopic data, the electrophysiologic data, and some as yet unconfirmed assumptions. For example, it is not clear whether the connection from DN to cortex is monosynaptic to the Purkinje cell or multisynaptic. The termination of all axons is schematicaly placed.

C. Brain Stem Neurons

Despite their lack of identity, the BS cells have been studied to establish their relationship with the cerebellar cortex. Antidromic activation has been obtained (Fig. 15a) after stimulation of the cortex. The BS neurons have been inhibited by stimulation of the cortex region in a manner analogous to the DN neurons. Bicuculline blocked both evoked inhibition (Fig. 15b) and inhibition caused by iontophoretic applicaton of GABA. However no consistent evidence has been found for any synaptic connections which these neurons may form in the cortical area.

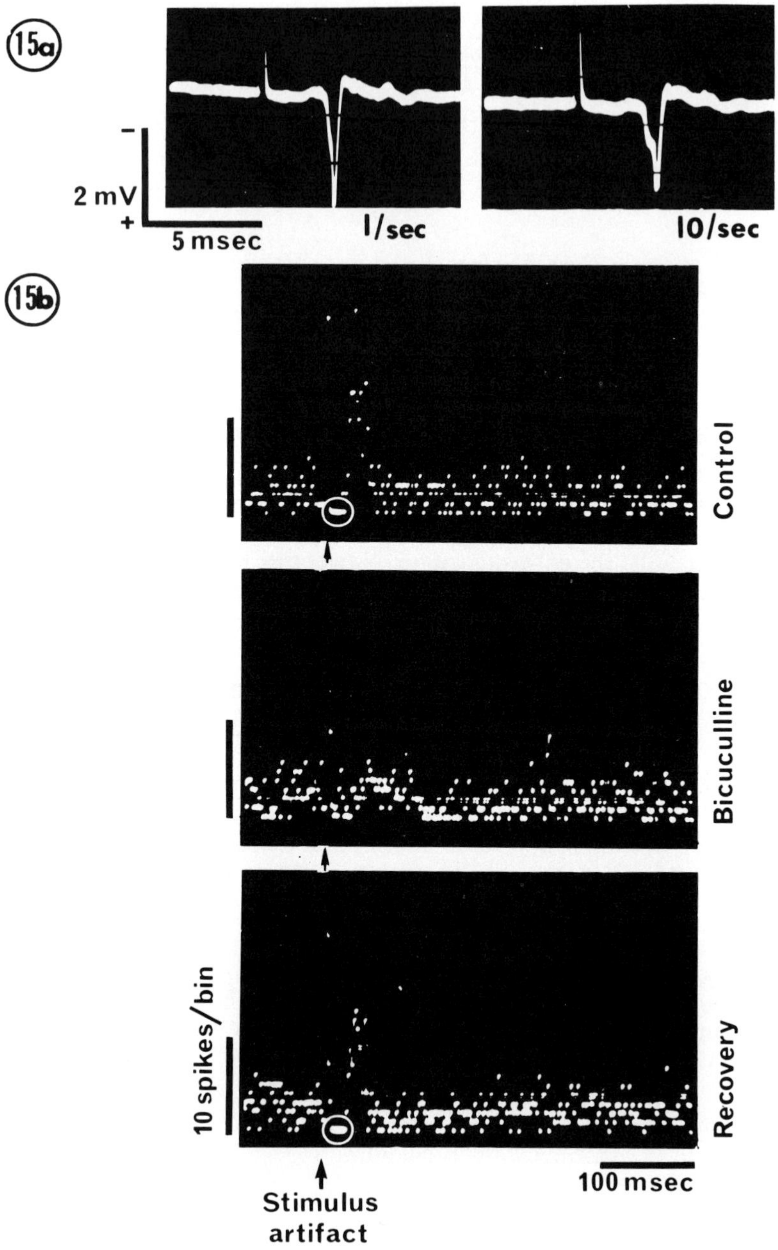

Legend: page 551

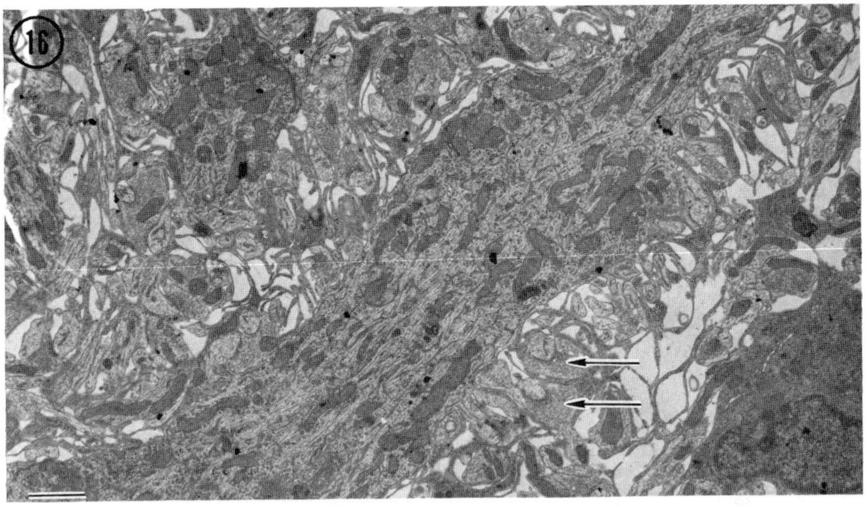

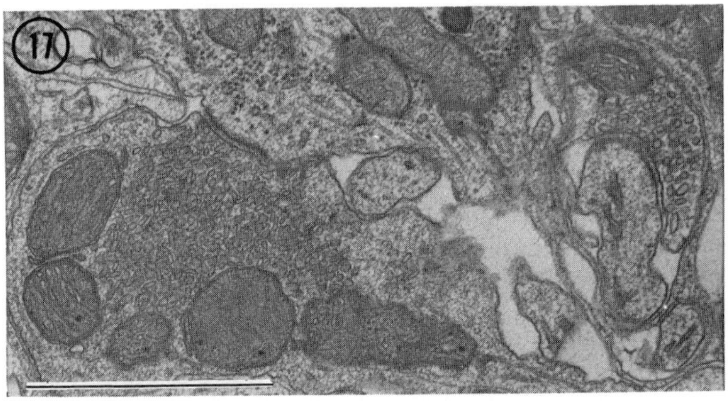

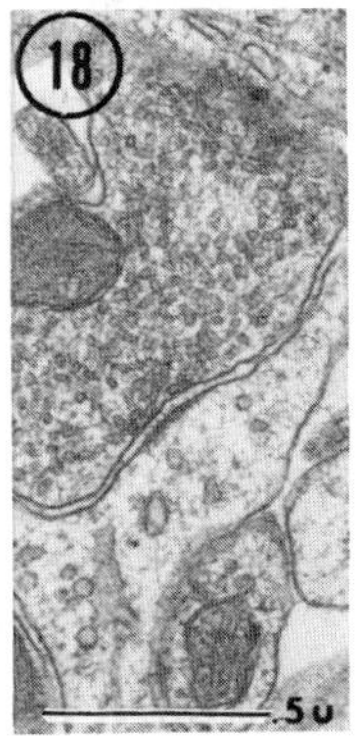

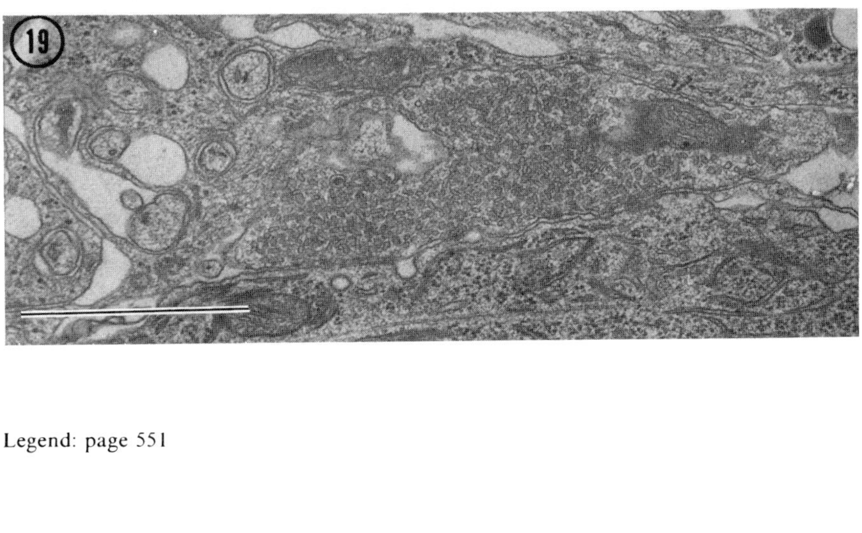

Legend: page 551

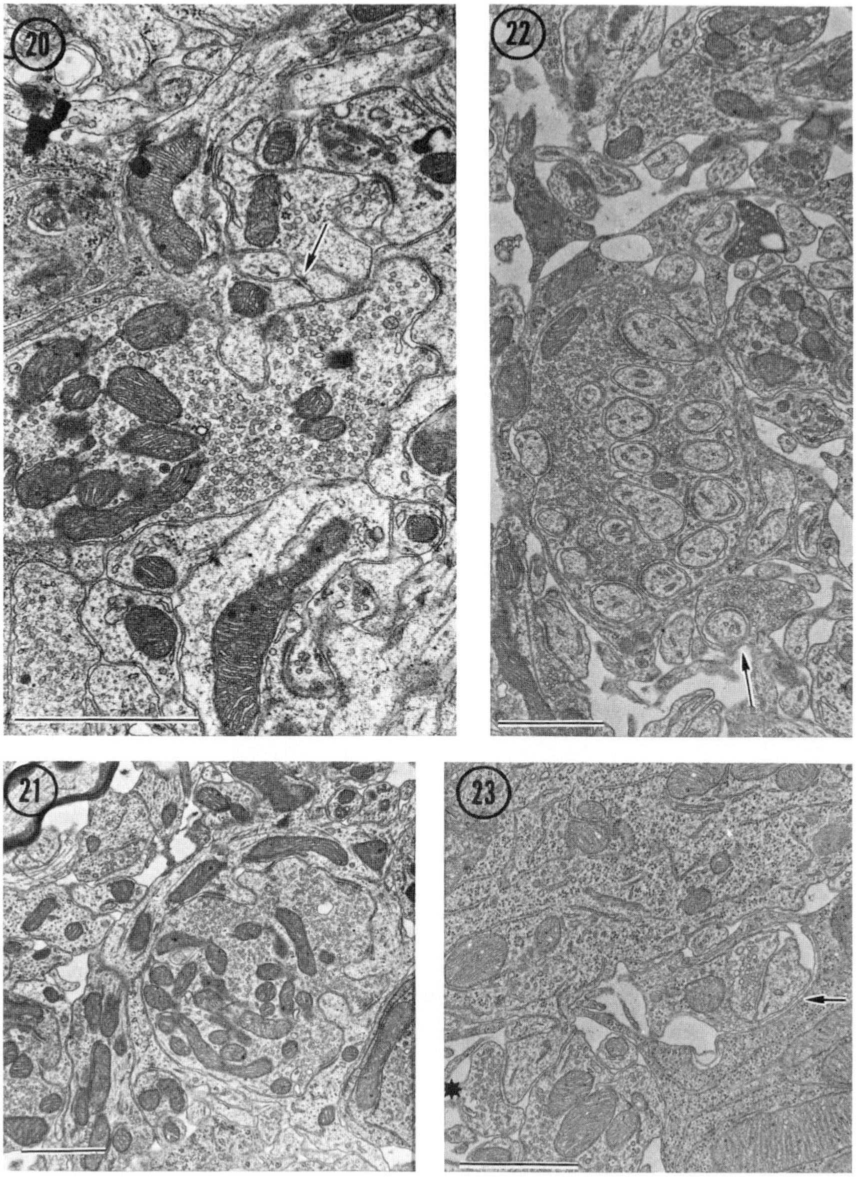
20
22
21
23

V. Electron Microscopy

The cultures were fixed in glutaraldehyde-formaldehyde and post osmicated using Millonig's buffer with 350 milliosmoles additional glucose. This fixation has revealed differences in the synaptic types present in these cultures.

The cortical region of the culture has been intensively studied to provide the evidence for the schema proposed in Fig. 14. Large dendrites have been identified and their size matches the size anticipated from the Golgi studies (Fig. 16). Many spines are seen adjacent to the dendrite and some are seen emerging. Some of the spines are contacted by terminals whose morphology matches that of parallel fibers. Other spines are without a presynaptic element in this plane of section. Recurrent collaterals have been identified by their pleomorphic vesicles and dark axoplasm (Fig. 17). Golgi terminals are present, with their small vesicles and the symmetric type contact (Fig. 18). A stellate terminal is seen forming a symmetric contact with the surface of a Purkinje neuron (Fig. 19). These terminals are characteristic and match those found in the intact animal (e.g., Palay and Chan-Palay, 1974).

There are terminals present in the cortical region which are reminiscent of mossy fibers (Figs. 20, 21): the terminal is large and forms multiple synaptic contacts around its periphery. Sometimes the post-synaptic elements are joined by desmosomal adhesions typical of granule cells (Fig. 20) and sometimes the post-synaptic element is indented in a "claw-like" manner (Fig. 21). This type of terminal has been reported repeatedly

FIG. 15a: Antidromic responses in a BS neuron after cerebellar cortical stimulation in the explant, 700 μ away (several superimposed sweeps).
FIG. 15b: Stimulation of the cerebellar cortical region inhibits the activity of BS neuron (white circle in upper record), followed by an excitatory response. This effect is blocked by bicuculline (8.5 μM) (middle record). The response returns after removal of the drug (lower record).The post-stimulus histograms were constructed from 200 oscilloscope sweeps (bin width was 2 msec).
FIG. 16: Large dendrite of Purkinje neuron: arrows indicate parallel fibre synapses. Bar = 1 μm.
FIG. 17: Recurrent collateral of Purkinje neuron. Bar = 1 μm.
FIG. 18: Golgi terminal. Bar = 0.5 μm.
FIG. 19: Stellate type terminal on the soma of a Purkinje neuron. Bar = 1 μm.
FIG. 20: Large terminal forming multiple synaptic contacts. Arrow points to desmosonal adhesion between two post-synaptic elements. Bar = 1 μm.
FIG. 21: Large terminal forming contacts with dendritic "claws". Bar = 1 μm.
FIG. 22: An hypertrophic parallel fiber terminal forming asymmetric synapses with many Purkinje dendritic spines. The usual 1:1 relationship is also shown (arrow). Bar = 1 μm.
FIG. 23: Terminal with flattened vesicles forming a contact with a spine emerging from a Purkinje neuron (*). The arrow indicates a normal type synapse with a similar spine. Bar = 1 μm.

(Hendelman, 1967; Wolf and Dubois-Dalcq, 1970; Kim, 1970; Aggerwal and Hendelman, 1976; Privat and Drian, 1975). Nonetheless, the origin of these terminals is still not clear. One of the possibilities is that these are terminals from the deep nuclear neurons, since the terminals are present in cultures which have DN neurons (Hendelman, 1972). This proposal still lacks conclusive experimental evidence.

In the neuropil, the parallel fibers were found to form multiple synaptic contacts with Purkinje dendritic spines (Fig. 22). This configuration has never been reported *in vivo* or in culture and represents an example of synaptic plasticity. Other types of terminals are present, such as the ones with flattened vesicles forming a contact with a spine (Fig. 23). It is possible that this is an abnormality of the culture or that this terminal represents the BS neurons, whose axons are known to enter the cortical region.

Conclusions

There seems little doubt that organotypic cultures do reproduce many of the patterns characteristic of the original tissue. In the case of cerebellar cultures, the cerebellar cortex-DN pathway is present as expected and a "new" pathway, DN to cerebellar cortex was reported before being demonstrated in the intact animal. The cerebellar cortex-DN connection seems to have the same neurotransmitter, GABA, as *in vivo.*

In addition to the organotypic features, there have been alterations to the cells. The morphology of the Purkinje neuron is different (Aggerwal, 1974) than *in vivo.* At the electron microscopic level, it is known that parallel fibers in culture form many more synapses with dendritic spines than normally (Aggerwal and Hendelman, 1975, 1976). It is known that the typical basket array is not formed and other examples of modifications and plasticity may yet be revealed.

At the outset, it was anticipated that the use of the model system would result in a correlation between the structure and the function. This has not yet been achieved. The best known pathway is that from cerebellar cortex to DN and it has not proven feasible as yet to map the terminations on the DN neurons. Likewise, it is still a mystery which terminals originate from the DN neurons and terminate in the cerebellar cortex. Other synapses and connections still are unknown.

The most unsettling problem is the consideration that abnormal connections form in such cultures. As yet, there is scant evidence that this occurs to any degree at all. However, the experimental model is now available to attempt to induce such errors, and to try to discover the factors that regulate synapse formation.

Addendum: Additional data on the electrophysiological activity of cerebellar cultures has appeared since this paper was submitted (Seil and Leiman, 1977.)

Acknowledgments

The authors are indebted to the devoted technical expertise of Brigitte Eiselt and Marianne Hegström. This work was supported by grants to W. J. Hendelman and K. C. Marshall from the Medical Research Council of Canada. Dr. Aggerwal is a fellow of the Medical Research Council and J. M. Wojtowicz is supported by the Ontario Mental Health Foundation.

References:

Aggerwal, A. S. 1974. The growth and development of Purkinje cells *in vitro*. Anat. Rec. 178:296.

Aggerwal, A. S. and W. J. Hendelman. 1975. Synapses on Purkinje somatic spines in organotypic cultures. Neurosci. Abstr. 1:815

Aggerwal, A. S. and W. J. Hendelman. 1976. Morphology, ultrastructure and synaptology of granule cells in organotypic cultures. Proceedings of the XXV annual conference of the Anatomical Society of India in Lucknow, India.

Allerand, C. D. 1971. Patterns of neuronal differentiation in developing cultures of neonatal mouse cerebellum: A living and silver impregnation study. J. Comp. Neur. 142:167-204.

Bloedel, J. R., H. Bantli and D. L. Tolbert. 1976. Cerebellar nucleocortical projection in the Rhesus monkey. Neuroscience abstracts 2, p. 106.

Bunge, R. P., R. Rees, P. Wood, H. Burton and C. P. Ko. 1974. Anatomical and physiological observations on synapses formed on isolated autonomic neurons in tissue culture. Brain Res. 66:401-412.

Calvet, M. C. and A. M. Lepault. 1975. *In vitro* Purkinje cell electrical behaviour related to tissular environment. Exp. Brain Res. 23:249-258.

Calvet, M. C., M. J. Drian and A. Privat. 1974. Spontaneous electrical patterns in cultured Purkinje cells grown with an antimitotic agent. Brain Res. 79:285-290.

Crain, S. M. 1976. *Neurophysiologic studies in tissue cultue.* Raven Press, New York.

Crain, S. M.and E. R. Peterson. 1974. Enhanced afferent synaptic functions in fetal mouse spinal cord-sensory ganglion explants following NGF-induced ganglion hyperophy. Brain Res. 79:145-152.

Gähwiler, B. H. 1975. The effects of GABA, picrotoxin and bicuculline on the spontaneous bioelectric activity of cultured cerebellar Purkinje cells. Brain Res. 99:85-95.

Gould, B. B. and A. M. Graybiel. 1976. Afferents to the cerebellar cortex in the cat: evidence for an intrinic pathway leading from the deep nuclei to the cortex. Brain Res. 110:601-611.

Hendelman, W. J. 1967. Electron microscope observations on tissue cultures of cerebellum. J. Cell. Biol. 35:169A.

Hendelman, W. J. 1972. The synaptology of cerebellar cultures of the vermis. Anat. Rec. 172:327.

Hendelman, W. J. and A. Aggerwal. 1974. Deep nuclear neurons of mouse in cerebellar organotypic cultures: a living and Golgi study. Proc. Can. Fed. Biol. Soc. 17:61.

Hendelman, W. J. and R. Rouf. 1974. The development of the Purkinje cell of the mouse: a Golgi analysis. Anat. Rec. 178:372.

Kim, S. U. 1970. Observations on cerebellar granule cells in tissue culture. A silver and electron mocroscopic study. Z. Zellforsch. 107:454-465.

Ko, C. P., H. Burton and R. P. Bunge. 1976. Synaptic transmission between rat spinal cord explants and dissociated superior cervical ganglion neurons in tissue culture. Brain Res. 117:437-460.

Leiman, A. L. and F. J. Seil. 1973. Spontaneous and evoked bioelectric activity in organized cerebellar tissue cultures. Exp. Neurol. 40:748-758.

Obata, K., K. Takeda and H. Shinoraki. 1970. Further study on pharmacological properties of the cerebellar-induced inhibition of Deiters' neurons. Exp. Brain Res. 11:327-342.

Olson, M. L. and R. P. Bunge. 1973. Anatomical observations on the specificity of synapse formation in tissue culture. Brain Res. 59:19-33.

Palay, S. L. and V. Chan-Palay. 1974. *Cerebellar Cortex. Cytology and Organization.* Springer-verlag New York, Heidelberg, Berlin.

Privat, A. and M. J. Drian. 1975. Specificity of the formation of the mossy fiber-granule cell synapses in the rat cerebellum. An *in vitro* study. Brain Res. 88:518-524.

Privat, A. and M. J. Drian. 1976. Postnatal maturation of rat Purkinje cells cultivated in the absence of two afferent systems: An ultrastructural study. J. Comp. Neurol. 166:201-244.

Schlapfer, W. T., A. M. Mamoon and C. A. Tobias. 1972. Spontaneous bioelectric activity of neurons in cerebellar cultures: evidence for synaptic interactions. Brain Res. 45:345-363.

Seil, F. J. 1972. Neuronal groups and fiber patterns in cerebellar tissue cultures. Brain Res. 42:33-51.

Seil, F. J. and R. M. Herndon. 1970. Cerebellar granule cells *in vitro*. A light and electron microscope study. J. Cell Biol. 45:212-220.

Seil, F. J. and A. L. Leiman. 1977. Spontaneous versus driven activity in intra cerebellar nuclei: A tissue culture study. Exp. Neurol. 54:110-127

Tolbert, D. L., H. Bantli and J. R. Bloedel. 1976. A cerebellar nucleo-cortical projection in the cat. Anat. Rec. 184:p 547.

Wojtowicz, J. M., K. C. Marshall and W. J. Hendelman. 1975. Electrical activity of cerebellar neurones in tissue culture. Canada Physiology. 6:69.

Wojtowicz, J. M., K. C. Marshall and W. J. Hendelman. 1976. Studies of synaptic connections in cerebellar tissue cultures. Neurosci. Abstr. 2:1031.

Wojtowicz, J. M., K. C. Marshall and W. J. Hendelman. 1977. Depression by magnesium ion of neuronal excitability in tissue cultures of central nervous system. Can. J. Physiol. and Pharmacol. (in press).

Wolf, M. K. 1964. Differentiation of neuronal types and synapses in myelinating cultures of mouse cerebellum. J. Cell Biol. 22:259-279.

Wolf, M. K. 1970. Anatomy of cultured mouse cerebellum. II. Organotypic migraton of granule cells demonstrated by silver impregnation of normal and mutant cultures. J. Comp. Neur. 140:281-297.

Wolf, M. K. and M. Dubois-Dalcq. 1970. Anatomy of cultured mouse cerebellum. I. Golgi and electron microscopic demonstration of granule cells, their afferent and efferent synapses. J. Comp. Neurol. 140:261-280.

CELL AND ORGANOTYPIC CULTURE STUDIES OF NEUROLOGICAL MUTATIONS AFFECTING STRUCTURAL DEVELOPMENT

MERRILL K. WOLF
Department of Anatomy
University of Massachusetts Medical School
Worcester, Massachusetts

I. Introduction

The neurological mutants of the mouse provide a powerful approach to the study of mammalian brain development and developmental disease. Of all the experimental mammals, *Mus musculus*, the house mouse, is first — and there is no good second — in our accumulation of formal genetic information and genetic engineering capabilities.

In 1976, an informal amalgamation of the gene lists of several major laboratories resulted in a list of about 647 mutant genes of the mouse, 68 of them newly described and provisionally accepted. Of these genes, 119, almost 20% of the total number, affect the neuromuscular system or organs of special sense. Description and experimental analysis of the disorder produced by a mutant gene provides the means to understand the function of the wild-type genetic material normally present at the mutant locus. The mouse has a representative mammalian brain, and information obtained in this way will be directly applicable to the development of other mammalian brains including that of man.

Culture experiments are uniquely suited to discriminate between intrinsic and extrinsic factors in the development and maintenance of the cells, tissues, or organs placed in culture. Richard L. Sidman, who has done so much to study these mutants himself and promote their study by others, is also widely experienced in tissue and organ culture research, having worked with Dame Honor Fell at the Strangeways Laboratory. Thus, it is natural that Sidman's laboratory at Harvard Medical School was prompt in bringing culture experimentation to bear on the study of mutants.

Two early studies on the eye will illustrate the potential of culture experiments in the analysis of a genetic disorder. The recessive mutation, retinal degeneration (rd/rd), produces degeneration of virtually all the retinal photoreceptors beginning about the tenth postnatal day (P 10), just as production of rod outer segment material is beginning. This might be due to an intrinsic defect in the genetic program for the development of the eye, or to an extrinsic influence, such as mutant animals elaborating a factor toxic to photoreceptors or failing to elaborate a trophic factor essential for these cells. Whole P8 to P10 mouse eyes were placed in organ culture in Trowell chambers (Sidman, 1963). By adding 20 μg/ml of cis-retinene to the culture medium, formation of rod outer segment material was obtained *in vitro* from normal mouse eyes, but not from rd/rd mouse eyes. This suggests that the defect in rd/rd is intrinsic to the eye, although this experimental design cannot be used to pursue the issue of which specific cell type is defective. Pink-eyed dilution (p/p), a pigment mutation, results in failure of melanophores to make normal amounts of melanin, so that pigmentation is reduced in the fur and almost absent from the pigment epithelium of the eye. When pigment epithelium from p/p rats and mice was cultured, the cells became strongly pigmented at rates largely controlled by the tyrosine concentration of the medium, suggesting that in this mutation an extrinsic influence, such as inadequate availability of tyrosine *in situ* for melanin synthesis, is responsible for the pigmentation defect in the eye (Sidman and Pearlstein, 1965).

Culture studies of mutations affecting the brain proper began at the same time and have followed two separate lines. The present reviewer, working first in Sidman's laboratory, later in his own laboratory at the University of Massachusetts Medical School, has used organotypic cultures of newborn cerebellum, which, if normal, will form abundant myelin in 2 to 3 weeks, to study the mutations which produce defective CNS myelination. Other mutations have also been studied in this way (Wolf, 1970; Wolf, 1974), but the studies of myelin-deficient mutations have been most fruitful and appear most promising at present. More recently, Anne Messer and Ekkhart Trenkner have been using dissociated cells from older immature cerebellum in plated culture (Messer) or reaggregated culture (Trenkner) to study two mutations which produce loss of cerebellar granule cells. In brief, the myelin-deficient mutant studies suggest that these genes produce intrinsic defects in the genetic program for myelin formation. The studies on the granule-cell mutations show that granule cells from affected animals survive normally *in vitro*, suggesting that their degeneration *in situ* is probably secondary to an abnormality in some other type of cell in the cerebellum. Both sets of studies however, have revealed unexpected features in the behavior of mutant tissue *in vitro* which do not conform to a simple "extrinsic vs. intrinsic" dichotomy, and which may be important guides to the modes of action of these genetic loci.

II. Mutants Deficient in CNS Myelination

The Catalogue of the Neurological Mutations of the Mouse (Sidman *et al.*, 1965) lists two myelin-deficient mutants: *quaking* (qk/qk), an autosomal recessive, and *jimpy* (jp/Y) a sex-linked recessive lethal. Subsequently, an additional sex-linked recessive lethal mutant of this type, *myelin synthesis defect* (jp^{msd}/Y) has been identified (Meier and MacPike, 1970). Direct tests for allelism of this new mutant with jp/Y are impossible because neither affected male lives long enough to produce viable sperm, but breeding data from a tetraparental chimera prepared with jp^{msd} and normal embryos provided strong inference that jp and jp^{msd} are allelic (Eicher and Hoppe, 1973). Two unpublished mutations being investigated at Cambridge (England) and at Sidman's laboratory appear to have basically similar pathology but involving different genetic loci than quaking and jimpy.

The common clinical features of all these mutants are a bobbing or bouncing tremor, which disappears at rest and is accentuated by excitement, and convulsions, which sooner or later are fatal. Some animals will become permanently paraplegic after a convulsion, apparently as a result of dorsal midline spinal cord hemorrhage. All other behavior seems to be normal.

The common pathological feature is a strikingly obvious deficit in CNS myelin. Gross inspection of a myelin-stained section of brain suggests that the white matter tracts, which are normal in position, size, and course, are entirely lacking in myelin. Microscopic examination at higher magnification is required to detect and quantitate the small amount of myelin which is formed. Peripheral nerve myelin appears normal, and examination of nerve roots shows that the myelin defect follows the boundary between the myelination domain of the PNS Schwann cell and the CNS oligodendrocyte. Homozygous qk/qk animals have the most CNS myelin, and the longest lifespan of the three; jp/Y have less CNS myelin, and a shorter lifespan; jp^{msd}/Y animals have the least myelin and the shortest lifespan.

Further examination of the morphological abnormalities in jp/Y (Meier *et al.*, 1974; Robain and Mandel, 1974; Skoff, 1976; and references cited therein) and qk/qk (Friedrich, 1974, 1975, and references cited therein) reveal some complicating features. In jp/Y white matter tracts, the presence of sudanophilic macrophages, coupled with evidence of abnormally active myelin lipid turnover, was originally taken as evidence of myelin breakdown. However, the culture studies described below do not support this idea, and Meier *et al.* found no ultrastructural evidence for it. As they point out, examination of histopathology in older animals is complicated by tissue damage secondary to convulsions and resulting anoxia. Axons in CNS fiber tracts of both jp/Y and qk/qk affected animals appear to be of smaller than normal diameter. The oligodendroglia of both mutants are abnormal in number: too few in jp/Y, too many in qk/qk. The myelin defect is not quantitatively uniform throughout the CNS, but is more severe in those fiber tracts originating from cell bodies in the forebrain than in those tracts originating from cell bodies in spinal cord. The cerebellum has not been quantitatively examined: qualitatively, the severity of hypomyelination in cerebellum resembles forebrain rather than spinal cord. Peripheral nerve myelin appears normal in comparison with the striking abnormality of CNS white matter, but at least in qk/qk (Samorajski *et al.*, 1970) it is not entirely normal: sciatic nerve shows reduced numbers of myelin lamellae and occasional large diameter unmyelinated axons not seen in littermate controls. Finally, Skoff (1976) has shown astrocytic hypertrophy and precocious development of gliafilaments in jp/Y optic nerve as early as the second postnatal day (P 2), thus preceding the appearance of oligodendroglia and the onset of myelination in normal optic nerve. Skoff argues that the astrocytic hypertrophy could directly inhibit oligodendroglial proliferation and myelin formation. It is also possible that it represents a reaction to some abnormality of the axon surface.

Biochemical studies have shown striking deficiencies of many structural lipids and enzymes of myelin metabolism in these mutants. To summarize the voluminous literature is beyond the scope of this volume and the competence of this reviewer: however, attention should be called to extensive work by Mandel and co-workers (for example, Mandel, 1972; Mandel *et al.*, 1972; Robain and Mandel, 1974, Sarlieve *et al.*, 1974), Bauman and co-workers (for example, Bauman *et al.*, 1971; Bauman *et al.*, 1973; Bourre *et al.*, 1973), Matthieu and co-workers (for example, Matthieu *et al.*, 1973, 1974a, 1974b), and no doubt many others. The problem is that many, perhaps all, of the observed deficits could well be the result rather than the cause of the striking deficit of myelin. There does not seem to be any one defective enzyme which can be directly correlated with any one of these defective genes (Roscoe O. Brady, personal communication).

We undertook to study myelination of brain tissue from myelin-deficient mutants and their littermate controls in organotypic culture, to see whether the diseases would be reproduced *in vitro*, by deficient myelination but normal development of neuronal networks, or would be modified in some way suggesting operation of extrinsic factors *in situ*. Cerebellum was chosen, because normal cerebellum myelinates well *in vitro*, mutant cerebellum is severely deficient in myelin *in situ*, and many specific features of the cerebellar neuronal network are well developed and have been well described in organotypic culture.

The culture system is discussed in other chapters of this volume (also see Wolf and Holden, 1969). It should be emphasized that tissue for organotypic culture must be taken from PO-P2 mice, several days before the onset of myelination *in situ*. Direct recognition of the affected animals is therefore impossible, and identification of animals with the mutant genotype requires the use of marker genes. We have elsewhere (Wolf, 1974) described the wasted time and uninterpretable data which resulted from attempting to study qk/qk without a marker. Closely linked semidominant marker genes are now available for all three myelin-deficient mutants.

Tabby (Ta), a sex-linked semidominant affecting hair and whisker development, provides the marker for jp and jp^{msd}. The stock of jp mice used for our published study of that mutant had Ta and jp on separate x-chromosomes. Ta+/+jp females had to be mated to Ta+/Y males to perpetuate the stock by producing daughters with the maternal genotype. The stocks of jp and jp^{msd} now available from the Jackson Laboratory have Tajp or Tajp^{msd} on the same chromosome, so that carrier females can be mated to any males; in practice, Fl hybrid male offspring of two inbred strains are used, to combine hybrid vigor with a modicum of genetic similarity. The frequency of genetic crossover, leading to misidentification of affected newborns, is 17%.

For qk, the T-locus provides a semidominant marker. Animals doubly heterozygous for quaking and for T^{2J} are mated to produce three types of embryos:

qk+/+ T^{2J} X qk+/+ T^{2J} (normal brains, shortened tails)

qk+/qk + (normal tails, affected)	qk+/+ T^{2J} (the parental genotype)	+T^{2J}/+T^{2J} (embryo grossly disorganized, dies *in utero*)

The shortening of the tail, though variable in amount, is always recognizable at birth. Since this stock is congenic with the inbred C57BL/6J strain, cultures from normal animals of this strain can legitimately be used as controls. This permits qk +/+ T^{2J} animals, which have somewhat impaired breeding performance, to be saved for breeding rather than expended for control cultures. The crossover frequency is only 1% for these two closely linked mutations.

The mutant jp was the first for which this essential genetic technology was available, and therefore it was the first mutant studied *in vitro* (Wolf and Holden, 1969). The first observation was that jp/Y cultures were profoundly deficient in myelin formation. The statistical summary of myelin formation by jimpy and control explants is given in Table I.

TABLE I

MYELIN FORMATION IN jp/Y EXPLANTS IN ORGANOTYPIC CULTURE, COMPARED TO LITTERMATE CONTROLS

	"Adequate" Myelin	Some Myelin	No Myelin Whatever
jp/Y explants	4	27	131
Littermate control explants (all genotypes lumped together)	218	93	27

From Wolf and Holden, 1969

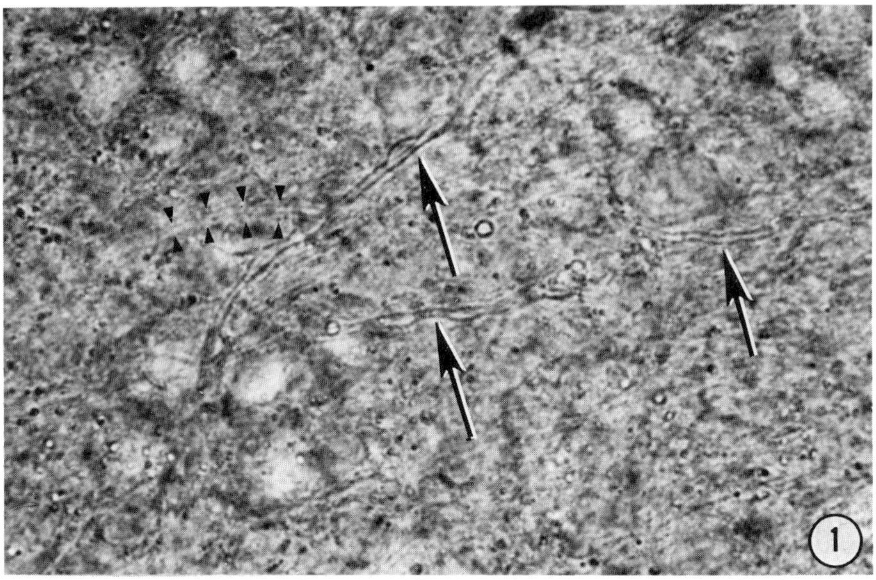

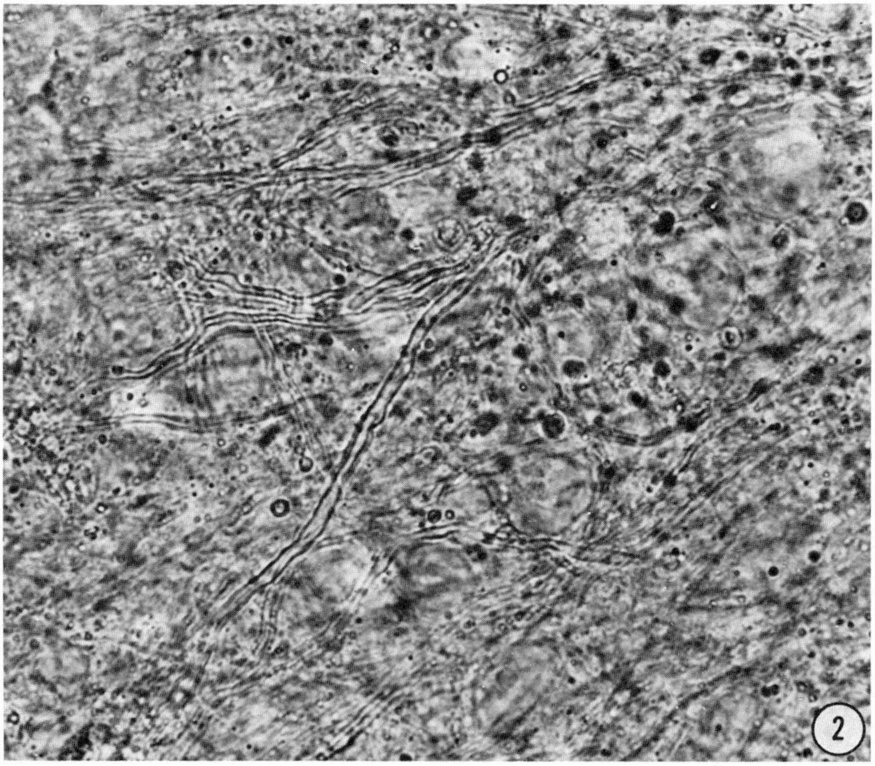

The arbitrary term "adequate" myelination was defined and used to take account of variable myelin formation by the control explants from the non-inbred mutant stock then available. Explants were counted instead of cultures to take account of experiments with explants of different genotypes on the same coverslip. Although some jp/Y explants produced some myelin, the deficiency in myelination compared to littermate control explants was consistent and categorical. Genetic crossovers were reliably detected by the fact that the amount of myelin formed fitted expectations for the wrong genotype, e.g., cultures from a Ta/Y male animal (presumably not jp) producing little or no myelin, or cultures from a non-Ta male animal (presumably jp) myelinating normally.

The next essential observation was that all other features of jp/Y cultures were normally developed. Holmes' method for neurofibrils provided ample confirmation of every neuroanatomical feature documented in normal cultures. This showed that the essence of the jp disease was reproduced *in vitro*, and therefore must represent an intrinsic defect in the myelin-forming system, axon plus oligodendrocyte.

Cultures of jp/Y cerebellum thus provide a sample of the disease on a coverslip, accessible to experiment and observation. Normal and jp/Y explants, when placed on the same coverslip, had no detectable influence on one another's development. This suggests that the deficiency involves cell-bound rather than diffusible factors, although it does not rule out a diffusible substance too unstable or present in too low a concentration to exert any effect at a distance of 3 to 5 mm. If the deficiency is cellular, one might expect heterozygous jp/+ females to present a mosaic of affected and normal cells due to random X-inactivation (Lyon, 1966). No such mosaicism has been detected, either *in situ* or *in vitro*. Either the affected cells are eliminated from a mosaic at a very early stage of embryogenesis, long before the expression of the disease, or the segment of the X chromosome carrying the jp locus does not undergo inactivation. Most significant was the test provided by the culture system of the idea that there is active ongoing destruction of myelin sheaths in jp/Y white matter. If such destruction were taking place, serial light microscopic observation of living cultures in Maximow assemblies should provide the ideal experimental tool

FIG. 1. Living myelinated axons in an organotypic culture of jp^{msd} cerebellum, 21 days *in vitro*. Arrows indicate myelinated axons, while a double row of arrowheads points to an axon which appears to have myelin at the limit of visibility. Similar "shadow myelin" fibers are seen elsewhere in the field. This field contains almost all the myelin or "shadow myelin" which was visible in this culture. Compare Fig. 2. About 900X.

FIG. 2. Living myelinated axons in an organotypic culture of cerebellum from a normal littermate, 21 days *in vitro*. Myelinated axons, too numerous and intertwined to count, can readily be identified without arrows or other visual aids. The myelin in this culture filled more than 10 microscopic fields of this size, at several levels of focus. About 900X.

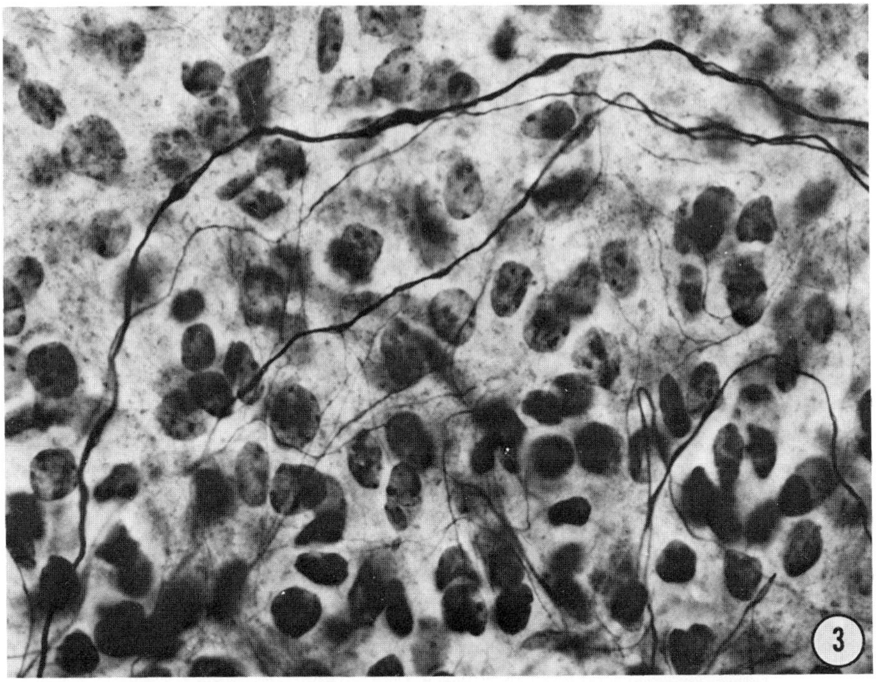
3

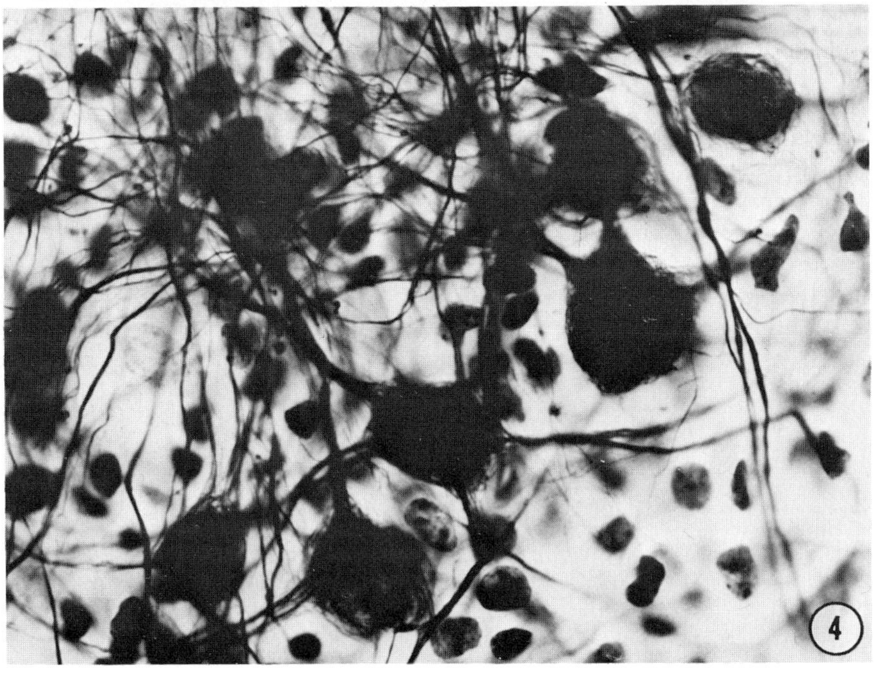
4

to document and describe it. No such destruction was observed, in hundreds of hours of observation. Those few myelin sheaths formed by jp/Y cultures were as stable as the much more numerous myelin sheaths in control cultures. Turnover of lipid must be taking place at a stage before incorporation of lipid into myelin.

Culture studies of qk/qk (Billings-Gagliardi *et al.*, 1976) and of jp^{msd} (unpublished) are more recent, and our accumulated experience, particularly with jp^{msd}, is not yet equal to our experience with jp. All three of these mutants show defective myelination *in vitro;* but their *relative* deficiency of myelination *in vitro* is paradoxical, for it does not follow the relative deficiency *in situ.* The deficiency of myelination, *in situ,* is least severe in qk/qk, more severe in jp/Y, most severe in jp^{msd}/Y. *In vitro,* this order is reversed. Cultures of jp^{msd}/Y develop *more* myelin cultures of jp/Y, while qk/qk cultures develop *no myelin whatever.*

TABLE II

COMPARISON OF MYELINATION OF qk/qk, jp/Y, AND jp^{msd}/Y CEREBELLUM *in situ* and *in vitro**

	Percentage of Normal Amount of Myelin Developed by Cerebellum *in situ*	Proportion of Cultures of Cerebellum Developing any Myelin Whatever
qk/qk	About 6%	0/115
jp/Y	Less than 1%	31/131
jp^{msd}/Y	Not yet counted: *much* less than 1%	24/52

*Data from Meier *et al.*, 1974; Friedrich, 1974; Wolf and Holden, 1969; and M. K. Wolf, unpublished observations.

Table II is a comparative summary of the numbers of qk/qk, jp/Y, and jp^{msd}/Y cultures which developed some myelin in culture, while Figures 1 and 2 are light micrographs of typical myelin in jp^{msd}/Y and littermate control cultures. The micrographs are not comparable in every respect, for the total amount of myelin in the control culture would cover more than ten times the area of Fig. 2 at several focal planes, while Fig. 1 shows almost all the myelin in that particular jp^{msd}/Y culture. Thus, the mutant myelin is

FIG. 3. Axons in the outgrowth of a silver impregnated organotypic culture of qk/qk cerebellum, 19 days *in vitro.* This culture contained no myelin or "shadow myelin" identifiable during life. Large, medium, and small diameter axons are well impregnated with silver and readily recognized. About 900X.

FIG. 4. Neuronal somas and dendrites in a thicker area of the same qk/qk culture are well impregnated and readily recognized. Numerous axons course through the field as well. About 900X.

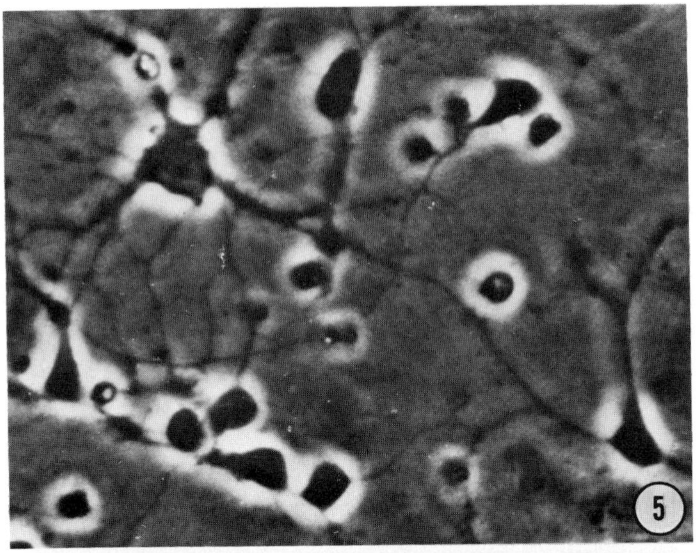

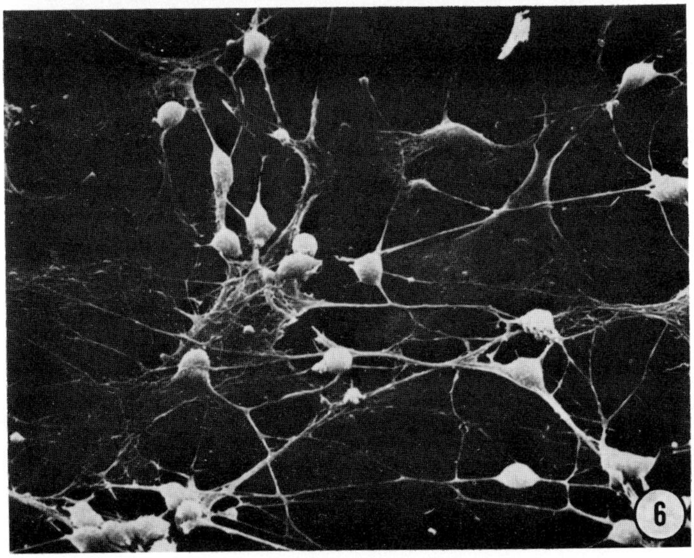

FIG. 5. Living, dissociated cerebellar cell culture from 7-day-old sg/sg cerebellum, 2 days *in vitro*, photographed with phase microscopy. Cell cultures from all genotypes tested look alike at this stage. However, sg/sg cultures maintain this appearance for 3 to 4 weeks *in vitro* better than cultures of other genotypes with most combinations of medium and plating surface. About 1100X. Reproduced by permission from Messer and Smith, 1977.

FIG. 6. Scanning electron micrograph of a dissociated cell culture of sg/sg cerebellum, 5 days *in vitro*. About 650X. Reproduced by permission from Messer and Smith, 1977.

clearly very abnormal. Myelin segments are too few in number and too small in diameter. However, as shown in Table II, this small amount of myelin occurs in a larger proportion of jp^{msd}/Y cultures than of jp/Y cultures. Furthermore, those jp^{msd}/Y cultures which develop myelin develop at least as much of it as jp/Y cultures, or perhaps even somewhat more. Particular attention is drawn to the faint parallel lines in Fig. 2 which appear to represent myelin segments so thin as to be at the threshold of visibility in the light microscope. This "shadow myelin" appears fairly often in jp^{msd}/Y cultures, and was not seen as often in jp/Y cultures. The idea that "shadow myelin" indeed represents myelin at the threshold of visibility, rather than totally unmyelinated axons, is reinforced by the fact that it was never seen in any qk/qk culture.

Interpretation of this finding is complicated by the fact that the strain of mice bearing jp, used for the earlier culture study, was quite different in the remainder of its genotype from the presently available strains bearing either jp or jp^{msd}. Furthermore, and very unfortunately, the presently available strains bearing jp and jp^{msd} are not entirely similar to each other. Transplantation of the two genes to completely comparable genetic backgrounds will be straighforward in principle but will require two to three years and a major commitment of animals and funds.

In the case of qk/qk, a picture comparable to Figs. 1 and 2 would show nothing at all. We have yet to find one segment of intact myelin, one bit of degenerating myelin, or even a myelin figure in over 100 qk/qk cultures examined to date. As shown in Table II, this is a striking discrepancy with the *in situ* observation that qk/qk white matter develops about 6% of the normal amount of myelin. This raises the question whether qk/qk cultures are simply exhibiting a non-specific degeneration, particularly since this appeared to be a problem in earlier work (Wolf, 1974) with a different qk/qk stock which did not have a marker gene. However, silver staining of qk/qk cultures from the present stock has already shown excellent development of large and small diameter axons (Fig. 3) and neuron somas and dendrites (Fig. 4) in cultures which had no myelin whatever during life. It seems clear that a series of silver impregnated qk/qk cultures numerically equal to the earlier series of jp/Y cultures will provide a similar illustration of every feature of normal neuronal development *in vitro.*

Two additional tentative observations may help provide a key. A few qk $+/+$ T^{2J} animals have been cultured. Myelination of qk $+/+$ T^{2J} cultures is not quite as good as that of normal, congenic C57BL/6J cultures. However, the T^{2J} gene is lethal to the early embryo in homozygous (T^{2J}/T^{2J}) state and might well have some non-specific deleterious effects on the development of CNS cultures even in the heterozygous state. The other observation is that qk/qk explants, in a few preliminary experiments, have shown a slight but distinct deleterious effect on the development of normal explants set up on the same coverslip. If this observation can be confirmed, it will suggest the possible operation of a diffusible factor, elaborated by qk/qk brain, which is toxic to developing myelin. Such a factor might be removed, to a significant but incomplete extent, by the circulation of blood *in vivo*, but be accumulated in the culture setting. This would explain the paradox of the disease seeming to be *more* severe *in vitro* than *in situ*, and more severe than the jp disease *in vitro*. Presumably jp/Y brain does not elaborate a toxic factor which is susceptible to accumulation in this way. Such a factor might also be elaborated in reduced amount by qk/+ brain, so that it is completely removed *in vivo* and only shows its effect *in vitro*.

A final observation has emerged from scanning electron microscopy (SEM), being carried out by Susan Billings-Gagliardi. It will be recalled that qk/qk white matter *in vivo* contains increased numbers of oligodendrocytes. Billings-Gagliardi has identified a type of small round cell, with numerous small narrow ruffles, which is extremely rare on the surface of normal cultures but quite common on the surface of qk/qk cultures. These cells seem to be categorically different in form from astrocytes, macrophages, or granule cell precursors: the latter cells in Billings-Gagliardi's SEM pictures look quite similar to the cells identified by Messer and Trenkner as immature granule cells in their cultures (see Figs. 5 and 6). The problem is that *both* the small round narrow-ruffled cells *and* the cells resembling granule cell precursors are present in increased numbers on the surfaces of some qk/qk cultures. These cultures must be compared, not only with normal ones, but also with cultures of the other myelin-deficient mutants. If the small round narrow-ruffled cells can be positively identified with the oligodendrocytes, this may contribute to a final settlement of the extended controversy about identification of granule cells versus oligodendrocytes *in vitro*.

III. Mutants Deficient in Granule Cells

Both staggerer (sg/sg) and weaver (wv/wv), the two mutants in this class, were described in the Catalogue of the Neurological Mutations of the Mouse (Sidman *et al.*, 1965); but much additional information has been obtained in subsequent years (Bignami and Dahl, 1974; Herrup and Mullen, 1976; Hirano and Dembitzer, 1975; Mallet *et al.*, 1976; Rakic and Sidman, 1973; Rezai and Yoon, 1972; Sidman, 1968, 1972; Sotelo and Changeux, 1974a, 1974b; Yoon, 1972, 1976). Both produce loss of almost all cerebellar granule cells, but have no apparent effect outside the cerebellum. In sg/sg, the granule cells multiply normally in the external granular layer, and migrate inward normally, but then disappear. This was originally thought to represent intrinsically programmed cell death. However, the Purkinje cells were found to be immature in their development at all stages of the granule cell migration; the Purkinje cells never develop the distal branchlet spines on which the granule cells normally synapse. In wv/wv, the granule cells never succeed in migrating inwards, but die in the external granular layer. The Bergmann glial fibers, along which normal granule cells migrate from external to internal granular layers, are markedly abnormal in wv/wv mice. Heterozygous wv/+ mice have an intermediate degree of Bergmann glial fiber abnormality and of granule cell loss. These observations have suggested that the granule cells of these two mutants are potentially normal, but die because of inability to complete an essential part of their developmental sequence, namely, formation of their principal efferent synapse, in sg/sg, and inward migration, in wv/wv. In both cases, however, other details of the abnormal development have been taken as evidence for a primary granule cell disorder (Sotelo and Changeux, 1974b; Yoon, 1972, 1976).

Anne Messer (Messer, 1975, 1977a, 1977b; Messer and Smith, 1977) has studied the properties of mutant granule cells in a monolayer culture system which affords them a considerable degree of isolation from other cell types. At postnatal day 7, mouse cerebellum is too old to survive in organotypic culture, but the external granule cells are more numerous than at birth. P7 cerebellum is dissociated by treatment with trypsin (1%) and deoxyribonuclease (0.1%) for 14 minutes followed by trituration in a pipette. After counting and checking dissociation, cells are plated at 1 - 1.5$\times$ 10^6 cells per 35 mm dish in 1.5 ml of a modification of Lasher and Zagon's high-K, high-glucose medium with 10% added serum. Fetal calf serum appears essential for initial attachment and outgrowth of processes, but after one week this must be changed to horse serum to obtain long-term survival of normal cultures. Cytosine arabinoside must be added after the first week to suppress division of non-neuronal cells. As Messer states, the entire system when grown with fetal calf serum is still somewhat variable. Excellent morphology and three or more weeks survival is shown by 30% of preparations, while 10% degenerate promptly and the remainder show intermediate performance. Figs. 5 and 6 show the appearance of Messer's cultures, by phase microscopy during life and SEM, respectively. Messer considers that at least 70-80% of the cells in these cultures are granule cells at some stage of development: her criteria include small size, rounded shape with 2 to 4 thin, straight, projecting processes, a "clockface" pattern of condensed chromatin on transmission electron microscopy, and failure to take up radioactive gamma-aminobutyric acid from the medium in the presence of a variety of other cells which do. All the cerebellar cortical neurons except the granule cells use GABA as transmitter and will take it up; glial cells, including the oligodendrocyte, are also reported to take up GABA.

Homozygous sg/sg and wv/wv animals were distinguished from littermate controls at P7 by direct inspection of their cerebella. Heterozygous sg/+ littermates were distinguished from +/+ ones by marker genes in the stock. Heterozygous wv/+ littermates were distinguished from +/+ by counts of dissociated cells, since wv/+ animals have an incomplete degree of the cerebellar abnormality. Both heterozygous and normal littermate cultures were compared with mutant cultures.

In brief, both sg/sg and wv/wv cultures showed survival of granule cells at least as good as control cultures, that is, for at least 3 weeks. Since cerebellum was taken for culture at 1 postnatal week, this reprents a total of 4 postnatal weeks of life, *in situ* plus *in vitro*. This shows that the massive death of granule cells, which is largely complete after 4 weeks of *in situ* life of the mutant animals, cannot be due to a genetic defect irreversibly programmed at postnatal day 7.

The cultures showed an additional, paradoxical property: under most culture conditions sg/sg cells survived *better* than their normal counterparts. The *best* cultures of control and mutant animals were quite similar, but sg/sg cultures showed optimum survival about 80% of the time, compared to about 30% for controls, and appear to show less clumping than controls. The difference between sg/sg and control cultures was maximum on untreated plastic surfaces, which promoted excessive clumping and early death of controls. Control cultures show less clumping and better survival on glass surfaces, did even better on polylysine-coated surfaces, and survived as well as mutant cultures (!) when plated on polylysine-coated surfaces with horse serum, rather than using fetal calf serum initially. (Horse serum could not be tested as initial feed with uncoated surfaces because it did not permit initial attachment of cells.)

Any explanation of this result requires some intrinsic or acquired difference between P7 normal and sg/sg granule cells. Indeed, the idea of such a difference is readily acceptable from the fact that the sg/sg cerebellum is already grossly abnormal in size and fissuration at P7, and, indeed, at P0 (birth). The nature of the difference, however, is obscure at present. Some of the available hypotheses involve the idea that the culture system arrests the development of the granule cells at an immature stage. This idea receives support from observations on wv/wv cultures. In Messer's culture system wv/wv cultures were essentially equivalent to controls, but were a bit more cellular after 10 days *in vitro*. The processes emerging from clumps of cells in wv/wv cultures tended not to form the fiber bundles that normal ones did. Messer suggested that another culture system might magnify these small differences between wv/wv and control cells. In fact, the culture system being explored by Trenkner (1976, and unpublished work) does magnify the difference.

Trenkner's objective has been to modify the treatment of dissociated cell suspensions so as to promote interactions of cells with one another rather than with the surface of the dish. Cell suspensions are prepared in a

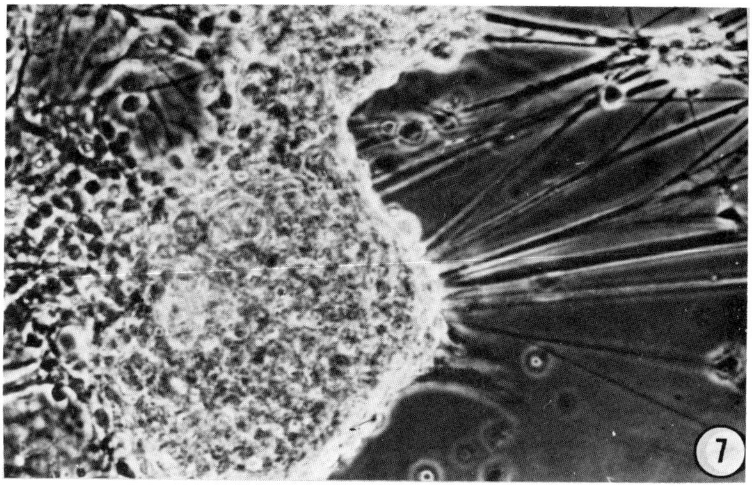

FIG. 7. Living culture of dissociated cells from 7-day-old cerebellum, partially reaggregated by the microwell technique. The bundles of processes extending from the clump of cells are characteristic of normal cerebellum in this system, but are not developed by wv/wv cerebellum. Phase microscopy: about 400X. Courtesy of Ekkhart Trenkner.

manner similar to Messer's, but 5 to 7 × 10^5 cells, suspended in 10 *microliters* of Eagle's Basal Medium with 10% horse serum, are placed in each microwell of a Falcon microtiter plastic plate. The ratio of tissue to medium; the proximity of tissue to a liquid-air interface; the choice of horse serum; the plastic surface; and the medium with normal rather than high potassium concentration; all differ from the optimum conditions for dispersed cell survival. The system shares some of the features of organotypic cultures or conventional shaker-flask reaggregate cultures, including promoting the formation of synapses, which have already been observed by Trenkner. The miniaturization of the system offers the intriguing possibility of making numerous replicate cultures from a single mouse cerebellum and of conserving scarce and costly biological materials which might be desirable for an experiment.

One of the regular morphological features of normal cerebellum in this culture system is the linkage of adjacent clumps of cells by bridges consisting of numerous closely applied cell processes. Granule cells migrate along these bridges (shown in Fig. 7) and along layers of non-neuronal cells which attach to the plastic. In wv/wv cultures, no bundles of bridging processes were formed; migration of granule cells was not observed, and large numbers of granule cells appeared to be degenerating. In cultures set up with delipidated rather than complete serum in the medium, wv/wv cells did *not* degenerate, and some bridging processes were formed. The absence of bridges and degeneration of cells could be reproduced by adding lipid extract of serum to the medium made with delipidated serum.

These preliminary results can readily be explained in terms of a glial defect in wv/wv. Presumably the specialized astrocytes, which make the Bergmann glial fibers *in situ*, are responsible in the cultures for providing the skeletons of the bundles of bridging processes and forming a monolayer of cells on the plastic, both of which serve as substrates for the granule cell migration. In wv/wv cultures, neither bridging bundles nor cell monolayers are normally developed; granule cells, being unable to migrate, die as they do *in situ*. Following this idea, one might suppose that wv/wv granule cells do as well as normal ones in Messer's cultures because the culture system tends to prevent them from entering a developmental sequence which they cannot complete, while sg/sg cells do even better than controls because their development is doubly retarded by the culture system and also by some influence acquired during their sojourn in a cerebellum which is already abnormal. The role of delipidation of the serum in reversing the wv/wv defect is another story: it may mark the beginning of a trail leading to the identification of a specific biochemical cell surface defect in this mutant.

Acknowledgments

The reviewer is grateful to Anne Messer and Ekkhart Trenkner for permission to reproduce published and unpublished photographs. The work of Susan Billings-Gagliardi and the reviewer has been supported by NIH grant No. NS-11425.

References

Baumann, N., J. M. Bourre, C. Jacque and S. Pollet. 1971. Genetic disorders of myelination. In: *Lipids, Malnutrition and the Developing Brain;* a Ciba Foundation Symposium. ASP (Elsevier, Excerpta Medica, North-Holland), Amsterdam.

Baumann, N. J. M. Bourre, C. Jacque and M. L. Harpin. 1973. Lipid composition of quaking mouse myelin: comparison with normal mouse myelin in the adult and during development. J. Neurochem. 20:753-759

Bignami, A. and D. Dahl. 1974. The development of Bergmann glia in mutant mice with cerebellar malformations: Reeler, staggerer and weaver. Immunofluorescence study with antibodies to the glial fibrillary acidic protein. J. Comp. Neur. 155:219-250.

Billings-Gagliardi, S., M. Suva and M. K. Wolf. 1976. Organotypic culture studies of myelin-deficient *quaking* mutant mice: a progress report. In Vitro 12:4, 321-322.

Bourre, J. M., S. Pollet, G. Chaix, D. Daudu and N. Baumann. 1973. In vitro study of fatty acids synthesized in brain microsomes of normal and quaking mice. Biochimie 55:1473-1479.

Eicher, E. M. and P. C. Hoppe. 1973. Use of chimeras to transmit lethal genes in the mouse and to demonstrate allelism of two x-linked male lethal genes, jp and msd. J. Exp. Zool. 183:181-184.

Friedrich, V. L. 1974. The myelin deficit in quaking mice. Brain Res. 82:168-172.

Friedrich, V. L. 1975. Hyperplasia of oligodendrocytes in quaking mice. Anat. Embryol. (Springer-Verlag) 147:259-271.

Herrup, K. and R. J. Mullen. 1976. Intrinsic Purkinje cell abnormalities in staggerer mutant mice revealed by analysis of a staggerer-normal chimera. Neuroscience Abstracts II: p.141.

Hirano, A. and H. M. Dembitzer. 1975. The fine structure of staggerer cerebellum. J. Neuropath. Exp. Neurol. 34:1-11.

Lyon, M. F. 1966. X-chromosome inactivation in mammals. Chapter 2 in: *Advances in Teratology*, Vol. I. D. H. M. Woollam, ed., Academic Press — Logos Press, New York and London. pp. 25-54.

Mallet, J., M. Huchet, R. Pougeouis and J. -P. Changeux. 1976. Anatomical, physiological and biochemical studies on the cerebellum from mutant mice. III. Protein differences associated with the weaver, staggerer and nervous mutations. Brain Res., 103:291-312.

Mandel, P. 1972. The control of myelin synthesis and inborn errors of metabolism. Adv. Exp. Med. Biol. 32:241-250.

Mandel, P., J. L. Nussbaum, N. M. Neskovic, L. L. Sarlieve and T. Kurihara. 1972. Regulation of myelinogenesis. In: *Advances in Enzyme Regulation* Vol. 10:101-118.

Matthieu, J. M., S. Widmer and N. Herschkowitz. 1973. Jimpy, an anomaly of myelin maturation. Biochemical study of myelination phases. Brain Res. 55:403-412.

Matthieu, J. M., R. D. Brady and R. H. Quarles. 1974. Anomalies of myelin-associated glycoproteins in "Quaking" mice. J. Neurochem. 22:291-296.

Matthieu, J. M., R. H. Quarles, H. F. DeWebster, R. H. Hogan and R. D. Brady. 1974. Characterization of the fraction obtained from the CNS of Jimpy mice by a procedure for myelin isolation. J. Neurochem. 23:517-523.

Meier, H. and A. D. MacPike. 1970. A neurological mutation (msd) of the mouse causing a deficiency of myelin synthesis. Experimental Brain Res. 10:512-525.

Meier, C., N. Herschkowitz and A. Bischoff. 1974. Morphological and biochemical observations in the jimpy spinal cord. Acta Neuropath. (Berlin) 27:349-362.

Messer, A. 1975. Monolayer cultures of cerebellar neurons from normal and neurological mutant mice. Neuroscience Abstracts I: 1245.

Messer, A. 1977a. The maintenance and identification of mouse cerebellar granule cells in monolayer culture. Brain Res. In press.

Messer, A. 1977b. Factors influencing monolayer cell culture morphology and survival of granule cells from wild-type and mutant mice. *ICN-UCLA Neurobiology Conference Proceedings.*

Messer, A. and D. M. Smith. 1977. *In vitro* behavior of granule cells from staggerer and weaver mutants of mice. Brain Res. In press.

Rakic, P. and R. L. Sidman. 1973. Sequence of developmental abnormalities leading to granule cell deficit in cerebellar cortex of weaver mutant mice. J. Comp. Neurol. 152:103-132.

Rezai, Z. and C. H. Yoon. 1972. Abnormal rate of granule cell migration in the cerebellum of "weaver" mutant mice. Develop. Biol. 29:17-26.

Robain, O. and P. Mandel. 1974. Etude quantitative de la myélinisation et de la croissance axonal dans le corps calleux et le cordon postérieur de la moëlle chez la souris Jimpy. Acta Neuropath. (Berlin) 29:293-309.

Samorajski, T., R. L. Friede and P. R. Reimer. 1970. Hypomyelination in the quaking mouse. A model for the analysis of disturbed myelin formation. J. Neuropath. Exp. Neurol. 29:507-523.

Sarlieve, L. L., N. M. Neskovic, G. Rebel and P. Mandel. 1974. PAPS-cerebroside sulphotransferase activity in developing brain of a neurological mutant of mouse (msd). Exp. Brain Res. 19:158-165.

Sidman, R. L. 1963. Organ-culture analysis of inherited retinal degeneration in rodents. In: *Symposium on organ culture: Studies of development, function, and disease.* National Cancer Institute Monograph no. 11. U.S. Department of Health, Education and Welfare, Public Health Service.

Sidman, R. L. 1968. Development of the interneuronal connections in brains of mutant mice. In: *Physiological and Biochemical Correlates of Nervous Integration.* F. D. Carlson, ed. Prentice-Hall, N.J. pp. 163-193.

Sidman, R. L. 1972. Cell interactions in developing mammalian nervous system. In: *Cell*

Interactions: Proceedings of the First Lepetit Colloquium. L. G. Silvestri, ed. North-Holland Publ. Co., Amsterdam, pp. 1-13.

Sidman, R. L. and R. Pearlstein. 1965. Pink-eyed dilation (p) gene in rodents: increased pigmentation in tissue culture. Develop. Biol. 12:93-116.

Sidman, R. L., P. W. Lane and M. M. Dickie. 1962. Staggerer, a new mutation in the mouse affecting the cerebellum. Science 137:610-612.

Sidman, R. L., M. G. Green and S. H. Appel. 1965. *Catalog of the Neurological Mutants of the Mouse*. Harvard University Press, Cambridge.

Skoff, R. P. 1976. Myelin deficit in the Jimpy mouse may be due to cellular abnormalities in astroglia. Nature (Lond.) 264:560-562.

Sotelo, C. and J. -P. Changeux. 1974a. Transsynaptic degeneration 'en cascade' in the cerebellar cortex of staggerer mutant mice. Brain Res. 67:519-526.

Sotelo, C. and J. P. Changeux. 1974b. Bergmann fibers and granular cell migration in the cerebullum of homozygous weaver mutant mouse. Brain Res., 77:484-491.

Trenkner, E., M. B. Hatten and R. L. Sidman. 1976. Histogenesis of normal and mutant mouse cerebellar cells in a micro culture system. Neuroscience Abstracts II: p. 1028.

Wolf, M. K. 1970. Anatomy of cultured mouse cerebellum. II. Organotypic migration of granule cells demonstrated by silver impregnation of normal and mutant cultures. J. Comp. Neurol., 140:281-298.

Wolf, M. K. 1974. Problems in culture analysis of neurological mutant disorders. In: *Methodological approaches to the study of brain maturation and its anomalies.* D. P. Purpura,and G. P. Reaser, eds. University Park Press, Baltimore, London, and Tokyo.

Wolf, M. K. and A. B. Holden. 1969. Tissue culture analysis of the inherited defect in central nervous system myelination in jimpy mice. J. Neuropath. Exp. Neurol. XXVIII:195-213.

Yoon, C. H. 1972. Developmental mechanism for changes in cerebellum of "staggerer" mouse, a neurological mutant of genetic origin. Neurology 22:473-754.

Yoon, C. H. 1976. Pleiotropic effect of the staggerer gene. Brain Res. 109:206-215.

NEUROSECRETION IN CULTURED SYSTEMS

DAVID PEARSON

Neuroendocrinology Section
Division of Neurosciences
City of Hope National Medical Center
Duarte, California

I. Introduction

Neurosecretion is the process that provides an interface between the two major communication systems which exist within animal organisms, neuronal and endocrine. Cells that are involved in this process can be appropriately described as biological transducers which are capable of converting neuronal signals into chemical outputs which are then distributed by humoral pathways to distant target tissues. Although all nervous cells have a chemical output, only this group of cells, the ultimate function of which is an electrical-to-chemical transduction, is called neurosecretory. These cells function not to chemically bridge the minute gap of a synapse so that elecrical transmission can continue, but rather as the final common neuroendocrine pathway.

Since the initial work of Ernst and Berta Scharrer and W. Bargmann which brought the highly suspect concept of neurosecretion into general acceptance several decades ago, dramatic strides have been made in the identification of neurosecretory systems, the determination of their hormone products, and the elucidation of their physiology. Comprehensive periodic reviews of this field are available as the published proceedings of international symposia on the subject (see First International Symposium on Neurosecretion, 1954; Bargmann *et al.*, 1958; Heller, 1962; Stutinsky, 1967; Bargmann and Scharrer, 1970; Knowles and Vollrath, 1974). Numerous other excellent reviews (Bern, 1966; Sachs, 1970; Cross *et al.*, 1975) and collected works (Walter, 1975) are available. In the face of the great burgeoning of discovery about the process of neurosecretion, a simple definition of the term which is sufficiently restrictive and consistently applicable has been strangely elusive. (For discussions of this problem see Knowles, 1974; Lederis, 1974). Definitions solely based on morphological or histochemical criteria have generally been inadequate. Definitions based on secretion of a hormone product into the circulation for delivery to target organs have been obscured by the discovery of ascending pathways from presumed neurosecretory cells which may form direct synapses with other neuronal tissues (Sterba, 1974). This is particularly interesting in light of the body of literature on behavioral effects of neurosecretory hormones (De Wied, 1969). For the purpose of the present discussion we will avoid the mounting complexities of a comprehensive definition by considering, primarily, the classical neurosecretory systems which include neuronal cells that synthesize a known hormone product and which secrete this product into the perivascular space from which it is carried to its sites of action by humoral pathways.

A schematic representation of a generalized neurosecretory cell is shown in Fig. 1. At least in the magnocellular neurosecretory system the secretory products are synthesized in the neuronal perikaryon using the normal protein synthetic mechanism of the cell. In some systems there is evidence that the hormones are synthesized as inactive precursors which are subsequently hydrolyzed to the active form (Sachs and Taketabake, 1964; Mendelson and Walter, 1976; Gainer *et al.*, 1977). Such an initial synthesis of inactive precursors seems to be a general feature of peptide hormone production since it is also seen in non-neuronal endocrine systems (Steiner, 1976). After ribosomal synthesis of the prohormones, they are packaged in the Golgi apparatus into discrete neurosecretory granules. In this form they are transported into the cell terminal via a fast axonal transport mechanism

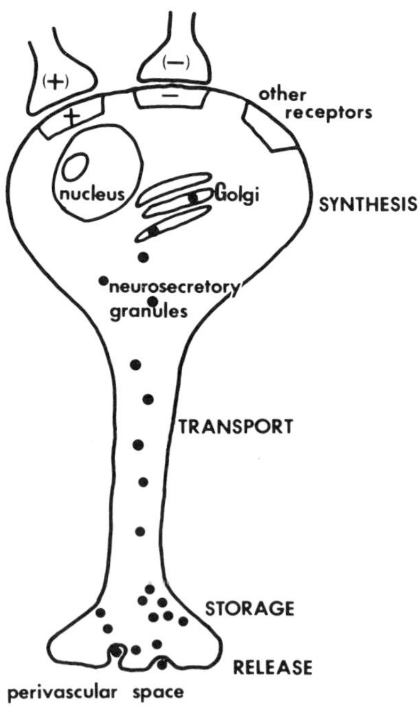

FIG. 1. Schematic representation of a generalized neurosecretory cell.

(Pickering and Jones, 1971). This is the site of storage and release in response to cell depolarization.

Thus, the process of neurosecretion can be divided into several discrete stages: synthesis (including the processing of precursor forms), transport (processing may also occur in this phase), storage, and release. The essential questions which have guided research on the process of neurosecretion are, what are the inter-relationships among these various stages, and what are the mechanisms by which they are controlled?

The number of workers who are utilizing long term *in vitro* systems to study various aspects of the neurosecretory process has increased considerably in recent years. Although the use of cell, tissue, and organ culture techniques to study neurosecretion is still definitely in the fledgling state, various cultured systems have been shown to retain the abilities to accomplish all of the individual stages in the neurosecretory process as defined above. I will present an overview of this work, and by holding description of specific experimental detail to a minimum I will attempt to give a broad coverage of the cultured neurosecretory systems which have been developed.

II. Studies on Hormone Biosynthesis in Culture

It is appropriate that vasopressin be the first neurosecretory hormone for which synthesis in culture was demonstrated. It is, after all, the product of the 'classical' neurosecretory system, the hypothalamoneurohypophysial system; one which has been studied much more extensively than any other. Biosynthesis of this hormone in organ culture was first demonstrated by Sachs and co-workers in 1970 (Sachs *et al.*, 1971) in an explant of the guinea pig anterior hypothalamus that had been maintained in culture for as long as two weeks. The explant included the supraoptic nuclei, but not the second hypothalamic site of synthesis of this hormone, the paraventricular nuclei. The explant was a large piece of tissue by organ culture standards. The generally accepted fact is that one cubic millimeter is the optimum size for a tissue explant to survive *in vitro* (Lumsden, 1968). This size would allow for adequate diffusion of nutrients and oxygen into the cells and adequate diffusion away of wastes. The neurosecretory explant of Sachs was at least one order of magnitude larger than this. The success obtained in this system was probably due to the fortuitous position of the neurons being studied. Since the neurosecretory cells in the supraoptic nuclei lie very close to the ventral surface of the hypothalmus they were maintained in good condition. Among the general population of cells in these cultures degeneration was, in fact, widespread.

The explant was maintained on a small piece of lens paper on a stainless steel grid supported 3 mm above the bottom of a plastic petri dish.

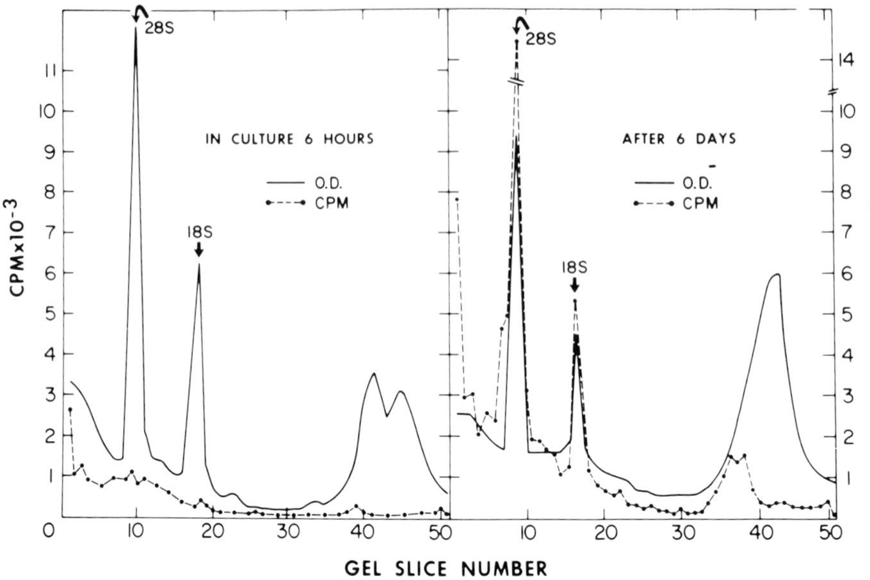

FIG. 2. Pattern of incorporation of ^{3}H-uridine into RNA in the hypothalamic segment of hypothalamoneurohypophysial organ cultures. Groups of 5 cultures were labeled with ^{3}H-uridine (40 μCi/ml) for 4 hours. The RNA was extracted and electrophoresed on 2.2% polyacrylamide gels.

The total volume of medium in the dish was 1.1 ml. The medium used was Dulbecco's Modified Eagle medium supplemented with Hepes, Tricine, a mixture of non-essential amino acids, glutamine, glucose, penicillin, and 10% fetal calf serum. In the initial studies on biosynthesis, explants were labeled for 8-20 hr. periods with rather large quantities of tritiated amino acids (100 μCi/ml each of proline, phenylalanine and tyrosine; specific radioactivity greater than 20 Ci/mmol). Since the finally isolated vasopressin contained only a few thousand counts per minute, rigorous standards of purification had to be met in order to establish unequivocally that biosynthesis of the hormone had, indeed, occurred. A series of chromatographic separations, including an elegant affinity chromatography step using neurophysin bound to sepharose, was used to purify vasopressin to constant specific radioactivity.

Subsequent to this initial work, the same culture techniques were applied to the entire hypothalamoneurohypophysial complex (HNC), that is, the hypothalmic explant described above with the neurohypophysis attached (Pearson *et al.*, 1975a). The presence of the intact neurosecretory system offered the potential for observation of the entire neurosecretory process *in vitro*.

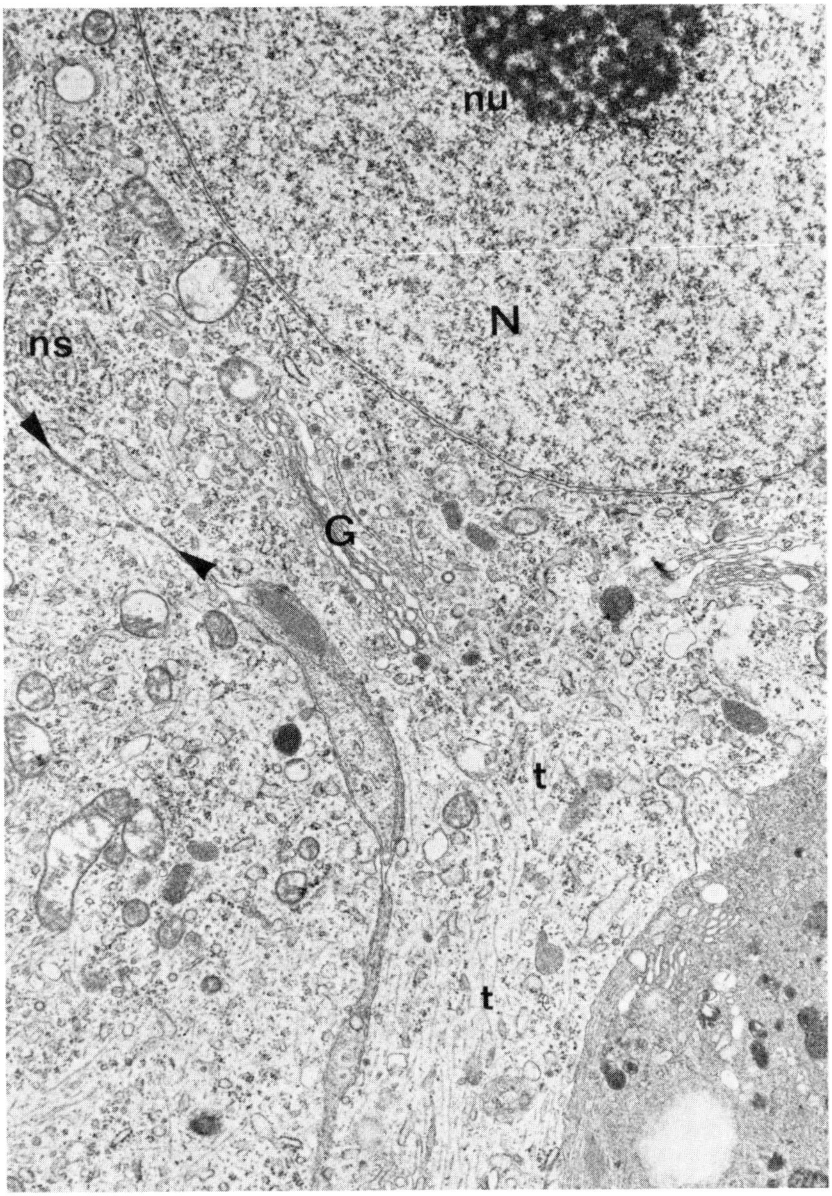

FIG. 3. Electron micrograph of neurosecretory cell. After 11 days in culture, the perikaryon is replete with organelles typical of a control preparation: Golgi apparatus (G), Nissl substance (ns), and neurotubules (t). Nucleus (N) is surrounded by nuclear membrane and contains a well-developed nucleolus (nu). Note intact plasmalemma (arrows) and the prominent process extending towards the lower part of the figure (×10,000) (Pearson *et al.*, 1975a).

Biosynthetic studies on the intact HNC showed that it had a somewhat greater capacity to synthesize the hormone than did the hypothalamic segment alone. It was observed that freshly explanted tissues had little or no hormone synthetic activity during the first few days in culture. Studies on the electrophoretic patterns of pulse labeled RNA produced during these early time periods showed that there was very little incorporation of tritiated uridine into cytoplasmic RNA species during this time, and, furthermore, that the incorporation which did occur, was present diffusely among the various molecular weight species. During the fifth through seventh days in culture incorporation of uridine into cytoplasmic RNAs increased and showed discrete peaks of radioactivity coincident with the major cytoplasmic species of RNA (see Fig. 2).

Although a significant amount of the incorporation observed under these conditions was due to biosynthesis in non-neuronal elements that began to grow out of the explants, inhibition of this growth with fluorodeoxyuracil did not prevent incorporation of labeled uridine into the normal cytoplasmic RNA species (Pearson *et al.*, 1975a). Furthermore, radioautography of tissues labeled during this period showed uridine incorporation into neurosecretory cells of the supraoptic nucleus. It was during this same period of culturing that the ability of the explants to synthesize vasopressin also showed an apparent recovery. As will be discussed below, other workers have observed a refractory period after explantation during which neurosecretory function could not be demonstrated (Marks *et al.*, 1973). This rather long period that is apparently necessary for the tissue to recover its functional integrity after the trauma of explantation, is a good demonstration of one of the unique values of organ and tissue culture systems as opposed to short term *in vitro* studies.

Parallel studies on the biosynthesis of vasopressin's carrier protein, neurophysin, were carried out in this system which showed that the neurophysin was synthesized in the cultured tissues with essentially the same kinetics as vasopressin (Sachs *et al.*, 1975). Table I shows the changes in incorporation into vasopressin as a function of time in culture.

In this system the morphological integrity of at least some of the neurosecretory neurons was maintained for at least two weeks in culture. Fig. 3 shows an electron micrograph of the neuronal perikaryon of a neurosecretory cell in an HNC organ culture that had been maintained for 15 days. The characteristic large nucleus with well-defined nucleolus can be seen along with a normal compliment of cytoplasmic organelles.

Thus, the ability of a cultured system to accomplish the first stage of the neurosecretory process, namely the biosynthesis of the active principle, was demonstrated. Since that time other hormone products have been shown to be synthesized in similar organ cultures. After incubation of hypothalamic organ cultures with tritiated amino acids, apparent synthesis of both thyrotropin-releasing hormone (TRH) and luteinizing hormone-releasing hormone (LHRH) was inferred from the fact that radioactivity co-migrated with the biological activity of these hormones in several chromatographic systems (Sachs *et al.*, 1974). TRH biosynthesis in organ culture of the guinea pig median eminence was definitively shown by McKelvy and coworkers (1975) by rigorous purification techniques. These workers observed that during a two-week culture period neuronal morphology showed progressive degeneration. Parallel loss of TRH biosynthetic ability was observed. In contrast, the ependymoglial elements retained morphological integrity.

TABLE I
VASOPRESSIN SYNTHESIS IN THE HYPOTHALAMONEUROHYPOPHYSIAL COMPLEX (HNC) IN ORGAN CULTURES

Days in Culture	Vasopressin Content (mU/Hypothalamus)	cpm in Vasopressin per Explant
0	34.6 ± 3.1	1010 ± 396
2	35.1 ± 4.7	281 ± 96
7	32.0 ± 1.9	986 ± 183
9	29.3 ± 4.8	902 ± 49
14	33.7 ± 5.1	605 ± 158
22	26.4	210

Guinea pig HNC explants were incubated with tritiated amino acids for 20 hours. The hypothalamic segments were then used for the isolation and purification of labeled hormone. Vasopressin content was determined by radioimmunoassay (±S.E.M.). (From Pearson *et al.*, 1975a)

The extent to which information on neurosecretory processes observed in tumor-derived cell lines can be extrapolated to the *in vivo* situation is unclear; nevertheless, useful information about bio-synthetic mechanisms may be gleaned from these sources. Grimm-Jörgensen and co-workers have studied the metabolism of TRH in two clonal cell lines derived from central nervous system tumors in the rat (Grimm-Jörgensen *et al.*, 1976). In both cell lines studied. TRH biosynthesis could be demonstrated during the logarithmic growth phase, but not during the stationary phase. Although loss of hormone due to the presence of a TRH-degrading enzyme is a problem in this tissue, the amount of the enzyme present is far less than observed in whole brain and hypothalamus.

Biosynthesis of vasopressin has also been demonstrated in a cell line derived from the anterior hypothalamus of fetal mice and transformed with simian virus 40 (De Vitry *et al.*, 1974). A homogeneous cell line has been isolated and maintained for at least 38 passages. Both neurophysin and vasopressin can be measured in homogenates of these cells, and their presence there has also been demonstrated by immunocytochemical techniques. This cell line has also been shown to incorporate labeled cysteine into proteins with an electrophoretic mobility identical with a neurophysin (De Vitry *et al.*, 1975). No oxytocin was detectable in this cell line.

III. Transport of Hormones to Storage and Release Sites

Using the cultured hypothalamoneurohypophysial complex as described above, transport of vasopressin from the neuronal perikarya in the hypothalamus to the storage and release sites in the posterior pituitary has been demonstrated (Pearson *et al.*, 1975b). Table II shows the amounts of labeled hormone isolated from the hypothalamus and in the posterior pituitary 22 hours after the addition of labeled amino acids to explants that had been maintained in culture for seven days. Under these conditions, at least 80% of the labeled hormone was still in the hypothalamus; however, significant amounts were isolated from the pituitary portion. By severing the neural stalk before addition of the labeled amino acids, it was demonstrated that the appearance of labeled hormone there was not due to re-uptake of labeled hormone that might have diffused to the pituitary via the medium. Under these conditions no counts appeared in vasopressin in the pituitary. If colchicine (1 μg/ml) was added to the medium at the time of addition of tracer, appearance of labeled hormone in the pituitary was similarly blocked (see Table II). It is interesting to note that electron micrographs of sections through the hypophysial stalk show no morphological differences between the control and colchicine treated tissues, and, furthermore, that the abundant appearance of intact microtubules in both of these preparations was indistinguishable from that observed in fresh tissue (Pearson *et al.*, 1975b).

IV. Controlled Release of Neurosecretory Hormones

The release of neurosecretory hormones from tissues maintained in organ culture in response to inputs designed to approximate physiological stimuli has been demonstrated by several groups of workers. Eggena and Polson (1974) have prepared a hypothalamoneurohypophysial explant from the toad similar to the one described above. In amphibian systems the action of vasotocin on the bladder is the primary mechanism for maintenance of the proper hydroosmotic balance, paralleling the antidiuretic action of vasopressin in the mammalian kidney. These workers found that a low and rather constant basal release of vasotocin during several days *in vitro* was increased dramatically when the osmolarity of the incubation medium was changed from 240 mOsmolar to 450 mOsmolar. By sectioning the preopticoneurohypophysial tract at the point of its emergence from the median eminence, they demonstrated that this effect was due to action on the hypothalamic segment, and not directly on the neurosecretory terminals in the pituitary.

TABLE II

TRANSPORT OF VASOPRESSIN IN THE HYPOTHALAMONEUROHYPOPHYSIAL COMPLEX (HNC) IN ORGAN CULTURES

	cpm in Vasopressin/Explant	
	Hypothalamus	Pituitary
Control	1728 ± 310	215 ± 92
Colchicine (1 μg/ml)	2941 ± 687	31 (1 expt.) 0 (2 expts.)
Pituitary Stalk Severed	1401 ± 116	0

Guinea pig HNC cultures maintained for 7 days were labeled for 22 hours with tritiated amino acids. Three experiments were performed in which each group contained 8 to 10 cultures. Colchicine was present only during labeling. In the stalk sectioned group the stalks were severed just prior to labeling. Grouped cultures in a given experiment were pooled. Hormone was isolated from the hypothalamic segment and the pituitary separately (±S.E.M.).

Recently, similar work has been performed in the hypothalamo-neurohypophysial complex taken from the rat. Sladek and Knigge (1976) have demonstrated that in organ cultures maintained for three to five days in 285 mOsmolar medium, dramatic stimulation of vasopressin release could be observed in response to very subtle manipulation of the ionic strength of the medium (an increase of only 15 mOsmolar, from 285 mOsmolar to 300 mOsmolar). This work strongly suggests that an osmoreceptor is present in the explant and that it retains its biological activity in culture.

In the same system Sladek and Knigge have demonstrated release of vasopressin in response to acetylcholine (Sladek and Knigge, 1976). *In vivo* electrophysiological studies have demonstrated that excitation of nicotinic type cholinergic receptors provides a major stimulatory input to the neurosecretory cells in the supraoptic nucleus. In the organ culture system release of vasopressin in response to acetylcholine has been shown to be concentration dependent from 10^{-8}M to 10^{-4}M. Furthermore, the nicotinic receptor blocking agent, hexamethonium, can completely prevent this release.

Related results have been obtained in a tissue culture system of a rather different nature. Marks and co-workers have developed a system which is capable of supporting the magnocellular neurons of the supraoptic nuclei taken from neonatal dogs. Supraoptic nuclei are minced and allowed to grown on coverslips for periods of two to three weeks. These workers have judged that death and degeneration of damaged cells is complete after about a week *in vitro* and that neurosecretory neurons remain with no efferent connections (Marks *et al.*, 1973). During the culture period the surviving cells project new axons. By penetrating individual cells in this preparation with a recording electrode, it was observed that most cells were quiescent, but that microsuperfusion with medium containing 10^{-6}M nicotine or glutamate elicited a continuous generation of action potentials. They went on to demonstrate that the spike frequency of this induced firing could be inhibited by addition of norepinephrine (10^{-5}M) or isoproterenol (10^{-5}M) but not by phenylephrine. The effect of isoproterenol could be reversed by equimolar amounts of propranolol. It is unfortunate that the small amount of tissue present in these preparations did not allow the measurement of hormone released by the various treatments. However, if the reasonable assumption is made that stimulus and secretion are directly coupled, these data are consistent with the observations made in the macro-organ culture system used by Sladek and Knigge.

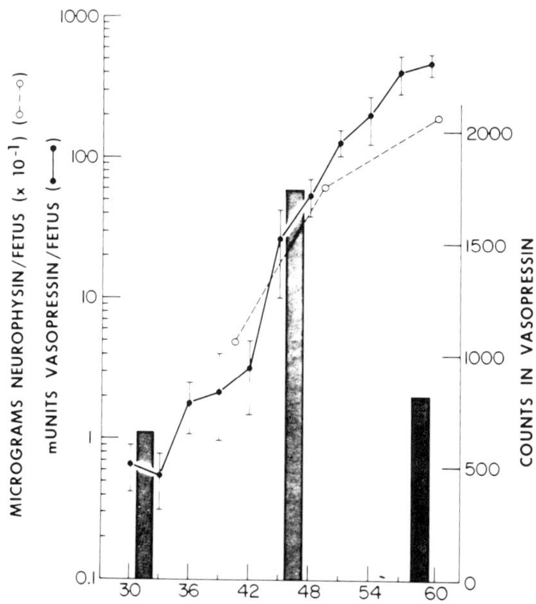

FIG. 4. Content of vasopressin (closed circles) and neurophysin (open circles) in the hypothalamoneurohypophysial complex of developing guinea pigs. The bars show the amount of ^{3}H-labeled amino acid incorporated into vasopressin in hypothalamoneurohypophysial organ cultures established from adult guinea pigs, which were labeled in medium conditioned by incubation with fetal hypothalamic tissue. Each incorporation experiment represents the results of 10 pooled organ cultures. Vasopressin content data are shown ± S.E.M. (from Pearson *et al.*, 1975c, copyright 1975 by the American Association for the Advancement of Science).

The presence of yet another distinct receptor activity on these cells was demonstrated by addition of angiotensin II to the superfusion medium (Sakai *et al.*, 1974). Spiking activity was elicited by this peptide in a concentration-dependent manner from 10^{-8}M to 10^{-4}M. This activity is blocked by the specific angiotensin II antagonist, cysteine-8-angiotensin II, which has no effect in spiking activity induced with nicotine.

V. Other Biological Responses to Effector Substances

As described above, many biological and pharmacological agents can induce neuronal firing and release of hormones in cultured systems. As yet, however, none of these treatments have been shown to alter the rate of hormone biosynthesis. While in the process of performing some studies on the developmental aspects of vasopressin biosynthesis, we discovered that a factor was present in the fetal guinea pig hypothalamus that was capable of stimulating the incorporation of labeled amino acids into vasopressin and neurophysin in cultured explants of hypothalamoneurohypophysial complexes taken from adult guinea pigs (Pearson *et al.*, 1975c). This increase in incorporation occurred in the absence of a general stimulation of incorporation into acid-precipitable proteins in these explants. The stimulatory factor was present only between the 40th and 55th day of gestation, which corresponds to the period during which the vasopressin and neurophysin content of the fetal tissues shows a dramatic increase (Fig. 4). Furthermore, the factor was not detectable in any other fetal tissue tested.

Another interesting biological activity of cultured neurosecretory tissues was demonstrated by Sachs and co-workers (Sachs *et al.*, 1974) in cells in the medial basal hypothalamus. Since this is the area of the hypothalamus that is primarily involved in the production of releasing and release-inhibiting hormones, it can be presumed to be a potential site of feedback control by circulating hormones. Sachs and co-workers demonstrated that in an explant taken from this area and maintained *in vitro* for 7 days, specific nuclear binding of estradiol-17-β occurred. More recently, Canick and co-workers (1977) have demonstrated that primary monolayer cultures established from fetal rat hypothalamus have the ability to form estrogens by aromatization of androgens. These investigators used cytosine arabinoside to inhibit the proliferation of non-neuronal elements, and they have found that the ability of the tissues to perform the aromatization is enhanced by this treatment.

VI. Other Neurosecretory Systems in Culture

In addition to the work described above in which cultured neurosecretory systems were shown to engage in one or more of the stages of the neurosecretory process, several other systems have been described in which neurosecretory neurons are maintained in culture but in which no specific neurosecretory functions have been demonstrated. Knigge and co-workers (1977) have recently described primary cultures of dispersed cells taken from the basal hypothalami of 10-12 day old rats. They found that neurons survived in this preparation for as long as three months, and that two general populations of neurons occurred. On the basis of immunocytochemical studies and fluorescence cytochemistry, populations of catecholamine containing neurons and luteinizing hormone-releasing hormone containing cells were seen.

Benda and co-workers (1975) used a similar preparation taken from fetal mouse hypothalamus to study developing structure and patterns of organization *in vitro*. They observed basal layers of astrocytes and ependymal cells, mature neurons, and cells which appeared to have the ultrastructural features of typical neurosecretory cells.

Vaccaro and co-workers (1976a) have described a method for maintaining primary monolayer cultures of cells from the hypothalami of fetal rats (utilized by Canick *et al.*, 1977 see above). A diverse population of surviving neurons in this preparation is shown by their demonstration that cultured cells have the ability to synthesize and accumulate acetylcholine and γ-aminobutyric acid but not dopamine or norepinephrine (Vaccaro *et al.*, 1976b), and that some cells contain vasopressin, neurophysin, or neurotensin (Vaccaro and Leeman, 1977).

Conclusions

The ultimate usefulness of cell, tissue, and organ cultures in unraveling the complexities of the neurosecretory process is yet to be demonstrated. The mass of work reviewed above describes systems which have been developed in recent years, many of which have been shown to retain in tissue culture the ability to engage in various specific stages of neurosecretion. Biosynthesis of vasopressin, neurophysin, LHRH, and TRH have been clearly shown. Transport of vasopressin from the hypothalamus to storage and release sites in the posterior pituitary has been demonstrated. Release of vasopressin in response to osmotic stimulation and to specific receptor interaction with acetylcholine has been shown, as have increased electrical activity of cultured neurosecretory cells in response to nicotinic agents and angiotensin II as well as inhibition by catecholamines (See Table III for summary). It is very encouraging that all of these processes, known to occur *in vivo* can be observed in cultured systems. New insights into the process of neurosecretion have not emerged from this work at the present time; however, given these early results and the unique experimental approaches that cultured systems allow, one must be optimistic about the future of such studies.

TABLE III

SUMMARY OF NEUROSECRETORY PHENOMENA OBSERVED IN CULTURED SYSTEMS

Observed Phenomenon	Preparation	References
Biosynthesis of vasopressin	Organ culture of guinea pig anterior hypothalamus	Sachs *et al.*,1971
Biosynthesis of vasopressin	Organ culture of guinea pig hypothalamoneurohypophysial complex	Pearson *et al.*, 1975(a)
Biosynthesis of neurophysin	Organ culture of guinea pig hypothalamoneurohypophysial complex	Sachs *et al.*, 1975
Biosynthesis of vasopressin	Cell line of transformed fetal mouse hypothalamic neurons	DeVitry *et al.*, 1975
Biosynthesis of TRH	Organ culture of guinea pig basal hypothalamus	McKelvy *et al.*, 1975; Sachs *et al.*, 1974
Biosynthesis of TRH	Clonal cell lines from rat CNS *tumor*	Grimm-Jörgensen *et al.*, 1976
Biosynthesis of LHRH	Organ culture of guinea pig basal hypothalamus	Sachs *et al.*, 1974
Transport of vasopressin to posterior pituitary	Organ culture of guinea pig hypothalamoneurohypophysial complex	Pearson *et al.*, 1975b
Release of vasotocin by osmotic stimulation	Organ culture of toad hypothalamoneurohypophysial complex	Eggena and Polson 1974
Release of vasopressin by osmotic stimulation	Organ culture of rat hypothalamoneurohypophysial complex	Sladek and Knigge, 1976
Release of vasopressin by acetylcholine	Organ culture of rat hypothalamoneurohypophysial complex	Sladek and Knigge, 1976
Stimulation of neuronal firing by nicotine	Tissue culture of neonatal dog supraoptic nucleus	Marks *et al.*, 1973
Inhibition of firing by catecholamines	Tissue culture of neonatal dog supraoptic nucleus	Marks *et al.*, 1973
Stimulation of firing by angiotension II	Tissue culture of neonatal dog supraoptic nucleus	Sakai *et al.*, 1974

References

Bargmann, W. and B. Scharrer, eds. 1970. V. *International Symposium on Neurosecretion.* Springer, Berlin-Heidelberg-New York.

Bargmann, W., B. Hanström, B. Scharrer and E. Scharrer, eds. 1958. II. *International Symposium on Neurosecretion.* Springer, Berlin-Göttingen-Heidelberg.

Benda, P., F. DeVitry, R. Picart and A. Tixier-Vidal. 1975. Dissociated cell cultures from fetal mouse hypothalamus: patterns of organization and ultrastructural features. Exp. Brain Res., 23:29-47.

Bern, H. A. 1966. On the production of hormones by neurones and the role of neurosecretion in neuroendocrine mechanisms. In: *Nervous and Hormonal Mechanisms of Integration.* Cambridge University Press, London. Vol. 20, 325-344.

Canick, J. A., D. E. Vaccaro K. J. Ryan, and S. E. Leeman. 1977. The aromatization of androgens by primary monolayer cultures of fetal rat hypothalamus. Endrocrinol., 100:250-253.

Cross, B., R. E. J. Dyball, R. G. Dyer, C. W. Jones, D. W. Lincoln, J. F. Morris and B. T. Pickering. 1975. Endocrine neurons. Recent Progr. in Hormone Res., 31:243-294.

DeVitry, F., M. Camier, P. Czernichow, P. Benda, P. Cohen and A. Tixier-Vidal. 1974. Establishment of a clone of mouse hypothalamic neurosecretory cells synthesizing neurophysin and vasopressin. Proc. Nat. Acad. Sci. U.S.A. 71:3575-3579.

DeVitry, F., M. Camier, P. Czernichow, P. Benda, P. Cohen and A. Tixier-Vidal. 1975. Characterization of a clonal strain of mouse hypothalamic cells synthesizing neurophysins and vasopressin. Ann. of N.Y. Acad. Sci., 248:64-79.

De Wied, D. 1969. Effects of peptides on behavior. In: *Frontiers in Neuroendocrinology* W. F. Ganong and L. Martin, eds. Oxford University Press, N. Y., pp. 97-144

Eggena, P., and A. X. Polson. 1974. Osmotic stimulation of vasotocin secretion by the toad's hypothalamo-neurohypophysial system *in vitro*. Endocrinol., 91:35-44.

First. International Symposium on Neurosecretion. 1954. Pubbl. Staz. Zool. Napoli. 24. Suppl. 1-98.

Gainer H., Y. Sarne and M. J. Brownstien. 1977. Conversion of a putative precursor during axonal transport. Science 195:1354-1356.

Grimm-Jörgensen, Y., S. E. Pfeiffer and J. F. McKelvy. 1976. Metabolism of thyrotropin releasing factor in two clonal cell lines of nervous system origin. Biochim. Biophys. Res. Commun. 70:167-173.

Heller, H. and R. B. Clark, eds. 1962. III. *International Symposium on Neurosecretion*. Mem. Soc. Endocrinol., No. 12. Academic Press, London-New York.

Knigge, K. M., G. Hoffman, D. E. Scott and J. R. Sladek, Jr. 1977. Identification of catecholamine and luteinizing hormone-releasing hormone containing neurons in primary cultures of dispersed cells of the basal hypothalamus. Brain Res. 120:393-405.

Knowles, F. 1974. Twenty years of neurosecretion. In: *Neurosecretion: the Final Neuroendocrine Pathway*. F. Knowles and L. Vollrath, eds. Springer, New York-Heidelberg-Berlin. pp. 3-14.

Knowles, F. and L. Vollrath, eds. 1974. VI. *International Symposium on Neurosecretion. Neurosecretion - The Final Neuroendocrine Pathway*. Springer, New York-Heidelberg-Berlin.

Lederis, K. 1974. Neurosecretion and the functional structure of the neurohypophysis. In: *Handbook of Physiology*, Vol. IV, Sec. 7, Part. 1, E. Knobil and W. H. Sawyer, eds. pp. 81-102.

Lumsden, C. E. 1968. Nervous tissue in culture. In: *The Structure and Function of Nervous Tissue*, Vol. I, G. Bourne, ed. Academic Press, New York. pp. 67-140.

Marks, B. H., K. K. Sakai, J. M. George and A. Koestner. 1973. Effects of catecholamines on supraoptic nucleus neurones in organ culture. In: *Frontiers in Catecholamine Research*. Pergamon Press, London. pp. 811-813.

McKelvy, J. F., M. Sheridan, G. Joseph, C. H. Phelps, and S. Perrie. 1975. Biosynthesis of thyrotropin-releasing hormones in organ cutures of the guinea pig median eminence. Endocrinol., 97:908-918.

Mendelson, I. S. and R. Walter. 1976. On the biosynthesis of putative neurophysin-vasopressin precursors in the hypothalamo-neurohypophysial gland of the guinea pig. In: *Hypothalamic Hormones, Second European Colloquium*. W. Voelter and D. Gupta, eds. Verlag Chemie, Weinheim. (in press).

Pearson, D., A. Shainberg, J. Osinchak, and H. Sachs. 1975(a). The hypothalamo-neurohypophysial complex in organ culture: morphologic and biochemical characteristics. Endocrinol. 96:982-993

Pearson, D., A. Shainberg, S. Malamed, and H. Sachs. 1975(b). The hypothalamo-neurohypophysial complex in organ culture: effects of metabolic inhibitors, biologic and pharmacologic agents. Endocrinol. 96:994-1003.

Pearson, D., R. Goodman, and H. Sachs. 1975(c). The fetal guinea pig hypothalamo-neurohypophysial complex in organ culture: presence of a stimulatory factor for vasopressin and neurophysin biosynthesis. Science 187: 1081-1082.

Pickering, B. T. and C. W. Jones. 1971. The biosynthesis and intraneuronal transport of neurohypophysial hormones. Mem Soc. Endocrin. 19:337-351.

Sachs, H. 1970. Neurosecretion. In: *Handbook of Neurochemistry*. A. Lajtha, ed. Plenum Press, New York, London. Vol. 4:373-428.

Sachs, H., and Y. Takatabake. 1964. Evidence for precursor in vasopressin biosynthesis. Endocrinol., 75:943-948.

Sachs, H., R. Goodman, J. Osinchak, and J. McKelvy. 1971. Supraoptic neurosecretory neurons of the guinea pig in organ culture. Biosynthesis of vasopressin and neurophysin. Proc. Nat. Acad. Sci. U.S.A. 68:2782-2786.

Sachs, H., D. Pearson, A. Shainberg, S. Shin, G. Bryce, S. Malamed and T. Mowles. 1974. Studies on the hypothalamo-neurohypophysial complex in organ cuture. In: *International Symposium on Recent Studies of Hypothalamic Function*. K. Lederis, ed. S. Karger Co., Basil. pp. 278-286.

Sachs, H., D. Pearson, and A. Neuredin. 1975. Guinea pig neurophysin: Isolation, developmental aspects, biosynthesis in organ culture. Ann. N. Y. Acad. Sci., 248:36-45.

Sakai, K. K., B. H. Marks, J. M. George, and A. Koestner. 1974. Specific angiotension II receptors in organ cultured canine supraoptic nucleus cells. Life Sci., 14:1337-1344.

Sladek, L., and K. M. Knigge. 1976. Vasopressin release in response to acetylcholine by organ cultured rat hypothalamic-neurohypophysial system. Abstracts of the Annual Meeting of the Soc. for Neurosci., p. 685.

Steiner, D. F. Peptide hormone precursors: biosynthesis, processing and significance. In: *Peptide Hormones*. J. A. Parsons, ed. University Park Press, Baltimore. pp. 49-64.

Sterba, G. 1974. Ascending neurosecretory pathways of the peptidergic type. In: *Neurosecretion — The Final Neuroendocrine Pathway*. F. Knowles and L. Vollrath, eds. Springer, New York-Heidelberg-Berlin. pp. 38-47.

Stubinsky, F., ed. 1967. IV. *International Symposium on Neurosecretion*. Springer, Berlin-Heidelberg-New York.

Vaccaro, D., A. Messer, D. M. K. Lam, S. E. Leeman. 1976(a). Primary monolayer cultures of rat hypothalamic cells. Fed. Proc., 35:1648.

Vaccaro, D., A. Messer, S. E. Leeman. 1976(b). Hypothalamic cells in primary monolayer culture. Annual meeting of Soc. for Neurosci., Abst. 1492.

Walter, R. 1975. Neurophysins: carriers of peptide hormones. Annals of N. Y. Acad. Sci., p. 248.

A PUTATIVE PRECURSOR TO NEUROPHYSINS AND NEUROHYPOPHYSEAL HORMONES*

IRWIN S. MENDELSON

Alvin Buckwold Centre
Department of Pediatrics
University of Saskatchewan
Saskatoon, Saskatchewan
Canada

RODERICH WALTER

Department of Physiology and Biophysics
University of Illinois Medical Center
Chicago, Illinois

Several lines of evidence suggest that arginine vasopressin (AVP) and neurophysin (Np) may be synthesized from a common precursor, e.g., 1) a concomitant biosynthesis of protein and the peptide (Sachs and Takabatake, 1964; Pickering *et al.*, 1975); 2) a puromycin-insensitive formation of both these molecules from presumed biosynthetic precursors (Pickering *et al.*, 1975); and 3) lack of a specific Np and AVP in rats homozygous for hereditary hypothalamic diabetes insipidus (Valtin, 1967; Valtin *et al.*, 1965; Burford *et al.*, 1971).

The work of Sachs and coworkers (D. Pearson, this volume) established that guinea pig hypothalamo-neorohypophyseal complex could be kept *in vitro* for weeks and retain its ability to synthesize Np and AVP; due to a slow synthesis, a 20 hour labeling period was, however, necessary to demonstrate incorporation of ^{35}S-cysteine into Np and AVP. To enhance formation of the protein and peptide we used hypothalamo-neurohypophyseal gland (HNG) preparations from guinea pigs, water deprived for 66 hours, in which the synthesis of Np was enhanced three-fold and that of AVP ten-fold; we studied the incorporation immediately after the removal of the preparation and used a 2-hour labeling period. Since preliminary pulse-chase experiments did not reveal rapid labeling of any particular protein, which could have been a precursor molecule, we used Tris-HCl (Tris(hydroxymethyl)aminomethane HCl), which inhibits the conversion of another precursor, proparathyroid hormone, to the active hormone (Chu *et al.*, 1974).

An 8M acid-urea polyacylamide gel (13.7%) was used to separate the protein from the peptide. Np and AVP standards migrated with R_f values

*This chapter was presented as part of the discussion of cell and tissue differentiation.

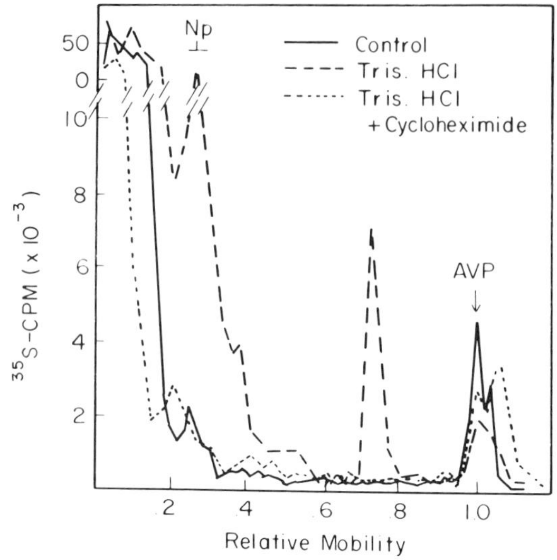

FIG. 1: Effects of Tris(hydroxymethyl)aminomethane-HC(Tris)HC1 (— — –), and Tris-HC1 plus cycloheximide (. . . .) on protein synthesis in hypothalano-neurohypophyseal gland (HNG) extracts studied by acid-urea polyacrylamide gel (13.7%) electrophoresis of HNG from guinea pig water-deprived for 66 hours. The control (———) shows the incorporation in the absence of Tris-HC1. The HNG preparations were incubated with ^{35}S-cysteine, 19 natural amino acids and Hank's medium at 36°C in an atmosphere of 95% O_2, 5% CO_2 for two hours. One external marker, ^{14}C-labeled AVP, was assigned an R_f value of 1.0 migrating toward the cathode on the right. A second external marker, guinea pig neurophysin (Np), was subject to electrophoresis in a parallel gel (R_f 0.3). From Mendelson and Walter, 1976.

of 0.3 and 1.0, respectively* (Fig. 1), and Np immunoreactive material was observed at R_f peaks of 0.3, 0.6, 0.7 and 0.8. The ^{35}S-cysteine labelled peak of R_f=1.0 was further identified as AVP by an antidiuretic bioassay. In the presence of Tris-HC1, a new ^{35}S-cysteine enriched peak appeared at R_f=0.8±0.1 and Tris-HC1 also caused a three-fold increase of immunoreactive Np-like material at this R_f. Sodium dodecyl sulphate polyacrylamide gel electrophoresis indicated that the peak at R_f=0.8±0.1 could be resolved into two major components with apparent molecular weights of 13,700 and 28,000 daltons; thus either of these polypeptides possess a higher molecular weight than the combined weight of Np(10,000 daltons) and AVP(1,000 daltons) (Mendelson and Walter, 1977).

The primary biosynthetic proteins translated from messenger RNA possess an N-terminal methionine residue (e.g., White *et al.*, 1973) and preparathyroid hormone has been found to be methionine rich in the

*The AVP was assigned an R_f value of 1.0 and the migration of the other compounds were expressed relative to this.

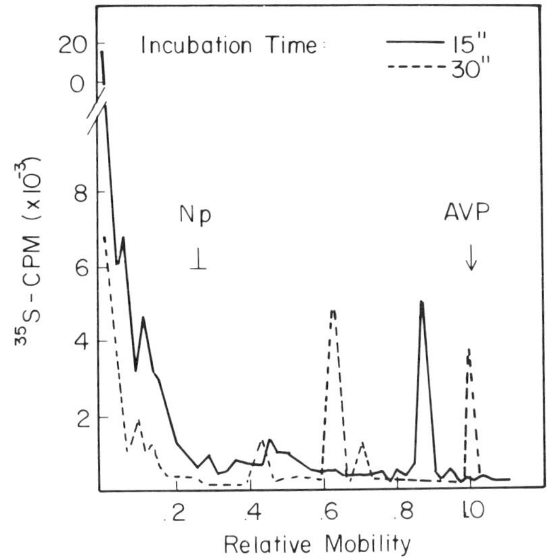

FIG. 2: Short pulses (15 and 30 seconds) of incorporation of ^{35}S-cysteine into the HNG preparation of water-deprived guinea pig. In this experiment extracts of two HNG were applied to the analytical gels. From Mendelson and Walter, 1976.

N-terminal region (Kemper *et al.*, 1976), while neither AVP nor Np protein of the guinea pig have methionine residues (Walter and Breslow, 1972). Repetition of the above 2 hour pulse experiment in the presence of Tris-HC1 was therefore carried out using ^{35}S-methionine instead of ^{35}S-cysteine in the hope of labeling the putative precursor. The R_f=0.8±0.1 peak was found to contain labeled methionine indicating that the material could represent an early translational product (Mendelson and Walter, 1977). A simultaneous incorporation of ^{3}H-labeled amino acids into the R_f=0.8±0.1 peak reveal the viability of the HNG preparation for *de novo* protein synthesis. In order to test whether the ^{35}S-cysteine/methionine enriched, neurophysin immunoreactive peak at R_f=0.8±0.1 was related to short-lived precursor(s) of AVP and/or Np, 15 and 30 second pulses with ^{35}S-cysteine were performed in the absence of Tris-HC1. It can be seen from Fig. 2. that only the peak of R_f of 0.8±0.1 is present after 15 seconds of labeling, while significant amounts of labeled AVP and a possible Np intermediate, are not detectable until after 30 seconds.

These experiments seem to be the first direct evidence for a 28,000 dalton precursor protein with a 13,700 dalton intermediate; ongoing research suggests that it may merit consideration as a precursor not only for Np and AVP but also for a second Np and oxytocin.

Acknowledgments

The work was supported by USPHS grant AM 18399.

References

Burford, G. D., C. W. Jones and B. T. Pickering. 1971. Tentative identification of a vasopressin-neurophysin and an oxytocin-neurophysin in the rat. Biochem. J. 124:809-813.

Chu, L. L. H., R. R. MacGregor, J. W. Hamilton and D. V. Cohn. 1974. Conversion of proparathyroid hormone: the use of amines as specific inhibitors. Endocrinology 95:1431-1438.

Kemper, B., J. F. Habener, M. D. Ernst, J. T. Potts and A. Rich. 1976. Pre-proparathyroid hormone: analysis of radioactive tryptic peptides and amino acid sequence. Biochem. 15:15-19.

Mendelson, I. S. and R. Walter. 1976 On the biosynthesis of putative neurophysin-vasopressin precursors in the hypothalamo-neurohypophyseal gland of the guinea pig. In: *Hypothalamic Hormones, Second European Colloquium, Tubingen.* W. Voelter and D. Gupta, eds. Verlag Chemie, Weinheim (in press).

Pickering, B. T., C. W. Jones, G. D. Burford, M. McPherson, R. W. Swann, P. F. Heap and J. F. Morris. 1975. The role of neurophysin proteins: suggestions from the study of their transport and turnover. Ann. N.Y. Acad. Sci. 248:15-35.

Sachs, H. and Y. Takabatake. 1964. Evidence for a precursor in vasopressin biosynthesis. Endocrinology 75:943-948.

Valtin, H. 1967. Hereditary hypothalamic diabetes insipidus in rats (Brattleboro Strain). Amer. J. Med. 42:814-827.

Valtin, H., W. H. Sawyer and H. W. Sokol. 1965. Neurohypophysial principles in rats homozygous and heterozygous for hypothalamic diabetes insipidus (Brattleboro strain). Endocrinology 77:701-706.

Walter, R. and E. Breslow. 1972. Methods of Isolation and identification of neurophysin protein. In: *Research Methods in Neurochemistry*, Vol. 2, N. Marks and R. Rodnight, eds. Plenum Press, New York. pp. 247-279.

White, A., P. Handler and E. L. Smith. 1973. *Principles of Biochemistry*, 5th Edition. McGraw-Hill Book Co., New York.

NEUROBLAST-GLIOBLAST INTERACTIONS: ECTO-ENZYMES

PAUL MANDEL, JAROSLAVA CIESIELSKI-TRESKA AND VLADIMIR STEFANOVIC

Centre de Neurochimie du CNRS
Strasbourg
France

I. Introduction

Cell culture provides an opportunity to investigate molecular aspects of nerve cell differentiation as well as molecular determinants of neural function in a defined microenvironment. Homogeneous cultures of neurons which divide synchronously and maintain differentiated properties would be particularly useful for this purpose. Unfortunately, so far, it has not been possible to obtain such neuronal cultures. Primary cultures of dissociated cells from embryonic nervous tissue are mixtures of different cell types, although several useful morphological and histochemical studies have been made. Glial primary cultures of new-born rats or mice seem to offer a good model of a rather homogeneous cell population. However these cells have a short life-span and undergo only a limited number of cell divisions. Biochemical investigations are only possible when microtechniques are available.

Clonal cell lines of nervous origin which exhibit a number of biochemical functions specific to nervous tissue may yield enough material for biochemical studies and represent a potentially useful system for the investigation of molecular aspects of nerve cell differentiation and function. Mouse neuroblastoma cells in culture undergo maturation illustrated by outgrowth of neurites (Seeds *et al.*, 1970), increase in enzymatic activities involved in neurotransmitter synthesis and degradation (Amano *et al.*, 1972; Richelson, 1973), increase in oxygen consumption (Nissen *et al.*, 1973) and production of action potentials after electrical or chemical stimulation (Nelson *et al.*, 1969; Harris and Dennis, 1970). Transformed cell lines may also display morphological structures and metabolic functions similar to neurons (Schubert, 1974).

Several clonal lines of glial origin synthetize the acidic protein S100 (Pfeiffer *et al.*, 1970; Schubert, 1974), have a high level of 2′,3′-cyclic nucleotide 3′-phosphohydrolase (EC 3.4.1.16), UDP-galactose:hydroxy fatty acid ceramide galactosyltransferase (EC 2.4.1.62) and PAPS-cerebroside sulphotransferase (EC 2.8.2.11) (Duch *et al.*, 1975), enzymatic activities involved in myelin synthesis, or in other functions assigned to glial cells.

It is obvious that clonal neural lines differ in many aspects from the untransformed neurons and glial cells. It has to be kept in mind that different clones present variable responses to the same effector or inducer, and that the properties of cells change with the number of passages. A biochemical heterogeneity is known to exist within cells of the same clone as shown by cytochemical methods (Ciesielski-Treska *et al.*, 1975b). Moreover, until now, tri- and tetrasialogangliosides specific for neuronal membranes could not be detected in clonal lines (Rebel *et al.*, 1973).

Thus there are several limitations for biochemical and genetic investigations on the clonal lines which have to be considered. This system, however, can be used for investigation of some clearly defined problems: effects of different effectors on cell differentiation, metabolic regulation of putative transmitters, drug testing, role of the plasma membrane in regulation of cell functions, genetic mapping, and so on. The present report summarizes the data obtained in recent years in our laboratory, concerning the study of surface membrane enzymes and their possible regulation during neural-glial interactions.

II. Ecto-Enzymes: General Aspects

The plasma membrane surrounding living cells mediates interactions between the cell and its external environment. Several membrane-bound enzymes have active sites which face the external medium rather than the cytoplasm, and these enzymes are referred to as "ecto-enzymes". A number of different lines of evidence suggest that the existence of ecto-enzymes probably is characteristic for most eukaryotic cells (Dunkley *et al.*, 1966; Agren *et al.*, 1971; De Pierre and Karnovsky, 1974a; Trams and Lauter, 1974). We have demonstrated that the active sites of the plasma membrane ATPase (ATP phosphohydrolase, EC 3.6.1.3), 5′-nucleotidase (5′-ribonucleotide phosphohydrolase, EC 3.1.3.5), inorganic pyrophosphatase (pyrophosphate phosphohydrolase, EC 3.6.1.1) and acetylcholinesterase (acetycholine ester hydrolase, EC 3.1.1.7) of nervous clonal lines in cultures are located at the external membrane surface. The comparison of enzyme activities in intact cells with cell homogenates permits the evaluation of the amount of ecto-membrane activity as compared to the intracellular activity.

These activities were not due to artifacts such as breakage of the cells during the incubation period since the cell integrity was checked before and after the incubation period. Slight proteolytic digestion was found to inactivate these enzymes, and they could also rapidly be inhibited by macromolecules which do not penetrate the cells. Lactate dehydrogenase activity, a cytoplasmic marker, was estimated in the incubation medium in order to establish possible loss of the cell cytosol constituents.

Verification that the products of ecto-enzymes were released to the extracellular medium was obtained by the procedure described by De Pierre and Karnovsky (1974a). For instance, in order to confirm ecto-membrane localization of ATPases, the intracellular pool of inorganic phosphate was labeled with ^{33}P. Labeled substrate, ^{32}P γATP, was supplied extracellularly and from the concentration of isotopes in the intra- and extracellular compartments it was evident that neither the substrate nor the product entered the cells.

Finally, the inhibition of ecto-enzyme activities by diazonium salt, a non-penetrating reagent which forms covalent bonds with many of the functional groups of proteins, also confirmed that the proteins involved in these activities must have functional groups located outside of the permeability barrier of the plasma membrane.

The experiments described below provide evidence that neuroblastoma and glial cell lines in cultures display a high level of ecto-ATPase, 5′-nucleotidase and ecto-pyrophosphatase activities. A detailed analysis of the kinetic properties, activation and substrate specificity of the enzymes, has already been performed (Stefanovic *et al.*, 1974, 1975a, 1975b, 1976a, 1976b, 1977). In addition, the acetyl- and butyrylcholinesterase activities have been measured and the activation of these enzymes by treatment with *Clostridium perfringens* sialidase has been demonstrated (Rosenberg *et al.*, 1975). In this paper, some properties of the ecto-ATPase and 5′-nucleotidase activities will be described. Clonal lines used in this study were normal hamster astroblasts NN (Shein *et al.*, 1970), C6 rat glioblastoma (Benda *et al.*, 1968) and mouse neuroblastoma lines N1E-115, N18 (Amano *et al.*, 1972) and M1 (Ciesielski-Treska *et al.*, 1975b).

TABLE I

Mg^{2+}- AND Ca^{2+}-DEPENDENT ATPASE ACTIVITIES AT THE EXTERNAL SURFACE OF NEUROBLASTOMA AND GLIAL CELLS IN CULTURE

Cell Line	Mg^{2+}-dependent ATPase	Ca^{2+}-dependent ATPase
C6	0.27±0.02	0.23±0.01
NN	0.35±0.02	0.33±0.01
M1	0.47±0.02	1.10±0.03
N18	1.20±0.04	2.38±0.05
N1E-115	1.49±0.03	3.25±0.04

NN and C6 glial cell lines and neuroblastoma clones were grown in Falcon flasks. For the study of enzymatic activity the cells were subcultured in Petri dishes and enzymatic activity was determined after 4 days. Activity is expressed as μmoles of inorganic phosphate liberated/mg cell protein/h, and each value is the mean of 5-6 experiments ± S.E.M. (from Stefanovic *et al.*, 1976a).

A. ATPase Activity at the External Surface of Neuroblastoma and Glial Cells

We found Mg^{2+}- and Ca^{2+}-dependent ATPase activities at the external surface of the N1E-115, N18 and M1 neuroblastoma clonal cell lines. The two glial cell lines, NN normal hamster astroblasts and C6 rat glioblastoma exhibited a very low Mg^{2+}- and Ca^{2+}-dependent ATPase activity (Table I). A different pattern of activation of neuroblastoma and glial cell ecto-ATPase by divalent cations is also evident. Mouse neuroblastoma clones had about twice as high activity with Ca^{2+} as with Mg^{2+}. Glial cell lines had higher Mg^{2+}-dependent ATPase activity. The enzyme is completely inactivated by 1 mM EDTA. The divalent cations Mn^{2+} and Co^{2+} activate the enzyme 74% and 46%, respectively, in hamster astroblasts, and about 50% in mouse neuroblasts (Table II).

The apparent Km values for Mg^{2+}- and Ca^{2+}-dependent ATPase in several lines have been determined. Apparent Km values (for ATP) for different neuroblastoma cell lines are from 0.053 to 0.191 mM for Mg^{2+}-ATPase and from 0.144 to 0.530 mM for Ca^{2+}-ATPase. The apparent Km values of ecto-ATPase in glial NN and C6 line are 0.036 to 0.065 mM for Mg^{2+}-ATPase and 0.027 to 0.043 mM for Ca^{2+}-ATPase.

TABLE II
EFFECT OF DIVALENT CATIONS ON ECTO-ATPASE ACTIVITY OF N1E-115 AND NN CELLS

Cation	ATPase activity	
	N1E-115	NN
Ca^{2+}	100[a]	100[b]
Mg^{2+}	55	134
Mn^{2+}	47	74
Co^{2+}	48	46
Cd^{2+}	28	34
Sr^{2+}	20	26
Cu^{2+}	10	34
Zn^{2+}	3	22

Divalent cations and ATP were present in a final concentration of 3 mM each. Values are the mean of 4-5 experiments ± S.E.M. Relative activity with Ca^{2+} is taken as 100. [a] 5.43±0.46 μmoles/mg cell protein/h. [b] 0.35±0.02 μmoles/mg cell protein/h (from Stefanovic *et al.*, 1976b).

The specificity of the enzyme system was examined by replacing ATP with other nucleoside tri-, di- and monophosphates. All nucleoside triphosphates tested were split. Nucleoside diphosphates were hydrolyzed at appreciable rates, while nucleoside monophosphates were not hydrolyzed under the same conditions. Competition studies have revealed that all of the nucleoside triphosphates are hydrolyzed by the same enzyme.

None of the sulfhydryl reagents, in the concentration tested, were inhibitory. A small inhibition ($p < 0.05$) of Mg^{2+}-ATPase by glutathione and cysteine was observed, while sodium fluoride, at a concentration of 1 mM, inhibited both ecto-ATPases. The enzymatic activity was not significantly changed by ouabain, dinitrophenol or oligomycin (Stefanovic *et al*, 1976b).

The properties exhibited by neuroblastoma and glial cell ecto-ATPases are similar to the properties of other mammalian ATPases. Their very broad substrate specificity is similar to the ecto-ATPases of human blood platelets (Chambers *et al.*, 1967), guinea pig leukocytes (De Pierre and Karnovsky, 1974b) and different normal and transformed cell lines in culture (Trams and Lauter, 1974). The lack of stimulation by dinitrophenol distinguishes ecto-ATPases of neuroblastoma and glial cells from mitochondrial ATPase.

B. 5′-Nucleotidase Activity at the External Surface of Neuroblastoma and Glial Cells

All neuroblastoma and glial cells tested have been shown to have

TABLE III
SUBSTRATE SPECIFICITY OF RAT C6 GLIOMA CELL 5′-NUCLEOTIDASE

	Nucleotidase activity	
Nucleotide	Ecto-	Internal
5′-AMP	4.23	2.57
5′-UMP	3.13	3.16
5′-CMP	2.34	3.03
5′-IMP	1.69	3.34
5′-GMP	1.52	2.70
5′-dAMP	1.10	2.57
3′-AMP	0	0
2′-AMP	0	0

Activity is expressed as μmoles of inorganic phosphate liberated/mg cell protein/h. Intact cells at the beginning of the stationary phase (approximately 10^6 cells) in monolayer culture were incubated in quadruplicate pH 7.4 and 37°C for 5 min in 1.5 ml of assay solution composed of 3 mM nucleotide, 130 mM NaCl, 1 mM $MgCl_2$, 5.5 mM glucose, and 30 mM Tris-HCl. Cell homogenates were prepared in distilled water at 0°C, reconstituted to the concentrations of nucleotide, glucose and salts, and incubated as described for intact cells. After incubation, inorganic phosphate was determined in the extracellular medium of intact cultures or in the supernatant fluid of homogenates. The differences between values obtained for whole cell homogenates and values for intact cell cultures are taken as internal cellular 5′-nucleotidase activities.

5′-nucleotidase activity at the external cell surface. Great variations in this activity have been found between different lines. The glioblastoma C6 line displays 5 to 20 times higher 5′-nucleotidase than other glial and neuroblastoma lines tested. The ecto-5′-nucleotidase rapidly hydrolyzed 5′-AMP. Other 5′-nucleotides were also hydrolyzed while 2′- or 3′-nucleotides were not attacked (Table III). The ecto-5′-nucleotidase obeyed Michaelis-Menten kinetics; the apparent Km for AMP in the C6 glioma cell line was 0.22 mM.

ATP and ITP powerfully repress the activity of ecto-5′-nucleotidase, and ADP behaves as a competitive inhibitor. Divalent cations were not needed for substantial activity. These findings indicate that the intact cell is capable of rapid hydrolysis of exogenous 5′-AMP, producing adenosine at the cell surface, at a rate which responds directly to extracellular AMP concentration and which can be controlled by extracellular ADP and ATP.

The assignment of a physiological role to the plasma membrane ecto-enzymes at present is still premature. There have been some indications of ecto-ATPase involvement in the permeability of cell membrane (Engelhardt, 1957), in nucleic acid metabolism by altering the level of nucleoside triphosphates (Manery *et al.,* 1968), and in cation transport (Henn *et al.,* 1972; Rosenblatt *et al.,* 1976). Differences between activities of Mg_{2+}-dependent ecto-ATPase in neoplastic and normal glial cells have been demonstrated (Agren *et al.,* 1971). The observed phenomenon was ascribed to the neoplastic transformation of cells, in particular of their membrane structure and function. 5′-Nucleotidase may make nucleosides which can penetrate the membranes and enter for the cell. Clearly further investigations are needed to establish the relationship between function of ecto-enzymes and membrane involvement in growth control and differentiation. In the following we will discuss the alteration of some ecto-enzymes as a function of cell interactions.

III. Ecto-Enzyme Activities in Coculture of Neuronal and Glial Clonal Lines

The importance of metabolic and functional coupling between neurons and glial cells in the nervous system is well established. However, the elucidation of mechanisms involved in the interaction of these cells is rather limited with regard to *in vivo* studies. Recently developed nervous system clonal lines provide a tool for the study of interactions between neuronal and glial cells in culture. The basic advantage of the *in vitro* approach in the study of neuronal-glial interactions lies in the possibility to manipulate and control the humoral and cellular environment of the cells. External surface enzymes seem most reliable for a study of neuron-glia interactions. Thus, we investigated ATPase, acetylcholinesterase, and 5′-nucleotidase activities in cocultures of neuroblastoma and glial cell lines as well as in the same lines reisolated from such cocultures. The effect of glial conditioned medium on neuroblastoma ectoATPase was also investigated.

TABLE IV
Mg^{2+}-AND Ca^{2+}-DEPENDENT ECTO-ATPASE ACTIVITIES OF NEUROBLASTOMA AND GLIAL CELLS IN CO-CULTURE

Co-cultures	Mg^{2+}-dependent ATPase	Ca^{2+}-dependent ATPase
M1 + NN	2.24 ± 0.03	3.71 ± 0.08
M1 + C6	0.43 ± 0.01	0.94 ± 0.02
N18 + NN	1.30 ± 0.04	3.02 ± 0.33
N115 + NN	1.59 ± 0.05	5.48 ± 0.23

Neuroblastoma and glial cells were co-cultured in Falcon Flasks for 10 days. For the study of enzymatic activity (μmoles/mg cell protein/h) they were subcultured in Petri dishes and used after 4 days. Each value is the mean of 5 experiments ± S.E.M. (from Stefanovic *et al.*, 1976a). For the activities of the individual lines, see Table I.

A. Ecto-ATPase Activities

An increase of both Mg^{2+}- and Ca^{2+}-dependent ecto-ATPase activities was observed in cocultures of neuroblastoma and glial cells (Table IV). Among several lines tested neuroblastoma M1 and astroblastoma NN cells in coculture were found to be the most potent in producing an increase in cell surface ATPase activity. In these cocultures Mg^{2+}- and Ca^{2+}-dependent ecto-ATPase activities were about 4 times higher, when compared with M1 line; compared with NN cell line, both ecto-ATPase activities increased 8-10 times.

The increase in Mg^{2+}- and Ca^{2+}-dependent ecto-ATPase activities in coculture of M1 and NN cells depends on the relative concentration of both cell types. A high increase in activity was observed at relatively low proportion of M1 cells in coculture with NN astroblasts (Fig. 1). The optimal increase in both ecto-ATPase activities was observed if equal numbers of M1 and NN cells were mixed.

The time course of the effect of cocultivation upon ecto-ATPase was investigated. For up to 4 days in coculture, values for ecto-ATPase activity were intermediate between those of the M1 and NN cell lines. On the 6th day, the activity exceeded the level of the M1 cell line. Cells in coculture for a longer period, exhibited much higher Mg^{2+}- and Ca^{2+}-dependent ATPase activities (Fig. 2).

To establish which type of cell in the cocultures is responsible for the increase of ecto-ATPase activities, neuroblastoma and glial cells were

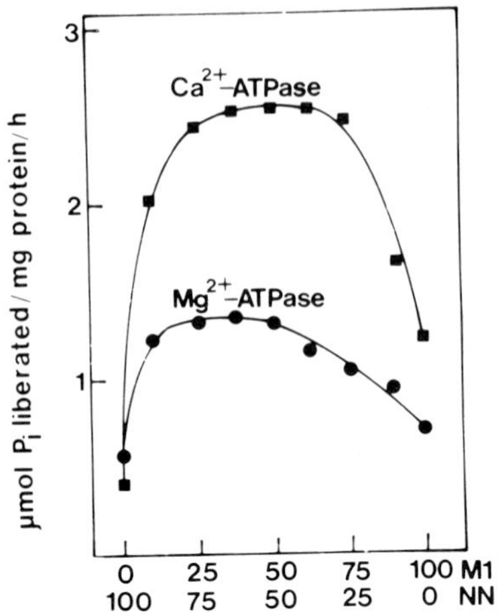

FIG. 1. Mg^{2+}- and Ca^{2+}-dependent ecto-ATPase activities of M1 and NN cell lines in coculture, shown as a function of the ratio between the cocultured cells. Cells were seeded in Petri dishes in the proportions indicated. Medium was changed after 2, 4 and 6 days and the enzymatic activity determined on the 7th day. (Closed circles), Mg^{2+}-ATPase. (Closed squares), Ca^{2+}-ATPase (from Stefanovic *et al.*, 1977).

reisolated. The procedure and characterization of reisolated cells have already been described (Stefanovic *et al.*, 1976a, 1977). Neuroblastoma M1 cells were isolated after 7 days and 2 months of coculture. The latter were extensively studied. Reisolated M1 cell line exhibited higher Mg^{2+}- and Ca^{2+}-dependent ecto-ATPase activities than the original M1 cell line. M1 cells isolated after a 7 day coculture had about 2-3 times higher ecto-ATPase activity; M1 cells reisolated after 2 months of coculture had 4-5 times higher ATPase activity than the original cell line. The

TABLE V

THE EFFECT OF SUBCULTIVATION ON ECTO-ATPase ACTIVITY OF REISOLATED NEUROBLASTOMA M1 CELLS

Number of replications	Mg^{2+}-dependent ATPase	Ca^{2+}-dependent ATPase
8	2.62 ± 0.18	5.73 ± 0.28
12	3.01 ± 0.06	5.72 ± 0.16
22	2.76 ± 0.08	5.87 ± 0.15
29	3.01 ± 0.08	5.65 ± 0.24

Activity is expressed as μmoles Pi liberated/mg cell protein/h. Cells were cultured in Petri dishes for 6 days. Medium was changed every other day. Each value is the mean of 4 experiments ± S.E.M. (from Stefanovic *et al.*, 1976a).

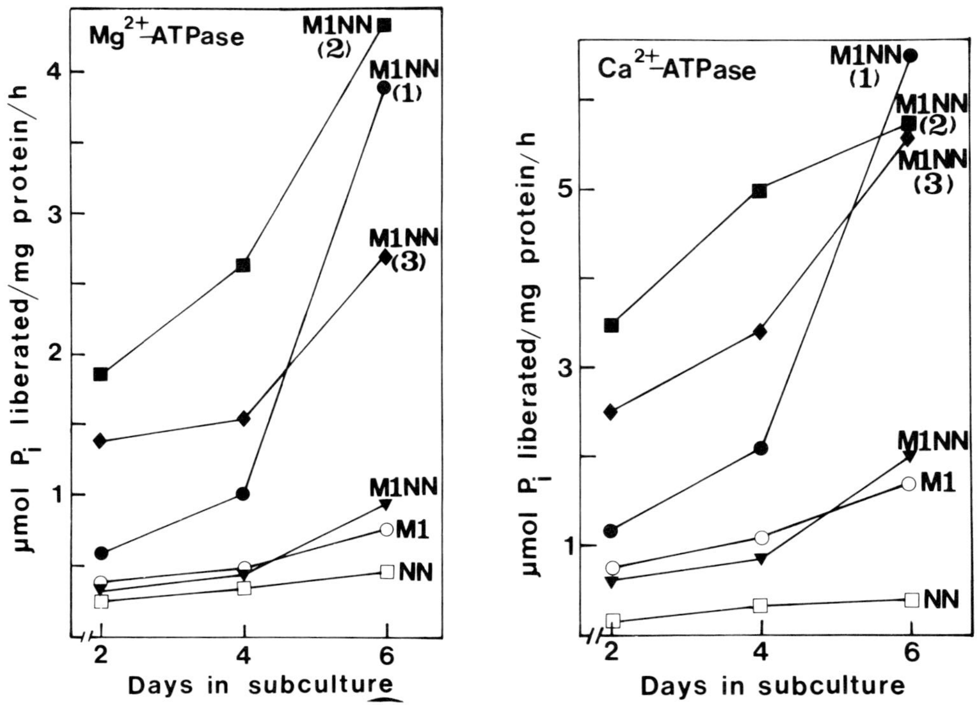

FIG. 2. Mg^{2+}- and Ca^{2+}-dependent ecto-ATPase activities of M1 and NN cell lines cocultured for different periods of time. Cocultures were prepared from equal parts of M1 and NN cells. Mg^{2+}-ATPase (A) and Ca^{2+}-ATPase (B) activities. (Open squares), NN hamster astroblasts. (Open circles), M1 neuroblastoma cell line. (Closed triangles), M1 and NN cells cocultured for the period indicated (2-6 days).(Closed circles), M1 and NN cells cocultured in Falcon flasks for 10 days and then subcultured in Petri dishes for the period indicated. (Closed squares), M1 and NN cells cocultured in Falcon flasks for 15 days and then subcultured in Petri dishes for the period indicated. (Closed diamonds), N1 and NN cells cocultured for 20 days in Falcon flasks, and then subcultured in Petri dishes for the period indicated.

ecto-ATPase activity of reisolated M1 cell line was monitored for about 20 replications and the typical points of such an experiment are presented in Table V. The ecto-ATPase activity did not change for more than 20 replications which represent about 250 cell generations after separation from the glial cells.

The possibility of removal of an inhibitor or the appearance of an activator of ATPase activity in reisolated M1 cells was further tested. The original M1 and reisolated M1 cell lines were cocultured for 6 days and the enzymatic activity determined in coculture and in the individual cell lines. The ATPase activities were strictly additive. Mg^{2+}- and Ca^{2+}-dependent ecto-ATPase activity of the coculture was higher than the theoretical activity by 8% and 10% respectively, which may be accounted for by the slightly shorter doubling time of reisolated M1 cells. These results indicate that diffusible inhibitors or activators of ecto-ATPase cannot account for the change in ecto-ATPase activity of M1 cells reisolated from the coculture with hamster astroblasts.

A kinetic study of ATPase activity was performed. Michaelis constants were determined for the original M1 and reisolated M1 cell lines, and the results are presented as double reciprocal plots in Fig. 3. Higher V_{max} and Km values in the reisolated M1 cell line as compared with the original cell line were found.

The following possibilities may be considered as an explanation of these findings: (1) an unknown change in the surface geometry of the cells, (2) a selection of a cell population with higher ecto-ATPase activities.

After 2 months in coculture of neuroblastoma M1 and astroblast NN cells, astroblast NN cells were reisolated. Reisolated NN cells exhibited higher activities of both Mg^{2+}- and Ca^{2+}-dependent ecto-ATPase than the original NN cell line. Reisolated NN cells in monolayer culture separated

TABLE VI
THE EFFECT OF SUBCULTIVATION ON ECTO-ATPase ACTIVITY OF REISOLATED NN ASTROBLAST CELLS

Number of replications	Mg^{2+}-dependent ATPase	Ca^{2+}-dependent ATPase
17	1.46 ± 0.04	1.17 ± 0.04
28	1.05 ± 0.07	0.92 ± 0.04
48	0.65 ± 0.05	0.48 ± 0.05

Activity is expressed as μmoles Pi liberated/mg cell protein/h and was determined on confluent cells. Each value is the mean of 6 experiments ± S.E.M. (from Stefanovic *et al.*, 1976a).

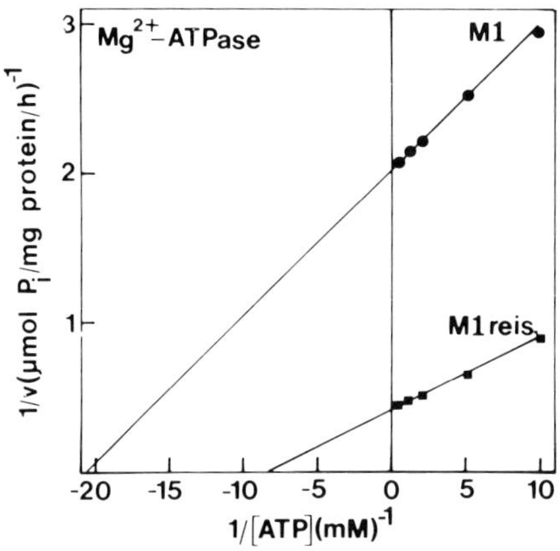

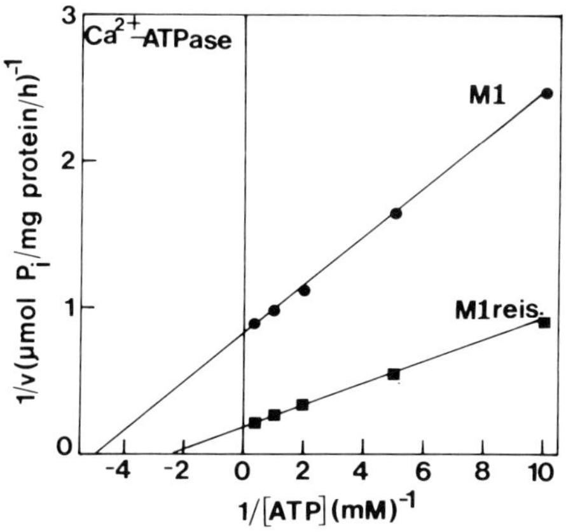

FIG. 3. Lineweaver-Burk plots of ecto-ATPase activity of M1 neuroblastoma cell line and M1 cell line reisolated after two months of coculture with NN hamster astroblasts. ATPase activity was determined in the presence of an ATP-regenerating system. Mg^{2+}-ATPase (A) and Ca^{2+}-ATPase (B) activities were determined on confluent cells at 5 different Mg^{2+}-ATP concentrations. (Closed circles), M1 neuroblastoma cell line. (Closed squares), M1 cell line reisolated from the coculture with NN hamster astroblasts (from Stefanovic *et al.*, 1977).

from neuroblastoma cells, changed their surface ATPase activity. After 48 passages both Mg^{2+}- and Ca^{2+}-dependent ATPase activities diminished but still remained higher than ATPase activity of the original NN clone (Table VI). The kinetic study showed that after 28 passages a decrease in V_{max} took place without a change in Km, followed by a decrease of both Km and V_{max} (Fig. 4).

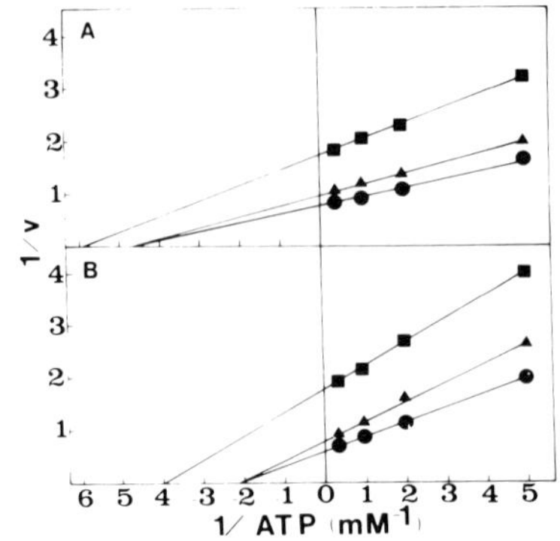

FIG. 4. Lineweaver-Burk plots of ATPase activity at the external surface of NN cells reisolated after cocultivation with M1 neuroblastoma clone. The effect of subcultivation. (Closed circles), 18 replications. (Closed triangles), 28 replications. (Closed squares), 48 replications. ATPase activity was determined in the presence of an ATP regenerating system. Mg^{2+}-ATP varied from 0.1 to 3.0 mM. (A) Mg^{2+}-dependent ATPase. (B) Ca^{2+}-dependent ATPase (from Stefanovic *et al.*, 1976b).

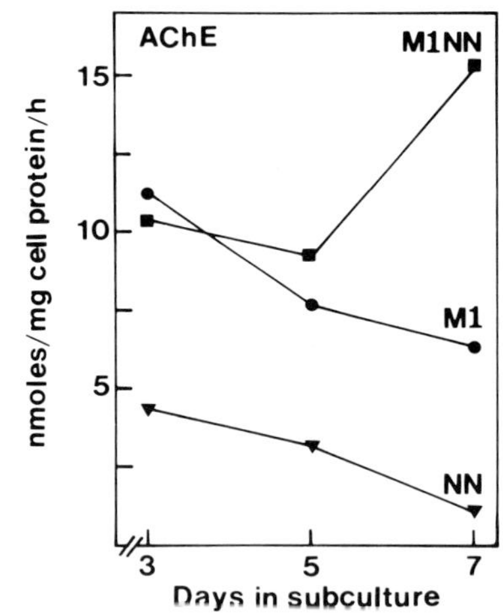

FIG. 5. Comparison of acetylcholinesterase activities of M1 and NN cell lines, and in cocultures of the two lines (M1, NN). The cocultures were prepared from equal parts of M1 and NN cells and enzymatic activity was measured on the 3rd, 5th and 7th day. Acetylcholinesterase was determined as described (Stefanovic, *et al.*, 1975b) and activity expressed as nmoles of acetylthiocholine hydrolyzed/mg total cell protein/h (from Ciesielski-Treska *et al.*, 1975a).

B. Acetylcholinesterase Activity

An increase in ecto-acetylcholinesterase activity was observed in the coculture of neuroblastoma M1 and astroblast NN cells (Fig. 5). After 7 days of coculture, acetylcholinesterase activity was respectively 7 and 2 times higher than the original NN and M1 cells. In contrast to the increase in acetylcholinesterase activity of M1 and NN cocultures, a decrease was detected in the coculture of M1 neuroblastoma and C6 glioblastoma. In addition, acetylcholinesterase activity of M1 cells reisolated after 2 months of coculture with NN astroblasts was about 8 times higher than that of the original M1 cells (Ciesielski-Treska *et al.*, 1975b).

C. 5′-Nucleotidase Activity

Table VII shows 5′-nucleotidase activity at the external surface of some neuroblastoma and glial cell lines in culture and in coculture of these cells. Relatively low 5′-nucleotidase activity was found at the external surface of all the neuroblastoma cell lines tested. NN astroblasts had a 5′-nucleotidase activity which was twice as high as that of the M1 cell line. 5′-Nucleotidase activity of cocultures was below the activity of NN cells. Reisolated M1 and NN cells had the same or lower 5′-nucleotidase activity than the original cell lines.

TABLE VII
5′-NUCLEOTIDASE AT THE EXTERNAL SURFACE OF NEUROBLASTOMA AND GLIAL CELL LINES IN CULTURE

Cell Line	5′-Nucleotidase activity
M1	0.065±0.003
N18	0.016±0.003
N1E-115	0.027±0.002
NN	0.146±0.005
M1 + NN	0.110±0.006
N18 + NN	0.068±0.002
N1E-115 + NN	0.109±0.006
M1 reisolated	0.074±0.005
NN reisolated	0.118±0.004

Activity is expressed as μmoles Pi liberated/mg cell protein/h. The cells were cultured in Falcon flasks for 10 days. For the study of enzymatic activity they were subcultured in Petri dishes and used after 4 days. Each value is the mean of 5 experiments ± S.E.M. (from Stefanovic *et al.*, 1976a).

The data presented show that nervous system cells in coculture undergo marked changes in their surface properties. Acetylcholinesterase and Mg^{2+} and Ca^{2+} dependent ecto ATPase activities of cocultured neuroblastoma and glial cells increased many fold over the level of the individual cell lines. Increased Mg^{2+}- and Ca^{2+}-dependent ecto-ATPase activities were also found in both glial and neuroblastoma cell lines reseparated from the coculture. After a certain number of generations, the ecto-ATPase activity of glial cells diminished but was still higher than in the original cell line. In contrast, high ecto-ATPase activities of neuroblastoma cells acquired in cocultures are maintained after many subcultures. In addition, the neuroblastoma cells have higher acetylcholinesterase activity than the original line. 5′-Nucleotidase activity at the external surface of cocultured NN and M1 cells or reisolated M1 and NN cells do not differ significantly from that of the original culture.

The data presented support the view that the nervous system cells exhibit a wide variety of functional interactions. In fact, communication from neurons to glia and vice-versa has been proposed, and different mechanisms discussed. Several authors have indicated that contact with glial cells enhances and maintains the differentiation of dissociated neurons in long-term cultures (Varon and Raiborn, 1969; Burnham *et al.*, 1972; Sensenbrenner and Mandel, 1974; Varon *et al.*, 1974). Choline acetyltransferase activity increased 100-fold and acetylcholine synthesis 100- to 1000-fold in cocultured dissociated sympathetic neurons with non-neuronal cells from sympathetic ganglia (Patterson and Chun, 1974).

Several lines of evidence suggest involvement of external cell surface ATPase activity in differentiated functions. Significant increase of ATPase activity takes place in the stationary phase of growth when some cells differentiate. Compared to the original line, the basal activity of adenyl cyclase and the increase of cyclic AMP evoked by adenosine are potentiated in neuroblastoma cells reisolated after coculture with glia cells. Thus tissue cultures afford an efficient tool for the clarification of the molecular mechanisms of the interaction of different types of nerve cells (Y. Kuroda, personal communication).

IV. The Effect of Astroblast Conditioned Medium on the Ecto-ATPase Activities of Neuroblastoma Cells

Experiments with a variety of tissue culture systems have indicated that cells release numerous factors to the medium which can influence other cell cultures either directly or by modifying the properties of the new medium (Peterson and Rubin, 1969; Halpern and Rubin, 1970). These factors are contained in so-called conditioned medium and have been shown to influence cellular aggregation (McClay and Moscona, 1974), growth (Rubin, 1966) and differentiation (Konigsberg, 1963). Glia conditioned medium was shown to induce a high degree of morphological differentiation of neuroblastoma cells (Monard *et al.*, 1973).

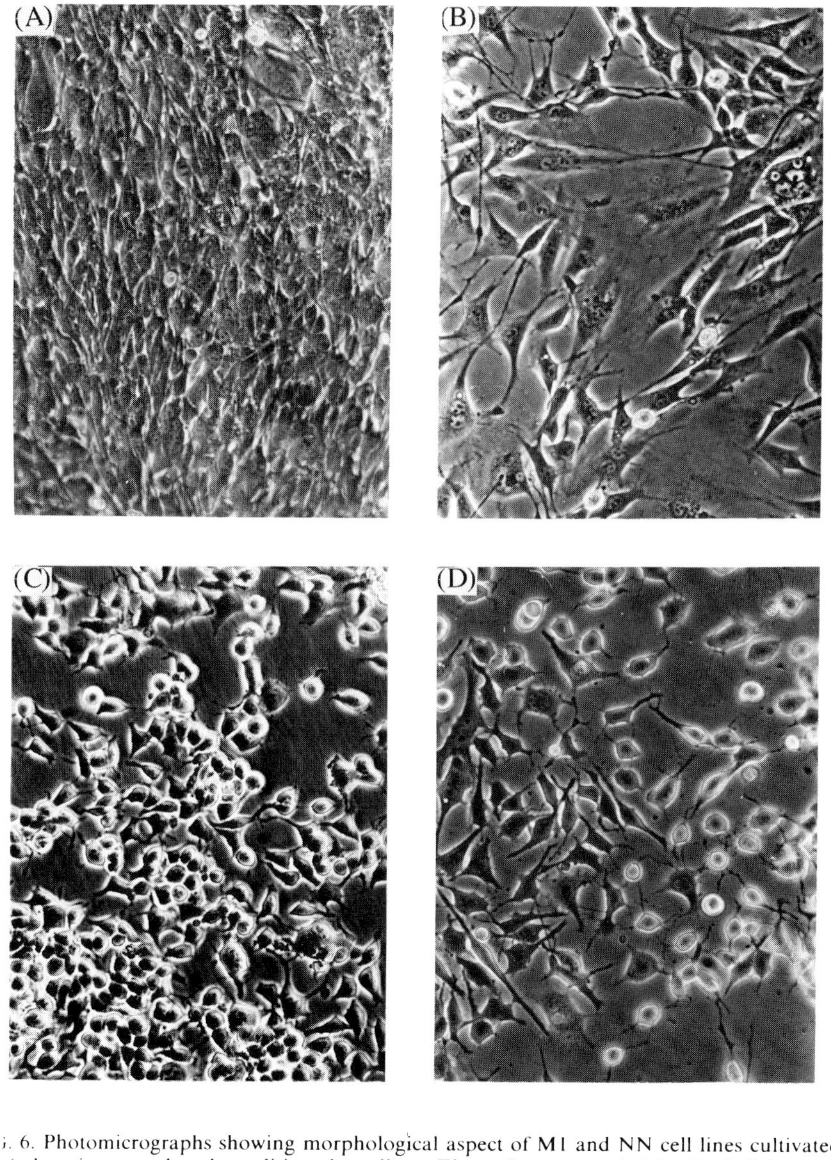

FIG. 6. Photomicrographs showing morphological aspect of M1 and NN cell lines cultivated for 6 days in normal and conditioned medium. The cells were seeded in Petri dishes and conditioned medium was added 18 h later. NN clone in normal medium (A), in the presence of 20% M1 neuroblastoma conditioned medium (B). M1 clone in normal medium (C), in the presence of 50% NN glia conditioned medium (D).

(A)

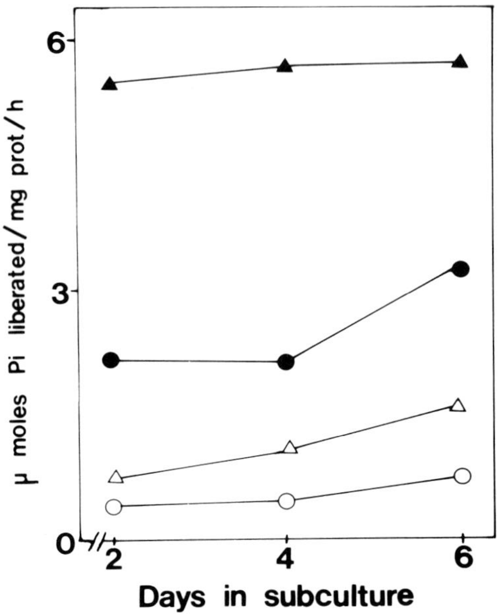
6
3
0
μ moles Pi liberated/mg prot/h
2
4
6
Days in subculture

(B)

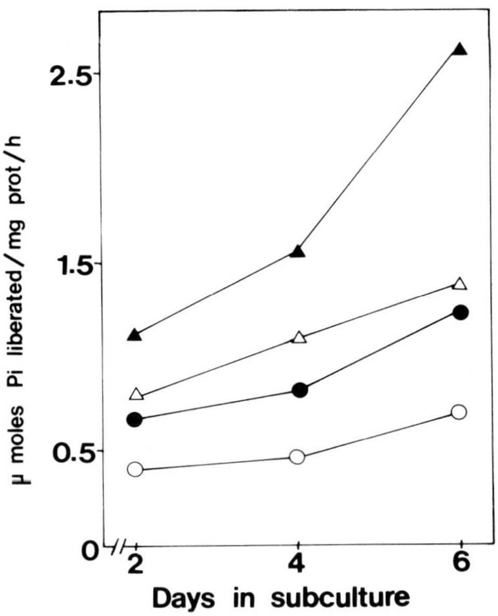
2.5
1.5
0.5
0
μ moles Pi liberated/mg prot/h
2
4
6
Days in subculture

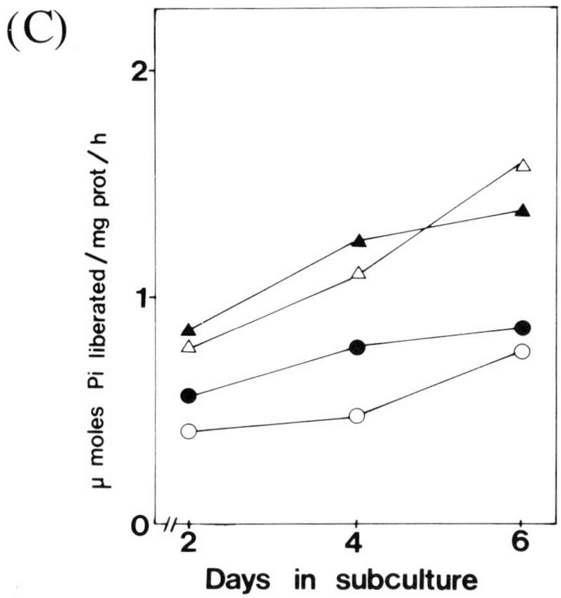

FIG. 7. Mg^{2+}- and Ca^{2+}-dependent ecto-ATPase activities of M1 cells cultivated in the presence of NN glia conditioned medium (A), C6 glia conditioned medium (B) and fibroblast conditioned medium (C). The M1 cells were maintained in the Falcon flasks in the presence of conditioned medium for 6 days, subcultured in Petri dishes and enzymatic activities were determined 2, 4 and 6 days thereafter. Each value represents the mean of 6-8 experiments. S.E.M. was less than 10%. M1 cells grown in normal medium: (Open circles), Mg^{2+}-ATPase; (Open triangles), Ca^{2+}-ATPase. M1 cells treated by conditioned medium: (Closed circles), Mg^{2+}-ATPase; (Closed triangles), Ca^{2+}-ATPase.

Neuroblastoma M1 cells were cultured in the presence of astroblast NN conditioned medium. The change in the morphological aspect of M1 cells is shown in Fig. 6. In contrast to the controls most of the neuroblastoma cells cultivated with glia conditioned medium extend expansions. On the other hand, M1 conditioned medium in a final concentraton of 20% inhibits growth of NN cells.

NN glia conditioned medium increases both ecto-ATPase activities of M1 neuroblastoma cells. In cultures of M1 neuroblastoma cells grown 13 days in NN conditioned medium the enzymatic activities of Mg^{2+}- and Ca^{2+}-dependent ecto-ATPase are 4 and 3 times higher, respectively, than the enzymatic activities of neuroblastoma cells cultivated in the absence of conditioned medium. Glioblastoma C6 and fibroblast conditioned media have very little effect on both enzymatic activities of M1 neuroblastoma cells (Fig. 7). After treatment with nerve growth factor or insulin, no significant changes in these activities were observed; moreover, after treatment with dibutyryl cyclic AMP the activities of Mg^{2+}- and Ca^{2+}-

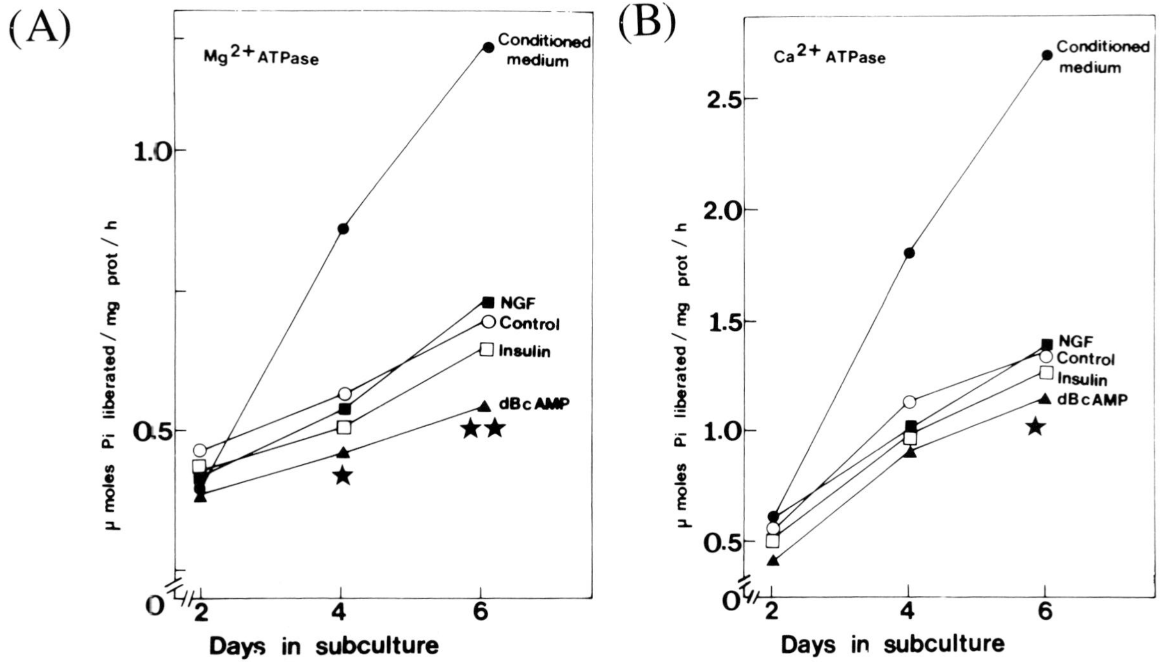

FIG. 8. Mg^{2+}- and Ca^{2+}-dependent ecto-ATPase activities of M1 cells cultured for the period of 6 days in the presence of NN glia conditioned medium or in normal medium with or without (control) addition of NGF, insulin or dibutyryl cyclic AMP. Mg^{2+}-ATPase (A) and Ca^{2+}-ATPase (B). Each value is the mean of 8-10 experiments. ★$p<0.05$.★★$p<0.02$.

dependent ecto-ATPases of neuroblastoma cells decreased despite the induced morphological differentiation (Fig. 8).

In order to determine whether the increase of ecto-ATPase activities evoked by the NN glia conditioned medium is reversible, normal medium was substituted after 2 or 3 weeks of conditioning. As shown in Table VIII, both ecto-ATPase activities decreased after withdrawal of NN glia conditioned medium and after 6 days reached similar (Mg^{2+}-ATPase) or lower (Ca^{2+}-ATPase) values than those of the control cells.

A kinetic analysis of the ATPase activities was performed. Michaelis constants were determined for M1 cells cultivated in normal medium and for M1 cells cultivated in NN glia conditioned medium, for 2 weeks, followed by 6 days of subculture either in NN glia conditioned medium, or in normal medium. As can be seen in Fig. 9, the highest V_{max} and Km values were found in M1 cells cultivated in NN glia conditioned medium. After 6 days of withdrawal of NN conditioned medium both V_{max} and Km decreased.

TABLE VIII
Mg^{2+}- AND Ca^{2+}-DEPENDENT ECTO-ATPase ACTIVITIES OF M1 NEUROBLASTOMA CELLS

Culture conditions	Mg^{2+}-ATPase	Ca^{2+}-ATPase
M1 control	0.76 ± 0.01	1.73 ± 0.04
M1 conditioned medium for 2 weeks+6 days	2.231 ± 0.039	5.744 ± 0.355
M1 conditioned medium for 2 weeks followed by 6 days in normal medium	0.627 ± 0.049	1.220 ± 0.046
M1 conditioned medium 3 weeks+6 days	2.168 ± 0.150	4.115 ± 0.558
M1 conditioned medium 3 weeks followed by 6 days in normal medium+6 days after withdrawal	0.996 ± 0.025	2,17 ± 0.147

Neuroblastoma cells were cultured in Falcon flasks for the period of 2 or 3 weeks in normal medium, or in NN glia conditioned medium. The subcultures were performed in the Petri dishes either in normal medium or in the conditioned medium. Enzymatic activities were determined 6 days later. Each value (μmoles/mg cell protein/h) represents the mean of 4-6 experiments ± S.E.M.

The data presented support the view that glial cells in cultures release a factor that produces increased Mg^{2+}- and Ca^{2+}- dependent ecto-ATPase

(A)

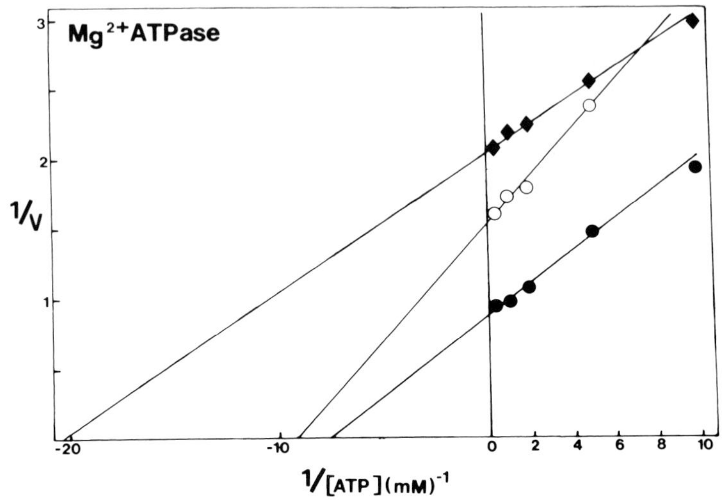

(B)

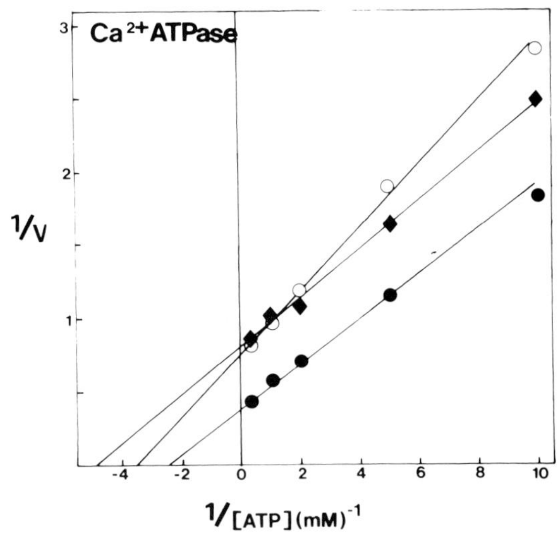

FIG. 9. Lineweaver-Burk plot of ATPase activities at the external surface of M1 cells cultivated in normal medium or in NN glia conditioned medium for a period of 2 weeks and thereafter subcultured either in the normal or in the NN glia conditioned medium. The enzymatic activities were determined 6 days after subculture. Mg^{2+}-ATPase (A) and Ca^{2+}-ATPase (B). (Closed diamonds), M1 cells cultivated in normal medium. (Closed circles), M1 cells cultivated in NN glia conditioned medium. (Open circles), M1 cells cultivated in NN glia conditioned medium followed by normal medium.

activities in neuroblastoma cells. However in contrast to the increase in ecto-ATPase activities observed in cocultures of M1 and NN cells the effect of glia conditioned medium is reversible. Therefore, it may be suggested either that actual cellular contacts are essential for the irreversible induction of the ATPases, or that more factors are involved, some of which are labile or not maintained in the conditioned medium in the absence of conditioning cells.

The lack of the effect of dibutyryl cyclic AMP, NGF or insulin on the ecto ATPase activities suggest that the effect of conditioned medium is more complex than the effect of isolated factors.

Experiments were undertaken to purify the factor(s) in NN glia conditioned medium responsible for the effects observed. Our preliminary results suggest that low molecular weight fraction of NN glia conditioned medium enhances morphological differentiation of M1 neuroblastoma cells and increases RNA and protein synthesis as well as the activities of both ecto-ATPases.

In summary, while cocultures of neuroblastoma and glial cells may provide a model for investigation of cell interactions, the study of the effect of conditioned medium may facilitate the isolation of molecules involved in these interactions.

References

Agren, G., J. Ponten, G. Ronquist and B. Westermark. 1971. Demonstration of an ATPase at the cell surface of intact normal and neoplastic human cells in culture. J. Cell. Physiol. 78:171-176.

Amano, T., E. Richelson and M. Nirenberg. 1972. Neurotransmitter synthesis by neuroblastoma clones. Proc. Natl. Acad. Sci. U.S.A. 69:258-263.

Benda, P., J. Lightbody, G. Sato, L. Levine and W. Sweet. 1968. Differentiated rat glial cell strain in tissue culture. Science 162:370-371.

Burnham, P., C. Raiborn and S. Varon. 1972. Replacement of nerve growth factor by ganglionic non-neuronal cells for the survival *in vitro* of dissociated ganglionic neurons. Proc. Natl. Acad. Sci. U.S.A. 69:3556-3560.

Chambers, D. A., E. W. Salzman and L. L. Neri. 1967. Characterization of "ecto-ATPase" of human blood platelets. Arch. Biochem. Biphys. 119:173-178.

Ciesielski-Treska, J., V. Stefanovic and P. Mandel. 1975a. Activité acétylcholinesterasique de cellules de neuroblastome et des cellules gliales en coculture. C. R. Acad. Sci. Paris, série D 281:1261-1264.

Ciesielski-Treska, J., S. Warter and P. Mandel. 1975b. Morphological, histo-chemical and chromosomal patterns of neuroblastoma parental and purine resistant lines. Neurobiology 5:382-392.

De Pierre, J. W. and M. L. Karnovsky. 1974a. Ecto-enzymes of the guinea pig polymorphonuclear leukocyte. I. Evidence for an ecto-adenosine monophosphatase, adenosine triphosphatase and ρ-nitriphenyl phosphatase. J. Biol. Chem. 249:7111-7120.

De Pierre, J. W. and M. L. Karnovsky. 1974b. Ecto-enzymes of the guinea pig polymorphonuclear leukocyte. II. Properties and suitability as markers for the plasma membrane. J. Biol. Chem. 249:7121-7129.

Duch, D., P. Mandel and H. Koprowski. 1975. Demonstration of enzymes related to myelogenesis in established human brain cell cultures. J. Neurol. Sci. 26:99-105.

Dunkley, C. R., J. F. Manery and E. E. Dryden. 1966. The conversion of ATP to IMP by muscle surface enzymes. J. Cell. Physiol. 68:241-248.

Engelhardt, W. A. 1957. Enzymes as structural elements of physiological mechanisms. Proc. Int. Symp. Enz. Chem. Tokyo and Kyoto 2:163-166.

Halpern, M. and M. Rubin. 1970. Proteins released from chick embryo fibroblasts in culture. I. Partial characterizaton of the proteins. Exp. Cell Res. 60:86-95.

Harris, A. J. and M. J. Dennis. 1970. Acetylcholine sensitivity and distribution on mouse neuroblastoma cells. Science 167:1253-1255.

Henn, F., H. Haljamäe and A. Hamberger. 1972. Glial cell function: active control of extracellular K^+ concentration. Brain Res. 43:437-443.

Konigsberg, I. R. 1963. Clonal analysis of myogenesis. Science 140:1273-1284.

Manery, J. F., J. R. Riordan and E. E. Dryden. 1968. Characteristics of nucleotide converting enzymes at the muscle surfaces with special reference to ion sensitivity. Canad. J. Physiol. Pharmacol. 46:537-547.

McClay, D. R. and A. A. Moscona. 1974. Purification of the specific cell aggregating factor from embryonic neural retina cells. Exp. Cell Res. 87:438-443.

Monard, D., F. Solomon, M. Rentsch and R. Gysin. 1973. Glia induced morphological differentiation in neuroblastoma cells. Proc. Natl. Acad. Sci. U.S.A. 70:1894-1897.

Nelson, P., W. Ruffner and M. Nirenberg. 1969. Neuronal tumor cells with excitable membranes grown *in vitro*. Proc. Natl. Acad. Sci. U.S.A. 64:1004-1010.

Nissen, C., J. Ciesielski-Treska, L. Hertz and P. Mandel. 1973. Regulation of oxygen consumption in neuroblastoma cells: effects of differentiation and of potassium. J. Neurochem. 20:1029-1035.

Patterson, P. H. and L. L. Y. Chun. 1974. The influence of non-neuronal cells on catecholamine and acetylcholine synthesis and accumulation in cultures of dissociated sympathetic neurons. Proc. Natl. Acad. Sci. U.S.A. 71:3607-3610.

Peterson, J. A. and H. Rubin. 1969. The exchange of phospholipids between cultured chick embryo fibroblasts and their growth medium. Exp. Cell Res. 58:365-378.

Pfeiffer, S. E., H. R. Herschman, J. Lightbody and G. Sato. 1970. Synthesis by a clonal line of rat glial cells of a protein unique to the nervous system. J. Cell. Physiol. 75:329-340.

Rebel, G., J. Ciesielski-Treska and P. Mandel. 1973. Etude des gangliosides d'un clone de cellules de neuroblastome. C. R. Acad. Sci. Paris, série D. 277:1193-1195.

Richelson, E. 1973. Regulation of tyrosine hydroxylase activity in mouse neuroblastoma clone NIE-115. J. Neurochem. 21:1139-1145.

Rosenberg, A., V. Stefanovic and P. Mandel. 1975. Acetyl- and butyrylcholinesterase of intact cultured neuroblastoma and glial cells and inactivation upon cellular desialylation. Feder. Proc. 34:244.

Rosenblatt, D. E., C. J. Lauter and E. G. Trams. 1976. Deficiency of a Ca^{2+}- ATPase in brains of seizure prone mice. J. Neurochem. 27:1299-1304.

Rubin, H. 1966. A substance in conditioned medium which enhances the growth of small numbers of chick embryo cells. Exp. Cell Res. 41:138-148.

Schubert, D. 1974. Induced differentiation of clonal rat nerve and glial cells. Neurobiology 4:376-387.

Seeds, N. W., A. G. Gilman, T. Amano and M. Nirenberg. 1970. Regulation of axon formation by clonal lines of a neural tumor. Proc. Natl. Acad. Sci. U.S.A. 66:160-167.

Sensenbrenner, M. and P. Mandel. 1974. Behaviour of neuroblasts in the presence of glial cells, fibroblasts and meningeal cells in culture. Exp. Cell Res 87:159-167.

Shein, H. M., A. Britva, H. H. Hess and D. J. Selkoe. 1970. Isolation of hamster brain astroglia by *in vitro* cultivation and subcutaneous growth, and content of cerebroside, ganglioside, RNA and DNA. Brain Res. 19:497-501.

Stefanovic, V., J. Ciesielski-Treska, A. Ebel and P. Mandel. 1974. Nucleoside triphosphatase activity at the external surface of neuroblastoma cells. Brain Res. 81:427-441.

Stefanovic, V., P. Mandel and A. Rosenberg. 1975a. Concanavalin inhibition of ecto-5′-nucleotidase of intact cultured C6 glioma cells. J. Biol. Chem. 250:7081-7083.

Stefanovic, V., P. Mandel and A. Rosenberg. 1975b. Activation of acetyl- and butyrylcholinesterase by enzymatic removal of sialic acid from intact neuroblastoma and astroblastoma cells in culture. Biochemistry 14:5257-5260.

Stefanovic V., J. Ciesielski-Treska, A. Ebel and P. Mandel. 1976a. Neuroblasts-glia interaction. The effect of co-cultivation upon ecto-ATPase activity of neuroblastoma and glioma cells. Exp. Cell Res. 98:191-203.

Stefanovic, V., M. Ledig and P. Mandel. 1976b. Divalent cation-activated ecto-nucleoside triphosphatase activity of nervous system cells in tissue culture. J. Neurochem. 27:799-805.

Stefanovic, V., J. Ciesielski-Treska and P. Mandel. 1977. Neuroblasts-glia interaction in tissue culture as evidenced by the study of ectoenzymes. Ecto-ATPase activity of mouse neuroblastoma cells. Brain Res. 122:313-323.

Trams, E. G. and C. J. Lauter. 1974. On the sidedness of plasma membrane enzymes. Biochim. Biophys. Acta. 345:180-197.

Varon, S. and C. W. Raiborn. 1969. Dissociation, fractionation and culture of embryonic brain cells. Brain Res. 12:180-199.

Varon, S., C. Raiborn and P. Burnham. 1974. Comparative effects of nerve growth factor and ganglionic non-neuronal cells on purified mouse ganglionic neurons in culture. J. Neurobiol. 5:355-371.

Nutrition

HORMONES AND GROWTH FACTORS IN CELL CULTURES: PROBLEMS AND PERSPECTIVES

J. MATHER AND G. SATO

Biology Department
University of California, San Diego
La Jolla, California

I. Introduction

The desirability of obtaining a completely defined medium for the growth of cells and organs of eukaryotic organisms was recognized early in the development of tissue culture (Lewis and Lewis, 1911). The complete definition of medium components awaited the knowledge of the role of essential amino acids, vitamins, salts, trace metals, and other nutritional requirements of cells and tissues in culture (Eagle, 1955a, b). With these requirements met in most synthetic media now available (Ham, 1965, Waymouth, 1972), it is possible to examine the role of serum, one of the last remaining undefined constituents of cell culture media. While some cell cultures can be maintained for short periods of time without serum, and others have been adapted for growth in the absence of serum (Evans *et al.*, 1956; Takaoka and Katsuta, 1971), until recently it has been impossible to completely replace serum without selecting, and thus altering, the cells growing in these conditions. Once a cell line has been selected to grow in the absence of serum, it becomes of course, impossible to study the role of serum in tissue culture, and therefore, to gain an understanding of the physiological role of the serum components.

It is now clear that one of the major roles of serum in tissue culture is to provide hormones needed by the cells for growth. We have been sucessful in growing three cell lines in serum-free media supplemented with various hormones and obtaining a growth rate equivalent to that in serum-supplemented medium (Hayashi and Sato, 1976; Hayashi *et al.*, 1977). With the availability of many differentiated cell lines retaining many of the physiological properties of their *in vivo* counterparts, and the ability to grow such cells in a completely defined medium, many problems such as hormone response, hormone production, growth control and differentiaton can be studied *in vitro* in a way previously not possible.

II. Materials and Methods.

Cells and Culture. Melanoma clone M2R was selected by P. Roberts, in our laboratory, from the B_{16} melanoma which arose spontaneously in a C57BL/6J mouse and has been carried as a transplantable tumor by the

Jackson Laboratory. HeLa was obtained from Dr. J. Holland, University of California, San Diego. GH_3 was developed by G. Sato (Buonassisi *et al.*, 1962). Stock cultures of all three cell lines were kept in a 1:1 mixture of Ham's F12 medium and Dulbecco's Modified Eagle's (DME) medium supplemented with 1.2 g/l sodium bicarbonate, 15 mM N-2-hydroxyethyl-piperazine-N′-2-ethane-sulfonic acid (Hepes), 192 units/ml penicillin, 200 ug/ml streptomycin and 25 μg/ml ampicillin, 5% (v/v) horse serum and 2.5% (v/v) fetal calf serum (FCS). The cells were grown on plastic tissue culture dishes in a humidified atmosphere of 95% air-5% CO_2 at 37° C. Cells were subcultured every 5-7 days.

Experimental Medium. A 1:1 mixture of Ham's F12 and DME supplemented with sodium bicarbonate, 15 mM Hepes, antibiotics and 3×10^{-8} M selenious acid (H_2SeO_3) was used. For the glycogen synthesis experiments, Ham's F12 medium was prepared without glucose (F12-glucose) and with the same additives as above.

Experimental Procedure. For each experiment, cells from exponentially growing stock cultures were trypsinized (0.1% (w/v) trypsin). The trypsin was inhibited with an equal volume (0.1% (w/v)) of soybean trypsin inhibitor. The cells were washed with the experimental medium, resuspended in medium and plated at the indicated inoculum under the various experimental conditions.

Cell numbers were obtained by removing the cells from the plates by trypsinization, neutralizing with an equal volume of serum-supplemented stock medium and counting, using a Coulter counter.

Glycogen Synthesis. Exponentially growing stock cultures were plated in serum-supplemented medium 4 days before use. Medium was changed to Ham's F12-glucose and the cells were kept without glucose for 5 hours. At time 0, ^{14}C-glucose (2 μCi/plate) and the indicated concentration of insulin were added. After one hour the medium was removed and the plates placed on ice. The cells were scraped off the plate and washed in F12-glucose. Glycogen synthesis was measured as described by Pflüger.

Hormones. Bovine crystalline insulin, human transferrin, progesterone and luteinizing hormone releasing factor (LRF) were obtained either from the Sigma Co., St. Louis, Missouri or Calbiochem, La Jolla, California. Follicle stimulating hormone (NIH.FSH.B1) was obtained from the National Institutes of Health Hormone Distribution Program, Bethesda, Maryland.

Plating Efficiency. Cells were prepared using the experimental procedure described above. The cells were plated in 100 mm dishes at an inoculum of 10^3 cells/plate. After incubation in the indicated media, the colonies were fixed on the plate in formalin and stained with crystal violet.

Conditioned Medium. To obtain conditioned medium, 10^6 cells were plated in a 100 mm tissue culture dish containing 10 ml of serum-free medium. After 2 days, the medium was removed, filtered through a Nalgene filter (0.2 nm pore size) and stored at 4° C until use (1-2 days).

III. Effects of Hormones and Growth Factors on Growth and Differentiation.

Growth was chosen as the criterion for success in replacing serum by hormones for several reasons. This assay is convenient and requires that the cells survive, undertake macromolecular synthesis (DNA, RNA, protein, etc.), and divide. A doubling time equivalent to that under favorable serum conditions would indicate that all of these functions were operating normally. The growth of clone M2R, developed from the B_{16} mouse melanoma, in serum and in hormone supplemented media is shown in Fig.

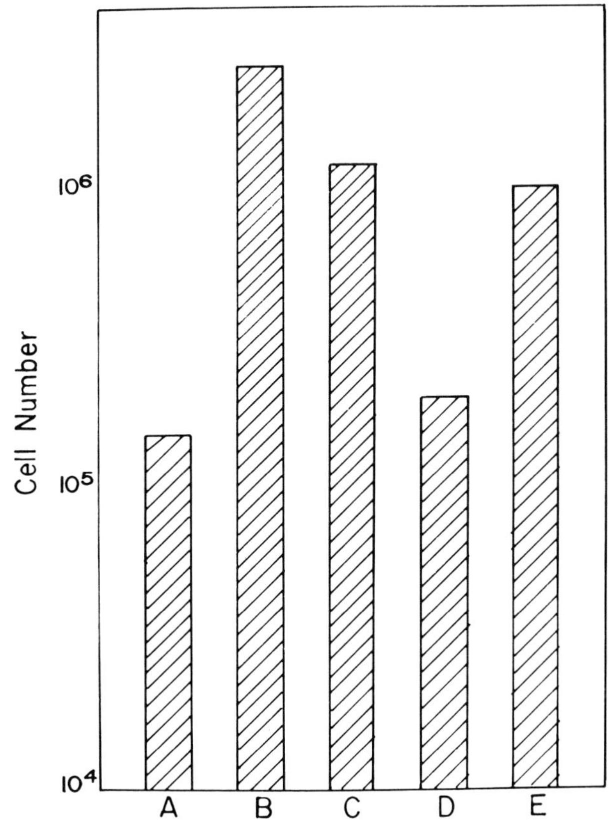

FIG. 1: Growth of melanoma M2R in serum-free medium with the addition of: (A) no hormones; (B) 5% (v/v) FCS; (C) 5 hormones (insulin, transferrin, progesterone, FSH, LRF); (D) 4 hormones above, omitting transferrin; (E) 4 hormones above, omitting FSH. Cells were plated at 4×10^4 cells/plate as described in Materials and Methods, and counted on day 4.

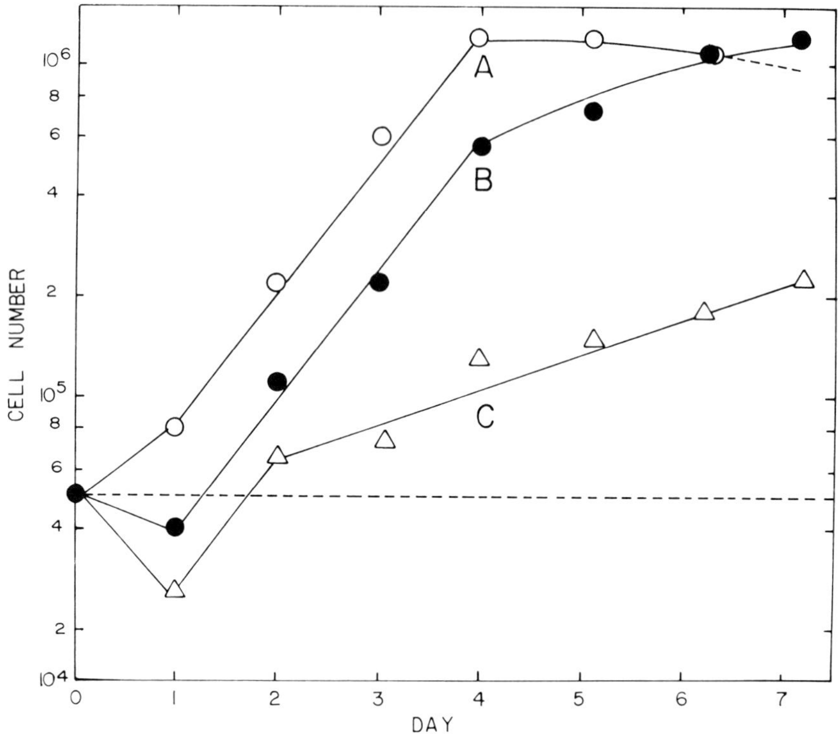

FIG. 2: Daily growth curve of melanoma M2R in serum-free medium (C), medium supplemented with 5 hormones (B) and supplemented with 5% (v/v) FCS (A). The doubling time in both serum and hormone supplemented media is 12 hours during logarithmic growth.

1. After 5 days the final cell number is slightly higher in media with serum than in media supplemented with insulin, transferrin, progesterone, FSH and LRF. That this difference is due to an increased sensitivity due to the trypsinization trauma and consequent lower survival, can be seen by comparing daily growth curves (Fig. 2). After recovery both the doubling time and the final density of the cells are identical under the two conditions. These hormonal conditions are thus necessary and sufficient to give growth equivalent to that seen in the presence of serum.

However, it should be emphasized that the five hormones mentioned above are not the only hormones or factors which may be growth-stimulatory. The present sets of hormone requirements were arrived at by testing the hormones shown in Table I. Some recently discovered factors were not then available for testing. It is also possible that there are growth factors in serum which have not yet been described. The serum-free system could well lead to the discovery of other such factors in much the same way that reduction of serum concentrations led to the discovery of FGF and OGF (Armelin, 1973; Armelin *et al.*, 1974; Gospodarowicz *et al.*, 1974). With this in mind, several possible sources of novel growth factors are being examined in this laboratory. A factor was purified from salivary gland extracts which is required for cell division in 3T3 cells (T. Nakamura, unpublished data). Several growth stimulatory factors have been isolated from blood meal, a commercial fertilizer. These fractions, active on GH_3 cells, are heat-stable, small molecular weight, basic proteins (T. Nakamura and G. Sato, unpublished results). One may be equivalent to one of the known factors found in plasma, somatomedin C, and can substitute for this

TABLE I
HORMONES AND FACTORS TESTED FOR GROWTH-STIMULATORY ACTIVITY

Hormone	Concentration
Luteinizing releasing factor (LRF)	10 ng/ml
Thyrotropin releasing factor (TRF)	10^{-9}M
Somatotropin releasing inhibitory factor (SRIF)	10 ng/ml
NIH-luteinizing hormone (LH)	0.5 μg/ml
NIH-follicle stimulating hormone (FSH)	10 ng/ml
Adrenocorticotropic hormone (ACTH)	5 μU/ml
Growth hormone (GH)	5 ng/ml*
Prolactin	10 ng/ml
Prostaglandin E_1 (PGE_1)	10 ng/ml
Prostaglandin $F_2\alpha$ ($PGF_2\alpha$)	10 ng/ml
Hydrocortisone (HC)	2×10^{-9}M
Estradiol	2×10^{-9}M
Progesterone	2×10^{-8}M
Aldosterone	50 pg/ml
Testosterone	2×10^{-9}M
Triiodothyronine (T_3)	3×10^{-10}M
Ceruloplasmin	5 μg/ml
Transferrin	5 μg/ml*
Gly-his-lys	0.2 μg/ml*
Somatomedin A(SMA)	1 μg/ml*
Somatomedin B(SMB)	1 μg/ml*
Somatomedin C(SMC)	1 μg/ml*
Insulin	5 μg/ml
Glucagon	0.5 μg/ml
Parathyroid hormone (PTH)	0.5 ng/ml
Calcitonin	0.1 ng/ml
Thyrotropin stimulating hormone (TSH)	5 ng/ml
Epidermal growth factor (EGF)	1 ng/ml*
Fibroblast growth factor (FGF)	1 ng/ml*
Ovarian growth factor (OGF)	1 ng/ml*

*concentration varies with purity of preparation

Table II
HORMONE REQUIREMENTS OF CELL LINES IN SERUM-FREE MEDIUM

Melanoma Clone M2R	HeLa	GH_3
Insulin* (5μg/ml)	Insulin* (5μg/ml)	Insulin* (5μg/ml)
Transferrin* (5μg/ml)	Transferrin* (5μg/ml)	Transferrin* (5μg/ml)
Progesterone†[1] (2×10^{-8}M)	Hydrocortisone† (3×10^{-8}M)	T_3†(10^{-11}M)
LRF† (10ng/ml)	EGF* (1ng/ml)	TRF† (1ng/ml)
NIH.FSH (0.4μg/ml)	NIH.LH (1μg/ml)	PTH† (0.5μg/ml)
		Blood meal [2]
		IEF-fractions
		pI 8.5
		pI 9.5
		pI 10.5

[1] can substitute testosterone (10^{-9}M)†
[2] can substitute somatomedin C* (1ng/ml) and FGF* (1ng/ml)
† = synthetic hormone
* = 95% purity

factor in the growth of GH^3 (see Table II). Further studies of this type may lead to the discovery of hormones and factors, which have not yet been defined but which may prove to be active on various cell types, including cell types which have, until now, been impossible to grow in the total absence of serum. Theoretically, it should be possible to obtain growth rates even greater than those found when serum is present since one is removing not only the growth promoting and indifferent hormones and factors in removing the serum, but also the inhibitory hormones. Likewise, one could also choose a set of growth-inhibitory hormones for each cell type.

It should be emphasized that the choice of medium is critical when working in serum-free conditions. An enriched medium, such as Ham's nutrient mixture F10 or F12 is necessary. Minimal media, such as Eagle's MEM, are not adequate to support the growth of most cells in the absence of serum.

If the hormone supplements can truly replace serum, then cells should be able to grow as well in hormone supplemented media not only in short-term experiments of 5-10 days but also for longer times. Fig. 3 shows melanoma cells maintained in the normal serum-supplemented stock medium compared with the same line maintained in hormone-supplemented medium for 6 months. For these cells, with a doubling time of 14 hours, this is over 300 generations. The hormone supplemented medium is thus adequate for prolonged cell culture, and provides all the requirements of the cells. The hormone response of these cells did not change after such a prolonged period of continued growth in hormone supplemented medium.

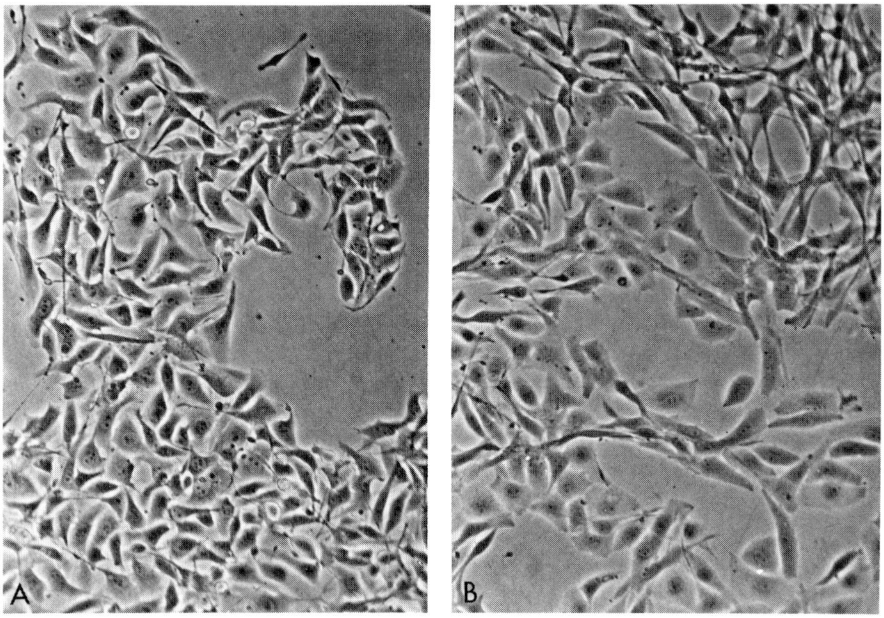

FIG. 3: Melanoma M2R cells: (A) stock culture and (B) cells maintained 6 months in hormone-supplemented medium.

The hormone requirements of three different cell types (GH_3, HeLa and melanoma) are shown in Table II. Concentrations of each hormone used and the purity of the preparations are noted. It should be noticed that insulin and transferrin are required by all three cell lines. These two factors are also required by all other cell types for which the hormone requirements have been partially worked out in our laboratory. These cell lines grow in media containing various hormones plus a low concentration of serum (0.1-1%) or some other partially purified substance such as fetuin. Such cell lines are C-6 glial cells, a teratoma cell line, a smooth muscle cell line and an ovarian cell line (G. Sato et al., unpublished data). It is not known if the insulin and transferrin are acting in the same manner in all cell lines. It was found that insulin is required by GH_3 cells for survival, and can thus be called a stringent requirement, but not by the other two cell types. Likewise, transferrin is a stringent requirement for melanoma (see Fig. 1) but not for HeLa or GH_3. Thus one can say there is a qualitative difference in the requirement for the same hormone by different cell types although it is not known if these are also quantitative differences.

It is also to be noted that the set of hormones required is different for each cell type and may be considered to be characteristic of the cell type. It may be possible to utilize such differences in hormone requirements to

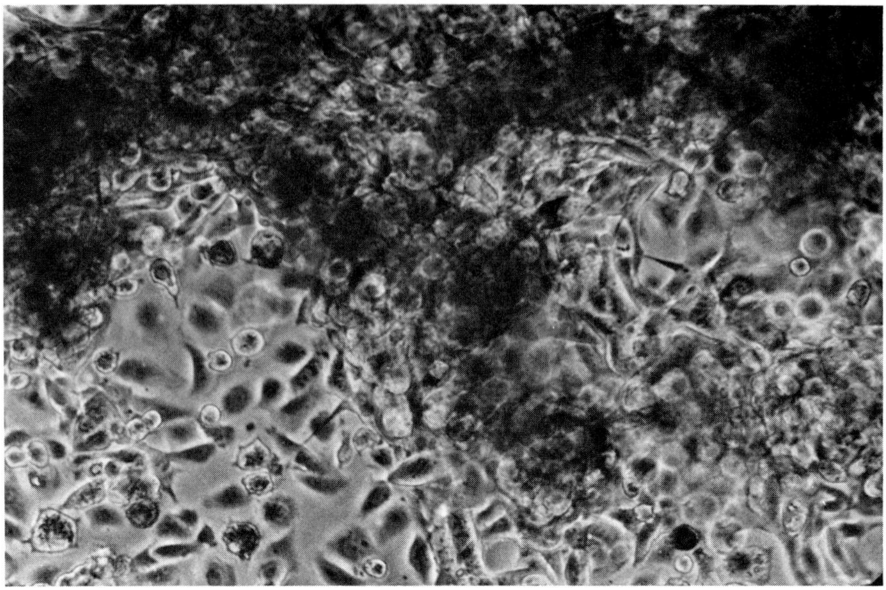

FIG. 4: Melanoma metastasis growing in hormone-supplemented medium. The lymph node was excised from an animal injected subcutaneously with 10^6 melanoma cells 3 weeks previously, minced and plated in the above medium. After 8 days, only melanoma cells survive.

select for a specific cell type in primary cultures of mixed cell populations. An example of such a selection is seen in Fig. 4. Melanoma cells were injected (10^6 cells/animal) subcutaneously. After three weeks the animal was killed and various organs such as spleen, lung and lymph nodes were minced and plated in serum-free medium supplemented with the hormone mixture required by melanoma cells for growth. After 8-10 days, the only surviving cells were the melanoma cells, which had metastasized into the organ plated. Thus one could select *in vitro* specifically for melanoma cells and delete the other cell types (lung, spleen, fibroblasts, etc.) in conditions optimal for the growth of the melanoma cells. One should also be able to use this as a method for selecting one normal cell type from an organ. This is presently being done in selecting Sertoli cells from disaggregated testis (J. Mather, unpublished results), and may be applicable to a broad range of other cell types.

Having defined the hormone requirements of a cell line one can then approach some basic questions in cell biology in a novel fashion. A primary goal is the elucidation of how these hormones act on the cell. What is the primary effect, site of action, specificity? An example of this type of study is the demonstration that at least one action of insulin in GH_3, HeLa (I. Hayashi, unpublished results) and melanoma is to increase glycogen

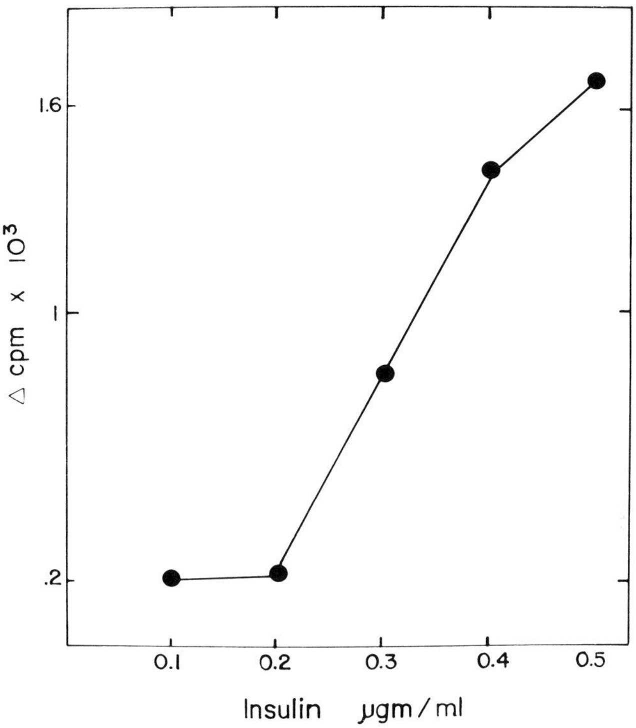

FIG. 5: Glycogen synthesis in response to insulin by M2R cells. Results are expressed as stimulation of incorporation over the basal (no added insulin) level. Basal level was 900 cpm.

synthesis (Fig. 5). Other possible effects such as an increase of glucose transport, or amino acid transport are being studied. The serum-free system offers a unique opportunity to study the effect of the interaction of hormones. One can study the effect of the presence of one or more additional hormones on the action of one hormone.

One can also study the effects of various hormones on the expression of a differentiated function of a cell line. It was found that GH_3 produces the same amount of prolactin when grown in the serum-free hormone supplemented media as in medium containing serum (Wu, unpublished data). The influence of the various hormones on the expression of this function is presently being studied. The relationship between hormones required for survival and growth and those required for the expression of the differentiated function, which may or may not be the same, can also be examined.

It has long been known that some cell types require feeder layers or conditioned media even in the presence of serum. The ability to use

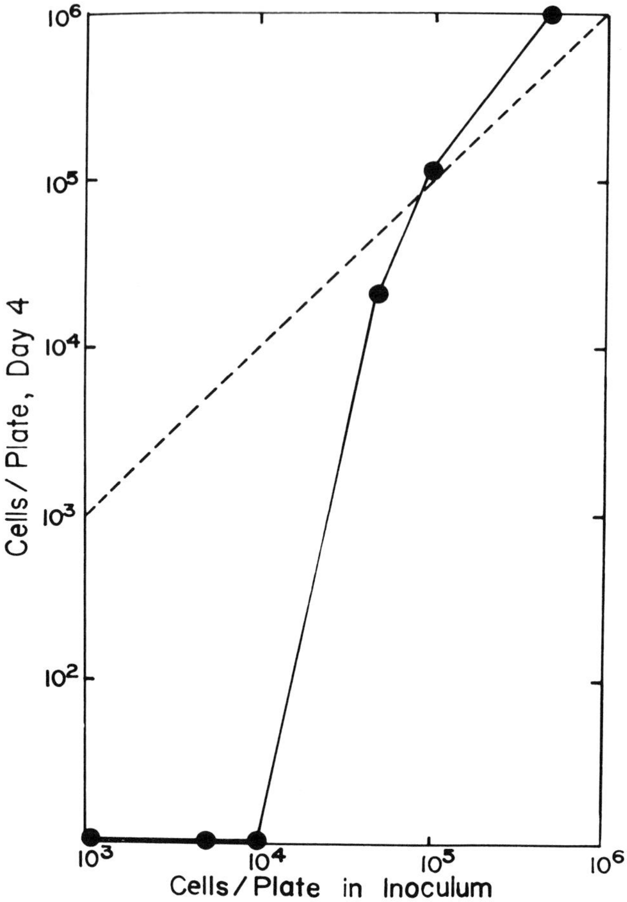

FIG. 6: Melanoma density dependent survival. Cells were plated at the indicated inoculum in a 60mm tissue culture dish containing 5 ml of serum-free medium. Surviving cells were counted on day 4 (black circles). Dotted line indicates theoretical 100% survival in the absence of cell multiplication.

hormone-supplemented media instead of adding serum makes it possible to study, and perhaps isolate, factors produced by cells and secreted into the media. The melanoma cell line M2R requires conditioned media for survival when grown in hormone-supplemented media at low plating densities (see Figs. 6, 7). Because these cells will grow in a completely defined medium, it is possible to attempt to isolate and characterize the factor or factors responsible for this conditioning effect. Such studies are presently in progress.

More detailed studies can also be carried out correlating the interrelationship of hormone growth requirements *in vitro* with the

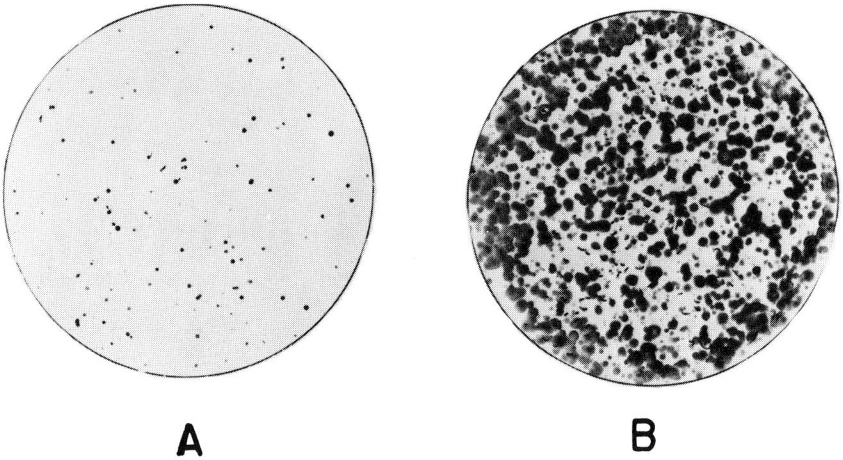

FIG. 7: Plating efficiency of M2R cells in hormone-supplemented medium (A) and hormone-supplemented conditioned medium (B). Cells were grown 7 days in the indicated media then changed to stock medium for 5 days before fixing and staining colonies.

requirements and responses of cells *in vivo*. The melanoma system is an example of one such system. Melanoma cell lines can be selected which metastasize in a high percentage of animals injected with the cells. The *in vitro* hormone requirements of such a line can be compared with those of a non-metastatic line. Differences could then be exploited to try to change the *in vivo* phenotype by manipulating the *in vitro* environment. Conversely one might study primary cultures of metastatic nodules from organs such as spleen, liver, lung and testis to see if the *in vivo* environment selects for, or causes, different hormone responses or requirements in a single cell line which could then be studied *in vitro*.

Finally, one can select *in vitro* for clonal cell lines which have lost the requirement for one of the hormones shown to be required by the parent line and thus study the processes by which the cell can carry out its functions without this hormone.

In summary, the ability to grow a variety of cell lines in hormone-supplemented serum-free media eliminates a number of technical difficulties inherent in earlier studies done in the presence of serum. More important, a range of new problems dealing with hormones can be approached, such as: the interaction of various hormones; the expression of differentiated cell functions and its dependence on hormones; correlation of hormone effects *in vitro* and *in vivo;* study of cell-produced factors; and selection and study of hormone-independent cell variants.

Acknowledgments

This research was supported by National Institutes of Health grant number 1-F22-CA00821-01 and grant number 5-R01-GM17019-07. We would like to thank Ms. M. A. Zurbach for assistance in preparation of the manuscript and Mr. L. Owashi for technical assistance.

References

Armelin, H. 1973. Pituitary extract and steroid hormones in control 3T3 cell growth. Proc. Nat. Acad. Sci. U.S.A. 70:2702-2706.

Armelin, H. A., K. Nishikawa and G. Sato. 1974. Control of mammalian cell growth in culture: the action of protein and steroid hormones as effector substances. In: *Control of Proliferation in Animal Cells.* B. Clarkson and R. Baserga, eds. Cold Spring Harbor Press, Cold Spring Harbor, New York. pp. 97-104.

Buonassisi, V., G. Sato and A. I. Cohen. 1962. Hormone-producing cultures of adrenal and pituitary tumor origin. Proc. Nat. Acad. Sci. U.S.A. 48:1184-1190.

Eagle, H. 1955a. The specific amino-acid requirements of a human carcinoma cell (strain HeLa) in tissue culture. J. Exp. Med. 102:37-48.

Eagle, H. 1955b. Nutrition needs of mammalian cells in tissue culture. Science, 122:501-504.

Evans, V. J., J. C. Bryant, M. C. Fioramonti, W. T. McQuilkin, K. K. Sanford and W. R. Earle. 1956. Studies of nutrient media for tissue cells *in vitro* I. A protein-free chemically defined medium for cultivation of strain L cells. Cancer Res. 16:77-86.

Gospodarowicz, D., K. L. Jones and G. Sato. 1974. Purification of a growth factor for ovarian cells from bovine pituitary glands. Proc. Nat. Acad. Sci. U.S.A. 71:2295-2299.

Ham, R. G. 1965. Clonal growth of mammalian cells in a chemically defined synthetic medium. Proc. Nat. Acad. Sci. U.S.A. 53:288-293.

Hayashi, I. and G. Sato. 1976. Replacement of serum by hormones permits growth of cells in a defined medium. Nature, (Lond.) 259:132-134.

Hayashi, I., S. Hutchings, J. Mather and G. Sato. 1977. Replacement of serum by hormones and its implication for primary cell culture. Experientia, in press.

Lewis, M. R. and W. H. Lewis. 1911. The cultivation of tissues from chick embryos in solutions of NaCl, $CaCl_2$, KCl and $NaHCO_3$. Anat. Rec. 5:277-293.

Takaoka, T. and H. Katsuta. 1971. Long-term cultivation of mammalian cell strains in protein- and lipid-free chemically defined synthetic media. Exptl. Cell Res. 67:295-304.

Waymouth, Charity. 1972. Construction of tissue culture media. In: *Growth Nutrition and Metabolism of Cells in Culture,* Vol. I. G. H. Rothblat, V. J. Cristofalo, eds. Academic Press, New York, N.Y. pp. 11-47.

NUTRITIONAL REQUIREMENTS OF CELLS IN CULTURE, WITH SPECIAL REFERENCE TO NEURAL CELLS

CHARITY WAYMOUTH

The Jackson Laboratory
Bar Harbor, Maine

I. Introduction

In spite of the fact that the first significant exercise in cell culture was the study by Ross Harrison (1906-1907) of nerve fiber development *in vitro*, the specific nutritional requirements of neural cells are, at the present time, no better defined than those of other specialized cell types. *The Bibliography of Tissue Culture, 1884-1950*, (Murray and Kopech, 1953) contains 34 pages of references on nerve cells in culture, but these reveal little attention to nerve cell *nutrition*. The succeeding quarter century has seen some advances in cell nutrition in general, but has only very recently begun to elucidate some specific needs of cells of the nervous system.

What I shall attempt — because I have no claim to be an expert in neural cell culture — is to review some of the advances that have been made in nutrition of cells, mostly of avian or mammalian origin. In the course of this review, I shall try to identify factors that are probably of general significance for all cells, including nerve cells, and also those that may apply specifically to neural cells.

If we examine what Ross Harrison did 70 years ago, we can identify his working premise, which was that the nutritional milieu for cells *in vitro* should imitate as closely as possible their environment *in vivo*. The use by Harrison of frog lymph for frog neuron development was based upon this idea. The work of Burrows — a student of Harrison — and of Carrel, with whom Burrows worked for many years, on the cultivation of embryonic chick cells in a coagulum of chicken plasma and chick embryonic extract, was based upon the same notion, although Carrel and Burrows regarded the plasma clot as a physical substratum for the cells, and the embryonic extract as the principal source of nutrients and "growth stimulants".

Another line of ancestry for culture media as we know them today derives from the work of the physiologists and pharmacologists, Ringer, Locke and Tyrode, who designed simple "physiological salt solutions" suitable for short-term survival and function of tissues and organs. These solutions, sometimes with added glucose, were intended by Ringer and

Locke for maintaining, in a functional and contracting state, preparations of heart and leg muscle; and by Tyrode for studying peristalsis in segments of intestine. Imitation of the ionic composition of the blood plasma was the basis of these solutions, which have been used as diluents for embryonic extracts and as components of various culture media that were otherwise composed of biological fluids.

Already by 1911, attempts to dispense with complex medium components, such as plasma, serum and embryonic extracts, and to define requirements for growth, metabolism and function, were initiated by Warren and Margaret Lewis (1911, 1912a). They varied the concentrations of NaCl, KCl, $CaCl_2$, $NaHCO_3$, with and without glucose, in media for several tissues from the chick embryo, and, in such simple media, observed neurite outgrowth from explants of intestine. In a study of nerve fiber outgrowth from chick intestine, Lewis and Lewis (1912b) used a series of Locke-glucose solutions (Locke, 1895, 1900), either alone or supplemented with peptones or chicken bouillon. In cultures of 2 days duration, extensive development of nerve plexus was seen in the salt-glucose solutions alone. We made up some of these solutions in my laboratory, in order to test their pHs and osmolalities and to compare them with those of more recent media used for nerve cell cultures. The solutions and these data are shown in Table I. The low bicarbonate solutions have a pH of 7.76 to 7.78, the high bicarbonate medium B has a pH of 7.95. The osmolalities ranged from 255 to 356 mOsm/kg.

TABLE I

MODIFIED LOCKE'S SOLUTIONS USED BY LEWIS AND LEWIS (1912a) FOR NERVE CELL CULTURE

Solution	NaCl	KCl	$CaCl_2 \cdot 2H_2O$	$NaHCO_3$	Dextrose	pH	Osmolality	Duration hours
A	5000	420	315	200	5000	7.78	255	10 to 26
B	9000	420	315	2000	2500	7.95	319	26
C	9000	240	315	200	2500	7.77	289	42
D	13500	240	315	200	2500	7.76	356	48
E	5000	240	315	200	15000	7.76	255	42

Concentrations of salts are in mg/l; pH and osmolality (mOsm/kg) measurements were made in the author's laboratory (pH with Corning model 10 pH meter, osmolality with Wescor model 5120 vapor pressure osmometer). Duration is that of Lewis' original cultures.

It should be remembered that, at this time, 65 years ago, none of the B vitamins had been characterized, several amino acids (e.g., threonine and methionine) were still unknown, and few hormones had yet been recognized or characterized. Beginning in the 1920s, the isolation and identification of many compounds of metabolic importance began. As these biologically important substances became chemically understood, and their roles *in vivo* explored, they were incorporated into some of the earliest serum-free artificial media. For example, Vogelaar and Erlichman (1933) and Baker and Ebeling (1938), in the late 1920s and early 1930s, included thyroxin and insulin in culture media, very soon after these hormones had been isolated. As the B vitamins were identified, they also were added to culture media. And by the late 1930s and 1940s, balanced salt solutions, designed especially for cells in culture rather than for physiological and pharmacological experiments, began to appear, e.g., those of Gey and Gey (1936), Earle (1943), and Hanks and Wallace (1949). Many of the defined media still in use today include salt mixtures based upon these standard saline solutions.

Successful defined culture media appear today with increasing frequency. The early examples were used for primary chick embryo cells, or for continuous cultures of heteroploid cells, such as the clone 929 (strain L) mouse line of Sanford *et al.* (1943). Media for epithelial cells are much more recent, and defined media for diploid cells are beginning to appear. The media that have been used for the culture of nerve cells have, with very few exceptions, been those designed for other uses. Table II shows examples of some of the most commonly used defined media. The list is by no means exhaustive, but is presented to illustrate that the choice of media is wide, and can be expected to widen as cell specific media become more sophisticated.

TABLE II
MOST COMMONLY USED DEFINED MEDIA

Medium Designation:	Used For:	Reference:
Medium 199	Primary chick embryo cells	Morgan *et al.* (1950)
Medium M150	Primary chick embryo cells	Morgan *et al.* (1955)
Basal medium	Human and mouse cell lines	Eagle (1955)
NCTC 109	Mouse cell lines	McQuilkin *et al.* (1957)
McCoy's 5A	Rat tumor cells	McCoy *et al.* (1959)
MB752/1	Mouse cell lines	Waymouth (1959)
Minimum essential medium	Mammalian cell lines	Eagle (1959)
CMRL 1066	Mouse cell lines	Parker (1961)
MD705/1	Mouse cell lines	Kitos *et al.* (1962)
L-15	Human and mouse cell lines	Leibovitz (1963)
F-10	Diploid human and hamster cells	Ham (1963a)
Nagle suspension medium	Mammalian cells in suspension	Nagle *et al.* (1963)

(Cont'd.)

(Cont'd.)

Medium Designation:	Used For:	Reference:
F-12	Diploid and established hamster and mouse cells	Ham (1965)
MAB87/3	Primary mouse cells	Gorham and Waymouth (1965)
NCTC 135 and other modifications of NCTC 109	Human, hamster, mouse and monkey cell lines	Price *et al.* (1966)
CMRL 1415	Primary mouse embryo cells	Healy and Parker (1966a)
RPMI 1640	Human normal and leukemic cells, hamster tumor cells	Moore *et al.* (1967)
Yamane autoclavable	Mammalian cells	Yamane *et al.* (1968)
Nagle autoclavable	Mammalian cell lines in suspension	Nagle (1968)
7C's	Rat tumor cells	Ling *et al.* (1968)
Birch and Pirt	Mouse LS cells in suspension	Birch and Pirt (1971)
Higuchi	Mouse clone 929, strain L cells	Higuchi (1970)
Nagle and Brown	Mammalian cells	Nagle and Brown (1971)
$HIWO_5$	Mouse clone 929, strain L cells in suspension	Morrison and Jenkin (1972)
F-12G	Fetal rat brain	Shapiro and Schrier (1973)
Higuchi and Robinson	Mammalian cell lines	Higuchi and Robinson (1973)
MPNL65/C	Mouse endocrine pancreas	Leiter *et al.* (1974)
Medium E	Rat liver cells	Williams and Gunn (1974)
Donta's medium	Human and rat glial cells	Donta (1975)
Takaoka and Katsuta	Mammalian cells	Takaoka and Katsuta (1975)
WO_{10}	Baby hamster kidney cells in suspension	Guskey and Jenkin (1976)
MCDB 102, 103	Diploid human fibroblasts	McKeehan *et al.* (1976)
MCDB 104, 201	Diploid human and chick embryo fibroblasts	McKeehan and Ham (1976)

II. Media for Growth or for Differentiation

Most of the early tissue culturists were fascinated by cell proliferation and, because fibroblasts proliferate readily, these cells became the most often used for nutritional studies. It was perhaps the neurobiologists, with their interest in differentiation, or in the reversible morphological changes that Paul Weiss (1939) named "modulation", who kept alive an interest in differentiation *in vitro*, until investigations were developed with functional epithelial cells, such as those of skin, liver, mammary gland, salivary gland, or the endocrine pancreas; with cardiac and skeletal muscle; and with immunologically active cells. As long as nutritional interest was confined to fibroblasts, and usually aneuploid fibroblasts at that, the nutritional problems were relatively simple, because no supplements such as hormones or specific growth factors appeared to be essential for the cells of most species. The design of media capable of eliciting specific cellular functions has recently begun to focus on the point, proposed a good many years ago (Waymouth, 1967), that each cell type is a special case, with receptors for hormones or other stimulants that enable the cell to perform its individual functions. Differentiation is generally defined as an irreversible capability to function. Some morphological manifestations such as neurite formation, often designated "differentiation", may be modulations. An example is the "differentiation" of mouse C 1300 neuroblastoma cells, observed by Schubert *et al.* (1971) to occur in serum-free and conditioned media. This "differentiation" is inversely proportional to the concentration of fetal bovine serum added. With increasing concentration of serum, growth in cell number occurs. Addition of serum to cultures already "differentiated" in a serum-free medium reverses the effect.

The rate of myelinization of rodent nerve cells *in vitro* can be controlled by metabolic means, i.e., retarded by inclusion of pyridoxine antagonists. The retardation can be reversed by addition of pyridoxine (Yonezawa *et al.*, 1969). Metabolic antagonists are also useful for differential cell selection, e.g. non-neuronal cells in mixed populations are retarded by inclusion of aminopterin (Peacock *et al.*, 1973). Differential selection has also been achieved by variation in levels of bicarbonate; removal of bicarbonate reduced the number of non-neuronal cells in cultures of rat sympathetic neurons (Mains and Patterson, 1973).

The type of medium used may affect the behavior of the culture, for example in terms of the growth, differentiation or function that occurs *in vitro*. Comparisons of conventional media for the expression of characteristic enzymes, as was done by Shapiro and Schrier (1973) in their study of choline acetyltransferase in fetal rat brain cells, demonstrated that whereas growth of these cells was better in Dulbecco's modified Eagle's medium (DMEM), the enzyme activity was greater in Ham's F-12G (in both cases supplemented with 10% FBS and with penicillin and streptomycin). Another effect of change of medium from DMEM to F-12G was noted by Peacock and Nelson (1973), whose media for mouse neuroblastoma also contained 0.4 μM of aminopterin to repress non-neuronal cells, namely a two-fold increase in hyperpolarization of the cells.

III. Sources of Variation in Culture Media

It is not always recognized that the most abundant component of culture media is water. Especially if one is attempting to define the conditions of culture precisely, the greatest care should be taken to use water of the highest quality. Electrical resistance is useful as a quality control measure for freedom from ionizable contaminants. In my laboratory, distilled water of not less than 2 mohms resistance is used. Deionized water may be ion free, but not necessarily free from organic contaminants. Not all water stills are equally efficient.

It should also be remembered that commercial chemicals have generally improved in quality in the last 25 to 50 years. This fact, and the availability of water of high purity, should remind us that trace substances that were probably present, e.g., in the early salt mixtures such as Tyrode's (1910) saline, are likely to be absent from the solution made up to his specifications today. Some of the adventitious contaminants of yesterday may indeed be important for cell nutrition. For example, the work of McKeehan *et al.* (1976) has shown the importance of trace elements in media for human diploid cell cultures.

The inclusion of biological fluids, such as serum, in culture media inevitably introduces biological variations into the total medium. Even when, as is usual in commercial practice, serum samples are pooled, the effects of age, sex, nutrition and physiological state result in batch variations. How wide these may be in *individual* fetal calf serum samples is shown by the following examples from the work of Olmsted (1967):

Component	Number of samples	Range of values	
Na^+	61	2680 to 4600	mg/l
Ca^{2+}	53	62 to 176	mg/l
Glucose	31	575 to 3189	mg/l
Cholesterol	26	177 to 592	mg/l
Total protein	79	34.1 to 65.6	g/l
Phospholipids	49	47.3 to 597.5	mg/l

A more recent study of *pooled* fetal bovine serum samples by Honn *et al.* (1975), using fewer (7) commercial samples, but examining many more parameters, also showed wide lot-to-lot variation. Examples of Honn's data are shown:

Component	Range of values	
Na^+	2357/2583	mg/l
Ca^{2+}	95 to 155	mg/l*
Glucose	2490 to 3820	mg/l
Cholesterol	290 to 1650	mg/l
Total protein	30.4 to 84.9	g/l
Testosterone	320 to 800	ng/l
Prolactin	1.5 to 136	ng/ml
Insulin	8.0 to 18.0	μU/ml
Growth hormone	4.1 to 167	ng/ml
Serum osmolarity	328 to 390	mOsm/l

*Original paper (Honn *et al.*, 1975) contains incorrect units. Personal communication, March 1977, from Kenneth V. Honn.

It is clear that, especially in cultures of hormone-dependent or hormone-sensitive cell types, variations due to inclusion of serum may profoundly affect the reproducibility of experiments.

In addition to chemical variations of this kind, hazards are presented by serum in the form of variable amounts of bacterial toxins (which may still be present after a previously contaminated serum has been filter-sterilized), and filterable microorganisms such as mycoplasmas and viruses. It is now well known that a great many continuous cell lines are contaminated with mycoplasmas, and that short-term cultures may also become so contaminated unless extreme care is taken to ensure that these organisms do not get into the cultures. Amongst the problems presented by mycoplasmal contamination is competition of the organisms, many strains of which are highly dependent on arginine, with the cultured cells for that nutrient.

Empirical testing of the effects of fetal bovine serum on a line of human fetal lung cells by Boone *et al.* (1972) revealed that about 10% of commercial serum lots contained bacteria and/or fungi, and 17/139 lots contained *Mycoplasma* sp. These authors compared eight lots of commercial serum with samples of serum collected under optimal conditions of sterility. The concentrations of free fatty acids in the commercial sera were significantly higher than in the specially processed sera, and there was a clear correlation between high free fatty acid content and poor cell growth.

Another study of fetal bovine serum from three different commercial sources on two human fibroblast lines (Milo *et al.* 1976) showed great differences in the abilities of these sera to support clonal growth. Ten samples of serum from one source, when used at 10 or 20% of the medium, were non-supportive of growth, whereas 10 samples from each of two other sources were adequate to excellent. In this study, a correlation was noted between steroid hormone content and the ability to support cell proliferation. The growth promoting sera were low in progesterone (40 to 100 pg/ml) and contained less than 100 pg/ml of 17 β estradiol. Such sera could be made non-supportive of growth by addition of a relatively large amount of progesterone (0.1 μg/ml). Sera that contained more than 1.0 ng/ml of progesterone were non-supportive, and could be made supportive by addition of 0.1 μg/ml of 17 β estradiol or of hydrocortisone. The poorest sera contained 1.0 to 1.5 mg/ml of progesterone.

Another potential source of variation in culture media is introduced when antibiotics are used. In the presence of even relatively small amounts of serum, antibiotic toxicity is a much lesser problem. However, in cultures in chemically defined media, we have noted significant reduction both in cell attachment to a glass or plastic substrate, and in cell growth (C. Waymouth and P. F. Ward, unpublished observations).

The procedures to which we subject culture media may contribute to changes in their nutritional value. For example, proteins, especially those that aggregate, may be retained on sterilizing filters. This type of selective loss may include loss of hormonally active compounds, e.g., it was noted by Lowy and Keighley (1968) that the hematopoietic hormone erythropoietin is strongly adsorbed on cellulose filters. At that time Lowy and Keighley recommended substitution of sterilizing filters made of pure silver. Polycarbonate (e.g. Nucleopore(R)) filters are generally less absorptive than cellulose ester filters, and are now used in our laboratory for most sterile filtration. One should also be aware of the effects of exposure of culture media to light, including room light, as described by Wang (1976), as well as the effects of exposing cells themselves to light (Parshad *et al.* 1977).

Medium osmolality is not only useful for checking the consistency of media from batch to batch (Waymouth, 1970) but is also a variable that may affect the growth and behavior of the cells placed in the medium. Little is known about optimal osmolalities for the various types of neural tissues, though Fisher *et al.* (1975) have noted that the numbers of neurons developing in cultures of chick spinal cord ganglia is maximal at 300 to 317 mOsm/kg, a level that occurred in their cultures by loss of water from a medium initially at 280 mOsm/kg. Media high in glucose appear to be commonly used for neural cell cultures (Murray *et al.* 1962) and the high osmolality medium L-15 of Leibovitz (1963), which has an osmolality of about 315 mOsm/kg (Waymouth, 1970), was found to work well for rat neurons by Mains and Patterson (1973). Preservation of cells during disaggregation may also be improved by a medium of high osmolality, as used for fetal rat brain (at 340 mOsm/kg) by Shapiro and Schrier (1973).

IV. Specific Nutrients

In 1972, in a review of nerve cell culture, Silberberg (1972) briefly surveyed the nutritional and metabolic needs of nerve tissues in culture. He noted the empirical, trial and error, state of the art at that time, reported on only a few variables, e.g., glucose requirements, alternate buffer systems to the conventional phosphate bicarbonate buffer, and noted the lack of systematic information on optimal pHs and osmolalities.

A. Potassium.

Amongst the nutrients or metabolic regulators that appear to have an unusually important role in cultures of nerve cells and of cardiac muscle (De Haan, 1967, 1970) is the potassium ion. Scott and Fisher (1970) reported a remarkable relationship between the number and survival of neurons produced in cultures of chick dorsal root ganglia in culture, and the

K^+ concentration. When the K^+ was raised from 6.8 mM, a sharp increase in number of neurons occurred at 40.8 mM. In cultures of dissociated rat cerebellum, a similar effect was observed by Lasher and Zagon (1972), but at a lower concentration of K^+ (24.5 mM). With 10 to 16 week human fetal neurons (Scott, 1971) the effect of K^+ was more rapid and was attained at a lower concentration (20.8 mM) than with chick ganglia. That these effects are specific for high K^+, rather than low Na^+, was shown by use of various osmotic adjuvants to maintain the media at constant osmolality of 295 m0sm/kg (Scott and Fisher, 1971). Potassium was also reported to stimulate oxygen uptake in cultures of chick glial cells (Hertz *et al.* 1973), although, with mouse neuroblastoma cells and a very high level of K^+ (50 mM), Nissen *et al.* (1973) observed inhibition of oxygen uptake in freshly explanted cells, but no effect on cells that had differentiated in culture.

B. Glucose.

High glucose was first shown to be important for myelination by Pomerat and Costero (1956), who used 2000 to 3000 mg/l (11.7 to 16.7 mM) for cat cerebellum. Murray *et al.* (1962) (cited by Silberberg, 1972) increased the level of glucose to 6000 mg/l (33.3 mM), which is of interest in the light of the very high glucose (83.3 mM) in Lewis and Lewis's solution E (Table I). The medium F-12G used by Shapiro and Schrier (1973) contains 26 mM glucose, instead of the 10 mM of the original Ham's F-12 (Ham, 1965).

C. Hormones.

1. Insulin. Not only glucose, but also high levels of insulin (0.1 to 0.3 I.U./ml) have been reported by Silberberg (1967) and Guillery *et al.* (1970) to promote myelination and to prolong the survival of nerve cells *in vitro.* Fiber outgrowth from sympathetic ganglia, on the other hand, is little affected by insulin, in spite of the reported similarity in structure and composition between insulin and the nerve growth factor (NGF) (Levi-Montalcini, 1966; Bradshaw and Young, 1975).

2. Hydrocortisone, which has the general property of stimulating ribosomal protein synthesis, has been noted, when used at 0.5 μM, to induce glycerolphosphate dehydrogenase (GPDH) in rat glioma (C6) and rat glia cultures (De Vellis and Kukes, 1973). A transient inhibition of cell proliferation in monolayer cultures of C6 cells lasting about 5 days, was observed by Grasso (1976) to be produced by 3 μM hydrocortisone. Recovery was such that the growth caught up with that of untreated controls by 7 days.

3. Thyroxin. In a medium of 50% horse serum or placental cord serum, with the 6000 mg/l of glucose that is favorable for myelination, addition of 3.0 μg/ml of thyroxin both accelerated myelinogenesis and prolonged the maintenance of myelination in cerebellar cultures (Hamburgh, 1966).

D. Nerve Growth Factor and Glial Factor.

The first of the target-cell-specific growth factors to be identified was the nerve growth factor. While other so-called growth factors have, as Baserga (1976) has pointed out, had a short half-life, the NGF has retained its validity, its chemical composition is now known, and it may properly be classified as a hormone or hormone-like molecule. The factor, which dramatically affects the outgrowth of neurites from explanted ganglia, is synthesized in tumors (Bueker, 1948; Levi-Montalcini and Hamburger, 1951), in snake venom (Cohen and Levi-Montalcini, 1956), and in especially large amounts in the adult male mouse submandibular glands (Cohen, 1960). It is found in other normal and tumor cells, including glia and gliomas (Bradshaw and Young, 1976). The apparent essentiality of the NGF for the survival and maintenance of sensory and sympathetic nerve cells *in vitro* (Levi-Montalcini and Angeletti, 1963), and its high activity e.g., at 0.1 μg/ml, (Levi-Montalcini and Angeletti, 1968), have made it an object of particular interest in the study of nerve cell nutrition and differentiation (Varon, 1975a).

Specific granules in the mouse submandibular gland are responsible for the synthesis of the NGF (Schwab *et al.* 1976), which consists of three sub-units (Varon *et al.* 1968). The β subunit, and the closely related 2.5 S subunit (Bradshaw and Young, 1976), exhibit the biological activity of NGF (Schwab *et al.* 1976), and the structure of the 2.5 S molecule has been determined to be that of a dimer of two identical polypeptide chains, each of molecular weight 13,259 (Angeletti *et al.* 1971, 1973), noncovalently linked. Its role *in vivo* appears to remain a mystery, but its functions *in vitro* are specific and significant.

Monard *et al.* (1975) have presented evidence that another factor, the glial factor, with properties, including target cell and immunological specificities that distinguish it from NGF, is responsible for differentiation of neuroblastoma cells.

V. Chemically Defined Media and Serum

Media selected for post-mitotic cells, such as neurons, where *function* is to be studied, should probably differ from media for neuroblasts or glial cells, that may be expected to *proliferate.* In the general field of cell and tissue culture, media for fibroblasts are generally selected for their ability to support proliferative growth. Media for epithelial cells may be selected or designed either for growth, or to enable functions to be expressed.

The design of media, and the roles played in them by particular nutrients have been reviewed by Waymouth (1965, 1972, 1974), by Higuchi (1973), and (for clonal growth) by Ham (1974a). Ham (1974b) has also drawn attention to the neglect of studies on the initiation of *primary cultures.* The volumes of Rothblat and Cristofalo (1972) form a useful introduction to the subject. I will here briefly survey some of the most recent and important advances. Sato (1973) has reviewed the culture of the nervous system, and Varon (1975b) the culture of neurons and glia.

It is still true that chemically defined media are for the most part not designed strictly for particular cells. "All purpose" media, supplemented with serum with the hope of thereby compensating for nutritional deficiencies, are much more widely used than defined media, created for the purpose at hand. It has been claimed that inclusion of proteins is essential for some cells. The realization is slowly spreading that serum proteins may have one or more roles contributing to cell nutrition. They may be utilized, after hydrolysis by cellular proteases, as sources of amino acids for the synthesis of cellular proteins. They may themselves contribute metabolic enzymes, or specific hormonal factors such as insulin or the NGF. They may act as carriers for active substances, such as vitamins (e.g. biotin), steroids, fatty acids or trace elements.

Fractionation of serum proteins has been done by many cell nutritionists. Several types of glycoprotein have been implicated in improving cell growth or attachment to substrates, of which fetuin (Pedersen, 1944) enjoyed some vogue as an attachment factor (Fisher *et al.* 1958; Lieberman and Ove, 1958) until it was shown (Lieberman *et al.* 1959) that more highly purified fetuin lacked this property.

The use of fetal bovine serum which these reports stimulated has, however, remained a seldom questioned tradition. In one report of nerve cell growth in a defined medium, Donta (1975) showed that functional glial cells from rat (C6 cells) and man (CHB_4), and an interstitial cell line from mouse (I_{10}), can be grown in a completely serum-free medium. These cells required adaptation to the synthetic medium, by decreasing the concentration of fetal bovine serum and use of 20 to 40% of conditioned medium.

Healy and Parker (1966b) demonstrated significant improvement in the growth of mouse embryo cells in medium CMRL 1415 by addition of macromolecular fractions from serum. The non-protein fractions were inactive. The activity of the protein fractions was associated primarily with α_2-macroglobulins and α_1-acid glycoproteins. The α_2-globulins could be replaced by non-protein macromolecules such as dextran or Ficoll, which alone were inert. The macroglobulins may perform carrier or protective functions, while it appears that the glycoproteins have nutritional or growth stimulating functions associated with them, either intrinsically or because they act as carriers or both.

Among the small molecules carried on proteins, the fatty acids received attention in the report of Ham (1963b) that, for clonal growth of Chinese hamster CHD-3 cells in medium F-10 + 10 μg/ml of linoleic acid improved the plating efficiency from 0.2% in unsupplemented medium to 83% of that achieved with 20 μg/ml of serum albumin. Ham's medium F-12 (Ham, 1965) and the more recent modifications from his laboratory MCDB 102 and 103 (McKeehan *et al.* 1976) and MCDB 201 (McKeehan and Ham, 1976) contain 0.08 μg/ml of linoleic acid, and MCDB 104 (McKeehan and Ham, 1976) contains only 22 ng/ml. Other evidence of the important role of fatty acids comes from the report that rat heart cells that fail to beat in lipid-free medium resume beating when linoleate (4×10^{-5}M), acetate (2.5×10^{-3}M) and palmitate (1×10^{-4}M) are added (Harary *et al.* 1966). More recently, Jenkin *et al.* (Morrison and Jenkin, 1972; Guskey and Jenkin, 1976) have identified oleate as one of the components of serum that will, for BHK-21/13 cells in serum-free medium supplemented with 2 mg/ml of fatty acid-free bovine serum albumin, enable at least 80 serial passages to be carried out. The concentration of oleate usually used was 10 μg/ml.

Ham and his colleagues have found that, by careful control of the relative proportions of the major components of a defined medium (Ham *et al.* 1977; McKeehan *et al.* 1977), and by addition of trace elements (McKeehan *et al.* 1976), most of the serum protein could be dispensed with in media for human diploid cells. Serum protein is supplied in media MCDB 102, 103, 104 and 201 in the form of a protein fraction from dialyzed fetal bovine serum (FBSP).

Part of the success in the use of 500 μg/ml or less of FBSP depends upon coating the substrate to which the cells attach with a basic polymer, such as polylysine or polyornithine (McKeehan and Ham, 1976); larger amounts of the serum protein are required for clonal growth without than with coated surfaces. The addition of selenium (5×10^{-9} to 3×10^{-8}M), silicon (5×10^{-7}M) and vanadium (5×10^{-11} or 5×10^{-9}M), provides these trace elements in amounts corresponding to those found in 5 to 20% of serum. The most recent report (McKeehan *et al.* 1977) has demonstrated growth of WI-38, MRC-5 and SW-567 human diploid cells in medium MCDB 104 with only 25 μg/ml of the FBSP.

Whereas Ham has emphasized the role of the trace elements in replacing serum, and has by this means reduced the requirement for serum protein to microgram per ml levels, Hayashi and Sato (1976) have examined the role of an array of about 25 major and minor hormones to be found in serum as replacements for whole serum for some epithelial cell lines (GH_3 rat pituitary cells, HeLa human carcinoma cells, and BHK cells) (See also Mather and Sato, this volume). The hormones examined are shown in

Table III. Combinations of several hormones replace serum, when added to medium F-12. This work demonstrates that hormone supplementation must be taken seriously, not only for the known hormone-dependent cells such as mammary gland, prostate, etc., but for other cell types also. We do not know whether traces of minor elements, such as vanadium or selenium may be carried along in the hormone fractions. Completely effective defined media should soon become the rule rather than the exception, when the proper proportions of major nutrients have been supplemented with fatty acids, steroids, hormones, and trace elements.

TABLE III
HORMONES AS REPLACEMENTS FOR WHOLE SERUM FOR SOME EPITHELIAL CELL LINES

Hormones		Cell Lines GH$_3$	BHK	HeLa
Triiodothyronine	T$_3$	3×10^{-10}M		3×10^{-10}M
Luteinizing hormone	LH		0.5 μg/ml	
Follicle stimulating hormone	FSH		10 mU/ml	
Corticotropin	ACTH		5 μU/ml	
Growth hormone	GH		5 ng/ml	
Prolactin			10 ng/ml	
Thyroid stimulating hormone	TSH		5 ng/ml	
Prostaglandin E$_1$			10 ng/ml	
Ceruloplasmin			5 μg/ml	5 μg/ml
Transferrin		5 μg/ml	5 μg/ml	5 μg/ml
Glycyl-histidyl-lysyl acetate			0.2 μg/ml	0.2 μg/ml
Somatomedin A		50 mU/ml	50 mU/ml	50 mU/ml
Insulin			50 ng/ml	50 ng/ml
Glucagon			50 ng/ml	50 ng/ml
Calcitonin			0.1 ng/ml	0.1 ng/ml
Thyroid releasing hormone	TRH	1×10^{-9}M	1×10^{-9}M	1×10^{-9}M
Somatostatin (Somatotropin releasing inhibitory factor)	SRIF		10 ng/ml	10 ng/ml
LH releasing hormone	LRH		10 ng/ml	10 ng/ml
Parathyroid hormone	PTH	0.5 ng/ml	0.5 ng/ml	0.5 ng/ml
Hydrocortisone acetate				2×10^{-9}M
17 β estradiol				2×10^{-9}M
Testosterone				2×10^{-9}M
Progesterone				2×10^{-8}M

(Modified from Hayashi and Sato, 1976)

Conclusions

Future media for nerve cells and tissues will be built by those who take careful note of the roles of (a) cation balance in maintaining membrane potentials, for cardiac muscle as well as for neurons; (b) fatty acids and trace elements for fibroblasts; and (c) the right selection of hormones for epithelial cells, but who still recognize that nerve cells require their own specific nutrition. Other variables not mentioned here need also to be considered, among them CO_2 and O_2, redox potential, and energy sources other than glucose. My plea to neurobiologists who undertake cell cultivation is that they should not settle for second best, by using media and environmental conditions designed for something else.

References:

Angeletti, R. H., M. A. Hermodson and R. A. Bradshaw. 1973. Amino acid sequences of mouse 2.5S nerve growth factor. II. Isolation and characterization of the thermolytic and peptic peptides and the complete covalent structure. Biochemistry 12: 100-115.

Angeletti, R. H., R. A. Bradshaw and R. D. Wade. 1971. Subunit structure and amino acid composition of mouse submaxillary gland nerve growth factor. Biochemistry 10: 463-469.

Baker, L. E. and A. H. Ebeling. 1938. Maintenance of fibroblasts in artificial and serumless media. Proc. Soc. Exp. Biol. Med. 39:291-294.

Baserga, R. 1976. Multiplication and division in mammalian cells. In: *Biochemistry of Disease.* Marcel Dekker, New York and Basel. Vol. 6, Chap. 10.

Birch, J. R. and S. J. Pirt. 1971. The quantitative glucose and mineral requirements of mouse LS (suspension) cells in chemically defined medium. J. Cell. Sci. 8:693-700.

Boone, C. W., N. Mantel, T. D. Caruso, E. Kazan and R. E. Stevenson. 1972. Quality control studies of fetal bovine serum used in tissue culture. In Vitro 7:174-189.

Bradshaw, R. A. and M. Young. 1976. Commentary. Nerve growth factor — recent developments and perspectives. Biochem. Pharmacol. 25:1445-1449.

Bueker, E. D. 1948. Implantation of tumors in the hind limb of the embryonic chick and the developmental response of the lumbosacral nervous system. Anat. Rec. 102:369-389.

Cohen, S. 1960. Purification of a nerve-growth promoting protein from the mouse salivary gland and its neurocytotoxic antiserum. Proc. Nat. Acad. Sci. USA. 46:302-311.

Cohen, S. and R. Levi-Montalcini. 1956. A nerve growth-stimulating factor isolated from snake venom. Proc. Nat. Acad. Sci. USA. 42:571-574.

DeHaan, R. L. 1967. Regulation of spontaneous activity and growth of embryonic chick heart cells in tissue culture. Devel. Biol. 16:216-249.

DeHaan, R. L. 1970. The potassium-sensitivity of isolated embryonic heart cells increases with development. Devel. Biol. 23:226-240.

De Vellis, J. and G. Kukes. 1973. Regulation of glial cell functions by hormones and ions. Tex. Rep. Biol. Med. 31:271-293.

Donta, S. T. 1975. Growth of functional glial cells in a serumless medium. Methods Cell Biol. 9:123-137.

Eagle, H. 1955. Nutrition needs of mammalian cells in tissue culture. Science 122:501-504.

Eagle, H. 1959. Amino acid metabolism in mammalian cell culture. Science 130:432-437.

Earle, W. R. 1943. Production of malignancy *in vitro.* IV. The mouse fibroblast cultures and changes seen in the living cells. J. Nat. Cancer. Inst. 4:165-212.

Fisher, H. W., T. T. Puck and G. Sato. 1958. Molecular growth requirements of single mammalian cells: the action of fetuin in promoting cell attachment to glass. Proc. Nat. Acad. Sci. USA. 44:4-10.

Fisher, K. R. S., S. Fedoroff and E. L. Wenger. 1975. Effect of osmotic pressure on neurogenesis in cultures of chick embryo spinal cords. In Vitro 11:329-337.

Gey, G. O. and M. K. Gey. 1936. The maintenance of human normal cells and tumor cells in continuous culture. I. Preliminary report: Cultivation of mesoblastic tumors and normal tissue and notes on methods of cultivation. Amer. J. Cancer 27:45-76.

Gorham, L. W. and C. Waymouth. 1965. Differentiation *in vitro* of embryonic cartilage and bone in a chemically defined medium. Proc. Soc. Exp. Biol. Med. 119:287-290.

Grasso, R. J. 1976. Transient inhibition of cell proliferation in rat glioma monolayer cultures by corisol. Cancer Res. 36:2408-2414.

Guillery, R. W., H. M. Sobkowicz and G. L. Scott. 1970. Relationship between glial and neuronal elements in the development of long-term cultures of the spinal cord of the fetal mouse. J. Comp. Neurol. 140:1-33.

Guskey, L. E. and H. M. Jenkin. 1976. The serial cultivation of suspended BHK-21/13 cells in serum-free Waymouth medium. Proc. Soc. Exp. Biol. Med. 151:221-224.

Ham, R. G. 1963a. An improved nutrient solution for diploid Chinese hamster and human cell lines. Exp. Cell Res. 29:515-526.

Ham, R. G. 1963b. Albumin replacement by fatty acids in clonal growth of mammalian cells. Science 140:802-803.

Ham, R. G. 1965. Clonal growth of mammalian cells in a chemically defined, synthetic medium. Proc. Nat. Acad. Sci. USA. 53:288-293.

Ham, R. G. 1974a. Unique requirements for clonal growth. J. Nat. Cancer Inst. 53:1459-1463.

Ham, R. G. 1974b. Nutritional requirements of primary cultures. A neglected problem of modern biology. In Vitro 10:119-129.

Ham, R. G., S. L. Hammond and L. L. Miller. 1977. Critical adjustment of cysteine and glutamine concentrations for improved clonal growth of WI-38 cells. In Vitro. (in press).

Hamburgh, M. 1966. Evidence for a direct effect of temperature and thyroid hormone on myelinogenesis *in vitro*. Devel. Biol. 13:15-30.

Hanks, J. H. and R. E. Wallace. 1949. Relation of oxygen and temperature in the preservation of tissues by refrigeration. Proc. Soc. Exp. Biol. Med. 71:196-200.

Harary, I., R. McCarl and B. Farley. 1966. Studies *in vitro* in single beating rat heart cells. IX. The restoration of beating by serum lipids and fatty acids. Biochim. Biophys. Acta 115:15-22.

Harrison, R. G. 1906-1907. Observations on the living developing nerve fiber. Proc. Soc. Exp. Biol. Med. 4:140-143.

Hayashi, I. and G. H. Sato. 1976. Replacement of serum by hormones permits growth of cells in a defined medium. Nature (Lond.) 259:132-134.

Healy, G. M. and R. C. Parker. 1966a. An improved chemically defined basal medium (CMRL-1415) for newly explanted mouse embryo cells. J. Cell Biol. 30:531-538.

Healy, G. M. and R. C. Parker. 1966b. Cultivation of mammalian cells in defined media with protein and nonprotein supplements. J. Cell Biol. 30:539-553.

Hertz, L., L. Dittman and P. Mandel. 1973. K^+ induced stimulation of oxygen uptake in cultured cerebral glial cells. Brain Res. 60:517-520.

Higuchi, K. 1970. An improved chemically defined culture medium for strain L mouse cells based on growth responses to graded levels of nutrients including iron and zinc. J. Cell. Physiol. 75:65-72.

Higuchi, K. 1973. Cultivation of animal cells in chemically defined media, a review. Adv. Appl. Microbiol. 16:111-136.

Higuchi, K. and R. C. Robinson. 1973. Studies in the cultivation of mammalian cell lines in a serum-free, chemically defined medium. In Vitro 9:144-121.

Honn, K. V., J. A. Singley and W. Chavin. 1975. Fetal bovine serum: a multivariate standard. Proc. Soc. Exp. Biol. Med. 149:344-347.

Kitos, P. A., R. Sinclair and C. Waymouth. 1962. Glutamine metabolism by animal cells growing in a synthetic medium. Exp. Cell Res. 27:307-316.

Lasher, R. S. and I. S. Zagon. 1972. The effect of potassium on neuronal differentiation in cultures of dissociated newborn rat cerebellum. Brain Res. 41:482-488.

Leibovitz, A. 1963. The growth and maintenance of tissue-cell cultures in free gas exchange with the atmosphere. Amer. J. Hyg. 78:173-180.

Leiter, E. H., D. L. Coleman and C. Waymouth. 1974. Cell culture of the endocrine pancreas of the mouse in chemically defined media. In Vitro 9:421-433.

Levi-Montalcini, R. 1966. The nerve growth factor: its mode of action on sensory and sympathetic nerve cells. Harvey Lects. 60:217-259.

Levi-Montalcini, R. and P. U. Angeletti. 1963. Essential role of the nerve growth factor in the survival and maintenance of dissociated sensory and sympathetic nerve cells in vitro. Devel. Biol. 7:653-659.

Levi-Montalcini, R. and P. U. Angeletti. 1968. Nerve growth factor. Physio. Rev. 48:534-569.

Levi-Montalcini, R. and V. Hamburger. 1951. Selective growth stimulating effects of mouse sarcoma on the sensory and sympathetic nervous system of the chick embryo. J. Exp. Zool. 116:321-361.

Lewis, M. R. and W. H. Lewis. 1911. The cultivation of tissues from chick embryos in solutions of NaC1, KC1 and $NaHCO_3$. Anat. Rec. 5:277-293.

Lewis, W. H. and M. R. Lewis. 1912a. The cultivation of chick tissues in media of known chemical composition. Anat. Rec. 6:207-211.

Lewis, W. H. and M. R. Lewis. 1912b. The cultivation of sympathetic nerves from the intestine of chick embryos in saline solutions. Anat. Rec. 6:7-31.

Lieberman, I., F. Lamy and P. Ove. 1959. Non-identity of fetuin and protein growth (flattening) factor. Science 129:43-44.

Lieberman, I. and P.Ove. 1958. A protein growth factor for mammalian cells. J. Biol. Chem. 233:637-642.

Ling, C. T., G. O. Gey and V. Richters. 1968. Chemically characterized concentrated corodies for continuous cell culture (the 7C's culture media). Exp. Cell Res. 52:469-489.

Locke, F. S. 1895. Artificial fluids as uninjurious as possible to animal tissues. Boston Med. Surg. J. 134:173.

Locke, F. S. 1900. Die Wirkung der Metalle das Blutplasmas und verschiedener Zucker auf des isoliertes Saügetierherz. Zentralbl. Physiol. 14:670.

Lowy, P. H. and G. Keighley. 1968. Inactivation of erythropoietin by Koshland's tryptophan reagent and by membrane filtration. Biochim. Biophys. Acta 160:413-419.

Mains, R. E. and P. H. Patterson. 1973. Primary cultures of dissociated sympathetic neurons. I. Establishment of long-term growth in culture and studies of differentiated properties. J. Cell Biol. 59:329–345.

McCoy, T. A., M. Maxwell and P. F. Kruse, 1959. Amino acid requirements of the Novikoff hepatoma *in vitro*. Proc. Soc. Exp. Biol. Med. 100:115–118.

McKeehan, W. L. and R. G. Ham. 1976. Stimulation of clonal growth of normal fibroblasts with substrata coated with basic polymer. J. Cell Biol. 71:727–734.

McKeehan, W. L., W. G. Hamilton and R. G. Ham. 1976. Selenium is an essential trace nutrient for growth of WI-38 diploid human fibroblasts. Proc. Nat. Acad. Sci. USA. 73:2023–2027.

McKeehan, W. L., K. A. McKeehan, S. L. Hammond and R. G. Ham. 1977. Improved medium for clonal growth of human diploid cells with limiting concentrations of serum protein. In Vitro 12: (in press).

McQuilkin, W. T., V. J. Evans and W. R. Earle. 1957. The adaptation of additional lines of NCTC 929 (Strain L) cells to chemically defined protein-free medium NCTC 109. J. Nat. Cancer Inst. 19:885–907.

Milo, G. E., W. B. Malarkey, J. E. Powell, J. R. Blakeslee and D. S. Yohn. 1976. Effects of steroid hormones in fetal bovine serum on plating and cloning of human cells in vitro. In Vitro 12: 23–30.

Monard, D., K. Stöckel, R. Goodman and H. Thoenen. 1975. Distinction between nerve growth factor and glial factor. Nature (Lond.) 258:444–445.

Moore, G. E., R. E. Gerner and H. A. Franklin. 1967. Culture of normal human leukocytes. J. Amer. Med. Assoc. 199:519–524.

Morgan, J. F., M. E. Campbell and H. J. Morton. 1955. The nutrition of animal tissues cultivated *in vitro*. I. A survey of natural materials as supplements to synthetic medium 199. J. Nat. Cancer Inst. 16:557–567.

Morgan, J. F., H. J. Morton and R. C. Parker. 1950. Nutrition of animal cells in tissue culture. I. Initial studies on a synthetic medium. Proc. Soc. Exp. Biol. Med. 73:1–8.

Morrison, S. J. and H. M. Jenkin. 1972. Growth of *Chlamydia psittaci* strain meningopneumonitis in mouse L cells cultivated in a defined medium in spinner cultures. In Vitro 8:94–100.

Murray, M. R. and G. Kopech. 1953. *The Bibliography of Tissue Culture, 1884-1950.* 2 vols. Academic Press, New York and London.

Murray, M. R., E. R. Peterson and R. P. Bunge. 1962. Cited by Silberberg (1972).

Nagle, S. C. 1968. Heat-stable chemically defined medium for growth of animal cells in suspension. Appl. Microbiol. 16:53-55.

Nagle, S. C. and B. L. Brown. 1971. An improved heat-stable glutamine-free chemically defined medium for growth of mammalian cells. J. Cell Physiol. 77:259-264.

Nagle, S. C., H. R. Tribble, R. E. Anderson and N. D. Gary. 1963. A chemically defined medium for growth of animal cells in suspension. Proc. Soc. Exp. Biol. Med. 112:340-344.

Nissen, C., J. Ciesielski-Treska, L. Hertz and P. Mandel. 1973. Regulation of oxygen consumption in neuroblastoma cells: Effects of differentiation and of potassium. J. Neurochem. 20:1029-1035.

Olmsted, C. A. 1967. A physico-chemical study of fetal calf serum used as a tissue culture nutrient correlated with biological tests for toxicity. Exp. Cell Res. 48:283-299.

Parker, R. C. 1961. *Methods of Tissue Culture.* 3rd Edition. Harper, New York, p. 77.

Parshad, R., K. K. Sanford, G. M. Jones, F. M. Price and W. G. Taylor. 1977. Oxygen and light effects on chromosomal aberrations in mouse cells in vitro. Exp. Cell Res. 104:199-205.

Peacock, J. H. and P. G. Nelson. 1973. Chemosensitivity of mouse neuroblastoma cells *in vitro.* J. Neurobiol. 4:363-374.

Peacock, J. H., P. G. Nelson and M. W. Goldstone. 1973. Electrophysiologic study of cultured neurons dissociated from spinal cords and dorsal root ganglia of fetal mice. Devel. Biol. 30:137-152.

Pedersen, K. O. 1944. Fetuin, a new globulin isolated from serum. Nature (Lond.) 154:575.

Pomerat, C. M. and I. Costero. 1956. Tissue cultures of cat cerebellum. Amer. J. Anat. 99:211-247.

Price, F. L., H. A. Kerr, W. F. Andersen, J. C. Bryant and V. J. Evans. 1966. Some *in vitro* requirements of cells of C3H mouse, Chinese hamster, green monkey, and human origin. J. Nat. Cancer Inst. 37:601-617.

Rothblat, G. H. and V. J. Cristofalo eds. 1972. *Growth, Nutrition and Metabolism of Cells in Culture.* 2 vols. Academic Press, New York and London.

Sanford, K. K., W. R. Earle and G. D. Likely. 1943. The growth *in vitro* of single isolated tissue cells. J. Nat. Cancer Inst. 9:229-246.

Sato, G. ed. 1973. *Tissue Culture of the Nervous System.* Plenum Press, New York and London.

Schubert, D., S. Humphreys, F. de Vitry and F. Jacob. 1971. Induced differentiation of a neuroblastoma. Devel. Biol. 25:514-546.

Schwab, M. E., K. Stöckel and H. Thoenen. 1976. Immunocytochemical localization of nerve growth factor (NGF) in the submandibular gland of adult mice by light and electron microscopy. Cell Tissue Res. 169:289-299.

Scott, B. S. 1971. Effect of potassium on neuron survival in cultures of dissociated human nervous tissue. Exp. Neurol. 30:297-308.

Scott, B. S. and K. C. Fisher. 1970. Potassium concentration and number of neurons in cultures of dissociated ganglia. Exp. Neurol. 27:16-22.

Scott, B. S. and K. C. Fisher. 1971. Effect of choline, high potassium and low sodium on the number of neurons in cultures of dissociated chick ganglia. Exp. Neurol. 31:183-188.

Shapiro, D. and B. K. Schrier. 1973. Cell cultures of fetal rat brain: growth and marker enzyme development. Exp. Cell Res. 77:239-247.

Silberberg, D. H. 1967. Phenylketonuria metabolites in cerebellum culture morphology. Arch. Neurol. 17:524-529.

Silberberg, D. H. 1972. Cultivation of nerve tissue. In: *Growth, Nutrition and Metabolism of Cells in Culture.* G. H. Rothblat and V. J. Cristofalo, eds. Academic Press, London and New York. pp. 131-167.

Takaoka, T. and H. Katsuta. 1975. An improved synthetic medium suitable for tissue culture of various mammalian cells. Japan. J. Exp. Med. 45:11-17.

Tyrode, M. V. 1910. The mode of action of some purgative salts. Arch. Int. Pharmacodyn. Ther. 20:205-223.

Varon, S. 1975a. Nerve growth factor and its mode of action. Exp. Neurol. 48(2):75-92.

Varon, S. 1975b. Neurons and glia in neural cultures. Exp. Neurol. 48(2):93-134.

Varon, S., J. Nomura and E. M. Shooter. 1968. Reversible dissociation of the mouse nerve growth factor protein into different subunits. Biochemistry 7:1296-1303.

Vogelaar, J. P. M. and E. Erlichman. 1933. A feeding solution for cultures of human fibroblasts. Amer. J. Cancer 18:28-48.

Wang, R. J. 1976. Effect of room fluorescent light on the deterioration of tissue culture media. In Vitro 12:19-22.

Waymouth, C. 1959. Rapid proliferation of sublines of NCTC clone 929 (strain L) mouse cells in a simple chemically defined medium (MB 752/1). J. Nat. Cancer Inst. 22:1003-1017.

Waymouth, C. 1965. Construction and use of synthetic media. In: *Cells and Tissues in Culture.* E. N. Willmer, ed. Academic Press, London and New York. pp. 99-142.

Waymouth, C. 1967. Somatic cells *in vitro:* Their relationship to progenitive cells and to artificial milieux. Nat. Cancer Inst. Monogr. 26:1-21.

Waymouth, C. 1970. Osmolality of mammalian blood and of media for culture of mammalian cells. In Vitro 6:109-127.

Waymouth, C. 1972. Construction of tissue culture media. In: *Growth, Nutrition and Metabolism of Cells in Culture.* G. H. Rothblat and V. J. Cristofalo, eds. Academic Press, New York and London. pp. 11-47.

Waymouth, C. 1974. "Feeding the baby" — designing the culture milieu to enhance cell stability. J. Nat. Cancer Inst. 53:1443-1448.

Weiss, P. 1939. *Principles of Development.* H. Holt and Co., New York.

Williams, G. M. and J. M. Gunn. 1974. Long-term cell culture of adult rat liver epithelial cells. Exp. Cell Res. 89:139-142.

Yamane, I., Y. Matsuya and K. Jimbo. 1968. An autoclavable powdered culture medium for mammalian cells. Proc. Soc. Exp. Biol. Med. 127:335-336.

Yonezawa, T., T. Mori and Y. Nakatani. 1969. Effects of pyridoxine deficiency in nervous tissue maintained *in vitro.* Ann. N.Y. Acad. Sci. 166:146-157.

Use of Cell Tissue and Organ Cultures in Neurobiology

CELL AND TISSUE DIFFERENTIATION IN CULTURES: DISCUSSION

*B. SCOTT**

Services to Developmentally Handicapped Persons
Surrey Place Centre
Toronto, Ontario
Canada

*F. T. OTERUELO**

Department of Anatomy
University of Saskatchewan
Saskatoon, Saskatchewan
Canada

I. Cell Lineages and Differentiation

The discussion on cell differentiation was initiated by Dr. Fedoroff who pointed out that in order to study the cell lineages and cell differentiation of the nervous system one must be aware of embryological events and must carefully select the cells or tissue to be studied in relation to the chronological program of developmental events. In order to facilitate the study of cell lineages a conceptual model consisting of three compartments has been devised (Fig. 1). The first two cell compartments, ventricular and subventricular, consist of proliferating cells and the third compartment consists of post mitotic differentiating and differentiated cells. The ventricular cell compartment is composed of stem cells which give rise to large, Golgi type I neurons and subventricular cells. The subventricular cells which are also stem cells, but which have somewhat limited capabilities, give rise to smaller neurons (Golgi type II) and to glial cells (astrocytes and oligodendrocytes). Chronologically, the astrocytes are formed first and their formation ceases before formation of oligodendrocytes stops. The difference between this model and that of renewing tissues such as hempoietic tissue is that as development progresses the proliferative cell compartments, consisting of stem cells, ventricular cells and subventricular cells, are virtually exhausted whereas in hemopoietic tissues the proliferative stem cell compartment remains constant throughout life. In the mature CNS only the end cell compartment composed of long living cells remains. In hemopoietic tissue the end cell compartment consists of short-lived cells.

*Drs. Scott and Oteruelo, as rapporteurs, have summarized and integrated the recorded discussion between the contributors to this volume on the above topic.

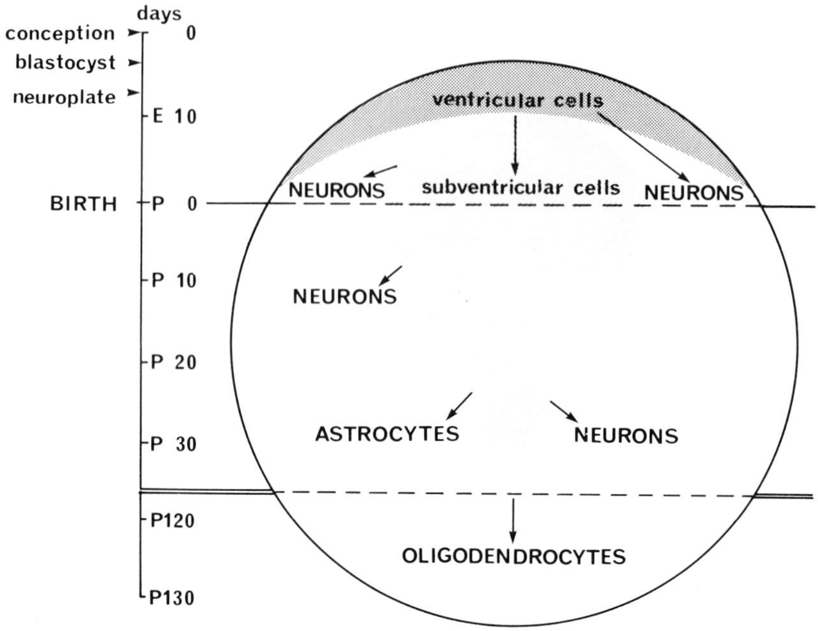

FIG. 1. A conceptual model of the developing nervous system, consisting of 3 compartments. (For details, see text). Time scale is that of mouse development. (S. Fedoroff, previously unpublished).

At present we know very little about detailed events within each compartment or the relationships between compartments, not to mention the mechanism which switches the cells into one or another lineage of differentiation and regulates the passage of cells through the compartments. Obviously, the cells passing through these compartments are subjected to cell interactions and signals varying in time and intensity. Therefore, what we can expect from cells in cultures will depend on how far they have progressed in the program, what cell interactions have been retained *in vitro* and whether or not the proper signals can be supplied at the right time and in the right way.

Dr. Carrato pointed out that in this model the microglia are missing. Microglia are not members of the functional neuronal-glial cell unit but nevertheless play an important role in the CNS. There is still argument about their origin and there has been no thorough work on their ultrastructure or their behaviour and identification in cultures.

Dr. Privat asked whether ependymal cells develop in cultures. Dr. Fedoroff replied that in cultures of subventricular cells, ciliated cells are observed but the cilia usually begin to appear after 10 to 14 days *in vitro*.

Dr. Varon described a conceptual "business approach" to cell

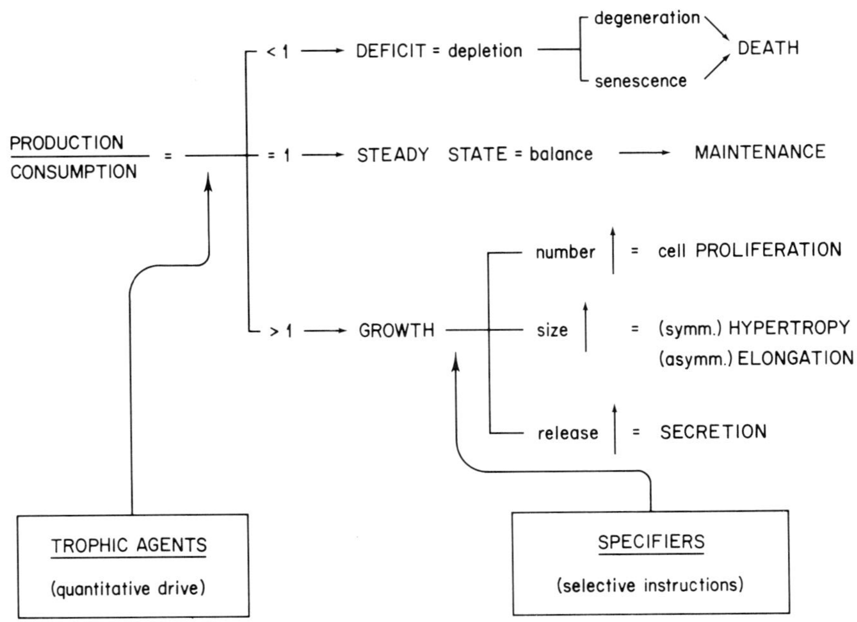

FIG. 2. A conceptual framework to help in the understanding of regeneration (and development) in neural tissue. (For details, see text). (S. Varon. 1977. Neural growth and regeneration: a cellular perspective. Exp. Neurol. 54:1-6).

development (Fig. 2) which might serve as a unifying framework for some of the problems concerning cell growth and differentiation, as well as cell interactions. The main idea in this "cellular bookkeeping" is the ratio of production to consumption. If this ratio is below one then a deficit occurs and the cell exhibits degeneration or senescence; if the ratio equals one then a steady state exists; if the ratio is greater, growth occurs. Dr. Varon proposed that the term trophic agent be restricted to those substances which interact with the cells by controlling this ratio of production to consumption, probably by regulating anabolic machinery. Growth could be expressed by the cell in several ways: by increasing the number of cells, i.e., proliferation; by increasing size, i.e., hypertrophy or elogation; and by excess product being exported or released in the form of a secretion. He suggested that the choice among these several modalities of growth was imposed by other factors, which he would prefer to call "specifying influences".

Using this approach Dr. Varon suggested that properties may develop as a function of age because some trophic agent changes in its availability to the target cell as time goes on in development. This change might be independent of the altered behavior of whatever properties, but it might also be the reason why there is an altered behavior, because different performances of the same cell might require different levels of availability of trophic agents in addition to specifying influences. (Cf. also Section IID.).

II. Cell Interactions

A. General Characteristics

Elaborating upon the approach illustrated in Fig. 2, Dr. Varon divided cellular interactions into several areas. As the first, he mentioned the influence of glial cells on neurons. For example, dorsal root ganglion explants show no response to exogenous nerve growth factor, whereas monolayers of dissociated ganglion cells are in desperate need of it. Cells are the same although, presumably, glial cells were selectively destroyed during the dissociation of the ganglia. The survival of dissociated ganglionic neurons, in the absence of exogenous nerve growth factors, is enhanced by adding increasing amounts of embryonic non-neurons (glial cells). This effect is abolished by exposing the non-neuronal cells to antibody to nerve growth factor and is a clear example of a trophic factor which is supplied from glial cells to neurons. It is a true trophic factor because, if supplied to the ganglion at a time when the neurons still proliferate, it will cause an increase in the number of cells by mitosis. The same molecule supplied at a later stage will cause growth by axonal elongation and, if supplied at an even later stage will cause a use of the excess product for secretory functional activity.

Dr. Varon continued to the influence of neurons on glial cells and mentioned that there is now ample evidence that the presence of neurons controls and regulates the proliferation of satellite cells in mouse dorsal root ganglia. There are also some interesting studies of the peripheral nervous system of the quaky mouse. This mutant is unable to form myelin and the defect has been pinned down to the Schwann cells which increase in number but somehow do not provide the signal to elongate and myelinate. Dr. Varon suggested that the primary effect might be in the cellular adhesiveness.

B. Growth of Neurites in Vitro

Directional growth of nerve fibers was discussed as possibly being a special case of neuronal/non-neuronal cell interaction. Dr. Hild mentioned the two possibilities that we know at present: growth towards a target (nerve-muscle cultures) and free growth of fiber without an apparent target. Dr. Varon mentioned the idea that trophic factors may influence the growth and guidance of fibers from an explant. For example, a nerve cell might need a certain trophic agent and receive enough to survive and to elongate but not to do its best; then, a gradient of that trophic factor encountered by the advancing growth cone could serve as a guidance mechanism. The growth cone is the most adhesive part of a neuron in culture and presumably also *in vivo*. If the growth cone in advancing does find a supply of trophic factor and if the trophic factor alters the adhesiveness, that growth cone may be encouraged to move, and it will continue to move better in the direction in which that trophic factor is available.

Dr. Crain added that there may be perhaps other guidance factors related to the geometry of the perikarya. He made reference to his earlier presentation regarding the remarkable asymmetry of formation of ventral roots emerging from isolated spinal cord explants in the absence of any target and to the fact that dorsal root fibers avoid the ventral cord tissues and pass preferentially into the dorsal cord tissue perhaps on account of a local avoidance factor.

Dr. Crain agreed with Dr. Varon that guidance by chemical gradients seems to be important and mentioned recent work in Sweden in which the possibilities for establishment of local gradients were increased by growing the nerve cells in a gel. Under these conditions there was much greater evidence of ordered growth of fibers both in relation to a target and in relation to localized sites of nerve growth factor deposit at a distance from the outgrowing fibers. Dr. Crain also drew attention to glial outgrowth that may be laying down carpets selectively, either facilitating the movement of dorsal root ganglion fibers and enhancing their penetration into the dorsal cord region, or veering away from the ventral cord region. In addition to reflecting chemical gradients this might conceivably also be due to cell contacts between the growth cones of the dorsal root fibers as they make initial contacts with this advanced carpet of different types of glial cells.

Dr. Seil supported Dr. Crain's concepts and added an example on the behavior of neurites in cultures of cerebellum. Vestibular neurons (which have lost their target in the brain stem) branch into all cortical areas of the explant where they are usually not found. Neurons from the trigeminal nucleus (which *in vivo* have peripheral targets) send all their processes out to the periphery of the culture suggesting that these cells possess some specific and inherent programs.

Dr. Messer commented that she could sometimes grow dissociated cerebellar cultures from staggerer mutant mice in which granule cells in the absence of any visible Purkinje cells with which they could make contact, did not degenerate. In fact, in her cultures they actually grew better than their normal counterparts. These studies thus did bring out clearly that we cannot use staggerer mutant mice as a model of programmed cellular degeneration if the cells fail to meet their target neurons.

Dr. Diamond elaborated on the concept that nerve sprouting may be controlled by the target but that other developmental mechanisms are needed to cause the migration of processes into certain areas. There are some boundaries, for example in the skin of the salamander, that hinder the migration of certain axons although other axons may go right across them. Dr. Diamond was of the opinion that this selective mechanism may actually operate on the non-neuronal cells and stop the advance of the non-neuronal cells. As a result of this the axons could only continue for a few hundred micra which is as much as they can advance unsupported.

The importance of studying the molecular basis for the different mechanisms of the cell interactions involved in axonal growth was pointed out by Dr. Mandel. The question of regulation at the molecular level was also raised by Dr. McDuffie, who mentioned that in hepatoma cells in culture, changes in expression of cell surface function seem to be totally related to the heparane sulfate of the cell. Alterations of these substances occur also in all of the mucopolysaccharidoses known to date, with accompanying mental retardation. Dr. McDuffie, therefore, urged the study of these substances in the culture situation.

C. Cell Aggregation and Cell Junctions

The challenge to study molecular mechanisms was also taken up by Dr. Garber who had developed a re-aggregation system of substantia nigra of the developing mouse mid-brain. These neurons can selectively be identified because of their catecholamine content. Garber and her co-workers found that these cells selectively associate and form a band, resembling substantia nigra, after four or five days of development in suspension cultures. They were now studying whether this recognition system can be disrupted by agents like haloperidol, which bind to the dopamine receptor, i.e., whether dopamine is a recognition molecule.

In another, more complex reaggregation system, i.e., telencephalon, Dr. Garber and Dr. Larramendi have recently observed a remarkably wide gap junction, which is about 40 nm in width and is filled with intercellular material. The possible importance of this junction in synaptogenesis is being studied.

Also Dr. Oteruelo presented work pertaining to the presence of characteristic cellular junctions of glial cells in culture. These junctions are present in 7-day, primary cultures from dissociated cells from cerebral hemispheres of new born DBA mice. The simplest type of junction occasionally found between these astrocytic cells is shown in Fig. 3a. This junction has some accumulation of an electron dense material at the membranes and no evidence of associated fibrillar or tubular material. More frequently, a more complex junction was observed, of the type shown in Figs. 3b and d. Fig. 3b is that of an interdigitated series of cellular processes demonstrating a junctional complex of a very characteristic type. An electron opaque material, filaments and tubules are associated at the junctional sites. What appears as a somewhat more organized junction is shown in Fig. 3c between two of these astrocytic cells. With high magnifications, Fig. 3d, it is possible to observe the jagged appearance of these junctional complexes and their association with bundles of thin filaments.

Dr. Hunt emphasized the importance of studying cell junctions and mentioned the fact that the topographical and functional commitments of the neural retina seems to be marked by the disappearance of gap junctions across the optic cup.

D. Synaptic Plasticity

Vigorous discussion was prompted by the question of synaptic plasticity in terms of neurotransmitter synthesis and the ability of the extracellular environment to affect the specificity of this synthesis. Dr. Hawrot described an extensive investigation of this question using cultures of sympathetic neurons from newborn rats. The method depended on an earlier observation by Patterson and Chung that non-neuronal cells had a marked influence on the sympathetic neurons causing them to produce high levels of acetylcholine. It was next observed that medium, conditioned by non-neuronal cells, could produce this effect without direct cell to cell

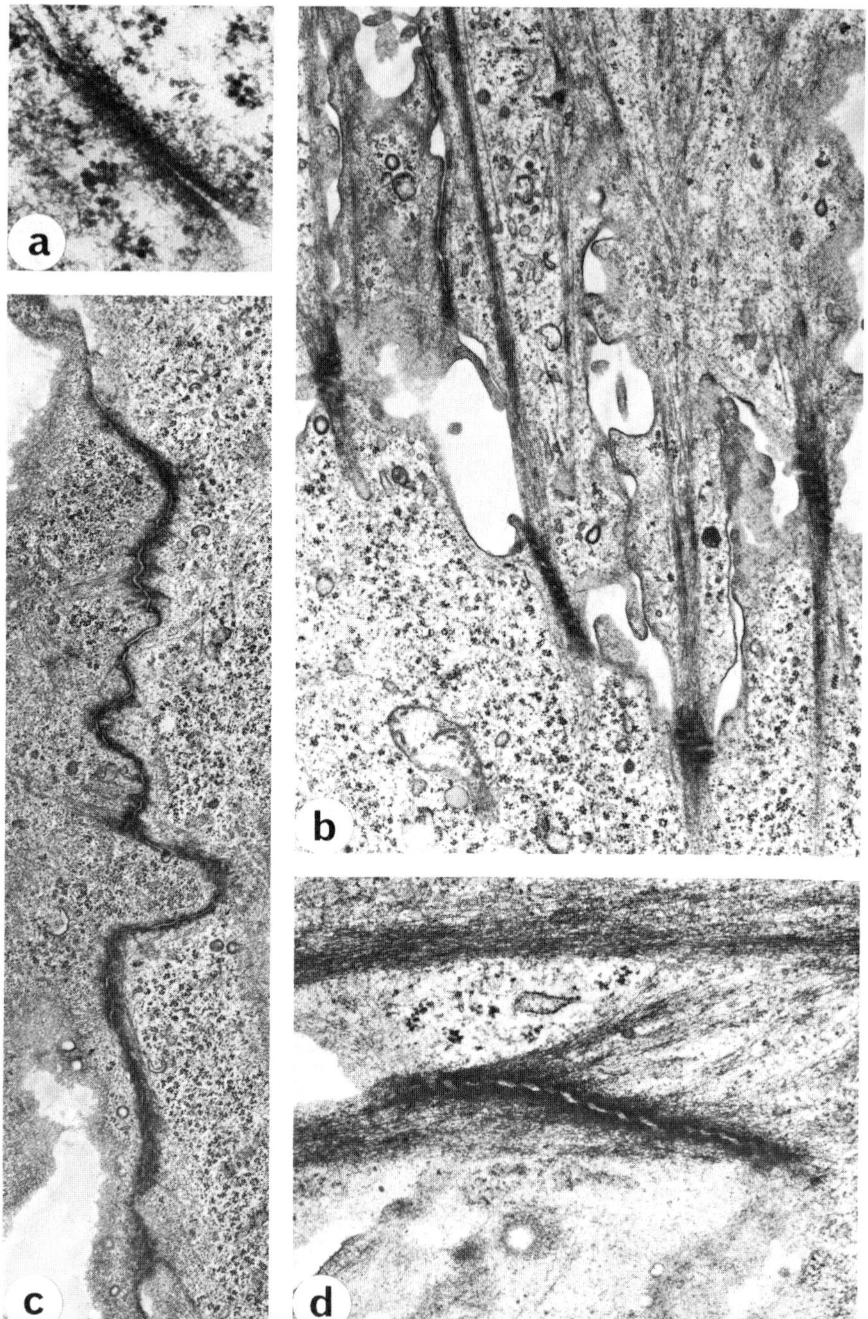
a
b
c
d

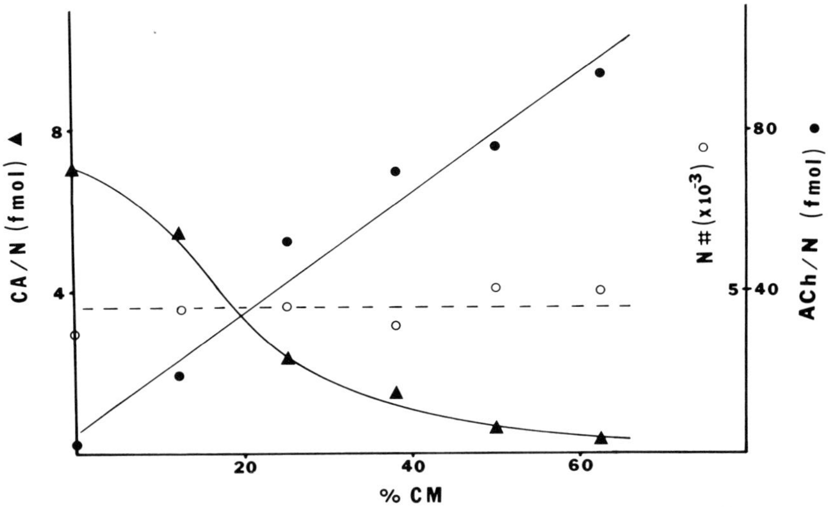

FIG. 4. The dependence of ACh and CA production and neuronal survival on CM concentration was determined on 20 day old cultures which had been treated with 10^{-5}M cytosine arabinoside on days 2 and 8 to virtually eliminate all nonneuronal cells. CM from confluent rat heart flasks was mixed with fresh medium at the concentration shown and added every 2 days, beginning at day 2. On day 20 neuronal somas were counted (open circles) and the cultures incubated with ^{3}H-choline and ^{3}H-tyrosine. Radioactive ACh (closed circles) and CA (closed triangle) were determined and the values are expressed on a per neuron basis. (P. H. Patterson and L. L. Y. Chun. 1977. The induction of acetylcholine synthesis in primary cultures of dissociated rat sympathetic neurons. I. Effects of conditioned media. Devel. Biol. 56:263-280).

contact. More recently Hawrot and Patterson examined the effect of adding various concentrations of conditioned medium and obtained the results illustrated in Fig. 4. In the absence of conditioned medium, they found the sympathetic neurons to have high levels of catecholamine synthesis and storage ability and very low levels of acetylcholine synthesis and storage. By increasing the amount of conditioned medium they forced a dramatic increase in the ability to synthesize and store acetylcholine and a corresponding decrease in catecholamine synthesis and storage. In this way they had a simple means of manipulating the culture conditions so that at high concentrations of conditioned medium, the sympathetic neurons are cholinergic and at low concentrations they are primarily adrenergic. From

FIG. 3. (a) Simple type of junction found in astrocytes in culture, ×46200. (b) Interdigitation of cellular processes with junctional complexes, ×12000. (c) Complex arrangement of a junction between two astrocytes, ×10100. (d) Serrated appearance of a junctional complex. Note filaments associated with the membranes at the junction, ×27800. (F. T. Oteruelo, previously unpublished).

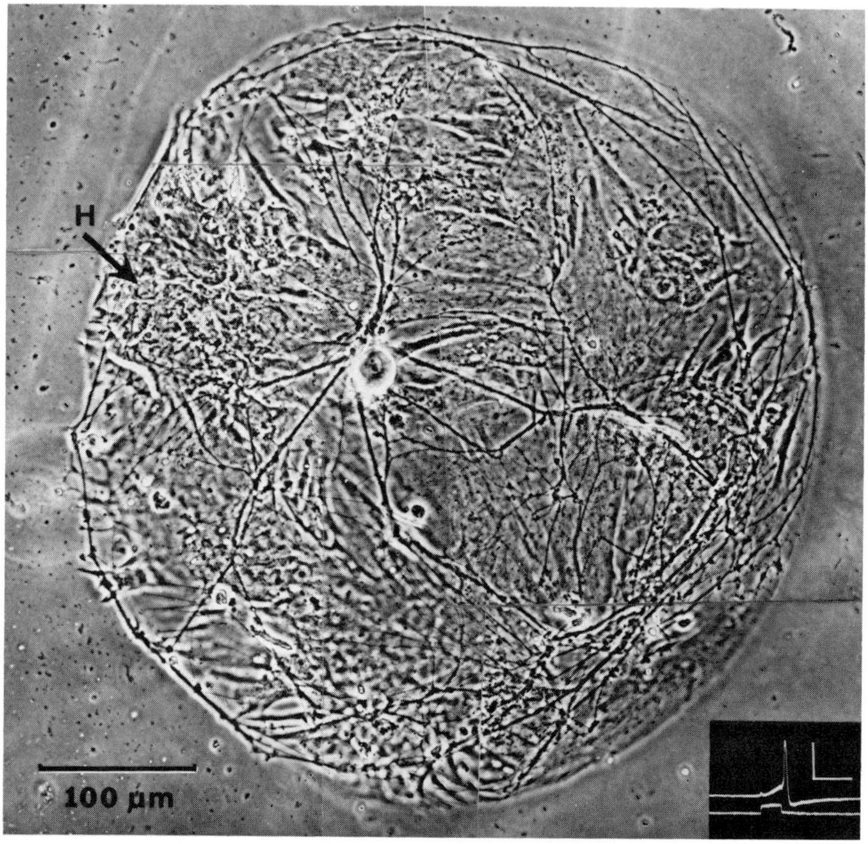

FIG. 5. A microculture containing a solitary neuron; arrow at H indicates a cluster of myocytes. Neuron at 19 days in culture. Inset shows impulse in this neuron (scales are 50 mv, 20 msec for y and x axis, respectively). (For details see text). (E. J. Furshpan, P. R. MacLeish, P. H. O'Lague and D. D. Potter. 1976. Chemical transmission between rat sympathetic neurons and cardiac myocytes developing in microcultures: Evidence for cholinergic, adrenergic, and dual neurons. Proc. Natl. Acad. Sci. U.S.A. 73:4225-4229).

these observations the question arose as to what was occurring at intermediate concentrations of conditioned medium where they had found the ability to synthesize and store both acetylcholine and catecholamine.

The method used to study this problem was to examine microcultures (Fig. 5) in which it was possible to grow a single sympathetic neuron on a monolayer of cardiac myocytes. Such a system was amenable to electrophysiological investigation because the field of innervation of the single neuron was restricted to an area of about 500 μm in diameter. Thus it was relatively easy to find electrical interactions between a given myocyte and the neuron. In these cultures three types of neurons were found. The first type, presumably cholinergic, consisted of neurons which, when stimulated, inhibited the beating of the myocytes and this inhibition was blocked by atropine. The second type, presumably adrenergic, consisted of neurons which, when stimulated, caused an increase in the beating rate of the cardiac myocyte and this effect was blocked by the β blocker, propranolol. The third type of neuron exhibited both properties. In stimulating these neurons one first observed an inhibition followed shortly thereafter by an acceleration in the beating rate of the cardiac myocyte. The initial hyperpolarization and inhibition of beating was blocked by atropine and the subsequent increase in beat rate was blocked by propranolol. It was concluded that such neurons have the capacity to synthesize and release both types of transmitters. Dr. Hawrot concluded by stating that the key question remaining was whether this duality in function was a transient developmental stage in these cultures after which the neurons would be restricted to a specific neurotransmitter or whether the duality was a permanent function due to the neurons receiving "a somewhat unusual situation of developmental cues so that they are receiving both inputs for producing adrenergic and cholinergic properties".

Other examples of the shifting from one type of synapse to another in culture and maybe also under certain conditions *in vivo* (Drs. Phillis, Lyser) were brought up. Dr. Crain mentioned the work of M. M. Bird and D. N. James in which dissociated spinal cord cells grown on conventional collagen coated coverslips showed no sign of dense core vesicles or histofluorescence. However, when grown on a feeder layer of liver cells the neurons did show catecholaminergic features. Other examples of dual function neurons were also mentioned. Dr. Diamond described the evidence that the Mauthner's neurons had both excitatory and inhibitory collaterals and Dr. Hunt mentioned that in aplysia some neurons use multiple transmitters and contain low levels of transmitter intracellularly in the cytoplasm which may not even be involved in neural transmission. Dr. Wood commented that a change in the transmitters used involves profound biochemical changes, i.e., alterations of both synthesis, release and re-uptake of the transmitter.

Dr. Varon saw the ability of neurons to shift from one neurotransmitter to another as a response of the neurons to the "specifying influence" mentioned earlier in the discussion. He warned against regarding the influence as a specific neuronal/glial interaction, because, although the first observation of this shift was made in sympathetic neuron cultures grown in the presence of large numbers of nonneuronal cells, presumably ganglionic glia, the fact that the response can occur using medium conditioned by exposure to heart or skeletal tissue indicates the origin of the influences can be peripheral.

From the observations of synaptic plasticity in terms of neurotransmitter synthesis and dual function, Dr. Hunt questioned whether "the notion of cholinergic and adrenergic neurons as principal categories of phenotypes is going to remain viable".

E. Induced Morphological Alteration in Cultures

Conceivably the alteration of cell morphology induced by cell extracts or other agents might again reflect a cell interaction and numerous questions were raised regarding the observations by Drs. Sensenbrenner and Lim of chemically induced alterations in the morphology of glial cells in culture. The most contentious issue was the specificity of the transformations induced by cyclic AMP, dibutyryl cyclic AMP and brain extract (GMF).

Dr. Westermark pointed out that if one treats human thyroid cells with either TSH or dibutyryl cyclic AMP one observes a picture which is indistinguishable from that observed by Dr. Sensenbrenner immediately after cyclic AMP treatment, that is, the thyroid cells transform into highly astrocyte-like cells. He also mentioned that the early changes observed by Dr. Lim were similar to those which occur when you change any resting cell population to a medium with high serum concentration.

Dr. Bornstein gave several other examples of alterations in cell morphology caused by changes in the environment. Glial cells in the outgrowth zone of fragment cultures formed a confluent epithelial monolayer close to the fragment but further out cells were less densely packed and occasionally assumed a neuron-like morphology with processes two or three mm long. Dr. Benitez has observed that a liver extract applied to cultured cells caused them to round up and develop very long processes, a transformation which was called neuralization at the time. Dr. Bornstein pointed out that to say such a transformation was a specific effect one would have to go a lot further. For example, there might be a similar effect of glial extract on liver cells. He described a similar morphological change when cultures were exposed to poly-vinylpyrrolidone (PVP). Neuroglial cells in a confluent sheet began to take up the PVP, gradually retract their processes, round up, detach themselves from the neuroepithelial sheet, put out pseudopodia and finally come to resemble closely normal microglia. All this occurred just by adding the plastic, PVP. Therefore, Dr. Bornstein cautioned against assuming a specific effect on the basis of a morphological transformation.

Both Dr. Sensenbrenner and Dr. Fedoroff had experienced that the developmental stage of the cultures, i.e., the time in culture (prepared from a fixed starting point, e.g., neonatal brain) determines which phenomena can be elicited. Thus, dBcAMP has to be added during a certain period of time to evoke a morphological differentiation, whereas addition to too young (Dr. Fedoroff) or too old (Dr. Sensenbrenner) cultures has no effect. According to Dr. Sensenbrenner also brain extract has to be added at certain stages. This was confirmed by Dr. Lim who added that repeated subculturing abolishes the effect of his maturation factor.

Dr. Mandel also addressed this problem of specificity by pointing out that before one can say that an effect is specific, one should know if the agent under consideration is just one of many, e.g., a phospholipase such as lysolecithin which can activate the cyclic GMP or cyclic AMP system, and if it is this activation that actually triggers the morphological transformation. He also pointed out that several methods are being used to produce cell differentiation and that we must be careful when studying different factors influencing cell differentiation. He reported, for example that when cell differentiation is produced with bromodeoxyuridine, a high proportion of thymidine in DNA is replaced by bromodeoxyuridine, producing a change in DNA in these cells and suggested that BrdU is perhaps not the best way to investigate cell differentiation.

Dr. Prasad commented that the presence of a dibutyryl cyclic AMP effect is not enough by itself to assign a role for a particular function to cyclic AMP; two or three more criteria must be fulfilled. The effect should also be reproducible by at least one stimulator of adenylate cyclase, e.g., prostaglandin and also at least one of the phosphodiesterases. Finally it should be shown that the increase in cyclic AMP depends on phosphorylation.

The question was raised as to the purity of the system being examined, that is, did the culture only contain glia or were the usually highly proliferative fibroblasts present. Several participants mentioned the similarity between the morphological alteration induced by dBcAMP or brain extract and the normal preparation of cells for mitosis and asked if mitotic inhibitors could mimic the effects of some of the factors used.

Regarding the chemical specificity of GMF, Dr. Lim stated that the induced morphological transformation could not be reproduced by PVP, concanavalin A, polylysine, polyglutamate, cyclic GMP or its analogues and many other substances. Dibutyryl cyclic AMP did mimic GMF but there were subtle morphological differences. Neither Drs. Lim nor

Sensenbrenner had tried mitotic inhibitors. The content of fibroblasts is low and GMF had only little effect on fibroblasts obtained from other organs.

Dr. Hild interpreted Dr. Lim's time lapse cinematographic studies as suggesting two effects, i.e., first a preparation for mitosis and then a differentiation. Also Dr. Privat pointed out that the phenomenon observed by Drs. Sensenbrenner and Lim is very similar to what happens in the *in vivo* situation when a cell leaves the subventricular layer to transform into an astrocyte; this occurs usually after mitosis when cells extend long processes. We can think that these cells are then submitted to some factors which last long enough to keep them in the form of elongated cells with numerous processes, abundance of filaments and glycogen. Dr. Varon liked the suggestion by Dr. Hild of considering two events or two factors involved. The first one, the early events that Dr. Lim described are very clearly a characteristic of the breaking of a contact inhibition of growth. This can be seen with fibroblasts, or with practically any confluent monolayer and with very many different agents whether they are from the serum or whether one uses proteases. So what attracted Dr. Varon to Dr. Lim's work was that there seems to be one or two rounds of mitosis and then mitosis stops. He continued that many people have raised the point that the real differentiation, the genetic imposition of a new program, may very well require a round or two of mitosis. After that something happens that makes the cell susceptible to other influences. Dr. Varon wondered whether it's possible to think that there are indeed two factors, maybe related, one which breaks the confluent inhibition of growth, generates first a synchronous mitotic activity which might involve the synchronous appearance of an adequate number of receptors to the second agent which then takes over and imposes on it the glial morphology.

III. Methodology

Dr. Fedoroff emphasized that for meaningful analysis of differentiation *in vitro* it is important to know the species and strain of animals and the specific stage of embryonic development at which the tissue or cells was obtained. Expressing developmental stages in terms of gestation time is misleading; two 7 day old embryos may actually be at different stages of development. The difference may be even greater when embryos of different animal strains are compared. It is important to specify whether tissue fragments or dissociated cells are used and whether the cells were obtained by dissociation with enzymes or by physical methods. Culture surfaces, plastics or substrates used must be specified since all these factors may act as selective pressures on particular cell types. It is also important to specify the precise regimen during culturing as well as the composition of the media, pH, osmolality, etc. In Dr. Fedoroff's laboratory it was observed that differentiation of glial cells can be altered by changing the type and concentration of serum in the medium. Dr. Lim had also noticed that increase of the serum concentration inhibited his glial maturation factor.

Dr. Bornstein added that there are many studies that have attempted to examine the types of serum and their effects on tissue cultures. He preferred human placental cord serum but in years gone by, many types of serum had been compared: horse serum, fetal calf serum, calf serum, bovine serum and so on. In his experience there were as many variations between serum from six different horses as between horse serum and calf serum and human placental cord serum. Drs. Hauw and Privat agreed on the importance of the serum quality and it was pointed out that the survival, especially of the granule cells in the cerebellum, might be critically dependent upon the quality of the serum. Dr. Messer had the experience that in media containing fetal calf serum normal granule cells undergo a variable, but substantial degree of clumping which eventually leads to a good deal of degeneration, whereas cells from the staggerer mutant undergo a lesser degree of clumping. Only by switching to horse serum, which completely changed the interaction and presumably the biochemical milieu, could she eliminate that surprising but discrete difference.

Dr. Pfeiffer mentioned the pronounced effect of minor alterations in the culturing regimen. For example, slight shifts towards acid or alkaline pH and alterations in feeding frequency had drastically altered the yield of thyrotropin releasing factor from a cell line. Furthermore, this factor is released from proliferating cells but not in stationary phase cultures.

Dr. Nelson commented briefly on hybridization. In one hybrid between fibroblasts and neuroblastoma an augmented choline acetyl transferase activity had been observed. Thus, despite the fact that neither parent has a high enzyme activity, one can get positive control or release of negative control by these hybridization procedures. Accordingly, the hybridization procedure seems to be a somewhat Monte Carlo method of bringing out a variety of features. It is obviously not completely chaotic but it is also not an averaging method or merely expressing characteristics of one parent or the other. There are positive interactions which can release potentialities of either parent that are not expressed in the original lines.

With respect to the applicability of tissue cultures in studies of the differentiation of the nervous system, Dr. Privat concluded that "if we leave some cells to express their full potential, we might happen to find something very different to what happens in the *in vivo* situation. I think we must keep on doing this kind of work and comparing our results and comparing very carefully the conditions under which the results have been obtained. . . We must try to vary these conditions under strict control, in order to explore all the possibilities of each cell, all the possible interactions of any cell with the other cells."

CULTURED NEURAL CELLS AND TISSUES AS MODELS OF THEIR IN VIVO COUNTERPARTS: DISCUSSION

*D. D. JOHNSON**

Department of Pharmacology
University of Saskatchewan
Saskatoon, Saskatchewan
Canada

I. Clonal Lines

A. Cell Characterization

In summarizing his view of the state of the art, Dr. Pfeiffer asked "what is required of investigators now?" He suggested that if we are going to continue to work with established lines, there are three things which are probably more important than others. One is to "clone" and preferably to "clone early" before experiments get too far along, because too often cultures that seem to be homogeneous change over a period of time.

The second point people seem to be remiss on, is in karyotyping. He suggested that it is no longer good enough to count the chromosomes since methods are available for determining how normal the karyotype is, not only in morphology at a gross level, but by doing banding patterns. He suggested that we could go a long way towards saying whether a particular line is normal or abnormal by doing banding patterns and he suspected that many clones are abnormal to some degree.

Dr. Hrushovetz elaborated further on the increase or decrease in the chromosomal number. Cell lines such as the HeLa and the "L" cell fibroblast lines have a varied number of chromosomes and usually one which has a higher number, called the modal number. Dr. Hrushovetz emphasized the fact that, at the beginning of experiments with cell lines, the distribution of the chromosome types should be known and this should also be determined at the end of the experiment. Obviously, genetically different nuclei will have genetically different physiological functions.

Regarding a change in the morphology of the chromosomes, if cells are

---

*Dr. Johnson, as rapporteur, has summarized and integrated the recorded discussion between the contributors to this volume on the above topic.

not analysed chromosome-wise at the beginning and end of experiments, cells may be selected when they go through a divisional stage, especially if experiments extend beyond a few hours.

Dr. Hrushovetz also pointed out that cells go through a cycle and that results determined at the molecular level will depend on the cell cycle stage. Incubation in a particular medium may select at certain specific stages and result in erroneous values. For example, if colchicine is used in the study of behaviour of cells in culture and DNA content is being determined, it will appear as if every cell had doubled its DNA content. There will be a 100% increase in DNA even though colchicine has no direct effect on DNA.

Dr. Hrushovetz finally spoke of cellular senescence. Embryonic cells are usually in phase one or early phase two; the adult tissue in an individual is in phase two and the tissue from an aged individual falls into phase three. Unless the number of cell divisions is known, cells from anywhere along this growth curve will be obtained and this could produce different values. The content of various enzymes and proteins varies with the different phases of a cell's life history.

Dr. Mandel commented that we must be aware of this basic problem, and that it is impossible to get synchronized cultures, or that we can get synchronization for only a very short period of time.

Dr. Pfeiffer continued his discussion of the problems in tissue culture and mentioned that also the problem of cell type designation is important. How are we really going to be able to say exactly what kind of cells we are working with? Neuroblastoma cells are neuroblastoma cells which are useful, but we really can't attach a name like granular cells or Purkinje cells to these. This is probably due to mixed cell types where one cell type is expressing both glial parameters and neuronal parameters or parameters of more than one glial cell. There are suggestions in the case of C-6 cells, that they are astrocytic and for a long time they were called astrocytes on the basis of the tumor from which they were derived having been termed an astrocyte tumor. On the other hand, it is one of the best lines for studying GPDH induction by hydrocortisone. Work by Dr. DeVellis (see J. DeVellis, this volume) has indicated this would suggest that they have oligodendrocyte characteristics. There is also the problem of the specificity of the assays. Is there really such a thing as an astrocyte specific marker or an oligodendrocyte specific marker?

B. Cell Biochemistry

The question of biochemical similarities and differences between cultured and non-cultured cells and tissues was opened by Dr. Mandel. He mentioned striking qualitative and quantitative differences in LDH activities between different clones. In a number of clones there was only one iso-enzyme of LDH, while in normal nerve cells there were five. Dr. Mandel proposed that for many clones, something very basic is missing regarding the enzymes involved in energy metabolism.

In normal nerve cells there is also an extremely high amount of adenine nucleotides. Dr. Mandel discussed the percentage distribution of free nucleotides in various cell lines. In N115 cells the quantity of adenine nucleotides was low compared to normal nerve cells. In proliferating neuroblastoma cells the content of adenine nucleotides was about 45% of total nucleotides, in differentiated cells about 53%, while in normal cells the content of adenine nucleotides was about 70% suggesting that, in this respect, cell lines are abnormal.

He pointed out that cell membranes from N115 cells contain about 30% of saturated fatty acids, 50% of monounsaturated fatty acids and a very low amount of polyunsaturated fatty acids. In plasma membranes and synaptic membranes isolated from *in vivo* systems there was about 40% of the so called hexanoic fatty acid which is a polyunsaturated fatty acid. If different polyunsaturated fatty acids were added to the culture medium, the quantity of monounsaturated fatty acid decreased and the quantity of polyunsaturated fatty acid increased, but normal values were not obtained. Dr. Mandel suggested that this was interesting from another point of view because it shows that we can change the membrane fluidity and that this is a rather important point to investigate. Thus the influence of membrane fluidity on the permeability to transmitters and their precursors could be investigated.

Some of the main components of plasma membranes or synaptic membranes are the gangliosides. When the distribution of the polysialogangliosides in different clones was investigated it was observed that there was a very high amount of monosialoganglioside, a low amount of disialoganglioside and no trisialogangliosides in all clones investigated. In the M1 clone the total amount of gangliosides is one tenth of that in brain tissue *in vivo*.

He reported that the content of GABA differs from one clone to another and that treatment with BrdU increased the GABA content in the M_5 line, but in the M_1 line there was no change. The absence of serum reduced GABA in both cell lines. Cyclic AMP and withdrawal of serum in different clonal lines did not produce striking effects except for a decrease on the withdrawal of serum. In the clonal line C6, a striking decrease of GABA occurred after the addition of cAMP. Coculturing of neuroblastoma and glial cells and subsequent reiosolation of the cells produced a decrease in the GABA content.

Dr. Pfeiffer pointed out that investigators should characterize cells both morphologically and biochemically. The presence or absence of a particular enzyme or protein is interesting, but it does not tell the whole story. The relative levels of a whole series of parameters is really what starts to characterize these cells. In addition, there are some problems of culture conditions in measuring these parameters. Dr. Rosenberg has shown, for example, the tremendous difference in specific activity occurring in certain enzymes as a function of growth in culture.

If a clonal line in culture is compared to a tumor, the level of enzyme activity may be vastly different. Some activities seem to be favoured in culture while others are favoured in the tumor. As a general rule plasma membrane antigens and activities seem to prefer the tumor configuration. This may be a matter of nutrition or of three-dimensional organization as it does not seem to be a genetic phenomenon.

Dr. Pfeiffer then discussed the continuing problems in tissue culture. One was the problem of regulation. He thought that it is a problem that we are going to continue to live with in tissue culture in general and in work with clonal established lines in particular.

Dr. Hertz questioned the value of using cell lines for quantitative biochemical studies. He suggested that if a certain parameter or function is measured in a glial cell line and it can be shown that primary cultures exert much more activity, there is little advantage of continuing to use the cell line. It may have been very important in establishing these functions, and remains so in the case of the neuroblastoma, where a corresponding primary culture is not available.

Dr. Rosenberg felt that the fidelity of translation from the culture system to the intact brain is a good one. He pointed out that, in his results, cultures which were most differentiated had the highest specific activity of choline-acetyltransferase and acetylcholinesterase. When considering the embryology of the developing brain, periependymal cells, which are uncommitted, divide a certain number of times as determined genetically and become committed to neuroblasts that are post mitotic. They then migrate out from the periependymal region to cortex to find their genetically determined position and morphologically differentiate and form synapses. It is then that the enzymes for chemical neurotransmission are induced. The highest specific activity is present in the most adult cortex; namely in the cortex where one has the greatest number of synapses after apical dendrites and spines have been produced. This is true for several enzymes related to neural transmitters in several brain areas.

C. Cell Functions

1. Transmitter metabolism. In continuation of his demonstration of induction of enzymes for chemical neurotransmission in differentiated cultures, Dr. Rosenberg mentioned that the clonal cell line of C6 that Dr. Schrier presented, had a very high affinity uptake for glutamate and formed

GABA. He suggested that this is a model system in a pure culture line to examine GABA synthesis which presumably also occurs in astrocytes in the intact brain.

Dr. Pfeiffer was not sure that cultures are useful for defining which cells in the central nervous system carry out a specific function. He suggested that if we find that a so-called astrocytoma cell in culture carries out a particular property or function, it does not mean that astrocytes in the normal brain perform that function. He suggested that this is not a useful thing to investigate in clonal cells, and that we are much better off to study this in the intact nervous system.

Dr. Mandel pointed out that several investigators have shown a high-affinity uptake of glutamate and aspartate. The usual metabolic inhibitors produce the expected abrupt blockade of aspartate and glutamate uptake. With regard to taurine, which is also taken up in neuronal clones and glial clones, Dr. Mandel reported that there was a competition by taurine analogs for this uptake mechanism and a very striking inhibition of uptake by some analogs. Dr. Mandel suggested that these data increase the likelihood that taurine is a neurotransmitter.

Dr. Wood suggested that one of the problems in using tissue cultures to study amino acid transmitters relates to the interrelationships of various compartments *in vivo* which perform the function as a whole. He indicated that the concentration of the neurotransmitter that is important is not the one in the glial cell or the nerve ending or the neuronal parikaryon, but that in the synaptic cleft. Obviously, although the concentrations in these other cells are not of primary importance, they are of extreme secondary importance. They are the concentrations that will determine the concentration of the transmitter in the synaptic cleft. Using pure cell cultures it is possible to look at these individual compartments to determine what's going on. If the concentration in the synaptic cleft is simply a summation of the different compartments, it would be reasonably easy, perhaps to get the overall picture. However, for the amino acid transmitters, a transmitter may be released, and then be taken up by another cell (glial cell) where it may be changed to some other compound which is then sent back into the nerve endings, which in turn converts it to the neurotransmitter. The integration and dependence of one compartment on the other is very difficult to study using cell cultures. The situation of the membrane people where they look at the functioning of the membrane, dissect and then study the purified constituents appears to be equivalent to the stage where we are at now using tissue culture methodology. We are looking at cell lines, but we must under very careful and known conditions, start putting these things back together to see if we can get them to function as a whole. The problems in tissue culture are much greater than those encountered by membranologists who have a simpler system.

Dr. Sellinger spoke of the value of using bulk prepared cells for the study of mechanisms of nerve cell function. His laboratory has always thought that it is equally feasible to get cell bodies directly from rats. Securing cell bodies from these animals enables the investigator to do a serial study of development and to look at all the parameters which a cell body exhibits. Obviously, transmitter phenomena cannot be studied in the cell body, however, he suggested that the enzymes producing these amines are synthesized in the polysomes. These polysomes are present in the cell body. He indicated that he isn't convinced that cell culture is the only methodology for studying neurons at the molecular level. The cell body isolation technique has been successful and has provided an entry into the specificity of neural function. Proteins can be isolated from nerve cell bodies from young or old rats. It is possible to do immunochemical studies with proteins obtained directly from the animal and his rationale for using this methodology is that it obviates the difficulties of producing many different cell lines.

Dr. Lin pointed out that both cell cultures and bulk separated cells can be very useful in studying the function and structural relationships of contractile and structural proteins which represent up to 20% of all nerve cell proteins.

For example, it is possible to isolate seven different forms of actin from whole brain. We can ask which form of actin is used for glial cell locomotion, or for axoplasmic transport or for the release of neurtransmitters. He suggested that it will only be possible to study these actins by isolating them from glial cell lines, neuroblastoma cells or bulk separated cells and not from a mixture of actins isolated from the whole brain.

As a further advantage of cloned lines in studies of transmitters and receptors, Dr. Rosenberg indicated that a hybrid line with opiate receptors is now available. He suggested that this is an ideal model for investigating the turnover of the opiate receptor in terms of its synthesis on the endoplasmic reticulum, delivery to the cell surface and its eventual endocytosis and degradation by lysosomes. This has provided an opportunity which is not accessible if one uses intact brain. Similarly, Dr. Rosenberg's laboratory is entering into the area of peptide neurotransmitters. One important peptide neurotransmitter is vaso-active intestinal polypeptide or VIP. There is a very high specific activity of VIP in neuroblastoma. They have investigated VIP activity regionally in the central nervous system and found it to be very high in the hypothalamus. These studies were influenced and were expedited by virtue of clonal cell lines. Dr. Rosenberg suggested that, in terms of neurotransmitter synthesis of a classic nature and of a peptide nature, clonal cell lines are indeed valuable systems.

Dr. Pfeiffer indicated that also the thyrotropin releasing factor may have a general brain function and perhaps acts as a neurotransmitter. Also, this factor has been found in neuronal cell lines from induced tumors in CDF rats, but only during the exponential growth phase.

2. Synapse formation. In the discussion on the use of cloned lines to study cell function, Dr. Nelson mentioned that the neuroblastoma is a rather chaotic cell in the sense that it is not particularly well organized in terms of a rigorous program of producing a given complicated set of functions; at one time or another all the functions that one may be interested in will, however, be expressed in these cells. The technical features of the system may allow dissection and isolation of these different steps. As long as the functions can be isolated this system may give an approach to analyzing the molecular basis for these functions. The notion that for a particular given function there's going to be some particular molecule directly related to that function is at the heart of this approach. Synapse formation is one of these functions. It is hoped that neuroblastoma cells will be useful for isolation of the various steps involved in the formation of specific synapses.

II. Primary Cultures

A. General Characteristics

Dr. Seil addressed the general topic of the validity of tissue culture as a model for the *in vivo* situation. He pointed out that with regard to the cultures of mouse cerebellum, if one looks at the overall organization there is a recognizable organization with the cortex, the white matter zone, the intercerebellar nuclei and the vestibular nuclei in their proper orientation. In these cultures there is evidence for the presence of all five neuronal types in the cortex.

Spontaneous electrical activity in the cortex from these cultures is usually of the phasic variety as opposed to the more regular variety. This type of activity appeared at the time of the appearance of a large number of synapses in the culture, at about eight days *in vitro.* If the cortex was stimulated, spikes occurred right after the stimulus and this was followed by an inhibitory pause of up to fifty milliseconds duration which in turn was followed by further spike discharge.

Using trains of stimuli and proper electrode orientation this was obtained with 12- and 18-day cultures. This correlated with the maturation of synapses in the culture. If fiber responses in the cortex were recorded, they developed initially from a deep to a superficial gradient just as in the animal. The pattern of excitation-inhibition-excitation is also like that found in the animal cerebellum.

It was possible to recognize the intercerebellar nuclei as focal points of converging Purkinje cell axons. Both large and small cells were seen as has been described in animals. If these cells were stimulated they were electrically excitable but no spontaneous activity occurred. Spontaneous activity was present in the vestibular nuclei.

Dr. Seil suggested that with this degree of organization at both the structural and functional level, that with some qualifications it is a reasonable model of the cerebellum *in vivo*.

Dr. Hauw, who cultured guinea pig and human cerebellum and did not remove the meninges, supported Seil's observations. After one month in culture the granule cells survived and the laminated structure of the cerebellar cortex was preserved with external and internal granule cell layers. As another example of architectural duplication in culture of an organ within the animal, Dr. Bornstein called attention to a paper that has just come out from Dr. Sobcowicz' laboratory in Wisconsin. In these cultures of the organ of Corti, the replication of the normal architectural relationship is absolutely astounding, the ganglion cells are sitting down at the bottom of the fragment and there's a limiting membrane and the hair cells are there, one after the other and there's one axon going up to each hair cell; it's just unbelievably beautiful; it's a lovely model.

Dr. Munkacsi indicated that certain experiments are more suitably carried out in tissue culture than in intact animals. For studies of the effect of hormones on certain cells the *in vivo* system may be too complex. He has administered hydrocortisone and thyroxin to primary explants of sympathetic ganglia to determine whether treatment with hormones would induce synaptogenesis. With thyroxin there was a very significant increase in synapse formation. After treatment with hydrocortisone there also seemed to be an increase in synapse formation. He emphasized the importance of establishing species and within-species variability. It is customary to say that there are no synapses in sympathetic ganglia before birth. He suggested that this is not quite true because if one looks at a large enough sample of animals there are instances when there are a significant number of synapses in the ganglia prior to birth.

Dr. Scott commented on his experiments in which he has counted neurons in culture to study neuron survival. Survival of neurons in cultures of 10-day chick embryo dorsal root ganglia decreases in an exponential fashion and levels off at a very low value of about 0.1%. In electrophysiological experiments these neurons appeared normal. More recently, he has used newborn mouse as the source of tissue and after 40 days in culture there was approximately 10% neuron survival. When he cultured cells from adult mouse dorsal root ganglia there was practically no neuron degeneration during the first 6 days; 97% of the neurons had survived and thereafter there was a very slow decrease in cell number levelling off at 65%. Therefore, adult cells when cultured under exactly the same conditions as the embryonic forms gave a much higher survival rate. Other workers in short term experiments obtained similar exponential decreases in neuron survival for embryonic material.

Dr. Scott commented that it would seem that in this classical sort of explant culture, the usual sort of thinning out phenomenon is due to neuron degeneration as well as neuron spreading. Dr. Scott suggested his data indicated that cultures of adult nervous tissue provide a much more valid and reliable culture system.

Dr. Privat mentioned that he has tried to grow primary cultures of cerebellum from 10-30 day old rats and had some success with rats up to 15 days of age. With 20-30 day old rats he was unable to keep the explants in good condition. It may be that the conditions were not optimal but they worked very well for cerebellar cells from newborn rats.

Dr. Hild raised the question why neurons can be explanted from stages of development when neuronal processes have already developed and why neuronal fractions can be obtained by mechanical separation when, in his cultures, degenerative processes begin immediately and the neuron dies if neuronal processes are severed even 350 μm from the perikaryon.

B. Morphology of Synapses in Vitro

The question arose whether the different types of synapses which have been observed *in situ* and which were reviewed by Dr. Phillis also occurred in culture. Dr. Hendelman reported that in organotypic cultures of cerebellum using recent fixation techniques he has seen plentiful evidence of both excitatory and inhibitory synapses. The presence of dense core vesicles in these cultures was not investigated.

Dr. Crain also has definitely observed the characteristic excitatory synapses in culture and some suggestion of the flattened vesicle type of inhibitory synapse. However, in many cases where there was electrophysiological evidence for inhibitory synapses, there were surprisingly few signs of the flattened vesicle type of synapse. Crain and co-workers are therefore tentatively considering that many of the synapses can look alike in terms of EM analysis and yet have both types of excitatory and inhibitory functions. They observed very few signs of electrotonic junctions in culture and no signs of the two way junctions mentioned by Phillis (J. W. Phillis, this volume). Dense core vesicles were very common in their cultures, being observed even in motor axon terminals in neuro-muscular junctions in culture. Many dense core vesicles were also seen in CNS neurons in culture. In this regard, Crain raised the question as to the origin of the many dense core vesicle synapses described by Dr. Sensenbrenner, in cultures of dissociated brain. Did these cultures contain brain stem, e.g., locus ceruleus or are these aberrant noradrenergic neurons which develop under the particular culture conditions from higher brain stem sources? Dr. Sensenbrenner replied that her cultures did not include any brain stem, and added that in long-term cultures (3-4 weeks), dense core vesicles occurred in synapses.

Dr. Nelson reported that in his laboratory a correlation of functional and fine structural characteristics of identified neuronal junctions in spinal cord culture has indicated very clearly that a wide variety of different synaptic morphologies occur. There were those with pre- and postsynaptic densities as well as the simpler kind of synaptic structures. There were endings with a good many dense core vesicles. With the fixatives used to date there hasn't emerged a fully developed flattened vesicle picture characteristic presumably of GABA-ergic connections but this might not be surprising in spinal cord cultures. Nelson also pointed out the possibility that the dense core vesicles arise from peptidergic neurons. He stressed that it was quite clear that there was variety of synaptic types both physiologically and morphologically and that he was trying to catalogue them to see what sort of detailed morphophysiologic correlation would emerge.

Dr. Nelson mentioned that it is possible to ask some detailed questions of mechanisms of central nervous system synaptic transmission in a satisfying, detailed and rigorous sort of way with dissociated normal cultures. The neuromuscular junction has served as an extremely valuable model for synaptic transmission, as Dr. Phillis pointed out, but other mechanisms undoubtedly exist. The next simplest system, the dorsal root ganglion is still a troublesome synapse. Why don't we know more about the basic nature of the synapse? The answer is that it's difficult to study even though it is the easiest one to get at in the central nervous system. His laboratory is presently focusing on the dorsal root ganglion synapse in dissociated preparations where they can see the location of the terminals, map out the chemosensitivity around these connections and get a more or less rigorous biophysical analysis of this particular synaptic connection in order to obtain as complete a picture as possible. He thought that this was a good expectation and that the answers are not going to be radically different from the intact situation. This general paradigm can be extended to a number of other situations where the analytical rigor will be useful and the extrapolation very likely quite reasonable.

Dr. Crain agreed with Dr. Nelson's remarks about potentialities and the value of utilizing primary cultures of dissociated neurons to focus on biophysical analysis of synaptic relationships between identified cells of the CNS and extension of the easier problems studied with nerve-muscle preparations. He would, however, extend this one step further in terms of encouraging concomitant investigations with primary cultures of explants, where it is possible to maintain interrelationships between neurons that may be irreversibly lost after dissociation. There may be whole areas of analysis relating to the kinds of problems that Dr. Hunt emphasized in regard to organization of arrays of neurons and the positional information that underlies these kinds of highly selective patterns of arrangements of neurons throughout the central nervous system. These might best be analyzed, at least at this stage of techniques, with judicious choice of intact explants in correlation with studies of isolated neurons. With improved techniques these kinds of interrelationships may be amenable to study by direct analysis after dissociation of the cells. Both techniques should be encouraged and may complement each other as analysis of these problems proceeds.

C. *Regeneration*

Dr. Varon mentioned that the problem of regeneration in neural lesions can certainly gain enormously from *in vitro* studies because it's the only place where we can determine what kind of agents might keep neurons alive, or the way that nerve growth factor and other factors influence neural regeneration. The conditions that promote elongation of axons and control the direction in which axons grow must be investigated. Dr. Varon suggested that we should attempt to study the problem of glial scar formation. There has been a recent report that aggregates of cerebellar cells develop a very normal looking network of glial fibres recognizable by GFA antibody, which after a few days undergo a degenerative condition accompanied by extensive gliosis. This did not happen in the presence of polylysine. He had recently reported that in monolayer cultures polylysine inhibits both mitosis and hypertrophy of glial cells. He cited this as an example of the kind of studies that may lead to a change in the idea that text books continue to claim, namely that central nervous system tissue does not regenerate.

Dr. Nathaniel commented that he had made lesions in the dorsal root of the rat spinal cord, and during the course of regeneration of the fibers, found an increase in collagen. There were very few fibroblasts present. A considerable number of collagen fibres appeared between the plasma membrane and the basement lamina of the Schwann cells. On the basis of morphology over a continuous period of study he concluded that the Schwann cells are producing the collagen.

III. Clinically Related Studies

Dr. Bornstein pointed out that there are a number of instances where tissue culture techniques are proving themselves to be of value in the examination of certain pathological processes known to occur in animals or in human beings. He has illustrated this for demyelinating disorders but there are many others. He is presently investigating factors that may affect the brain in uremia. In the intact animal these factors are very difficult to separate. It is perfectly obvious that when you have a tissue culture system it is separate from the rest of the animal so that those other tissues and organs present in the animal are not interfering with the examination of whatever particular factor you are looking at in culture. Thus, for uremia, for alcohol, or for other toxic factors one can examine various metabolites for their specific action on these cells. This is not possible in the animal.

Dr. Bornstein reported that his laboratory was examining industrial toxins in culture and they were able to examine various modifications of these toxins which are affecting patients, particularly in terms of peripheral neuropathies. Using tissue culture it is possible to examine these toxins to determine which is the active agent.

He suggested that cultured tissues can be used to examine many parameters of pathological processes, and in return some pathological processes give us insight into normal processes. This has been true very frequently in classical pathology and he thought that this is also holding true in the case of tissue culture.

Following up what Dr. Bornstein had said, Dr. Rosenberg said that the use of neuroblastoma cells is very effective as a model system for evaluating the Guillain-Barré syndrome, an acute form of polyneuropathy, which has received a great deal of publicity as it pertains to the Swine Flu vaccination program in the United States. Ten patients with Guillain-Barré syndrome had IGg and IGn immunoglobulin species in their serum that were cytotoxic to neuroblastoma cultures. The IGg and IGn moieties bind to cell surface neuroblastma receptors. There was induction of patching and capping and occurrence of cytolysis. This provides an immunologic means for investigating a very specific and very common form of polyneuropathy. In addition, patients with chronic progressive motor neuropathy have also been examined and have lower titers of similar kinds of IGg and IGn antibodies.

Dr. Rosenberg continued that the use of neuroblastoma in screening of chemotherapeutic agents, specifically against neuroblastoma, has been extremely effective. Dr. Prasad (K. N. Prasad, this volume) had alluded to the fact that patients had been treated with papaverine and chemotherapeutic agents as a direct extension of neuroblastoma culture studies. He wondered whether the transformation from ganglioneuroma to medulloblastoma might occur because of loss of regulation of phosphodiesterase and asked if Dr. Prasad could give further information in terms of chemotherapeutic intervention specifically with agents that inhibit phosphodiesterase.

Dr. Prasad replied that the clinical trial with papaverine is being carried out in Memorial Hospital in New York on terminal patients with neuroblastoma. With the addition of papaverine, 17 out of 19 patients have been clinically free of disease between 12 and 24 months. This is the first time in this type of patient that a clinical cure has been achieved. During treatment, when a biopsy was taken, there were signs of transformation to ganglioneuroma. This was not seen in earlier studies with cytotoxic drugs alone. Papaverine thus plays a kind of strategic role in this kind of treatment (see also K. N. Prasad, this volume).

As another example of practical clinically relevant use of cell and tissue cultures, Dr. Bornstein referred to the experiments by Dr. Wolf in which he has examined genetic traits and developed nerve cultures from animals which have genetic disorganizations. Using this methodology one can study endogenous factors which are causes of disease.

Dr. Messer elaborated on this question and defined three levels at which mutants can be studied, i.e., 1) the primary gene defect, 2) the intermediate signals along the way between the primary gene defect and the altered phenotypic expression and 3) the immediate phenomenology. Dr. Messer noted that such experiments are even becoming useful in the study of neurological disease. Since obviously brain tissue cannot be obtained from the patients, fibroblast cultures, especially, have been used to study the gene defect. Such studies have shown a defect in guanosine metabolism in Lesch-Nyhan disease, a defect in oxidative metabolism in Friedreich's ataxia and an altered cell density at confluency in fibroblasts from patients suffering from Huntington's chorea.

With respect to the second level, i.e., the intermediate signals, Dr. Messer emphasized, like Dr. Wolf, the importance of a marker gene.

Conclusions

Dr. Hild questioned the need of repeating in tissue culture what we already know is happening in the nervous system or what can be assessed much better with other methods. He mentioned, as an example, studies of neurosecretion and finished by stating that there is no need of excising the hypothalamus and making a caricature preparation of the whole system in tissue culture when it will be much simpler and exact to study it in the intact animal.

Dr. Messer answered Dr. Hild's comments by remarking on the particularly complex problems of hypothalamic interactions, e.g., hormonal influence of neuronal development, interactions between neurotransmitters, hormones and releasing factors, etc. The best way to study these variables, she said, is in culture, by removing the cells from a more complex system. Along the same lines Dr. Bornstein emphasized the opportunity of obtaining unique data from some discrete immunological techniques in culture. Also Dr. Hunt disagreed with the statements made by Dr. Hild and said that he believes that interactions of precursors are much better studied in culture.

Dr. Mather noted that cultured cells and tissues in general do not only use medium constituents but they do also release certain factors. These factors can be studied much easier *in vitro* than *in vivo.* Also Dr. Nelson referred to the usefulness of tissue culture for the study of factors which produce one effect or another. Dr. Nelson thought that tissue culture systems were ideal for studying this problem. The specificity of the effect may be questionable, but a conditioned medium paradigm does allow the possibility of isolating the particular molecules that are producing these effects and analyzing the nature of the mechanism. He predicted that there is likely to be a large number of these factors and that they all should be interesting. The tissue culture system should be a necessary and crucial methodologic feature in such studies. He emphasized, however, also that the pros and cons for *in vivo* and *in vitro* studies always have to be weighed with extreme care. Dr. Pearson mentioned that in a culture system a steady-state input of a certain labeled amino acid can be maintained in a preparation in a stable condition. After a given period the hormone can be isolated from the tissue and the medium and that is probably the only way in which the rates of synthesis can be measured.

With respect to the choice of one or another culture system, Dr. Schrier suggested that the Heisenberg Uncertainty Principle be applied to tissue culture. When you look at nature, by the fact that you are looking at it, you are disturbing it and, therefore it is not exactly what it was when you started. He suggested that when we become stranded on which simplified system is better, a lot of time is wasted. The fact is that with simplification we lose normality. It isn't that all normality is lost, we simply have to know how much is lost. When we select a system and then try to organize questions around that system, we are going in the wrong direction. The proper way to approach something is to say, "Yes, I would like to know about this particular function," in our case of the nervous system, and then look for the best system to answer this question in a clean fashion. Clean answers are obtained out of a clean system, be it very simple, and then we must go to something more complex and see if it gives the same answer. When using simplified systems such as C6 cells or neuroblastoma cells, if you can demonstrate that nature is capable of doing something in a clean fashion, it means that nature, in a more complex system, is still capable of doing that. It is therefore important. Whether it's the same cell type or not is an academic thing which can be handled at a later date.

Dr. Pfeiffer commented that cultures do have a use as model systems for studying a specific function or a specific parameter which is known to exist *in vivo.* An example of this might be transmitter metabolism. There is already good evidence from *in vivo* work that neurotransmitter metabolism in glial cells is probably important and interesting. In a clonal system, Dr. Schrier has shown that there are very definite advantages of having an open system into which you can add or remove things very readily.

Dr. Hunt did not quite agree with Dr. Pfeiffer and suggested one major and one minor contribution that cell lines can provide. Cell lines can be employed as starting points for analytical biochemistry, but rather than studying known functions, to Dr. Hunt they offered the prospect of identifying as yet unknown molecular components of "glia" or "neurons" with the ultimate aim of raising antibodies to these components and localizing and analyzing these functionally in primary cultures of neurons and glia now that these are becoming available. He proposed that the principal thing that cell lines can do is to provide masses of tissue to find things fast.

Dr. Hunt suggested that the second contribution of cell lines involved the discovery of minor molecular components which are present in extremely small amounts. The power of genetics, which is blind to quantitative magnitudes of things and selects on the basis of cruciality, aids in the identification and biochemical studies of minor components. Thus cell lines may reveal potential components that play a minor role. For example, if sub-clones can be found in which a particular receptor disappears, they can be analyzed from the point of view of looking for insertion proteins or proteins that would bring a receptor into the membrane. Other minor components involved in post-translational control and packaging of receptors, transmitters, vesical assembly etc., also may be discovered.

Dr. Varon stated that it was time to stop arguing about which method is "the" method for studying nerve cells. He suggested that the "bulk people" are excellent producers of material which may very well be dead, but which is available in large amounts for very special investigations. People working with clones provide massive amounts of live cells. Clonal cells provide opportunities for genetic analysis, hybridization, and gene regulation that normal cells do not. Investigators (including himself) who use primary cell cultures prefer to think that they are looking at the real thing. The problem is that they are not. They are going beyond the level of normal neural tissue. Instead of trying to prove how well the *in vitro* systems mimic the *in vivo* situation we should be examining what neural cells can do, not necessarily what they do *in vivo*. The problem is that we must distinguish between capabilities of neural cells and what functions actually exist in *in vivo* systems. It is often thought that the nervous system in the adult is a very rigid system, but it is not. It has to become adult in the first place and during development interactions change continually. Adulthood does not mean fixation, but a different rate of development. Somehow we age in the neural system too, so there is evolution. We cannot forget that there is also pathology of the nervous system. Dr. Varon concluded that the only way to understand these problems is to know the limits of the capabilities of the cells; define all the circumstances under which these capabilities are expressed or inhibited and then perhaps offer, as a byproduct, a description of how the normal system works.

Dr. Bornstein echoed what Dr. Varon had said; that all manner of tissue culture has its advantages and disadvantages and we have to learn to choose the particular model system that will best answer the questions. Also, Dr. Wood indicated that tissue culture systems are going to play a very important role as they have already. Dr. Hunt commented that cell and tissue culture experiments should always be performed with the aim of ultimately going back to primary cultures to study the fully known function. Dr. Hunt expressed hope that in the next five years more attempts will be made in cataloguing parameters in cell lines but that then people must return to normal neurons and glia and try to characterize where they are and what they do.

In his concluding remarks, Dr. Mandel indicated that, as would be expected from such an audience, there is general agreement with the idea of using tissue culture as a tool for investigating neural cell differentiation and cell function. He concluded that in this field like any other, the main problem is to ask the right questions with the right system; to know what the system can give, not to ask of it more than it can give, and to be critical.

His last point was that, if we want to understand how the central nervous system works, it is obvious that whatever is discovered in tissue culture must also be demonstrated *in vivo*.

Index

C

D

E

F

G

H

N

T

U

V

W

X

A
B 7
C 8
D 9
E 0
F 1
G 2
H 3
I 4
J 5